Springer Proceedings in Complexity

Springer Proceedings in Complexity publishes proceedings from scholarly meetings on all topics relating to the interdisciplinary studies of complex systems science. Springer welcomes book ideas from authors. The series is indexed in Scopus.

Proposals must include the following:

- name, place and date of the scientific meeting
- a link to the committees (local organization, international advisors etc.)
- scientific description of the meeting
- list of invited/plenary speakers
- an estimate of the planned proceedings book parameters (number of pages/articles, requested number of bulk copies, submission deadline)

Submit your proposals to: Hisako.Niko@springer.com

Santo Banerjee · Asit Saha
Editors

Nonlinear Dynamics and Applications

Proceedings of the ICNDA 2022

Volume 2

Springer

Editors
Santo Banerjee
Department of Mathematics
Politecnico di Torino
Torino, Italy

Asit Saha
Sikkim Manipal Institute of Technology
Sikkim Manipal University
East-Sikkim, India

ISSN 2213-8684 ISSN 2213-8692 (electronic)
Springer Proceedings in Complexity
ISBN 978-3-030-99794-6 ISBN 978-3-030-99792-2 (eBook)
https://doi.org/10.1007/978-3-030-99792-2

This Springer imprint is published by the registered company Springer Nature Switzerland AG
The registered company address is: Gewerbestrasse 11, 6330 Cham, Switzerland

Contents

Graphs, Networks and Communication

Fractional System and Applications

Artificial Intelligence, Internet of Things and Smart Learning

Contributors

Abdikian A. Department of Physics, Malayer University, Malayer, 65719-95863, Iran

Ajith K. M. National Institute of Technology Karnataka, Mangaluru, India

Akleylek Sedat Ondokuz Mayıs University, Samsun, Turkey

Alam Shariful Indian Institute of Engineering Science and Technology, Shibpur, Howrah, West Bengal, India

Ali Demir Mehmet Computer Engineering Department, Kocaeli University, Kocaeli, Turkey

Amrutha R. Department of Physics, KCG College of Technology, Chennai, India

Amrutha S. V. Department of Physics, National Institute of Technology Karnataka, Surathkal, Mangalore, Karnataka, India

Ananthan Rohit Mechanical Engineering, Indian Institute of Information Technology, Design and Manufacturing, Kancheepuram, Chennai, India

Anh Mai The Vinh University, Vinh city, Nghean, Vietnam

Anshu Department of Mathematics, BITS Pilani, Pilani, Rajasthan, India

Ashwani School of Basic Sciences, Indian Institute of Technology Mandi, Mandi, Himachal Pradesh, India

Aslam Humaira Department of Mathematics, Adamas University, Kolkata, India

Atteya A. Department of Physics, Faculty of Science, Alexandria University, Alexandria, Egypt

Bandyopadhyay Anup Department of Mathematics, Jadavpur University, Kolkata, India

Banerjee Saayan Centre for Engineering Analysis and Design, Combat Vehicles R&D Establishment, DRDO, New Delhi, India

Barman Dipesh Indian Institute of Engineering Science and Technology, Shibpur, Howrah, West Bengal, India

Baroi Juhi School of Sciences, Christ (Deemed to Be University) Delhi NCR, Ghaziabad, India

Bhardwaj S. B. Department of Physics, SUS Govt. College Matak-Majri, Karnal, India

Bishi Binaya K. Department of Mathematics, Lovely Professional University, Jalandhar, Phagwara, Panjab, India;
Department of Mathematical Sciences, University of Zululand, Kwa-Dlangezwa, South Africa

Biswas Santanu Department of Mathematics, Adamas University, Kolkata, India; Department of Mathematics, Jadavpur University, Kolkata, India

Borah Joydeep Department of Mathematics, D. D. R. College, Chabua, AS, India

Borah Rupjyoti Department of Mathematics, Dibrugarh University, Dibrugarh, AS, India

Boral Rony Department of Physics, Indian Institute of Technology Guwahati, Guwahati, Assam, India

Büyüksaraçoğlu Sakallı Fatma Trakya University, Edirne, Turkey

Chakrabarty Siddhartha P. Indian Institute of Technology Guwahati, Guwahati, Assam, India

Chand Fakir Department of Physics, Kurukshetra University, Kurukshetra, India

Chanda Koustav L&T Infotech, Ranaghat, India

Chandra Swarniv Jadavpur University, Kolkata, India;
Department of Physics, Govt. General Degree College at Kushmandi, Dakshin Dinajpur, India;
Institute of Natural Sciences and Applied Technology, Kolkata, India

Chandrashekhar G. Department of Mathematics, Osmania University, Hyderabad, TG, India

Chatterjee Prasanta Department of Mathematics, Siksha Bhavana, Visva-Bharati, Santiniketan, Santiniketan, West Bengal, India

Chaube Mithilesh Kumar Dr. SPM IIIT Naya Raipur, C.G., Raipur, India

Chaudhuri Kripasindhu Department of Mathematics, Jadavpur University, Kolkata, India

Chukwuma Victor Iheanacho Institute of Radioelectronics, Riga Technical University, Riga, Latvia

Chuong Le Van Vinh University, Vinh city, Nghean, Vietnam

Cuong Ngo Tri Nam Systemtec JSC, Hanoi, Vietnam

Das Ankush Department of Electrical Engineering, JIS College of Engineering, Kalyani, West Bengal, India

Das Apurba Narayan Department of Mathematics, Alipurduar University, Alipurduar, West Bengal, India

Das Chinmay Department of Mathematics, Govt. General Degree College at Kushmandi, Dakshin Dinajpur, India;
Institute of Natural Sciences and Applied Technology, Kolkata, India

Das Debabrata Cooch Behar Panchanan Barma University, Cooch Behar, India

Das Rituparna Department of Mathematics, Pandu College, Guwahati, India

Das Satyajit Department of Mathematics, Adamas University, Kolkata, India

Das Susmita Cooch Behar Panchanan Barma University, Cooch Behar, India

Deb Biswajit Department of Mathematics, Sikkim Manipal Institute of Technology, Sikkim Manipal University, East-Sikkim, India

Dehingia Hirak Jyoti Department of Mathematics, Dibrugarh University, Dibrugarh, Assam, India

Deka P. N. Department of Mathematics, Dibrugarh University, Dibrugarh, Assam, India

Devesh M. School of Mechanical Engineering, Vellore Institute of Technology, Chennai, India

Dey Mondal Sukarna Department of Mathematics, Dr. B.C. Roy Engineering College, MAKAUT, Kolkata, West Bengal, India

Dey Debasish Department of Mathematics, Dibrugarh University, Dibrugarh, AS, India

Dey Dipankar Depatment of Computer Science and Technology, Global Institute of Science and Technology, Haldia, WB, India

Dhar Subham Cooch Behar Panchanan Barma University, Cooch Behar, WB, India

Donnarumma Davide Università di Torino, Torino, Italy

Doss M. Arockia Department of Chemistry, St. Joseph University, Dimapur, India

Dubey Balram Department of Mathematics, BITS Pilani, Pilani Campus, Pilani, Rajasthan, India

Dutta Dharitree Department of Physics, Anandaram Dhekial Phookan College, Nagaon, Assam, India

Fataf Nur Aisyah Binti Abdul Cyber Security Centre, National Defence University of Malaysia (NDUM), Kuala Lumpur, Malaysia

Félix-Beltrán O. Faculty of Electronics Sciences, Benemérita Universidad Autónoma de Puebla, Puebla, Mexico

Gandhi Rishu Department of Mathematics, Birla Institute of Technology and Science, Pilani, Rajasthan, India

Gangele Rajkumar Department of Mathematics and Statistics, Dr. Harisingh Gour Vishwavidyalaya, Sagar, Madhya Pradesh, India

Gangwar Sanu Kumar Department of Physics, Indian Institute of Technology, Guwahati, Assam, India

Gangwar Sonali Department of Physics, Indian Institute of Technology, Guwahati, Assam, India

García López Álvaro Universidad Rey Juan Carlos, Madrid, Spain

Gautam Anjan Department of Mathematics, Sikkim Manipal Institute of Technology, Sikkim Manipal University, East-Sikkim, India

Gautam Manjeet Singh Department of Physics, Government College, Jind, India

Ghosh Aditya Department of Mathematics, Adamas University, Kolkata, India

Ghosh Basudev Jadavpur University, Kolkata, India

Ghosh S. S. Indian Institute of Geomagnetism, Mumbai, India

Ghosh Sabyasachi Indian Institute of Technology Bhilai, Sejbahar, Raipur, Chhattisgarh, India

Gomathy G. Department of Mathematics, School of Advanced Sciences, Vellore Institute of Technology, Vellore, Tamilnadu, India

Goswami Damodar Prasad Asutosh Mookerjee Memorial Institute, Sivotosh Mookerjee Science Centre, Kolkata, India

Goswami Jyotirmoy Jadavpur University, Kolkata, India

Goswami K. S. Centre of Plasma Physics—Institute for Plasma Research, Assam, India

Goswami Pinkimani Department of Mathematics, University of Science and Technology Meghalaya, Ri-Bhoi, ML, India

Gowrisankar A. Department of Mathematics, School of Advanced Sciences, Vellore Institute of Technology, Vellore, Tamil Nadu, India

Gupta Ashvini Department of Mathematics, BITS Pilani, Pilani Campus, Pilani, Rajasthan, India

Gupta Nishi National Institute of Technology Silchar, Silchar, Assam, India

Hassan M. K. Department of Physics, University of Dhaka, Dhaka, Bangladesh

Ipatovs Aleksandrs Institute of Radioelectronics, Riga Technical University, Riga, Latvia

Jana Chandan Department of Applied Mathematics with Oceanology and Computer Programming, Vidyasagar University, Midnapore, West Bengal, India

Jana Soovoojeet Department of Mathematics, Ramsaday College, Amta, Howrah, India

Jeppu Yogananda Honeywell Technology Solutions, Hyderabad, Telangana, India

Joseph Subin P. Government Engineering College, Wayanad, Thalapuzha, Kerala, India

Joshi Sudhir University of Petroleum & Energy Studies, Dehradun, Uttarakhand, India

Kairi Rishi Raj Department of Mathematics, Cooch Behar Panchanan Barma University, Cooch Behar, India

Kalita Pankaj School of Energy Science and Engineering, Indian Institute of Technology Guwahati, Guwahati, India

Kar T. K. Department of Mathematics, Indian Institute of Engineering Science and Technology, Shibpur, Howrah, India

Karakoc Seydi Battal Gazi Department of Mathematics, Faculty of Science and Art, Nevsehir Haci Bektas Veli University, Nevsehir, Turkey

Karmakar Mithun Department of CSE, CIT Kokrajhar, Kokrajhar, Assam, India

Kaur Manveet Department of Physics, Guru Nanak Dev University, Amritsar, India

Kaur Rajneet Department of Physics, Guru Nanak Dev University, Amritsar, India

Kavitha L. Department of Physics, School of Basic and Applied Sciences, Central University of Tamil Nadu, Thiruvarur, Tamil Nadu, India;
The Abdus Salam International Centre for Theoretical Physics, Trieste, Italy

Khan Aftab Department of Electrical Engineering, JIS College of Engineering, Kalyani, West Bengal, India

Khandale K. Y. Department of Physics, Shivaji University, Kolhapur, India

Khanduri Umesh Department of Mathematics, Birla Institute of Technology and Science, Pilani, Rajasthan, India

Khare Shivani Department of Mathematics and Statistics, Dr. Harisingh Gour Vishwavidyalaya, Sagar, Madhya Pradesh, India

Khasim Ahamed National Institute of Technology Karnataka, Mangaluru, India

Khatua Anupam Department of Mathematics, Indian Institute of Engineering Science and Technology, Shibpur, Howrah, India

Kodi Raghunath Department of Humanities and Sciences (Mathematics), Bheema Institute of Technology and Science, Adoni, AP, India

Kostić Srđan Geology Department, Jaroslav Černi Water Institute, Belgrade, Serbia

Kulkarni Vinayak School of Energy Science and Engineering, Indian Institute of Technology Guwahati, Guwahati, India

Kumar Bose Goutam Department of Mechanical Engineering, Haldia Institute of Technology, Haldia, West Bengal, India

Kumar Dey Pabitra Department of Computer Applications, Dr. B.C. Roy Engineering College, MAKAUT, Durgapur, West Bengal, India

Kumar Jha Saket Department of Electronics and Communication Engineering, Sikkim Manipal Institute of Technology, Sikkim Manipal University, Majitar, Sikkim, India

Kumar Maiti Dilip Department of Applied Mathematics with Oceanology and Computer Programming, Vidyasagar University, Midnapore, West Bengal, India

Kumar Mishra Mohit Department of Electronics and Communication Engineering, Manipal University, Jaipur, India

Kumar Sarkar Ashis Department of Mathematics, Jadavpur University, Kolkata, India

Kumar Abhishek Department of Electrical Engineering, Manipal University Jaipur, Jaipur, Rajasthan, India;
Department of Aerospace Engineering, University of Petroleum and Energy Studies, Dehradun, India

Kumar Ajay Department of Mathematics and Humanities, Sardar Vallabhbhai National Institute of Technology, Surat, Gujarat, India

Kumar Anand Department of Physics, Chaudhary Ranbir Singh University, Jind, India

Kumar Ankit Department of Mathematics, BITS Pilani, Pilani, Rajasthan, India

Kumar Ayush Department of Electronics and Communication Engineering, Sikkim Manipal Institute of Technology, Sikkim Manipal University, Majitar, Sikkim, India

Kumar B. Rushi Department of Mathematics, SAS, Vellore Institution of Technology, Vellore, Tamil Nadu, India

Kumar Bhagwan Department of Mathematics and Statistics, Dr. Harisingh Gour Vishwavidyalaya, Sagar, Madhya Pradesh, India

Kumar Hitender Department of Physics, Government College for Women, Gharaunda, India

Kumar Santosh Dr. SPM IIIT Naya Raipur, C.G., Raipur, India

Kumar Vijil Department of Mathematics and Computing, Indian Institute of Technology (ISM), Dhanbad, Jharkhand, India

Kumar Vinod Department of Physics, Chaudhary Devi Lal University, Sirsa, India

Kumari Agin Department of Mathematics, Chaudhary Bansi Lal University, Bhiwani, India

Kurt Pehlivanoğlu Meltem Kocaeli University, Kocaeli, Turkey

Kushvah Badam Singh Department of Mathematics and Computing, Indian Institute of Technology (ISM), Dhanbad, Jharkhand, India

Lavanya C. Department of Physics, Periyar University, Salem, Tamil Nadu, India

Lepse Pratik V. Department of Mathematics, Lovely Professional University, Jalandhar, Phagwara, Panjab, India

Mahata Animesh Mahadevnagar High School, Maheshtala, Kolkata, West Bengal, India

Majee Suvankar Department of Mathematics, Indian Institute of Engineering Science and Technology, Shibpur, Howrah, India

Makrariya Akshara School of Advanced Science-Mathematics, VIT Bhopal University, Bhopal, Madhya Pradesh, India

Malakar Bidhan Department of Electrical Engineering, JIS College of Engineering, Kalyani, West Bengal, India

Malgol Amit National Institute of Technology Calicut, Kattangal, Kerala, India

Mali Prakash Chandra Department of Mathematics, Jadavpur University, Kolkata, India

Malik Anand Department of Physics, Chaudhary Ranbir Singh University, Jind, India

Mamun Miah M. Department of Mathematics, Khulna University of Engineering & Technology, Khulna, Bangladesh

Mani Rajan M. S. Department of Physics, University College of Engineering, Anna University, Ramanathapuram, India

Mani Awadesh Condensed Matter Physics Division, Indira Gandhi Centre for Atomic Research, Kalpakkam, Tamil Nadu, India

Manigandan R. School of Mechanical Engineering, Vellore Institute of Technology, Chennai, India

Maqbul Md. National Institute of Technology Silchar, Silchar, Assam, India

Mathur Kunwer Singh Department of Mathematics and Statistics, Dr. Harisingh Gour Vishwavidyalaya, Sagar, Madhya Pradesh, India

Medellín-Neri N. Faculty of Electronics Sciences, Benemérita Universidad Autónoma de Puebla, Puebla, Mexico

Meher Ramakanta Department of Mathematics and Humanities, Sardar Vallabhbhai National Institute of Technology, Surat, Gujarat, India

Meiyazhagan J. Department of Nonlinear Dynamics, Bharathidasan University, Tiruchirappalli, Tamil Nadu, India

Micheal David Raj Division of Mathematics, School of Advanced Sciences, Vellore Institute of Technology, Chennai, Tamil Nadu, India

Mishra Pankaj K. Department of Physics, Indian Institute of Technology Guwahati, Guwahati, Assam, India

Mishra Subodha Department of Physics, Siksha 'O' Anusandhan, Deemed to be University, Bhubaneswar, 751030, India

Mondal Biswajit Raja N.L. Khan Women's College (Autonomous), Midnapore, West Bengal, India

Mondal Dipankar Indian Institute of Technology Guwahati, Guwahati, Assam, India

Mondal Kajal Kumar Department of Mathematics, Cooch Behar Panchanan Barma University, Cooch Behar, West Bengal, India

Mondal Supriya MMI College of Nursing, Pachpedi Naka, Raipur, Chhattisgarh, India;
VY Hospital, Adjacent to Kamal Vihar (Sector 12), Raipur, Chhattisgarh, India

Mortuja Md. Golam Dr. SPM IIIT Naya Raipur, C.G., Raipur, India

Mukherjee Supriya Department of Mathematics, Gurudas College, Kolkata, West Bengal, India

Munoz-Pacheco J. M. Faculty of Electronics Sciences, Benemérita Universidad Autónoma de Puebla, Puebla, Mexico

Murmu Tanushree Department of Mathematics, Jadavpur University, Kolkata, India

Murthy B. S. N. Department of Mathematics, Aditya College of Engineering and Technology, Surampalem, Andhra Pradesh, India

Murugesan Regan Vel Tech Rangarajan Dr. Sagunthala R & D Institute of Science and Technology, Chennai, Tamil Nadu, India

Nag Amitava Department of CSE, CIT Kokrajhar, Kokrajhar, Assam, India

Nagaram Nagadevi Bala Vel Tech Rangarajan Dr. Sagunthala R & D Institute of Science and Technology, Chennai, Tamil Nadu, India

Natarajan C. Department of Mathematics, Srinivasa Ramanujan Centre, SASTRA Deemed University, Kumbakonam, India

Nath Ghosh Dipendra Controller of Examinations, Kazi Nazrul University, Asansol, West Bengal, India

Nath Surojit Department of Electrical Engineering, JIS College of Engineering, Kalyani, West Bengal, India

Nayak Sima Department of Mechanical Engineering, Indian Institute of Technology Guwahati, Guwahati, India

Nikam P. P. Department of Physics, Devchand College, Arjunnagar, Kolhapur, India

Oligo Jubert B. College of Teacher Education, Nueva Vizcaya State University, Bayombong, Philippines

Ozah Jintu Dibrugarh University, Dibrugarh, Assam, India

Pain Pritam Department of Mechanical Engineering, Haldia Institute of Technology, Haldia, West Bengal, India

Pal Chahal Rishi Department of Physics, Chaudhary Bansi Lal University, Bhiwani, India

Pal Charli Chinmayee Department of Physics, Siksha 'O' Anusandhan, Deemed to be University, Bhubaneswar, 751030, India

Pal Debjit Department of Electrical Engineering, JIS College of Engineering, Kalyani, West Bengal, India

Pal Nikhil Department of Mathematics, Siksha-Bhavana, Visva-Bharati University, Santiniketan, India

Palit Aniruddha Department of Mathematics, Surya Sen Mahavidyalaya, Siliguri, India

Panigoro Hasan S. Department of Mathematics, State University of Gorontalo, Bone, Bolango, Indonesia

Patidar Neelam School of Advanced Science-Mathematics, VIT Bhopal University, Bhopal, Madhya Pradesh, India

Patil S. D. Department of Physics, Devchand College, Arjunnagar, Kolhapur, Maharashtra, India

Patil S. S. Department of Physics, Vivekanand College, Kolhapur, India

Patra Maiti Atasi Directorate of Distance Education, Vidyasagar University, Midnapore, West Bengal, India

Pattanayak Solanki Department of Computer Science, Haldia Institute of Management, Haldia, WB, India

Paul Anindya Department of Mathematics, Cooch Behar Panchanan Barma University, Cooch Behar, West Bengal, India

Paul Debraj Department of Electrical Engineering, JIS College of Engineering, Kalyani, West Bengal, India

Paul Kamakhya North Eastern Hill University, Shillong, ML, India

Paul Niranjan Department of Mathematics, Siksha Bhavana, Visva-Bharati, Santiniketan, Santiniketan, West Bengal, India

Paul Subrata Department of Mathematics, Arambagh Govt. Polytechnic, Arambagh, West Bengal, India

Pavithra T. Department of Physics, Central University of Tamil Nadu, Thiruvarur, Tamil Nadu, India

Pawar V. S. Department of Physics, Rajer Ramrao Mahavidyalay, Jath, Maharashtra, India

Pikulins Dmitrijs Institute of Radioelectronics, Riga Technical University, Riga, Latvia

Pio Ferreira Claudia São Paulo State University (UNESP), Botucatu, Brazil

Poddar Nanda Cooch Behar Panchanan Barma University, Cooch Behar, WB, India

Poonam Department of Mathematics, BITS Pilani, Pilani, Rajasthan, India

Prabhu A. Department of Physics, Periyar University, Salem, India

Prabhu Department of Chemistry, Periyar University, Salem, Tamil Nadu, India

Pradhan Barsha Department of Mathematics, Sikkim Manipal Institute of Technology, Sikkim Manipal University, Majitar, Rangpo, East-Sikkim, India

Prakash Om Department of Aerospace Engineering, University of Petroleum and Energy Studies, Dehradun, Uttarakhand, India

Pramanik Sukanta Department of Statistics, Siliguri College, North Bengal University, Siliguri, West Bengal, India

Prasad Punam Kumari Department of Mathematics, Sikkim Manipal Institute of Technology, Sikkim Manipal University, East-Sikkim, India

Priyanka T. M. C. Department of Mathematics, School of Advanced Sciences, Vellore Institute of Technology, Vellore, Tamil Nadu, India

Punacha Shreyas Department of Physics, National Institute of Technology Karnataka, Surathkal, Mangalore, Karnataka, India

Raghavi K. Department of Physics, School of Basic and Applied Sciences, Central University of Tamil Nadu, Thiruvarur, India

Rahmi Emli Department of Mathematics, State University of Gorontalo, Bone, Bolango, Indonesia

Rajan M. S. Mani Anna University, University College of Engineering, Ramanathapuram, India

Rao Shiva Dibrugarh University, Dibrugarh, Assam, India

Rasappan Suresh University of Technology and Applied Sciences- Ibri, Sultanate of Oman, Ibri, Oman

Raut Santanu Department of Mathematics, Mathabhanga College, Coochbehar, India

Ravisankar R. Department of Physics, Indian Institute of Technology, Guwahati, Assam, India

Ravuri Mohana Ramana Department of Basic Science and Humanities (Mathematics), Narasaraopeta Engineering College, Narasaraopeta, AP, India

Ray Rajendra K. School of Basic Sciences, Indian Institute of Technology Mandi, Mandi, Himachal Pradesh, India

Rehman Attiq ul Department of Mathematical Sciences, BGSB University, Rajouri, India

Remi T. Department of Physics, Farook College University of Calicut, Kozhikode, Kerala, India

Rino Nelson N. Mechanical Engineering, Indian Institute of Information Technology, Design and Manufacturing, Kancheepuram, Chennai, India

Rout Anil Kumar School of Energy Science and Engineering, Indian Institute of Technology Guwahati, Guwahati, India

Roy Antara Asansol Institute of Engineering and Management-Polytechnic, Asansol, India

Roy Ashim Department of Mathematics, Alipurduar Univeristy, Alipurduar, India

Roy Banamali Department of Mathematics, Bangabasi Evening College, Kolkata, West Bengal, India

Roy Jyotirmoy Indian Institute of Engineering Science and Technology, Shibpur, Howrah, West Bengal, India

Roy Subrata Department of Mathematics, Cooch Behar Panchanan Barma University, Cooch Behar, India

Rushi Kumar B. Department of Mathematics, School of Advanced Sciences, Vellore Institute of Technology, Vellore, Tamilnadu, India

Saha Abhijit Department of Mechanical Engineering, Haldia Institute of Technology, Haldia, West Bengal, India

Saha Ashesh National Institute of Technology Calicut, Kattangal, Kerala, India

Saha Dipan Advanced Centre for Nonlinear and Complex Phenomena, Kolkata, India

Saha Gourab Cooch Behar Panchanan Barma University, Cooch Behar, India

Saha Sandip Department of Mathematics, Madanapalle Institute of Technology & Science, Madanapalle, Andhra Pradesh, India;
School of Advance Sciences (SAS) Mathematics, Vellore Institute of Technology Chennai, Chennai, Tamilnadu, India

Sahoo Niranjan Department of Mechanical Engineering, School of Energy Science and Engineering, Indian Institute of Technology Guwahati, Guwahati, India

Sahu P. K. Department of Mathematics, Government Shyama Prasad Mukharjee College, Sitapur, Chhattisgarh, India

Saikia Banashree Department of Mathematics, Dibrugarh University, Dibrugarh, Assam, India

Saini Nareshpal Singh Department of Physics, Guru Nanak Dev University, Amritsar, India

Sajan Department of Mathematics, BITS Pilani, Pilani, Rajasthan, India

Samanta Sabyasachi Department of Information Technology, Haldia Institute of Technology, Haldia, WB, India

Sapkota Niraj Department of Mathematics, Sikkim Manipal Institute of Technology, Sikkim Manipal University, Sikkim, India

Sardar Sankirtan Department of Mathematics, Guru Ghasidas Vishwavidyalaya, Bilaspur, India

Sarkar Jit Jadavpur University, Kolkata, India

Sarkar Kshirod Raja N.L. Khan Women's College (Autonomous), Midnapore, West Bengal, India

Sarkar Sudipta Department of Mathematics, Heritage Institute of Technology, Anandapur, Kolkata, West Bengal, India

Sarkar Swarup Department of Physics, Indian Institute of Technology Guwahati, Guwahati, Assam, India

Sarkar Tanay Department of Mathematics, Jadavpur University, Kolkata, India

Sarma Ridip Department of Mathematics, Assam Don Bosco University, Tapesia, Sonapur, Assam, India

Sasmal Sourav Kumar Department of Mathematics, BITS Pilani, Pilani, Rajasthan, India

Savapondit Santonu Department of Mathematics, Sikkim Manipal Institute of Technology, Sikkim Manipal University, Sikkim, India

Sebastian Anupama Department of Physics, National Institute of Technology Karnataka, Surathkal, Mangalore, Karnataka, India

Sekhar Roy Deb Department of Electrical Engineering, JIS College of Engineering, Kalyani, West Bengal, India

Sengupta Debajyoti Department of Electrical Engineering, JIS College of Engineering, Kalyani, West Bengal, India

Senthilvelan M. Department of Nonlinear Dynamics, Bharathidasan University, Tiruchirappalli, Tamil Nadu, India

Shajahan T. K. Department of Physics, National Institute of Technology Karnataka, Surathkal, Mangalore, Karnataka, India

Shanmugavelan S. Department of Mathematics, Srinivasa Ramanujan Centre, SASTRA Deemed University, Kumbakonam, India

Sharma Bhupendra K. Department of Mathematics, Birla Institute of Technology and Science, Pilani, Rajasthan, India

Sharry Physics Department, Guru Nanak Dev University, Amritsar, India

Shibin Raj C. P. National Institute of Technology Calicut, Kattangal, Kerala, India

Shilpi Physics Department, Guru Nanak Dev University, Amritsar, India

Shukla Manoj Kumar Lovely Professional University, Phagwara, Punjab, India

Sibeesh Puthiyapurayil Department of Physics, National Institute of Technology Karnataka Surathkal, Mangalore, India

Singh Chandraketu School of Sciences, Christ (Deemed to Be University) Delhi NCR, Ghaziabad, India

Singh Tamang Jitendra Department of Electronics and Communication Engineering, Sikkim Manipal Institute of Technology, Sikkim Manipal University, Majitar, Sikkim, India

Singhal Abhinav School of Sciences, Christ (Deemed to Be University) Delhi NCR, Ghaziabad, India

Singh Kuldeep Department of Physics, Guru Nanak Dev University, Amritsar, India;
Department of Mathematics, Khalifa University of Science and Technology, Abu Dhabi, UAE

Singh Madan Mohan Department of Basic Sciences & Social Sciences, North Eastern Hill University, Shillong, ML, India

Singh Ram Department of Mathematical Sciences, BGSB University, Rajouri, India

Singh Ram Mehar Department of Physics, Chaudhary Devi Lal University, Sirsa, India

Singh Ritesh Manipal University Jaipur, Jaipur, Rajasthan, India;
University of Petroleum & Energy Studies, Dehradun, Uttarakhand, India

Singla Sunidhi Department of Physics, Guru Nanak Dev University, Amritsar, India

Sinha Kaushik Department of Electrical Engineering, JIS College of Engineering, Kalyani, West Bengal, India

Sinha Sudipta Burdwan Raj College, University of Burdwan, Burdwan, India

Slathia Geetika Department of Physics, Guru Nanak Dev University, Amritsar, India

Solo Peter Department of Chemistry, St. Joseph University, Dimapur, India;
Department of Chemistry, St. Joseph's College (Autonomous), Jakhama, India

Sonal Deepa Department of Computer Science, V.K.S. University, Arrah, India

Sowmiya C. Department of Mathematics, SAS, Vellore Institution of Technology, Vellore, Tamil Nadu, India

Srinivas M A S Department of Mathematics, Jawaharlal Nehru Technological University, Hyderabad, Telangana, India

Srinivas M N Department of Mathematics, School of Advanced Sciences, Vellore Institute of Technology, Vellore, Tamilnadu, India

Subha P. A. Department of Physics, Farook College University of Calicut, Kozhikode, Kerala, India

Subramanian K. Department of Physics, SRM Institute of Science and Technology, Ramapuram Campus, Chennai, India

Subramanian Venkatesh National Institute of Technology Karnataka, Mangaluru, India

Sucu Derya Yildirim Faculty of Science and Art, Department of Mathematics, Nevsehir Haci Bektas Veli University, Nevsehir, Turkey

Sunny Geo Department of Physics, School of Basic and Applied Sciences, Central University of Tamil Nadu, Thiruvarur, India

Takale M. V. Department of Physics, Shivaji University, Kolhapur, Maharashtra, India

Takale P. T. Department of Physics, Shivaji University, Kolhapur, India

Talukdar Pranabendra Department of Mathematics, Alipurduar University, Alipurduar, West Bengal, India

Tamang Jharna Department of Mathematics, Sikkim Manipal Institute of Technology, Sikkim Manipal University, Majitar, East-Sikkim, India;
Department of Mathematics, Sikkim Alpine University, Kamrang, Namchi, South-Sikkim, India

Thilakavathy J. Department of Science and Humanities, Jerusalem College of Engineering, Chennai, India

Tiwari Rakhi Department of Mathematics, Babasaheb Bhimrao Ambedkar Bihar University, Muzaffarpur, India

Tiwari Rakshit Indian Institute of Technology Guwahati, Guwahati-781039, Assam, India

Tjukovs Sergejs Institute of Radioelectronics, Riga Technical University, Riga, Latvia

Tolga Sakallı Muharrem Trakya University, Edirne, Turkey

Urunkar T. U. Department of Physics, Shivaji University, Kolhapur, India

Valderama Julius S. College of Arts and Sciences, Nueva Vizcaya State University, Bayombong, Philippines

Varghese Steffy Sara Space and Planetrary Science Center, Khalifa University, Abu Dhabi, UAE

Varunkumar M. Department of Basic Sciences and Humanities, GMR Institute of Technology, Rajam, Andhra Pradesh, India;
Department of Mathematics, School of Advanced Sciences, VIT-AP University, Amaravati, Andhra Pradesh, India

Vasović Nebojša Department of Applied Mathematics and Informatics, Faculty of Mining and Geology, University of Belgrade, Belgrade, Serbia

Veni Saravana Department of Physics, Amirta College of Engineering and Technology, Erachakulam Campus, Nagercoil, India

Venkatalaxmi A. Department of Mathematics, Osmania University, Hyderabad, TG, India

Venturino Ezio Università di Torino, Torino, Italy

Zambrano-Serrano E. Facultad de Ingeniería Mecánica y Eléctrica, Universidad Autónoma de Nuevo León, San Nicolás de los Garza, Nuevo León, Mexico

Mathematical Modeling: Trends and Applications

Role of Additional Food in a Delayed Eco-Epidemiological Model with the Fear-Effect

Chandan Jana, Dilip Kumar Maiti, and Atasi Patra Maiti

Abstract This paper proposes an eco-epidemiological system with saturated incidence kinetics and a generalised Holling type-response function. Logistically growing prey species are partitioned into susceptible and infected prey. Predator-induced fear among prey populations suppresses the logistic growth rate and incidence rate. Additional food is supplied to predators to support them. Time-delay is incorporated to transform susceptible prey into infected prey. We derive the conditions for the existence, permanence, stability of all feasible equilibrium and the occurrence of Hopf bifurcation. Optimal control strategies are used for disease controlling by supplying additional food. Numerically, we verify analytical results and exhibit the system's dynamicity. Predator-induced fear lowers their size and switches an unstable system into a stable one. Also, an appropriate additional food supply to predators protects them from extinction and controls prey's infection.

Keywords Eco-epidemic · Fear · Additional food · Delay · Disease control · Chaos

1 Introduction

Eco-epidemiology is the intermixing of two biological fields: ecology which studies about population dynamics, and epidemiology which studies about infectious diseases in biological communities. In an epidemiological system, disease transmits from infected populations to susceptible ones due to their mutual coexistence and interaction. There are several disease transmission functions (incidence rate): the

C. Jana · D. Kumar Maiti (✉)
Department of Applied Mathematics with Oceanology and Computer Programming, Vidyasagar University, Midnapore 721102, West Bengal, India
e-mail: d_iitkgp@yahoo.com

A. Patra Maiti
Directorate of Distance Education, Vidyasagar University, Midnapore 721102, West Bengal, India

S. Banerjee and A. Saha (eds.), *Nonlinear Dynamics and Applications*,
Springer Proceedings in Complexity,
https://doi.org/10.1007/978-3-030-99792-2_60

law of mass action (βSI) [2], saturated incidence $\left(\frac{\beta SI}{1+\alpha I}\right)$ [9], and standard incidence $\left(\frac{\beta SI}{S+I}\right)$[6], etc. In ecology the prey-predator correlation is described by the term 'functional responses' which is the intake rate of a predator in unit time as a function of food density. Over the last decades, researchers have studied various 'functional response', like Holling type-II [7], Holling type-III [1], Crowley-Martin [5] to demonstrate prey-predator interaction.

Some recent studies [10, 11] established that due to fear of predation, prey individuals enhance their vigilance, give up their favourite food zone and habitation, migrate to a relatively low-risk region for foraging, and control their reproductive strategies accordingly. These survival strategies lower their birth rate, mutual contact and consequently decrease the infection rate. Samaddar et al. [10] explored the influence of predator-imposed fear in the prey-predator system associated with additional food to predators. They investigated that the fear together with additional food play an essential role in persisting a stable coexistence ecosystem.

Through literature reviews, to best of our knowledge, an eco-epidemiological system with (i) saturated incidence rate, (ii) Holling type-II functional response, (iii) predator-imposed fear among preys, (iv) incidence delay and (v) additional food supply to predators has not been studied yet. The main objectives of the present work are:

- To investigate how does the force of infection drive the population dynamicity?
- To illuminate the contribution of fear-effects to species survival and improvement of ecosystem stability?
- To observe whether additional food supply to predators can sustain their existence and control the disease of prey species?

This paper is arranged as follows: In Sect. 2, we formulate both non-delayed and delayed models. The well-posedness of the model is verified in Sect. 3. We analyse local stability and Hopf-bifurcation in Sect. 4. An attempt is made to control the disease in Sect. 5. In Sect. 6, we validate analytical results and investigate the system's dynamicity through numerical simulation. At last, the conclusion and significance of the work are presented in Sect. 7.

2 Mathematical Model Formulation

Some basic assumptions are taken into account to formulate the model.

- The model consists three subpopulations: susceptible prey ($S(t)$), infected prey ($I(t)$) and predator ($P(t)$) at any time t.
- Disease transmits from infected preys to susceptible individuals due to their coexistence by saturated incidence rate $\frac{\beta SI}{1+\alpha I}$ where β represents the infection force and α is the effect of inhibition.
- Prey-predator interaction is governed by a generalised Holling type II functional response.

Table 1 Units and description of the used symbols

Parameters	Biological meaning	Unit
r	Birth rate of prey population	Time^{-1}
k	Environmental carrying capacity for total prey	Biomass
k_1, k_2	Cost of fear	Biomass^{-1}
α	Inhibition effect	Time^{-1}
β	Force of infection	Time^{-1}
m_1	Rate of predation of susceptible prey	Time^{-1}
m_2	Rate of predation of infected prey	Time^{-1}
a_1, a_2	Handling time	Time^{-1}
e_1, e_2 and e_3	Conversion efficiency of P on S, I and Λ, respectively	Constant
Λ	Additional food	Constant
φ	Quality of additional food	Constant
n	Quantity of additional food	Biomass
d_1	Death rate of infected prey	Time^{-1}
d_2	Death rate of predator	Time^{-1}

- Due to fear of predation, fear among prey populations suppresses the logistic growth rate and incidence rate.
- The non-reproducing additional food, proportionate to the density of predators, is provided to predators at constant rate to survive them from their extinction.

Considering all these biological factors, proposed model is expressed as follows:

$$\frac{dS}{dt} = \frac{rS}{1+k_1P}\left(1-\frac{S+I}{k}\right) - \frac{\beta SI}{(1+\alpha I)(1+k_2P)} - \frac{m_1SP}{1+a_1S+a_2I+\varphi n\Lambda}$$
$$= \bar{f_1}(S, I, P), \tag{1}$$

$$\frac{dI}{dt} = \frac{\beta SI}{(1+\alpha I)(1+k_2P)} - \frac{m_2IP}{1+a_1S+a_2I+\varphi n\Lambda} - d_1I = \bar{f_2}(S, I, P), \tag{2}$$

$$\frac{dP}{dt} = \frac{(e_1m_1S+e_2m_2I+e_3n\Lambda)\,P}{1+a_1S+a_2I+\varphi n\Lambda} - d_2P = \bar{f_3}(S, I, P) \tag{3}$$

where $r, k, n, a_1, a_2, d_1, d_2, e_1, e_2, e_3, k_1, k_2, m_1, m_2, \alpha, \beta, \varphi$ and Λ are positive constants (ref. Table 1) and $S(0) = s_0 > 0$, $I(0) = i_0 > 0$, $P(0) = p_0 > 0$. A schematic representation of interaction among S, I and P is presented in Fig. 1.

Let, τ be the time lag to transform the susceptible individual into an infected one until the disease spreads to his body up to a certain level. Then, the delayed system is governed by the following equations:

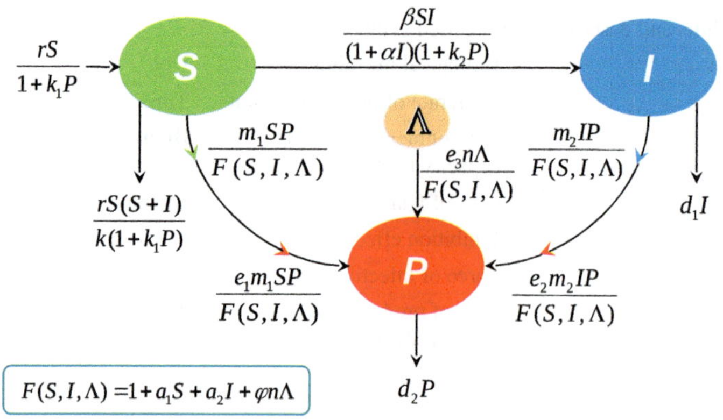

Fig. 1 Schematic view of the dynamical interactions of S, I and P in the propose model

$$\frac{dS}{dt} = \frac{rS}{1+k_1P}\left(1-\frac{S+I}{k}\right) - \frac{\beta SI(t-\tau)}{\{1+\alpha I(t-\tau)\}(1+k_2P)} - \frac{m_1SP}{1+a_1S+a_2I+\varphi n\Lambda} = f_1(S, I, P), \tag{4}$$

$$\frac{dI}{dt} = \frac{\beta SI(t-\tau)}{\{1+\alpha I(t-\tau)\}(1+k_2P)} - \frac{m_2IP}{1+a_1S+a_2I+\varphi n\Lambda} - d_1I = f_2(S, I, P), \tag{5}$$

$$\frac{dP}{dt} = \frac{(e_1m_1S+e_2m_2I+e_3n\Lambda)P}{1+a_1S+a_2I+\varphi n\Lambda} - d_2P = f_3(S, I, P) \tag{6}$$

where $S(0) = \vartheta_1(\xi) > 0$, $I(0) = \vartheta_2(\xi) > 0$, $P(0) = \vartheta_3(\xi) > 0$, $\xi \in [-\tau, 0]$ and $\vartheta_j \in C\left([-\tau, 0] \to \mathbb{R}^+\right)$ for j=1, 2, 3.

3 Basic Mathematical Results

Theorem 1 *The system (Eqs. 1–3) is invariant in the positive octant of* $\mathbb{R}_3$ *(i.e.,* $\mathbb{R}_3^+$*).*

Theorem 2 *All the solutions* (S, I, P) *of the system (Eqs. 1–3), starting from* $\mathbb{R}_3^+$*, are uniformly bounded in the area* $\Delta = \{(S, I, P) \in \mathbb{R}_3^+ : 0 < S+I \le k, 0 < P \le \frac{\phi}{\eta} - k\}$*, where* $\phi = \frac{rk}{4}\left(1+\frac{\eta}{r}\right)^2$ *and* $0 < \eta \le min\left\{d_1, d_2 - \frac{e_3}{\varphi}\right\} > 0$*, provided* $d_2 > \frac{e_3}{\varphi}$.

4 Equilibrium Points and Stability Analysis

4.1 Equilibrium Points

The stagnant state $\left(\bar{f}_i(S, I, P) = 0 \text{ for } i = 1, 2, 3\right)$ of the system (Eqs. 1–3) yield the following ecologically meaningful equilibria:

(i) Trivial equilibria $E_0(0, 0, 0)$ which always exists.

(ii) Disease-free and predator-free equilibrium $E_1(k, 0, 0)$ exists if $\beta \leq d_1$ and $\frac{e_1 m_1 k}{1+a_1 k+\varphi n \Lambda} \leq d_2$.

(iii) Disease-free equilibria $E_2(\hat{S}, 0, \hat{P})$ where $\hat{S} = \frac{d_2(1+n\varphi\Lambda)-e_3 n\Lambda}{e_1 m_1 - a_1 d_2}$ (provided $e_1 m_1 > a_1 d_2$ and $d_2(1 + n\varphi\Lambda) > e_3 n\Lambda$) and $\hat{P}$ is the positive root of the following equation: $k k_1 m_1 P^2 + k m_1 P - \left(k - \hat{S}\right)\left(1 + a_1 \hat{S} + n\varphi\Lambda\right) = 0$. Since $k > \hat{S}$, using Descartes' rule of signs, this equation must have unique positive root of P, say, $\hat{P}$.

(iv) Predator-free equilibria $E_3(S', I', 0)$ where $S' = \frac{d_1(1+\alpha I')}{\beta}$ and I' is obtained from the following equation: $(d_1\alpha + \beta)r\alpha I^2 + \{r(d_1\alpha + \beta) + \beta^2 k - \alpha r(\beta k - d_1)\}I - (\beta k - d_1) = 0$. For the existence of E_3, we must have $\beta k > d_1$. Using Descartes' rule of signs, it is deduced that this equation possesses unique positive root of I, say, I'.

(v) Coexistence equilibria $E_4(S^*, I^*, P^*)$: (I^*, P^*) is the intersection point of null-clines: $\Phi_1(I, P) = \frac{r}{1+k_1 P}\left(1 - \frac{f(I)+I}{k}\right) - \frac{m_1 P}{1+a_1 f(I)+a_2 I+\varphi n\Lambda} - \frac{\beta I}{(1+\alpha I)(1+k_2 P)}$ and $\Phi_2(I, P) = \frac{\beta f(I)}{(1+\alpha I)(1+k_2 P)} - \frac{m_2 P}{1+a_1 f(I)+a_2 I+\varphi n\Lambda} - d_1$. Here, $S = \frac{(a_2 d_2 - e_2 m_2)I}{e_1 m_1 - a_1 d_2} + \frac{d_2+(d_2\varphi - e_3)n\Lambda}{e_1 m_1 - a_1 d_2} = f(I)$, and consequently one can compute S^* using the value of I^*.

4.2 Stability Analysis

The characteristic of the delayed system (Eqs. (4–6)) at E_4 is

$$\lambda^3 + M_1\lambda^2 + M_2\lambda + M_3 + e^{-\lambda\tau}\left(N_1\lambda^2 + N_2\lambda + N_3\right) = 0 \tag{7}$$

where $M_1 = -(m_{11} + m_{22})$, $M_2 = m_{11}m_{22} - m_{12}m_{21} - m_{13}m_{31} - m_{23}m_{32}$, $M_3 = m_{11}m_{23}m_{32} - m_{13}m_{21}m_{32} + m_{13}m_{22}m_{31} - m_{12}m_{23}m_{31}$, $N_1 = -n_{22}$, $N_2 = m_{11}n_{22} - m_{21}n_{12}$ and $N_3 = m_{31}(m_{13}n_{22} - m_{23}n_{12})$.

Non-delay system ($\tau = 0$): The characteristic Eq. (7) becomes $\lambda^3 + L_1\lambda^2 + L_2\lambda + L_3 = 0$ where $L_i = M_i + N_i$ $(i = 1, 2, 3)$. By applying Routh Hurwitz stability criterion, we provide the stability condition at E_4 in the Theorem 3.

Theorem 3 *The interior equilibria E_4 locally asymptotically stable (LAS) if the conditions are fulfilled: (i) $L_1 > 0$, $L_3 > 0$ and (ii) $L_1L_2 - L_3 > 0$.*

Lemma 1 *Trivial equilibrium E_0 is always saddle.*

Lemma 2 *Axial equilibria E_1 and predator-free equilibria E_3 are always locally asymptotic stable.*

Lemma 3 *Disease-free equilibria $E_2(\hat{S}, 0, \hat{P})$ is LAS if $r\hat{S}(1 + a_1\hat{S} + \varphi n\Lambda) > a_1 m_1 k\hat{S}\hat{P}(1 + k_1\hat{P})$.*

Delayed system($\tau > 0$): Here, we derive the critical value of delay parameter (τ) at which delayed system (Eqs. (4–6)) switch its dynamical behaviour.

Theorem 4 *The delayed system is asymptotically stable around E_4 for $\tau < \tau_c$ and undergoes through Hopf bifurcation at $\tau = \tau_c$ which is given by*

$$\tau_c = \frac{1}{\hat{\omega}} \arccos\left(\frac{L_4L_6 + L_5L_7}{L_4^2 + L_5^2}\right) + \frac{2j\pi}{\hat{\omega}}, \quad j = 0, 1, 2.... \tag{8}$$

where $L_4 = N_1\hat{\omega}^2 - N_3$, $L_5 = -N_2\hat{\omega}$, $L_6 = M_3 - M_1\hat{\omega}^2$ and $L_7 = M_2\hat{\omega} - \hat{\omega}^3$.

4.3 Hopf Bifurcation

Theorem 5 *Taking the force of infection (β) as bifurcation parameter, non-delay system undergoes through Hopf bifurcation under the following conditions:*

(i) $L_1 > 0$ *and* $L_3 > 0$ *at* $\beta = \beta_c$,
(ii) $L_1L_2 - L_3 = 0$ *at* $\beta = \beta_c$ *(for pair of purely imaginary eigenvalues),*
(iii) $[L_1(\beta_c)L_2(\beta_c)]' \neq L_3'(\beta_c)$,

where L_i's ($i = 1, 2, 3$) are mentioned formerly.

5 Implementation of Optimal Control to Disease

We apply control on the quality (φ) and quantity (n) of additional food to minimize the infection and finally to eradicate diseases from system. Due to seasonal variation of contact rate, we consider φ and n as time dependent. Let us consider objective functional J as

$$J = \min_{\varphi,n} \int_{t_0}^{t_f} \{I + \Upsilon_1\varphi^2(t) + \Upsilon_2 n^2(t)\}dt \tag{9}$$

subject to the Eqs. (1–3) and the parameters t_0 and t_f are beginning and end time, respectively. We have to optimize J. Here, Υ_1 and Υ_2 are weights related with controls φ and n, respectively. For the optimal control (φ^*, n^*), we have $J(\varphi^*, n^*) = \min_{\varphi,n\in\Delta} J(\varphi, n)$ where $\Delta = \{(\varphi, n) : 0 \leq \varphi(t) \leq M_\varphi,\ 0 \leq n(t) \leq M_n,\ t \in [t_0, t_f]\}$

is the measurable set. M_φ and M_n represent the respective upper bound of controls φ and n, respectively.

To solve the optimal control problem (Eq. (9)), let us consider the Lagrangian $L = I + \Upsilon_1\varphi^2 + \Upsilon_2 n^2$ which is to be minimized. Then, the Hamiltonian of the system is $H = L + \gamma_1\frac{dS}{dt} + \gamma_2\frac{dI}{dt} + \gamma_3\frac{dP}{dt}$ where the adjoint variables γ_i $(i = 1, 2, 3)$ can be computed by solving the system of Eq. (10):

$$\dot{\gamma}_1 = -\frac{\partial H}{\partial S}, \ \dot{\gamma}_2 = -\frac{\partial H}{\partial I} \text{ and } \dot{\gamma}_3 = -\frac{\partial H}{\partial P} \tag{10}$$

satisfying transversality conditions: $\gamma_i(t_f) = 0$ for $i = 1, 2, 3$. Let $\bar{\gamma}_1, \bar{\gamma}_2, \bar{\gamma}_3$ be the solutions of the system (Eq. 10) and $(\bar{S}, \bar{I}, \bar{P})$ as optimum value of (S, I, P).

Theorem 6 *[3, 4] There is an optimal control* (φ^*, n^*) *such that* $J(I(t), \varphi^*(t), n^*(t)) = \min_{\varphi,n} J(I(t), \varphi(t), n(t))$ *subject to the Eqs. (1–3).*

We use Pontryagin's Maximum Principle [8, 9] to prove the Theorem 6.

Theorem 7 *Over the region* Δ, *values of optimal control pair* (φ^*, n^*) *which minimizes* J *is given by* $\varphi^* = \max\{0, \min(\hat{\varphi}, M_\varphi)\}$ *and* $n^* = \max\{0, \min(\hat{n}, M_n)\}$ *where* $\hat{\varphi}$ *and* $\hat{n}$ *are provided later.*

Proof According to the optimality condition, we have $\frac{\partial H}{\partial \varphi} = 0$ when n is fixed and $\frac{\partial H}{\partial n} = 0$ when φ is fixed. These equations yield least positive real roots, say, $\hat{\varphi}$ and $\hat{n}$. Again, $0 \le \varphi(t) \le M_\varphi$ and $0 \le n(t) \le M_n$ for $t \in [t_0, t_f]$. So, we have $\varphi^* = \max\{0, \min(\hat{\varphi}, M_\varphi)\}$ and $n^* = \max\{0, \min(\hat{n}, M_n)\}$, and consequently the optimized J.

6 Numerical Simulation

We perform numerical simulation to validate the analytical results and observe dynamical behaviour of the system. A set of parameters' values are considered, based on idea related to the sensitivity of parameters, as $Z = \{r = 1,\ k = 5,\ a_1 = 1,\ a_2 = 0.9,\ d_1 = 0.15,\ d_2 = 0.1,\ e_1 = 0.5,\ e_2 = 0.4,\ e_3 = 0.15,\ k_1 = 0.15,\ k_2 = 0.02,\ m_1 = 0.4,\ m_2 = 0.5, \alpha = 0.6,\ \beta = 2,\ \varphi = 1,\ n = 0.65,\ L = 1\}$.

6.1 Non-delay System

Evidently, for the set Z two nullclines $\Phi_1(I, P) = 0$ and $\Phi_2(I, P) = 0$ intersect uniquely at $(0.4517, 0.5754)$ in the IP-plane (ref. Fig. 2a). Then, we have $S^* = 0.1788$, i.e. the coexistence equilibria $E_4(0.1788, 0.4517, 0.5754)$ exists (ref. Fig. 2b). At E_4, we have the eigenvalues as $-0.0297 + 0.4295i$, $-0.0297 - 0.4295i$,

-0.0029, i.e., real part of all the eigenvalues are negative. Moreover, all the quantities $L_1 = 0.0560$, $L_2 = 0.0008$ and $L_1 L_2 - L_3 = 0.0112$ are positive at E_4 (i.e., all the criterias of Theorem 3 are fulfilled). Therefore, E_4 is asymptotically stable for non-delay system which is illuminated in Fig. 2d.

In Fig. 3, an attempt is made to illuminate system behaviour with the variation of force of infection β considering three cases: (a–c) absence of fear ($k_1 = 0, k_2 = 0$) and additional food ($\Lambda = 0$), (d–f) absence of fear ($k_1 = 0, k_2 = 0$) and the presence of additional food ($\Lambda = 1$), and (g–i) presence of fear ($k_1 = 0.15, k_2 = 0.02$) and additional food ($\Lambda = 1$). For these three situations, corresponding Hopf-points are $\beta_c = 1.35,\ 2.1$ and 1.7, respectively, i.e., at those points system's stability switches from unstable to stable one. It is seen that in the absence of fear ($k_1 = 0, k_2 = 0$) and additional food ($\Lambda = 0$), predator extinct beyond $\beta = 1.55$. But, in the presence of additional food (ref. Fig. 3f), predators extinct beyond $\beta = 2.4$. Biologically speaking, an additional food supply to predators can survive themselves from its extinction

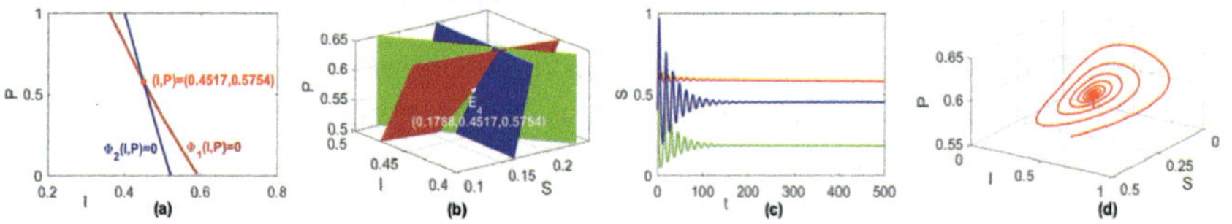

Fig. 2 **a** Existence of the intersection point of the nullclines $\Phi_1(I, P) = 0$ and $\Phi_2(I, P) = 0$; **b** existence of unique $E_4(0.1788, 0.4517, 0.5754)$; **c** time histories of populations; **d** stable focus for the parameter set Z of system (Eqs. 1–3)

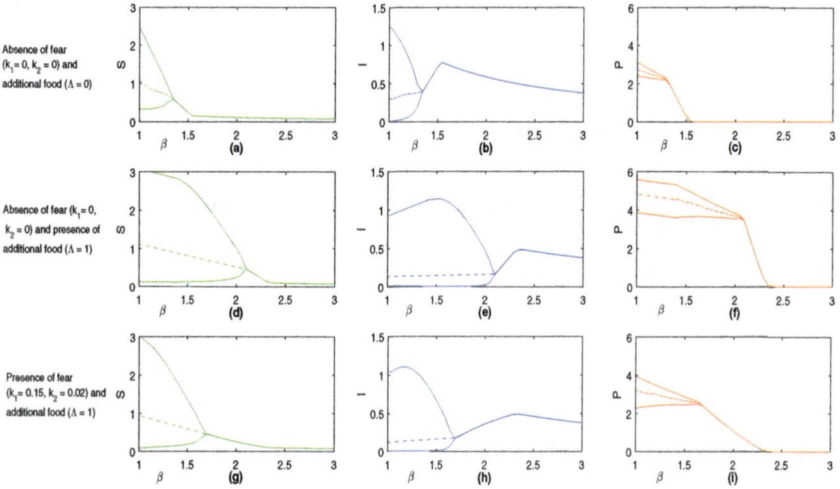

Fig. 3 Bifurcation diagram against force of infection β for the parameter set Z. **a–c** in the absence of fear ($k_1 = 0, k_2 = 0$) and additional food ($\Lambda = 0$), **d–f** in the absence of fear ($k_1 = 0, k_2 = 0$) and the presence of additional food ($\Lambda = 1$), and **g–i** in the presence of fear ($k_1 = 0.15, k_2 = 0.02$) and additional food ($\Lambda = 1$)

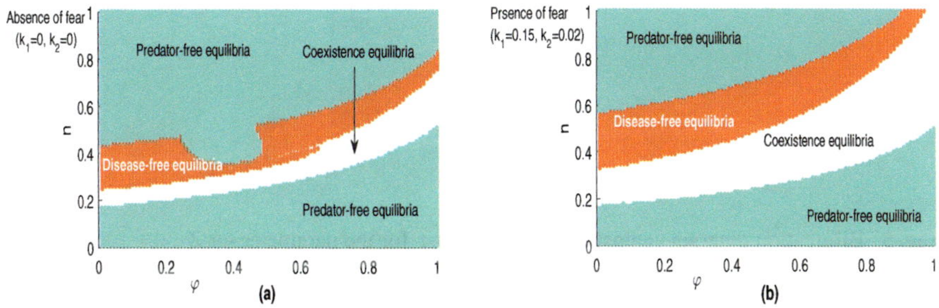

Fig. 4 Existence of various equilibria in $\varphi\eta$-plane

on a large scale of β, relatively. Further, incorporating fear-effect (ref. Fig. 3g–i), it can be seen that system becomes stable on a relatively large scale of β. Hence, fear-effect is efficient to make a sustainable stable ecosystem. It may be mentioned that a certain level of fear is fruitful to survive all the populations. But, a level of fear may eradicate predators from the system. Evidently, infected populations increase up to a certain level of increasing β. After that, all the populations decrease noticeably with a further increase in β. This is biologically reasonable because the density of the I-population attains a maximum value at a threshold β. Thereafter, the system declines to increase the density of infected population as S-populations decrease with increasing β.

In Fig. 4, we present various equilibria in φn-plane in the presence or absence of fear. One may observe that both quality (φ) and quantity (n) of additional food must be a reasonable amount for the existence of all the populations; otherwise, predators may become extinct from the system. Also, by supplying suitable quality (φ) and quantity (n) of additional food, we make a system disease-free as seen in Fig. 4. We use Pontryagin's Maximum Principle to find the optimum value of (φ, n) for disease controlling by supplying additional food. Moreover, the presence of the fear-effect increase the region of existence of interior equilibria.

6.2 Delayed System

Taking τ as a bifurcation parameter, the dynamical behaviour of the delayed system is explored in Fig. 5 with the help of the bifurcation diagram. The delayed system is stable for $\tau < 0.72$. For $\tau > 0.72$, the delayed system's exhibits complex dynamics, discussed in Fig. 5. It is seen that when τ crosses the critical values, steady-state values separated into maximum and minimum of the periodic oscillatory solution, and coexistence steady-state becomes unstable, i.e., Hopf bifurcation occurs at point of separation. For a fixed τ, the occurrence of too many local maxima/ minima indicates that the system performs chaotic behaviour. However, during the chaotic situation, populations take successive values randomly occupying a fixed range of correspond-

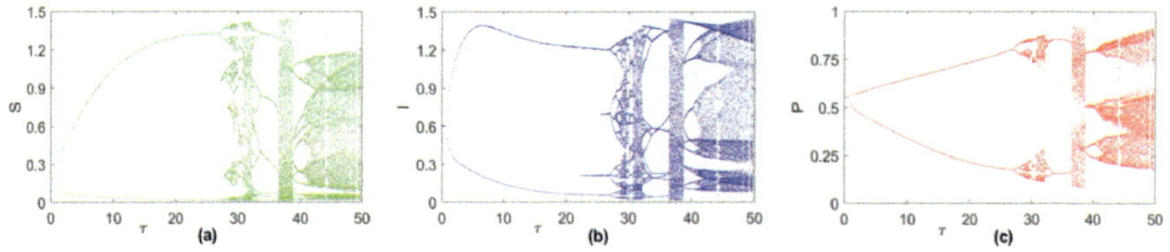

Fig. 5 Bifurcation diagram against delay parameter τ for the parameter set Z

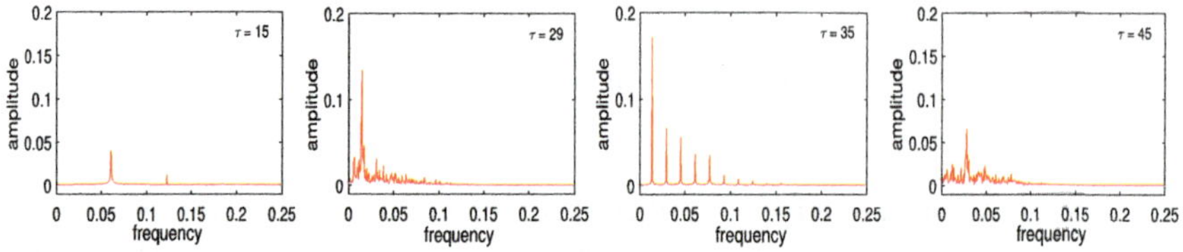

Fig. 6 Spectra of instantaneous S- population for several fixed τ for the parameter set Z

ing populations , similar to a situation of $sin(\frac{1}{x})$ as x gets closer to zero. It is seen that the system is periodic for $\tau \in [0.72, 27.04] \cup [33.57, 36.75] \cup [39.25, 40.53]$, otherwise chaotic. This complexity in the system's dynamics can be verified from Fig. 6. A finite number of the spectral peak corresponds to periodic behaviour, while no definite spectral peak exposes system's chaotic behaviour.

7 Conclusion

Here, we have proposed and analysed an eco-epidemiological system with infection among prey species, and additional food is provided to predators. Fear-effect is incorporated as a reducing factor in logistic growth of prey species and the disease transmission function. Also, we consider incidence delay to make the system biologically more realistic.

The numerical analyses is executed extensively to investigate the potentiality that cost of fears and additional supplied food can play a pivotal role in the system. It has been observed that a suitable choice of additional food may eradicate the disease from the system and help the predators to survive. Also, an adequate level of fear transforms an unstable system into a stable one. The system exhibits rich dynamics in the presence of incidence delay. Hence, a reasonable amount of fear-effect and additional food is essential to survive all the populations and to make a sustainable stable ecosystem. Our studies reveal a non-chemical approach to control disease in an eco-epidemiological system.

Acknowledgements Author Dilip K. Maiti acknowledges supports given by the DST-FIST, INDIA (Sanction Order No.: SR/FST/MS-1/2018/21(C) dated-13/12/2019) for upgradation of research facility at the departmental level.

References

1. Banerjee, R., Das, P., Mukherjee, D.: Global dynamics of a holling type-iii two prey-one predator discrete model with optimal harvest strategy. Nonlinear Dyn. **99**(4), 3285–3300 (2020)
2. Greenhalgh, D., Khan, Q.J., Al-Kharousi, F.A.: Eco-epidemiological model with fatal disease in the prey. Nonlinear Anal. Real World Appl. **53**, 103072 (2020)
3. Kar, T., Ghorai, A., Jana, S.: Dynamics of pest and its predator model with disease in the pest and optimal use of pesticide. J. Theoret. Biol. **310**, 187–198, 103072 (2012)
4. Lukes, D.L.: Differential Equations: Classical to Controlled (1982)
5. Maiti, A.P., Jana, C., Maiti, D.K.: A delayed eco-epidemiological model with nonlinear incidence rate and crowley-martin functional response for infected prey and predator. Nonlinear Dyn. **98**(2), 1137–1167, 103072 (2019)
6. Maji, C., Kesh, D., Mukherjee, D.: The effect of predator density dependent transmission rate in an eco-epidemic model. Differ. Equ. Dyn. Syst. **28**(2), 479–493, 103072 (2020)
7. Panday, P., Samanta, S., Pal, N., Chattopadhyay, J.: Delay induced multiple stability switch and chaos in a predator-prey model with fear effect. Math. Comput. Simul. **172**, 134–158, 103072 (2020)
8. Pontryagin, L.S.: Mathematical Theory of Optimal Processes. CRC Press (1987)
9. Sahoo, B.: Role of additional food in eco-epidemiological system with disease in the prey. Appl. Math. Comput. **259**, 61–79, 103072 (2015)
10. Samaddar, S., Dhar, M., Bhattacharya, P.: Effect of fear on prey–predator dynamics: exploring the role of prey refuge and additional food. Chaos Interdiscip. J. Nonlinear Sci. **30**(6), 063129 (2020)
11. Sha, A., Samanta, S., Martcheva, M., Chattopadhyay, J.: Backward bifurcation, oscillations and chaos in an eco-epidemiological model with fear effect. J. Biolog. Dyn. **13**(1), 301–327, 103072 (2019)

Impact of Predator Induced Fear in a Toxic Marine Environment Considering Toxin Dependent Mortality Rate

Dipesh Barman, Jyotirmoy Roy, and Shariful Alam

Abstract The concentration of harmful toxins in the marine ecosystem coming from different external sources is increasing day by day and becomes a serious threat to living organisms. On the other hand, in the field of predator-prey interactions, one main aspect which has been neglected for decades is predator-induced fear to prey species that affects reproduction rate of prey individuals. Keeping in mind both the factors, a mathematical model has been proposed by incorporating predator-induced fear in that toxic ecosystem. It has been observed that the system undergoes Hopf-bifurcation with respect to both the parameters associated with fear factor and toxicity. Also, there is a complex relationship between these two parameters and fear factor plays a significant role in predator extinction. All the analytical findings have been verified through numerical simulations by considering appropriate hypothetical parameter sets.

Keywords Fear effect · Toxicity · Hopf-bifurcation · Population density

1 Introduction

Till decades one of the most ignoring fact in predator-prey interactions is to avoid the indirect effect of predator induced fear to prey species. Most of the researchers formulated predator-prey model by taking into consideration the direct predation of prey species and accordingly analyze the system dynamics. But, in 2011, Zanette et al. [16] made an experiment to song sparrow by supplying their predator's vocal cues to some of the song sparrow population while the remaining others do not experience any vocal cues of their predators. However, it is observed that, all the song sparrow populations who suffered from predator cues have shown a lessen activity in reproduction by 40% as compared to the population who did not receive

D. Barman (✉) · J. Roy · S. Alam
Indian Institute of Engineering Science and Technology, Shibpur, Howrah 711103, West Bengal, India
e-mail: dipeshrs2018@gmail.com

S. Banerjee and A. Saha (eds.), *Nonlinear Dynamics and Applications*,
Springer Proceedings in Complexity,
https://doi.org/10.1007/978-3-030-99792-2_61

any disturbance of predator's sound. Influencing by this phenomenon, Wang et al. [14] formulated a mathematical model and observed that predator induced fear has an immense impact on system dynamics. Later on many researchers [1–3, 13, 15, 17] modified predator-prey model by considering this fear effect and analyzes them accordingly (for more interest one may read the references therein).

Marine pollution happens due to the entry of harmful chemicals, metals, etc., from different point and non point sources into the sea water. The vast majority of marine contamination comes from land. Air contamination is likewise a contributing component via carting away iron, sulfur, nitrogen, silicon, carbonic acid, pesticides or residue particles into the ocean [5]. The point source pollution includes entry of harmful particles from a easily identified source while non point source contamination describes pollutants coming from agricultural activities, wind-blown debris, and dust. At the point when pesticides are joined into the marine environment, they immediately become consumed into marine food webs and as a consequence various types of illness such as disease, mutation, tissue related problems, decrease in reproduction, etc., begin to occur in marine life [7, 9, 11, 12]. So, due to toxicity many marine species suffered from increasing mortality rate. Not only this toxin harms marine species, it also has an immense impact on the higher tropic level in food web including sea birds and human being who consume this species [6].

In recent days, a lot of attention by researchers was paid to address the impact of toxicants in environment. In this regard, Hallam and Clark [8] studied the first order kinetics of a population in the presence of toxicity. After that, Hallam and De Luna [9] investigated a model considering toxins from both the environmental and food chain pathways. Later on, Chattopadhyay [4] performed a study on two species competing with each other and observed that all the populations persist with the help of toxicants. Again, Huang et al. [10] demonstrated through a research study of a prey-predator model with the influence of environmental toxicants where both prey and predator are exposed to the toxicants simultaneously. However, in this study, we have made an attempt to realize the effect of toxic substance in marine environment in the presence of predator induced fear. Here, we have made an attempt in exploring the role of predator exerted fear on prey population where both the species suffer from an increased mortality rate due to toxicity of water in marine ecosystem. The motivation of this study includes to explore the role of fear effect and toxicity in system dynamics in the presence of external toxin sources. Not only this thing, how toxicity and fear level are related to each other in system dynamics is also an interest of this study. This article has been organized in several sections such as: Sect.2 describes about the formation of model system (1); Sect.3 verifies the well-posedness of system (1) while Sect.4 manages to analyse local stability around different equilibrium points. The findings of this system have been obtained in Sect. 5 with the help of MATLAB & MATCONT. Finally, this article ends with Sect. 6 as conclusion.

2 Model Formulation

Here, we have formulated a predator-prey model in (1) by considering predator induced fear to prey population in a toxic marine ecosystem as follows:

$$\begin{cases} \dfrac{dx}{dt} = \dfrac{rx}{1+fy} - d_1 x - mx^2 - \dfrac{\alpha xy}{1+bx} - \dfrac{e\gamma T}{1+e\gamma T}x, \\ \dfrac{dy}{dt} = \dfrac{\beta xy}{1+bx} - d_2 y - \dfrac{e\gamma T}{1+e\gamma T}y, \\ \dfrac{dT}{dt} = A - aT - \gamma(x+y)T, \end{cases} \tag{1}$$

with initial conditions

$$x(0) > 0, \;\; y(0) > 0, T(0) > 0, \tag{2}$$

where x, y and T respectively denote the density of prey population, predator population and harmful toxin concentration at any time t. The above model (1) has been constructed based on some assumptions such as

(i) In the absence of predator population, prey individuals grow logistically with growth rate r and natural mortality rate d_1. Furthermore, prey individuals engage into a clash among themselves at a rate m for food resources.

(ii) Predator population consumes prey individuals according to Holling type -II functional response at a rate α and predator species get benefited by reproducing new offsprings at a rate β from this food consumption. They die naturally at a rate d_2.

(iii) The birth rate of prey species reduces due to the fear exerted from predator species according to a function $\phi(f, y) = \dfrac{1}{1+fy}$ as proposed by Wang et al. [14]. One can read the detailed biological assumptions for constructing this fear function described in [14].

(iv) The harmful toxins for both species are coming from various types of man made external sources like industries, households, pesticides used in agriculture, etc. at a constant rate A. It is assumed that toxin declines naturally or toxic materials in marine system have been removed at a rate a due to different types of government initiatives or awareness. Furthermore, it is assumed that toxin concentration reduces due to the interaction between toxin and both the species at a rate γ.

(v) It is further assumed that due to the interaction with toxin, both the species die out from the system and this death rate depends on the concentration of toxin and strength in toxin, i.e., toxicity e. We have proposed that both the species die according to the function $\Psi(e, T) = \dfrac{e\gamma T}{1 + e\gamma T}$. Clearly, it is to be noted that, both species do not die in the absence of toxin as $\Psi(e, T) = 0$; level of toxin, i.e., toxicity may influence the death rate because for $e = 0$, $\Psi(e, T)$ becomes zero. Interestingly, $\dfrac{\partial\Psi}{\partial e} > 0$ and $\dfrac{\partial\Psi}{\partial T} > 0$ indicates that both population suffers a higher death rate with the increase of both the toxicity level e and toxin concentration T.

3 Well-posedness

3.1 Positivity

Theorem 1 *Every solution of system* (1) *starting from IC* (2) *remains positive for any time* $t > 0$.

Proof The proof is very much straight forward and hence omitted.

3.2 Boundedness

Theorem 2 *Every solution of system* (1) *starting from IC* (2) *is always bounded.*

Proof The proof is very much straight forward and hence omitted.

4 Stability Analysis

4.1 Fixed Points with Their Existence Conditions

Model system (1) has three fixed or equilibrium points, namely

(i) Axial equilibrium point $E_1\left(0, 0, \dfrac{A}{a}\right)$ always exists;

(ii) Planar equilibrium point $E_2\left(\bar{x}, 0, \dfrac{A}{a + \gamma\bar{x}}\right)$ exists for $Ae\gamma < (r - d_1)(a + Ae\gamma)$ and $\bar{x}$ has to be extracted from the underlying equation

$$m\gamma x^2 + \{m(a + Ae\gamma) - \gamma(r - d_1)x\} + Ae\gamma - (r - d_1)(a + Ae\gamma) = 0.$$

(iii) Interior equilibrium point $E^*\ (x^*, y^*, T^*)$ where

$$x^* = \frac{e\gamma T^* + d_2(1+e\gamma T^*)}{(1+e\gamma T^*)(\beta - bd_2) - be\gamma T^*} \text{ exists for } \beta - bd_2 > \frac{Abe\gamma}{Ae\gamma + a + \gamma(x^*+y^*)},$$
$$T^* = \frac{A}{a+\gamma(x^*+y^*)},$$

and y^* has to be calculated from

$$\alpha f y^2 - \{f(d_2 - d_1 - mx^*)(1+bx^*) - \alpha - \beta f x^*\}y + \beta x^* - r - (d_2 - d_1 - mx^*)(1+bx^*) = 0,$$

under the restriction $\beta x^* + (mx^* + d_1 - d_2)(1+bx^*) < r$.

4.2 Local Stability Analysis (LAS)

In this subsection, our main interest is to explore the restrictions under which system (1) remains close enough to the corresponding fixed points under a slight given perturbation and this job has been performed with the help of the following theorems.

Theorem 3 *System* (1) *exhibits LAS behavior around axial equilibrium point* $E_1\left(0, 0, \frac{A}{a}\right)$ *for* $r < d_1 + \frac{Ae\gamma}{a+Ae\gamma}$.

Proof Eigenvalues of Jacobian matrix computed at axial equilibrium point $E_1\left(0, 0, \frac{A}{a}\right)$ are

$$\lambda_1 = r - d_1 - \frac{Aea\gamma}{a+Aea\gamma} < 0 \text{ for } r < d_1 + \frac{Aea\gamma}{a+Aea\gamma},\ \lambda_2 = -d_2 - \frac{Ae\gamma}{a+Ae\gamma} < 0,\ \lambda_3 = -a < 0.$$

Hence the result.

Theorem 4 *System* (1) *exhibits LAS behavior around planar equilibrium point* $E_2\left(\bar{x}, 0, \frac{A}{a+\gamma\bar{x}}\right)$ *for*

(i) $\frac{\beta\bar{x}}{1+b\bar{x}} < d_2 + \frac{e\gamma\bar{T}}{1+e\gamma\bar{T}}$,

(ii) $r < a + (\gamma + 2m)\bar{x} + d_1 + \frac{e\gamma\bar{T}}{1+e\gamma\bar{T}}$,

(iii) $(a+\gamma\bar{x})\left(r - d_1 - 2m\bar{x} - \frac{e\gamma\bar{T}}{1+e\gamma\bar{T}}\right) + \frac{e\gamma^2\bar{x}\bar{T}}{(1+e\gamma\bar{T})^2} < 0.$

Proof One eigenvalue of Jacobian matrix computed at planar equilibrium point $E_2\left(\bar{x}, 0, \frac{A}{a+\gamma\bar{x}}\right)$ is

$$\lambda_1 = \frac{\beta \bar{x}}{1+b\bar{x}} - d_2 - \frac{e\gamma \bar{T}}{1+e\gamma \bar{T}} < 0 \text{ for } \frac{\beta \bar{x}}{1+b\bar{x}} < d_2 + \frac{e\gamma \bar{T}}{1+e\gamma \bar{T}},$$

and other two eigenvalues has to be obtained from

$$\lambda^2 - \left(r - d_1 - 2m\bar{x} - \frac{e\gamma \bar{T}}{1+e\gamma \bar{T}} - a - \gamma \bar{x}\right)\lambda - \frac{e\gamma^2 \bar{x}\bar{T}}{(1+e\gamma \bar{T})^2} - (a+\gamma \bar{x})\left(r - d_1 - 2m\bar{x} - \frac{e\gamma \bar{T}}{1+e\gamma \bar{T}}\right) = 0.$$

The above equation will have two negative real roots for conditions (ii) and (iii). Hence the result.

Theorem 5 *System* (1) *displays LAS behavior close to interior equilibrium point* $E^*(x^*, y^*, T^*)$ *if* $\xi_1 > 0$, $\xi_3 > 0$ *and* $\xi_1 \xi_2 > \xi_3$ *hold.*

Proof Characteristic equation of the Jacobian matrix computed at interior equilibrium point $E^*(x^*, y^*, T^*)$ is given by

$$\lambda^3 + \xi_1 \lambda^2 + \xi_2 \lambda + \xi_3 = 0, \tag{3}$$

where

$$\begin{aligned}
\xi_1 =& - J_{11}^* - J_{22}^* - J_{33}^*,\\
\xi_2 =& J_{11}^* J_{22}^* + J_{11}^* J_{33}^* + J_{22}^* J_{33}^* - J_{23}^* J_{32}^* - J_{12}^* J_{21}^* + J_{13}^* J_{21}^*,\\
\xi_3 =& J_{12}^* J_{21}^* J_{33}^* - J_{12}^* J_{31}^* J_{23}^* - J_{13}^* J_{21}^* J_{33}^* + J_{13}^* J_{23}^* J_{31}^* - J_{11}^* J_{22}^* J_{33}^* + J_{11}^* J_{23}^* J_{32}^*
\end{aligned}$$

and

$$\begin{aligned}
J_{11}^* &= \frac{r}{1+fy^*} - d_1 - 2mx^* - \frac{\alpha y^*}{(1+bx^*)^2} - \frac{e\gamma T^*}{1+e\gamma T^*}, \quad J_{12}^* = -\frac{rfx^*}{(1+fy^*)^2} - \frac{\alpha x^*}{1+bx^*},\\
J_{13}^* &= -\frac{e\gamma x^*}{(1+e\gamma T^*)^2}, \quad J_{21}^* = \frac{\beta y^*}{(1+bx^*)^2}, \quad J_{22} = \frac{\beta x^*}{1+bx^*} - d_2 - \frac{e\gamma T^*}{1+e\gamma T^*}, \quad J_{23} = -\frac{e\gamma y^*}{(1+e\gamma T^*)^2},\\
J_{31}^* &= -\gamma T^*, \quad J_{32}^* = -\gamma T^*, \quad J_{33}^* = -a - \gamma(x^* + y^*).
\end{aligned}$$

From Routh Hurtwitz criteria, equation (3) have negative root or roots having negative real part if $\xi_1 > 0$, $\xi_3 > 0$ and $\xi_1 \xi_2 > \xi_3$ hold. Hence the theorem.

4.3 Existence of Hopf-bifurcation

Here, we are going to explore the Hopf-bifurcation existence conditions of the model system (1) around the interior equilibrium point $E^*(x^*, y^*, T^*)$ w.r.t the fear effect f.

Theorem 6 *The system* (1) *undergoes Hopf-bifurcation around the interior equilibrium point* $E^*(x^*, y^*, T^*)$ *w.r.t the fear effect* f *if* f *exceeds the threshold value*

$f = f^*$. The necessary and sufficient conditions for occurring Hopf-bifurcation at $f = f^*$ of the model system (1) *are*

(i) $\eta_1(f^*)\eta_2(f^*) - \eta_3(f^*) = 0$,

(ii) The transversality condition $\dfrac{d}{df}[Re(\lambda(f))]_{f=f^*} \neq 0$.

Proof The characteristic equation (3) at $f = f^*$ becomes

$$(\lambda^2 + \eta_2)(\lambda + \eta_1) = 0, \tag{4}$$

i.e., $\lambda_1 = i\sqrt{\eta_2}$, $\lambda_2 = -i\sqrt{\eta_2}$ and $\lambda_3 = -\eta_1$.

For $f \in (f^* - \mu, f^* + \mu)$, where μ is a sufficiently small positive quantity and the general root can be taken as follows

$$\begin{aligned}\lambda_1(f) &= \theta_1(f) + i\theta_2(f),\\ \lambda_2(f) &= \theta_1(f) - i\theta_2(f),\\ \lambda_3(f) &= -\eta_1.\end{aligned}$$

Now, let us try to figure out the restrictions for which the transversality condition is satisfied.

Substituting $\lambda_1(f) = \theta_1(f) + i\,\theta_2(f)$ into (4) and taking the derivative w.r.t f, we get

$$\begin{aligned}E(f)\theta_1'(f) - F(f)\theta_2'(f) + G(f) &= 0,\\ F(f)\theta_1'(f) + E(f)\theta_2'(f) + H(f) &= 0,\end{aligned}$$

where

$$\begin{aligned}E(f) &= 3\theta_1^2(f) + 2\eta_1(f)\theta_1(f) + \eta_2(f) - 3\theta_2^2(f),\\ F(f) &= 6\theta_1(f)\theta_2(f) + 2\eta_1(f)\theta_2(f),\\ G(f) &= \theta_1^2(f)\eta_1'(f) + \eta_2'(f)\theta_1(f) + \eta_3'(f) - \eta_1'(f)\theta_2^2(f),\\ H(f) &= 2\theta_1(f)\theta_2(f)\eta_1'(f) + \eta_2'(f)\theta_2(f).\end{aligned}$$

Since $\theta_1(f^*) = 0$, $\theta_2(f^*) = \sqrt{\eta_2(f^*)}$, so

$$\begin{aligned}E(f^*) &= -2\eta_2(f^*),\\ F(f^*) &= 2\eta_1(f^*)\sqrt{\eta_2(f^*)},\\ G(f^*) &= \eta_3'(f^*) - \eta_1'(f^*)\eta_2(f^*),\\ H(f^*) &= \eta_2'(f^*)\sqrt{\eta_2(f^*)}.\end{aligned}$$

Therefore,

$$
\begin{aligned}
&\frac{d}{df}[Re(\lambda(f))]_{f=f^*}\\
&= -\frac{F(f^*)H(f^*) + E(f^*)G(f^*)}{E^2(f^*) + F^2(f^*)}\\
&= -\frac{2\eta_1(f^*)\sqrt{\eta_2(f^*)}\eta_2'(f^*)\sqrt{\eta_2(f^*)} + (-2\eta_2(f^*))\{\eta_3'(f^*) - \eta_1'(f^*)\eta_2(f^*)\}}{(-2\eta_2(f^*))^2 + \{(2\eta_1(f^*)\sqrt{\eta_2(f^*)}\}^2}\\
&= -\frac{\eta_1(f^*)\eta_2'(f^*) - \eta_3'(f^*) + \eta_1'(f^*)\eta_2(f^*)}{2\{\eta_2(f^*) + \eta_1(f^*)\}^2}\\
&\neq 0, \qquad \text{if } \eta_3'(f^*) \neq \eta_1(f^*)\eta_2'(f^*) + \eta_1'(f^*)\eta_2(f^*) \text{ and } \lambda_3(f^*) = -\eta_1(f^*) \neq 0.
\end{aligned}
$$

Hence the theorem.

5 Numerical Simulation

In this section, we are going to verify all the analytical findings as discussed before by using the softwares MATLAB and MATCONT. So, at first, we have to choose the parameter set as follows:

$$
\begin{aligned}
&r = 0.5,\ f = 0.01,\ m = 0.1,\ \alpha = 0.5,\ \beta = 0.4,\ b = 2,\ e = 1.16,\ A = 0.6,\ a = 0.05,\\
&\gamma = 0.07,\ d_1 = 0.01,\ d_2 = 0.02. \qquad (5)
\end{aligned}
$$

To check the influence of predator generated fear in system dynamics, we have considered periodic behavior of system dynamics by choosing $e = 0.5$ and varied fear level in different ranges. We have observed that fear level f plays a contributory role in the context of stability through a super-critical Hopf-bifurcation (as the first lyapunov co-efficient is a negative quantity). It is observed that system remains in periodic mode until f passes away its threshold value $f = f^* = 2.8$ as shown in Fig. 1. Ecologically, it signifies that with the increase in fear level, predator's food resources decrease which consequently makes a balance in system dynamics to exhibit stable behavior. Now, we want to explore the impact of toxin in system dynamics through the parameter strength in toxin e. Similarly, here also we have noticed that e has a huge impact in system dynamics in terms of stability. The system exhibits unstable behavior for lower level of toxicity ($e \leq 1.16$) and becomes stable for higher level of toxicity ($e > 1.16$) as displayed in Fig. 2. It is to be noted that with increasing value of e, both prey and predator population suffer from higher mortality rate and hence both population density decline which somehow makes a balance in the system to exhibit stable behavior. Now, we have plotted population biomass with respect to f to check its impact in population density in Fig. 3. From Fig. 3, we observed that both prey density and toxin concentration initially increases and then saturates as f increases. The prey population saturates after increasing its

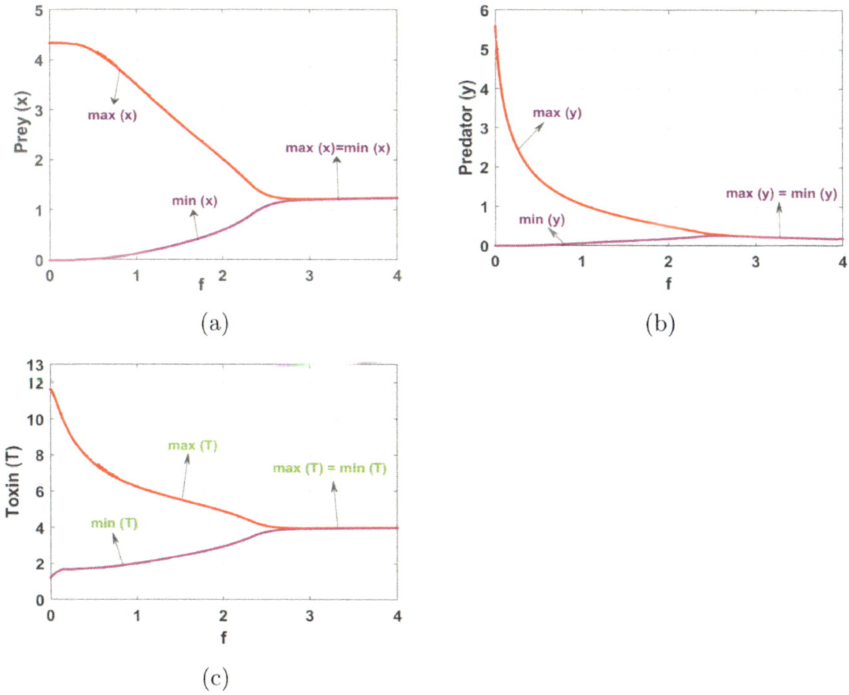

Fig. 1 Bifurcation diagram w.r.t. predator induced fear f in $[0, 4]$ by considering $e = 0.5$ and rest other parameters are kept as fixed. All the three figures jointly describe that the system (1) remains in periodic manner as long as f does not exceed $f = f^* = 2.8$. But, as soon as f crosses the threshold value $f = f^* = 2.8$, the system instantly becomes stable by removing the periodic oscillations of the solution trajectory

density because of controlled population growth rate. But, strangely, as f increases, predator population goes to extinction. The reason behind it may be interpreted as the reduction in food sources for predator species due to the decline in prey's growth rate caused by increasing level of fear.

However, one question comes to our mind that, is there any relation between fear level f and toxicity e ? To answer this question we have plotted a two parametric bifurcation diagram (see Fig. 4) around the co-existence steady state and observed that both f and e are inversely related with each other. The reason behind such observation may be interpreted as prey species are less frightened from predation due to the illness of predator species caused from consuming more toxic foods. From Fig. 4, it is noticed that low level of fear f is required to obtain the Hopf-bifurcation curve with the increasing level of toxicity e. It explains that prey species is less sensitive in perceiving the predation risk with increasing toxicity e in that ecosystem.

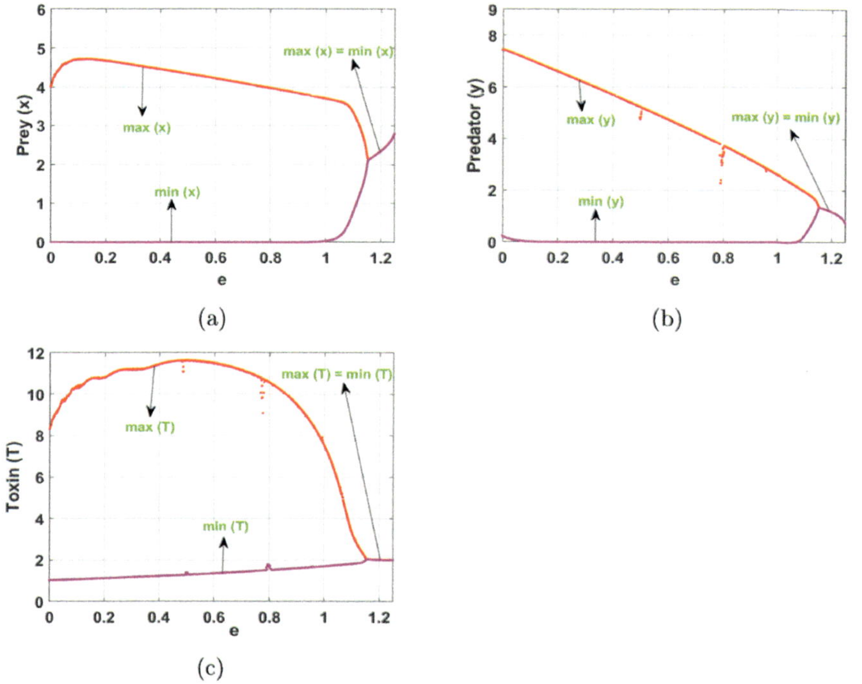

Fig. 2 Bifurcation diagram w.r.t. strength of toxin e in [0, 1.25] by considering $f = 0.01$ and rest other parameters are kept as fixed of system (1). All the three figures jointly describe that the system remains in periodic manner as long as e does not surpass $e = e^* = 1.16$. But, as soon as e passes the threshold value $e = e^* = 1.16$, the system quickly becomes stable by eliminating the periodic oscillations of the solution trajectory

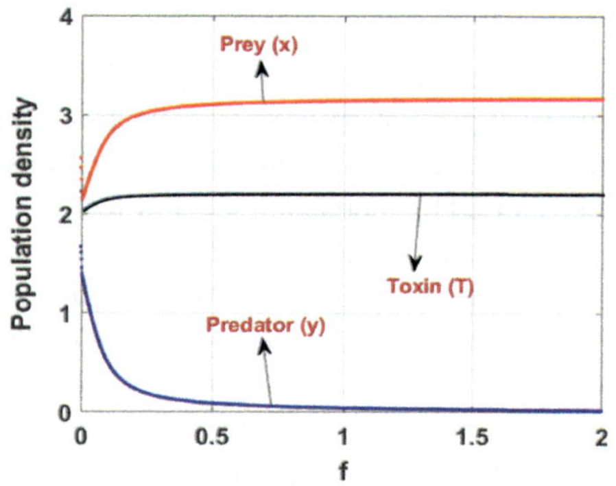

Fig. 3 Plot of population density under the variation of predator incited fear f by choosing $e = 1.16$. It is noticed that both prey density and toxin concentration initially increases and then saturates as f increases. But, interestingly, as f increases, predator population goes to extinction

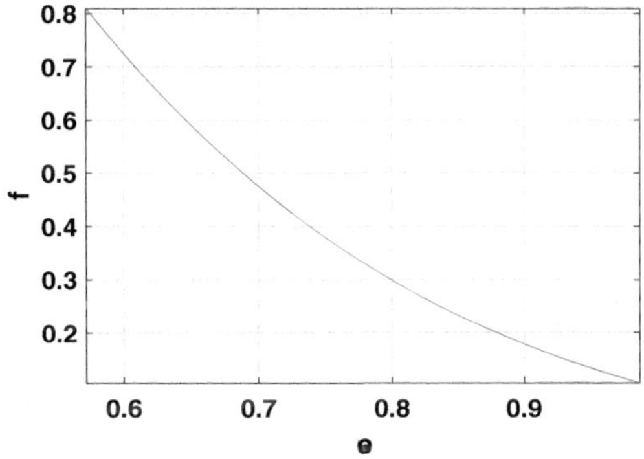

Fig. 4 Two dimensional projection of Hopf-bifurcation curve in f vs e parametric plane revealing that the existence of inverse linear relationship between f and e. It is clear that need of higher level of predator induced fear is necessary to obtain the critical line of Hopf-bifurcation curve for decaying level in strength of toxin e

6 Conclusion

In this article, a predator-prey model has been proposed and analysed in a toxic marine environment by considering predator induced fear to prey's birth rate and toxin dependent death rate for both species. The well-posedness of model system (1) along with stability analysis has been performed analytically. It is noticed that both fear factor and toxicity play a crucial role in controlling the system dynamics in a complex way. Model system (1) undergoes Hopf-bifurcation w.r.t. both these parameters. Predator produced fear has a stable impact on system dynamics for increasing level of fear in the presence of toxin. Strangely, for lower level of toxicity the system exhibits periodic solution whereas for higher level of toxicity it shows stable behavior by eliminating the periodic oscillations. The lower the toxicity level, the higher the periodic oscillations which is quite interesting. Apart from controlling the stability of system dynamics w.r.t. fear factor, it has an immense impact on population density. Although, fear factor does not influence prey biomass and toxin concentration significantly, it horribly forces the predator species in extinction due to lack of nourishing food resources. This model can be further extended by taking into consideration various toxin related factors which can be left as future work to the interested audience.

References

1. Barman, D., Roy, J., Alam, S.: Trade-off between fear level induced by predator and infection rate among prey species. J. Appl. Math. Comput. **64**(1), 635–663 (2020)
2. Barman, D., Roy, J., Alam, S.: Dynamical behaviour of an infected predator-prey model with fear effect. Iranian J. Sci. Technol. Trans. A: Sci. **45**(1), 309–325 (2021)

3. Barman, D., Roy, J., Alrabaiah, H., Panja, P., Mondal, S.P., Alam, S.: Impact of predator incited fear and prey refuge in a fractional order prey predator model. Chaos Solitons Fractals **142**, 110420 (2021)
4. Chattopadhyay, J.: Effect of toxic substances on a two-species competitive system. Ecological Model **84**(1–3), 287–289, 110420 (1996)
5. Duce, R.A., Galloway, J.N., Liss, P.S., et al.: The impacts of atmospheric deposition to the ocean on marine ecosystems and climate. World Meteorol. Organ. (WMO) Bull. **58**(1), 61 (2009)
6. Espinoza, R.: Chemical waste that impact on aquatic life or water quality. https://blog.idrenvironmental.com/chemical-waste-that-impact-on-aquatic-life-or-water-quality. Accessed: 18 July 2021
7. Griffitt, R.J., Luo, J., Gao, J., Bonzongo, J.C., Barber, D.S.: Effects of particle composition and species on toxicity of metallic nanomaterials in aquatic organisms. Environ. Toxicol. Chem. An Int. J. **27**(9), 1972–1978, 110420 (2008)
8. Hallam, T., Clark, C., Lassiter, R.: Effects of toxicants on populations: a qualitative approach i. equilibrium environmental exposure. Ecological Model. **18**(3–4), 291–304 (1983)
9. Hallam, T., De Luna, J.: Effects of toxicants on populations: a qualitative: approach iii. environmental and food chain pathways. J. Theoret. Biol. **109**(3), 411–429 (1984)
10. Huang, Q., Wang, H., Lewis, M.A.: The impact of environmental toxins on predator-prey dynamics. J. Theoret. Biol. **378**, 12–30, 110420 (2015)
11. Kahru, A., Savolainen, K.: Potential hazard of nanoparticles: from properties to biological and environmental effects. Toxicology **2**(269), 89–91 (2010)
12. Rana, S., Samanta, S., Bhattacharya, S., Al-Khaled, K., Goswami, A., Chattopadhyay, J.: The effect of nanoparticles on plankton dynamics: a mathematical model. Biosystems **127**, 28–41, 110420 (2015)
13. Roy, J., Barman, D., Alam, S.: Role of fear in a predator-prey system with ratio-dependent functional response in deterministic and stochastic environment. Biosystems **197**, 104176 (2020)
14. Wang, X., Zanette, L., Zou, X.: Modelling the fear effect in predator-prey interactions. J. Math. Biol. **73**(5), 1179–1204, 110420 (2016)
15. Wang, X., Zou, X.: Modeling the fear effect in predator-prey interactions with adaptive avoidance of predators. Bull. Math, Biol. **79**(6), 1325–1359, 110420 (2017)
16. Zanette, L.Y., White, A.F., Allen, M.C., Clinchy, M.: Perceived predation risk reduces the number of offspring songbirds produce per year. Science **334**(6061), 1398–1401 (2011)
17. Zhang, H., Cai, Y., Fu, S., Wang, W.: Impact of the fear effect in a prey-predator model incorporating a prey refuge. Appl. Math. Comput. **356**, 328–337, 110420 (2019)

Stability Analysis of the Leslie-Gower Model with the Effects of Harvesting and Prey Herd Behaviour

Md. Golam Mortuja, Mithilesh Kumar Chaube, and Santosh Kumar

Abstract Prey herd behaviour has been studied using a modified Leslie-Gower model, including harvesting in both populations. Three separate fixed points can be seen in the model. Local stability theory has been used to investigate the fixed point's stability. The stability of the interior fixed point under a parametric condition is investigated. Few numerical simulations are run to verify the results.

Keywords Predator-prey system · Prey group defense · Fixed points · Local stability · Environmental sustainability

1 Introduction

Since the classical Lotka-Volterra model [1, 2], which was developed by Lotka and Volterra, many researchers have become interested in such concerns [3–5], and they have approached the topic from several perspectives yielding numerous important conclusions [6–8]. Such as the authors in [9] analyzed the bifurcations of a Leslie type predator-prey model with Holling type-III functional response. Lin et al. discussed the multitype bi-stability using the population model in [10]. In [11] the authors analyzed the model using Allee effect. Specifically, the authors in [12] looked at the Leslie-Gower predator-prey model, which is given by:

$$\begin{cases} \frac{dX}{dt} &= rX\left(1-\frac{X}{P}\right)-\frac{\alpha_1 XY}{N_1+X}, \\ \frac{dY}{dt} &= SY\left(1-\frac{\alpha_2 XY}{N_2+X}\right) \end{cases} \tag{1}$$

where X is the density of prey, and Y is the density of predators. The authors have studied the above model, and by using the Lyapunov function, they analyzed the

Md. G. Mortuja (✉) · M. K. Chaube · S. Kumar
Dr. SPM IIIT Naya Raipur, C.G., Raipur, India
e-mail: mdgolam@iiitnr.edu.in

S. Banerjee and A. Saha (eds.), *Nonlinear Dynamics and Applications*,
Springer Proceedings in Complexity,
https://doi.org/10.1007/978-3-030-99792-2_62

stability of the fixed points. In [13] the authors examined the global stability of Leslie-Gower model with feedback controls.

Harvesting is an excellent strategy for people [14, 15] to keep predator and prey populations in balance so that the population can continue to grow appropriately and offer economic benefits [16, 17]. In research, prey harvesting is generally seen merely to control population numbers. The Leslie-Gower predator-prey systems with constant prey harvesting were examined by the authors in [18]. Later, the Leslie-Gower predator-prey model with nonlinear prey harvesting was explored by the authors in [19]. In a Leslie-Gower model, Xie et al. [20] investigated the impact of harvesting and looked at the super-critical Hopf bifurcation that leads to a stable limit cycle. In a discrete modified LG model, Anuraj et al. [21] investigated many co-dimension 1 and 2 bifurcations. A diffusive LG model with Allee effects and mutual predator interface was developed and analyzed by Tiwari & Raw [22]. The dynamics of a stochastic modified LG model with time delay and prey harvest were investigated in [23]. We must consider the harvesting in the prey population and the predator for ecological balance and good economic development. In the present paper, proportional harvesting is regarded in both populations.

Some prey populations show herd behavior, in which predator and prey interact around the perimeter of the prey species, resulting in the predator's hunting rate of prey that varies from that predicted by conventional models. In the ocean, a fish's rate of collecting zooplankton is more significant than a fish's rate of capturing phytoplankton. In this case, the phytoplankton is acting in a herd-like way. That kind of interaction cannot be fully described by Holling-type functional responses. To comprehend the prey population's herd behavior, Ajraldi et al. [24] utilized the square root of the prey density, such that there is an interaction between both species in the prey herd behavior. The paper's main objective is to analyze the dynamics of the Leslie-Gower model with harvesting in both populations considering prey herd behaviour.

1.1 Mathematical Modeling

The modified Leslie-Gower model [12] with harvesting in both population considering prey herd is as follows (after scaling the parameter and variables):

$$\begin{cases} \frac{dX}{dt} = X(1-X) - \frac{\alpha\sqrt{X}Y}{m+\sqrt{X}} - \gamma x, \\ \frac{dY}{dt} = kY\left(1 - \frac{dY}{m+\sqrt{X}}\right) - \delta Y \end{cases} \tag{2}$$

The Jacobian matrix of the above system at origin is undefined. For that the transformation $X = x^2$, $Y = y$ has been applied on the above system:

$$\begin{cases} \frac{dx}{dt} = \frac{1}{2}\left[x\left(1 - x^2\right) - \frac{\alpha y}{m+x} - \gamma x\right], \\ \frac{dy}{dt} = ky\left(1 - \frac{dy}{m+x}\right) - \delta y \end{cases} \tag{3}$$

2 Fixed Points and Their Stability

In this section, the existence and stability of the fixed point of the system have been studied. For the background of biology it is considered that the fixed points all are non negative.

2.1 *Existence of Fixed Points*

To find the fixed points of the system need to solve the following equations:

$$\begin{cases} x(1 - x^2) - \gamma x - \frac{\alpha y}{1+mx} = 0, \\ ky(1 - \frac{dy}{m+x}) - \delta y = 0, \end{cases} \tag{4}$$

The following theorem is about the fixed points of the system and their existence.

Theorem 1 *The system have three fixed points which are:*

(i) $E_0 = (0, 0)$, *the population free fixed point.*
(ii) $E_1 = (x_1, 0)$, *the predator fixed point, if* $\gamma < 1$.
(iii) $E_2 = (x_2, y_2)$, *the interior fixed point.*

Proof By solving the equations it is easy to see that $(0, 0)$ is a fixed point of the system. Now if $y = 0$ then $1 - x^2 - \gamma = 0$. As we considered that the fixed point are non negative, the predator free fixed point will be $(\sqrt{1-\gamma}, 0)$ if $\gamma < 1$ from $1 - x^2 - \gamma = 0$. Now, the interior fixed point E_2 is exists satisfying the equations:

$$\begin{cases} x_2(1 - x_2^2) - \gamma x_2 = \frac{\alpha}{d}(1 - \frac{k}{\delta}), \\ y_2 = \frac{1}{d}(m + x_2)(1 - \frac{k}{\delta}), \end{cases} \tag{5}$$

2.2 *Stability of the Fixed Points*

The stability of the fixed point has been analyzed in this subsection. The Jacobian matrix of the system at (x, y):

$$J(x, y) = \begin{pmatrix} \frac{1}{2}\left[1 - 3x^2 + \frac{\alpha y}{(m+x)^2} - \gamma\right] & -\frac{1}{2}\left(\frac{\alpha}{m+x}\right) \\ \frac{dky^2}{(m+x)^2} & k - \frac{2dy}{m+x} - \delta \end{pmatrix}.$$

Theorem 2 *The fixed point E_0 is:*

(i) unstable if $1 > \gamma, k > \delta$,
(ii) stable if $1 < \gamma, k < \delta$,
(iii) saddle if $1 > \gamma, k < \delta$ *or* $1 < \gamma, k > \delta$,
(iv) non-hyperbolic if either $1 = \gamma$ *or* $k = \delta$.

Proof The real eigenvalues of the Jacobian matrix $J(x, y)$ at E_0 are $\frac{1}{2}(1 - \gamma), k - \delta$. It is easy to see that if $1 > \gamma, k > \delta$ then the eigenvalues are positive implies E_0 is unstable node. If $1 < \gamma, k < \delta$ then the eigenvalues are negative implies E_0 is stable node. If $1 > \gamma, k < \delta$ or $1 < \gamma, k > \delta$, then the eigenvalues are opposite in sign implies E_0 is saddle. If either $1 = \gamma$ or $k = \delta$ then one of the eigenvalues is zero implies E_0 is non-hyperbolic.

Theorem 3 *The fixed point E_1 is:*

(i) stable if $k < \delta$,
(ii) saddle if $k > \delta$,
(iv) non-hyperbolic if $k = \delta$.

Proof The real eigenvalues of the Jacobian matrix $J(x, y)$ at E_1 are $-x_1^2, k - \delta$. First eigenvalue is always positive as $x_1 = \sqrt{1 - \gamma} > 0$ since $\gamma < 1$. It is easy to see that if $k < \delta$ then the eigenvalues are negative implies E_1 is stable node. If $k > \delta$, then the eigenvalues are opposite in sign implies E_1 is saddle. If $k = \delta$ then the second eigenvalue is zero implies E_1 is non-hyperbolic.

Theorem 4 *The fixed point E_2 is stable if $T < 0$, $D > 0$.*

Proof The characteristic equation of the Jacobian matrix J at (x_2, y_2) is given by $\mu^2 - T\mu + D = 0$, where

$$\begin{cases} T = \dfrac{1}{2}\left[1 - 3x_2^2 + \dfrac{\alpha y}{(m + x_2)^2} - \gamma\right] + \dfrac{dy_2}{m + x_2}, \\ D = \dfrac{1}{2}\left[1 - 3x_2^2 + \dfrac{\alpha y}{(m + x_2)^2} - \gamma\right]\left(\dfrac{dy_2}{m + x_2}\right) + -\dfrac{1}{2}\left(\dfrac{\alpha}{m + x}\right)\left(\dfrac{dky^2}{(m + x)^2}\right). \end{cases} \tag{6}$$

Now if $T < 0$, $D > 0$ then the Jacobian matrix J has two eigenvalues having negative real parts. Hence the theorem.

3 Discussion with Numerical Examples

By introducing the parameter values the following results have been identified:

For $\gamma = 0.5337, \alpha = 0.005, d = 0.14, m = 0.8, \delta = 0.5$, if $k = 0.8 > \delta$ then the axial fixed point E_1 is saddle (Fig. 1b) and if $k = 0.45 < \delta$ then the axial fixed point E_1 is stable (Fig. 1a). For $\alpha = 0.005, d = 0.14, m = 0.8, \delta = 0.5, k = 0.8$, if $\gamma = 0.15$ then the interior fixed point E_2 is stable (Fig. 1c). For $\gamma = 0.1, \alpha = 0.05, d = 0.14, m = 0.6, \delta = 0.5$ we plot the time series with respect to prey and predator by changing the parameter value $k = 0.8 > \delta$ and $k = 0.45 < \delta$. There are three equilibrium states, as we can see. The first is $E_0(0, 0)$, which means there are no members of either species present. The second is $E_1(x_1, 0)$, $x_1 = \sqrt{1-\gamma}$, in which the predators are not present and the preys are at their maximum sustainable number x_1. Both populations are present in the third state. Using the above mentioned parameters values, we see from the eigenvalues that E_1 is a strictly stable node for $k < \delta$, and is a saddle and therefore unstable for any $k > \delta$. At this point we have a rather complete picture of the equilibria of the system and the stability. For $k < \delta$, the only stable equilibrium state is the all prey populations state at the maximum sustainable population $x_1 = \sqrt{1-\gamma}$ (Fig. 2).

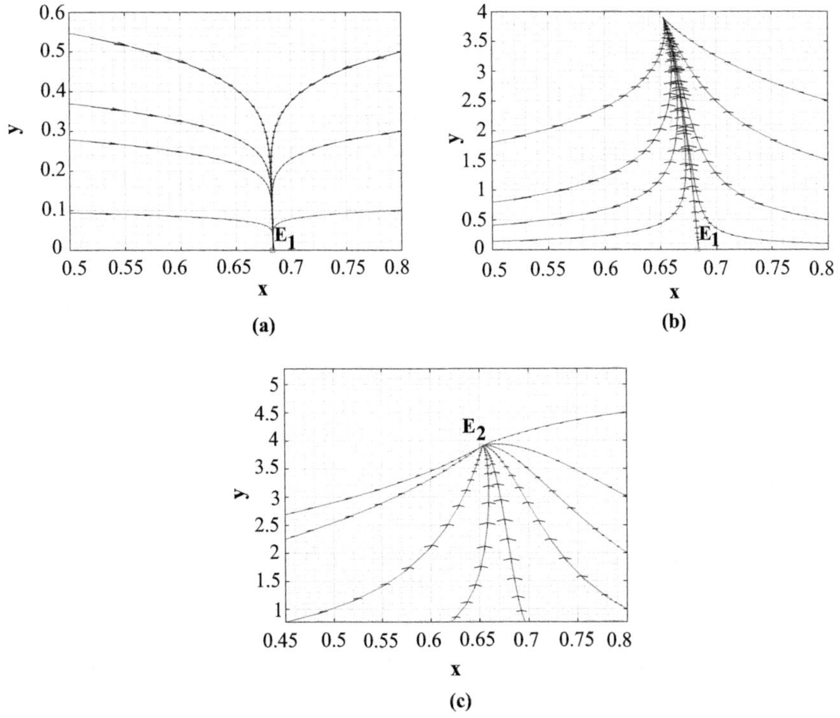

Fig. 1 **a** E_1 is stable, **b** E_1 is saddle, **c** E_2 is stable

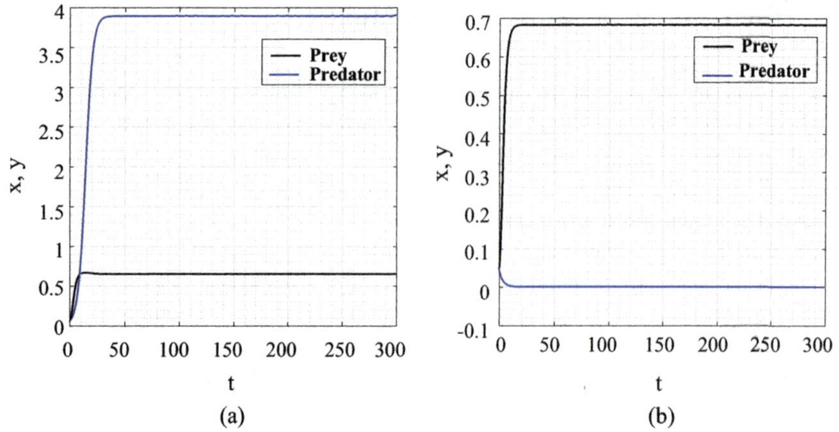

Fig. 2 Time series **a** $k > \delta$, **b** $k < \delta$

4 Conclusion

A modified Leslie-Gower model with harvesting in both populations has been developed considering prey herd behaviour. The model exhibits three different fixed points population-free, predator-free and interior fixed point. The stability of the fixed point has been analyzed by local stability theory. If $1 < \gamma, k < \delta$, the origin is stable, otherwise unstable, or non-hyperbolic. Predator-free fixed point is stable if $1 < \gamma, k < \delta$. It is examined that the interior fixed point is stable under a parametric condition. We also include some numerical simulations to support our analytical findings and conclusions. As a result, our research could help to ensure long-term environmental sustainability. In a pond or natural ocean system, huge fishes eat small fish as their main source of nutrition. To defend their predator, certain fish gather. As a result, the findings of the article could be useful in fisheries management. Management will be informed of the rate at which little fish species (preys) must be taken to maintain ecological balance based on the findings of this study.

References

1. Lotka, A.J.: Elements of Physical Biology. Williams & Wilkins (1925)
2. Volterra, V.: Principes de biologie mathématique. Acta biotheoretica **3**(1), 1–36 (1937)
3. Britton, N.F.: Essential Mathematical Biology. Springer, Berlin, Germany (2013)
4. Murray, J.D.: Mathematical Biology I: An Introduction. Springer, Berlin, Germany (2002)
5. Murray, J.D.: Mathematical Biology: Spatial Models and Biomedical Applications. Springer, Berlin, Germany (2002)
6. Brauer, F., Castillo-Chavez, C.: Mathematical Models in Population Biology and Epidemiology. Springer, Berlin, Germany (2011)

7. Yang, W., Li, Y.: Dynamics of a diffusive predator-prey model with modified Leslie-Gower and Holling-type III schemes. Comput. Math. Appl. **65**(11), 1727–1737 (2013)
8. Wu, F. Jiao, Y.: Stability and Hopf bifurcation of a predator-prey model. Boundary Value Problems (2019)
9. Huang, J., Ruan, S., Song, J.: Bifurcations in a predator prey system of Leslie type with generalized Holling type III functional response. J. Differ. Equ. **257**(6), 1721–1752 (2014)
10. Lin, G., Ji, J., Wang, L., Yu, J.: Multitype bistability and long transients in a delayed spruce budworm population model. J. Differ. Equ. **283**, 263–289 (2021)
11. Mart-Jeraldo, N., Aguirre, P.: Allee effect acting on the prey species in a Leslie-Gower predation model. Nonlinear Anal. Real World Appl. **45**, 895–917 (2019)
12. Aziz-Aloui, M., Daher Okiye, M.: Boundedness and global stability for a predator-prey model with modified Leslie-Gower and holling-type II schemes. Appl. Math. Lett. **16**, 1069–1075 (2003)
13. Chen, L., Chen, F.: Global stability of a Leslie-Gower predator-prey model with feedback controls. Appl. Math. Lett. **22**(9), 1330–1334 (2009)
14. Huang, J., Gong, Y., Gong, Y., Ruan, S.: Bifurcation analysis in a predator-prey model with constant-yield predator harvesting. Discrete Contin. Dyn. Syst. B **18**(8), 2101–2121 (2013)
15. Chakraborty, S., Pal, S., Bairagi, N.: Predator-prey interaction with harvesting: mathematical study with biological ramifications. Appl. Math. Model. **36**(9), 4044–4059 (2012)
16. Lenzini, P., Rebaza, J.: Nonconstant predator harvesting on ratio-dependent predator-prey models. Appl. Math. Sci. **4**, 791–803 (2010)
17. Ali, J., Shivaji, R., Wampler, K.: Population models with diffusion, strong Allee effect and constant yield harvesting. J. Math. Anal. Appl. **352**(2), 907–913 (2009)
18. Zhu, C., Lan, K.: Phase portraits. Hopf bifurcations and limit cycles of Leslie-Gower predator-prey systems with harvesting rates. Discrete Contin. Dyn. Syst. Ser. B **14**, 289–306 (2010)
19. Gupta, R.P., Chandra, P.: Bifurcation analysis of modified Leslie-Gower predator-prey model with Michaelis-Menten type prey harvesting. J. Math. Anal. Appl. **398**(1), 278–295 (2013)
20. Xie, J., Liu, H., Luo, D.: The effects of harvesting on the dynamics of a Leslie-Gower model. Discrete Dyn. Nature Soc. (2021)
21. Singh, A., Malik, P.: Bifurcations in a modified Leslie-Gower predator-prey discrete model with Michaelis-Menten prey harvesting. J. Appl. Math. Comput., 1–32 (2021)
22. Tiwari, B., Raw, S.N.: Dynamics of Leslie-Gower model with double Allee effect on prey and mutual interference among predators. Nonlinear Dyn. **103**(1), 1229–1257 (2021)
23. Liu, Y., Liu, M., Xu, X.: Dynamics analysis of stochastic modified Leslie-Gower model with time-delay and Michaelis-Menten type prey harvest. J. Appl. Math. Comput., 1–28 (2021)
24. Ajraldi, V., Pittavino, M., Venturino, E.: Modeling herd behavior in population systems. Nonlinear Anal. Real World Appl. **12**(4), 2319–2338 (2011)

Modeling the Symbiotic Interactions Between *Wolbachia* and Insect Species

Davide Donnarumma, Claudia Pio Ferreira, and Ezio Venturino

Abstract A sex-structured mathematical model is proposed to address the interactions among *Wolbachia* and *Aedes* mosquitoes. Several features associated with the infection that impacts mosquito phenotype like the cytoplasmic incompatibility, sex ratio biased to females, and maternal inheritance are considered. The analysis of the model shows the presence of three equilibria: the infection-free point, which is attainable only for a narrow range of initial conditions, the point where all individuals are infected, which however arises only in a very particular situation, namely the full vertical transmission of the bacterium, and the endemic equilibrium. Thresholds for the stability of the coexistence equilibrium are obtained, and they involve a relation among parameters and the initial prevalence of the infected mosquito in the population. As expected, increasing the sex ratio biased to females promotes the fixation of the infection on the population at high values.

Keywords Ordinary differential equations · Stability analysis · Thresholds

1 Introduction

Wolbachia is a common type of Gram-negative bacteria that infects about 60% of all arthropods, but is harmless for animals and humans. These obligated intracellular parasites are harbored mainly in the reproductive organs of the insects [20], but also in the legs and guts [6]. It is known that *Wolbachia* infection of the common mosquito *Culex pipiens* alters its reproduction in diverse ways that favor its invasion

D. Donnarumma · E. Venturino (✉)
Università di Torino, 120123 Torino, Italy
e-mail: ezio.venturino@unito.it

D. Donnarumma
e-mail: davide.donnarumma@edu.unito.it

C. Pio Ferreira
São Paulo State University (UNESP), Botucatu 18618-970, Brazil
e-mail: claudia.pio@unesp.br

S. Banerjee and A. Saha (eds.), *Nonlinear Dynamics and Applications*,
Springer Proceedings in Complexity,
https://doi.org/10.1007/978-3-030-99792-2_63

of wild non-infected populations. For example, infected males cannot reproduce with uninfected females because of cytoplasmic incompatibility (CI). Female-biased sex ratio can burst the size of the infected population and can be achieved by killing part of the males at the larval stage, or by the feminization of males (males develop as sterile females), or by parthenogenesis (infected females generate offsprings without mating with males) [9]. Fitness cost associated with carrying the bacteria (like an increase in the mortality rate) can be overcome by the CI and maternal inheritance (infected females can transmit vertically the *Wolbachia* bacteria to their offsprings) [14].

Two types of mosquitoes have attained the limelight in recent years, being particularly harmful. They are currently spreading northwards (and southwards) from tropical regions, carrying with them the viruses of formerly unknown diseases at temperate latitudes, such as Chikungunya, Dengue, Zika [20]. They are the *Aedes albopictus* and the *Aedes aegypti* mosquitoes. Ways of controlling their spread vary from larvicide and insecticide spraying, through mosquito surveillance, and population education programs with focusing attention on their living surroundings, for example, by removing standing water which may serve as a breeding site for the mosquito [22].

A different way of fighting mosquitoes, that does not involve genetically modified insects, is based on the observation that *Aedes aegypti* does not harbor the *Wolbachia* bacteria. This fact has been exploited to devise a biological control program, for which *Aedes aegypti* are inoculated with *Wolbachia*. Thus, males infected in the laboratory are subsequently released into the environment, and their mating with wild females is unsuccessful, producing eggs that do not hatch, reducing the size of the next generation population. The release however should be monitored and repeated in time, as *Wolbachia* dies with the host insect [21].

Results on the effectiveness of these programs have been reported in several countries. Besides the reduction of the mosquito population, the presence of *Wolbachia* interferes with arboviruses by reducing its infection into mosquitoes [17]. Therefore, geographic areas where the presence of the infected mosquito is high are expected to report less prevalence of arboviruses on the human population. For instance, in Indonesia, the release of *Aedes aegypti* inoculated by the wMel strain of *Wolbachia* in several randomly-selected geographical locations were proven to offer protection efficacy (by reducing the incidence of dengue disease) against the four serotypes of dengue in 77.1% of the people taking part in the experiment [19]. As a consequence of the bacterium introgression in the mosquito population, the need for hospital treatment was reduced for the virologically-confirmed dengue patients living in the areas subject to the treatment. The symptomatic cases as well as the hospitalizations were significantly lower than in other areas where *Wolbachia* release had not been implemented.

In Brazil, deployments of the wMel strain of *Wolbachia* in selected geographic areas were monitored by keeping track of the monthly reported human cases of Dengue, Chikungunya, and Zika, against the data coming from an untreated region. The bacterium release was associated with a 69% reduction in Dengue incidence, in a little more than a half reduction of the incidence for Chikungunya and a little more than a third of the Zika cases. Thus, a significant benefit can result for the health system by the presence of bacterium-affected mosquitoes in urban areas, independently of the possible spatial heterogeneity [13].

In Italy, a similar technique is being used against *Aedes albopictus* [4]. These mosquitoes have been inoculated with the *Wolbachia* naturally present in *Culex pipiens*, and then reared in the laboratory to produce males that are incompatible with the wild *Aedes albopictus* females. Again, these males are subsequently released into the environment, leading to mating that produce infertile eggs.

From the mathematical point of view, several models have been proposed to study aspects of this new symbiotic interaction. For instance, [1] showed that the cytoplasmic incompatibility does not guarantee establishment of the *Wolbachia*-infected mosquitoes if the imperfect maternal transmission is considered. For this, the optimal *Wolbachia* release problem was studied in different scenarios, where the decision makers either aim for replacement or co-existence of both populations. As in other works [5, 16], the viability of the technique relies on the number of infected mosquito released, and on and periodicity of the releasing. In [2, 8] a delay differential model was derived with the aim of studying the colonization and persistence of the *Wolbachia*-transinfected *Aedes aegypti* mosquito. The conditions of existence for each equilibrium of the model were obtained analytically. The persistence of both infected and wild populations was explored in the context of mosquito's fitness, host-symbiont interaction, and temperature change. It was shown that the increase of the delay, which represents the development time, can promote, through Hopf bifurcation, stability switch towards instability for the nonzero equilibria. A two-sex mosquito model, that accounts for multiple pregnant states and the aquatic-life stage, is used to compare the effectiveness of different integrated mitigation strategies to establish the *Wolbachia* infection on the population [15]. The proposed model presents a subcritical bifurcation indicating that even when $R_0 < 1$, there can still exist a stable endemic equilibrium. Also, it argues that mitigation strategies to reduce the population of wild uninfected mosquitoes before releasing numerous *Wolbachia*-infected mosquitoes could improve the establishment of the infection in the population. Coupling mosquito population with human population, other works assessed the reduction of dengue transmission by the fact that a *Wolbachia* infection is established in the mosquito population [7, 10, 23]. In this case, the susceptibility of *Wolbachia*-carrying mosquito to dengue infection is explored under different scenarios that take in consideration variation in mosquito fitness, maternal inheritance, and cytoplasmic incompatibility inherent to each *Wolbachia* strain.

In this paper, we develop a model to study the introduction and persistence of the *Wolbachia* bacteria into the mosquito population, partitioning the mosquito population among sexes and infection status. Only adult mosquitoes were considered. The paper is organized as follows. In the next section, we formulate the model. Trajectories are shown to be bounded in Sect. 3, the equilibria are studied in Sect. 4, and the simulations for the endemic case are reported in Sect. 4.3. A thorough investigation of the equilibria in terms of the model parameter variations concludes the paper.

2 Model Setup

Let F represent the susceptible female mosquitoes, i.e. those not infected with the *Wolbachia* bacterium, I the *Wolbachia*-infected female mosquitoes, M the susceptible male mosquitoes and U the *Wolbachia*-infected male mosquitoes. Now, reproduction success is determined by the presence of the *Wolbachia*, in the sense that reproduction after mating of an infected male with any female occurs at a much lower rate than for a couple of healthy individuals, and furthermore an infected female that mates a healthy male has very, very low chances of success, if at all. In other words, the outcomes are schematically represented in Table 1.

The ordinary differential system is given by

$$
\begin{aligned}
\frac{dF}{dt} &= \frac{1}{2} r \frac{M}{M+U} F + p r_1 (1-\rho) I + p k r \frac{U}{U+M} F - mF - aF(F+M+I+U), \\
\frac{dI}{dt} &= p r_1 \rho I - \mu I - b I (F+M+I+U), \\
\frac{dM}{dt} &= \frac{1}{2} r \frac{M}{M+U} F + q r_1 (1-\rho) I + q k r \frac{U}{U+M} F - mM - cM(F+M+I+U), \\
\frac{dU}{dt} &= q r_1 \rho I - \mu U - gU(F+M+I+U).
\end{aligned} \tag{1}
$$

Note that in order to make sense both mathematically and biologically, in this model we assume

$$U + M \neq 0. \tag{2}$$

The other parameters of the model are as follows. In all (susceptible) compartments, the natural mortality m explicitly appears. Also, in the last terms of all the equations, we allow for possible intraspecific competition among the four classes, at rate a for the susceptible females, b for the infected ones, c for susceptible males and g for the infected ones.

The first equation describes the evolution of susceptible female mosquitoes. They are born by mating with a susceptible male, first term with reproduction (oviposition) rate r, assuming that the offsprings split evenly among the two sexes with the same ratio (1/2). A similar term is indeed found in the susceptible male's equation. The

Table 1 Reproduction rates given by all possible mating combinations. The wild populations are denoted by F and M, and the *Wolbachia*-infected one by I and U. Here $r_1 < r$ and $k << 1$, possibly $k = 0$

Male	Female	
	F	I
M	r	r_1
U	kr	r_1

second term accounts for the mating of an infected female, which may occur with a healthy male, with probability $M(M+U)^{-1}$, or with an infected one, with probability $U(M+U)^{-1}$. The resulting reproduction however has a lower rate $r_1 < r$ than for a healthy-healthy mating, and furthermore we allow for possible vertical transmission of the *Wolbachia*. Thus, if $\rho < 1$ represents the vertical transmission of the infection, a portion $1-\rho$ of the offspring are born healthy. In case of infection, we also account for the possibility of *Wolbachia* presence altering the sex ratio of the offspring: p is the fraction of female offspring, while $q = 1 - p$ represents the male fraction [3]. To sum up, the second term represents the healthy females generated by mating infected females with a male. The third term models the mating of a susceptible female with an infected male at a rate r, which has a probability of success $k << 1$, if not $k = 0$. The offspring have an altered sex ratio so that we find p healthy females.

The second equation for the infected females contains the recruitment term coming from the mating of infected females at a reduced rate r_1 with any kind of male. The sex ratio is altered, p for the female offspring, and the bacteria are vertically transmitted with probability ρ which generates infected individuals. The second term contains the natural plus infection-induced mortality $\mu > m$.

Susceptible males dynamics is represented in the third equation, with reproduction and mortality terms that are identical to those of the healthy females. However, here, the offsprings coming from infected females have the sex ratio altered by the fraction q.

The infected males, fourth equation, again are recruited from infected females with reduced reproduction rate r_1 and with a fraction q of sex ratio and probability of vertical bacteria transmission ρ. They are subject to natural plus infection-related mortality μ, the very same as for the females.

The Jacobian of (1), needed for the stability analysis, turns out to be:

$$J = \begin{pmatrix} J_{1,1} & pr_1(1-\rho) - aF & r\frac{FU(\frac{1}{2}-pk)}{(M+U)^2} - aF & -r\frac{FM(\frac{1}{2}-pk)}{(M+U)^2} - aF \\ -bI & J_{2,2} & -bI & -bI \\ r\frac{\frac{1}{2}M+qkU}{M+U} - cM & qr_1(1-\rho) - cM & J_{3,3} & -r\frac{MF(\frac{1}{2}-qk)}{(M+U)^2} - cM \\ -gU & qr_1\rho - gU & -gU & J_{4,4} \end{pmatrix}, \tag{3}$$

with

$$\begin{aligned} J_{1,1} &= \frac{1}{2}r\frac{M}{M+U} + pkr\frac{U}{U+M} - m - a(F+M+I+U) - aF, \\ J_{2,2} &= pr_1\rho - \mu - b(F+M+I+U) - bI, \\ J_{3,3} &= \frac{1}{2}r\frac{U}{(M+U)^2}F - qkr\frac{U}{(M+U)^2}F - m - c(F+M+I+U) - cM, \\ J_{4,4} &= -\mu - g(F+M+I+U) - gU. \end{aligned}$$

3 Boundedness

First, note that

$$p + q = 1. \tag{4}$$

Let $Z = F + I + M + U$ denote the total mosquito population. Summing the equations in (1), for $\eta > 0$, recalling (4) we obtain

$$\begin{aligned}\frac{dZ}{dt} + \eta Z = {} & r\frac{M + kU}{M + U}F + r_1 (1 - \rho) I - m(F + M) - \mu(U + I) + r_1 \rho I \\ & -(aF + bI + cM + gU)Z + \eta Z \\ \le {} & (r + \eta)F + (r_1 + \eta)I - aF^2 - bI^2 - cM^2 - gU^2 \\ \le {} & (r + \eta)F + (r_1 + \eta)I - aF^2 - bI^2 \\ \le {} & F_* + I_* = Z_*, \quad \text{with} \quad F_* = \frac{(r + \eta)^2}{4a}, \quad \text{and} \quad I_* = \frac{(r_1 + \eta)^2}{4b}.\end{aligned}$$

From the resulting differential inequality for Z,

$$\frac{dZ}{dt} + \eta Z \le Z_*,$$

we get

$$Z(t) \le e^{-\eta t}\left[Z(0) - \frac{Z_*}{\eta}\right]\frac{Z_*}{\eta} \le \max\left\{Z(0), \frac{Z_*}{\eta}\right\} = Z_+.$$

In view of the fact that all populations are nonnegative, we finally find

$$F(t), I(t), M(t), U(t) \le Z_+$$

i.e. the solution trajectories of (1) are bounded.

4 Equilibria

Here, we are looking for points $(\bar{F}, \bar{I}, \bar{M}, \bar{U})$ that solve the equilibrim equations of the system (1).

4.1 The Infection-Free Case

In this case we have $\bar{I} = 0$, $\bar{U} = 0$ and the admissible equilibrium turns out to be

$$E_1 = \left(\frac{2arm - 4am^2 + 4cm^2 - 4cmr + cr^2}{2a\,(2ma + ar - 2cm + cr)}, 0, \frac{r\,(r - 2m)}{2\,(2ma + ar - 2cm + cr)}, 0 \right).$$

To achieve feasibility, we need to ensure that the following inequalities hold true:

$$\bar{F} = \frac{cr^2 - 2mr\,(2c - a) - 4m^2\,(a - c)}{2a[r\,(a + c) - 2m\,(c - a)]} > 0, \quad \bar{M} = \frac{r\,(r - 2m)}{2u[r\,(a + c) - 2m\,(c - a)]} > 0.$$

Now, the denominator of both fractions is nonnegative for

$$r > 2m\frac{c - a}{c + a}. \tag{5}$$

Note that the above condition holds unconditionally in case $c < a$. To have the numerator of $\bar{F}$ nonnegative we need either one of the following inequalities to be satisfied

$$r < 2m\frac{c - a}{c}, \quad r > 2m, \tag{6}$$

with the proviso that for $c < a$ the former is not satisfied, since all model's parameters are positive. Now, to have the numerator of $\bar{M}$ nonnegative we need

$$r > 2m, \tag{7}$$

and combining (5), (6) and (7), E_1 is feasible for

$$r > 2m. \tag{8}$$

Thus, this equilibrium feasibility depends only on the birth and death rates. Following the same argument it is easy to see that if

$$r < 2m\frac{c - a}{c + a}$$

we cannot have both $\bar{F}$ and $\bar{M}$ nonnegative.

Figure 1 shows the total population at equilibrium E_1 as a function of m and r for $a = c = 1$. The initial conditions are the following: $M_0 = F_0 = 1$, $U_0 = I_0 = 0$. The other parameters are arbitrary, since the equilibrium does not depend on them. Assuming that the condition $r > 2m$ is satisfied, either the increase of r or decrease of m or both promote the increase of $\bar{Z} = \bar{F} + \bar{M}$.

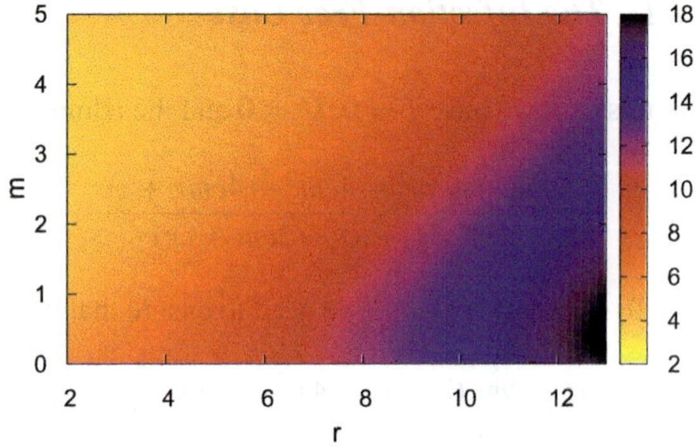

Fig. 1 The figure shows the total healthy population $\bar{Z} = \bar{F} + \bar{M}$ (here $\bar{I} = \bar{U} = 0$) at the infection-free equilibrium E_1 as function of r and m for $a = 1, c = 1$

Stability of this equilibrium can be assessed in view of the fact that the characteristic equation factorizes into the product of the two minors of the following matrices:

$$A = \begin{pmatrix} \frac{r}{2} - m - a\left(\bar{F} + \bar{M}\right) - a\bar{F} & -a\bar{F} \\ \frac{r}{2} - c\bar{M} & -m - c\left(\bar{F} + \bar{M}\right) - c\bar{M} \end{pmatrix},$$

$$B = \begin{pmatrix} \frac{\rho r_1}{2} - \mu - b\left(\bar{F} + \bar{M}\right) & 0 \\ \frac{\rho r_1}{2} & -\mu - g\left(\bar{F} + \bar{M}\right) \end{pmatrix}.$$

The Routh-Hurwitz conditions for A give

$$\mathrm{tr}\,(A) = \frac{r}{2} - m - a\left(\bar{F} + \bar{M}\right) - a\bar{F} - m - c\left(\bar{F} + \bar{M}\right) - c\bar{M} < 0$$

which explicitly becomes

$$\frac{r^2\left(c^2 + 3\,ac\right) - r\left(c^2 + ac - a^2\right) - 4\,m^2\left(ac - c^2\right)}{2a\left(r\left(c + a\right) - 2m\left(c - a\right)\right)} > 0. \tag{9}$$

The numerator is positive for $r < r_-$ or $r > r_+$, with

$$r_\pm = 2m\frac{-a^2 + ac + c^2 \pm \sqrt{a^4 - 2\,a^3c + 2\,a^2c^2}}{c\left(3\,a + c\right)}.$$

The denominator is positive for $r > r_3$, with

$$r_3 = 2m\frac{c - a}{c + a}.$$

Note that $r_\pm, r_3 < 2m$. Therefore, (9) holds if the following condition is satisfied,

$$r > \max\{r_\pm, r_3\}.$$

The second Routh-Hurwitz condition for A is

$$\det(A) = \left[\frac{r}{2} - m - a\left(\bar{F} + \bar{M}\right) - a\bar{F}\right]\left[-m - c\left(\bar{F} + \bar{M}\right) - c\bar{M}\right] + a\bar{F}\left[\frac{r}{2} - c\bar{M}\right]$$

and explicitly it yields

$$\frac{4\,am^2 - 2\,amr - 4\,cm^2 + 4\,cmr - cr^2}{4\,a} < 0$$

which is satisfied for $r < 2m(c-a)c^{-1}$ or for $r > 2m$.

Combining these result with (8), these stability conditions for E_1 coming from A hold unconditionally whenever the equilibrium is feasible.

For the minor B we have

$$\mathrm{tr}\,(B) = \frac{\rho\,r_1}{2} - \mu - b\left(\bar{F} + \bar{M}\right) - \mu - g\left(\bar{F} + \bar{M}\right) < 0$$

which explicitly gives the first stability condition

$$(\rho\,r_1 - 4\,\mu)\,a + (b+g)\,(2m - r) < 0. \tag{10}$$

Further,

$$\det(B) = \left[\frac{\rho\,r_1}{2} - \mu - b\left(\bar{F} + \bar{M}\right)\right]\left[\mu + g\left(\bar{F} + \bar{M}\right)\right] < 0.$$

from which we obtain the second stability condition

$$b\,(r - 2m) > a\,(\rho\,r_1 - 2\,\mu)\,. \tag{11}$$

In conclusion, E_1 is stable if both (10) and (11) hold.

Figure 2c shows the total mosquito population $\bar{Z} = \bar{F} + \bar{M}$ at equilibrium E_1 as a function of the parameters a and c. The influence of the latter is scant, while the population size grows fast for very small values of a, the intraspecific competition rate affecting the healthy females. Figure 2a shows the decrease in the different populations when a increases. The females disappear faster than the males, as it should be expected, since a influences directly their population. Figure 2b shows the populations behavior when c increases. The females grow, males drop, while the total population remains constant. For $c < a$ males exceed the females. In all sets of parameters, the initial condition is the same one used in Fig. 1; recall that the other parameters are chosen arbitrarily.

In conclusion, feasibility for E_1 requires the oviposition rate for healthy mosquitoes to be at least twice as much as their mortality rate. Stability hinges on (10) and (11); in particular E_1 can be destabilized by the suitable introduction of bacteriological infection, as these conditions contain parameters related to the *Wolbachia* action.

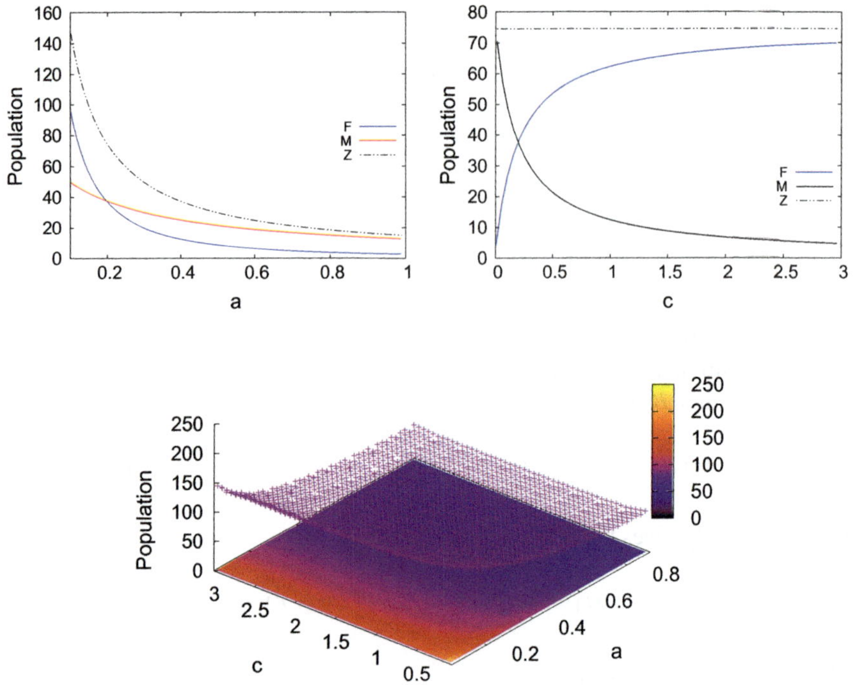

Fig. 2 In **a** top right, Female F, males M, and the total population $Z = F + M$ at the disease-free equilibrium E_1 as function of the females intraspecific competition rate a with $c = 0.2$, and in **b** top left, F, M, and Z populations at the disease-free equilibrium E_1 as function of the males intraspecific competition rate c, with $a = 0.2$, and in **c** bottom, the total healthy population Z at the disease-free equilibrium E_1 as function of the intraspecific competition rates a for females and c for males, with $r = 30$, $m = 0.1$. In both (**a**) and (**b**) the parameters r and m are fixed as in (**c**)

4.2 Persistence of the *Wolbachia*-Infected Population and Extinction of the Wild Population

Here we consider the equilibrium with $\bar{F} = \bar{M} = 0$, but this situation could arise only in the full vertical transmission case $\rho = 1$. The corresponding equilibrium is

$$E_2 = \left(0,\ \frac{2br_1\mu - 4b\mu^2 + 4g\mu^2 - 4g\mu r_1 + gr_1^2}{2b\left(2\mu b + br_1 - 2g\mu + gr_1\right)},\ 0,\ \frac{r_1\left(r_1 - 2\mu\right)}{2\left(2\mu b + br_1 - 2g\mu + gr_1\right)}\right).$$

The analysis follows very closely the one of E_1, so that we just mention the final results. Feasibility holds in case

$$r_1 > 2\mu, \tag{12}$$

which concerns only reproduction rate and mortality of the infected mosquitoes and is independent of their intraspecific competition rates b and g. For stability, the Jacobian again factorizes and ultimately E_2 is stable whenever feasible.

Similarly, the dependence of the population values on b and g indicates that increasing the intraspecific competition b among infected females implies a sharp reduction in the total (infected) population $\bar{Z} = \bar{I} + \bar{U}$ (as here $\bar{F} = \bar{M} = 0$), while the latter is scantly affected by changes in g, although the male to female ratio is much reduced.

In summary, this equilibrium would be beneficial, because all the insects would be *Wolbachia*-affected. Thus, in case of a bacterium with perfect vertical transmission rate, it would always be possible to attain a stable endemic state at which all individuals are infected and the population is lower, using considerations already seen for equilibrium E_1, due to the bacteria-induced reduced reproductive capacity and higher mortality.

4.3 Coexistence with Endemic Infection

This equilibrium is investigated numerically. To this end, we use information obtained from the literature [12], such as the vertical bacterial transmission that is taken to be almost perfect, $\rho \simeq 99\%$.

The reduction of adult longevity is about $p \simeq 21\%$, thus from

$$\mu = m \frac{p}{100 - p},$$

we find

$$\mu \simeq 1.27\, m. \tag{13}$$

Egg survival is about $\simeq 83 - 96\%$, which combined with their reduction due to the action of the bacterium, estimated to be about $\simeq 15\%$ [8], implies that the reduced reproduction rate is

$$0.7055\, r < r_1 < 0.8160\, r. \tag{14}$$

Finally, following [18], we assume that *Wolbachia* prevents the successful mating of an infected male with a susceptible female, thereby setting

$$k \simeq 0. \tag{15}$$

For the remaining parameters, we fix the following hypothetical values

$$\begin{aligned} &r = 150, \quad r_1 = 120, \quad m = 20, \quad \mu = 25.4, \quad \rho = 0.99, \\ &k = 0.01, \quad a = 1, \quad b = 1, \quad c = 1, \quad g = 1, \quad p = q = 0.5. \end{aligned} \tag{16}$$

The initial conditions are

$$F_0 = I_0 = M_0 = U_0 = 1. \tag{17}$$

Solving numerically system (1), three cases of interest arise:

(i) bacteria-free equilibrium (the equilibrium E_1 which was already discussed in Sect. 4.1), with $\bar{F}_1 = 27.5$, $\bar{M}_1 = 27.5$ and of course $\bar{I}_1 = \bar{U}_1 = 0$ (observe that the condition $r > 2m$ is satisfied for this chosen parameter set). The Jacobian eigenvalues are all negative, -55.0, -75.0, -80.4 and -21.0, so it is asymptotically stable.
(ii) two coexistence points:
    (a) $E_3 = (\bar{F}_3, \bar{I}_3, \bar{M}_3, \bar{U}_3)$ with

$$\bar{F}_3 = 0.2, \quad \bar{I}_3 = 16.8, \quad \bar{M}_3 = 0.2, \quad \bar{U}_3 = 16.8.$$

    The eigenvalues of the Jacobian matrix are all negative, -34.0, -51.6, -54.0, -59.4. Thus, this point is asymptotically stable; and
    (b) $E_4 = (\bar{F}_4, \bar{I}_4, \bar{M}_4, \bar{U}_4)$ where

$$\bar{F}_4 = 12.1, \quad \bar{I}_4 = 4.9, \quad \bar{M}_4 = 12.1, \quad \bar{U}_4 = 4.9.$$

    The eigenvalues of the Jacobian matrix in this case are: -34.0, -54.0, -59.4 and 14.9. Because the last one is positive, this equilibrium is unstable.

Note that the total mosquitoes populations are $\bar{Z} = \bar{F} + \bar{M} = 55$ at E_1 and $\bar{Z} = \bar{F} + \bar{I} + \bar{M} + \bar{U} = 34$ at E_4, with almost all (99%) mosquitoes infected. Thus, in addition to render the insects infected, *Wolbachia* reduces also their population size, in agreement with the field studies [4, 6, 13, 19].

The unstable equilibrium E_4 gives us tips about the invasion threshold; it says that the initial prevalence of the infection must exceed 0.285, i.e.,

$$I_0 + U_0 > 0.285\,(F_0 + I_0 + M_0 + U_0)\,. \tag{18}$$

In Figure 3 we show the bistability situation that arises in case (ii). We plot the system trajectories in the projection of the phase space onto the plane that contains the infected populations $I + U$ versus the healthy ones $F + M$. The initial points are in red, in blue the equilibria. The one near the vertical axis represents the endemic case E_3, while the one over the horizontal axis is the disease-free point E_1. In the former, as stated above, the healthy population is scant, and the level of the infected population at the endemic equilibrium $\bar{I} + \bar{U}$, and consequently the resulting total population at this point, is much lower than the total (healthy) population $\bar{Z} = \bar{F} + \bar{M}$ at the disease-free equilibrium E_1. We also report both healthy and infected populations of females against males, in Fig. 4. Because these are projections of the trajectories of a four dimensional system onto planes, there are some apparent intersections, not

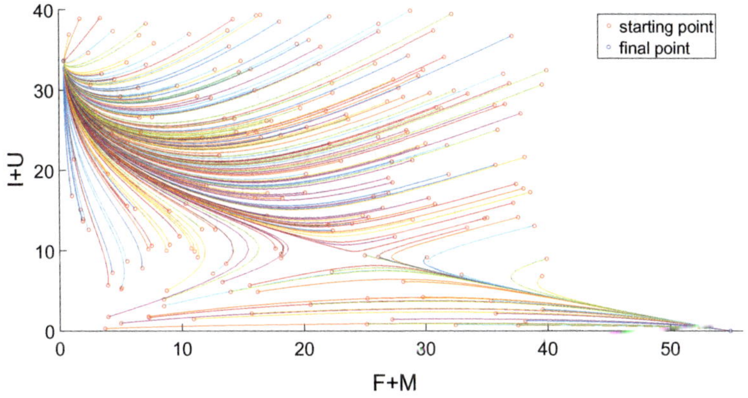

Fig. 3 Projections of the system trajectories on the infected $I + U$ versus healthy $F + M$ total populations, with randomly generated initial conditions (red circles). The trajectories tend to either one of the equilibria (blue circles), the endemic point near the vertical axis and the disease-free over the horizontal one. Here, the initial conditions are always chosen so that $I_0 = U_0$ and $M_0 = F_0$

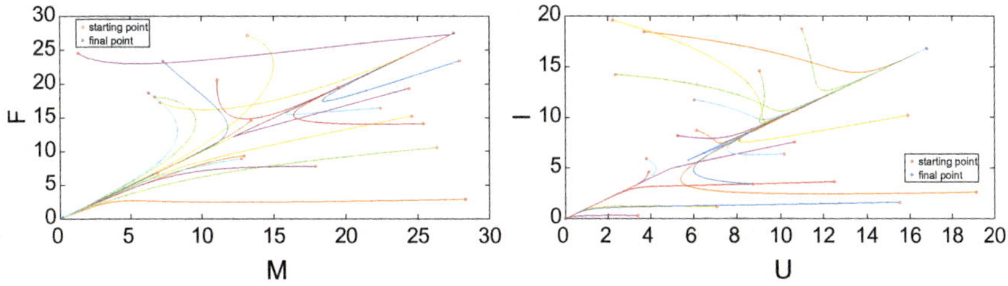

Fig. 4 The same simulations of Fig. 3, with the same parameters and with randomly generated initial conditions for all the populations, with trajectories projected onto the healthy plane $M - F$, left, and onto the infected plane $U - I$, right. In these simulations, conditions $I_0 = U_0$ and $M_0 = F_0$ do not hold

occurring in the four dimensional phase space. Note also that in these simulations taking $p = q = 0.5$, as well as $a = c$ and $b = g$, implies that the equilibria are found on the bisectrix, as the size of male and female populations are the same. Figure 3 suggests the presence of a separatrix in the four dimensional population space. Its position and shape depends on the choice of the parameters, here (16). Note that in particular, condition (17) has been set so that the bacterium can settle endemically in the population.

5 Numerical Simulations

In this section, we will focus on the case of persistence of both populations of mosquitoes, the infected and uninfected ones. Therefore, all simulations were done using the baseline parameters. Also, the initial subpopulation values are set to 1, so that $I_0 + U_0 = 2$, $F_0 + I_0 + M_0 + U_0 = 4$, and their ratio is $0.5 > 0.285$, ensuring the endemicity of the bacterium. This sets up a proxy for the study of the coexistence equilibrium as a function of some parameters.

5.1 *Oviposition*

Figure 5a shows the total mosquito population $Z = F + I + M + U$ as function of the reproduction parameters r and r_1. For low values of both oviposition rates, the population is much reduced. For higher values of r and low r_1 the population grows linearly, with the healthy portion prevailing on the infected one, as can be seen in Fig. 5a and b. When instead r_1 grows and r is small in comparison, the bacterium establishes endemically in the mosquito population. In such case, the total population (on the right of Fig. 5a) grows in a slower way and attains lower values than those found for the healthy one, on the left of the Fig. 5a. This is in agreement with the fact that the introduction of *Wolbachia*-carrying mosquitoes, when successful, can decrease the total mosquitoes population. The sharp jump that separates the two surfaces represents the reduction of mosquitoes due to the *Wolbachia* presence. Its position clearly depends on the model parameters.

5.2 *Mortality*

For high values of the mortality rates ($2m > r$, $2\mu > r_1$) the population dies out, which is apparent in the front part of the Fig. 6a, which is rotated, with the origin in the back. For values of m and μ becoming lower, the population increases linearly, as expected. Again, on the left of Fig. 6a we find the bacteria-free case, as the *Wolbachia*-

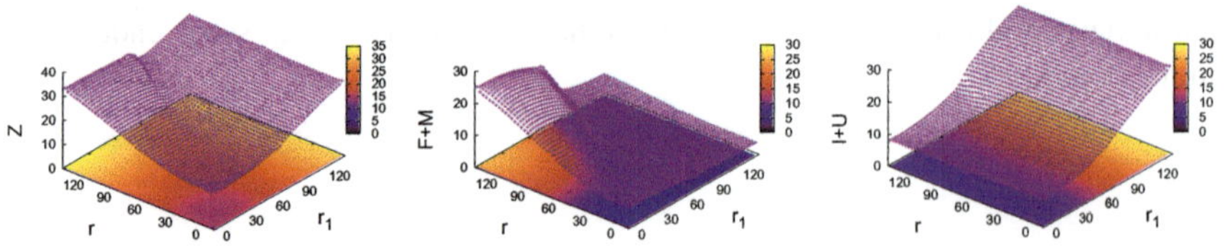

Fig. 5 Mosquito populations at equilibrium as a function of the oviposition rates r and r_1. In **a** left, the total mosquito population $Z = F + I + M + U$, in **b** center, the healthy mosquito population $F + M$, and in **c** right, the infected mosquito population $I + U$

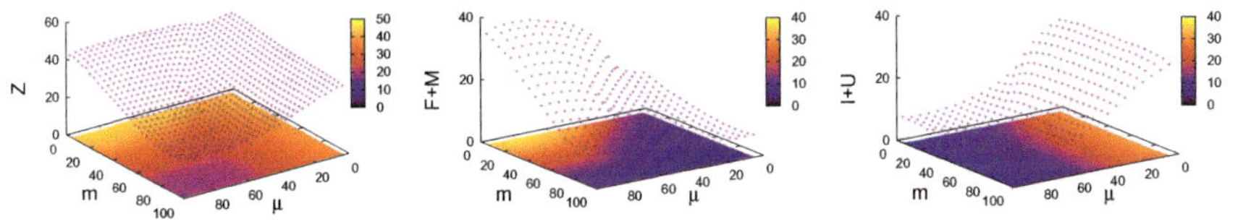

Fig. 6 Mosquito population at equilibrium in terms of the mortality rates m and μ. In **a** left, the total mosquito population $Z = F + I + M + U$; note that this frame is rotated with respect to most of the other ones, the origin being located at the far end of the frame, in **b** center, the healthy mosquito population $F + M$, and in **c** right, the infected mosquito population $I + U$

induced mortality μ is much higher than the natural mortality m; on the right of Fig. 6a instead the system settles to the endemic situation, because here the bacterium has a reduced impact on the single individual. This situation corresponds to the phenomenon observed in various natural situations, where diseases with high killing rates are of course lethal for the individual, but in this way they have fewer chances to spread in the whole community, with a thereby resulting smaller impact for the population as a whole. For better understanding of these remarks, see Fig. 6b and c for the total healthy and infected subpopulations.

5.3 *Vertical Transmission*

Figure 7 shows the healthy, infected and total mosquito populations as function of the *Wolbachia* vertical transmission rate ρ, in the meaningful range $\rho \in [0.7, 1]$, because for lower values the equilibrium becomes bacterium-free. The infected population is clearly seen to rise with higher values of the transmission coefficient. Therefore, it is important to choose a strain of *Wolbachia* that optimizes the transmission of the bacteria from mother to its offspring.

5.4 *Mating Prevention*

The role of *Wolbachia* as preventing agent for mating reproduction success among an infected male and a healthy female transpires from Fig. 8. In view of the field findings, [18], only very small values of $k << 1$ are meaningful, and in such case the healthy population settles at very low levels, as it should be expected. Comparing Figs. 8 and 7 we observe that the coexistence equilibrium is more sensitive to the parameter ρ that is related to bacteria inheritance than to k that accounts for the cytoplasmic incompatibility.

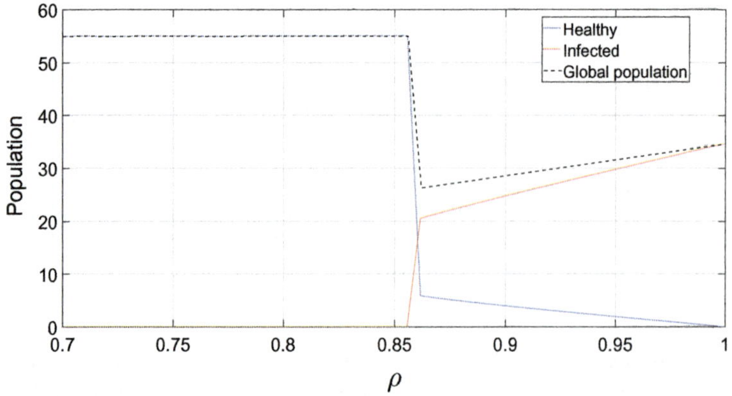

Fig. 7 Mosquito populations at equilibrium in terms of the vertical transmission rate ρ. Blue line: the total healthy population $F + M$; red line: the total infected population $I + U$; black line: the total mosquito population $Z = F + I + M + U$

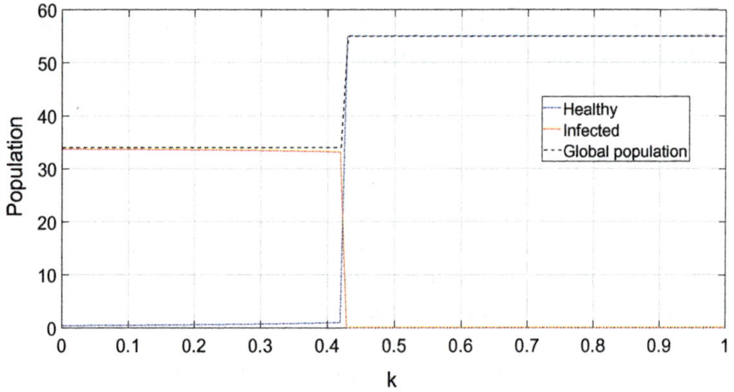

Fig. 8 Mosquito populations at equilibrium in terms of the bacterium-induced mating prevention action k. Blue line: the total healthy population $F + M$; red line: the total infected population $I + U$; black line: the total mosquito population $Z = F + I + M + U$

5.5 *Intraspecific Competition*

In Fig. 9a, where again the origin is at the far end, for decreasing and low values of b, the intraspecific competition experienced by the infected females entails a large increase in the mosquito level, independently of the intraspecific competition experienced by the healthy females a. The increase is due to *Wolbachia* endemic presence in the population, see Fig. 9b and c. Large values of b coupled with low values of a determine also a sharp population increase, in this case, being due to the bacterium eradication.

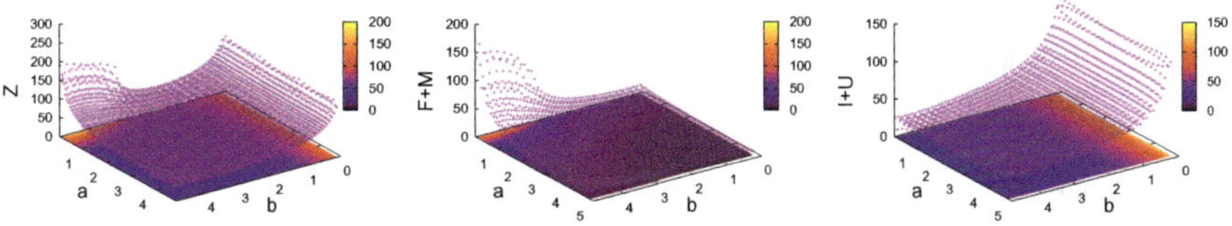

Fig. 9 Equilibrium in terms of the healthy and infected females intraspecific competitions a and b. In **a** left, the total mosquito population $Z = F + I + M + U$, in **b** center, the healthy mosquito population $F + M$, and in **c** right, the infected mosquito population $I + U$

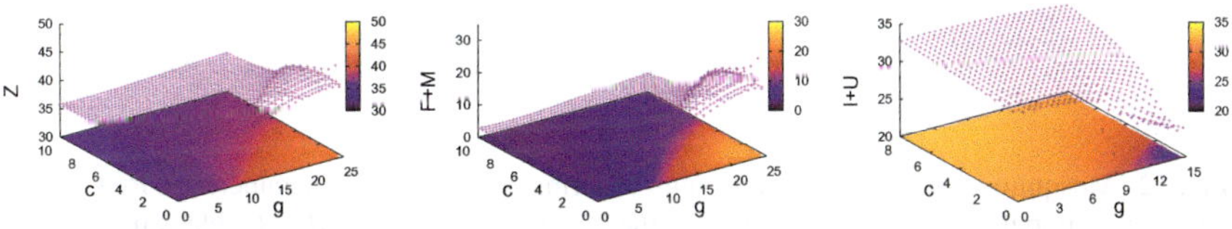

Fig. 10 Equilibrium in terms of the healthy and infected males intraspecific competitions c and g. In **a** left, the total mosquito population $Z = F + I + M + U$, in **b** center, the healthy mosquito population $F + M$, and in **c** right, the infected mosquito population $I + U$

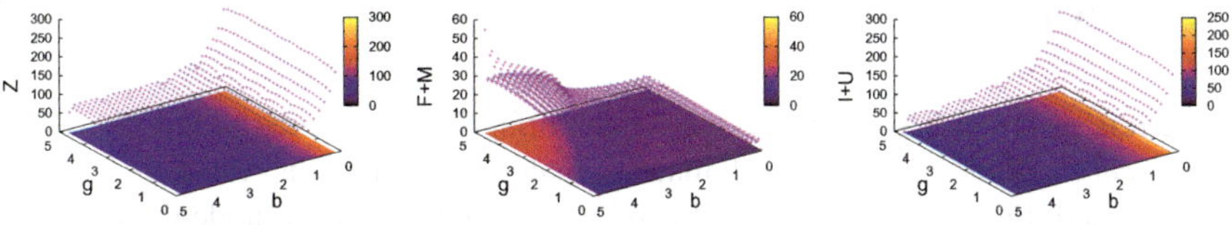

Fig. 11 Equilibrium in terms of the infected females and males intraspecific competitions b and g. In **a** left, the total mosquito population $Z = F + I + M + U$; note that the figure is rotated and the origin is located in the right corner; in **b** center, the healthy mosquito population $F + M$; and in **c** right, the infected mosquito population $I + U$

A sharp increase in the mosquito population level is found in the right corner of Fig. 10a, i.e. for low values of healthy males intraspecific competition c and large ones for their infected counterparts, g. Again, the sudden surge is due to the disappearance of the bacterium from the population, compare Fig. 10b and c.

As it has been discovered in the left frame of Fig. 9, also in case of changes in the infected subpopulations intraspecific coefficients b and g a jump is observed, Fig. 11, when large values of both coefficients produce higher equilibrium value, where the infection is eradicated. On the other hand, small values of the female competition b induce the endemic equilibrium, independently of the value that a attains, see the central and right frames in Fig. 11.

Finally, the healthy female and male intraspecific parameters a and c do not alter the coexistence equilibrium. And, in order to increase its presence in the mosquito population, the bacteria must increase male's mortality.

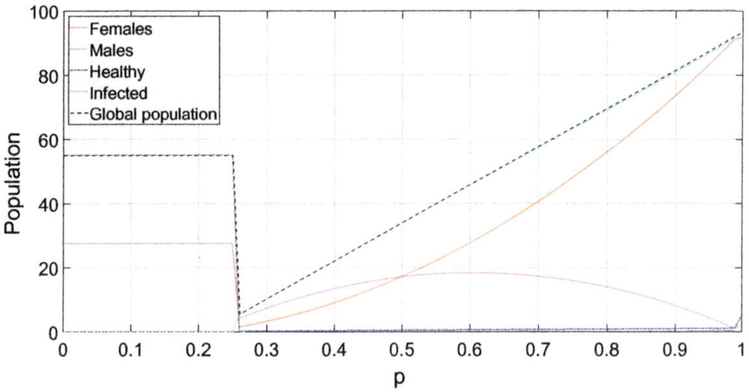

Fig. 12 Mosquito populations at equilibrium in terms of the sex ratio parameter p. Red line: the total female population $F + I$; magenta line: the total male population $M + U$; black line: the total mosquito population $Z = F + I + M + U$; cyan line: the total infected population $I + U$ blue line: the total healthy population $F + M$

5.6 *Sex Ratio*

Here it is seen that a low value of the female offsprings newborns p, induced by the *Wolbachia* action, entails a low female to male subpopulations' ratio, as it should be expected, Fig. 12. At a critical value of $p \simeq 0.25$ the females suddenly disappear, to rebound to higher values for increasing values of p, up to 90% when $p = 1$. A corresponding linear increase of the total population is also observed. The males instead in the interval of $p \in [0.25, 1]$ experience an increase to the peak at $p \simeq 0.6$, to decline afterwards up to a tenth of the total population when $p = 1$. If we consider the infected population, it is absent for $p < 0.25$, and correspondingly the total population is healthy. For $p > 0.25$ the infected population starts to increase, with increasing p, attaining 90% of the population for $p = 1$, and a corresponding decrease of the healthy individuals occurs. Thus, as it is expected, sex relation bias to female ($p > 0.5$) increases the probability of the *Wolbachia* infection's fixation in the population, especially when the bacteria vertical transmission rate is high.

6 Conclusion

The introduction of *Wolbachia*-infected mosquito into uninfected wild populations of *Aedes* mosquito can reduce or halt the transmission of some arboviruses. This occurs because disease transmission depends on the ratio between vector and human populations, and the *Wolbachia*-carrying mosquitoes have reduced vectorial capacity when challenged with dengue virus. The establishment of a *Wolbachia*-carrying population in an area where a *Wolbachia*-free population is endemic depends on several parameters like mosquitoes fitness, maternal inheritance and cytoplasmic incompatibility. Therefore, using a sex-structured epidemiological model, we discussed populations competition and persistence under several scenarios. The proposed model has four equilibria: (i) extinction of mosquito population, (ii) extinction of infected population and persistence of uninfected one, (iii) extinction of uninfected population and persistence of infected one, and (iv) coexistence of both populations. First, the persistence of the population (infected or uninfected one) occurs only for $r > 2m$ or, $r_1 > 2\mu$ which means that reproduction rates overcome mortality rates. Besides, case (iii) is possible only when maternal inheritance is perfect. Moreover, coexistence is achieved given that the initial infection prevalence is above a threshold value (given by Eq. 18) which depends on model parameters. Finally, female-biased sex ratio (see Fig. 12), or high maternal inheritance (see Fig. 7), or high cytoplasmic incompatibility (see Fig. 8) promote the fixation of the *Wolbachia*-infection on the mosquito population. Overall, the results obtained here are in agreement with the ones described in the literature.

References

1. Adekunle, A.I., Meehan, M.T., McBryde, E.S.: Mathematical analysis of a *Wolbachia* invasive model with imperfect maternal transmission and loss of *Wolbachia* infection. Infect. Dis. Model. **4**, 265–285 (2019)
2. Benedito, A.S., Ferreira, C.P., Adimy, M.: Modeling the dynamics of *Wolbachia*-infected and uninfected *Aedes aegypti* populations by delay differential equations. Math. Model. Nat. Phenom. **15**, 76 (2020)
3. Biedler, J.K., Tu, Z.: Sex determination in mosquitoes. Adv. Insect Physiol. **51**, 37–66 (2016)
4. Calvitti, M.: Studio della simbiosi tra *Aedes Albopictus* e *Wolbachia Pipientis* in rapporto allo sviluppo di tecniche di lotta basate sulla incompatibilità citoplasmatica. (Symbiosis study between *Aedes Albopictus* and *Wolbachia Pipientis* in relatioship with fighting techniques development based on citoplasmatic incompatibility) Università degli Studi della Tuscia di Viterbo (2008). https://dspace.unitus.it/handle/2067/1966?mode=full&locale=it
5. Campo-Duarte, D.E., Vasilieva, O., Cardona-Salgado, D., Svinin, M.: Optimal control approach for establishing wMelPop *Wolbachia* infection among wild *Aedes aegypti* populations. J. Math. Biol. **76**(7), 1907–1950 (2018)
6. Ding, H., Yeo, H., Puniamoorthy, N.: *Wolbachia* infection in wild mosquitoes (Diptera: Culicidae): implications for transmission modes and host-endosymbiont associations in Singapore. Parasites Vectors **13**, 612 (2020). https://doi.org/10.1186/s13071-020-04466-8
7. Dorigatti, I., McCormack, C., Nedjati-Gilani, G., Ferguson, N.M.: Using *Wolbachia* for dengue control: insights from modelling. Trends Parasitol. **34**(2), 102–113 (2018)
8. Ferreira, C.P.: *Aedes aegypti* and *Wolbachia* interaction: population persistence in an environment changing. Theor. Ecol. **13**, 137–148 (2020). https://doi.org/10.1007/s12080-019-00435-9
9. Fallon, A.M.: Growth and maintenance of *Wolbachia* in insect cell lines. Insects **12**(8), 706 (2021)
10. King, J.G., Souto-Maior, C., Sartori, L.M., Maciel-de-Freitas, R., Gomes, M.G.M.: Variation in *Wolbachia* effects on *Aedes* mosquitoes as a determinant of invasiveness and vectorial capacity. Nat. Commun. **9**(1), 1–8 (2018)
11. Medici, A.: Ricerche di campo per lo sviluppo della tecnologia del maschio sterile nella lotta ad *Aedes Albopictus*. (Field research for the development of sterile males in the fight against *Aedes Albopictus*) Università di Bologna (2013). http://amsdottorato.unibo.it/5386/

12. Moretti, R., Yen, P.S., Houé, V., Lampazzi, E., Desiderio, A., Failloux, A.B., Calvitti, M.: Combining *Wolbachia*-induced sterility and virus protection to fight *Aedes albopictus*-borne viruses. PLOS Negl. Trop. Dis. (2018). https://doi.org/10.1371/journal.pntd.0006626
13. Pinto, S.B., Riback, T.I.S., Sylvestre, G., Costa, G., Peixoto, J., Dias, F.B.S., Tanamas, S.K., Simmons, C.P., Dufault, S.M., Ryan, P.A., O'Neil, S.L., Muzzi, F.C., Kutcher, S., Montgomery, J., Green, B.R., Smithyman, R., Eppinghaus, A., Saraceni, V., Durovni, B., Anders, K.L., Moreira, L.A.: Effectiveness of *Wolbachia*-infected mosquito deployments in reducing the incidence of dengue and other *Aedes*-borne diseases in Niterói, Brazil: A quasi-experimental study. PLOS Negl. Tropic. Dis. (2021). https://doi.org/10.1371/journal.pntd.0009556
14. Pocquet, N., et al.: Assessment of fitness and vector competence of a New Caledonia w Mel *Aedes aegypti* strain before field-release. PLoS Negl. Trop. Dis. **15**(9), e0009752 (2021)
15. Qu, Z., Xue, L., Hyman, J.M.: Modeling the transmission of *Wolbachia* in mosquitoes for controlling mosquito-borne diseases. SIAM J. Appl. Math. **78**(2), 826–852 (2018)
16. Rafikov, M., Meza, M.E.M., Ferruzzo Correa, D.P., Wyse, A.P.: Controlling *Aedes aegypti* populations by limited Wolbachia-based strategies in a seasonal environment. Math. Methods Appl. Sci. **42**(17), 5736–5745 (2019)
17. Reyes, J.I., Suzuki, Y., Carvajal, T., Muñoz, M.N., Watanabe, K.: Intracellular interactions between arboviruses and *Wolbachia* in *Aedes aegypti*. Front. Cell. Infect. Microbiol. 11, 690087 (2021). https://doi.org/10.3389/fcimb.2021.690087
18. Ross, P.A., Wiwatanaratanabutr, I., Axford, J.K., White, V.L., Endersby-Harshman, N.M., Hoffman, A.A.: *Wolbachia* infections in *Aedes aegypti* differ markedly in their response to cyclical heat stress. PLoS Pathog **13**(1), e1006006 (2017). https://doi.org/10.1371/journal.ppat.1006006. https://journals.plos.org/plospathogens/article?id=10.1371/journal.ppat.1006006
19. Utarini, A., Indriani, C., Ahmad, R.A., Tantowijoyo, W., Arguni, E., Ansari, R., Supriyati, E., Wardana, D.S., Meitika, Y., Ernesia, I., Nurhayati, I., Prabowo, E., Andari, B., Green, B.R., Hodgson, L., Cutcher, Z., Rancès, E., Ryan, P.A., O'Neil, S.L., Dufault, S.M., Tanamas, S.K., Jewell, N.P., Anders, K.L., Simmons, C.P.: Efficacy of *Wolbachia*-infected mosquito deployments for the control of dengue. New Engl. J. Med. (2021) https://www.nejm.org/doi/10.1056/NEJMoa2030243
20. Centers for Disease Control and Prevention. https://www.cdc.gov/mosquitoes/mosquito-control/community/sit/wolbachia.html
21. World Mosquito Program. https://www.worldmosquitoprogram.org/en/work/wolbachia-method/how-it-works
22. Yang, H.M., Ferreira, C.P.: Assessing the effects of vector control on dengue transmission. Appl. Math. Comput. **198**(1), 401–413 (2008)
23. Zhang, H., Lui, R.: Releasing *Wolbachia*-infected *Aedes aegypti* to prevent the spread of dengue virus: A mathematical study. Infect. Dis. Model. **5**, 142–160 (2020)

Effect of Nonlinear Harvesting on a Fractional-Order Predator-Prey Model

Kshirod Sarkar and Biswajit Mondal

Abstract In this paper, we have introduced nonlinear harvesting to study a fractional-order predator-prey model with Holling-II response. Existence of multiple equilibrium points of the model and their stability under different conditions are analyzed. Global stability analysis of the co-existence equilibrium point has examined by considering suitable Lyapunov function. we have observed that fractional-order model is more viable for the memory-based system than integer-order model. Our model exhibits Hopf bifurcation with respect to fractional-order of the derivative. Furthermore, we have investigated period of doubling bifurcation with respect to some other parameters. Numerical simulations have performed to verify the theoretical results using Adams-Bashforth-Moulton type scheme.

Keywords Fractional-order · Nonlinear harvesting · Stability analysis · Hopf bifurcation

1 Introduction

For the basic needs, harvesting of biological resources are rigorously practiced for fishing and management of wildlife etc. The study of dynamical systems with harvesting is another important research topic in population dynamics. It is observed that the nonlinear harvesting is more reasonable when the number of harvesting of species is very large. Furthermore, It is quite realistic to implement linear harvesting when prey population is low. In linear harvesting, the harvesting function $h(x) = qEx$, increases proportionally as harvesting species x increases, which is unrealistic in the

K. Sarkar (✉) · B. Mondal
Raja N.L. Khan Women's College (Autonomous), Vidysagar University Road, Midnapore 721102, West Bengal, India
e-mail: kshirodsarkar8@gmail.com; ks_mathematics@rnlkwc.ac.in

B. Mondal
e-mail: bm-mathematics@rnlkwc.ac.in

S. Banerjee and A. Saha (eds.), *Nonlinear Dynamics and Applications*,
Springer Proceedings in Complexity,
https://doi.org/10.1007/978-3-030-99792-2_64

biological sense. But these unreasonable features can be fixed by applying nonlinear type harvesting.

Many researchers have considered nonlinear harvesting in their models. Hu and Cao [7] has incorporated a nonlinear type prey harvesting in Leslie-Gower model and have discussed different types of bifurcation analysis. A bioeconomic predator-prey model is studied by [9] and they consider nonlinear harvesting for prey only. Also [6] have discussed the predator-prey model with nonlinear harvesting in predator population and established that nonlinear harvesting is more realistic for large population. Stability and bifurcation analysis in a prey-predator model with group defence as well as nonlinear harvesting are studied by [8]. Effect of nonlinear harvesting in a prey-predator model with square root functional response are discussed by [14]. Furthermore, a notable research considering nonlinear type harvesting have examined by [5]. Stability and bifurcation analysis of a prey-predator model with Holling-IV functional response and nonlinear harvesting are examined by [16]. Recently, [11] have examined the dynamical effects of nonlinear harvesting in prey species with Holling-II predation in their model and have considered the harvesting function $h(x) = \frac{hu}{h+u}$ where h is positive constants. But the form of this type nonlinear harvesting function is a particular case rather than the general case. The general form of nonlinear harvesting function is $f(u) = \frac{qEu}{m_1E+m_2u}$. Here, q denotes the harvesting coefficient, E denotes the harvesting effort and m_1, m_2 are positive constants. We have incorporated nonlinear harvesting function $\frac{qEu}{m_1E+m_2u}$ in our model

$$\begin{aligned}\frac{du}{d\tau} &= ru\left(1-\frac{u}{k}\right) - \frac{buv}{c+u} - \frac{qEu}{m_1E+m_2u}\\ \frac{dv}{d\tau} &= \frac{euv}{c+u} - dv,\end{aligned} \tag{1}$$

where $u(0) > 0$, $v(0) > 0$. Here r and k are intrinsic growth rate and carrying capacity of prey species respectively. Parameter b is the consumption rate per unit time, c is the half saturation constant, e is the conversion factor per unit time by per predator and d is the death rate of predator species. To reduce free and floating parameters, we have applied non-dimensional scheme. Thus we take some set of transformation as follows: $x = \frac{u}{k}$, $y = \frac{bv}{k}$, $t = r\tau$, $\beta = \frac{1}{r}$, $\alpha = \frac{c}{k}$, $h = \frac{qE}{m_2rk}$, $m = \frac{m_1E}{m_2k}$, $\beta_1 = \frac{e}{r}$ and $\delta = \frac{d}{r}$. Then we have obtained a dynamical model:

$$\begin{aligned}\frac{dx}{dt} &= x(1-x) - \frac{\beta xy}{\alpha+x} - \frac{hx}{m+x}\\ \frac{dy}{dt} &= \frac{\beta_1 xy}{\alpha+x} - \delta y,\end{aligned} \tag{2}$$

Fractional order differential equations manifests greater degrees of freedom and some realistic results in complex dynamical model. Fractional order differential equation is more appropriate to analyze biological system as it contains the memory kernel. In recent years, many researches have been done in fractional-order model due to its ability to execute a better approximation of nonlinear dynamics. Now consider $n-1<\mu<n$ and $n \in N$, then the Caputo type fractional derivative of order $\mu>0$ is defined as [2]:

$$D_t^{\mu} f(t)=\frac{1}{\Gamma(n-\mu)} \int_{t_0}^{t} \frac{f^n(s)}{(t-s)^{\mu+1-n}} ds,$$

where $f(t)$ is a function of order μ and $f(t) \in C^n([t_0, \infty), \mathrm{R})$. There are very few articles which have discussed fractional-order prey-predator model with nonlinear harvesting. Fractional order Leslie-Gower model with nonlinear harvesting is studied by [4]. Further, authors [12] have considered a fractional-order Leslie-Gower model with nonlinear harvesting and discussed the optimal control strategies. Here we consider the model 2 with fractional order differential equation as follows:

$$\begin{aligned}
\frac{d^{\mu}x}{dt^{\mu}} &= x(1-x)-\frac{\beta xy}{\alpha+x}-\frac{hx}{m+x}\\
\frac{d^{\mu}y}{dt^{\mu}} &= \frac{\beta_1 xy}{\alpha+x}-\delta y,
\end{aligned} \tag{3}$$

where $x(0)>0$, $y(0)>0$ and $0<\mu\leq 1$.

The arrangement of remaining portion of the paper is given as follows: existence of different equilibrium points and their local stability are examined in Sect. 2. The global stability and Hopf bifurcation are discussed in Sects. 3 and 4 respectively. Numerical simulations have been organized in Sect. 5 and finally, conclusions have discussed in Sect. 6.

2 Equilibrium Points and Local Stability

The trivial equilibrium point $E_0=(0,0)$ is always exist. This model has at most two predator free equilibrium(PFE) points, namely it is $E_L=(x_L, 0)$ and $E_R=(x_R, 0)$ and it depends on the equation $g(x)=1-x-\frac{h}{m+x}=0$. Then roots of the equation are x_L and x_R, where $x_L=\frac{1}{2}\left(1-m-\sqrt{(1-m)^2-4(h-m)}\right)$ and $x_R=\frac{1}{2}\left(1-m+\sqrt{(1-m)^2-4(h-m)}\right)$. Conditions for the existence of two PFE points is $m<h<m_0=\left(\frac{1+m}{2}\right)^2$, where $m<1$. In this case obviously, $g'(x_L)>0$ and $g'(x_R)<0$. Only axial PFE point $E_R=(x_R, 0)$ exist for $h\leq m$. If $h=m_0$, these two PFE points collides each other at the point $\left(\frac{1-m}{2}, 0\right)$ and then $g'\left(\frac{1-m}{2}\right)=0$. No axial equilibrium point exist for $h>m_0$. Co-existing equilibrium point is obtained at $E_*=\left(x^*, y^*=\left(\frac{\alpha+x^*}{\beta}\right)g(x^*)\right)$, where $x^*=\frac{\delta\alpha}{\beta_1-\delta}$, which is meaningful only when $\beta_1>\delta$. Hence from the above studies, we have the following results:

(i) If $m<h<m_0$, system has two PFE points E_R and E_L. Interior equilibrium point E_* exists if $x_L<x^*<x_R$.

(ii) If $m \geq h$, system has one PFE point E_R. Interior equilibrium point E_* exist if $0 < x^* < x_R$.

(iii) If $h = m_0$, two PFE points E_R and E_R collide with each other. No interior equilibrium point exists when $h \geq m_0$.

The Jacobian matrix at any point (x, y) of the system 3 is

$$J(x, y) = \begin{pmatrix} 1 - 2x - \frac{\beta\alpha y}{(\alpha+x)^2} - \frac{hm}{(m+x)^2} & -\frac{\beta x}{\alpha+x} \\ \frac{\beta_1 \alpha y}{(\alpha+x)^2} & \frac{\beta_1 x}{\alpha+x} - \delta \end{pmatrix}. \quad (4)$$

2.1 $m < h < m_0$

Theorem 1 *(a) The trivial equilibrium point E_0 is locally asymptotically stable (LAS).*

(b) Both the equilibrium points E_L and E_R are saddle point if $x_L < x^* < x_R$.

(c) The equilibrium point E_L is unstable, while equilibrium point E_R is a saddle point if $0 < x^* < x_L$.

(d) The equilibrium point E_L is saddle point, while equilibrium point E_R is LAS if $x_R < x^*$.

(e) If $T(J(E_*)) \leq 0$, then the equilibrium point E_* of the system 3 is LAS.

(f) If $T(J(E_*)) > 0$ and $T^2(J(E_*)) < 4D(J(E_*))$, then the equilibrium point E_* is LAS for μ if $\sqrt{\mid T^2(J(E_*)) - 4D(J(E_*)) \mid} > T(J(E_*))tan\left(\frac{\mu\pi}{2}\right)$.

Proof We have analysed the local stability of the different equilibrium points according to the results of [13, 15]. (a) We have obtained the eigen values of the matrix $J(E_0)$. It is observed that the eigenvalues of $\lambda_1 < 0$ if $h > m$ and $\lambda_2 < 0$, consequently we get $\mid arg(\lambda_1) \mid = \pi > \frac{\mu\pi}{2}$. Hence according to Matignon's condition, E_0 is LAS when $h > m$.

(b) From Eq. 4, we have obtained the eigenvalues of the Jacobian matrix $J(E_L)$ as $\lambda_1 = x_L g'(x_L)$ and $\lambda_2 = \frac{\beta_1 x_L}{\alpha+x_L} - \delta$. For $x_L < x^* < x_R$ it gives $g'(x_L) > 0$ and $\frac{\beta_1 x_L}{\alpha+x_L} < \delta$, consequently we have $\lambda_1 > 0$ and $\lambda_2 < 0$. Thus E_L is saddle point. The eigenvalues of the matrix $J(E_R)$ are obtained as $\lambda_1 = x_R g'(x_R)$ and $\lambda_2 = \frac{\beta_1 x_R}{\alpha+x_R} - \delta$. But in this case we have $\lambda_1 < 0$ and $\lambda_2 > 0$, as $g'(x_R) < 0$ and $\frac{\beta_1 x_R}{\alpha+x_R} > \delta$. Hence E_R is also a saddle point.

(c) For the case $0 < x^* < x_L$, it is obvious that eigenvalue $\lambda_1 > 0$ of $J(E_L)$ as $g'(x_L) > 0$ and eigenvalue $\lambda_2 > 0$ as $\frac{\beta_1 x_L}{\alpha+x_L} - \delta > 0$. So the condition $\mid arg(\lambda_i) \mid > \frac{\mu\pi}{2}$, for $i = 1, 2$ does not satisfy.

Hence in this case E_L is unstable. Now eigenvalue of the matrix $J(E_R)$ are such that $\lambda_1 = x_R g'(x_R)$ and $\lambda_2 = \frac{\beta_1 x_R}{\alpha+x_R} - \delta$. Similarly in this case we obtained as $\lambda_1 < 0$ and $\lambda_2 > 0$. Therefore E_R is a saddle point.

(d) If $x_R < x^*$, then eigenvalues of $J(E_L)$ is such that $\lambda_1 > 0$ and $\lambda_2 < 0$. Hence E_L is saddle point. Now eigenvalues of $J(E_R)$ is such that $\lambda_1 < 0$ and $\lambda_2 < 0$. Therefore, using Matignon's condition of stability we have E_R is LAS.

(e) In this case coexistence equilibrium point E_* exist when $x_L < x^* < x_R$. The characteristic equation of $J(E_*)$ is given by $\lambda^2 - T(J(E_*))\lambda + D(J(E_*)) = 0$. Then it gives $T(J(E_*)) = 1 - 2x^* - \frac{\beta\alpha y^*}{(\alpha+x^*)^2} - \frac{hm}{(m+x^*)^2}$ and $D(J(E_*)) = \frac{\beta\alpha\beta_1 x^* y^*}{(\alpha+x^*)^3} > 0$. Now if $T(J(E_*)) = 0$, then $J(E_*)$ has two purely imaginary complex conjugate eigenvalues λ_1 and λ_2 such that $| arg(\lambda_{1,2}) |= \frac{\pi}{2}$. Hence the equilibrium point E_* is stable. If $T(J(E_*)) < 0$, then also the eigenvalues of $J(E_*)$ are satisfied the Matignon's condition and hence E_* is LAS.

(f) If $T(J(E_*)) > 0$ and $T^2(J(E_*)) < 4D(J(E_*))$, then $J(E_*)$ has a pair of complex conjugate eigenvalues λ_1 and λ_2 with positive real parts. From the given condition it is obvious that $Im(\lambda_1) > Re(\lambda_1)\tan(\frac{\mu\pi}{2})$ and $-Im(\lambda_2) > Re(\lambda_2)\tan(\frac{\mu\pi}{2})$. After some calculation it implies that $| arg(\lambda_i) |> \frac{\mu\pi}{2}$, for $i = 1, 2$. Hence E_* is LAS for $\mu \in (0, \mu^*]$, where the critical value μ^* is given by $\mu^* = \frac{2}{\pi}\tan^{-1}\left(\sqrt{| T^2(J(E_*)) - 4D(J(E_*)) |}/T(J(E_*))\right)$.

2.2 $h \leq m$

Theorem 2 *(a) The trivial equilibrium point E_0 is saddle point when $h < m$ and stable when $h = m$.*
(b) Only PFE point E_R is saddle point if $0 < x^* < x_R$ and LAS if $x_R < x^*$.
(c) Interior equilibrium point E_* of the system 3 is LAS if $T(J(E_*)) \leq 0$.
(d) If $T(J(E_*)) > 0$ and $T^2(J(E_*)) < 4D(J(E_*))$, then the equilibrium point E_* is LAS for μ if $\sqrt{| T^2(J(E_*)) - 4D(J(E_*)) |} > T(J(E_*))tan\left(\frac{\mu\pi}{2}\right)$.

Proof (a) The eigenvalues of $J(E_0)$ are $\lambda_1 = 1 - \frac{h}{m}$ and $\lambda_2 = -\delta < 0$. Now for $h < m$, we have $\lambda_1 > 0$ and hence E_0 is saddle point for the case $h < m$.

(b) In this case only one PFE point E_R exist. Now we obtain the eigenvalues of $J(E_R)$ as $\lambda_1 = x_R g'(x_R)$ and $\lambda_2 = \frac{\beta_1 x_R}{\alpha+x_R} - \delta$. For $0 < x^* < x_R$, it gives $g'(x_R) < 0$ and $\frac{\beta_1 x_R}{\alpha+x_R} > \delta$. Consequently we have $\lambda_1 < 0$ and $\lambda_2 > 0$. Thus E_R is saddle point. But when $x_R < x^*$, we have $\lambda_1 < 0$ and $\lambda_2 < 0$. Hence in this case E_R is LAS, since it satisfies the Matignon's condition. In this case coexistence equilibrium point E_* exist when $0 < x^* < x_R$. Local stability analysis around E_* of the Theorem 2c and d are same as the Theorem 1e and f respectively.

2.3 $h \geq m_0$

It is clear that when $g'(x_L) = 0$, then E_L coincide with E_R. Thus in this case saddle-node bifurcation occurs at $h = m_0$, when $m \neq 1$. the only equilibrium point E_0 exists when $h \geq m_0$. Eigen values of $J(E_0)$ are $\lambda_1 = 1 - \frac{h}{m} < 0$ and $\lambda_2 = -\delta < 0$ and hence E_0 is LAS when $h \geq m_0$.

3 Global Stability

Theorem 3 *The co-existence equilibrium point E_* of the dynamical model 3 is globally asymptotically stable if it is locally asymptotically stable.*

Proof For $E_* = (x^*, y^*)$, we have $1 - x^* - \frac{\beta y^*}{\alpha+x^*} - \frac{h}{m+x^*} = 0$ and $\frac{\beta_1 x^*}{\alpha+x^*} - \delta = 0$. Consider a function $L_1(x, y) = \left(x - x^* - x^* log\frac{x}{x^*}\right) + a_1\left(y - y^* - y^* log\frac{y}{y^*}\right)$. where $a_1 > 0$ is a constant. Now applying μ fractional order derivative of $L_1(x, y)$ and apply the Lemma 3.1 of [17]. After some calculations it follows that $\frac{d^\mu L_1}{dt^\mu} \leq \left(\frac{\beta y^*}{(\alpha+x)(\alpha+x^*)} + \frac{h}{(m+x)(m+x^*)} - 1\right)(x - x^*)^2 + (a_1\alpha\beta_1 - \beta x^* - \beta\alpha)\frac{(x-x^*)(y-y^*)}{(\alpha+x)(\alpha+x^*)}$.

Now choosing the constant $a_1 = \frac{\beta(\alpha+x^*)}{\alpha\beta_1}$ and after some calculations it has $\frac{d^\mu L_1}{dt^\mu} \leq \left(\frac{\beta y^*}{\alpha(\alpha+x^*)} + \frac{h}{m(m+x^*)} - 1\right)(x - x^*)^2$. Now consider the two functions $L_2(x, y) = \frac{1}{2}[(x - x^*) + \frac{\beta}{\beta_1}(y - y^*)]^2$ and $L_3(x, y) = [(x - x^*) + \frac{\beta}{\beta_1}(y - y^*)]$.

Then take μ fractional-order derivative of $L_3(x, y)$ and using linearity property we get that $\frac{d^\mu L_3}{dt^\mu} = -x(x - x^*) + \frac{hx(x-x^*)}{(m+x)(m+x^*)} + \frac{\beta y^*(x-x^*)}{\alpha+x^*} - \frac{\beta x^*(y-y^*)}{\alpha+x^*}$. Again applying μ fractional-order derivative in $L_2(x, y)$ we have

$$\frac{d^\mu L_2}{dt^\mu} < \left(\frac{h}{m+x^*} + \frac{\beta y^*}{\alpha+x^*} + \frac{\beta y^*}{\beta_1}\right)(x-x^*)^2 - \frac{\beta^2 x^*(y-y^*)^2}{\beta_1(\alpha+x^*)} + \left(\frac{\beta h}{\beta_1(m+x^*)} + \frac{\beta^2 y^*}{\beta_1(\alpha+x^*)} - \frac{\beta x^*}{\beta_1} - \frac{\beta x^*}{\alpha+x^*}\right)(x-x^*)(y-y^*). \quad (5)$$

It is noted that $\left(\frac{\beta h}{\beta_1(m+x^*)} + \frac{\beta^2 y^*}{\beta_1(\alpha+x^*)} - \frac{\beta x^*}{\beta_1} - \frac{\beta x^*}{\alpha+x^*}\right)(x - x^*)(y - y^*) \leq P(x - x^*)^2 + \frac{\beta^2 x^*(y-y^*)^2}{2\beta_1(\alpha+x^*)}$, where $P = \frac{\left(\frac{\beta h}{\beta_1(m+x^*)} + \frac{\beta^2 y^*}{\beta_1(\alpha+x^*)} - \frac{\beta x^*}{\alpha+x^*} - \frac{\beta x^*}{\beta_1}\right)^2}{2\beta^2 x^*/\beta_1(\alpha+x^*)}$. Thus from Eq. 5 we have $\frac{d^\mu L_2}{dt^\mu} \leq \left(\frac{h}{m+x^*} + \frac{\beta y^*}{\alpha+x^*} + \frac{\beta y^*}{\beta_1} + P\right)(x - x^*)^2 - \frac{\beta^2 x^*(y-y^*)^2}{2\beta_1(\alpha+x^*)}$.

Now consider the Lyapunov function $L(x, y) = L_1(x, y) + a_2 L_2(x, y)$, where $a_2 > 0$ is a positive constant. Therefore taking μ order derivative of $L(x, y)$ and after simplification it has

$$\frac{d^\mu L}{dt^\mu} \leq -\left(\left(1 - \frac{\beta y^*}{\alpha(\alpha+x^*)} - \frac{h}{m(m+x^*)}\right) - a_2\left(\frac{\beta y^*}{\alpha+x^*} + \frac{h}{m+x^*} + \frac{\beta y^*}{\beta_1} + P\right)\right) \times (x-x^*)^2 - \frac{a_2\beta^2 x^*(y-y^*)^2}{2\beta_1(\alpha+x^*)}.$$

Now we can choose constant a_2 in a suitable way such that $\left(1 - \frac{\beta y^*}{\alpha(\alpha+x^*)} - \frac{h}{m(m+x^*)}\right) - a_2\left(\frac{\beta y^*}{\alpha+x^*} + \frac{h}{m+x^*} + \frac{\beta y^*}{\beta_1} + P\right) = \frac{1}{2}\left(1 - \frac{\beta y^*}{\alpha(\alpha+x^*)} - \frac{h}{m(m+x^*)}\right)$. Thus it gives us $\left(1 - \frac{\beta y^*}{\alpha(\alpha+x^*)} - \frac{h}{m(m+x^*)}\right) = 2a_2\left(\frac{\beta y^*}{\alpha+x^*} + \frac{h}{m+x^*} + \frac{\beta y^*}{\beta_1} + P\right) > 0$. Then we get that $\frac{d^\mu L}{dt^\mu} \leq -\frac{1}{2}\left(1 - \frac{\beta y^*}{\alpha(\alpha+x^*)} - \frac{h}{m(m+x^*)}\right)(x - x^*)^2 - \frac{a_2\beta^2 x^*(y-y^*)^2}{2\beta_1(\alpha+x^*)}$. Therefore we have

$\frac{d^{\mu}L}{dt^{\mu}} < 0,\ for\ (x, y) \neq (x^*, y^*)$ and $\frac{d^{\mu}L}{dt^{\mu}} = 0\ for\ (x, y) = (x^*, y^*)$. Hence E_* of the system 3 is GAS.

4 Hopf Bifurcation

Theorem 4 *The necessary and sufficient conditions for the dynamical system 3 exhibits Hopf bifurcation through E_* if the parameter μ crosses the critical value $\mu^* = \frac{2}{\pi} tan^{-1}\left(\frac{\sqrt{|T^2(J(E_*)) - 4D(J(E_*))|}}{T(J(E_*))}\right)$, where $T(J(E_*)) > 0$ and $T^2(J(E_*)) < 4D(J(E_*))$.*

Proof From the given conditions it is clear that the eigenvalues of the Jacobian matrix $J(E_*)$ are $\lambda_{1,2} = \phi \pm i\omega$ where we have $\phi = \frac{1}{2}T(J(E_*)) > 0$ and also $\omega = \frac{1}{2}\sqrt{|\ T^2(J(E_*)) - 4D(J(E_*))\ |}$. Now it has $G_{1,2}(\mu^*) = \frac{\mu^*\pi}{2} - |\ arg(\lambda_i)\ | = \frac{\mu^*\pi}{2} - tan^{-1}\ |\ \omega/\Phi\ | = tan^{-1}\ |\ \omega/\Phi\ | - tan^{-1}\ |\ \omega/\Phi\ | = 0$ and obviously $\frac{\partial G_{1,2}}{\partial \mu}\ |_{\mu=\mu^*} = \frac{\pi}{2} \neq 0$. Therefore, according to Sect. (2.2) of [1] Hopf bifurcation occurs if the bifurcation parameter μ crosses the value $\mu = \mu^*$.

5 Numerical Simulation

Here we have performed some numerical computations to check the feasibility of our model. In this fractional-order model, a generalized Adams-Bashforth-Moulton type scheme is used accordingly by [3, 10]. We have observed that the parameters h and m have a major biological impact on the model. In Fig. 1, we have drawn

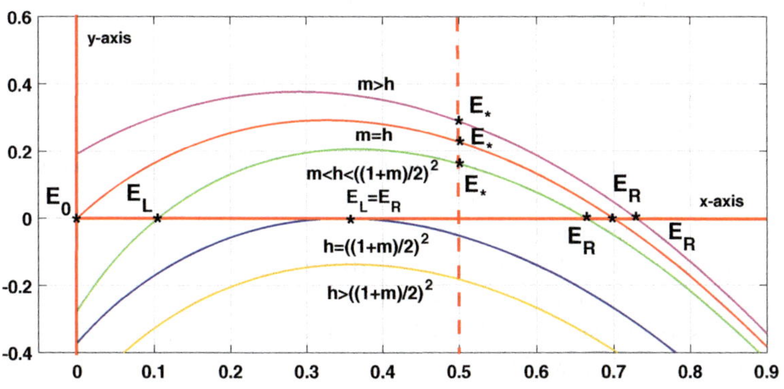

Fig. 1 The existence of various PFE equilibrium points depends on the parameters h and m. X-nullcline and y-nullcline for different values of parameters that are fixed except h and m are obtained

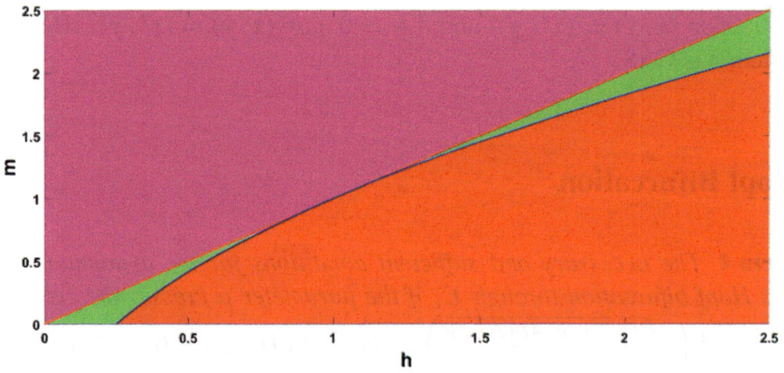

Fig. 2 No interior, as well as PFE point exists in the red shaded region. The interior equilibrium point and two PFE points exist in the green shaded region. Whereas interior equilibrium point and one PFE point exist in the cyan region

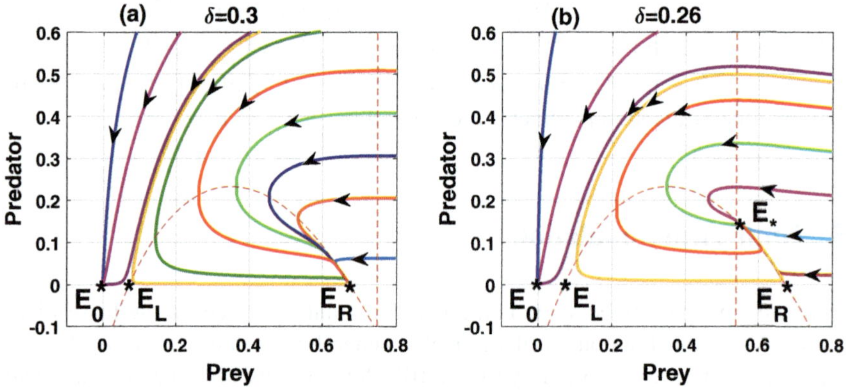

Fig. 3 Phase portraits are obtained for $\mu = 0.99, \alpha = 0.5, \beta = 0.55, \beta_1 = 0.5, h = 0.3, m = 0.25$. **a** no interior equilibrium exist, E_R is LAS, E_L is saddle point and E_0 is LAS. **b** Interior equilibrium point E_* is nodal sink, E_R and E_L are saddle point and E_0 is LAS

x-nullcline and y-nullcline where we consider different values of h, m along with $\beta = 0.55$, $\beta_1 = 0.5$, $\alpha = 0.5$, $\delta = 0.25$.

The Existence of the interior equilibrium and PFE points depends on two parameters h and m, which are shown in Fig. 2. First we check the case $m < h < m_0$ and consider the parameters $\beta = 0.55$, $\beta_1 = 0.5$, $\alpha = 0.5$, $h = 0.3$, $m = 0.25$ and $\delta = 0.3$. In this case two PFE points $E_R = (0.67604, 0)$ and $E_L = (0.07396, 0)$ are exist and no interior equilibrium exist as the condition $x_L < x^* < x_R$ does not follows. It is observed that E_0 is LAS as $h > m$, E_R is a nodal sink and E_L is a saddle point as $x^* > x_R$ and drawn in Fig. 3a. Now consider $\delta = 0.26$ and other parameters are the same. Then the system has two PFE points and interior equilibrium points at $E_* = (0.54167, 0.15035)$ which is a nodal sink as $T(J(E_*)) = -0.2411 < 0$ and depicted in Fig. 3b.

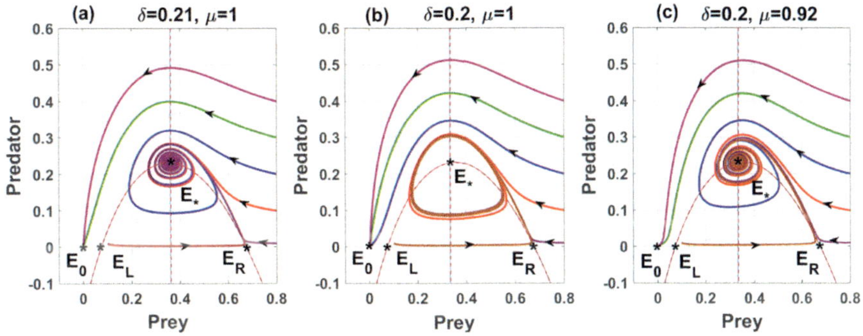

Fig. 4 Phase planes are drawn at $\beta = 0.55$, $\beta_1 = 0.5$, $\alpha = 0.5$, $h = 0.3$ and $m = 0.25$

Consider $\delta = 0.21$ and other parameters are same. Then the Phase portraits given by Fig. 4a shows that system 3 has interior equilibrium at $E_* = (0.36207, 0.23165)$. We found that E_* is a spiral sink and E_0 is LAS. Here E_L and E_R are saddle points. Now take $\delta = 0.2$ with the other parameters are same. We have seen that from Fig. 4b, coexistence equilibrium point $E_* = (0.34746, 0.23176)$ is unstable as $T(J(E_*) = 0.0062 > 0$. But if we consider $\mu = 0.92$, system 3 becomes LAS around the E_* and displayed in Fig. 4c. Now consider $\delta = 0.1$ and other parameters are fixed. Then $E_* = (0.125, 0.085227)$ is unstable and two PFE points are saddle points, which is shown in Fig. 5a. When $\delta = 0.05$, no interior equilibrium exists. We examine that E_L is a nodal source and E_R is a saddle point because $0 < x^* < x_L$ and depicted in Fig. 5b.

To check the dynamics for the case $h < m$, we have considered $\beta = 0.55$, $\beta_1 = 0.5$, $\alpha = 0.5$, $h = 0.3$, $m = 0.35$ and $\delta = 0.3$. Here, no coexistence equilibrium exist as $x^* > x_R$, one PFE point E_R exist and it is LAS and trivial equilibrium E_0 is a saddle point as $h < m$ and shown in Fig. 6a. Now consider $\delta = 0.25$ along with the same parameters value. It has examined that E_* exist as $x^* > x_R$ and it is

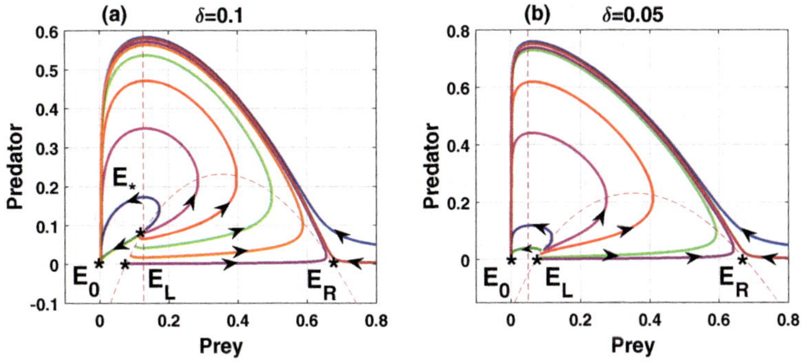

Fig. 5 Phase planes are drawn at $\mu = 0.99$, $\beta = 0.55$, $\alpha = 0.5$, $\beta_1 = 0.5$, $h = 0.3$ and $m = 0.25$

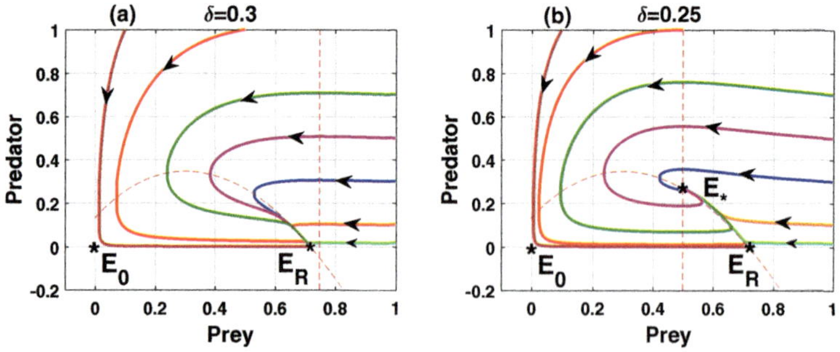

Fig. 6 Phase planes are drawn with the values of $\mu = 1$, $\beta = 0.55$, $\alpha = 0.5$, $\beta_1 = 0.5$, $h = 0.3$ and $m = 0.35$

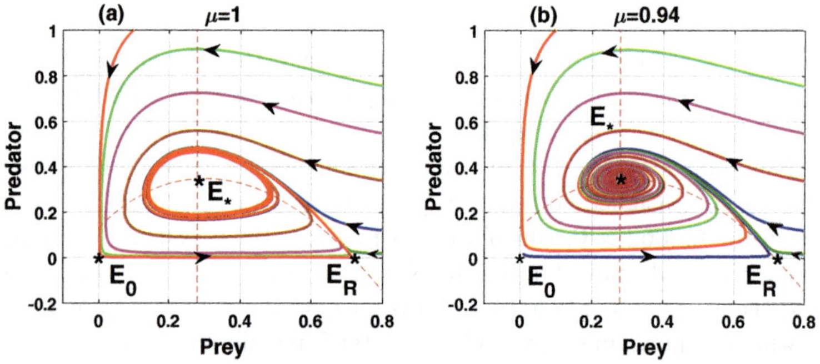

Fig. 7 Phase planes are drawn at $\beta = 0.55$, $\alpha = 0.5$, $\beta_1 = 0.5$, $h = 0.3$, $m = 0.35$ and $\delta = 0.18$

spiral sink. Beside these, E_0 and E_R are saddle point. Now consider $\delta = 0.18$ along with the same parameters value. We have obtained for integer-order system that the coexistence equilibrium E_* is a unstable and E_0, E_R are saddle points. But the system 3 with $\mu = 0.94$ becomes LAS around E_* as $\mu < \mu^* = 0.9655$, obtained from the Theorem 4 and shown in Fig. 7.

With the same parameters value we have obtained a figure with various regions of influence in (h, μ)-plane in Fig. 8. For $h = 0.3$, we have obtained $\mu^* = 0.9655$ and which is again verified in this figure. To check the effect δ and h in the system 3, we have obtained some bifurcation diagrams and have displayed in Figs. 9 and 10. We have observed that for $\mu = 1$ model 3 exhibits Hopf bifurcation around E_* at $\delta = 0.2331$. It is observed from Fig. 9 that predator species can be extinct if $\delta > 0.4985$. The bifurcation diagram given by the Fig. 10 shows that extinction of both species can happen if $h > 0.81$.

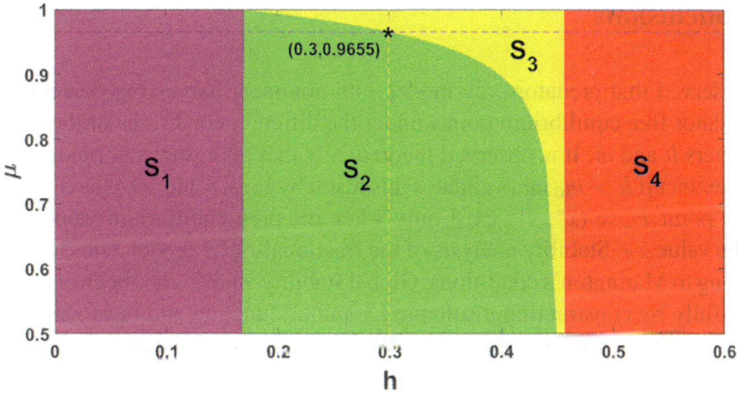

Fig. 8 We have obtained various regions in (h, μ)-plane for the parameters value $\beta = 0.55$, $\alpha = 0.5$, $m = 0.35$, $\beta_1 = 0.5$ and $\delta = 0.18$. Equilibrium point E_* is LAS in S_1 for integer-order as well as fractional-order system, LAS only for fractional-order system in S_2, unstable in S_3 and no co-existence equilibrium point exists in S_4

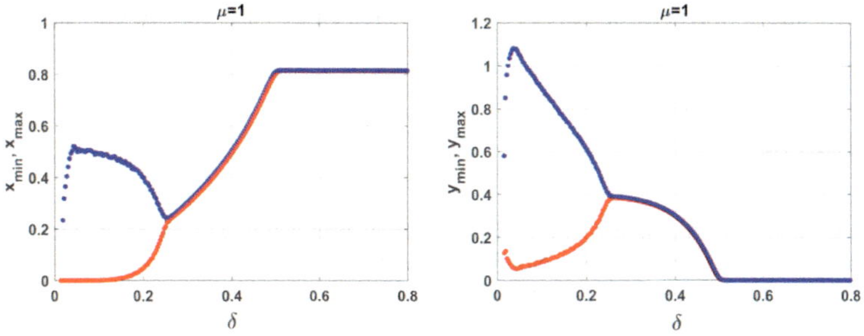

Fig. 9 Double period bifurcation occurs for δ when $\beta = 0.9$, $\alpha = 0.5$, $h = 0.3$, $m = 0.8$ and $\beta_1 = 0.8$. Hopf bifurcation occurs at the critical value $\delta = 0.2331$. When $\mu = 1$, model 3 becomes unstable around E_* when $0 \leq \delta < 0.2331$, LAS when $0.2331 < \delta < 0.5$ and extinction of predator species can happen for $\delta > 0.5$

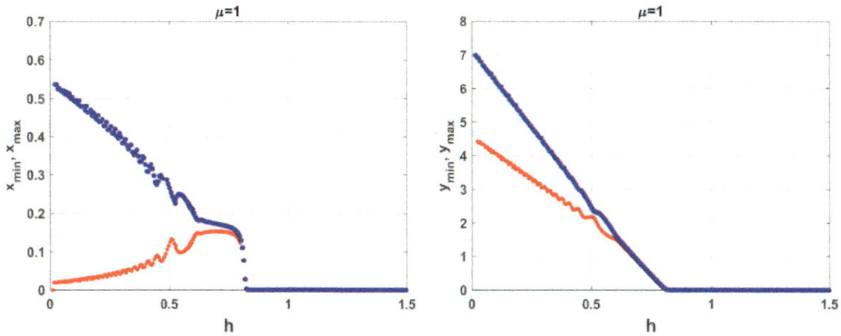

Fig. 10 Double period bifurcation when $\mu = 1$ and other parameters are considered as $\beta = 0.9$, $\alpha = 0.5$, $m = 0.8$, $\beta_1 = 0.8$ and $\delta = 0.2$

6 Conclusions

It is observed that predator-prey model with nonlinear harvesting possesses multiple predator free equilibrium points under the different conditions of the harvesting parameters h and m. It is observed theoretically that both prey and predator species can be extinct if $h > m_0$ and validates this result by Figs. 8 and 10. Coexistence equilibrium point $E_* = (x^*, y^*)$ exist only when the prey equilibrium value x^* is less than the value x_R. Stability analysis of the fractional-order model 3 has been studied according to Matington's conditions. Global stability analysis has been demonstrated successfully by constructing a suitable Lyapunov function and have verified it that E_* is globally asymptotically stable. We have discussed the occurrence of Hopf bifurcation for the fractional-order μ and have examined it theoretically and numerically. Also we have studied that Hopf bifurcation of the system 3 when $\mu = 1$ for the parameters δ and h occurs around the coexistence equilibrium point and have discussed in numerical simulation. It is shown that a larger death rate of predators can be extinct predator species. We have examined the influence of the fractional-order μ on the fractional-order system and it is shown that an unstable integer order system becomes stable in fractional-order system which obeys the theoretical results.

References

1. Abdelouahab, M.S., Hamri, N.E., Wang, J.: Hopf bifurcation and chaos in fractional-order modified hybrid optical system. Nonlinear Dyn. **69**(1), 275–284 (2012)
2. Diethelm, K., Ford, N.J.: Analysis of fractional differential equations. J. Math. Anal. Appl. **265**(2), 229–248 (2002)
3. Diethelm, K., Ford, N.J., Freed, A.D.: A predictor-corrector approach for the numerical solution of fractional differential equations. Nonlinear Dyn. **29**(1), 3–22 (2002)
4. Ghaziani, R.K., Alidousti, J., Eshkaftaki, A.B.: Stability and dynamics of a fractional order Leslie-Gower prey-predator model. Appl. Math. Model. **40**, 2075–2086 (2016)
5. Gupta, R., Banerjee, M., Chandra, P.: Period doubling cascades of prey- predator model with nonlinear harvesting and control of over exploitation through taxation. Commun. Nonlinear Sci. Numer. Simul. **19**(7), 2382–2405 (2014)
6. Gupta, R., Chandra, P., Banerjee, M.: Dynamical complexity of a prey- predator model with nonlinear predator harvesting. Discret. Contin. Dyn. Syst. B **20**(2), 423–443 (2015)
7. Hu, D., Cao, H.: Stability and bifurcation analysis in a predator-prey system with michaelis-menten type predator harvesting. Nonlinear Anal. Real World Appl. **33**, 58–82 (2017)
8. Kumar, S., Kharbanda, H.: Chaotic behavior of predator-prey model with group defense and non-linear harvesting in prey. Chaos Solitons Fractals **119**, 19–28 (2019)
9. Li, M., Chen, B., Ye, H.: A bioeconomic differential algebraic predator-prey model with nonlinear prey harvesting. Appl. Math. Model. **42**, 17–28 (2017)
10. Li, C., Tao, C.: On the fractional adams method. Comput. Math. Appl. **58**(8), 1573–1588 (2009)
11. Lv, Y., Pei, Y., Wang, Y.: Bifurcations and simulations of two predator-prey models with nonlinear harvesting. Chaos Solitons Fractals **120**, 158–170 (2019)
12. Ma, L., Liu, B.: Dynamic analysis and optimal control of a fractional order singular Leslie-Gower prey-predator model. Acta Math. Sci. **40**, 1525–1552 (2020)
13. Matignon, D.: Stability results for fractional differential equations with applications to control processing. In: Computational Engineering in Systems Applications, vol. 120, pp 963–968. Citeseer (1996)

14. Mortuja, M.G., Chaube, M.K., Kumar, S.: Dynamic analysis of a predator-prey system with nonlinear prey harvesting and square root functional response. Chaos Solitons Fractals **148**, 111071 (2021)
15. Petráš, I.: Fractional-order nonlinear systems: modeling, analysis and simulation. Springer Science & Business Media (2011)
16. Shang, Z., Qiao, Y., Duan, L., Miao, J.: Stability and bifurcation analysis in a nonlinear harvested predator-prey model with simplified holling type IV functional response. Int. J. Bifurc. Chaos **30**, 2050205 (2020)
17. Vargas-De-León, C.: Volterra-type Lyapunov functions for fractional-order epidemic systems. Commun. Nonlinear Sci. Numer. Simul. **24**(1–3), 75–85 (2015)

A Numerical Application of Collocation Method for Solving KdV-Lax Equation

Seydi Battal Gazi Karakoc and Derya Yildirim Sucu

Abstract In this paper, a mathematical model representing the numerical solutions of Lax equation which is version of generalized fifth-order nonlinear KdV equation (fKdV) is studied. Collocation method with septic B-splines has been used for the model problem. Using a powerful Fourier series analysis of the linearized scheme, the numerical results have been shown to be unconditionally stable. L_2 and L_∞ error norms are calculated for single solutions to show practicality and robustness of proposed scheme. The obtained numerical results are shown in the table. Also, all simulations are shown to illustrate the numerical behavior of a single soliton. Present results show that the method provides highly accurate solutions. Therefore, the current scheme will be useful for other nonlinear scientific problems.

Keywords Lax equation · Finite element method · Collocation

1 Introduction

Fifth-order KdV-type equation in its general form is given by

$$u_t + \alpha u^2 u_x + \beta u_x u_{xx} + \gamma u u_{xxx} + u_{xxxxx} = 0. \tag{1}$$

Here α, β and γ are arbitrary positive parameters. These parameters strongly change the properties of the equation.This type of fifth-order Eq. (1) is the universal model for the study of shallow water waves with surface tension and has many physical applications a wide range of areas. Many versions of fKdV equation can be generated using different values of these parameters. For example, the following KdV-Lax equation

S. B. G. Karakoc · D. Y. Sucu (✉)
Faculty of Science and Art, Department of Mathematics, Nevsehir Haci Bektas Veli University, Nevsehir 50300, Turkey
e-mail: deryasucu@aksaray.edu.tr

S. B. G. Karakoc
e-mail: sbgkarakoc@nevsehir.edu.tr

S. Banerjee and A. Saha (eds.), *Nonlinear Dynamics and Applications*,
Springer Proceedings in Complexity,
https://doi.org/10.1007/978-3-030-99792-2_65

$$u_t + 30u^2u_x + 30u_xu_{xx} + 10uu_{xxx} + u_{xxxxx} = 0, \tag{2}$$

with $\alpha = 30$, $\beta = 30$, and $\gamma = 10$.

In this study, Lax equation, which is one of the fKdV equation, will be discussed. Many applications of various methods for the all forms of the fKdV equation can be found in the literature, but our scheme has not been implemented before. Because of the great importance of fKdV equation in nonlinear equations, many scientists obtained analytical and numerical solutions. In the literature, one can find out that the equation was solved with several methods, among others; Adomian [1], extended tanh [2], Haar wavelet sorting [3], Hirota's bilinear [4], auto-Bäcklund and Hirota transform [5], inverse scattering transform [6].

In addition, Inan and Ugurlu applied exp-function method for fifth-order KdV equation [7]. Bilige et al. [8] proposed an extended simplest equation method to search for full traveling wave solutions for various forms of the fifth-order KdV equation. The existence and stability of traveling waves of the fifth order KdV equation are investigated for a general class of nonlinearity satisfying power-like scaling relationships at [9]. A numerical approach based on the Homotopy perturbation transform method (HPTM) was applied to obtain exact and approximate solutions of nonlinear fifth-order KdV equations to study magneto-acoustic waves in the plasma at [10]. Travelling wave solutions were found for the generalized nonlinear fifth-order Korteweg-de Vries (KdV) equations using the direct algebraic method at [11]. Seventh order Lax equation is analyzed by Darvishi et al. with pseudospectral method [12].

The main form of our study can be briefly stated as follows: In Sect. 2, the septic B-spline approach is shown and the solution of the KdV-Lax equation by the finite element method is proposed. The stability analysis of the method is discussed in Sect. 3. In Sect. 4, numerical applications and their results are shown in table and graphs. In the final, a brief conclusion is given on the method presented in Sect. 5.

2 Septic B-Spline Approximation

In this study, we are interested in the numerical solutions of the KdV-Lax equation, whose bidirectional generalisation is given below:

$$u_t + 30u^2u_x + 30u_xu_{xx} + 10uu_{xxx} + u_{xxxxx} = 0, \tag{3}$$

with initial and boundary conditions

$$u(x,0) = f(x), \quad a \le x \le b, \tag{4}$$

$$u(a,t) = 0, \quad u_x(a,t) = 0, \tag{5}$$

$$u(b,t) = 0, \quad u_x(b,t) = 0, \quad t > 0. \tag{6}$$

KdV-Lax equation is searched into the boundary conditions $u \to 0$ while $x \to \pm\infty$, x and t which generally denote time and space, respectively. The equation is a member of the completely integrable hierarchy of higher-order KdV equations [13]. To start the procedure, our first task to solve the initial-boundary value problem given in Eqs. (3)–(6) numerically is to separate the solution domain.The septic B-spline functions $\{\phi_{-3}(x), \phi_{-2}(x), \ldots, \phi_{N+3}(x)\}$, at the nodes x_m are described on the solution zone $[a,\ b]$ in [14].

Now, we continue the numerical treatment, which we will apply using the septic B-spline collocation finite element method, by generating an approximate solution for the equation system. We find the numerical approximation solution $u_N(x,t)$ in the following form,

$$u_N(x,t) = \sum_{m=-3}^{N+3} \phi_m(x)\sigma_m(t). \tag{7}$$

Applying the following transformation

$$h\xi = x - x_m, \quad 0 \le \xi \le 1 \tag{8}$$

to the specific finite region $[x_m, x_{m+1}]$ is planned to more easily practicable region [0, 1] [15]. Thus, septic B-splines depending on variable ξ over the finite element [0, 1] are described as:

$$\begin{aligned}
\phi_{m-3} &= 1 - 7\xi + 21\xi^2 - 35\xi^3 + 35\xi^4 - 21\xi^5 + 7\xi^6 - \xi^7,\\
\phi_{m-2} &= 120 - 392\xi + 504\xi^2 - 280\xi^3 + 84\xi^5 - 42\xi^6 + 7\xi^7,\\
\phi_{m-1} &= 1191 - 1715\xi + 315\xi^2 + 665\xi^3 - 315\xi^4 - 105\xi^5 + 105\xi^6 - 21\xi^7,\\
\phi_m &= 2416 - 1680\xi + 560\xi^4 - 140\xi^6 + 35\xi^7,\\
\phi_{m+1} &= 1191 + 1715\xi + 315\xi^2 - 665\xi^3 - 315\xi^4 + 105\xi^5 + 105\xi^6 - 35\xi^7,\\
\phi_{m+2} &= 120 + 392\xi + 504\xi^2 + 280\xi^3 - 84\xi^5 - 42\xi^6 + 21\xi^7,\\
\phi_{m+3} &= 1 + 7\xi + 21\xi^2 + 35\xi^3 + 35\xi^4 + 21\xi^5 + 7\xi^6 - \xi^7,\\
\phi_{m+4} &= \xi^7.
\end{aligned} \tag{9}$$

The values of u_m,and its derivatives at the knots are calculated from using Eq. (7) and septic B-splines (9) in terms of element parameters σ_m in following form

$$\begin{aligned}
u_N(x_m,t) &= \sigma_{m-3} + 120\sigma_{m-2} + 1191\sigma_{m-1} + 2416\sigma_m + 1191\sigma_{m+1} + 120\sigma_{m+2} + \sigma_{m+3},\\
u'_m &= \tfrac{7}{h}(-\sigma_{m-3} - 56\sigma_{m-2} - 245\sigma_{m-1} + 245\sigma_{m+1} + 56\sigma_{m+2} + \sigma_{m+3}),\\
u''_m &= \tfrac{42}{h^2}(\sigma_{m-3} + 24\sigma_{m-2} + 15\sigma_{m-1} - 80\sigma_m + 15\sigma_{m+1} + 24\sigma_{m+2} + \sigma_{m+3}),\\
u'''_m &= \tfrac{210}{h^3}(-\sigma_{m-3} - 8\sigma_{m-2} + 19\sigma_{m-1} - 19\sigma_{m+1} + 8\sigma_{m+2} + \sigma_{m+3}),\\
u^{iv}_m &= \tfrac{840}{h^4}(\sigma_{m-3} - 9\sigma_{m-1} + 16\sigma_m - 9\sigma_{m+1} + \sigma_{m+3}),\\
u^{v}_m &= \tfrac{2520}{h^5}(-\sigma_{m-3} + 4\sigma_{m-2} - 5\sigma_{m-1} + 5\sigma_{m+1} - 4\sigma_{m+2} + \sigma_{m+3}).
\end{aligned} \tag{10}$$

Now, using the (7) and (10) into Eq. (2), following general form equation is reached for the linearization technique:

$$\dot{\sigma}_{m-3}+120\dot{\sigma}_{m-2}+1191\dot{\sigma}_{m-1}+2416\dot{\sigma}_{m}+1191\dot{\sigma}_{m+1}+120\dot{\sigma}_{m+2}+\dot{\sigma}_{m+3}$$
$$+(30Z_{m1}+20Z_{m2})\tfrac{7}{h}(-\sigma_{m-3}-56\sigma_{m-2}-245\sigma_{m-1}+245\sigma_{m+1}+56\sigma_{m+2}+\sigma_{m+3})$$
$$+10Z_{m3}\tfrac{210}{h^3}(-\sigma_{m-3}-8\sigma_{m-2}+19\sigma_{m-1}-19\sigma_{m+1}+8\sigma_{m+2}+\sigma_{m+3})$$
$$+\tfrac{2520}{h^5}(-\sigma_{m-3}+4\sigma_{m-2}-5\sigma_{m-1}+5\sigma_{m+1}-4\sigma_{m+2}+\sigma_{m+3})=0, \tag{11}$$

where $\dot{\sigma}=\frac{d\sigma}{dt}$ and

$$Z_{m1}=u^2=(\sigma_{m-3}+120\sigma_{m-2}+1191\sigma_{m-1}+2416\sigma_{m}+1191\sigma_{m+1}+120\sigma_{m+2}+\sigma_{m+3})^2,$$
$$Z_{m2}=u_{xx}=\tfrac{42}{h^2}(\sigma_{m-3}+24\sigma_{m-2}+15\sigma_{m-1}-80\sigma_{m}+15\sigma_{m+1}+24\sigma_{m+2}+\sigma_{m+3}),$$
$$Z_{m3}=u=\sigma_{m-3}+120\sigma_{m-2}+1191\sigma_{m-1}+2416\sigma_{m}+1191\sigma_{m+1}+120\sigma_{m+2}+\sigma_{m+3}.$$

Let's discretize for time parameters σ_i's according to the Crank-Nicolson formula and it is separated using forward finite difference approximation for its spatial variables and their derivatives $\dot{\sigma}$'s in the following form in Eq. (11):

$$\sigma_i=\frac{\sigma_i^{n+1}+\sigma_i^n}{2},\ \dot{\sigma}_i=\frac{\sigma_i^{n+1}-\sigma_i^n}{\Delta t}. \tag{12}$$

Thus, the above operation allows us to derive a recursion relationship between two time levels based on the parameters δ_i^{n+1} and δ_i^n for as:

$$\begin{aligned}&\lambda_1\sigma_{m-3}^{n+1}+\lambda_2\sigma_{m-2}^{n+1}+\lambda_3\sigma_{m-1}^{n+1}+\lambda_4\sigma_{m}^{n+1}+\lambda_5\sigma_{m+1}^{n+1}+\lambda_6\sigma_{m+2}^{n+1}+\lambda_7\sigma_{m+3}^{n+1}\\&=\lambda_7\sigma_{m-3}^{n}+\lambda_6\sigma_{m-2}^{n}+\lambda_5\sigma_{m-1}^{n}+\lambda_4\sigma_{m}^{n}+\lambda_3\sigma_{m+1}^{n}+\lambda_2\sigma_{m+2}^{n}+\lambda_1\sigma_{m+3}^{n},\end{aligned} \tag{13}$$

where

$$\begin{aligned}&\lambda_1=[1-E-T-M],\\&\lambda_2=[120-56E-8T+4M],\\&\lambda_3=[1191-245E+19T-5M],\\&\lambda_4=[2416],\\&\lambda_5=[1191+245E-19T+5M],\\&\lambda_6=[120+56E+8T-4M],\\&\lambda_7=[1+E+T+M],\\&E=\tfrac{a}{2}\Delta t,\quad T=\tfrac{b}{2}\Delta t,\quad M=\tfrac{2520}{2h^5}\Delta t,\\&a=[30Z_{m1}+20Z_{m2}],\\&b=[\tfrac{2100}{h^3}Z_{m3}].\end{aligned} \tag{14}$$

If we take a look at the algebraic system (13) we obtained above, the number of linear equations are less than the number of unknown coefficients, that is, the system involves of $(N+1)$ equation $(N+7)$ unknown time dependent parameters [16]. The simplest way to find a unique solution is to remove six unknowns $\sigma_{-3},\sigma_{-2},\sigma_{-1},\ldots,\sigma_{N+1},\sigma_{N+2}$, and σ_{N+3} from the system. This procedure is applied using the boundary conditions with the values of u and after eliminating unknowns, a matrix system of $(N+1)$ linear equations with $(N+1)$ unknown parameters $d^n=(\sigma_0,\sigma_1,\ldots,\sigma_N)^T$ are obtained in the form of the following matrix-vector system:

$$R\mathbf{d}^{\mathbf{n+1}}=S\mathbf{d}^{\mathbf{n}}. \tag{15}$$

3 Stability Analysis

In this part, we explain through the Von-Neumann theory to demonstrate stability of the linearized numerical algorithm. To show the stability analysis, the fKdV equation was linearized by supposing that the quantities u^2, u_{xx} and u in the nonlinear terms $u^2 u_x$, $u_x u_{xx}$ and $u u_{xxx}$ are locally constant, respectively. Growth factor ξ of a characteristic Fourier mode is identified as:

$$\sigma_m^n = \xi^n e^{imkh}, \tag{16}$$

here $i = \sqrt{-1}$, h is the element greatness and k is the mode number, is obtained from the linear stability analysis of the algorithm. Putting the equality (16) into the iterative system (13), which gives the growth factor

$$\xi = \frac{\rho_1 - i\rho_2}{\rho_1 + i\rho_2}, \tag{17}$$

where

$$\begin{aligned} \rho_1 &= 2\cos(3kh) + 240\cos(2kh) + 2382\cos(kh) + 2416, \\ \rho_2 &= (2M + 2T + 2E)\sin(3kh). \end{aligned} \tag{18}$$

$|\xi| = 1$ is obtained when we take the modulus of Eq. (17). In this way, we demonstrate that scheme (13) is unconditionally stable under the present conditions.

4 Numerical Applications and Discussions

In this section, we exemplify our method improved in Sect. 2 to the KdV-Lax equation for different parameters of the time and space division. To check the sensibility and reliability of the presented method, the following L_2 and L_∞ error norms will be used respectively:

$$L_2 = \left\| u^{exact} - u_N \right\|_2 \simeq \sqrt{h \sum_{j=1}^{N} \left| u_j^{exact} - (u_N)_j \right|^2}, \tag{19}$$

$$L_\infty = \left\| u^{exact} - u_N \right\|_\infty \simeq \max_j \left| u_j^{exact} - (u_N)_j \right|, \quad j = 1, 2, \ldots, N. \tag{20}$$

Lax equation has an exact solution of the form

$$u(x,t) = 2k^2 \left(2 - 3\tanh^2\left[k\left(x - 56k^4 t - x_0\right)\right]\right), \tag{21}$$

where k and x_0 are arbitrary real numbers. We will consider the Lax equation with the boundary-initial condition which is

Table 1 Error norms for $k = 0.01$, $\Delta t = 0.0004$ and various values of h

	$\Delta t = 0.0004$, $h = 0.5$		$\Delta t = 0.0001$, $h = 0.05$	
t	L_2	L_∞	L_2	L_∞
0.1	0.0007476185	0.0003786723	0.0006605026	0.0003870472
0.2	0.0008112957	0.0003797258	0.0006965400	0.0004162210
0.3	0.0008167458	0.0003793303	0.0007209887	0.0003999905
0.4	0.0008542324	0.0003784037	0.0007876002	0.0004253867
0.5	0.0008354053	0.0003899616	0.0008298723	0.0005542919
0.6	0.0008850945	0.0003787136	0.0009135698	0.0009698522
0.7	0.0008491875	0.0003800426	0.0011118805	0.0022608287
0.8	0.0008693723	0.0003949290	0.0019634349	0.0059890208
0.9	0.0009128289	0.0003794148	0.0048874414	0.0170258202
1.0	0.0008974475	0.0003791528	0.0140005587	0.0499546306

$$u(x, 0) = 2k^2 \tanh^2 [k(x - x_0)], \tag{22}$$

where $u \to 0$ as $x \to \pm\infty$. We let $x_0 = 0$ and $k = 0.01$ over the interval $x \in [-20, 20]$, to present numerical solutions.

The algorithm has been performed in the calculation range $[-20, 20]$ and up to time $t = 1$. It was recorded that the solitary wave has amplitude $A = 0.0002$ at $x = 0$ and at the initial moment of $t = 0$. In simulation calculations, typical values we use are $\Delta t = 0.0004$; 0.0001 with $h = 0.5$ and 0.05. Values of the L_2 and L_∞ error norms are displayed in Table 1. It has been observed that the calculated values of the error norms are found to be adequately small. It is evident that the minimum L_∞ error norm 3.786723×10^{-4} with the parameters $\Delta t = 0.0004$ and $h = 0.5$. These errors do not change much over time. In addition, it can be seen that the real solutions and the numerical solutions are in good agreement and the method is efficient when the values of the error norms are seen from the table. If the Fig. 1 is examined, we can clearly see that the three dimensional states of the bell shaped solitary wave solutions produced from $t = 0$ to $t = 1$. It can be said that amplitude and shape are preserved as time passes. On the other hand, numerical error distribution is also plotted at time $t = 1$ for different values of h and Δt in Fig. 1.

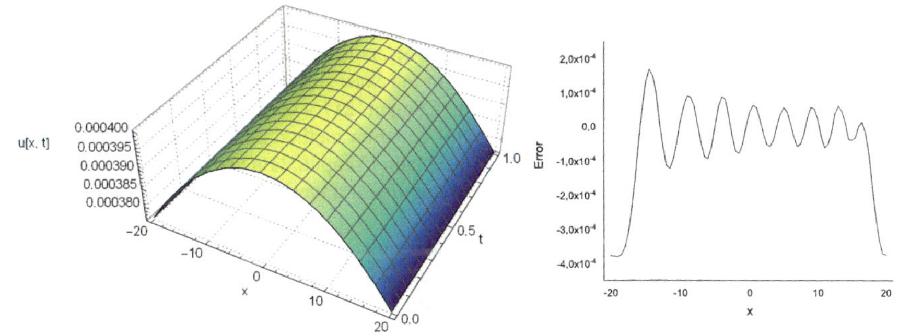

Fig. 1 Motion of single solitary wave and the error distributions at $t = 1$ for the parameters $\Delta t = 0.0004$ and $h = 0.5$

5 Conclusion

In this study, numerical solutions of Lax equation, which is a fifth-order KdV equation, are investigated by considering some fixed selection initial and boundary conditions. In this trajectory, a combination of the collocation method on the finite element approach is used to construct the numerical scheme of the equation. Septic B-splines have been chosen as the interpolation functions of this application. We have shown that our linearized scheme is unconditionally stable. In order to perform numerical experiments, the algorithm is studied with a single solitary wave motion whose analytical solution is known. The performance and validity of the numerical scheme was measured by calculating both L_2 and L_∞ error norms. All experiments were supported by figures and table. The sampled results confirm that our error norms are good enough as required. It may be concluded that the method used here is powerful, efficient and confidental technique for solving a wide class of nonlinear evolution equations.

References

1. Kaya, D.: An explicit and numerical solutions of some fifth-order KdV equation by decomposition method. Appl. Math. Comput. **144**, 353–363 (2003)
2. Wazwaz, A.M.: The extended tanh method for new solitons solutions for many forms of the fifth-order KdV equations. Appl. Math. Comput. **184**(2), 1002–1014 (2007)
3. Saleem, S., Hussain, M.Z.: Numerical solution of nonlinear fifth-order KdV-type partial differential equations via Haar wavelet. Int. J. Appl. Comput. Math. **6**(6), 1–16 (2020)
4. Wazwaz, A.M.: N-soliton solutions for the combined KdV-CDG equation and the KdV-Lax equation. Appl. Math. Comput. **203**(1), 402–407 (2008)
5. Lei, Y., Fajiang, Z., Yinghai, W.: The homogeneous balance method, Lax pair, Hirota transformation and a general fifth-order KdV equation. Chaos Solitons Fractals **13**(2), 337–340 (2002)

6. Ablowitz, M.J., Ablowitz, M.A., Clarkson, P.A., Clarkson, P.A.: Solitons, Nonlinear Evolution Equations and Inverse Scattering. Cambridge University Press (1991)
7. Inan, I.E., Ugurlu, Y.: Exp-function method for the exact solutions of fifth order KdV equation and modified Burgers equation. Appl. Math. Comput. **217**(4), 1294–1299 (2010)
8. Bilige, S., Chaolu, T.: An extended simplest equation method and its application to several forms of the fifth-order KdV equation. Appl. Math. Comput. **216**(11), 3146–3153 (2010)
9. Esfahani, A., Levandosky, S.: Existence and stability of traveling waves of the fifth-order KdV equation. Phys. D **421**, 132872 (2021)
10. Goswami, A., Singh, J., Kumar, D.: Numerical simulation of fifth order KdV equations occurring in magneto-acoustic waves. Ain Shams Eng. J. **9**(4), 2265–2273 (2018)
11. Seadawy, A.R., Lu, D., Yue, C.: Travelling wave solutions of the generalized nonlinear fifth-order KdV water wave equations and its stability. J, Taibah Univ. Sci. **11**(4), 623–633 (2017)
12. Darvishi, M.T., Khani, F., Kheybari, S.: A numerical solution of the Lax's 7th-order KdV equation by Pseudospectral method and Darvishi's preconditioning. Int. J. Contemp. Math. Sci **2**(22), 1097–1106 (2007)
13. Lax, P.D.: Integrals of nonlinear equations of evolution and solitary waves. Commun. Pure Appl. Math. **21**(5), 467–490 (1968)
14. Prenter, P.M.: Splines and Variational Methods. Wiley, New York (1975)
15. Karakoc, S.B.G., Saha, A., Sucu, D.: A novel implementation of Petrov-Galerkin method to shallow water solitary wave pattern and superperiodic traveling wave and its multistability: Generalized Korteweg-de Vries equation. Chin. J. Phys. **68**, 605–617 (2020)
16. Karakoc, S.B.G., Omrani, K., Sucu, D.: Numerical investigations of shallow water waves via generalized equal width (GEW) equation. Appl. Numer. Math. **162**, 249–264 (2020)

Influence of Suspension Lock on the Four-Station Military Recovery Vehicle with Trailing Arm Suspension During Crane Operation

M. Devesh, R. Manigandan, and Saayan Banerjee

Abstract The present work is focused on the development of a non-linear dynamic mathematical model of a four-station military recovery vehicle with trailing arm hydro-gas suspension (HSU) during crane operations over flat terrain. The influence of the crane payload non-linear motion on the trailing arm dynamic behavior is brought out in the dynamic model. The model additionally contains non-linearities due to the penalty contact phenomenon, which are associated with the HSU rebound/bump-stoppers, suspension locks or between the dummy masses and ground. Second-order coupled governing non-linear differential equations of motion are formulated for 13 degrees of freedom of the vehicle, namely, sprung mass bounce, pitch and roll, angular motions of the 4 unspring masses, crane payload angular motions in the longitudinal and lateral directions as well as bounce motion of the 4 dummy masses. The equations of motion are coded and solved in MATLAB. The maximum pay-load lifting capacity could be determined by modeling the dynamic influence of the suspension locks on the recovery vehicle static equilibrium configuration, which is an essential pre-requisite before deciding upon the vehicle moving speed with the crane payload. This model is novel, generic and would provide deep insight into the development of a recovery vehicle simulator.

Keywords Military recovery vehicle · Suspension lock · Trailing arm · Hydro-gas suspension · Non-linear dynamics

---

M. Devesh · R. Manigandan
School of Mechanical Engineering, Vellore Institute of Technology, Chennai, India
e-mail: m.devesh2016@vitstudent.ac.in

R. Manigandan
e-mail: manigandan.r2016@vitstudent.ac.in

S. Banerjee (✉)
Centre for Engineering Analysis and Design, Combat Vehicles R&D Establishment, DRDO, New Delhi, India
e-mail: saayanbanerjee.cvrde@gov.in

S. Banerjee and A. Saha (eds.), *Nonlinear Dynamics and Applications*,
Springer Proceedings in Complexity,
https://doi.org/10.1007/978-3-030-99792-2_66

1 Introduction

Military recovery vehicle cranes are typically used to lift and carry required payloads from the battlefield to the nearest base workshop. The payload inertias can directly influence the vehicle stability, which necessitates a detailed study to determine the maximum payload capacity with the trailing arm hydro-gas suspension configuration. The comparative dynamic analysis of a military recovery vehicle during crane operations with linear suspension over flat terrain was discussed by Nikhil [1]. However, in Nikhil [1], linear vertical spring-mass system was considered to determine the maximum crane payload which would yield definite differences in sprung mass responses when compared to that with trailing arm suspension. The effects of the integrated ride and cornering dynamics of a military vehicle on the weapon responses were brought out by Banerjee et al. [2]. A non-linear mathematical model of a single station with hydro-gas trailing arm suspension was developed by Banerjee et al. [3]. However, in Banerjee et al. [2] and Banerjee et al. [3], the dynamic effects from sprung mass large pitch and roll angular motions were not considered in the governing equations. Moreover, in Banerjee et al. [2] and Banerjee et al. [3], contact algorithms were not incorporated into the governing equations to simulate the stoppers or loss of ground contact. The meshed gear profile penalty contact formulation is established accurately by Xiufeng and Yabin [4]. A sprung mass non-linear pitch dynamics mathematical model with a trailing arm torsion bar suspension system was developed by Devesh et al. [5]. However, in Devesh et al. [5], an in-plane dynamic model with two degrees of freedom is only considered to determine the sprung mass dynamic responses.

It is noteworthy that extensive research has been undertaken in the field of military vehicle dynamics. However, the present study brings out the integrated dynamic influence of crane payload inertias and suspension locks with their associated penalty contact formulation on the military recovery vehicle static settlement configuration through a detailed non-linear dynamic mathematical model, which has not been reported in the literature to date. The road-holding phenomenon of the unsprang masses under such integrated dynamic influence is additionally simulated through penalty contact formulation in the governing equations, which has also not been brought out in literature to date. The dynamic effects of sprung mass large pitch and roll angular motions are additionally considered in the present mathematical model. This dynamic model is very novel and generic. It would establish a stepping stone towards deciding upon the vehicle moving speed along with the crane payload and development of a recovery vehicle simulator.

2 Description of the Four-Station Military Recovery Vehicle Dynamic Model

The second-order coupled governing differential equations of motion consists of 13 degrees of freedom, namely, sprung mass bounce, pitch, roll about its CG, left (LH_i) and right (RH_i) unsprung mass angular motions from the rebound ($i = 1$ to 2), crane payload angular motions in the longitudinal and lateral directions with respect to the vertical axis and as well as left (LH_i) and right (RH_i) dummy mass bounce motions ($i = 1$ to 2). The hydro-gas suspension rebound, bump-stoppers and suspension locks are modelled by using penalty contact formulation with high magnitudes of torsional stiffness and damping. The loss of wheel-to-ground contact under high pay-load magnitudes is also simulated through the penalty contact formulation by the introduction of suitable dummy masses. Figure 1 represents the coordinates for different degrees of freedom and other vehicle parameters. Figure 2 describes the associated free body diagram with a detailed representation of forces and moments

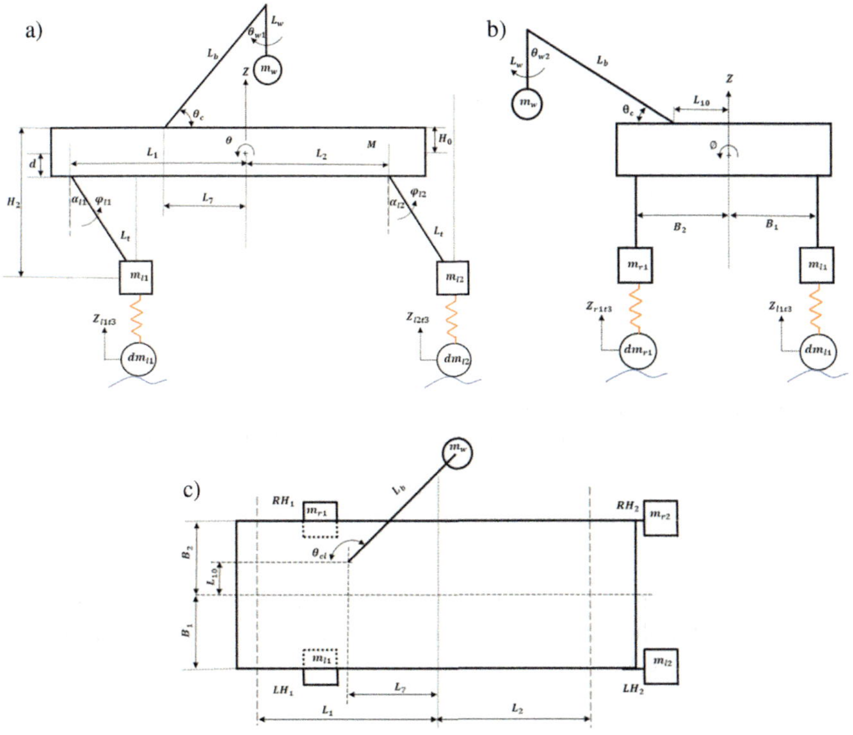

Fig. 1 Vibration model of the military recovery vehicle with crane payload- **a** Side View of the vehicle, **b** Front View of the vehicle, **c.** Top View of the vehicle

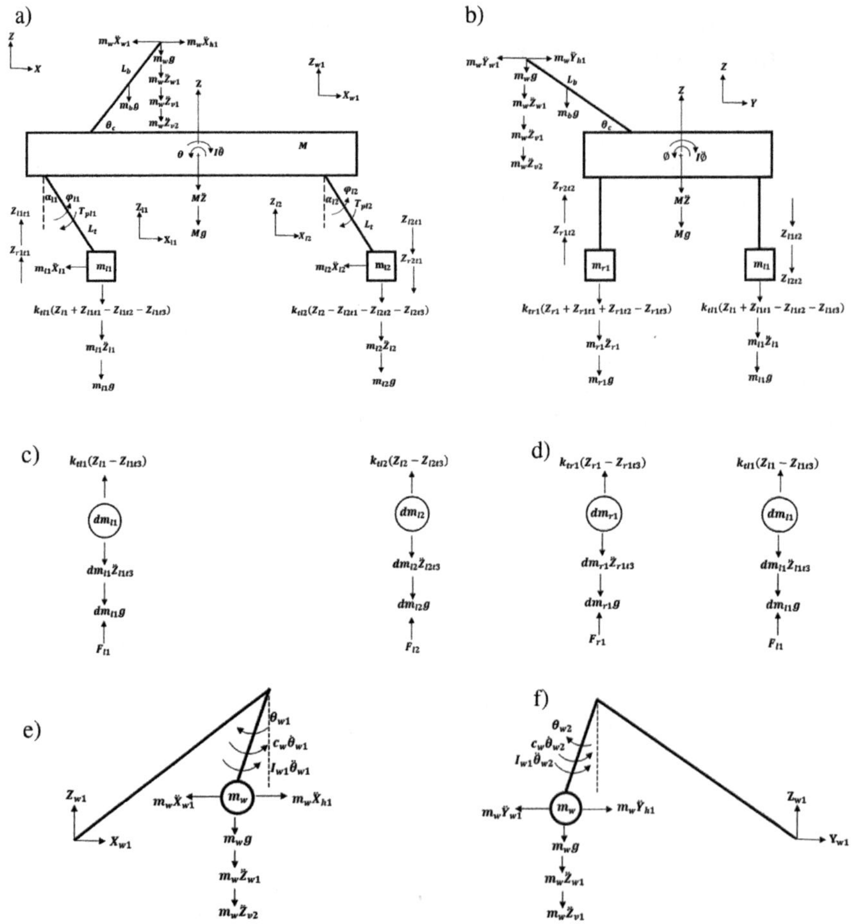

Fig. 2 Free body diagram of four-station military recovery vehicle with crane payload- **a** Sprung mass bounce & pitch angular motion and angular motion of unsprung masses, **b** Sprung mass roll motion, **c** Bounce motion of the dummy masses from side view, **d** Bounce motion of the dummy masses from front view, **e** Crane payload angular motion in the longitudinal direction, **f** Crane payload angular motion in the lateral direction

on the crane, sprung and unsprung masses. The description of different variables which are used in the free body diagram and their corresponding magnitudes are highlighted in Table 1.

Table 1 Description and magnitudes of the variables

Variable	Definition	Magnitude
M	Sprung mass	20,000 kg
I	Pitch moment inertia of sprung mass about the CG	170,000 kgm^2
J	Roll moment inertia of sprung mass about the CG	17,000 kgm^2
m_{li}, m_{ri}	LH_i& RH_i $(i = 1, 2)$ unsprung masses	500 kg
k_{tli},k_{tri}	LH_i& RH_i $(i = 1, 2)$ road-wheel & track pad vertical stiffness	8000 kN/m
c	Viscous damping coefficient along the direction of actuator piston motion	400 kNs/m
L_1	Longitudinal dist. from sprung mass CG to front suspension pivotal points	3.3 m
L_2	Longitudinal dist. from sprung mass CG to rear suspension pivotal points	2.7 m
B_1	Lateral dist. from sprung mass CG to left suspension stations	1.45 m
B_2	Lateral dist. from sprung mass CG to right suspension stations	1.55 m
L_7	Longitudinal dist. from sprung mass CG to crane boom mounting location	2.8 m
$L\ 10$	Lateral dist. from sprung mass CG to crane boom mounting location	1 m
L_t	Axle arm length	0.5 m
H_2	Vertical dist. from unsprung mass CG to crane boom mounting location	2 m
H_0	Vertical dist. from sprung mass CG to crane boom mounting location	1 m
d	Vertical dist. from sprung mass CG to the suspension pivotal points	0.25 m
α_{li},α_{ri}	LH_i& RH_i $(i = 1, 2)$ suspension axle arm rebound angles	42^0
m_b	Mass of the crane boom	1100 kg
c_w	Torsional viscous damping coefficient about the payload hinge point	100 kNms/rad
L_b	Length of the crane boom	5.4 m
L_w	Length of the payload string	3 m
θ_c	Crane boom inclination with respect to longitudinal axis (in XZ plane)	67^0
θ_{cl}	Crane boom inclination with respect to longitudinal axis (in XY plane)	135^0
dm_{li}, dm_{ri}	LH_i& RH_i $(i = 1, 2)$ dummy masses	1 kg
k_{sli},k_{sri}	Contact stiffness between the dummy masses and ground	10^5 kN/m
c_{sli},c_{sri}	Contact damping between dummy masses and ground	1 kNs/m
k_{ub}	Contact stiffness of the bump stopper	10^5 kNm/rad
k_{rb}	Contact stiffness of the rebound stopper	10^5 kNm/rad

(continued)

Table 1 (continued)

Variable	Definition	Magnitude
c_{ub}	Contact damping coefficient of the bump stopper	1 kNms/rad
c_{rb}	Contact damping coefficient of the rebound stopper	1 kNms/rad
m_w	Crane payload mass including the mass of the string and hook (Variable magnitude)	
φ_s	Bump stopper/Suspension lock angle with respect to rebound position (Variable magnitude)	

3 Equations of Motion of the Four-Station Military Vehicle

The second order coupled governing non-linear differential equations of motion are derived by referring to the free body diagram in Fig. 2. The non-linear bounce equation of the sprung mass, which is measured at its CG, is written as

$$\begin{aligned} M\ddot{Z}+P_{l1}+P_{l2}+P_{r1}+P_{r2}+k_{tl1}Z_{tl1}+k_{tl2}Z_{tl2}+k_{tr1}Z_{tr1}+k_{tr2}Z_{tr2} \\ +M_1g+m_w\ddot{Z}_{w1}+m_w\ddot{Z}_{v1}+m_w\ddot{Z}_{v2}=0 \end{aligned} \tag{1}$$

where $M\ddot{Z}$ is the sprung mass vertical inertia, P_{li} & P_{ri} are the vertical inertias of the LH$_i$ & RH$_i$ (i = 1 to 2) unsprung masses, respectively, $k_{tli}Z_{tli}$ & $k_{tri}Z_{tri}$ are restoring forces from the road wheel springs of the LH$_i$ and RH$_i$ (i = 1 to 2) stations, respectively, M_1g is the force due to self-weight of the sprung and unsprung masses as well as crane boom and payload, $m_w\ddot{Z}_{w1}$, $m_w\ddot{Z}_{v1}$ and $m_w\ddot{Z}_{v2}$ are the crane payload vertical inertias due to sprung mass bounce, pitch and roll motions as well as due to its own angular motion in the longitudinal and lateral directions, respectively. The effects of sprung mass large pitch and roll angular motions along with its inertia coupling with the unsprung mass and crane payload angular motion, are highlighted in the subsequent equations.

In Eq. (1), $P_{li}=m_{li}\ddot{Z}_{li}$, $P_{ri}=m_{ri}\ddot{Z}_{ri}$, where

$$Z_{li}=Z+L_t\cos\alpha_{li}-L_t\cos(\alpha_{li}+\varphi_{li}+\theta)+d-d\cos\theta+pL_i\sin\theta+B_1\sin\varnothing \tag{2}$$

$$Z_{ri}=Z+L_t\cos\alpha_{ri}-L_t\cos(\alpha_{ri}+\varphi_{ri}+\theta)+d-d\cos\theta+pL_i\sin\theta-B_2\sin\varnothing \tag{3}$$

($p=-1$ for $i=1$ and $p=+1$ for $i=2$).

The LH$_i$ and RH$_i$ road-wheel spring vertical displacements Z_{tli} and Z_{tri} ($i=1$ to 2), respectively, which are obtained by accounting for the load transfer effects due to crane payload inertias in both longitudinal and lateral directions, are described by

$$Z_{tli}=\left(Z_{li}+pZ_{lit1}-Z_{lit2}-Z_{lit3}\right) \tag{4}$$

($p = +1$ for $i = 1$ and $p = -1$ for $i = 2$)

$$Z_{tri} = (Z_{ri} + pZ_{rit1} + Z_{rit2} - Z_{rit3}) \tag{5}$$

($p = +1$ or $i = 1$ and $p = -1$ for $i = 2$)

where Z_{lit1} & $Z_{rit1}(i = 1 \text{ to } 2)$ are components of the LH_i & $\text{RH}_i(i = 1 \text{ to } 2)$ road-wheel spring vertical displacements, respectively, due to load transfer effects from the crane payload longitudinal inertias and Z_{lit2} & $Z_{rit2}(i = 1 \text{ to } 2)$ are components of the LH_i & $\text{RH}_i(i = 1 \text{ to } 2)$ road-wheel spring vertical displacements, respectively, due to load transfer effects from the crane payload lateral inertias.

$$Z_{jit1} = \frac{(L_{ji}B_q)\left(m_w\ddot{X}_{h1} - m_w\ddot{X}_{w1}\right)\{(H_2 - H_0)\cos\theta + Z\}}{(B_1 + B_2)(k_{tj1}L_{j1}^2 + k_{tj2}L_2^2)} \tag{6}$$

($j = l$ or r; $q = 2$ for $j = l$; $q = 1$ for $j = r$)

$$Z_{jit2} = \frac{B_q\left(m_w\ddot{Y}_{h1} - m_w\ddot{Y}_{w1}\right)\{(H_2 - H_0)\cos\theta + Z\}}{2(k_{tl1}B_1^2\cos\varnothing + k_{tr1}B_2^2\cos\varnothing)} \tag{7}$$

($j = l$ or r; $q = 1$ for $j = l$; $q = 2$ for $j = r$)

In Eqs. (1), (6) and (7),

$$L_{j1} = L_1\cos\theta - L_t\sin\left(\alpha_{j1} + \varphi_{j1} + \theta\right) - d\sin\theta \tag{8}$$

$$L_{j2} = L_2\cos\theta + L_t\sin\left(\alpha_{j2} + \varphi_{j2} + \theta\right) + d\sin\theta \tag{9}$$

where $j = l$ or r

$$M_1 = (M + m_{l1} + m_{l2} + m_{r1} + m_{r2} + m_b + m_w) \tag{10}$$

$$Z_{w1} = Z - L_7\sin\theta - L_{10}\sin\varnothing \tag{11}$$

$$Z_{v1} = L_w - L_w\cos\theta_{w1} \tag{12}$$

$$Z_{v2} = L_w - L_w\cos\theta_{w2} \tag{13}$$

The sprung mass non-linear pitch equation of motion about its CG, is written as

$$\begin{aligned} I\ddot{\theta} &- P_{l1}L_{l1} + P_{l2}L_{l2} - P_{r1}L_{r1} + P_{r2}L_{r2} + H_{l1}L_{l3} + H_{l2}L_{l4} + H_{r1}L_{r3} \\ &+ H_{r2}L_{r4} - k_{tl1}Z_{tl1}L_{l1} + k_{tl2}Z_{tl2}L_{l2} - k_{tr1}Z_{tr1}L_{r1} + k_{tr2}Z_{tr2}L_{r2} \\ &- m_{l1}gL_{l1} + m_{l2}gL_{l2} - m_{r1}gL_{r1} + m_{r2}gL_{r2} - \left(m_wg + m_w\ddot{Z}_{w1} + m_w\ddot{Z}_{v1} + m_w\ddot{Z}_{v2}\right) \end{aligned}$$

$$U_{c1} + (m_w\ddot{X}_{h1} - m_w\ddot{X}_{w1})V_c - m_b g U_{c2} = 0 \quad (14)$$

In Eq. (14), $H_{li} = m_{li}\ddot{X}_{li}, H_{ri} = m_{ri}\ddot{X}_{ri}$, where

$$X_{li} = -pL_i - L_t\sin\alpha_{li} + pL_i\cos\theta + d\sin\theta + L_t\sin(\alpha_{li} + \varphi_{li} + \theta) \quad (15)$$

$$X_{ri} = -pL_i - L_t\sin\alpha_{ri} + pL_i\cos\theta + d\sin\theta + L_t\sin(\alpha_{ri} + \varphi_{ri} + \theta) \quad (16)$$

($p = -1$ for $i = 1$ and $p = +1$ for $i = 2$)

$$X_{h1} = L_w\sin\theta_{w1} \quad (17)$$

$$X_{w1} = L_7 - L_7\cos\theta \quad (18)$$

$$L_{j3} = L_t\cos(\alpha_{j1} + \varphi_{j1} + \theta) + d\cos\theta + L_1\sin\theta \quad (19)$$

$$L_{j4} = L_t\cos(\alpha_{j2} + \varphi_{j2} + \theta) + d\cos\theta - L_2\sin\theta \quad (j = l \text{ or } r) \quad (20)$$

$$U_{c1} = L_b\cos(\theta_c - \theta - \varnothing)\cos\theta_{cl} + L_7\cos\theta + H_0\sin\theta \quad (21)$$

$$V_c = L_b\sin(\theta_c - \theta - \varnothing) - L_7\sin\theta + H_0\cos\theta \quad (22)$$

$$U_{c2} = (L_b/2)\cos(\theta_c - \theta - \varnothing)\cos\theta_{cl} + L_7\cos\theta + H_0\sin\theta \quad (23)$$

The second order non-linear roll equation of the sprung mass with reference to the CG of the vehicle is written as

$$\begin{aligned} J\ddot{\varnothing} &+ P_{l1}B_1\cos\varnothing + P_{l2}B_1\cos\varnothing - P_{r1}B_2\cos\varnothing - P_{r2}B_2\cos\varnothing + k_{tl1}Z_{tl1}B_1\cos\varnothing \\ &+ k_{tl2}Z_{tl2}B_1\cos\varnothing - k_{tr1}Z_{tr1}B_2\cos\varnothing - k_{tr2}Z_{tr2}B_2\cos\varnothing + m_{l1}gB_1\cos\varnothing + m_{l2}gB_1\cos\varnothing \\ &- m_{r1}gB_2\cos\varnothing - m_{r2}gB_2\cos\varnothing - (m_w\ddot{Z}_{w1} + m_w\ddot{Z}_{v1} + m_w\ddot{Z}_{v2} + m_w g)U_{d1} \\ &+ (M_w\ddot{Y}_{h1} - M_w\ddot{Y}_{w1})V_d - m_b g U_{d2} = 0 \end{aligned} \quad (24)$$

where $J\ddot{\varnothing}$ is the sprung mass roll moment of inertia about its CG, $m_w\ddot{Y}_{h1}$ &$m_w\ddot{Y}_{w1}$ are the crane payload inertias in the lateral direction due its own angular motion as well as due to coupling with the sprung mass large roll angular motion, respectively.

In Eq. (24), $Y_{w1} = L_{10} - L_{10}\cos\varnothing$, $Y_{h1} = L_w\sin\theta_{w2}$

$$U_{d1} = L_b\cos(\theta_c - \theta - \varnothing)\sin\theta_{cl} + L_{10}\cos\varnothing + H_0\sin\varnothing \quad (25)$$

$$V_d = L_b \sin(\theta_c - \theta - \varnothing) + L_{10} \sin\varnothing + H_0 \cos\varnothing \tag{26}$$

$$U_{d2} = (L_b/2)\cos(\theta_c - \theta - \varnothing)\sin\theta_{cl} + L_{10}\cos\varnothing + H_0 \sin\varnothing \tag{27}$$

The governing non-linear equations of motion, which represent the angular motions of LH$_i$ & RH$_i$ ($i = 1$ to 2) unsprung masses about their respective pivotal points, are written as.

$$\begin{aligned} &P_{li} L_t \sin(\alpha_{li} + \varphi_{li} + \theta) + H_{li} L_t \cos(\alpha_{li} + \varphi_{li} + \theta) + m_{li} g L_t \sin(\alpha_{li} + \varphi_{li} + \theta) \\ &+ k_{tli} Z_{tli} \sin(\alpha_{li} + \varphi_{li} + \theta) + T_{pli} = 0 \end{aligned} \tag{28}$$

$$\begin{aligned} &P_{ri} L_t \sin(\alpha_{ri} + \varphi_{ri} + \theta) + H_{ri} L_t \cos(\alpha_{ri} + \varphi_{ri} + \theta) + m_{ri} g L_t \sin(\alpha_{ri} + \varphi_{ri} + \theta) \\ &+ k_{tri} Z_{tri} L_t \sin(\alpha_{ri} + \varphi_{ri} + \theta) + T_{pri} = 0 \end{aligned} \tag{29}$$

where T_{pli} & T_{pri} are the sum of moments about pivotal points of the LH$_i$ & RH$_i$ ($i = 1$ to 2) unsprung masses due to the gas restoring force & damping force on the actuator piston as well as due to bump-stopper/rebound stopper contact forces. It may be noted that the bump-stopper or rebound stopper contact forces would act only if contact is established with the suspension axle arm during its angular motion. The hydro-gas suspension kinematics, stiffness & damping properties are arrived at by referring to [3].

$$T_{pji} = \begin{cases} T_{ji}\{\varphi_{ji}\} + c\dot{x}_{ji}\{\varphi_{ji}\} L_0;\ 0 \le \varphi_{ji} \le \varphi_s,\ j = l \text{or} r, \\ \left(T_{ji}\{0\} - k_{rb}(\varphi_{ji})^2\right) + \left(c\dot{x}_{ji}\{0\} L_0 + c_{rb}(\dot{\varphi}_{ji})\right);\ \varphi_{ji} < 0, \\ \left(T_{ji}\{\varphi_s\} + k_{ub}(\varphi_{ji} - \varphi_s)^2\right) + \left(c\dot{x}_{ji}\{\varphi_s\} L_0 + c_{ub}(\dot{\varphi}_{ji})\right);\ \varphi_{ji} > \varphi_s \end{cases} \tag{30}$$

The governing second order non-linear equations, which represent vertical motion of the LH$_i$ & RH$_i$ ($i = 1$ to 2) dummy masses, are written as

$$dm_{li} \ddot{Z}_{lit3} - k_{tli}(Z_{li} - Z_{lit3}) + dm_{li} g - F_{li} = 0 \tag{31}$$

$$dm_{ri} \ddot{Z}_{rit3} - k_{tri}(Z_{ri} - Z_{rit3}) + dm_{ri} g - F_{ri} = 0 \tag{32}$$

where $dm_{li}\ddot{Z}_{lit3}$ & $dm_{ri}\ddot{Z}_{rit3}$ are the LH$_i$ & RH$_i$ ($i = 1$ to 2) dummy mass vertical inertia forces, $k_{tli}(Z_{li} - Z_{lit3})$ & $k_{tri}(Z_{ri} - Z_{rit3})$ are the LH$_i$ & RH$_i$ ($i = 1$ to 2) road-wheel spring restoring forces, $dm_{li}g$ & $dm_{ri}g$ represent self-weight of the LH$_i$ & RH$_i$ ($i = 1$ to 2) dummy masses, F_{li} & F_{ri} are the ground contact forces on the LH$_i$ & RH$_i$ ($i = 1$ to 2) dummy masses such that

$$F_{ji} = \begin{cases} k_{sji}(Z_{jit3})^2 - c_{sji}(\dot{Z}_{jit3}); & Z_{jit3} \leq 0, j = lorr, \\ 0; & Z_{jit3} > 0 \end{cases} \tag{33}$$

The crane pay-load angular motion about its hinge point in the longitudinal and lateral directions are written as

$$\begin{aligned} I_{w1}\ddot{\theta}_{w1} + m_w g L_w \sin\theta_{w1} - m_w \ddot{X}_{w1} L_w \cos\theta_{w1} + c_w \dot{\theta}_{w1} \\ + m_w \ddot{Z}_{w1} L_w \sin\theta_{w1} + m_w \ddot{Z}_{v2} L_w \sin\theta_{w1} = 0 \end{aligned} \tag{34}$$

$$\begin{aligned} I_{w1}\ddot{\theta}_{w2} + m_w g L_w \sin\theta_{w2} - m_w \ddot{Y}_{w1} L_w \cos\theta_{w2} + c_w \dot{\theta}_{w2} \\ + m_w \ddot{Z}_{w1} L_w \sin\theta_{w2} + m_w \ddot{Z}_{v1} L_w \sin\theta_{w2} = 0 \end{aligned} \tag{35}$$

where $I_{w1}\ddot{\theta}_{w1}$ & $I_{w1}\ddot{\theta}_{w2}$ are the crane payload rotational inertia about its hinge point in the longitudinal and lateral directions, respectively, $c_w\dot{\theta}_{w1}$ & $c_w\dot{\theta}_{w2}$ are the viscous torsional damping moments about the crane payload hinge point in the longitudinal and lateral directions, respectively, $m_w \ddot{X}_{w1} L_w \cos\theta_{w1}$ & $m_w \ddot{Y}_{w1} L_w \cos\theta_{w2}$ are the moments about the hinge point due to the payload horizontal inertia by virtue of its coupling with the sprung mass large pitch & roll motions, respectively, $m_w \ddot{Z}_{w1} L_w \sin\theta_{w1}$ & $m_w \ddot{Z}_{w1} L_w \sin\theta_{w2}$ are the moments about the hinge point due to the payload vertical inertia by virtue of its coupling with the sprung mass degrees of freedom.

4 Static Equilibrium Response Comparison with and Without Suspension Lock Over Flat Terrain

Equations (1) to (35) are coded in MATLAB and solved by using similar technique as that described in [3]. The magnitudes of various fixed parameters are described in Table 1. The four-station military vehicle is considered to have bump-stopper (φ_s = 60°) at every suspension station with which the dynamic effects of variation in the crane payload (m_w) from 4.5 t, 8.5 t, 12.5 t & 14.5 t on the sprung mass roll angle and RH_1 & LH_2 wheel reactions are observed initially. Thereafter, suspension locks, which are similar in functionality to the usual bump-stoppers, were added to each station in order to limit the wheel travel and sprung mass roll as well as to reduce the suspension loads. In this regard, the suspension lock angle φ_s is reduced to 35° from 60°. It may be noted that since the crane payload acts on the vehicle right side and that too on one corner, RH_1 & LH_2 wheel stations encounter the extreme loading conditions. Therefore, the transient dynamic and subsequent static equilibrium response comparison (in terms of sprung mass roll angle and LH_2 & RH_1 vertical wheel reactions) with the suspension lock and usual bump-stopper over flat terrain under different crane-payload conditions, is brought out from the non-linear dynamic mathematical model of the military vehicle.

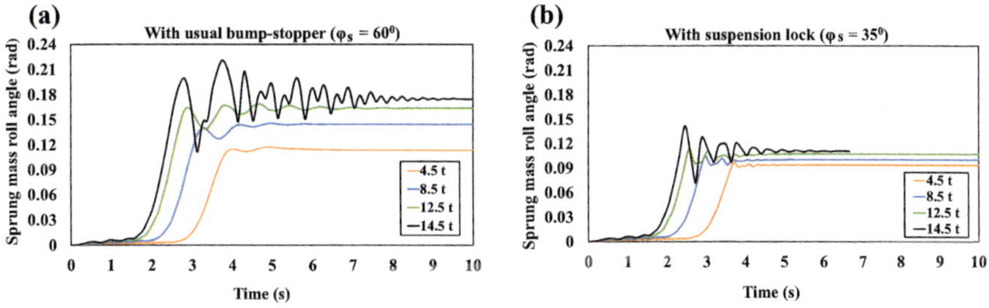

Fig. 3 **a** Sprung mass roll under different payloads when $\varphi_s = 60°$, **b** Sprung mass roll under different payloads when $\varphi_s = 35°$

The HSU rebound gas charging pressure is considered to be 95 bar in all the stations for all the load cases. Figure 3a and b highlight the time dependent sprung mass roll angular displacement variation with usual bump-stopper at 60° and suspension lock at 35°, respectively. Figure 4a and b represent the vertical reaction load variation on the RH_1 suspension station and Fig. 5a and b highlight the vertical reaction load variation on the LH_2suspension station with usual bump-stopper at 60° and suspension lock at 35°, respectively. Table 2 indicates the comparative final static

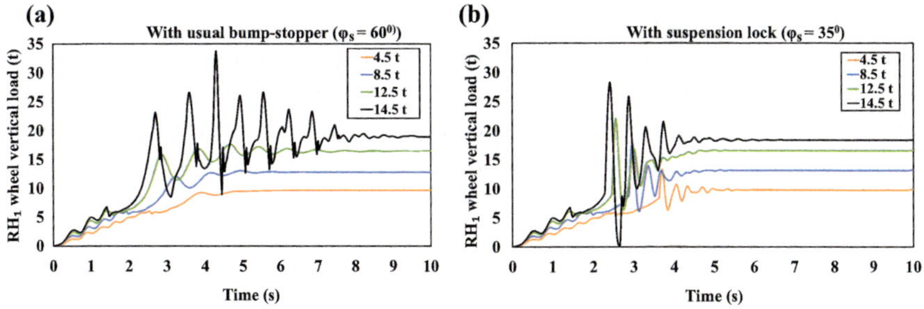

Fig. 4 **a** RH_1 wheel vertical load under different payloads when $\varphi_s = 60°$, **b** RH_1 wheel vertical load under different payloads when $\varphi_s = 35°$

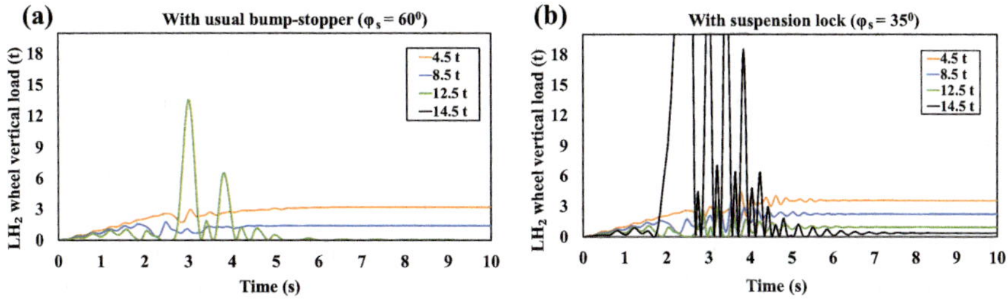

Fig. 5 **a** LH_2 wheel vertical load under different payloads when $\varphi_s = 60°$, **b** LH_2 wheel vertical load under different payloads when φ_s =35°

Table 2 Final static equilibrium configuration of the four-station military vehicle with the crane payload

Crane payload (t)	Sprung mass roll (rad)		RH_1 wheel vertical reaction (t)		LH_2 wheel vertical reaction (t)	
	$\varphi_s = 60°$	$\varphi_s = 35°$	$\varphi_s = 60°$	$\varphi_s = 35°$	$\varphi_s = 60°$	$\varphi_s = 35°$
4.5	0.114	0.093	9.65	9.84	3.23	3.59
8.5	0.144	0.1	12.76	13.13	1.42	2.25
12.5	0.164	0.106	16.47	16.57	0.06	0.96
14.5	0.174	0.11	18.34	18.99	0	0.37

equilibrium response in terms of the sprung mass roll, RH_1 & LH_2 vertical wheel reaction loads. It is observed from Fig. 3a and b and Table 2 that the sprung mass roll angular displacement reduces further with the suspension lock when compared to that with the usual bump-stopper under all the loading conditions. Therefore, as the sprung mass roll is reduced, the wheel travel also gets limited. Moreover, as the crane payload increases, there is more reduction in the sprung mass roll displacement. Therefore, it clearly reveals the advantage of implementing the suspension lock at a lesser angle by reducing the roll over tendency under given payloads. This is evident from the LH_2 wheel vertical reaction load which tends to become zero (i.e., tends to lose ground contact) without the suspension lock. It is observed from Figs. 4a and b, 5a and b and from Table 2 that there is a marginal increase in the RH_1 and LH_2 wheel vertical reaction loads with implementation of the suspension lock. However, it may be noted that with addition of the suspension lock at a lesser angle, the vertical reaction load is shared by both the lock and HSU. This reduces the overall suspension loads with addition of the lock. It is estimated from Table 2 that the safe crane payload for the given vehicle & suspension configuration is 12.5 t by considering a load factor of 1.5.

5 Conclusion

The non-linear mathematical model of the four-station military vehicle with trailing arm HSU and crane hanging payload is developed to simulate the dynamic influence of suspension lock on the vehicle dynamic behavior and final static equilibrium configuration. The trailing arm dynamic behavior, inertia coupling effects between the sprung & unsprung masses as well as hanging payload, rebound/bump-stopper and ground contact phenomenon is brought out in the mathematical model. The safe crane payload can also be estimated from the model. With the suspension lock at a relatively lesser angle, sprung mass roll and overall suspension load reduction are observed. Therefore, the implementation of suspension lock ascertains vehicle stability. The model is a prerequisite for deciding upon the allowable vehicle speed during movement with the crane payload. The model is very novel & generic and provides deep insight into the development of the recovery vehicle simulator.

Acknowledgements The authors would like to thank Shri Balamurugan V, Director, CVRDE & Shri Murugesan R, Addl. Director (CEAD) for extending all required facilities to carry out the research work, Shri S. Pazhanikumar, Sc 'G' & Shri Arun Kumar Deokar, Sc 'C' from SV group and Shri N.S. Sekar, Sc 'F' (Retd.) from RG group. The authors express their gratitude to Dr. V. Balamurugan, Addl. Director (AP) & Prof. R. Krishna Kumar, IIT (M). The authors would also like to thank Shri Abu Bakr Azam, Research Scholar, Nanyang Technological University for his noteworthy assistance.

References

1. Mankar, N.A.: Comparative dynamic analysis of a military recovery vehicle during crane operations with linear suspension. M. Tech Thesis, DIAT (Pune), India (2019)
2. Banerjee, S., Balamurugan, V., Krishna, KR.: Effect of integrated ride and cornering dynamics of a military vehicle on the weapon responses. In: Proceedings of The Institution of Mechanical Engineers, Part K: Journal of Multi-body Dynamics, 232.4 pp. 536–554 (2018)
3. Banerjee, S., Balamurugan, V., Krishna, KR.: Ride dynamics mathematical model for a single station representation of tracked vehicle. J. Terramech. **53**, 47–58 (2014)
4. Li, X., Wang, Y.: Analysis of mixed model in gear transmission based on ADAMS. Chin. J. Mech. Eng. **25**(5), 968–973 (2012)
5. Devesh, M., Manigandan, R., Banerjee, S., Mailan, L.B.: Development of sprung mass non-linear mathematical model of pitch dynamics with trailing arm suspension. Vibroeng Proc 37, 1–6 (2021)

One-Dimensional Steady State Heat Conduction Equation with and Without Source Term by FVM

Neelam Patidar and Akshara Makrariya

Abstract A One-dimensional (1D) steady-state heat conduction equation with and without source term is presented in this paper. The temperature distribution in a body is determined by the model of the heat equation based on some physical assumptions. The heat equation is solved by finite volume method (FVM). By using the Matlab software, a numerical simulation of the raised examples was investigated. The results of the heat equation with and without source term are well compared and it is found that the temperature distribution of 1D steady-state heat equation with source term is parabolic whereas the temperature distribution without source term is linear. The results concluded that the numerical solutions perfectly matched the exact solutions as expected.

Keywords Heat equation · Heat conduction equation · Source term · Finite volume method (FVM) · Partial differential equation (PDE) · One-dimensional (1D)

1 Introduction

The heat equation is a crucial PDE that defines how heat (or temperature variation) is distributed in a particular location over time. Mathematical analysis, numerical computations, and experiments are all made easier by the heat equation. It's also extremely practical: engineers must ensure that engines do not melt and computer chips do not overheat [1]. We must examine the concepts in depth because of these and

VIT Bhopal University

N. Patidar (✉) · A. Makrariya
School of Advanced Science-Mathematics, VIT Bhopal University, Bhopal, Madhya Pradesh, India
e-mail: neelampatidar1994@gmail.com

A. Makrariya
e-mail: akshara.makrariya@vitbhopal.ac.in

S. Banerjee and A. Saha (eds.), *Nonlinear Dynamics and Applications*,
Springer Proceedings in Complexity,
https://doi.org/10.1007/978-3-030-99792-2_67

many other real-world applications of the heat equation [2, 3]. The control volume approach has become a prominent fluid flow solution procedure over the last two decades.

Chai et al. [4] presented a finite volume method (FVM) to capture collimated beam. To demonstrate the FVM's capabilities, it has been used in two- and three-dimensional enclosures with transparent, emitting, absorbing and anisotropically scattering media. All The obtained results have been compared with other published results. And, it was conclude that the FVM is accurate and efficient.

Li et al. [5] proposed a new FVM for cylindrical heat conduction issues. The problem was taken based on a local analytical solution. The novel approach's computation results are compared to those of the traditional second-order FVM. The developed method for cylindrical heat conduction issues is more accurate than previous methods. The obtained results reveal that the novel method takes much less computation effort than traditional methods to achieve the same degree of accuracy.

Belghazi et al. [6] presented an analytical approach of unsteady-state heat conduction. The study has done for two-layered material based on moving Gaussian laser. The homogeneous part of the heat equation was solved by the separation of variables method. This model can also be used to calculate the thermal contact resistance between layers.

A method for the solution of heat conduction problems with phase-changing and movable boundary conditions has been developed by Juan [7]. The problem has been solved by the element-free Galerkin method [8, 9]. According to the results of this approach, it can effectively deal with the marked nonlinearity of fusion latent heat release in heat transfer problems with phase change. Numerical results are obtained accurate and stable by its straightforward implementation.

A numerical simulation of non-Fourier heat conduction in fins under periodic boundary conditions has been done by Liu et al. [10]. To evaluate the non-fourier heat conduction, the Lattice Boltzmann approach was used [11]. The study analyzed the effect of frequency, shape and relaxation time of the base temperature oscillations on heat transfer efficiency [12].

We have looked at 1D heat conduction in a steady state. Steady means the temperatures do not change with time and hence, the heat flow does not change over time. 1D means that the temperature is determined by a single dimension [13].

Heat Conduction: The thermal conductivity is defined by Fourier's law. According to Fourier's law, the area at right angles to the gradient through which the heat flows is proportional to the negative gradient of temperature and the time rate of heat transfer. Fourier's law is also known as the law of heat conduction [14]. The aim of this study is to derive a 1D heat equation with and without source term, and its solution using the FVM along with its numerical analysis using MATLAB.

Finite volume method: The FVM is a numerical method that converts PDEs into discrete algebraic equations. The PDEs express conservation laws across differential volumes. Discrete algebraic equations can be solved over finite volumes [15, 16]. To obtain the values of the dependent variables for each element, the system of these equations was solved. The resulting system of equations usually includes fluxes that enter the finite volume faces, hence flux calculations are crucial in FVM [17, 18]. Interpolation and gradient methods are also used in this process [19].

2 Solution of 1D Steady-State Heat Conduction Equation by FVM

The 1D steady-state heat conduction equation with the source term that we are using in this study is created using the given formula

$$\frac{d}{dx}\left(k\frac{dT}{dx}\right) + S = 0$$

where 'T' is the rod's temperature. The temperature boundary of values at A and B are specified. Thermal diffusivity is denoted by the letter k and the source term is S.

2.1 Grid Generation

The discretization of the domain into discrete control volumes is the first stage in the finite control volume, as shown in Fig. 1.

In this figure, the number of nodal points is placed between A and B. Here are three nodal points. We are calculating the equation at node P. The finite control volume is represented by the highlighted area. P is the center of control volume and W and E are the neighboring nodes of the node P to its west and east respectively. The w represents the interface between W-P and e represents the interface between P-E. Grid size is denoted by δx. For this problem, we have assumed a uniform grid [15, 20].

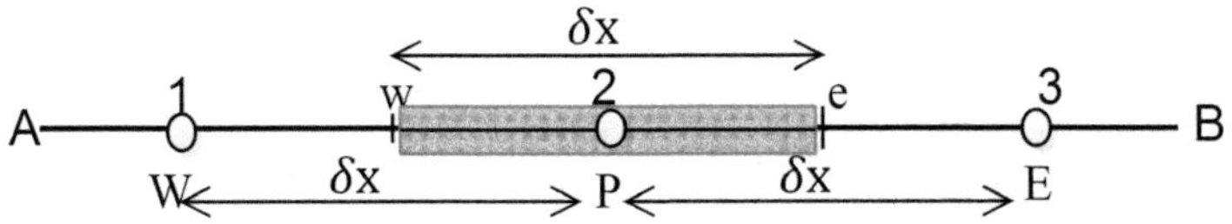

Fig. 1 discretization of the domain

2.2 Discretization

Over the control volume, integrate the governing equation as follows:

$$\int_{\Delta V} \frac{d}{dx}\left(\alpha \frac{dT}{dx}\right) dV + \int_{\Delta V} S dV = 0$$

The cross-sectional area of the control volume is A. ΔV is the control volume. The average value of the source S over the control volume is S. Hence, $dv = \text{Adx}$ and $\Delta V = \text{A}\delta\text{x}$. since the heat generation rate is uniform, So the equation becomes

$$\int_{w}^{e} \frac{d}{dx}\left(\alpha A \frac{dT}{dx}\right) dx + S \int_{\Delta V} dV = 0$$

$$\left[\left(\alpha A \frac{dT}{dx}\right)_e - \left(\alpha A \frac{dT}{dx}\right)_w\right] + S\Delta V = 0$$

The above equation states that the generation of temperature is equal to the diffusive flux of temperature leaving the east face of the control volume minus the diffusive flux of temperature entering the west face. As we can observed that, the flow is conserved within the control volume [21]. To derive useful forms of the discretized equation, the interface diffusion coefficient and temperature gradient at east and west face are required. Linear approximations seem to be the obvious and this is the easiest method of computing interface values and gradients. This practice is called a central differencing scheme [22].

$$\alpha_w = \frac{\alpha_W + \alpha_P}{2} \quad , \alpha_e = \frac{\alpha_E + \alpha_P}{2}$$

And the diffusive flux terms are calculated as follows:

$$\left(\alpha A \frac{dT}{dx}\right)_e = \alpha_e A_e \left(\frac{T_E - T_P}{\delta x_{PE}}\right)$$

$$\left(\alpha A \frac{dT}{dx}\right)_w = \alpha_w A_w \frac{T_P - T_W}{\delta x_{WP}}$$

The finite volume method approximates the source term as

$$S\Delta V = S_u + S_p T_P$$

The final equation can be rearranged as

$$\left(\alpha_e A_e \cdot \frac{T_E - T_P}{\delta x_{PE}}\right) - \left(\alpha_w A_w \cdot \frac{T_P - T_W}{\delta x_{WP}}\right) + (S_u + S_p T_P) = 0$$

This can be arranged as

$$\left(\frac{\alpha_e A_e}{\delta x_{PE}} + \frac{\alpha_w A_w}{\delta x_{WP}} - S_p\right) T_P = \left(\frac{\alpha_w A_w}{\delta x_{WP}}\right) T_W + \left(\frac{\alpha_e A_e}{\delta x_{PE}}\right) T_E + S_u$$

Identifying the coefficient of T_P, T_W, T_E as a_W, a_P, a_E and rearranging the equation as under

$$a_P T_P = a_W T_W + a_E T_E + S_u$$

where

$$a_W = \frac{a_w A_W}{\delta x_{WP}}, \quad a_P = \frac{\alpha_e A_e}{\delta x_{PE}}, \quad a_E = a_W + a_P - S_p$$

3 Numerical Solution

A textbook exam is taken "An Introduction to Computational Fluid Dynamics" by Versteeg and Malasekara [15]: Consider the problem of heat conduction with and without source term in an insulated rod whose ends are kept at a constant temperature of 100 °C and 500 °C respectively. Calculate the temperature distribution in a road at a steady state. The cross-sectional area A is 10×10^{-3} m^2 and the thermal conductivity is 1000 W/m/K.

3.1 The Heat Conduction Equation with Source Term

Under uniform heat generation $S = 1000\text{K}$, the equation for the governing process is

$$\frac{d}{dx}\left(K * \frac{dT}{dx}\right) + S = 0$$

Integrating above the control volume dV and discretization of the above equation will yield

$$K \cdot A \cdot \left(\frac{T_E - T_P}{dx}\right) - K \cdot A \cdot \left(\frac{T_p - T_W}{dx}\right) = SAdx \frac{2K \cdot A}{dx} \cdot T_P = \frac{K \cdot A}{dx} \cdot T_E + \frac{K \cdot A}{dx} \cdot T_W + SAdx$$

The above equation represents the solution of temperature at the inner nodes. The boundary conditions for this domain are $T_1 = 100\,°\text{C}$ (left end) and $T_2 = 500\,°\text{C}$ (right end). At the left control volume, the node associated with the left boundary is

$$\frac{2KA}{dx} \cdot T_P = \frac{KA}{dx} \cdot T_A + \frac{KA}{dx} \cdot T_E + SAdx$$

And, the node associated with the right boundary is

$$\frac{2KA}{dx} \cdot T_P = \frac{KA}{dx} \cdot T_B + \frac{KA}{dx} \cdot T_W + SAdx$$

3.2 The Heat Conduction Equation Without Source Term

$$\frac{d}{dx}\left(K * \frac{dT}{dx}\right) = 0$$

Integrating above control volume dV and discretization of the above equation will yield

$$KA \cdot \left(\frac{T_E - T_P}{dx}\right) - KA \cdot \left(\frac{T_p - T_W}{dx}\right) = 0$$

$$\frac{2KA}{dx} \cdot T_P = \frac{KA}{dx} \cdot T_E + \frac{KA}{dx} \cdot T_W$$

The above equation represents the solution of temperature at the inner nodes. The node associated with the left boundary

$$\frac{2KA}{dx} \cdot T_P = \frac{KA}{dx} \cdot T_A + \frac{KA}{dx} \cdot T_E$$

The node associated with the right boundary

$$\frac{2KA}{dx} \cdot T_P = \frac{KA}{dx} \cdot T_B + \frac{KA}{dx} \cdot T_W$$

To verify the accuracy, the final equation is performed in MATLAB on 5, 10, and 25 nodes. The numerical solution is compared to the exact solution. The exact solution is contained in the textbook [15].

4 Results and Discussion

The temperature distributions were analyzed at 5, 10 and 25 nodes to validate the results. From the results, it was found that numerical accuracy increases as we increase the number of nodes. The following outputs were obtained from the study.

The temperature distribution of the rod is shown in Fig. 2. In this figure, the straight line shows the temperature distribution of a rod obtained by numerical method and the dotted line represents the exact temperature distribution. The temperature of a

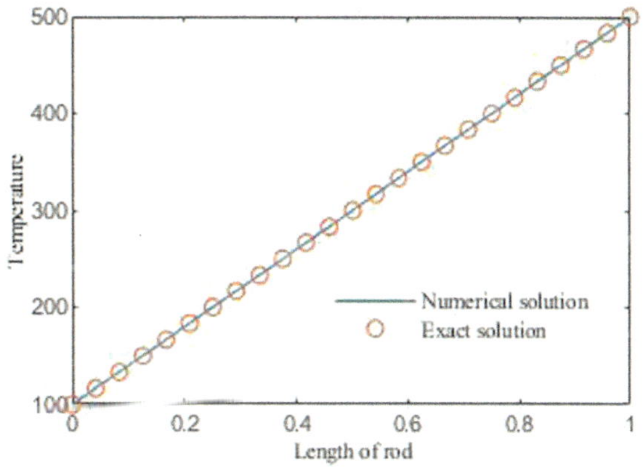

Fig. 2 Temperature distribution of rod

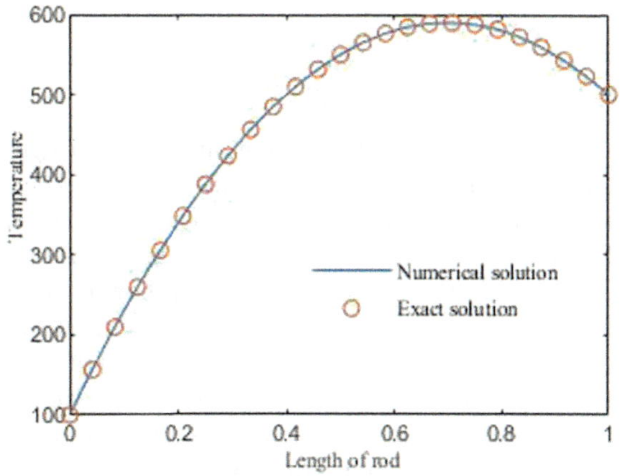

Fig. 3 Temperature distribution of rod on adding external source

rod increases with increasing the length of rod. The temperature start increasing from 100 °C then reached at 500 °C. The temperature increases linearly.

Figure 3 shows the distribution of temperature of the rod on adding an external heat source. In this figure, the straight line shows the temperature distribution of a rod obtained by FVM and the dotted line represents the exact temperature distribution. The temperature of a rod increases as increasing the length of the rod. The temperature starts increasing from 100 °C and reached at 590 °C then again starts decreasing and reached at 500 °C. The temperature distribution was obtained parabolically.

Figure 4 shows the comparison between one-dimensional heat conduction equations with and without source term. It was obtained that the temperature distribution of a rod without an external heat source is linear. On adding the external heat source in a rod, the temperature distribution gets changed and becomes parabolic. From all these results it was observed that the external source affects the temperature distribution of a rod.

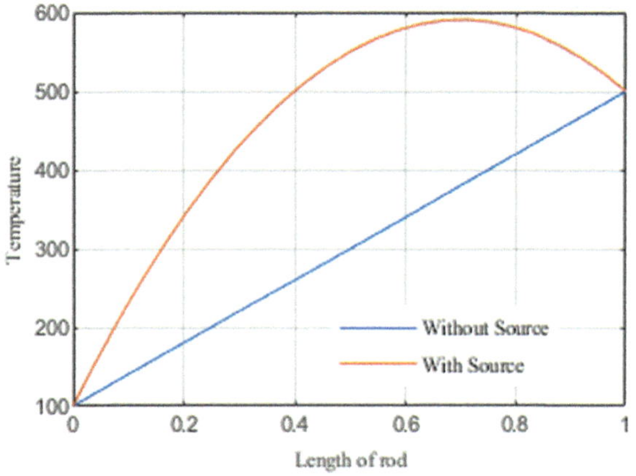

Fig. 4 Comparison between heat equations with and without source term

5 Conclusion

A one-dimensional heat equation can be used to represent many physical phenomena that are connected to temperature distribution, as demonstrated in this study. Analytical solutions are sometimes insufficient for understanding the behavior of solutions. Due to this, we may rely on numerical solutions to find out more about the inherent problems. In this study, we have seen how to obtain and solve a 1D steady-state heat equation with and without a source term. It was observed that the temperature distribution of 1D steady-state heat equation with source term is parabolic whereas the temperature distribution without source term is linear. The numerical solutions were found to be similar to the exact solutions, as expected. Furthermore, by using MATLAB programming, we have provided a real comprehension of the example mentioned in the study.

References

1. Tipler, PA., Mosca, G.: Physics for Scientists and Engineers. Macmillan (2007)
2. Asmar, N.H.: Partial Differential Equations with Fourier Series and Boundary Value Problems. Courier Dover Publications (2016)
3. Abdisa, L.T.: One Dimensional Heat Equation and its Solution by the Methods of Separation of Variables, Fourier Series and Fourier Transform (2021)
4. Chai, J.C., Lee, H.S., Patankar, S.V.: Finite volume method for radiation heat transfer. J. Thermophys. Heat Transf. **8**(3), 419–425 (1994)
5. Li, W., Yu, B., Wang, X., Wang, P., Sun, S.: A finite volume method for cylindrical heat conduction problems based on local analytical solution. Int. J. Heat Mass Transf. **55**(21–22), 5570–5582 (2012)
6. Belghazi, H., El Ganaoui, M., Labbe, J.-C.: Analytical solution of unsteady heat conduction in a two-layered material in imperfect contact subjected to a moving heat source. Int. J. Thermal Sci. **49**(2), 311–318 (2010)

7. Álvarez-Hostos, J.C., Gutierrez-Zambrano, E.A., Salazar-Bove, J.C., Puchi-Cabrera, E.S., Bencomo, A.D.: Solving heat conduction problems with phase-change under the heat source term approach and the element-free Galerkin formulation. Int. Commun. Heat Mass Transf. **108**, 104321 (2019)
8. Hostos, J.C., Fachinotti, V.D., Piña, A.J., Bencomo, A.D., Cabrera, E.S.: Implementation of standard penalty procedures for the solution of incompressible Navier-Stokes equations, employing the element-free Galerkin method. Eng. Anal. Boundary Elem. **96**, 36–54 (2018)
9. Ling, J., Yang, D.S., Zhai, B.W., Zhao, Z.H., Chen, T.Y., Xu, Z.X., Gong, H.: Solving the single-domain transient heat conduction with heat source problem by virtual boundary meshfree Galerkin method. Int. J. Heat Mass Transf. **1**(115), 361–7 (2017)
10. Liu, Y., Li, L., Zhang, Y.: Numerical simulation of non-Fourier heat conduction in fins by lattice Boltzmann method. Appl. Therm. Eng. **166**, 114670 (2020)
11. Bamdad, K., Ashorynejad, H.R.: Inverse analysis of a rectangular fin using the lattice Boltzmann method. Energy Convers. Manage. **97**, 290–7 (2015)
12. Wankhade, P.A., Kundu, B., Das, R.: Establishment of non-Fourier heat conduction model for an accurate transient thermal response in wet fins. Int. J. Heat Mass Transf. **126**, 911–23 (2018)
13. Date, A.W.: Introduction to Computational Fluid Dynamics. Cambridge University Press, Cambridge (2005)
14. Wang, L.: Generalized Fourier law. Int. J. Heat Mass Transfer **37**(17), 2627–2634 (1994)
15. Versteeg, H.K., Malalasekera, W.: An Introduction to Computational Fluid Dynamics: The Finite Volume Method [OpenFOAM], vol. M (2015)
16. Moukalled, F., Mangani, L., Darwish, M.: The Finite Volume Method in Computational Fluid Dynamics, vol. 113. Springer, Berlin (2016)
17. Akhtar, F.: Introduction and application of finite volume method (FVM) for 1-D linear heat conduction equation. Heat Equ 1–17 (2021)
18. Banerjee, S.: Mathematical Modeling- Models. CRC Press, Analysis and Applications (2014)
19. An Introduction to Computational Fluid Dynamics: The Finite Volume Method. [OpenFOAM] (2015)
20. Zienkiewicz, O.C., Taylor, R.L., Nithiarasu, P., Zhu, J.Z.: The Finite Element Method, vol. 3. McGraw-Hill, London (1977)
21. Barth, T., Ohlberger, M.: Finite volume methods: foundation and analysis. Encyclop. Comput. Mech. **1**(15), 1–57 (2004)
22. Akhtar, F.: Solving 1-D steady state heat conduction equation using finite volume method (FVM) M, 1–15 (2021)
23. Makrariya, A., Adlakha, N.: Two-dimensional finite element model to study temperature distribution in peripheral regions of extended spherical human organs involving uniformly perfused tumors. Int. J. Biomath. **8**(06), 1550074 (2015)
24. Makrariya, A., Pardasani, K.R.: Finite element method to study the thermal effect of cyst and malignant timpr in women's breast during menstrual cycle under cold environment. Adv. Appl. Math. Sci. **18**(1), 29–43 (2018)

Travelling and Solitary Wave Solutions of (2+1)-Dimensional Nonlinear Evolution Equations by Using Khater Method

Ram Mehar Singh, S. B. Bhardwaj, Anand Malik, Vinod Kumar, and Fakir Chand

Abstract The most of the physical systems are nonlinear by nature which can be represented by various nonlinear partial differential equations. Here, we present a simple technique say Khater method to find the Travelling and Solitary Wave solutions of (2+1)-dimensional nonlinear evoluation equations. This method is a very powerful tool for obtaining the exact solutions of various nonlinear differential equations. In this study, modified Korteweg- de Vries-Zakharov-Kuznetsov (mKdV-ZK) equation is taken as an example of nonlinear evolution equation which is used in astrophysics to study various space phenomena, dynamics of plasma etc.

Keywords mKdV-ZK equation · Khater method · Traveling wave solutions · Solitary wave solutions

1 Introduction

The concept of solitary wave was introduced by Zabusky and Kruskal in 1965 in their well known experiment on KdV- equation . It gain interest of researchers working in areas of nonlinear dynamics. Various phenomena such as prolongation structures, space curves , gauge-equivalence, Lie-algebraic properties , singularity structures are related to the concept of solitons and can be described by different nonlinear evolution equations(NLEEs) [1–3]. The investigation of travelling and solitary wave solutions of these equations is a major component of the

R. M. Singh (✉) · V. Kumar
Department of Physics, Chaudhary Devi Lal University, Sirsa 125055, India
e-mail: dixit_rammehar@yahoo.co.in; dixitrammehar@cdlu.ac.in

S. B. Bhardwaj
Department of Physics, SUS Govt. College Matak-Majri, Karnal 132041, India

A. Malik
Department of Physics, Chaudhary Ranbir Singh University, Jind 126102, India

F. Chand
Department of Physics, Kurukshetra University, Kurukshetra 136119, India

S. Banerjee and A. Saha (eds.), *Nonlinear Dynamics and Applications*,
Springer Proceedings in Complexity,
https://doi.org/10.1007/978-3-030-99792-2_68

research that play a significant role in the description of various nonlinear systems. The exact solutions of such equations yield lot of information about the system concerned [4, 5] which can be obtained by reducing the nonlinear partial differential equations(NLPDEs) to associated ordinary differential equations by using the ansatz $g(x, y, t) = g(\xi), \quad \xi = x + y - ct$. In past, several efforts have been made by mathematicians and physicists to find the exact solutions of the nonlinear partial differential equations by employing various methods like extended modified auxiliary equation mapping method, exp-function method, ansatz method, trial equation method, extended direct algebaric method, auxiliary equation method, $e^{(-i\phi\xi)}$-expansion method, extended tanh-function method, Kudryashov and modified Kudryashov methods, improved $tan(\frac{\phi}{2})$-expansion method, $(\frac{G'}{G})$-expansion method , novel $(\frac{G'}{G})$-expansion method, improved $(\frac{G'}{G})$-expansion method etc. [6–12].

Now, we introduce a new method for solving various NLEEs i.e. Khater method which is one of the few general methods available for solving various NLEEs. There are lot of NLEEs to describe various nonlinear systems but in the present study, we have computed the travelling and solitary wave solutions of (2+1)-dimensional mKdV-ZK equation [13–15]which is an important class of NLEEs arising in fluid dynamics, plasma physics,Bose-Einstein condensate, shallow water waves, nonlinear optics astrophysics, quantum optic, hydrodynamic and mathematical physics to study nonlinear physical phenomena.The organization of the paper is as follow : The basic formulation of the Khater method is discussed in Sect. 2. Under the elegance of Khater method, exact solitonic solutions of (2+1)-dimensional mKdV-ZK equation and their graphical representations are given in Sect. 3. Finally concluding remarks are addressed in Sect. 4.

2 Khater Method

Consider the nonlinear evolution equation as :

$$P(u, D_x^\eta q, D_y^\eta q, D_z^\eta q, D_t^\eta q, D_x^\eta D_y^\eta q, D_x^\eta D_t^\eta q, D_y^\eta D_t^\eta q, ...) = 0, \tag{1}$$

where P is a polynomial in q(x,y,t) and its partial derivatives .

Step 1. Consider the transformation

$$g(x, y, t) = g(\xi), \quad \xi = x + y - c_1 t, \tag{2}$$

where c_1 is the constant, then Eq. (2) transformed in the following ODE:

$$Q(q, q', q'', q''', ...) = 0 \tag{3}$$

where Q is a polynomial in $q(\xi)$ and its various derivatives.

Step 2. Consider solution of Eq. (3) as

$$q(\xi) = \sum_{i=0}^{N} a_i a^{ig(\xi)}, \tag{4}$$

where a, a_i, are constants to be calculated, such that $a_N \neq 0$ and $g(\xi)$ satisfies the following differential equation :

$$g'(\xi) = \frac{1}{ln(a)}\left(\eta a^{-g(\xi)} + \rho + \delta a^{g(\xi)}\right) \tag{5}$$

Step 3. Determine the positive integer N in Eq. (5) by balancing the highest order derivatives and the nonlinear terms.

Step 4. Inserting Eq. (4) along Eq. (5) into Eq. (3) and rationalization of the resultant expression, we met a set of algebraic equations, which can be solved by symbolic computation to get the values of a_i and (η, ρ, δ).

The solutions of Eq. (5):

When $(\rho^2 - \eta\delta < 0 \;\&\; \sigma = 0)$.

$$a^{f(\xi)} = \left[\frac{-\rho}{\delta} + \frac{\sqrt{-(\rho^2 - \eta\delta)}}{\delta} tan\left(\frac{\sqrt{-(\rho^2 - \eta\delta)}}{2}\xi\right)\right] \tag{6}$$

or

$$a^{f(\xi)} = \left[\frac{-\rho}{\delta} + \frac{\sqrt{-(\rho^2 - \eta\delta)}}{\delta} cot\left(\frac{\sqrt{-(\rho^2 - \eta\delta)}}{2}\xi\right)\right] \tag{7}$$

When $(\rho^2 + \eta\delta > 0 \;\&\; \delta \neq 0)$.

$$a^{f(\xi)} = \left[\frac{-\rho}{\delta} - \frac{\sqrt{(\rho^2 - \eta\delta)}}{\delta} tanh\left(\frac{\sqrt{(\rho^2 - \eta\delta)}}{2}\xi\right)\right] \tag{8}$$

or

$$a^{f(\xi)} = \left[\frac{-\rho}{\delta} - \frac{\sqrt{-(\rho^2 - \eta\delta)}}{\delta} coth\left(\frac{\sqrt{(\rho^2 - \eta\delta)}}{2}\xi\right)\right] \tag{9}$$

When $(\rho^2 + \eta^2 > 0 \;\&\; \delta \neq 0 \;\&\; \delta = -\eta)$.

$$a^{f(\xi)} = \left[\frac{\rho}{\eta} + \frac{\sqrt{(\rho^2 + \eta^2)}}{\eta} tanh\left(\frac{\sqrt{(\rho^2 + \eta^2)}}{2}\xi\right)\right], \tag{10}$$

or

$$a^{f(\xi)} = \left[\frac{\rho}{\eta} + \frac{\sqrt{(\rho^2 + \eta^2)}}{\eta} coth\left(\frac{\sqrt{(\rho^2 + \eta^2)}}{2}\xi\right)\right]. \tag{11}$$

When $(\rho^2+\eta^2<0\ \&\ \delta\neq 0\ \&\ \delta=-\eta)$.

$$a^{f(\xi)}=\left[\frac{\rho}{\eta}+\frac{\sqrt{-(\rho^2+\eta^2)}}{\eta}tan\left(\frac{\sqrt{-(\rho^2+\eta^2)}}{2}\xi\right)\right], \tag{12}$$

or

$$a^{f(\xi)}=\left[\frac{\rho}{\eta}+\frac{\sqrt{-(\rho^2+\eta^2)}}{\eta}cot\left(\frac{\sqrt{-(\rho^2+\eta^2)}}{2}\xi\right)\right]. \tag{13}$$

When $(\rho^2-\eta^2<0\ \&\ \delta=\eta)$.

$$a^{f(\xi)}=\left[\frac{-\rho}{\eta}+\frac{\sqrt{-(\rho^2-\eta^2)}}{\eta}tan\left(\frac{\sqrt{-(\rho^2-\eta^2)}}{2}\xi\right)\right], \tag{14}$$

$$a^{f(\xi)}=\left[\frac{-\rho}{\eta}+\frac{\sqrt{-(\rho^2-\eta^2)}}{\eta}cot\left(\frac{\sqrt{-(\rho^2-\eta^2)}}{2}\xi\right)\right]. \tag{15}$$

When $(\rho^2-\eta^2>0\ \&\ \delta=\eta)$.

$$a^{f(\xi)}=\left[\frac{-\rho}{\eta}+\frac{\sqrt{(\rho^2+\eta^2)}}{\eta}tanh\left(\frac{\sqrt{(\rho^2+\eta^2)}}{2}\xi\right)\right], \tag{16}$$

$$a^{f(\xi)}=\left[\frac{-\rho}{\eta}+\frac{\sqrt{(\rho^2+\eta^2)}}{\eta}coth\left(\frac{\sqrt{(\rho^2+\eta^2)}}{2}\xi\right)\right]. \tag{17}$$

When $(\boldsymbol{\eta=\delta=0})$.

$$a^{f(\xi)}=\left[\frac{-(1+e^{2\rho\xi})\pm\sqrt{2(e^{4\rho\xi}+1)}}{e^{2\rho\xi}-1}\right] \tag{18}$$

or

$$a^{f(\xi)}=\left[\frac{-(1+e^{2\rho\xi})\pm\sqrt{e^{4\rho\xi}+6e^{2\rho\xi}+1}}{2e^{2\rho\xi}}\right] \tag{19}$$

When $(\boldsymbol{\rho^2=\eta\delta})$.

$$a^{f(\xi)}=\left[\frac{-\eta(\rho\xi+2)}{\rho^2\xi}\right] \tag{20}$$

When $(\boldsymbol{\rho=k,\ \eta=2k,\ \delta=0})$.

$$a^{f(\xi)}=\left[e^{k\xi}-1\right]. \tag{21}$$

When (ρ=k, δ=2k, η=0).

$$a^{f(\xi)} = \left[\frac{e^{k\xi}}{1-e^{k\xi}}\right]. \tag{22}$$

When ($2\rho = \eta + \delta$).

$$q(x, y, t) = -(c_1 + \eta + \delta) - 4\delta\left[\frac{1-\eta e^{\frac{1}{2}(\eta-\delta)(x+y-c_1 t)}}{1-\delta e^{\frac{1}{2}(\eta-\delta)(x+y-c_1 t)}}\right], \tag{23}$$

or

$$q(x, y, t) = -(c_1 + \eta + \delta) - 4\delta\left[\frac{1+\eta e^{\frac{1}{2}(\eta-\delta)(x+y-c_1 t)}}{-1-\delta e^{\frac{1}{2}(\eta-\delta)(x+y-c_1 t)}}\right]. \tag{24}$$

When (−2ρ=η+δ).

$$a^{f(\xi)} = \left[\frac{\eta e^{1/2(\eta-\delta)\xi} + \eta}{\delta e^{1/2(\eta-\delta)\xi} + \delta}\right]. \tag{25}$$

When (η=0).

$$a^{f(\xi)} = \left[\frac{\rho e^{\rho\xi}}{1+\frac{\delta}{2}e^{\rho\xi}}\right] \tag{26}$$

When (ρ = η = δ ≠0).

$$a^{f(\xi)} = \left[\frac{-(\eta\xi+2)}{\eta\xi}\right]. \tag{27}$$

When (ρ = η =0).

$$a^{f(\xi)} = \left[\frac{-2}{\delta\xi}\right]. \tag{28}$$

When (β = 0, η = δ).

$$a^{f(\xi)} = \left[tan\left(\frac{\eta\xi + C}{2}\right)\right]. \tag{29}$$

When (δ = 0).

$$a^{f(\xi)} = \left[e^{\beta\xi} - \frac{\eta}{2\rho}\right]. \tag{30}$$

where C is arbitrary constant.

Step 5. Implying these values and the solutions of Eq. (5) into Eq. (4), one obtains the exact solutions of Eq. (1).

3 Example

In this section, we employ Khater method to obtain travelling and solitary wave solutions of the (2+1)-dimensional mKdV-ZK equation [13–15] given by

$$q_t + \alpha q^2 q_x + (q_{xx} + q_{yy})_x = 0, \tag{31}$$

Using the transformation(2) in Eq. (31), we get

$$6q'' + \alpha q^3 - c_1 q = 0, \tag{32}$$

by using the principle of homogenity, we have $N = 1$ then Eq. (4) becomes

$$q(\xi) = a_0 + a_1 a^{g(\xi)}, \tag{33}$$

Inserting Eq. (33) and its derivative into Eq. (32) and then after rationalization, one obtain a system of algebraic equations on solving by symbolic computation

$$a^{0g} : 6a_1\eta\rho + \alpha a_0^3 - ca_0 = 0, \tag{34}$$

$$a^{1g} : 6a_1(\rho^2 + 2\eta\rho) + 3\alpha a_0^2 a_1 - ca_1 = 0 \tag{35}$$

$$a^{2g} : 18a_1\delta\rho + 3\alpha a_0 a_1^2 = 0 \tag{36}$$

$$a^{3g} : 12a_1\delta^2 + \alpha a_1^3 = 0 \tag{37}$$

After solving these equations, we get

$$a_1 = 2\delta\sqrt{\frac{-3}{\alpha}} = 2\delta m, \quad a_0 = -\frac{-3\rho}{\alpha m} = \sqrt{\frac{c_1 - 12\eta\delta}{\alpha}}. \tag{38}$$

So that, the exact traveling wave solution of equation (31) be in the form:

$$q(x, y, t) = \sqrt{\frac{c_1 - 12\eta\delta}{\alpha}} + 2\delta m \ \ a^{g(\xi)}. \tag{39}$$

Thus, solitary wave solutions for different papametric conditions are When ($\rho^2 - \eta\delta < 0$ & $\delta \neq 0$).

$$q(x, y, t) = -\sqrt{\frac{c_1 - 12\eta\delta}{\alpha}} + 2\sqrt{\frac{3(\rho^2 - \eta\delta)}{\alpha}} \ \ tan\left(\frac{\sqrt{-(\rho^2 - \eta\delta)}}{2}(x + y - c_1 t)\right), \tag{40}$$

or

$$q(x, y, t) = -\sqrt{\frac{c_1 - 12\eta\delta}{\alpha}} + 2\sqrt{\frac{3(\rho^2 - \eta\delta)}{\alpha}}\ cot\left(\frac{\sqrt{-(\rho^2 - \eta\delta)}}{2}(x + y - c_1 t)\right), \tag{41}$$

When $(\rho^2 + \eta\delta > 0\ \&\ \delta \neq 0)$.

$$q(x, y, t) = -\sqrt{\frac{c_1 - 12\eta\delta}{\alpha}} + 2\sqrt{\frac{-3(\rho^2 - \eta\delta)}{\alpha}}\ tan\left(\frac{\sqrt{(\rho^2 - \eta\delta)}}{2}(x + y - c_1 t)\right), \tag{42}$$

or

$$q(x, y, t) = -\sqrt{\frac{c_1 - 12\eta\delta}{\alpha}} + 2\sqrt{\frac{-3(\rho^2 - \eta\delta)}{\alpha}}\ cot\left(\frac{\sqrt{(\rho^2 - \eta\delta)}}{2}(x + y - c_1 t)\right), \tag{43}$$

When $(\rho^2 + \eta^2 > 0\ \&\ \delta \neq 0\ \&\ \delta = -\eta)$.

$$q(x, y, t) = -\sqrt{\frac{c_1 - 12\eta\delta}{\alpha}} - 2\sqrt{\frac{-3(\rho^2 + \eta^2)}{\alpha}}\ tan\left(\frac{\sqrt{(\rho^2 + \eta^2)}}{2}(x + y - c_1 t)\right), \tag{44}$$

or

$$q(x, y, t) = -\sqrt{\frac{c_1 - 12\eta\delta}{\alpha}} - 2\sqrt{\frac{-3(\rho^2 + \eta^2)}{\alpha}}\ cot\left(\frac{\sqrt{(\rho^2 + \eta^2)}}{2}(x + y - c_1 t)\right), \tag{45}$$

When $(\rho^2 + \eta^2 < 0\ \&\ \delta \neq 0\ \&\ \delta = -\eta)$.

$$q(x, y, t) = -\sqrt{\frac{c_1 - 12\eta\delta}{\alpha}} - 2\sqrt{\frac{3(\rho^2 + \eta^2)}{\alpha}}\ tan\left(\frac{\sqrt{-(\rho^2 + \eta^2)}}{2}(x + y - c_1 t)\right), \tag{46}$$

or

$$q(x, y, t) = -\sqrt{\frac{c_1 - 12\eta\delta}{\alpha}} - 2\sqrt{\frac{3(\rho^2 + \eta^2)}{\alpha}}\ cot\left(\frac{\sqrt{-(\rho^2 + \eta^2)}}{2}(x + y - c_1 t)\right), \tag{47}$$

When $(\rho^2 - \eta^2 < 0\ \ \&\ \delta = \eta)$.

$$q(x, y, t) = \sqrt{\frac{c_1 - 12\eta\delta}{\alpha}} + 2\sqrt{\frac{3(\rho^2 - \eta^2)}{\alpha}}\ tan\left(\frac{\sqrt{(-\rho^2 - \eta^2)}}{2}(x + y - c_1 t)\right), \tag{48}$$

or

$$q(x, y, t) = \sqrt{\frac{c_1 - 12\eta\delta}{\alpha}} + 2\sqrt{\frac{3(\rho^2 - \eta^2)}{\alpha}}\ cot\left(\frac{\sqrt{(-\rho^2 - \eta^2)}}{2}(x + y - c_1 t)\right), \tag{49}$$

When $(\rho^2 - \eta^2 > 0\ \ \&\ \delta = \eta)$.

$$q(x,y,t)=\sqrt{\frac{c_1-12\eta\delta}{\alpha}}+2\sqrt{\frac{-3(\rho^2-\eta^2)}{\alpha}}\ tanh\left(\frac{\sqrt{(\rho^2-\eta^2)}}{2}(x+y-c_1t)\right), \tag{50}$$

or

$$q(x,y,t)=\sqrt{\frac{c_1-12\eta\delta}{\alpha}}+2\sqrt{\frac{-3(\rho^2-\eta^2)}{\alpha}}\ coth\left(\frac{\sqrt{(\rho^2-\eta^2)}}{2}(x+y-c_1t)\right), \tag{51}$$

When $(\rho^2=\eta\delta)$.

$$q(x,y,t)=\sqrt{\frac{c_1-12\rho^2}{\alpha}}-\left[\frac{2\sqrt{-3}(\rho(x+y-c_1t)+2)}{\sqrt{\alpha}(x+y-c_1t)}\right] \tag{52}$$

When $(\rho=k,\delta=2k,\eta=0)$.

$$q(x,y,t)=\sqrt{\frac{c_1}{\alpha}}+4k\sqrt{\frac{-3}{\alpha}}\left[\frac{e^{k(x+y-c_1t)}}{1-e^{k(x+y-c_1t)}}\right]. \tag{53}$$

When $(2\rho=\eta+\delta)$.

$$q(x,y,t)=\sqrt{\frac{c_1-12\eta\delta}{\alpha}}+2\delta\sqrt{\frac{-3}{\alpha}}\left[\frac{1-\eta e^{\frac{1}{2}(\eta-\delta)(x+y-c_1t)}}{1-\delta e^{\frac{1}{2}(\eta-\delta)(x+y-c_1t)}}\right], \tag{54}$$

or

$$q(x,y,t)=\sqrt{\frac{c_1-12\eta\delta}{\alpha}}+2\delta\sqrt{\frac{-3}{\alpha}}\left[\frac{1+\eta e^{\frac{1}{2}(\eta-\delta)(x+y-c_1t)}}{-1-\delta e^{\frac{1}{2}(\eta-\delta)(x+y-c_1t)}}\right]. \tag{55}$$

When $(-2\rho=\eta+\delta)$.

$$q(x,y,t)=\sqrt{\frac{c_1-12\eta\delta}{\alpha}}+2\delta\sqrt{\frac{-3}{\alpha}}\left[\frac{\eta+e^{\frac{1}{2}(\eta-\delta)(x+y-c_1t)}}{\delta+e^{\frac{1}{2}(\eta-\delta)(x+y-c_1t)}}\right], \tag{56}$$

When $(\eta=0)$

$$q(x,y,t)=\sqrt{\frac{c_1}{\alpha}}+2\delta\sqrt{\frac{-3}{\alpha}}\left[\frac{\rho e^{\rho(x+y-c_1t)}}{1+\frac{\delta}{2}e^{\rho(x+y-c_1t)}}\right] \tag{57}$$

When $(\rho=\eta=\delta\neq 0)$.

$$q(x,y,z,t)=\sqrt{\frac{c_1-12\eta^2}{\alpha}}+2\eta\sqrt{\frac{-3}{\alpha}}\left[\frac{-(\eta(x+y-c_1t)+2)}{\eta(x+y-c_1t)}\right]. \tag{58}$$

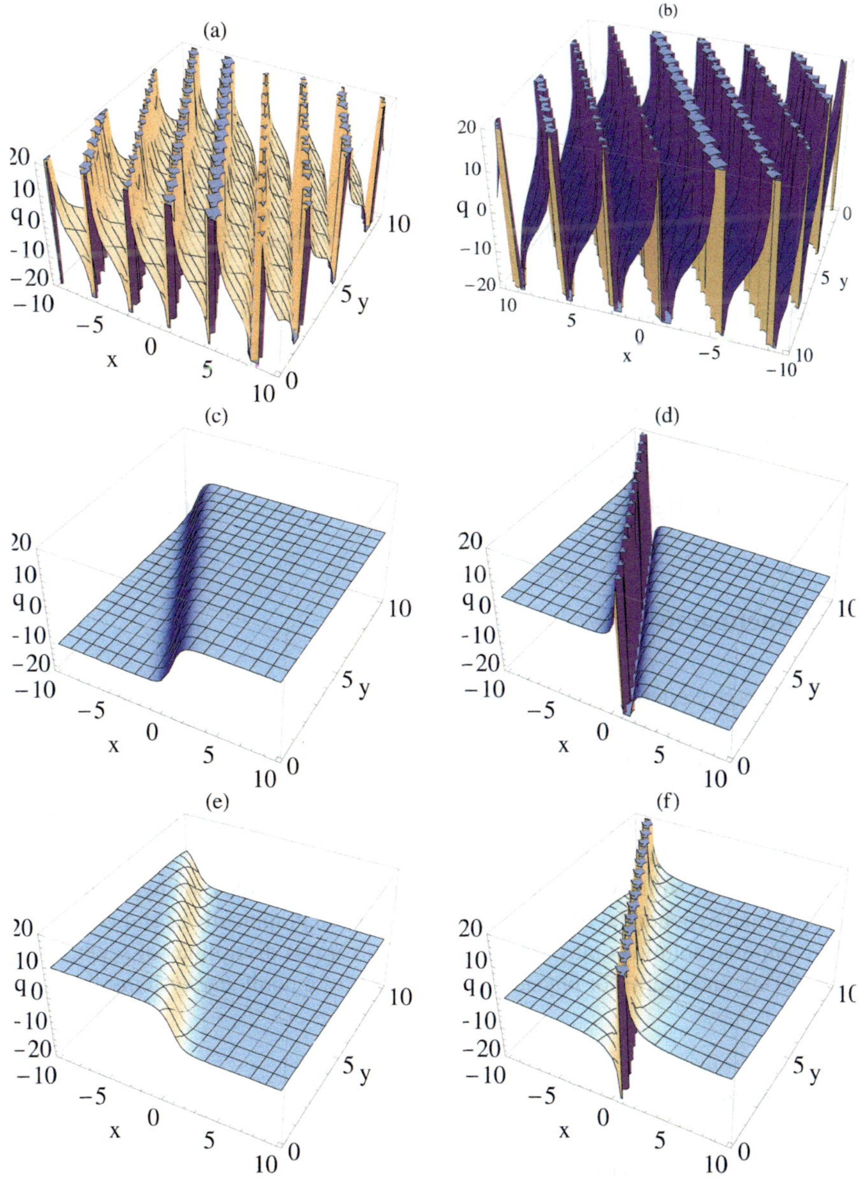

Fig. 1 Solitary wave solution of Eqs. (41, 44–46, 50, 56) with $\alpha = -1, \delta = \eta = t = c_1 = 1$ & $\rho = 2$

When ($\rho = \eta = 0$).

$$q(x, y, z, t) = -\frac{4\sqrt{-3}}{\sqrt{\alpha}(x + y - c_1 t)}. \tag{59}$$

When ($\rho = 0, \eta = \delta$).

$$q(x, y, t) = 2\eta\sqrt{\frac{-3}{\alpha}} tan\left(\frac{\eta(x + y + z - c_1 t) + C}{2}\right). \tag{60}$$

Graphical representation of solutions: We have obtains the travelling and solitary wave solutions of mKdV-ZK equation for various case in term of unknown parameters . The solitary wave solutions can be derived from the travelling wave solutions for the specific value of parameters . Some plots of solitary waves for suitable values of unknown parameters are shown in Fig. 1 . For $\alpha = -1$, $\delta = \eta = t = c_1 = 1$ & $\rho = 2$. Equations (41), (44), (45), (46) , (50) and (56) represents the periodic waves, Kink wave, dark and bright solitons, soliton like travelling wave and bright solitons etc.

4 Conclusions

In this paper, we have obtained the solitary and travelling wave solutions of the (2+1)-dimensional mKdV-ZK equation by Khater method using the symbolic computation. Various higher order polynomials arising in mathematical physics, high energy physics, fluid dynamics, geochemistry and chemical kinematics can be solved by this method. It is seen Khater method provide an effective that powerful technique for solving various nonlinear evaluation equations used in area of science and engineering.The solitary wave solutions computed in this study are hyperbolic, trignometric and exponential forms , which are discussed graphically as shown in Fig. 1. The computed solutions play significant role in various disciplines of sciences and engineering and useful for physical interpretation of a nonlinear systems also [13].

References

1. Wazwaz, A.-M.: The tanh method for traveling wave solutions of nonlinear equations. Appl. Math. Comput. **154**(3), 713–723 (2004)
2. Xu, G.: An elliptic equation method and its applications in nonlinear evolution equations. Chaos, Solitons Fractals **29**(4), 942–947 (2006)
3. Seadawy, A.R.: Travelling-wave solutions of a weakly nonlinear two-dimensional higherorder Kadomtsev-Petviashvili dynamical equation for dispersive shallow-water waves. Eur. Phys. J. Plus **29**, 132 (2017)
4. Seadawy, A.R., Lu, D.: Bright and dark solitary wave soliton solutions for the generalized higher 4 order nonlinear Schrödinger equation and its stability. Results Phys. **6**(590), 3 (2016)

5. Kim, H., Sakthivel, R.: New exact traveling wave solutions of some nonlinear higher dimensional physical models. Rep. Math. Phys. **70**(1), 39–50 (2012)
6. Selima, E.S., Seadawy, A.R., Yao, X.: The nonlinear dispersive Davey-Stewartson system for surface waves propagation in shallow water and its stability. Eur. Phys. J. Plus **131**(425), (2016)
7. Islam, M.T., Akbar, M.A., Azad, M., et al.: A Rational (G/G)-expansion method and its application to the modified KdV-Burgers equation and the (2+1)-dimensional Boussinesq equation. Nonlinear Study **6**(4), 1–11 (2015)
8. Khater, M.M., Seadawy, A.R.: D. Lu Elliptic and solitary wave solutions for Bogoyavlenskii equations system, couple Boiti-Leon-Pempinelli equations system and Time-fractional Cahn-Allen equation, Results in physics **7**, 2325–2333 (2017)
9. Alam, M.N., Akbar, M.A., Hoque, M.F.: Exact travelling wave solutions of the (3+1)-dimensional mKdV-ZK equation and the (1+1)-dimensional compound KdVB equation using the new approach of generalized (G/GG/G)-expansion method. Pramana **83**(3), 317–329 (2014)
10. Biswas, A., Triki, H., Hayat, T., Aldossary, O.M.: 1-Soliton solution of the generalized Burgers equation with generalized evolution. Appl. Math. Comput. **217**(24), 10289–10294 (2011)
11. Kuo, C.-K.: The new exact solitary and multi-soliton solutions for the (2+1)-dimensional Zakharov-Kuznetsov equation. Comput. Math. Appl. **75**(8), 2851–2857 (2018)
12. Bekir, A., Cevikel, A.C., Güner, Ö., San, S.: Bright and dark soliton solutions of the (2 + 1)-dimensional evolution equations. Math. Model. Anal. **19**(1), 118–126 (2014)
13. Alam, M.N., Hafez, M., Akbar, M.A., et al.: Exact traveling wave solutions to the (3 + 1)-dimensional mKdV-ZK and the (2+ 1)-dimensional Burgers equations via exp (- Φ (η))-expansion method. Alex. Eng. J. **54**(3), 635–644 (2015)
14. Aslan, İ: Generalized solitary and periodic wave solutions to a (2 + 1)-dimensional Zakharov-Kuznetsov equation. Appl. Math. Comput. **217**(4), 1421–1429 (2010)
15. Wang, G., Kara, A.: A (2+ 1)-dimensional KdV equation and mKdV equation: symmetries, group invariant solutions and conservation laws. Phys. Lett. A **383**(8), 728–731 (2019)

Cosmological Models for Bianchi Type-I Space-Time in Lyra Geometry

Pratik V. Lepse and Binaya K. Bishi

Abstract The main purpose of this manuscript is to investigate the Bianchi type-I dark energy cosmological models in the framework of Lyra geometry. The modified Einstein's field equations is derived for Lyra geometry and obtained the exact solutions. In order to obtained the exact solutions volumetric expansion law is used. As per the Exponential and Power-law expansion, we have discussed the two cosmological models. Several physical parameters are obtained for both the models and discuss its physical importance following the observational data.

Keywords Bianchi type-I · Cosmological constant · Lyra geometry

1 Introduction

The concept of dark energy should be explored for the better understanding of the universe. The cosmologist, scientist and astronomers believed that dark energy is a kind of a repulsive force which acts as an antigravity and responsible for accelerated expansion of the universe, is termed as dark energy. The experiments, particularly WMAP (Wilkiinson Microwave Anisotropic Probe) and satellite experiment [1] concludes that, our universe consists of three major components namely dark energy nearly 73%, dark matter 23% and usual matter is about 4% [1–3]. The supernova project and HIGH-Z Supernova team reveals that the universe is expanding with acceleration. For closed universe, Hubble parameter H, the matter energy density parameter Ω_M and the dark energy density parameter Ω_Λ, predicted by Tegmark [4] are near about $H \approx 0.32$, $\Omega_m \approx 0.23$ and $\Omega_\Lambda \approx 1.17$. For flat cosmological model,

P. V. Lepse (✉) · B. K. Bishi
Department of Mathematics, Lovely Professional University, Jalandhar, Phagwara 144401, Panjab, India
e-mail: pratiklepse124@gmail.com

B. K. Bishi
Department of Mathematical Sciences, University of Zululand, Kwa-Dlangezwa 3886, South Africa

S. Banerjee and A. Saha (eds.), *Nonlinear Dynamics and Applications*, Springer Proceedings in Complexity,
https://doi.org/10.1007/978-3-030-99792-2_69

the cosmological observation [5, 6] suggested the existence of a positive cosmological term-Λ with magnitude $\Lambda \approx 10^{-123}$ and with $\Omega_m = 0.3$, $\Omega_\Lambda = 0.7$ in the accelerating universe.

The dark energy is characterised through the equation of state (EoS) parameter $p = \omega\rho$ in the universe. Here ω may be a constant or a function of cosmic time. The cosmic time dependent $\omega(t)$ is revive from experimental data. The analysis is performed on the experimental data in ordered to determine $\omega(t)$ [11, 12]. This parameter $\omega(t)$ have been calculated with some reasoning leading to simple parameterization of the dependence character of ω by many authors [13–16]. In view of galaxy clustering statistics [4] and SNe Ia data with CMBR, ω is approximated as $\omega \approx -0.977$. The above discussed results are consistent with both time dependent and time independent equation of state parameter ω. The different values of ω corresponds to the different type of universe. The universe is classified with the different value of ω as [17, 18]

$$\omega = \begin{cases} -1: & \text{Vacuum fluid universe} \\ 0: & \text{Dust fluid univesrse} \\ \frac{1}{3}: & \text{Radiation fluid universe} \\ 1: & \text{Stiff fluid universe} \end{cases}$$

The variable ω as $\omega(t)$ or $\omega(z)$, z denotes redshift is investigated by Jimenez [19] and Das et al. [20]. The quintessence models, $\omega > -1$ and phantom model, $\omega < -1$ give rise to dependent $\omega(t)$ [21–24]. Several forms of $\omega(t)$ can be found in the literature, which are involved in investigating dark energy [25–28].

After the formulation of General Relativity, Einstein devoted his entire life in the search of the theory that takes into account both gravitation and electromagnetism. He was not happy with the field equations $R_{ij} - \frac{1}{2}Rg_{ij} = -kT_{ij}$ of General Relativity. Since the right hand side is not geometrical in nature although the left hand side is. Further, the solution of the field equations are not free from singularities. A total field theory should not give rise to singularities. There have been various attempts to unify the gravitation and electromagnetism either by considering non-Riemannian geometry of four dimensions or by considering Riemannian spaces of higher dimensions.

Lyra [29] adduced Riemannian geometry by incorporating a gauge function in to the manifold, which is in fact structure less. In such case the geometry naturally gives rise to cosmological constant. This bears the remarkable resemblance to Wely's geometry. The Einstein's field equations (EFEs) based on Lyra's manifold in normal gauge is expressed as (Take $c = 1$ and $8\pi G = 1$)

$$R_{ij} - \frac{1}{2}Rg_{ij} + \frac{3}{2}\phi_i\phi_k - \frac{3}{4}g_{ik}\phi_m\phi^m = -kT_{ij}. \tag{1}$$

In which ϕ_i is the displacement vector defined through the gauge function β. The displacement vector field is considered as

$$\phi_i = (0, 0, 0, \beta(t)). \tag{2}$$

In this paper, we have studied the exponential-law expansion model and power-law expansion model in Bianchi type-I universe in view of Lyra geometry. We distributed our work in section wise like in Sect. 2. Bianchi type-I metric is presented and field equations for Lyra's manifold are derived. In Sect. 3 the solution of the Bianchi type-I cosmological model is constructed under exponential and power-law expansion. Conclusive remarks of the work is described in Sect. 4.

2 Field Equations for Bianchi Type-I Line Element

let us consider the Bianchi type-I line element of the form

$$ds^2 = -dt^2 + A^2dx^2 + B^2dy^2 + c^2dz^2. \tag{3}$$

Here A, B and C are function of t only. The energy momentum tensor T_i^j is considered as

$$T_j^i = (\rho + p)v_i v^j + pg_i^j \tag{4}$$

Here ρ and p represents the proper energy density and pressure respectively. The quantity θ is the scalar of expansion which is given by,

$$\theta = \nu^i_{|i} \tag{5}$$

and v^i satisfies the relation

$$g_{ij}v^i v^j = -1. \tag{6}$$

we assume that coordinates to be co-moving, so that

$$v^1 = v^2 = v^3 = 0, v^4 = 1. \tag{7}$$

Equation (4) yield

$$\begin{aligned} T_1^1 &= p_x = \omega_x \rho = \omega\rho, \\ T_2^2 &= p_y = \omega_y \rho = (\omega + \gamma)\rho, \\ T_3^3 &= p_z = \omega_z \rho = (\omega + \delta)\rho, \\ T_4^4 &= p_x = -\rho. \end{aligned} \tag{8}$$

where p_x , p_y , p_z and ω_x , ω_y , ω_z indicates the directional pressure and equation of state (EoS) parameters along x, y and z axes respectively. The EoS parameter of the fluid, which is deviation free is denoted by ω. We have parameterized the deviation from isotropy by setting $\omega_x = \omega$ and then introducing skewness parameter γ and δ that are the deviation from ω along y and z axes, which are $\omega_y = (\omega + \gamma)$ and $\omega_z = (\omega + \delta)$ respectively.

The field Eq. (1) in Lyra's geometry, for the metric (2), using Eqs. (2) and (4) takes the form

$$\frac{B_{44}}{B} + \frac{C_{44}}{C} + \frac{B_4C_4}{BC} + \frac{3}{4}\beta^2 = -\omega\rho \tag{9}$$

$$\frac{A_{44}}{A} + \frac{B_{44}}{B} + \frac{A_4B_4}{AB} + \frac{3}{4}\beta^2 = -(\omega + \gamma)\rho \tag{10}$$

$$\frac{A_{44}}{A} + \frac{C_{44}}{C} + \frac{A_4C_4}{Ac} + \frac{3}{4}\beta^2 = -(\omega + \delta)\rho \tag{11}$$

$$\frac{A_4B_4}{AB} + \frac{B_4C_4}{BC} + \frac{A_4C_4}{AC} - \frac{3}{4}\beta^2 = \rho \tag{12}$$

where $A_4 = \frac{dA}{dt}$,$A_{44} = \frac{d^2A}{dt^2}$.

The energy conservation $T^{j}_{i,j} = 0$ leads to

$$\frac{3}{2}\beta\beta_4 + \frac{3}{2}\beta^2\left(\frac{A_4}{A} + \frac{B_4}{B} + \frac{C_4}{C}\right) = 0 \tag{13}$$

The spatial volume V for model and the average scale factor are defined as

$$V = a^3 = ABC \tag{14}$$

In terms of scale factor or metric potentials, the Hubble parameter may be define as

$$H = \frac{a_4}{a} = \frac{1}{3}\left(\frac{A_4}{A} + \frac{B_4}{B} + \frac{C_4}{C}\right) \tag{15}$$

The deceleration parameter, q is expressed as

$$q = -\frac{aa_{44}}{a_4^2} \tag{16}$$

The following physical parameters are defined as

$$\text{Scalar expansion} = \theta = \frac{A_4}{A} + \frac{B_4}{B} + \frac{C_4}{C} \tag{17}$$

$$\text{Shear scalar} = \sigma^2 = \frac{1}{2}\sum_{i=1}^{3}\left(H_i^2 - \frac{1}{3}\theta^2\right) \tag{18}$$

$$\text{Average anisotropy parameter} = A_m = \frac{1}{3}\sum_{i=1}^{3}\left(\frac{\Delta H_i}{H}\right)^2 \tag{19}$$

where $\Delta H_i = H_i - H$, $i = 1, 2, 3$. H_i represents the directional Hubble parameter H along x, y and z axes respectively.

3 Solutions of Field Equation

The field Eqs. (9) to (12) are the four differential equations with eight unknown parameters A, B, C, β, ρ, ω, δ and γ. To make the system complete, we need the additional four conditions. Special law of variation for generalized Hubble's parameter is used as the first condition, which provides a constant value of deceleration parameter. The mean Hubble parameter H and the average scale factor are related through the expression as

$$H = la^{-n} \tag{20}$$

From which we write (using Eq. (15))

$$a_4 = la^{-n+1} \tag{21}$$

$$a_{44} = -l^2(n-1)a^{-2n+1} \tag{22}$$

Using Eqs. (21) and (22) in the Eq. (16) we have

$$q = n - 1 \tag{23}$$

The inflation of a model is determined by the sign of q. q has positive sign for $n > 1$ corresponding to "standard" decelerating model whereas it has negative sign for $0 \leq n < 1$ indicating accelerating model [30]. Integration of Eq. (21), leads to

$$a = \begin{cases} (nlt + \alpha_1)^{\frac{1}{n}}, & \text{for } n \neq 0 \\ \alpha_2 e^{lt}, & \text{for } n = 0 \end{cases} \tag{24}$$

where α_1 and α_2 are constants and which come out from the integration. Thus, the law (20) generate volumetric expansion in the form of power-law (PL) and exponential-law (EL)(One can see from Eq. (24)). We assume the second condition as $\sigma_1^1 \propto \theta$, which leads

$$A = (BC)^m \tag{25}$$

where $m > 0$. As a third assumption, we take $\gamma = \delta$ that means the deviation from ω along y and z axes are equal. From the differential Eqs. (10) and (11), we deduced

$$\frac{B}{C} = \alpha_3 \exp\left[\alpha_4 \int (ABC)^{-1} dt\right] \tag{26}$$

where α_3 and α_4 are integration constants. In view of the parameter n, for $n = 0$ and $n > 0$ corresponds to the exponential-law (EL) and power-law (PL) respectively.

3.1 Model in Exponential-Law expansion

In this case, lengthy but straight forward calculation leads to the metric potentials in the following forms:

$$\begin{aligned} A(t) &= \alpha_2^{\frac{3m}{m+1}} \exp\left(\frac{3ml}{m+1} t\right) \\ B(t) &= \sqrt{\alpha_3}\alpha_2^{\frac{3}{2(m+1)}} \exp\left[\frac{3lt}{2(m+1)} - \frac{\alpha_4}{6l\alpha_2^3} e^{-3lt}\right] \\ C(t) &= \frac{\alpha_2^{\frac{3}{2(m+1)}}}{\sqrt{\alpha_3}} \exp\left[\frac{3lt}{2(m+1)} + \frac{\alpha_4}{6l\alpha_2^3} e^{-3lt}\right] \end{aligned} \tag{27}$$

Thus in EL expansion (24), the required metric is

$$\begin{aligned} ds^2 = -dt^2 + \alpha_2^{\frac{6m}{m+1}} \exp\left(\frac{6mlt}{m+1}\right) dx^2 + \alpha_3 \alpha_2^{\frac{3}{(m+1)}} \exp\left[\frac{3lt}{(m+1)} - \frac{\alpha_4}{3l\alpha_2^3} e^{-3lt}\right] dy^2 \\ + \frac{\alpha^{\frac{3}{(m+1)}}}{\alpha_3} \exp\left(\frac{3lt}{m+1} + \frac{\alpha_4}{3l\alpha_2^3} e^{-3lt}\right) dz^2 \end{aligned} \tag{28}$$

At early time, the scale factor is constant but it increases with time and reaches infinity at the late epoch of time. It shows that at first, the universe starts with a constant volume and expands exponentially to infinity. Let us calculate all physical quantities $\beta(t)$, H_1, H_2, H_3, θ, σ, A_m, ρ, ω, γ and δ for exponential law. When Eq. (15) is applied to Eq. (13), the gauge function $\beta(t)$ takes the form

$$\beta(t) = \frac{N}{\alpha_2^3 e^{3lt}}, \tag{29}$$

where N is constant of integration. At initial stage of time, gauge function $\beta(t)$ return constant value and it is decreasing exponentially with increasing in time at later time. So that gauge function $\beta(t)$ of the model is goes over to the model of general relativity (Pradhan et al. [35]). The directional Hubble parameter H along x, y and z axes are given by

$$H_1 = \frac{A_4}{A} = \frac{3ml}{m+1} \tag{30}$$

$$H_2 = \frac{B_4}{B} = \frac{3l}{2(m+1)} + \frac{\alpha_4}{2\alpha_2^3}e^{-3lt} = \frac{3l}{2(m+1)} + \frac{\alpha_4}{2}\left(\frac{\beta}{N}\right) \tag{31}$$

$$H_3 = \frac{C_4}{C} = \frac{3l}{2(m+1)} - \frac{\alpha_4}{2\alpha_2^3}e^{-3lt} = \frac{3l}{2(m+1)} - \frac{\alpha_4}{2}\left(\frac{\beta}{N}\right) \tag{32}$$

It is noticed from the above equations that, H_i $(i = 2, 3)$ depends upon gauge function $\beta(t)$. When $\beta(t) = 0$, this shows the nature of directional Hubble parameter and they are constant at early stage as well as lateral stage of time t. The following physical parameters for the model (28) are deduced as

$$\theta = 3H = 3l \tag{33}$$

$$\sigma^2 = \frac{3l^2(2m-1)^2}{4(m+1)^2} + \frac{\alpha_4^2}{4\alpha_2^6}e^{-6lt} = \frac{3l^2(2m-1)^2}{4(m+1)^2} + \frac{\alpha_4^2}{4N^2}\beta^2 \tag{34}$$

$$A_m = \frac{(2m-1)^2}{2(m+1)^2} + \frac{\alpha_4^2}{6l^2\alpha_2^6}e^{-6lt} = \frac{(2m-1)^2}{2(m+1)^2} + \frac{\alpha_4^2}{6l^2N^2}\beta^2 \tag{35}$$

The expansion scalar θ exhibit constant value for whole range of time t. This show uniform exponential expansion of the model. The shear of the model is depending upon gauge function $\beta(t)$. Initially it has constant value and goes on decreasing with increasing time t. When $t \to \infty$, $\sigma^2 \to \frac{(2m-1)^2}{2(m+1)^2}$ and $A_m \to \frac{(2m-1)^2}{2(m+1)^2}$. Further, it is noticed that at late time shear and average anisotropy parameter vanishes for $m = \frac{1}{2}$. Using the Eqs. (27) and (29), in differential Eq. (12), we obtained ρ as

$$\begin{aligned}\rho &= \frac{9l^2(4m+1)}{4(m+1)^2} - \frac{3N^2}{4\alpha_2^4}e^{-4lt} - \frac{\alpha_4^2}{4\alpha_2^6}e^{-6lt} \\ &= \frac{9l^2(4m+1)}{4(m+1)^2} - \frac{3}{4}\beta^2 - \frac{\alpha_4^2}{4N^2}\beta^2\end{aligned} \tag{36}$$

The energy density depends upon the gauge function $\beta(t)$. At initial time, the energy density has a constant value $9(4m+1)\alpha_2^6l^2 - (m+1)^2(\alpha_4^2 + 3\alpha_2^2N^2)$. The positivity of energy density leads to the constraint on m such that $\frac{4m+1}{(m+1)^2} > \frac{(\alpha_4^2+3\alpha_2^2N^2)}{9\alpha_2^6l^2}$. When $t \to \infty$, the energy density $\rho \to \frac{9l^2(4m+1)}{4(m+1)^2}$. With the help of Eqs. (9) and (12), the EoS parameter ω has been calculated as

$$\omega = \frac{\left(27\, l^2 {\alpha_2}^6 e^{6lt} + {\alpha_4}^2 (m+1)^2\right) e^{4lt} + 3\, N^2 {\alpha_2}^2 e^{6lt} (m+1)^2}{\left(-9\, {\alpha_2}^6 (4\, m+1)\, l^2 e^{6lt} + {\alpha_4}^2 (m+1)^2\right) e^{4lt} + 3\, N^2 {\alpha_2}^2 e^{6lt} (m+1)^2}$$
$$= \frac{-27N^2 l^2 - 3N^3 (m+1)^2 \beta^2 - \alpha_4^2 (m+1)^2 \beta^2}{9\, (4\, m+1)\, l^2 N^2 - 3\, N^3\, (m+1)^2\, \beta^2 - {\alpha_4}^2\, (m+1)^2\, \beta^2} \tag{37}$$

In this exponential-law of expansion, the EoS parameter ω depends upon gauge function $\beta(t)$. In early time of the universe, the ω takes a constant value equal to $\frac{27\, \alpha 2^6 l^2 + (m+1)^2 \left(3\, N^2 \alpha 2^2 + \alpha 4^2\right)}{(m+1)^2 \left(3\, N^2 \alpha 2^2 + \alpha 4^2\right) - 9\, (4\, m+1) \alpha 2^6 l^2}$. In late time, it approaches $-\frac{3}{4m+1}$. We write the values of the skewness parameter γ and δ, using the Eqs. (9), (10), (29) and (36) as follows:

$$\gamma = \delta = \frac{18\, (m+1)\, e^{4lt}\, (2\, m-1)\, e^{6lt} {\alpha_2}^6 l^2}{3\, {\alpha_2}^2 \left(-3\, {\alpha_2}^4 l^2\, (4\, m+1)\, e^{4lt} + N^2\, (m+1)^2\right) e^{6lt} + (m+1)^2\, e^{4lt} {\alpha_4}^2}$$
$$= \frac{18\, (m+1)\, N^2\, (2\, m-1)\, l^2}{{\alpha_4}^2\, (m+1)^2\, \beta^2 + 3\, N^2\, (m+1)^2\, \beta^2 - 9\, (4\, m+1)\, l^2 N^2} \tag{38}$$

For the value of time $t = \tau_0$ given by

$$\tau_0 = \frac{1}{2l} \log \left[\frac{2\alpha_4^2 (m+1)^2 \Sigma_0}{\alpha_2^2 (\Sigma_0^2 - 2N^2 (m+1)^2 \Sigma_0 + 4N^4 (m+1)^4)} \right], \tag{39}$$

which leads to equation of state parameter $\omega = 0$. So that dusty universe at time $t = \tau_0$ given by Eq. (39). Again, it is to be noted that when time t lies in open interval $\tau_1 < t < \tau_2$ with

$$\tau_1 = \frac{1}{2l} \ln \left[\frac{178 (m+1)^2 \alpha_4^2 \Sigma_1}{\alpha_2^2 (\Sigma_1^2 - 178 (m+1)^2 N^2 \Sigma_1 + 31684 (m+1)^4 N^4)} \right] \tag{40}$$

$$\tau_2 = \frac{1}{2l} \log \left[\frac{6 (m+1)^2 \alpha_4^2 \Sigma_2}{\alpha_2^2 (\Sigma_2^2 + 6N^2 (m+1)^2) \Sigma_2 + 36 (m+1)^4 N^4} \right] \tag{41}$$

the equation of state parameter ω lies in the range $-1.67 < \omega < -0.62$, which consistent and in good agreement with the limiting observational results [31]. Where

$$\Sigma_0 = \sqrt[3]{-8 \left(-\frac{3\sqrt{3}{\alpha_4}^2 l}{2} \sqrt{4\, N^6\, (m+1)^2 + 27\, {\alpha_4}^4 l^2} + N^6\, (m+1)^2 + \frac{27\, {\alpha_4}^4 l^2}{2} \right) (m+1)^4}$$

$$\Sigma_1 = \sqrt[3]{\begin{array}{l} -\left\{ -\frac{2\, l \alpha 4^2}{89} \sqrt{-\left(44589\, (m+1)^2\, N^6 - 251001\, \alpha 4^4 l^2 \left(m - \frac{133}{668}\right)\right) \left(m - \frac{133}{668}\right)} \right. \\ \left. -\frac{1002 \alpha 4^4 l^2}{89} \left(m - \frac{133}{668}\right) + (m+1)^2 N^6 \right\} 5639752\, (m+1)^4 \end{array}}$$

$$\Sigma_2 = \sqrt[3]{\begin{array}{l} -216(m+1)^4\left\{-\frac{2\alpha_4^2 l}{3}\sqrt{-\left(31\,(m+1)^2 N^6 - \frac{961\,\alpha_4{}^4 l^2}{9}\left(m-\frac{119}{124}\right)\right)\left(m-\frac{119}{124}\right)} \right. \\ \left. +N^6 m^2 + \left(2N^6 - \frac{62\,\alpha_4{}^4 l^2}{9}\right)m + N^6 + \frac{119\,\alpha_4{}^4 l^2}{18}\right\} \end{array}}$$

Further for the value of time $t = \tau$, given by

$$\tau = \frac{1}{2l}\ln\left[\frac{2\alpha_4^2(m+1)^2\Sigma}{\alpha_2^2(\Sigma^2 - 2N^2(m+1)^2\Sigma + 4N^4(m+1)^4)}\right] \tag{42}$$

where

$$\Sigma = \sqrt[3]{\begin{array}{l} -8(m+1)^4\left\{-\frac{3\alpha_4{}^2 l}{2}\sqrt{\left(9\,(2m-1)\,\alpha_4{}^4 l^2 - 4\,N^6\,(m+1)^2\right)(2m-1)} \right. \\ \left. +N^6\,(m+1)^2 - 9\,l^2\left(m-\frac{1}{2}\right)\alpha_4{}^4\right\} \end{array}}$$

The values of ω comes out to be -1. For flat model, the matter energy density Ω_M and dark energy Ω_Λ, satisfies

$$\Omega_M + \Omega_\Lambda = 1 \tag{43}$$

where

$$\Omega_M = \frac{\rho}{3H^2} \quad \text{and} \quad \Omega_\Lambda = \frac{\Lambda}{3H^2} \tag{44}$$

Thus, Eqs. (43) and (44) gives

$$\rho + \Lambda = 3H^2$$

and then using Eqs. (33) and (36), we write the expressions of Λ as

$$\begin{aligned} \Lambda &= 3l^2 + \frac{3N^2}{4\alpha_2^4}e^{-4lt} + \frac{\alpha_4^2}{4\alpha_2^6}e^{-6lt} - \frac{9l^2(4m+1)}{4(m+1)^2} \\ &= 3l^2 + \frac{3}{4}\beta^2 + \frac{\alpha_4^2}{4N^2}\beta^2 - \frac{9l^2(4m+1)}{4(m+1)^2} \end{aligned} \tag{45}$$

From Eq. (45), one can noticed that cosmological constant Λ takes a positive constant value $\frac{3\,l^2(2\,m-1)^2\alpha 2^6+(m+1)^2(3\,N^2\alpha 2^2+\alpha 4^2)}{4\alpha 2^6(m+1)^2}$ at early stage of the universe and approaches to a positive value $3l^2\left[1-\frac{3(4m+1)}{4(m+1)^2}\right]$ at late time for all values of m and l. From Eqs. (33), (36) and (45), we have

$$\Omega_M = \frac{3(4m+1)}{4(m+1)^2} - \frac{N^2}{4l^2\alpha_2^4}e^{-4lt} - \frac{\alpha_4^2}{12l^2\alpha_2^6}e^{-6lt}$$
$$= \frac{3(4m+1)}{4(m+1)^2} - \frac{1}{4l^2}\beta^2 - \frac{\alpha_4^2}{12l^2N^2}\beta^2 \quad (46)$$

$$\Omega_\Lambda = \frac{N^2}{4l^2\alpha_2^4}e^{-4lt} + \frac{\alpha_4^2}{12l^2\alpha_2^6}e^{-6lt} - \frac{3(4m+1)}{4(m+1)^2} + 1$$
$$= \frac{1}{4l^2}\beta^2 + \frac{\alpha_4^2}{12l^2N^2}\beta^2 - \frac{3(4m+1)}{4(m+1)^2} + 1 \quad (47)$$

From Eq. (46), we have observed that the matter energy density Ω_m takes the value $\frac{3(4m+1)}{4(m+1)^2} - \left[\frac{N^2}{4l^2\alpha_2^4} + \frac{\alpha_4^2}{12l^2\alpha_2^6}\right]$ at initial time. At late time, Ω_m is positive valued with the value $\frac{3(4m+1)}{4(m+1)^2}$. Further Eq. (47) indicates that, the dark energy Ω_Λ is positive valued at initial and late time for all the values of parameters involved in the expression. Ω_Λ is a decreasing function of cosmic time and approaches to $1 - \frac{3(4m+1)}{4(m+1)^2}$ when $t \to \infty$.

3.2 Model in Power-Law Expansion

We solved the differential Eqs. (9), (10), by using the power-law expansion ($n \neq 0$) (24) and Eqs. (25) and (26), we arrived at

$$A(t) = (nlt + \alpha_1)^{\frac{3m}{n(m+1)}}$$
$$B(t) = \sqrt{\alpha_3}(nlt + \alpha_1)^{\frac{3}{2n(m+1)}} exp\left[\frac{\alpha_4}{2l(n-3)}(nlt + \alpha_1)^{\frac{n-3}{n}}\right] \quad (48)$$
$$C(t) = \frac{1}{\sqrt{\alpha_3}}(nlt + \alpha_1)^{\frac{3}{2n(n-3)}} exp\left[-\frac{\alpha_4}{2l(m+1)}(nlt + \alpha_1)^{\frac{n-3}{n}}\right]$$

Thus the metric (3) reduced to ($for\ n \neq 0$), for power law expansion

$$ds^2 = -dt^2 + (nlt + \alpha_1)^{\frac{6m}{n(m+1)}}dx^2 + \alpha_3(nlt + \alpha_1)^{\frac{3}{n(m+1)}} exp\left[\frac{\alpha_4}{2l(n-3)}(nlt + \alpha_1)^{\frac{n-3}{n}}\right]dy^2$$
$$+ \frac{1}{\alpha_3}(nlt + \alpha_1)^{\frac{3}{n(m+1)}} exp\left[-\frac{\alpha_4}{2l(n-3)}(nlt + \alpha_1)^{\frac{n-3}{n}}\right]dz^2 \quad (49)$$

At early time, the scale factor is constant but it increases with time and reaches infinity at the late epoch of time. It shows that at first, the universe starts with a constant volume and expands rapidly to infinity. Let us calculate all physical quantities $\beta(t)$, H_1, H_2, H_3, θ, σ, A_m, ρ, ω, γ and δ for power law. When Eq. (15) is applied to Eq. (13), the gauge function $\beta(t)$ takes the form for $n \neq 0$ as,

$$\beta(t) = N(nlt + \alpha_1)^{-\frac{3}{n}} \tag{50}$$

where N is constant of integration. At initial stage of time, gauge function $\beta(t)$ return constant value and it is decreasing with evolution of time. The directional Hubble parameters H_i along x, y and z axes are given by

$$H_1 = \frac{A_4}{A} = \frac{3ml}{m+1}\frac{1}{(nlt+\alpha_1)} \tag{51}$$

$$H_2 = \frac{B_4}{B} = \frac{1}{2}\left[\frac{3l}{(1+m)(\alpha_1+nlt)} + \alpha_4(\alpha_1+nlt)^{\frac{-3}{n}}\right] \tag{52}$$

$$H_3 = \frac{C_4}{C} = \frac{1}{2}\left[\frac{3l}{(1+m)(\alpha_1+nlt)} - \alpha_4(\alpha_1+nlt)^{\frac{-3}{n}}\right] \tag{53}$$

The physical parameters for the model (49) are deduced as

$$\theta = 3H = 3l(nlt+\alpha_1)^{-1} \tag{54}$$

$$\begin{aligned}\sigma^2 &= \frac{1}{4}\left[\frac{3l^2(1-2m)^2}{(1+m)^2(\alpha_1+nlt)^2} + {\alpha_4}^2\beta(t)^2N^{-2}\right]\\ &= \frac{1}{4}\left[\frac{3l^2(1-2m)^2}{(1+m)^2(\alpha_1+nlt)^2} + {\alpha_4}^2(\alpha_1+nlt)^{\frac{-6}{n}}\right]\end{aligned} \tag{55}$$

$$\begin{aligned}A_M &= \frac{1}{6}\left[\frac{3(1-2m)^2}{(1+m)^2} + \frac{{\alpha_4}^2\beta(t)^2N^{-2}(\alpha_1+nlt)^2}{l^2}\right]\\ &= \frac{1}{6}\left[\frac{3(1-2m)^2}{(1+m)^2} + \frac{{\alpha_4}^2(\alpha_1+nlt)^{\frac{2n-6}{n}}}{l^2}\right]\end{aligned} \tag{56}$$

The expansion scalar θ is a decreasing function of cosmic time t. It exhibit a constant value initially and approaching to zero with the evolution of cosmic time. The shear and average anisotropy parameters of the model is depending upon the gauge function $\beta(t)$. Both the parameters are decreasing function of cosmic time t. Initially σ^2 has constant value and approaching towards zero at late time. The anisotropy parameter A_M has constant value initially and $A_M \to \infty$ and $A_M \to \frac{(1-2m)^2}{2(1+m)^2}$ for $n > 3$ and $0 < n < 3$ respectively. Using the Eqs. (48) and (50), in differential Eq. (12), we obtained ρ as

$$\begin{aligned}\rho &= \frac{1}{4}\left[\frac{9l^2(1+4m)}{(1+m)^2(\alpha_1+nlt)^2} - {\alpha_4}^2\beta(t)^2N^{-2} - 3\beta(t)^2\right]\\ &= \frac{1}{4}\left[\frac{9l^2(1+4m)}{(1+m)^2(\alpha_1+nlt)^2} - (\alpha_1+nlt)^{\frac{-6}{n}}({\alpha_4}^2 + 3N^2(\alpha_1+nlt)^{\frac{2}{n}})\right]\end{aligned} \tag{57}$$

The energy density depends upon the gauge function $\beta(t)$. At initial time, the energy density has a constant value $\frac{1}{4}\left[\frac{9l^2(1+4m)}{{\alpha_1}^2(1+m)^2} - \alpha_1{}^{\frac{-6}{n}}({\alpha_4}^2 + 3N^2\alpha_1{}^{\frac{2}{n}})\right]$. The positivity of energy density leads to the constraint on m such that $\frac{4m+1}{(m+1)^2} > \frac{4\alpha_1{}^{\frac{2n-6}{n}}({\alpha_4}^2+3N^2\alpha_1{}^{\frac{2}{n}})}{9l^2}$. When $t \to \infty$, the energy density $\rho \to \frac{9l^2(4m+1)}{{\alpha_1}^2(m+1)^2}$. Using the Eqs. (48) and (57) in Eq. (9), the EoS parameter ω have been calculated as

$$
\begin{aligned}
\omega &= \frac{\frac{3l^2(4(1+m)n-9)}{(1+m)^2(\alpha_1+nlt)^2} - {\alpha_4}^2\beta(t)^2N^{-2} - 3\beta(t)^2}{\frac{9l^2(1+4m)}{(1+m)^2(\alpha_1+nlt)^2} - {\alpha_4}^2\beta(t)^2N^{-2} - 3\beta(t)^2} \\
&= \frac{\frac{3l^2(4(1+m)n-9)}{(1+m)^2(\alpha_1+nlt)^2} - (\alpha_1+nlt)^{\frac{-6}{n}}({\alpha_4}^2 + 3N^2(\alpha_1+nlt)^{\frac{2}{n}})}{\frac{9l^2(1+4m)}{(1+m)^2(\alpha_1+nlt)^2} - (\alpha_1+nlt)^{\frac{-6}{n}}({\alpha_4}^2 + 3N^2(\alpha_1+nlt)^{\frac{2}{n}})}
\end{aligned}
\tag{58}
$$

In this power-law of expansion, the EoS parameter ω depends upon gauge function $\beta(t)$ from (50). In early time of the universe, the ω takes a constant value equal to $\frac{\frac{3l^2(-9+4(1+m)n)}{{\alpha_1}^2(1+m)^2} - \alpha_1{}^{\frac{-6}{n}}({\alpha_4}^2)+3\alpha_1{}^{\frac{2}{n}}N^2}{\frac{9l^2(1+4m)}{{\alpha_1}^2}(1+m)^2 - \alpha_1{}^{\frac{-6}{n}}({\alpha_4}^2)+3\alpha_1{}^{\frac{2}{n}}N^2}$. In late time, it approaches $\frac{4n(1+m)-3}{1+4m}$ for $0 < n < 2$. We write the values of the skewness parameter γ and δ, using the Eqs. (50), (57) and (58) in (10) as follows

$$
\begin{aligned}
\gamma = \delta &= \frac{4\left[\frac{9l^2(1+4m)}{4(1+m)^2(\alpha_1+nlt)^2} + \frac{3l^2(3+2m(3-2m(n-3)-3n)-2n)}{4(1+m)^2(\alpha_1+nlt)^2} + \frac{\frac{3l^2(4(1+m)n-9)}{(1+m)^2(\alpha_1+nlt)^2} - {\alpha_4}^2\beta(t)^2N^{-2} - 3\beta(t)^2}{\frac{9l^2(1+4m)}{(1+m)^2(\alpha_1+nlt)^2} - {\alpha_4}^2\beta(t)^2N^{-2} - 3\beta(t)^2}\right]}{\frac{9l^2(1+4m)}{(1+m)^2(\alpha_1+nlt)^2} - {\alpha_4}^2\beta(t)^2N^{-2} - 3\beta(t)^2} \\
&= \frac{4\left[\frac{9l^2(1+4m)}{4(1+m)^2(\alpha_1+nlt)^2} + \frac{3l^2(3+2m(3-2m(n-3)-3n)-2n)}{4(1+m)^2(\alpha_1+nlt)^2} + \frac{\frac{3l^2(4(1+m)n-9)}{(1+m)^2(\alpha_1+nlt)^2} - (\alpha_1+nlt)^{\frac{-6}{n}}({\alpha_4}^2+3N^2(\alpha_1+nlt)^{\frac{2}{n}})}{\frac{9l^2(1+4m)}{(1+m)^2(\alpha_1+nlt)^2} - (\alpha_1+nlt)^{\frac{-6}{n}}({\alpha_4}^2+3N^2(\alpha_1+nlt)^{\frac{2}{n}})}\right]}{\frac{9l^2(1+4m)}{(1+m)^2(\alpha_1+nlt)^2} - (\alpha_1+nlt)^{\frac{-6}{n}}({\alpha_4}^2 + 3N^2(\alpha_1+nlt)^{\frac{2}{n}})}
\end{aligned}
\tag{59}
$$

For the value of time $t = \tau_0$ given by

$$
\tau_0 = \frac{1}{nl}\left[\frac{l}{m+1}\sqrt{\frac{3[4n(m+1)-9]}{\alpha_4^2 + 3N^2(\alpha_1+nlt)^{\frac{2}{n}}}}\right]^{\frac{n}{n-3}} - \frac{\alpha_1}{nl}
\tag{60}
$$

the equation of state parameter $\omega = 0$. So that we have dusty universe at time $t = \tau_0$ given by Eq. (60). It is to be noted that when time t lies in open interval $\tau_1 < t < \tau_2$, where

$$\tau_1 = \frac{1}{nl}\left[\frac{1.06l}{m+1}\sqrt{\frac{[4n(m+1)+4(5m-1)]}{\alpha_4^2+3N^2(\alpha_1+nlt)^{\frac{2}{n}}}}\right]^{\frac{n}{n-3}} - \frac{\alpha_1}{nl} \tag{61}$$

$$\tau_2 = \frac{1}{nl}\left[\frac{1.36l}{m+1}\sqrt{\frac{[4n(m+1)+7.44m-7.1]}{\alpha_4^2+3N^2(\alpha_1+nlt)^{\frac{2}{n}}}}\right]^{\frac{n}{n-3}} - \frac{\alpha_1}{nl} \tag{62}$$

then the equation of state parameter ω lies in the range $-1.67 < \omega < -0.62$, which consistent and in good agreement with the limiting observational results [31]. When the domain of time t is $\tau_3 < t < \tau_4$, in which

$$\tau_3 = \frac{1}{nl}\left[\frac{1.13l}{m+1}\sqrt{\frac{[4n(m+1)+15.96m-5.01]}{\alpha_4^2+3N^2(\alpha_1+nlt)^{\frac{2}{n}}}}\right]^{\frac{n}{n-3}} - \frac{\alpha_1}{nl} \tag{63}$$

$$\tau_4 = \frac{1}{nl}\left[\frac{1.29l}{m+1}\sqrt{\frac{[4n(m+1)9.48m-6.63]}{\alpha_4^2+3N^2(\alpha_1+nlt)^{\frac{2}{n}}}}\right]^{\frac{n}{n-3}} - \frac{\alpha_1}{nl} \tag{64}$$

then we observed the values of ω as $-1.33 < \omega < -0.79$ which coincides with the values obtained from observational results [32]. Also it is noticed that when t lies in the open interval $\tau_5 < t < \tau_6$ where

$$\tau_5 = \frac{1}{nl}\left[\frac{1.10l}{m+1}\sqrt{\frac{[4n(m+1)+17.28m-4.68]}{\alpha_4^2+3N^2(\alpha_1+nlt)^{\frac{2}{n}}}}\right]^{\frac{n}{n-3}} - \frac{\alpha_1}{nl} \tag{65}$$

$$\tau_6 = \frac{1}{nl}\left[\frac{1.25l}{m+1}\sqrt{\frac{[4n(m+1)+11.04m-6.24]}{\alpha_4^2+3N^2(\alpha_1+nlt)^{\frac{2}{n}}}}\right]^{\frac{n}{n-3}} - \frac{\alpha_1}{nl} \tag{66}$$

then the values of ω are found to be $-1.44 < \omega < -0.92$ which are very much consistent with the values of ω of latest observational results in 2009 [33, 34]. Further for the value of time $t = \tau$, given by

$$\tau = \frac{1}{nl}\left[\frac{1.22l}{m+1}\sqrt{\frac{[4n(m+1)+12m-6]}{\alpha_4^2+3N^2(\alpha_1+nlt)^{\frac{2}{n}}}}\right]^{\frac{n}{n-3}} - \frac{\alpha_1}{nl}, \tag{67}$$

the values of ω comes out to be -1. The geometrical and physical behaviour of the dark energy model is to be discussed on the values of EoS parameter and the gauge function $\beta(t)$. The gauge function $\beta(t)$ is appeared in all physical quantity like $H, \sigma, A_M, \rho, \omega, \delta, \gamma$. If the constant of integration $N = 0$ then $\beta(t) = 0$ from Eq. (50). Thus for $N = 0$, the physical quantity ρ, ω and time t in our model goes over to the result of Pradhan et al. [35]. If $N \neq 0$, then the gauge function $\beta(t)$ play a role in our models. Thus in this note, an attempt has been made to generalize the model of Pradhan et al. [35], in Lyra's geometry. For flat model (in the absence of curvature), the matter energy density Ω_M and dark energy Ω_Λ, satisfies the relation

$$\Omega_M + \Omega_\Lambda = 1 \tag{68}$$

where

$$\begin{aligned}\Omega_M &= \frac{\rho}{3H^2}\\ \Omega_\Lambda &= \frac{\Lambda}{3H^2}.\end{aligned} \tag{69}$$

Thus, Eqs. (68) and (69) gives

$$\rho + \Lambda = 3H^2$$

and then using Eqs. (54) and (57), we write the values of Λ as

$$\begin{aligned}\Lambda &= \frac{1}{4}\left[\frac{3l^2(3+4m+4m^2)}{(1+m)^2(\alpha_1+nlt)^2} + {\alpha_4}^2\beta(t)^2N^{-2} + 3\beta(t)^2\right]\\ &= \frac{1}{4}\left[\frac{3l^2(3+4m+4m^2)}{(1+m)^2(\alpha_1+nlt)^2} + (\alpha_1+nlt)^{\frac{-6}{n}}({\alpha_4}^2 + 3N^2(\alpha_1+nlt)^{\frac{2}{n}})\right]\end{aligned} \tag{70}$$

From Eq. (70), one can noticed that cosmological constant Λ takes a positive constant value $\frac{1}{4}\left[\frac{3l^2(3+4m+4m^2)}{(1+m)^2\alpha_1^2} + \alpha_1^{\frac{-6}{n}}({\alpha_4}^2 + 3N^2\alpha_1^{\frac{2}{n}})\right]$ at early stage of the universe and approaches to zero at late time for all the parameters involved in the expression (70). The cosmological term Λ is a decreasing function of cosmic time t. From Eqs. (54), (57) and (70), we write the values of the matter energy density $\Omega_M = \frac{\rho}{3H^2}$ and dark energy $\Omega_\Lambda = \frac{\Lambda}{3H^2}$ as,

$$\Omega_M = \frac{3}{4}\frac{(4m+1)}{(m+1)^2} - \frac{(\alpha_1+nlt)^{2-\frac{6}{n}}({\alpha_4}^2 + 3N^2(\alpha_1+nlt)^{\frac{2}{n}})}{12l^2} \tag{71}$$

$$\Omega_\Lambda = \frac{1}{4}\frac{(4m^2-4m+1)}{(m+1)^2} + \frac{(\alpha_1+nlt)^{2-\frac{6}{n}}({\alpha_4}^2 + 3N^2(\alpha_1+nlt)^{\frac{2}{n}})}{12l^2} \tag{72}$$

Here we noticed that both the matter energy density Ω_M and dark energy Ω_Λ start evolve with a constant value. Further, $\Omega_M \to \frac{3}{4}\frac{(4m+1)}{(m+1)^2}$ and $\Omega_\Lambda \to \frac{1}{4}\frac{(4m^2-4m+1)}{(m+1)^2}$ when $t \to \infty$ for $0 < n < 2$. The Ω_Λ is a decreasing function of cosmic time t.

4 Concluding Remarks

In this manuscript, we have studied the dark energy cosmological models for Bianchi type-I universe in view of Lyra geometry. We have analysed two cosmological models namely PL model and EL model. The outcomes obtained from these two models are presented below:

- In both the models, the energy density ρ starts with a constant value initially and approaches to same value $\frac{9l^2(4m+1)}{4(m+1)^2}$ at late time for $\alpha_1 = \pm 2$.
- The cosmological constant Λ is depending on the gauge function and in terms of cosmic time, the qualitative behaviour of Λ is decreasing in nature for both the discussed models. In case of exponential model and power law model, cosmological constant Λ approaches to $3l^2\left[1 - \frac{3(4m+1)}{4(m+1)^2}\right]$ and zero at late time respectively.
- In both the models, the dark energy Ω_Λ is a decreasing function of cosmic time t. In PL $(n = 0)$and EL $(0 < n < 2)$ models $\Omega_\Lambda \to 1 - \frac{3(4m+1)}{4(m+1)^2}$ and $\Omega_\Lambda \to \frac{(2m-1)^2}{4(m+1)^2}$ respectively at late time.
- The results discussed in Lyra's geometry reduces to the results of general relativity for $N = 0$.

References

1. Bennett, C.L., Halpern, M., Hinshaw, G.: First-year wilkinson microwave anisotropy probe (WMAP) observations: preliminary maps and basic results. Astrophys. J. Suppl. Ser. ED-148, 1–7 (2003)
2. Spergel, D.N., Verde, L., Peiris, H.V., Komatsu, E., Nolta, M.R., Bennett, C.L., Halpern, M., Hinshaw, G., Jarosik, N., Kogut, A., Limon, M.: First-year Wilkinson microwave anisotropy probe (WMAP) observations: determination of cosmological parameters. Astrophys. J. Suppl. Ser. ED-148(1), 175–179 (2003)
3. Tegmark, M., Strauss, M.A. Blanton, M.R., Abazajian, K., Dodelson, S., Sandvik, H., Wang, X., Weinberg, D.H., Zehavi, I., Bahcall, N.A., Hoyle, F.: Cosmological parameters from SDSS and WMAP. Phys. Rev. D ED-69(10), 103501 (2004)
4. Tegmark, M., Strauss, M.A., Blanton, M.R., Abazajian, K., Sandvik, H., Wang, X., Weinberg, D.H. Zehavi, I., Bahcall, N.A., Hoyle, F.: Cosmological parameters from SDSS and WMAP. Phys. Rev. D ED-69(10), 103501 (2004)
5. Garnavich, P.M., Kirshner, R.P. Challis, P., Tonry, J., Gilliland, R.L., Smith, R.C., Clocchiatti, A., Diercks, A., Filippenko, A.V., Hamuy, M., Hogan, C.J.: Constraints on cosmological models from Hubble Space Telescope observations of high-z supernovae. Astrophys. J. Lett. ED-493(2), L53 (1998)
6. Garnavich, P.M., Jha, S., Challis, P., Clocchiatti, A., Diercks, A., Filippenko, A.V., Gilliland, R.L., Hogan, C.J., Kirshner, R.P., Leibundgut, B., Phillips, M.M.: Supernova limits on the cosmic equation of state. Astrophys. J. ED-509(1), 74 (1998)
7. Perlmutter, S., Gabi, S., Goldhaber, G., Goobar, A., Groom, D.E., Hook, I.M. Kim, A.G., Kim, M.Y., Lee, J.C., Pain, R., Pennypacker, C.R.: Measurements of the cosmological parameters Ω and Λ from the first seven supernovae at $z \geq 0.35$. Astrophys. J. ED-483(2), 565 (1997)
8. Perlmutter, S., Aldering, G., Della Valle, M., Deustua, S., Ellis, R.S., Fabbro, S., Fruchter, A., Goldhaber, G., Groom, D.E., Hook, I.M., Kim, A.G.: Discovery of a supernova explosion at half the age of the Universe. Nature ED-391(6662), 51 (1998)

9. Perlmutter, S., Aldering, G., Goldhaber, G., Knop, R.A., Nugent, P., Castro, P.G., Deustua, S., Fabbro, S., Goobar, A., Groom, D.E., Hook, I.M.: Measurements of Ω and Λ from 42 high-redshift supernovae. Astrophys. J. ED-517(2), 565 (1999)
10. Sahni, V., Starobinsky, A.: Reconstructing dark energy. Int. J. Modern Phys. D ED- 15(12), 2105–2132 (2006)
11. Sahni, V., Shafieloo, A., Starobinsky, A.A.: Two new diagnostics of dark energy. Phys. Rev. D ED-78(10), 103502 (2008)
12. Huterer, D., Turner, M.S.: Probing dark energy: Methods and strategies. Phys. Rev. D ED-64(12), 123527 (2001)
13. Weller, J., Albrecht, A.: Future supernovae observations as a probe of dark energy. Phys. Rev. D ED-65(10), 103512 (2002)
14. Linden, S., Virey, J.M.: Test of the Chevallier-Polarski-Linder parametrization for rapid dark energy equation of state transitions. Phys. Rev. D ED-78(2), 023526 (2008)
15. Krauss, L.M., Jones-Smithand Huterer, D.K.: Dark energy, a cosmological constant, and type Ia supernovae. New J. Phys. ED-9(5), 141 (2007)
16. Usmani, A.A., Ghosh, P.P., Mukhopadhyay, U., Ray, P.C., Ray, S.: The dark energy equation of state. Monthly Notices Royal Astronom. Soc. Lett. ED- 386(1), L92-L95 (2008)
17. Kujat, J., Linn, A.M., Scherrer, R.J., Weinberg, D.H.: Prospects for determining the equation of state of the dark energy: what can be learned from multiple observables?' Astrophys. J. ED-572(1), 1 (2002)
18. Bartelmann, M., Dolag, K., Perrotta, F., Baccigalupi, C., Moscardini, L., Meneghetti, M., Tormen, G.: Evolution of dark-matter haloes in a variety of dark-energy cosmologies. New Astron. Rev. ED- 49(2–6), 199-203, (2005)
19. Jimenez, R.: The value of the equation of state of dark energy. New Astron. Rev. ED-47(8–10), 761–767 (2003)
20. Das, A., Gupta, S., Saini, T.D., Kar, S.: Cosmology with decaying tachyon matter. Phys. Rev. D ED-72(4), 043528 (2005)
21. Turner, M.S., White, M.: CDM models with a smooth component. Phys. Rev. D ED- 56(8), R4439 (1997)
22. Caldwell, R.R., Dave, R., Steinhardt, P.J.: Cosmological imprint of an energy component with general equation of state. Phys. Rev. Lett. ED-80(8), 1582 (1998)
23. Liddle, A.R., Scherrer, R.J.: Classification of scalar field potentials with cosmological scaling solutions. Phys. Rev. D ED-59(2), 023509 (1998)
24. Steinhardt, P.J., Wang, L., Zlatev, I.: Cosmological tracking solutions. Phys. Rev. D ED-59(12), 123504 (1999)
25. Rahaman, F., Bhui, B.C., Bhui, B.: Cosmological model with a viscous fluid in a Kaluza-Klein metric. Astrophys. Space Sci. ED-301(1–4), 47–49 (2006)
26. Mukhopadhyay, U., Ray, S., DChoudhury, S.: Λ-CDM universe: a phenomenological approach with many possibilities. Int. J. Modern Phys. D ED- 17(02), 301–309 (2008)
27. Ray, S., Rahaman, F., Mukhopadhyay, U., Sarkar, R.: Variable equation of state for generalized dark energy model. Int. J. Theoret. Phys. ED-50(9), 2687–2696 (2011)
28. Akarsu, Ö, Kılınç, C.B.: Bianchi type III models with anisotropic dark energy. General Relat. Gravit. ED-42(4), 763–775 (2010)
29. Lyra, G.: Ubereine Modifikation der Riemannschen Geometrie, Math.Z. ED-54(1), 52–54 (1957)
30. Vishwakarma, V.: A study of angular size-redshift relation for models in which Λ decays as the energy density. Classical Quant. Grav. ED- 17(18), 3833 (2000)
31. Knop, R.A., Aldering, G., Amanullah, R., Astier, P., Blanc, G., Burns, M.S. Conley, A., Deustua, S.E., Doi, M., Ellis, R., Fabbro, S.: New constraints on Ω_M, Ω_Λ, and ω from an independent set of 11 high-redshift supernovae observed with the Hubble Space Telescope. Astrophys. J. ED-598(1), 102 (2003)
32. Tegmark, M., Blanton, M.R. Strauss, M.A., Hoyle, F. Schlegel, D. Scoccimarro, R., Vogeley, M.S., Weinberg, D.H., Zehavi, I., Berlind, A., Budavari, T.: The three-dimensional power spectrum of galaxies from the sloan digital sky survey. Astrophys. J. ED-606(2), 702 (2004)

33. Hinshaw, G., Weiland, J.L., Hill, R.S., Odegard, N., Larson, D., Bennett, C.L., Dunkley, J., Gold, B., .Greason, M.R., Jarosik, N., Komatsu, E.: Five-year wilkinson microwave anisotropy probe observations: data processing, sky maps, and basic results. Astrophys. J. Suppl. Ser. ED-180(2), 225 (2009)
34. Komatsu, E., Dunkley, J., Nolta, M.R., Bennett, C.L., Gold, B., Hinshaw, G., Osik, J.N., Larson, D., Limon, M., Page, L., Spergel, D.N.: Five-year wilkinson microwave anisotropy probe observations: cosmological interpretation. Astrophys. J. Suppl. Ser. ED-180(2), 330 (2009)
35. Pradhan, A., Amirhashchi, H., Saha, B.: Bianchi type-I anisotropic dark energy model with constant deceleration parameter. Int. J. Theoreti. Phys. ED-50(9), 2923–2938 (2011)

A Non-linear Model of a Fishery Resource for Analyzing the Effects of Toxic Substances

Sudipta Sarkar, Tanushree Murmu, Ashis Kumar Sarkar, and Kripasindhu Chaudhuri

Abstract The goal of the proposed model is to investigate and analyze the qualitative behaviour of predator-prey fishery resource in an aquatic ecosystem by a non-linear mathematical model in which prey and predator species are contaminated by the toxic substances released by each of the species. In this model the species are subjected to bio-economic combined harvesting and obey the logistic growth rate function. Bio-economic harvesting of prey-predator species in presence of harmful toxic substances released by them is analyzed here by using modified catch rate function. Boundedness of the proposed model is examined here. Biological and bionomic steady states of the proposed model are derived. The conditions for local behaviour, instability and global behaviour of the steady states are exhibited in this paper. Optimal harvesting policy with the help of Pontryagin's maximal principle and finally, numerical exmples are illustrated to verify theoretical observations obtained from proposed model.

Keywords Fishery resource · Bioeconomic combined harvesting · Stability · Steady state · Optimal equilibrium · Toxicity · Net revenue

1 Introduction

Sustainable resources fishery, forestry, wild life etc. are vital origins of food and other necessary commodities in human life. These resources play a salient role for existence and advancement of biological populations. For these renewable resources management, our aim is to maximize the current value of advantages obtained from

---

Supported by organization x.

S. Sarkar (✉)
Department of Mathematics, Heritage Institute of Technology, Anandapur, Kolkata 700107, West Bengal, India
e-mail: sudipta.jumath@gmail.com

T. Murmu · A. Kumar Sarkar · K. Chaudhuri
Department of Mathematics, Jadavpur University, Kolkata 700032, India

S. Banerjee and A. Saha (eds.), *Nonlinear Dynamics and Applications*,
Springer Proceedings in Complexity,
https://doi.org/10.1007/978-3-030-99792-2_70

these sustainable resources and their proper preservation so that the extinction of these resources can be preserved. One of the most vital problems in an aquatic ecological system is the consequence of toxic elements. The growth of the fish and other aquatic organisms is highly influenced by these toxic substances. During the last few decades, so many researchers have carried out their investigations regarding fishery resource management. Clark [2, 3], Meserton-Gibbons [10, 11] have pursued their valuable analysis based on mathematical modeling of harvesting of fishery. Excessive and unregulated harvesting is not only the cause of depletion of fish species but also the consequence of toxic substances, the competitiveness of inter and intra species among predator-prey resources etc. are reasons. Chattopadhyay [1], Kar and Chaudhuri [7], Kar et al. [8], Dubey and Hussain [4] and Mukhopadhyay et al. [12] have conducted their mathematical analysis based on non-linear mathematical modeling of predator-prey fishery resource keeping the salient focus in toxicity. Our main focus is to find out the optimal control so that maximal value of the benefits obtained from this predator-prey fishery resource in presence of toxic substances by using a modified catch rate function preserving the extinction of both species. Further, the concept of Maynard [9] for competing fishery model subjected to commercially exploitation and toxicity was developed by Kar and Chaudhuri [7]. Ghosh et al. [5] developed a non-linear model to show the consequences of toxic elements on predator-prey fishery resource. Haque and Sarwardi [6], Pal et al. [13] developed a non-linear model to exhibit the effects of toxic elements for an aquatic system. A non-linear model was introduced by Sarkar et al. [15] using a modified catch rate function.

Our current model is arranged in the following manner: In Sect. 2: problem construction, Sect. 3: equilibria of the proposed dynamical system are exhibited. In Sect. 4: qualitative behaviour of the proposed model are analyzed. Finally, optimal harvesting policy with the help of Pontryagin's maximal principle, numerical results as well as interpretations of our proposed model have been shown in the consecutive sections.

2 Model Formulation

After incorporation of toxic affects, the dynamical system of prey-predator competing fishery resource model is as follows:

$$\begin{aligned} \frac{dx}{dt} &= rx\left(1-\frac{x}{K}\right) - \alpha xy - \gamma_1 x^3 y - \frac{q_1 E x}{b_1 + E}, \\ \frac{dy}{dt} &= sy\left(1-\frac{y}{L}\right) - \beta xy - \gamma_2 x y^2 - \frac{q_2 E y}{b_2 + E}. \end{aligned} \tag{1}$$

Here, $x = x(t)$, $y = y(t)$ denote prey and predator population density at time t respectively. r, s, α, β, K, L denote maximum growth rate of prey and predator fish species, coefficients of competition and environmental carrying capacities of prey and predator fish species respectively.

Here, γ_1,γ_2 represent the toxicity co-efficient of prey and predator fish populations. The parameters assumed in this paper are all positive constants. The term $\gamma_1 x^3$ is treated as one kind of response function of the predator fish to the prey population and it originates from the toxic substances by the predator species to put off the prey population from sharing the common resources. Here, the terms $\frac{d(\gamma_1 x^3)}{dx} = 3\gamma_1 x^2 > 0$ and $\frac{d^2(\gamma_1 x^3)}{dx} = 6\gamma_1 x > 0$ are positive. So, increasing growth rate of the toxic elements is observed as the biomass of prey-predator populations s are increased. Here, q_1, q_2 denote the coefficients of catchability of both fish species respectively and b_1, b_2 are all positive constants.

In the beginning of fishery resurce models, the catch rate function is usually taken in the form $h = qEx$ based on the CPUE (catch-per-unit-effort) hypothesis [2]. Later on, it is revised in the form as $h = \frac{qEx}{bE+lx}$.

It is assumed that the fisherman search randomly in a given area effectively which is a function of the effort level to harvest the fish resource by the fisherman. We rename this concept as a searching efficiency for the area of discovery. The capture rate of fish resource dependent on how effectively(efficiently) the effort level is used in presence of other fisherman. On the basis of the above hypothesis accordingly, we modify the catch rate function as a function of the fish resource population being captured for different effort levels in the form $h = \frac{qEx}{b+E}$, where E, q, b denote the harvesting effort, the catchability coefficient and a positive constant respectively.

3 The Equilibria and Their Feasibility

There are four equilibria of the dynamical system of Eqs. (1) which are $E_0(0,0)$, $E_1(x_1, 0)$, $E_2(0, y_2)$ & $E^*(x^*, y^*)$, where $x_1 = \frac{K}{r}\left(r - \frac{q_1 E}{b_1+E}\right)$, $y_2 = \frac{L}{s}\left(s - \frac{q_2 E}{b_2+E}\right)$, $y^* = \frac{L(s-\beta x^* - \frac{q_2 E}{b_2+E})}{s+L\gamma_2 x^*}$ and we get the value of x^* from the following cubic equation

$$A_1 {x_1^*}^3 + A_2 {x_1^*}^2 + A_3 x_1^* + A_4 = 0, \tag{2}$$

where $A_1 = \gamma_1 L\beta$, $A_2 = \frac{-r}{K}L\gamma_2 - \gamma_1 Ls + \frac{L\gamma_1 q_2 E}{b_2+E}$, $A_3 = \frac{-rs}{K} + rL\gamma_2 + \alpha L\beta - \frac{q_1 EL\gamma_2}{b_1+E}$ and $A_4 = rs - \alpha Ls + \frac{q_2 E\alpha L}{b_2+E} - \frac{q_1 Es}{b_1+E}$.

It is observed that the steady state E_0 is always feasible, E_1 is feasible if $r > \frac{q_1 E}{b_1+E}$ and E_2 exists if $s > \frac{q_2 E}{b_2+E}$.

If $D < 0$, then there exists at least one positive root say x^* of the Eq. (3). So, the existence condition of $E^*(x^*, y^*)$ is $s - \beta x^* - \frac{q_2 E}{b_2+E} > 0$ and $s > \frac{\alpha L q_2 E}{\alpha L(b_2+E) - r(b_2+E) + \frac{q_1 E(b_2+E)}{(b_1+E)}}$.

4 Qualitative Analysis

4.1 Boundedness

In the system (1), the solutions are uniformly bounded.

Proof

To prove the boundedness of the system let us choose a function $\omega = x + y$.

$$\begin{aligned}\frac{d\omega}{dt} + \zeta\omega &= rx\left(1 - \frac{x}{K}\right) + sy\left(1 - \frac{y}{L}\right) - (\alpha + \beta)\,xy - xy(\gamma_1 x^2 + \gamma_2 y)r \\ &\quad - E\left(\frac{q_1 x}{b_1 + E} + \frac{q_2 y}{b_2 + E}\right) + \zeta\,(x + y) \\ &< \frac{K}{4r}\left(r + \zeta - \frac{Eq_1}{b_1 + E}\right)^2 + \frac{L}{4s}\left(s + \zeta - \frac{Eq_2}{b_2 + E}\right)^2 = \eta.\end{aligned}$$

Thus, we get $0 < \omega(x, y) < \frac{\eta}{\zeta}(1 - \exp^{-\zeta t}) + \omega(0)\exp^{-\zeta t} < \max\left\{\frac{\eta}{\zeta}, \omega(0)\right\}$. Therefore, the solutions are uniformly bounded in

$$\mathbf{R}_{xy} = \left\{(x, y) \in \Re^2_+ : \omega(x, y) \le \frac{\eta}{\zeta} + \varepsilon \text{ for any } \varepsilon > 0\right\}.$$

4.2 Local Behaviour of the Equilibria

Let J_n be the variational matrix of the steady states E_n where $n = 0, 1, 2$. The eigenvalues of J_0 are $r - \frac{q_1 E}{b_1+E}$ and $s - \frac{q_2 E}{b_2+E}$. So, $E_0(0, 0)$ is stable node if $E > \max\left(\frac{b_1 r}{q_1 - r}, \frac{b_2 s}{q_2 - s}\right)$ and unstable if $E < \min\left(\frac{b_1 r}{q_1 - r}, \frac{b_2 s}{q_2 - s}\right)$.

The eigen values of J_1 are $\lambda_1 = \frac{-r x_1}{K}$ and $\lambda_2 = s - \beta x_1 - \frac{q_2 E}{b_2+E}$. Obviously, $\lambda_1 < 0$ and hence $E_1(x_1, 0)$ is a stable node if $\lambda_2 < 0$ which imply that $E > \frac{b_2(s - \beta x_1)}{q_2 - s + \beta x_1}$. It can be shown that $E_1(x_1, 0)$ is saddle if $E < \frac{b_2(s - \beta x_1)}{q_2 - s + \beta x_1}$.

The eigen values of J_2 are $\lambda_1 = r - \alpha y_2 - \frac{q_1 E}{b_1+E}$ and $\lambda_2 = \frac{-s y_2}{L}$. It is observed that $\lambda_2 < 0$. Hence $E_2(0, y_2)$ is a stable node when $\lambda_1 < 0$ i.e., if $E > \frac{b_1(r - \alpha y_2)}{q_1 - r + \alpha y_2}$ and saddle point if $E < \frac{b_1(r - \alpha y_2)}{q_1 - r + \alpha y_2}$.

The variational matrix of (1) around E^* is as follows:

$$J^* = \begin{pmatrix} \frac{-r_1 x^*}{K} - 2\gamma_1 x^{*2} y^* & -\alpha x^* - \gamma_1 x^{*3} \\ -\beta y^* - \gamma_2 y^{*2} & \frac{-s y^*}{L} - \gamma_2 x^* y^* \end{pmatrix}$$

Here, $trace(J^*) < 0$. So, for the local stability of the system of Eq. (1) around E^*, we have to show that $det(J^*)$ is strictly positive.

$$det(J^*) = x^* y^* \left[\frac{rs}{KL} - \alpha\beta + \gamma_2 \left(\frac{q_1 E}{b_1 + E} - r \right) - \beta\gamma_1 x^{*2} + 2x^* \left(\gamma_1 \gamma_2 x^* y^* + \frac{s\gamma_1 y^*}{L} + \frac{r\gamma_2}{K} \right) \right]. \tag{3}$$

So, $det(J^*) > 0$ if $x^{*2} < \frac{1}{\beta\gamma_1} \left[\frac{rs}{KL} + \gamma_2 \left(\frac{q_1 E}{b_1+E} - r \right) - \alpha\beta \right]$.

4.3 Non-existence of Periodic Solution

The system of Eq. (1) can be written in the form $\dot{Y} = G(Y)$, where $Y = (x, y)$ and $G = (G_1, G_2)$. Here, $G_1, G_2 \in C^\infty(\mathbf{R})$, $G_1 = rx\left(1 - \frac{x}{K}\right) - \alpha xy - \gamma_1 x^3 y - \frac{q_1 E x}{b_1+E}$ and $G_2 = y\left(1 - \frac{y}{L}\right) - \beta xy - \gamma_2 x y^2 - \frac{q_2 E y}{b_2+E}$. Let us consider a function $F(x, y) = \frac{1}{xy}$. Then $F(x, y) > 0$ for $(x, y) \in \mathbf{R}_{xy}$.

$$\nabla.(F\mathbf{G}) = -\left[\frac{r}{Ky} + 2\gamma_1 x + \frac{s}{Lx} + \frac{\gamma_2 y}{x} \right] < 0 \ \forall (x, y) \in \mathbf{R}_{xy}. \tag{4}$$

It follows that $\nabla.(F\mathbf{G}) < 0$ always. So, by Bendixson-Dulac's criterion, there does not exist any periodic orbit for the proposed model.

4.4 Global Stability Analysis

For global stability analysis of the dynamical system (1), a Lyapunov function is constructed as follows:

$$V(x, y) = \left[\left(x - x^* \log \frac{x}{x^*} \right) + \left(y - y^* \log \frac{y}{y^*} \right) \right]. \tag{5}$$

Here, V is positive definite $\forall (x, y) \in \Re^2_+$. Now,

$$\frac{dV}{dt} = -\left[P\left(x - x^*\right)^2 + R\left(x - x^*\right)\left(y - y^*\right) + Q\left(y - y^*\right)^2 \right], \tag{6}$$

where $P = \frac{r}{K} + \gamma_1 y\left(x + x^*\right) > 0$, $Q = \frac{s}{L} + \gamma_2 x > 0$, $R = \alpha + \beta + \gamma_1 x^{*2} + \gamma_2 y^* > 0$.

Now, $4PQ - R^2 = 4\left\{\frac{r}{K} + \gamma_1 y(x + x^*)\right\}\left\{\frac{s}{L} + \gamma_2 x\right\} - \left\{\alpha + \beta + \gamma_1 x^{*2} + \gamma_2 y^*\right\}^2 > 0$, if $4rs > KL\{\alpha + \beta + \gamma_1 x^* + \gamma_2\}^2$.

So, $\frac{dV}{dt} = 0$ at $E^*(x^*, y^*)$ and $\frac{dV}{dt} < 0$ at all the points other than $E^*(x^*, y^*)$. Using Lyapunov- LaSalle's invariance principle, we have shown that $E^*(x^*, y^*)$ is globally asymptotically stable under certain conditions.

4.5 Bionomic Steady States

When the total revenue (TR) achieved by selling the harvested biomass is equal to the total cost (TC) for the harvesting, then the bionomic equilibrium is obtained. Then the net revenue at any time is as follows:

$$\pi(x, y, E, t) = \left(\frac{p_1 q_1 x}{b_1 + E} + \frac{p_2 q_2 y}{b_2 + E} - C\right) E, \tag{7}$$

where C=constant cost for fishing per unit effort, p_1=constant price per unit prey biomass, p_2= constant price per unit predator biomass. To make our calculation simple, the cost of harvesting per unit effort is considered as constant.

Now, from equation $\dot{x} = 0$, we get $E = \frac{\alpha y b_1 + \gamma_1 x^2 y b_1 - r\left(1 - \frac{x}{K}\right) b_1}{r\left(1 - \frac{x}{K}\right) - \alpha y - \gamma_1 x^2 y - q_1}$.

Thus, E is positive, when $\alpha y + \gamma_1 x^2 y < r\left(1 - \frac{x}{K}\right) < \alpha y + \gamma_1 x^2 y + q_1$.

Similarly, from the equation $\dot{y} = 0$ we get $E = \frac{\beta b_2 x + \gamma_2 x y b_2 - s\left(1 - \frac{y}{L}\right) b_2}{s\left(1 - \frac{y}{L}\right) - \beta x - \gamma_2 x y - q_2}$.

So, E is positive when $\beta x + \gamma_2 x y < s\left(1 - \frac{y}{L}\right) < \beta x + \gamma_2 x y + q_2$. Therefore, the interior equilibrium point exists on the curve

$$\frac{\alpha y b_1 + \gamma_1 x^2 y b_1 - r\left(1 - \frac{x}{K}\right) b_1}{r\left(1 - \frac{x}{K}\right) - \alpha y - \gamma_1 x^2 y - q_1} = \frac{\beta b_2 x + \gamma_2 x y - s\left(1 - \frac{y}{L}\right) b_2}{s\left(1 - \frac{y}{L}\right) - \beta x - \gamma_2 x y - q_2}, \tag{8}$$

where $0 \leq x \leq K$ and $0 \leq y \leq L$.

We can find the bionomic steady states of the open access fishery model using Eq. (8) and condition $\pi = TR - TC = 0$, which gives the result $\left(\frac{p_1 q_1 x}{b_1 + E} + \frac{p_2 q_2 y}{b_2 + E} - C\right) = 0$.

5 Optimal Harvesting Policy

Our salient task is to fix the optimal adjustment between the current and future harvests values. The current value I of revenues is as follows:

$$I = \int_0^\infty \pi(x, y, E, t) e^{-\delta t} dt. \tag{9}$$

Here, $\pi(x, y, E, t) = \frac{p_1 q_1 x E}{b_1+E} + \frac{p_2 q_2 y E}{b_2+E} - CE$ and δ denotes the instant annual discount rate, C is the fishing cost per unit effort, p_1, p_2 are prices per unit prey and predator biomass respectively. Our main focus is to maximize I along with the Eq. (1) with the help of Pontryagin's Maximum Principle [14]. Here, $0 \leq E(t) \leq E_{max}$ and control set $V_t = [0, E_{max}]$ where E_{max} is the possible maximum value of the effort of harvesting.

For our problem, the Hamiltonian is given by:

$$H = \left(\frac{p_1 q_1 x}{b_1 + E} + \frac{p_2 q_2 y}{b_2 + E} - C\right) E e^{-\delta t} + \lambda_1 \left[rx\left(1 - \frac{x}{K}\right) - \alpha xy - \gamma_1 x^3 y - \frac{q_1 E x}{b_1 + E}\right]$$
$$+ \lambda_2 \left[sy\left(1 - \frac{y}{L}\right) - \beta xy - \gamma_2 xy^2 - \frac{q_2 E y}{b_2 + E}\right]. \quad (10)$$

Here, λ_1, λ_2 are adjoint variables. For finding an optimal solution, we are taking

$$E = \frac{\alpha y b_1 + \gamma_1 x^2 y b_1 - r\left(1 - \frac{x}{K}\right) b_1}{r\left(1 - \frac{x}{K}\right) - \alpha y - \gamma_1 x^2 y - q_1} = \frac{\beta b_2 x + \gamma_2 x y b_2 - s\left(1 - \frac{y}{L}\right) b_2}{s\left(1 - \frac{y}{L}\right) - \beta x - \gamma_2 xy - q_2}. \quad (11)$$

Now, from the two adjoint equations ($\frac{d\lambda_1}{dt} = -\frac{\partial H}{\partial x}$ and $\frac{d\lambda_2}{dt} = -\frac{\partial H}{\partial y}$), we get

$$\frac{d^2\lambda_1}{dt^2} - \left(\frac{rx}{K} + 2\gamma_1 x^2 y + \frac{sy}{L} + xy\gamma_2\right)\frac{d\lambda_1}{dt}$$
$$+ \left[\left(\frac{rx}{K} + 2\gamma_1 x^2 y\right)\left(\frac{sy}{L} + xy\gamma_2\right) - \left(\beta y + \gamma_2 y^2\right)\left(\alpha x + \gamma_1 x^3\right)\right]\lambda_1 = M_1 e^{-\delta t}, \quad (12)$$

where $M_1 = \left(\frac{p_1 q_1 E\delta}{b_1+E}\right) + \left(\frac{sy}{L} + xy\gamma_2\right)\frac{p_1 q_1 E}{b_1+E} - \left(\beta y + \gamma_2 y^2\right)\frac{p_2 q_2 E}{b_2+E}$.

The auxiliary equation for the Eq. (12) is as follows:

$$\mu^2 - \left(\frac{rx}{K} + 2\gamma_1 x^2 y + \frac{sy}{L} + xy\gamma_2\right)\mu$$
$$+ \left[\left(\frac{rx}{K} + 2\gamma_1 x^2 y\right)\left(\frac{sy}{L} + xy\gamma_2\right) - \left(\beta y + \gamma_2 y^2\right)\left(\alpha x + \gamma_1 x^3\right)\right] = 0. \quad (13)$$

Then, complete solution of the Eq. (13) is as follows:

$$\lambda_1(t) = A_1 e^{\mu_1(t)} + A_2 e^{\mu_2(t)} + \left(\frac{M_1}{N_1}\right) e^{-\delta t}. \quad (14)$$

Here, $N_1 = \left[\delta^2 - \delta\left(\frac{rx}{K} + 2\gamma_1 x^2 y + \frac{sy}{L} + xy\gamma_2\right) + \left(\frac{rx}{K} + 2\gamma_1 x^2 y\right)\left(\frac{sy}{L} + xy\gamma_2\right) - \left(\beta y + \gamma_2 y^2\right)\left(\alpha x + \gamma_1 x^3\right)\right] \neq 0$. It is true that λ_1 is bounded if $A_1 = A_2 = 0$.

So, we have $\lambda_1(t) = \left(\frac{M_1}{N_1}\right)e^{-\delta t}$ and $\lambda_2(t) = \left(\frac{M_2}{N_1}\right)e^{-\delta t}$, where $M_2 = \frac{p_2 q_2 E\delta}{b_2+E} + \left(\frac{rx}{K} + 2\gamma_1 x^2 y\right)\frac{p_2 q_2 E}{b_2+E} - \left(\alpha x + \gamma_1 x^3\right)\frac{p_1 q_1 E}{b_1+E}$.

At $t \to \infty$, $\lambda_i(t)e^{\delta t}$, $i = 1, 2$ become constant if they satisfy the transversality condition.

Now, the equation $\frac{\partial H}{\partial E} = 0$ along with $\lambda_1(t)$ and $\lambda_2(t)$ gives the following:

$$\left(p_1 - \frac{M_1}{N_1}\right)\frac{q_1 b_1 x}{(b_1+E)^2} + \left(p_2 - \frac{M_2}{N_1}\right)\frac{q_2 b_2 y}{(b_2+E)^2} = C. \tag{15}$$

Using the Eqs. (11) and (15), we get the optimal equilibrium of the populations i.e., $x = x_\delta$, $y = y_\delta$.

At $\delta \to \infty$, Eq. (15) gives the result $\frac{p_1 q_1 b_1 x}{(b_1+E)^2} + \frac{p_2 q_2 b_2 y}{(b_2+E)^2} = C$ which implies $\frac{\partial \pi}{\partial E} = 0$. Thus, the economic rent is totally vanished. Then, from the Eqs. (11) and (15), we get the following:

$$\frac{\partial \pi}{\partial E} = \frac{M_1}{N_1}\frac{q_1 b_1 x}{(b_1+E)^2} + \frac{M_2}{N_1}\frac{q_2 b_2 y}{(b_2+E)^2}. \tag{16}$$

Since M_1 and M_2 are $o(\delta)$ where N_1 is $o(\delta^2)$, it is found that $\frac{\partial \pi}{\partial E}$ is $o(\delta^{-1})$. So, $\frac{\partial \pi}{\partial E}$ is gradually decreasing function with $\delta(\geq 0)$.

Therefore, it is concluded that $\frac{\partial \pi}{\partial E}$ attains it's maximum value at $\delta = 0$.

6 Numerical Simulations

Here, we verify the analytical findings numerically using MATLAB-2016a and Maple-18. For this purpose, a set of parameter values have been taken as follows: r=3.5, k=300, $\alpha = 0.02$, $q_1 = 0.9$, $E = 10$, $b_1 = 0.5$, $\gamma_1 = 0.0002$, s = 1.1, l = 100, $\beta = 0.001$, $\gamma_2 = 0.0001$, $q_2 = 0.7$, $b_2 = 0.3$. For the same set of parameter values, it is shown that: (i)$E_0(0, 0)$ is unstable, (ii)$E_1(226.53, 0)$ is stable node, (iii)$E_2(0, 38.22)$ is stable node, (iv)$E^*(16.8679, 31.8064)$ is both locally and globally asymptotically stable node.

From Figs. 1 and 2, it is seen that the system possesses an interior equilibrium point $E^* = (16.8679, 31.8064)$ and around that point, the system (1) is globally asymptotically stable.

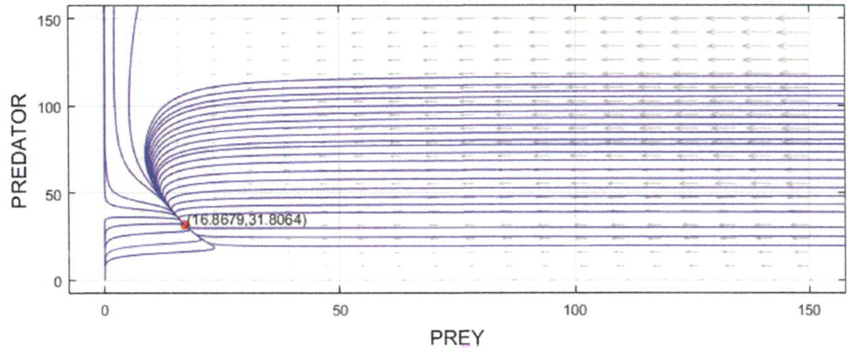

Fig. 1 Globally stable steady state of the fishery resource model with various initial values

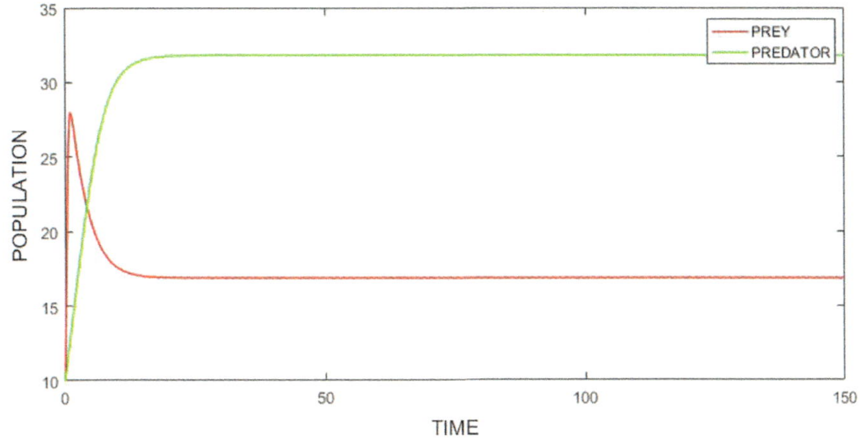

Fig. 2 Globally stable steady state of the fishery resource model with various initial values

It is found that the bionomic equilibrium i.e., $(x_\infty, y_\infty) = (16.5346, 33.2731)$ (Fig. 3) occurs for the same parameter values $p_1 = 10$, $p_2 = 10$, C = 50 and $E = \frac{\alpha y b_1 + \gamma_1 x^2 y b_1 - r(1-\frac{x}{k})b_1}{r(1-\frac{x}{k}) - \alpha y - \gamma_1 x^2 y - q_1}$.

Without the effect of toxicity i.e., $(\gamma_1 = \gamma_2 = 0)$, we found $(x_\infty, y_\infty) = (190.84, 21.24)$ (Fig. 4).

From the figures, it is concluded that the population density of prey $(x_\infty = 190.84)$ at which the bionomic equilibrium occurs without toxicity is greater than the biomass density of prey $(x_\infty = 16.5346)$ with toxicity. On the other hand, the population density of predator $(y_\infty = 21.24)$ at which the bionomic equilibrium occurs without toxicity is lower than the population density of predator $(y_\infty = 33.2731)$ with toxicity.

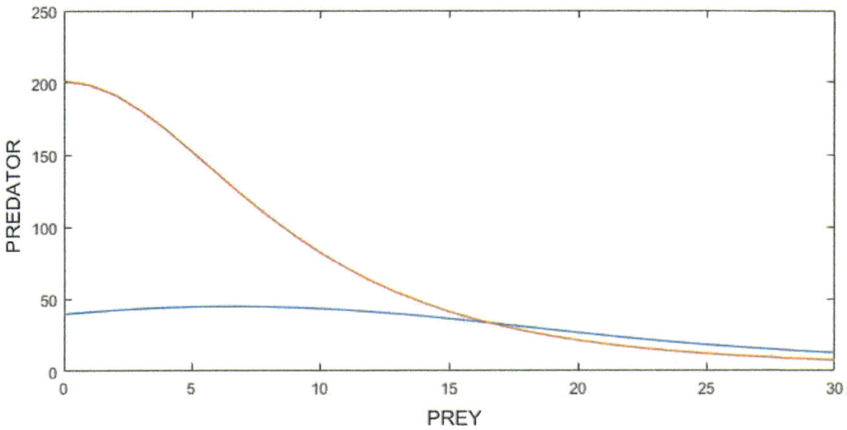

Fig. 3 Bionomic equilibrium of the fishery model when toxicity $\neq 0$

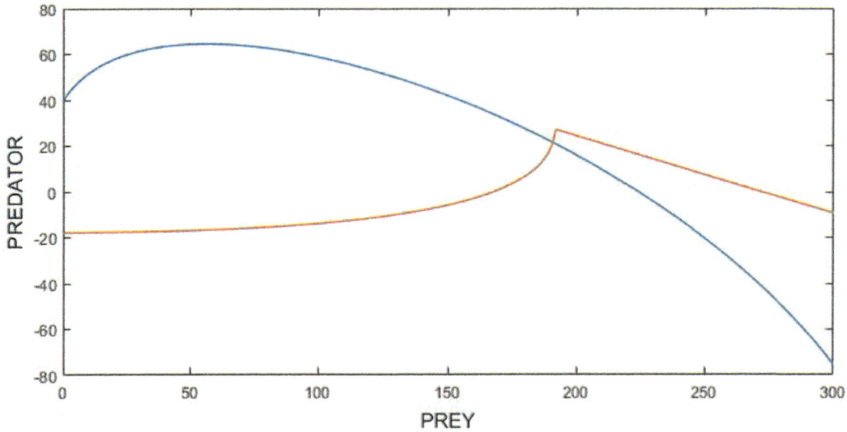

Fig. 4 Bionomic equilibrium of the fishery model when toxicity $= 0$

7 Conclusions

On a note to conclude, it can be prudently placed that this paper, has attempted to discuss the consequences of toxic elements released by both species in a competing predator-prey fishery model, where both the species are subjected to harvesting with a modified catch rate function. An effort has been taken to examine both the local and global stabilities. An endeavour, in this article has also been taken to delineate the existence of bionomic equilibrium in presence of toxicity along with the task of portrayal of the state of the bionomic equilibrium in absence of toxicity. The numerical examples adopted for the present purpose, suggest that the severity of releasing toxins by predator-prey species may change the qualitative nature of the proposed model, explicating the fact that toxicity may cause extinction of any one of the species. It has been observed that if the toxicity is considered, the population level of the bionomic equilibria steady states for the first species quickly decreases, whereas in the second species, it slowly increases. Thus, to put this in a nutshell, it can be said that the moderate increase of toxic substances come from both the predator-prey species, have severe impact on both and would finally lead to annihilation. The optimal harvesting policy, has also been discussed, as another key component of the paper. The policy, which has been employed, applying the Pontryagin maximum principle, has revealed the maximization of net revenue . The investigation has also ascertained that the shadow price remains constant with respect to time, as the transversality condition is satisfied by the optimal equations at infinity. The model has led us to derive the fact that the maximization of net revenue is led by the zero-discounting rate while the completion dissipation of net revenues is led by the infinite discounting rate.

References

1. Chattopadhyay, J.: Effect of toxic substances on a two species competitive system. Ecol. Model **84**, 287–289 (1997)
2. Clark, C.W.: Mathematical Bioeconomics: the Optimal Management of Renewable Resources. Wiley, New York (1976)
3. Clark, C.W.: Bioeconomic Modeling and Fisheries Management. Wiley, New York (1985)
4. Dubey, B., Hussain, J.: A model for the allelopathic effect on two competing fish species. Ecol. Model **129**, 195–207 (2000)
5. Ghosh, M., Chandra, P., Sinha, P.: A mathematical model to study the effects of toxic chemicals on a prey-predator type fishery. J. Biol. Syst. **10**(2), 97–105 (2002)
6. Haque, M.: Sarwadi: effect of toxicity on a harvested fishery model. Model. Earth Syst. Environ. **2**, 122 (2016)
7. Kar, T.K., Chaudhuri, K.S.: On non-selective harvesting of two competing fish species in the presence of toxicity. Ecol. Model. **161**, 125–137 (2003)
8. Kar, T.K., Pahari, U.K., Chaudhuri, K.S.: Management of a prey -predator fishery based on continuous fishinf effort. J. Biol. Syst. **12**(3), 301–313 (2004)
9. Maynard, J.M.: Models in Ecology. xii, 146. University Press, New York (1974)
10. Meserton-Gibbons, M.: On the optimal policy for the combined harvesting of predator and prey. Nat. Res. Model **3**, 63–90 (1988)
11. Meserton-Gibbons, M.: A technique for finding optimal two species harvesting policies. Ecol. Model **92**, 235–244 (1996)
12. Mukhopadhyay, A., Chattopadhyay, J., Tapaswi, P.K.: A delay differential equations model of plankton alleopathy. Maths. Biosci. **149**, 167–189 (1998)
13. Pal, D., Mahapatra, G.S., Mahato, S.K., Samanta, G.P.: A mathematical model of a prey-predsator type fishery in the presence of toxicity with fuzzy optimal harvesting with fuzzy optimal harvesting. J. Appl. Math. Inf. **38**, 13–36 (2020)
14. Pontryagin, L.S., Boltyanskii, V.S., Gamkrelidze, R.V., Mishchencko, E.F.: The Mathematical Theory of Optimal Processes. Wiley, New York (1962)
15. Sarkar, S., Sarkar, A., Chaudhuri, K.S.: Modeling of single species fishery resource harvesting with modified catch rate function. Bull. Cal. Math. Soc. **112**(6), 512–524 (2020)

Analysis for the Impact of HIV Transmission Dynamics in Heterosexuality and Homosexuality

Regan Murugesan, Suresh Rasappan, and Nagadevi Bala Nagaram

Abstract The purpose of this paper is to examine the aspects of the mathematical analysis of HIV transmission through homo and heterosexual. A system of differential equations is designed for the transmission for homo and heterosexual. Equilibrium and interior equilibrium points are identified. Bifurcation analysis forms an important tool in this study. The asymptotic mean square stability criterion is derived for non-deterministic situations. The non-deterministic analysis has been performed around the interior equilibrium point. Numerical simulation is carried out and it supports the theoretical result. Derivation of a mathematical model is the outcome of this study.

Keywords HIV transmission · Homosexual · Heterosexual · Bifurcation

1 Introduction

Mathematical modeling comes in handy to describe and analyze a real life situation. The Human Immunodeficiency Virus (HIV) turned into diagnosable in 1980s [1]. It is a type of lentivirus. It continues to be a noteworthy worldwide medical problem [2]. According to 2017 statistics, 36.9 million individuals live with HIV and most strikingly around 25% of them do not know that they have the virus [3]. 940,000

Supported by Vel Tech University, Avadi, Chennai, Tamil Nadu, India.

R. Murugesan (✉) · N. B. Nagaram
Vel Tech Rangarajan Dr. Sagunthala R & D Institute of Science and Technology, # 42 Avadi- Vel Tech Road, Avadi, Chennai 600062, Tamil Nadu, India
e-mail: mreganprof@gmail.com

S. Rasappan
University of Technology and Applied Sciences- Ibri, Sultanate of Oman, Ibri, Oman

S. Banerjee and A. Saha (eds.), *Nonlinear Dynamics and Applications*,
Springer Proceedings in Complexity,
https://doi.org/10.1007/978-3-030-99792-2_71

people have died due to AIDS related illnesses. The homosexual and heterosexual form the major root cause for the transmission of HIV. In India, 2.1 million people live with HIV and especially 86% of people are affected by HIV through sexual transmission. It has been the third biggest HIV epidemic around the world [3].

The majority among HIV people are located in low- and middle-income countries [3]. Sex laborers, men who have intercourse with men (MSM), individuals who infuse drugs (PWID), Hijiras/transgender individuals, migrant workers and truck drivers are the major key affected HIV population in India. There is no solution to cure the diseases till now but controlling the diseases is possible. Controlling the HIV transmission in the external i.e., person to person through sex, needles etc., is more important than the internal of the human body i.e., after presence of HIV in human body cells. Prevention and public awareness are the major instruments to control the HIV infection.

NACO [3] is in charge of defining policy and executing programs for the hindrance and control of the HIV prevalence in India. The main motive of the NACO is to reduce the annual new HIV infections through HIV treatment, education, care and support for those at high risk of HIV.

AIDS is one of the universal menaces to humanity, so most of the mathematicians have been evincing interest to take a look at the transmission of HIV/AIDS and its dynamics by using the mathematical model. Sex structured models are bringing out the better understanding the associated nuances of the disease. Many mathematicians have developed various models without considering the sex structure wherein they have mainly focused on the dynamics of the disease by considering suitable systems of nonlinear differential equations. However, inclusion of sex structure in a mathematical model would make it more realistic.

It would be worthwhile to refer to some of the related works of the previous researchers. Abu and Emeje analyzed the strategies to control HIV/AIDS with mathematical models [4]. The objective of their research is the identification of the effect of condom use and antiretroviral therapy separately and the combination of both as control strategies. Miao et al proposed an SIR model for a stochastic system and investigated the transmission vertically i.e., mother to child and vaccination [5]. The threshold dynamics conditions are explored. They have incorporated less noise and large noise in their model. Their model helps to control the epidemic of infective disease. Bushnaq et al built a biological model to examine the existence and stability of HIV/AIDS infection employing fractional order derivative [6].

The present study focuses attention on the prevention of HIV affected people by undertaking mathematical analysis. Its main focus is on HIV education and counselling support. Due to continuous monitoring (feedback from HIV prevention programmes) of key affected people, the annual threshold level may be stabilized. Our results may also support development in the direction of 90-90-90 objectives in India.

2 System Description

A mathematical model is one way of describing the system using mathematical concept or any mathematical tools. HIV/AIDS is a challenging illness in medical sector. Sex laborers, men who have intercourse with men (MSM), individuals who infuse drugs (PWID), Hijras/transgender individuals, migrant workers and truck drivers are identified as key transmitters. A system of differential equation was designed for HIV transmission [7, 8] under the category homo and heterosexuality. In the present study, a mathematical model is developed based on the following general assumptions:

The sexually active population is divided into five sectors respectively as susceptible female, infected female, susceptible male, infected male and AIDS affected individuals The population sizes of the genders are initially assumed respectively as λ_1 and λ_2.The direct transmission from female-to-female phenomenon is considered. The age factor is ignored. The natural mortality rate μ is considered to be equal in all the sectors. All possibilities of HIV transmission through the factors are considered.

The following Fig. 1 represents the flow diagram of HIV transmission. The Mathematical model for the HIV transmission is developed as

$$
\begin{aligned}
\dot{x}_1 &= (\lambda_1 + \gamma_1 - \mu - \alpha_1 - \psi_2 - \Delta_1)x_1 + \eta_1 x_2 + y_1 + \kappa_1 y_2 \\
\dot{x}_2 &= \alpha_1 x_1 + (\gamma_2 - \eta_1 - \kappa_2 - \mu - \alpha_2 - \phi_1)x_2 + \Delta_2 y_1 + \phi_2 y_2 \\
\dot{y}_1 &= \psi_2 x_1 + \kappa_2 x_2 + (\lambda_2 + \delta_1 - \psi_1 - \Delta_2 - \beta_1 - \mu)y_1 + \eta_2 y_2 \\
\dot{y}_2 &= \Delta_1 x_1 + \phi_1 x_2 + \beta_1 y_1 + (\delta_2 - \eta_2 - \kappa_1 - \phi_2 - \beta_2 - \mu)y_2 \\
\dot{z} &= \alpha_2 x_2 + \beta_2 y_2 - (\mu + r_1)z
\end{aligned}
\tag{1}
$$

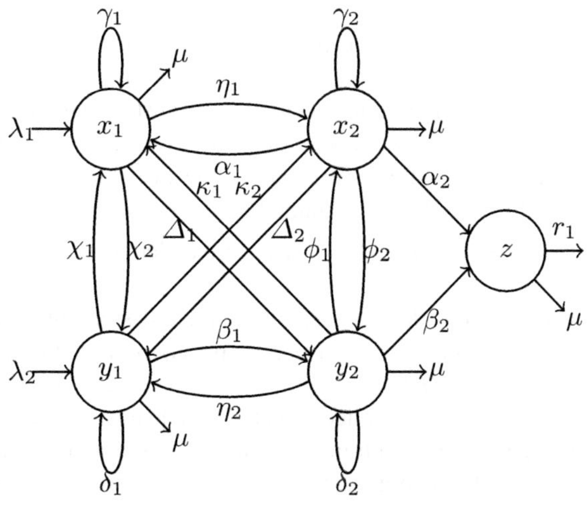

Fig. 1 Flow diagram of HIV transmission

The modified Mathematical model is

$$\begin{aligned}\dot{x}_1 &= \xi_1 x_1 + \eta_1 x_2 + \psi_1 y_1 + \kappa_1 y_2 \\ \dot{x}_2 &= \alpha_1 x_1 + \xi_2 x_2 + \Delta_2 y_1 + \phi_2 y_2 \\ \dot{y}_1 &= \psi_2 x_1 + \kappa_2 x_2 + \xi_3 y_1 + \eta_2 y_2 \\ \dot{y}_2 &= \Delta_1 x_1 + \phi_1 x_2 + \beta_1 y_1 + \xi_4 y_2 \\ \dot{z} &= \alpha_2 x_2 + \beta_2 y_2 - \xi_5 z\end{aligned} \tag{2}$$

where x_1 identifies the population suppressed under susceptible females, x_2 represents the individual population of infected females , y_1 identifies the individual population suppressed under susceptible males, y_2 represents the individual population of infected males, z represents the AIDS affected population , η_1 and α_1 are the rates of transmission from x_1 to x_2 and x_2 to x_1,β_1 and η_2 are the rates of transmission from y_1 to y_2 and y_2 to y_1, $\chi_1$and χ_2 are the rates of transmission from y_1 to x_1 and x_1 to y_1, ϕ_1 and ϕ_2 are the rates of transmission from y_2 to x_2 and x_2 to y_2, Δ_1 and κ_1 are the rates of transmission from x_1 to y_2 and y_2 to x_1, Δ_2 and κ_2 are the rates of transmission from x_2 to y_1 and y_1 to x_2, γ_1 and γ_2 are the rates of transmission from x_1 to x_1 and x_2 to x_2, δ_1 and δ_2 are the rates of transmission from y_1 to y_1 and y_2 to y_2, α_2 and β_2 are the rates of transmission from x_2 to z and y_2 to z, μ and r_1 are the natural death rate at all states and death rate due to AIDS concerning z, respectively.

3 HIV Impact, Equilibrioception and Stability

3.1 Equilibrioception Points

The equilibrioception points arising from the HIV impact are obtained with the fulfillment of the condition $\dot{x_1} = \dot{x_2} = \dot{y_1} = \dot{y_2} = \dot{z} = 0$. The equilibrioception is estimated at individual sectors of the system which is carried out in five cases.

Case 1: The HIV transmission rate from susceptible female is assumed to be zero, i.e., $\dot{x_1} = 0$. After computation, the possible equilibrioception points of the susceptible females are obtained as $(-\frac{(\kappa_1 y_2+\psi_1 y_1)}{\xi_1}, 0, 0, 0, 0)$, $(-\frac{(\eta_1 x_2+\psi_1 y_1)}{\xi_1}, 0, 0, 0, 0)$, $(-\frac{(\eta_1 x_2+\kappa_1 y_2)}{\xi_1}, 0, 0, 0, 0)$. These are called as equilibrioception points of susceptible females.

Case 2: Consider the case of the transmission rate from infected females taking the value zero i.e., $\dot{x_2} = 0$. The possible equilibrioception points of the infected females are obtained as $(0, -\frac{(\Delta_2 y_1+\phi_2 y_2)}{\xi_2}, 0, 0, 0)$, $(0, -\frac{(\alpha_1 x_1+\phi_2 y_2)}{\xi_2}, 0, 0, 0)$, $(0, -\frac{(\alpha_1 x_1+\Delta_2 y_1)}{\xi_2}, 0, 0, 0)$. They are called infected female equilibrioception points.

Case 3: The rate of transmission from susceptible males is taken as zero, i.e., $\dot{y_1} = 0$. The equilibrioception points in this case are obtained as $(0, 0, -\frac{(\kappa_2 x_2+\eta_2 y_2)}{\xi_3}, 0, 0)$, $(0, 0, -\frac{(\psi_2 x_1+\eta_2 y_2)}{\xi_3}, 0, 0)$ and $(0, 0, -\frac{(\psi_2 x_1+\kappa_2 x_2)}{\xi_3}, 0, 0)$. They are called susceptible male equilibrioception.

Case 4: The infected male equilibrioception points are determined when $\dot{y_2} = 0$. They are $(0, 0, 0, -\frac{(\Delta_1 x_1 + \phi_1 x_2)}{\xi_4}, 0)$, $(0, 0, 0, -\frac{(\beta_1 y_1 + \Delta_1 x_1)}{\xi_4}, 0)$, $(0, 0, 0, -\frac{(\beta_1 y_1 + \phi_1 x_2)}{\xi_4}, 0)$
Case 5: Finally, when the transmission rate of AIDS affected people is considered as zero, i,e., $\dot{z_h} = 0$, the equilibrioception points are derived as $(0, 0, 0, 0, -\frac{\beta_2 y_2}{\xi_5})$ and $(0, 0, 0, 0, -\frac{\beta_2 x_2}{\xi_5})$.

3.2 *Endemic Equilibrioception*

The endemic equilibrium point is got as $E^*\left(x_1^*, x_2^*, y_1^*, y_2^*, x_1^*, z^*\right)$ where $x_1^* = -(\eta_1 x_2 + \kappa_1 y_2 + \psi_1 y_1)/\xi_1$, $x_2{}^* = -(\alpha_1 x_1 + \Delta_2 y_1 + \phi_2 y_2)/\xi_2$, $y_1{}^* = -(\psi_2 x_1 + \kappa_2 x_2 + \eta_2 y_2)/\xi_3$, $y_2{}^* = -(\beta_1 y_1 + \Delta_1 x_1 + \phi_1 x_2)/\xi_4$ and $z^* = (\alpha_2 x_2 + \beta_2 y_2)/x_{i5}$. It is seen that HIV transmission persists or exists.

4 The Stability of the HIV Impaction

Theorem 1 *The endemic equilibrioception point E^* is asymptotically stable globally, if* $(x_1{}^* - x_1) = \xi_1 + (\eta_1 x_2 + \kappa_1 y_2 + \psi_1 y_1)/x_1$, $(x_2{}^* - x_2) = \xi_2 + (\alpha_1 x_1 + {}_2 y_1 + \phi_2 y_2)/x_2$, $(y_1{}^* - y_1) = \xi_3 + (\psi_2 x_1 + \kappa_2 x_2 + \eta_2 y_2)/y_1$, $(y_2{}^* - y_2) = \xi_4 + (\beta_1 y_1 + {}_1 x_1 + \phi_1 x_2)/y_2$ *and* $(z^* - z) = (\alpha_2 x_2 + \beta_2 y_2)/z - \xi_5$.

Proof The stability of HIV impaction of the model is obtained by Lyapunov function. It is HIV transmitting function. $x_1{}^* - x_1$, $x_2{}^* - x_2$, $y_1{}^* - y_1$, $y_2{}^* - y_2$ and $z^* - z$ are acting as HIV disease spreading reducers. The stability of the model is depending on the disease reducers. These reducers are considered around the endemic equilibrioception.

Consider the Lyapunov function

$$\begin{aligned} V &= (x_1 - x_1{}^* - x_1{}^* \log(x_1 / x_1{}^*)) + (x_2 - x_2{}^* - x_2{}^* \log(x_2 / x_2{}^*)) \\ &\quad + (y_1 - y_1{}^* - y_1{}^* \log(y_1 / y_1{}^*)) \\ &\quad + (y_2 - y_2{}^* - y_2{}^* log(y_2 / y_2{}^*)) + (z - z^* - z^* log(z/z^*)) \\ \dot{V} &= (x_1 - x_1{}^*)\dot{x}_1 / x_1 + (x_i s_2 - x_2{}^*)\dot{x}_2 / x_2 + (y_1 - y_1{}^*)\dot{y}_1 / y_1 \\ &\quad + (y_2 - y_2{}^*)\dot{y}_2 / y_2 + (z - z^*)\dot{z}/z. \end{aligned}$$

By applying endemic equilibrioception point, we obtain

$$
\begin{aligned}
\dot{V} &= (x_1 - x_1{}^*)(\xi_1 + (\eta_1 x_2 + \kappa_1 y_2 + \psi_1 y_1)/x_1) \\
&\quad + (x_2 - x_2{}^*)(\xi_2 + (\alpha_1 x_1 + {}_2 y_1 + \phi_2 y_2)/x_2) \\
&\quad + (y_1 - y_1{}^*)(\xi_3 + (\psi_2 x_1 + \kappa_2 x_2 + \eta_2 y_2)/y_1) \\
&\quad + (y_2 - y_2{}^*)(\xi_4 + (\beta_1 y_1 + {}_1 x_1 + \phi_1 x_2)/y_2) \\
&\quad + (z - z^*)((\alpha_2 x_2 + \beta_2 y_2)/z - \xi_5) \\
\dot{V} &= -(x_1 - x_1{}^*)^2 - (x_2 - x_2{}^*)^2 - (y_1 - y_1{}^*)^2 - (y_2 - y_2{}^*)^2 - (z - z^*)^2 \\
\dot{V} &< 0
\end{aligned}
$$

which is negative definite. Hence by LaSalle's invariance principle, the HIV transmission dynamic (1) is globally asymptotically stable at endemic equilibrioception.

5 The Non-deterministic Model

The non-deterministic model is considered for the homo and heterosexual population in a HIV transmission. In this model, the perturbations are permitted into the factors x_1, x_2, y_1, y_2 and z around the endemic equilibrioception E* for the situation when it is feasible and regionally asymptotically steady. Local steadiness of E^* is implied via the existence conditions of E^*. So in the model (1), it is assumed that the stochastic disturbances of the factors around their values at E^* are of white noise type, which are relative to the distances of x_1, x_2, y_1, y_2 and z from the values of x_1*, x_2*, y_1*, y_2* and z^*. The Stochastic differential equation of the HIV is

$$
\begin{aligned}
dx_1 &= [\xi_1 x_1 + \eta_1 x_2 + \psi_1 y_1 + \kappa_1 y_2]dt + \sigma_1[x_1 - x_1*]dw_1(t) \\
dx_2 &= [\alpha_1 x_1 + \xi_2 x_2 + {}_2 y_1 + \phi_2 y_2]dt + \sigma 2[x_2 - x_2*]dw_2(t) \\
dy_1 &= [\psi_2 x_1 + \kappa_2 x_2 + \xi_3 y_1 + \eta_2 y_2]dt + \sigma 3[y_1 - y_1*]dw_3(t) \\
dy_2 &= [{}_1 x_1 + \phi_1 x_2 + \beta_1 y_1 + \xi_4 y_2]dt + \sigma 4[y_2 - y_2*]dw_4(t) \\
dz &= [\alpha_2 x_2 + \beta_2 y_2 - \xi_5 z]dt + \sigma_4[z - z^*]dw_5(t)
\end{aligned}
\tag{3}
$$

where $\sigma_i, i = 1, 2, 3, 4, 5$ are real constants and $w_i(t), i = 1, 2, 3, 4, 5$ are independent from each other by standard Wiener processes [9].

6 Stability of the Non-deterministic Model

The non-deterministic model (3) can be centered at its endemic equilibrioception E^* positively by the change of variables $u_1 = (x_1 - x_1^*)$, $u_2 = (x_2 - x_2^*)$, $u_3 = (y_1 - y_1^*)$, $u_4 = (y_2 - y_2^*)$, $u_5 = (z - z^*)$. The non-deterministic differential equations around the endemic equilibrioception E^* are linearized and taken in the form

$$du(t) = f(u(t))dt + g(u(t))dw(t) \tag{4}$$

where $u(t) = [u_1(t)u_2(t)u_3(t)u_4(t)u_5(t)]^T$

$$f(u(t)) = \begin{bmatrix} \xi_1 & 0 & 0 & 0 & 0 \\ 0 & \xi_2 & 0 & 0 & 0 \\ 0 & 0 & \xi_3 & 0 & 0 \\ 0 & 0 & 0 & \xi_4 & 0 \\ 0 & 0 & 0 & 0 & \xi_5 \end{bmatrix} u(t) \tag{5}$$

$$g(u(t)) = \begin{bmatrix} \sigma_1 u_1 & 0 & 0 & 0 & 0 \\ 0 & \sigma_2 u_2 & 0 & 0 & 0 \\ 0 & 0 & \sigma_3 u_3 & 0 & 0 \\ 0 & 0 & 0 & \sigma_4 u_4 & 0 \\ 0 & 0 & 0 & 0 & \sigma_5 u_5 \end{bmatrix} \tag{6}$$

In (4) the positive interior equilibrium E^* corresponds to the trivial solution $u(t) = 0$. Let U be the set $U = \{(t \geq t_0) \times R^n, t_0 \in R^+\}$.

Thus, $V \in {C_2}^0(U)$ is two times constantly differentiable function with respect to u and a continuous function with respect to t. With reference to Afanasev et al. [10], the Eq. (4) implies the following Itö differential equation

$$LV(t,u) = \frac{\partial V(t,u)}{\partial t} + f^T(u)\frac{\partial V(t,u)}{\partial t}\frac{1}{2}\text{trace}\left[g^T(u)\frac{\partial^2 V(t,u)}{\partial u^2}g(u)\right]$$

where $\frac{\partial V}{\partial u} = Col\left(\frac{\partial V}{\partial u_1}, \frac{\partial V}{\partial u_2}, \frac{\partial V}{\partial u_3}, \frac{\partial V}{\partial u_4}, \frac{\partial V}{\partial u_5}\right)$ and $\frac{\partial^2 V(t,u)}{\partial u^2} = \frac{\partial^2 V}{\partial u_j \partial u_i}, i, j = 1, 2, 3, 4.$

Remarks

Suppose the function $V \in {C_2}^0(U)$ exists as above. Then if the inequalities

$$M_1|u|^q \leq V(t,u) \leq M_2|u|^q, \tag{7}$$

$$LV(t,u) \leq -M_3|u|^q, M_i > 0, q > 0 \tag{8}$$

hold, then the trivial solution of (4) is exponentially $q-$stable for $t \geq 0$. When $q = 2$ the trivial solution due to (7) and (8) is globally asymptotically steady in probability.

Theorem 2 *If* ${\sigma_1}^2 = -4\xi_1$, ${\sigma_2}^2 = -4\xi_2$, ${\sigma_3}^2 = -4\xi_3$, ${\sigma_4}^2 = -4\xi_4$ *and* ${\sigma_5}^2 = -4\xi_5$, *then the solution with the zero of* (4) *is asymptotically mean square steady.*

Proof Let us assume the Lyapunov function

$$V(u) = \frac{1}{2}[w_1{u_1}^2 + w_2u_2 + w_3{u_3}^2 + w_4{u_4}^2 + w_5{u_5}^2]$$

where $w_i, i = 1, 2, 3, 4, 5$ are real positive constants to be chosen. It is straightforward to check that the inequalities (7) and (8) hold with $q = 2$. Now the Itö Process gives

$$V(u) = w_1(\xi_1u_1)u_1 + w_2(\xi_2u_2)u_2 + w_3(\xi_3u_3)u_3 + w_4(\xi_4u_4)u_4 + w_5(\xi_5u_5)u_5 + \frac{1}{2}\text{trace}\left[g^T(u)\frac{\partial^2 V(t,u)}{\partial u^2}g(u)\right]$$

$$\begin{aligned} V(u) &= w_1(\xi_1u_1)u_1 + w_2(\xi_2u_2)u_2 + w_3(\xi_3u_3)u_3 + w_4(\xi_4u_4)u_4 + w_5(\xi_5u_5)u_5 \\ &\quad + \frac{1}{2}[{\sigma_1}^2w_1{u_1}^2 + {\sigma_2}^2w_2{u_2}^2 + {\sigma_3}^2w_3{u_3}^2 + {\sigma_4}^2w4{u_4}^2 + {\sigma_5}^2w_5{u_5}^2] \\ &= w_1(\xi_1 + {\sigma_1}^2){u_1}^2 + w_2(\xi_2 + {\sigma_2}^2){u_2}^2 + w_3(\xi_3 + {\sigma_3}^2){u_3}^2 \\ &\quad + w_4(\xi_4 + {\sigma_4}^2){u_4}^2 + w_5(\xi_5 + {\sigma_5}^2){u_5}^2 \end{aligned}$$

$$V(u) = -\xi_1w_1{u_1}^2 - \xi_2w_2{u_2}^2 - \xi_3w_3{u_3}^2 - \xi_4w_4{u_4}^2 - \xi_5w_5{u_5}^2$$

Hence the system is asymptotically mean square stable.

7 Numerical Simulation for Deterministic Model

A simulation was carried out by taking the initial conditions $x_1 = 5.013,\ y_1 = 4.218,\ z = 1.298,\ x_2 = 4.558,\ y_2 = 9.383$. and the parametric values

$\xi_1 = 0.2345, \eta_1 = 0.3837, \psi_1 = 0.2312, \varphi_1 = 0.2345, \xi_2 = 0.4834, \eta_2 = 0.7845, \psi_2 = 0.7543, \varphi_2 = 0.4567, \xi_3 = 0.1987, \alpha_1 = 0.1976, \beta_1 = 0.9109, 1 = 0.1209, \xi_4 = 0.2987, \alpha_2 = 0.7652, \beta_2 = 0.2109, 2 = 0.5768, \xi_5 = 0.2345, \kappa_1 = 0.9872, \kappa_2 = 0.6092$

As a result of this simulation process between the five states, it is observed that the dissemination of HIV in a homo and heterosexual affected populace is stabilized at the origin. The result is depicted in Fig. 2.

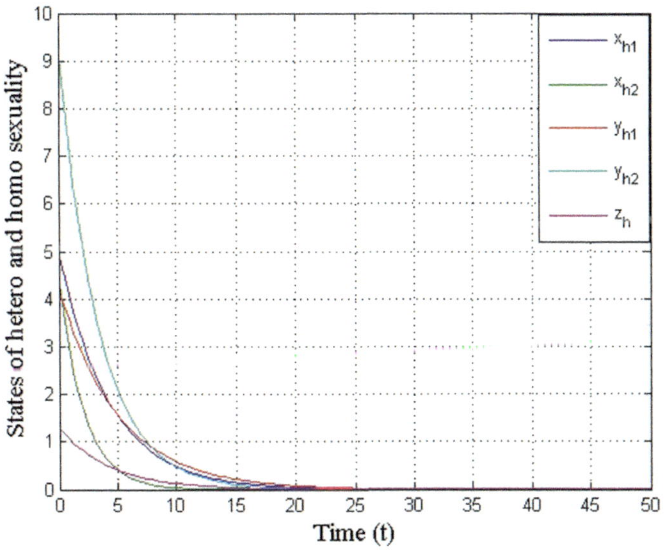

Fig. 2 Analysis of deterministic system for dissemination of HIV among homosexual and heterosexual affected population

8 Numerical Example for Non-deterministic Model

A simulation was carried out by taking the initial conditions $x_1 = 15.013$, $y_1 = 17.218$, $z = 5.298$, $x2 = 18.558$, $y2 = 15.383$ and the parametric values $\xi_1 = 0.5645$, $\eta_1 = 0.7247$, $\psi_1 = 0.6542$, $\varphi_1 = 0.7250$, $\xi_2 = 0.5234$, $\eta_2 = 0.6845$, $\psi_2 = 0.3543$, $\varphi_2 = 0.6567$, $\xi_3 = 0.7987$, $\alpha_1 = 0.2376$, $\beta_1 = 0.7609$, $1 = 0.6209$, $\xi_4 = 0.1257$, $\alpha_2 = 0.7651$, $\beta_2 = 0.8654$, $2 = 0.2768$, $\xi_5 = 0.8765$, $\kappa_1 = 0.1872$, $\kappa_2 = 0.1092$.

The noise band in Wiener process is taken as the closed interval [0, 1]. With the above values, the following observations are made.

When the value of σ is in the range between 0.1 and 0.49, susceptible males and infected females are affected more by HIV. When the value of σ is in the range between 0.5 and 1, the persons in all the five categories namely susceptible female, infected female, susceptible male, infected male and AIDS affected people are impacted by the HIV transmission of homo and heterosexual population. The simulation result for the non-deterministic model is furnished in Fig. 3.

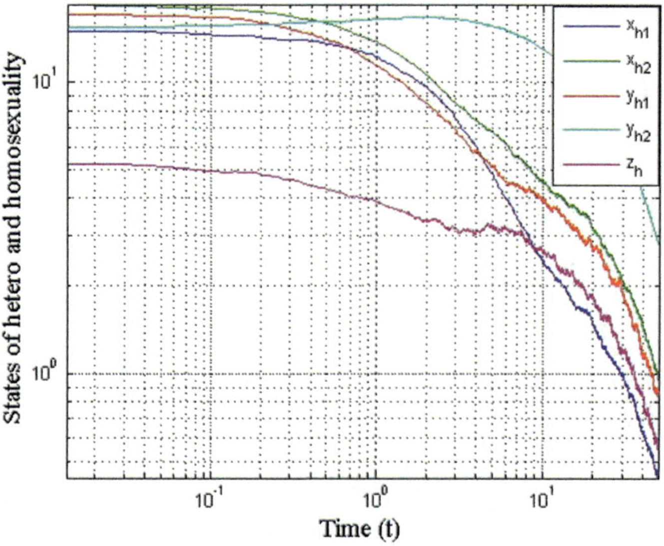

Fig. 3 Non-deterministic nature of homosexual and heterosexual affected population

9 Hyperbolic Equilibria and Bifurcation Analysis

9.1 Hyperbolic Equilibria

The Jacobian matrix of the model (1) is

$$J = \begin{bmatrix} \xi_1 & \eta_1 & \psi_1 & \kappa_1 & 0 \\ \alpha_1 & \xi_2 & \Delta_2 & \phi_2 & 0 \\ \psi_2 & \kappa_2 & \xi_3 & \eta_2 & 0 \\ \Delta_1 & \phi_1 & \beta_1 & \xi_4 & 0 \\ 0 & \alpha_2 & 0 & \beta_2 & -\xi_5 \end{bmatrix} \tag{9}$$

The characteristic equation of the Jacobian matrix J is

$$\lambda^5 - a\lambda^4 - b\lambda^3 - c\lambda^2 - d\lambda - E = 0 \tag{10}$$

In this analysis, If $E \neq 0$, then there exists a hyperbolic equilibrium point for the system. This equilibrium point is robust. Therefore the system of equations is structurally stable and when $E = 0$, there exists a non-hyperbolic equilibrium point.

9.2 Bifurcation

If $\eta_1, \psi_1, \kappa_1, \alpha_2, \beta_1$ and β_2 are negative, or $\eta_1, \psi_1, \kappa_1, \beta_1 = 0$ or negative, $\alpha_2 = 0$ and $\beta_2 = 0$, then J has a pair of imaginary eigen values. Consequently, the stability of the system may change or disappear or split into many equilibrium points. This indicates the existence of bifurcation. If $\xi_1 = \xi_4 = \xi_5 = 0$, then J has zero eigenvalue and a pair of imaginary eigen values. In this case also, bifurcation exists.

10 Conclusion

The mathematical analysis of the problem of HIV transmission through homo and heterosexual investigated by designing a new mathematical model. A non-deterministic model is investigated and its qualitative properties are analyzed. Hyperbolic equilibrium points and bifurcation analysis strengthen the qualitative properties of the proposed model. The hyperbolic equilibrium points give the robustness of the proposed model. Numerical computations concerning the stability analysis of the proposed model using the MATLAB software are presented. From the bifurcation analysis, it is observed that the disease-free equilibrium condition is not possible if the individual can suppress AIDS and the mortality rate or mortality span may undergo change due to continuous prevention and treatment. As a result of this study it is seen that, due to the severity of the HIV transmission, the endemic equilibrioception may change, disappear or split into many endemic equilibrioception.

Acknowledgements The authors are thankful to the Reviewer for the comments and suggestions towards the improvement of the paper.

References

1. Dietz, K.: On the transmission dynamics of HIV. Math. Biosci. **90**(1–2), 397–414 (1988)
2. Kremer, M., Morcom, C.: The effect of changing sexual activity on HIV prevalence. Math. Biosci. **151**(1), 99–122 (1998)
3. Global information and education on HIV and AIDS. https://www.avert.org/global-HIV-and-AIDS-statistics (2019)
4. Abu, O., Emeje, M.A.: Comparative analysis of HIV/AIDS control strategies with mathematical models. J. Scient. Eng. Res. **3**, 442–448 (2016)
5. Miao, A., Zhang, J., Zhang, T., Pradeep, B.G.: Threshold dynamics of a stochastic model with vertical transmission and vaccination. Comput. Math. Methods Med., 1–10 (2017)
6. Bushnaq, S., Khan, S.A., Shah, K., Zaman, G.: Mathematical analysis of HIV/AIDS infection model with Caputo-Fabrizio fractional derivative. Cogent Math. Stat. **5**(1), 1432521 (2018)
7. May, R.M., Anderson, R.M., McLean, A.R.: Possible demographic consequences of HIV/AIDS epidemics II. Assuming HIV infection does not necessarily lead to AIDS. In: Mathematical Approaches to Problems in Resource Management and Epidemiology (Ithaca, NY, 1987), Lecture Notes in Biomathematics, Springer, Berlin, 81, 220–248 (1989)

8. Sani, A., Kroese, D.P., Pollett, P.K.: Stochastic models for the spread of HIV in a mobile heterosexual population. Math. Biosci. **208**(1), 98–124 (2007)
9. Mukherjee, D.: Stability analysis of a stochastic model for prey-predator system with disease in the prey. Nonlinear Anal. Model. Control **8**(2), 83–92 (2003)
10. Afanas'ev, V.N., Kolmanowski, V.B., Nosov, V.R.: Mathematical Theory of Global Systems Design. Kluwer Academic, Dordrecht (1996)

Exact Traveling Wave Solutions to General FitzHugh-Nagumo Equation

Subin P. Joseph

Abstract A general FitzHugh–Nagumo equation is considered in this paper and new traveling wave exact solutions for this nonlinear partial differential equation are derived. This nonlinear reaction-diffusion equation models several evolution equations such as the transmission of nerve impulses and the evolutionary rescue in the case of spatial invasions. Since the equation is highly nonlinear, the exact solutions are computed using certain ansatz forms and using a computer algebra system.

Keywords FitzHugh–Nagumo equation · Nonlinear evolution equation · Exact solutions

1 Introduction

In this paper we consider the cubic nonlinear evolutionary equation

$$\frac{\partial u}{\partial t} - \eta \frac{\partial^2 u}{\partial x^2} = \alpha_1 u + \alpha_2 u^2 + \alpha_3 u^3 \tag{1}$$

where $u = u(x, t)$ and η, α_1, α_2 and α_3 are real parameters. This equation is the generalization of FitzHugh– Nagumo equation [11, 12] given by

$$\frac{\partial u}{\partial t} - \frac{\partial^2 u}{\partial x^2} = u(1-u)(u-\rho), \tag{2}$$

with $\alpha_1 = -\rho, \alpha_2 = (1+\rho)$, $\alpha_3 = -1$, $\eta = 1$ and $0 < \rho < 1$. Equation (2) has many applications in the fields of astrophysics, fluid mechanics, exciting electronic circuit theory, population genetics, heart electrical waves, chemical chemistry, trans-

Supported by TEQIP-II Four Funds, Government Engineering College, Wayanad.

S. P. Joseph (✉)
Government Engineering College, Wayanad, Thalapuzha, Kerala, India
e-mail: subinpj@gecwyd.ac.in

S. Banerjee and A. Saha (eds.), *Nonlinear Dynamics and Applications*,
Springer Proceedings in Complexity,
https://doi.org/10.1007/978-3-030-99792-2_72

mission of nerve impulses and the evolutionary rescue in the case of spatial invasions etc. [1, 2, 4–7, 10, 13–15, 17–19].

The theory which investigate evolutionary rescue in the case of well-mixed populations that are declining due to an environmental shift and spatial invasions are presented in [19] . The prevention of population extinction by adaptation is termed as evolutionary rescue [3, 8]. If $u(x, t)$ is the relative invader frequency, then its rate of change is

$$\frac{\partial u}{\partial t} - \eta \frac{\partial^2 u}{\partial x^2} = \omega u(1-u), \tag{3}$$

where η and ω are the species diffusion constant and the frequency independent fitness advantage for the invader respectively. Similarly, when the fitness of the resident is linearly decreasing with respect to the frequency of the invader, then

$$\frac{\partial u}{\partial t} - \eta \frac{\partial^2 u}{\partial x^2} = \omega u^2(1-u). \tag{4}$$

The Eqs. (3) and (4) are special cases of the general Eq. (1). Putting $\alpha_1 = \omega, \alpha_2 = -\omega$, $\alpha_3 = 0$ and putting $\alpha_1 = 0, \alpha_2 = \omega, \alpha_3 = -\omega$ in Eq. (1), we can recover these equations respectively.

Since the evolutionary equations given above are having many applications in different fields such as natural computations, exact solutions are necessary for the better analysis of the situation in hand as well as checking the accuracy of approximate solutions. But the available exact solutions are rare for these nonlinear evolutionary equations [5–7]. Nonlinear partial differential equations are very difficult to solve in general to find exact solutions. Several methods are developed to find exact solutions for nonlinear equations that appear in mathematical physics. But the efficient and widely used method to solve these equations are various ansatz methods, such as tanh method, sech method, sinh-cosh method, Exp-function method, the Jacobi elliptic function expansion method and(G'/G)-expansion method and several variants of these methods [9, 16, 20, 21]. In this paper also we apply ansatz method to find new traveling wave exact solutions to the general evolution Eq. (1). Since the computations are much involved, we need to depend on any of the powerful computer algebra system in performing the required computations. In the next section we explain the algorithm used to compute the solutions. Several new exact solutions for the general evolution equation are derived in the third section. The paper is concluded in the last section with a discussion.

2 Method

We need to find the traveling wave solutions to the evolution Eq. (1). First of all, the traveling wave transformation $u(x, t) = v(\xi)$ is used to convert this equation in to an ordinary differential equation, where $\xi = dx + ct$ and d and c are traveling wave parameters. The resulting ordinary differential equation is then given by

$$d^2\eta v''(\xi) - cv'(\xi) + \alpha_1 v(\xi) + \alpha_2 v(\xi)^2 + \alpha_3 v(\xi)^3 = 0. \tag{5}$$

Now, to derive the exact solutions, we assume that the solutions to the above ordinary differential equation exist in the form

$$v(\xi) = a_0 + \frac{a_1 + a_2 g_1(\xi) + a_3 h_1(\xi)}{b_1 + b_2 g_2(\xi) + b_3 h_2(\xi)} \tag{6}$$

where the functions g_1, g_2, h_1 and h_2 and the parameters a_i's and b_i's are to be determined later. Then, $v(\xi)$ given by Eq. (6) is substituted in to the ordinary differential Eq. (5). Then we obtain a system of nonlinear algebraic equations corresponding to this differential equation by using any of the computational algebra system. This system of equations is then solved to determine the value of the parameters. Using these parametric values we can determine the exact solutions to Eq. (1). We use different trial functions to derive new exact solutions in the following section.

3 Exact Solutions

To obtain the first family of solutions, we select the trial function $g_1(\xi) = \text{sech}^2(\xi)$, $g_2(\xi) = \text{csch}^2(\xi)$, $h_1(\xi) = \tanh(\xi)$ and $h_2(\xi) = \coth(\xi)$. Then Eq. (6) becomes

$$v(\xi) = a_0 + \frac{a_1 + a_2 \text{sech}^2(\xi) + a_3 \text{csch}^2(\xi)}{b_1 + b_2 \tanh(\xi) + b_3 \coth(\xi)}. \tag{7}$$

Substituting this value of $v(\xi)$ in Eq. (5), we get a system of thirteen nonlinear algebraic equations given in appendix. Solving this system of algebraic equations, the following sets of different solutions are obtained.

SET-I:

$$\begin{aligned} &c = \pm\frac{\alpha_2\sqrt{\alpha_2^2 - 4\alpha_1\alpha_3}}{4\alpha_3}, \quad d = -\frac{\sqrt{4\alpha_1\alpha_3 - \alpha_2^2}}{2\sqrt{2}\sqrt{\alpha_3}\sqrt{\eta}}, \quad b_1 = -\frac{a_1\alpha_2}{2\alpha_1}, \quad a_0 = 0 \\ &b_2 = \mp\frac{a_1\sqrt{\alpha_2^2 - 4\alpha_1\alpha_3}}{2\alpha_1}, \quad b_3 = 0, \quad a_2 = \frac{1}{4}a_1\left(\frac{\alpha_2^2}{\alpha_1\alpha_3} - 4\right), \quad a_3 = 0. \end{aligned} \tag{8}$$

SET-II:

$$\begin{aligned} &c = \pm\frac{\alpha_2\sqrt{\alpha_2^2 - 4\alpha_1\alpha_3}}{4\alpha_3}, \quad d = -\frac{\sqrt{4\alpha_1\alpha_3 - \alpha_2^2}}{2\sqrt{2}\sqrt{\alpha_3}\sqrt{\eta}}, \quad b_1 = -\frac{a_1\alpha_2}{2\alpha_1}, \quad a_0 = 0 \\ &b_2 = 0, \quad b_3 = \mp\frac{a_1\sqrt{\alpha_2^2 - 4\alpha_1\alpha_3}}{2\alpha_1}, \quad a_2 = 0, \quad a_3 = a_1\left(1 - \frac{\alpha_2^2}{4\alpha_1\alpha_3}\right). \end{aligned} \tag{9}$$

SET-III:

$$\begin{aligned}
&c=\pm\frac{\alpha_2\sqrt{\alpha_2^2-4\alpha_1\alpha_3}}{8\alpha_3},\quad d=-\frac{\sqrt{4\alpha_1\alpha_3-\alpha_2^2}}{4\sqrt{2}\sqrt{\alpha_3}\sqrt{\eta}},\quad b_1=-\frac{a_1\alpha_2}{2\alpha_1},\quad a_0=0\\
&b_2=\mp\frac{a_1\sqrt{\alpha_2^2-4\alpha_1\alpha_3}}{4\alpha_1},\quad b_3=\mp\frac{a_1\sqrt{\alpha_2^2-4\alpha_1\alpha_3}}{4\alpha_1},\\
&a_2=\frac{1}{16}a_1\left(\frac{\alpha_2^2}{\alpha_1\alpha_3}-4\right),a_3=\frac{1}{16}a_1\left(4-\frac{\alpha_2^2}{\alpha_1\alpha_3}\right).
\end{aligned}\tag{10}$$

SET-IV:

$$\begin{aligned}
&c=\pm\frac{\alpha_2\sqrt{\alpha_2^2-4\alpha_1\alpha_3}}{8\alpha_3},\quad d=-\frac{\sqrt{4\alpha_1\alpha_3-\alpha_2^2}}{4\sqrt{2}\sqrt{\alpha_3}\sqrt{\eta}},\quad b_1=-\frac{a_1\alpha_2}{2\alpha_1},\quad a_0=0\\
&b_2=\mp\frac{a_1\sqrt{\alpha_2^2-4\alpha_1\alpha_3}}{4\alpha_1},\quad b_3=\mp\frac{a_1\sqrt{\alpha_2^2-4\alpha_1\alpha_3}}{4\alpha_1},\quad a_2=0,\quad a_3=0.
\end{aligned}\tag{11}$$

SET-V:

$$\begin{aligned}
&c=\pm\frac{\alpha_2\sqrt{\alpha_2^2-4\alpha_1\alpha_3}}{4\alpha_3},\quad d=\frac{\sqrt{4\alpha_1\alpha_3-\alpha_2^2}}{2\sqrt{2}\sqrt{\alpha_3}\sqrt{\eta}},\quad b_1=-\frac{a_1\alpha_2}{2\alpha_1},\quad a_0=0\\
&b_2=0,\quad b_3=\mp\frac{a_1\sqrt{\alpha_2^2-4\alpha_1\alpha_3}}{2\alpha_1},\quad a_2=0,\quad a_3=0.
\end{aligned}\tag{12}$$

SET-VI:

$$\begin{aligned}
&c=\pm\frac{\alpha_2\sqrt{\alpha_2^2-4\alpha_1\alpha_3}}{4\alpha_3},\quad d=\frac{\sqrt{4\alpha_1\alpha_3-\alpha_2^2}}{2\sqrt{2}\sqrt{\alpha_3}\sqrt{\eta}},\quad b_1=-\frac{a_1\alpha_2}{2\alpha_1},\quad a_0=0\\
&b_2=\mp\frac{a_1\sqrt{\alpha_2^2-4\alpha_1\alpha_3}}{2\alpha_1},\quad b_3=0,\quad a_2=0,\quad a_3=0.
\end{aligned}\tag{13}$$

Corresponding to each set of above solutions, the exact solutions for the FitzHugh–Nagumo equation are given below respectively.

$$v_{1,2}(x,t)=-\frac{\text{sech}^2(ct+dx)\left(\alpha_2^2+4\alpha_1\alpha_3\sinh^2(ct+dx)\right)}{2\alpha_3\left(\alpha_2\pm\sqrt{\alpha_2^2-4\alpha_1\alpha_3}\tanh(ct+dx)\right)},\tag{14}$$

$$v_{3,4}(x,t)=\frac{\alpha_2^2\text{csch}^2(ct+dx)-4\alpha_1\alpha_3\coth^2(ct+dx)}{2\alpha_3\left(\alpha_2\pm\sqrt{\alpha_2^2-4\alpha_1\alpha_3}\coth(ct+dx)\right)},\tag{15}$$

$$v_{5,6}(x,t) = \frac{\operatorname{csch}(ct+dx)\operatorname{sech}(ct+dx)\left(\alpha_2^2 - 4\alpha_1\alpha_3\cosh^2(2(ct+dx))\right)}{4\alpha_3\left(\alpha_2\sinh(2(ct+dx)) \pm \sqrt{\alpha_2^2 - 4\alpha_1\alpha_3}\cosh(2(ct+dx))\right)}, \quad (16)$$

$$v_{7,8}(x,t) = -\frac{2\alpha_1\sinh(2(ct+dx))}{\alpha_2\sinh(2(ct+dx)) \pm \sqrt{\alpha_2^2 - 4\alpha_1\alpha_3}\cosh(2(ct+dx))}, \quad (17)$$

$$v_{9,10}(x,t) = -\frac{2\alpha_1}{\alpha_2 \pm \sqrt{\alpha_2^2 - 4\alpha_1\alpha_3}\coth(ct+dx)}, \quad (18)$$

$$v_{11,12}(x,t) = -\frac{2\alpha_1}{\alpha_2 \pm \sqrt{\alpha_2^2 - 4\alpha_1\alpha_3}\tanh(ct+dx)}, \quad (19)$$

where the values of c and d are given by Eqs. (8)–(13). Graphical representation of the first solution $u_1(x,t)$ is given in Fig. 1.

Now, we will derive second family of solutions by letting $g_1(\xi) = \coth(\xi)$, $g_2(\xi) = 0$, $h_1(\xi) = \coth^2(\xi)$ and $h_2(\xi) = 0$. Then Eq. (6) becomes

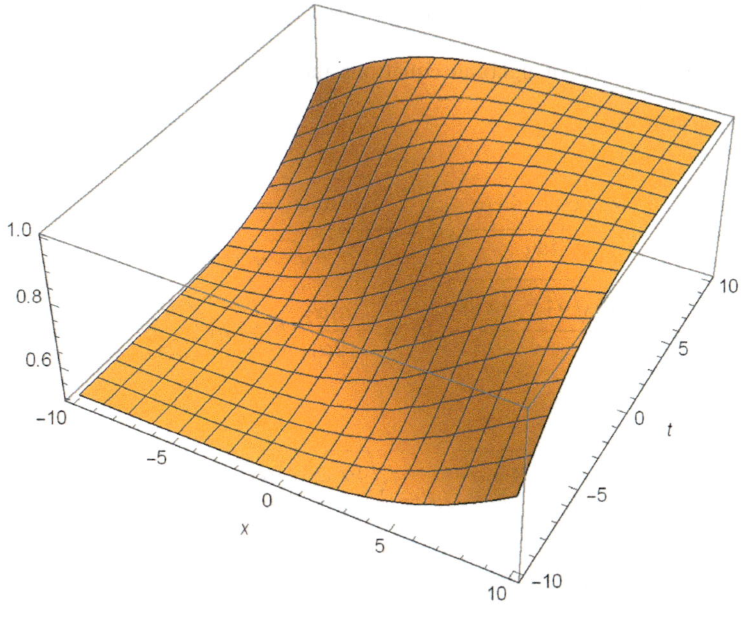

Fig. 1 Graphical representation of the exact solution $v_1(x,t)$ given by Eq. (14) of the FitzHugh–Nagumo equation, where $\eta = 1, \alpha_1 = -\frac{1}{2}, \alpha_2 = \frac{3}{2}, \alpha_3 = -1$

$$v(\xi) = a_0 + \frac{a_1 + a_2 \coth(\xi)}{b_1 + b_2 \coth^2(\xi)}. \tag{20}$$

Substituting this value of $v(\xi)$ in Eq. (5), we get another system of nonlinear algebraic equations given by

$$\begin{aligned}
&a_2 b_1^2 c = \alpha_3 a_1^3 + a_1^2 b_1 (3a_0\alpha_3 + \alpha_2) + a_0 b_1^3 (a_0 (a_0\alpha_3 + \alpha_2) + \alpha_1) \\
&+ a_1 b_1 \left(b_1 \left(3\alpha_3 a_0^2 + 2\alpha_2 a_0 + \alpha_1\right) - 2b_2 d^2\eta\right), \\
&2a_1 (b_1 - b_2) b_2 c = \alpha_3 a_2^3 + 2a_2 \left(b_2 b_1 \left(2a_0\alpha_2 + 3a_0^2\alpha_3 + \alpha_1 + 6d^2\eta\right)\right. \\
&\left.+ b_2 \left(a_1 (3a_0\alpha_3 + \alpha_2) + b_2 d^2\eta\right) + b_1^2 d^2\eta\right), \\
&b_2 \left(a_0 b_2 (a_0 (a_0\alpha_3 + \alpha_2) + \alpha_1) - a_2 c + 2a_1 d^2\eta\right) = 0, \\
&b_2 \left(2a_1 b_2 c + a_2 \left(b_2 \left(-2a_0\alpha_2 - 3a_0^2\alpha_3 - \alpha_1 + 2d^2\eta\right) + 6b_1 d^2\eta\right)\right) = 0, \\
&b_2 \left(a_2^2 (3a_0\alpha_3 + \alpha_2) + 3a_0 b_1 b_2 (a_0 (a_0\alpha_3 + \alpha_2) + \alpha_1)\right. \\
&\left.+ a_2 b_2 c + a_1 \left(b_2 \left(2a_0\alpha_2 + 3a_0^2\alpha_3 + \alpha_1 - 8d^2\eta\right) - 6b_1 d^2\eta\right)\right) = 0, \\
&2a_1 b_1 b_2 c + a_2 \left(3\alpha_3 a_1^2 + b_1^2 \left(2a_0\alpha_2 + 3a_0^2\alpha_3 + \alpha_1 - 2d^2\eta\right)\right. \\
&\left.+ b_1 \left(2a_1 (3a_0\alpha_3 + \alpha_2) - 6b_2 d^2\eta\right)\right) a_2^2 (3a_1\alpha_3 + b_1 (3a_0\alpha_3 + \alpha_2)) \\
&+ a_2 b_1^2 c + b_2 \left(a_1^2 (3a_0\alpha_3 + \alpha_2) + 3a_0 b_1^2 (a_0 (a_0\alpha_3 + \alpha_2) + \alpha_1)\right. \\
&\left.+ 2a_1 \left(b_1 \left(2a_0\alpha_2 + 3a_0^2\alpha_3 + \alpha_1 + 4d^2\eta\right) + 3b_2 d^2\eta\right)\right) = 0.
\end{aligned} \tag{21}$$

Solving this system of algebraic equations, the following sets of different solutions are obtained.
SET-I:

$$\begin{aligned}
c &= \pm\frac{\alpha_2\sqrt{\alpha_2^2 - 4\alpha_1\alpha_3}}{4\alpha_3}, d = -\frac{\sqrt{4\alpha_1\alpha_3 - \alpha_2^2}}{2\sqrt{2}\sqrt{\alpha_3}\sqrt{\eta}}, \\
a_2 &= \mp\frac{a_1\sqrt{\alpha_2^2 - 4\alpha_1\alpha_3}}{\alpha_2}, b_1 = -\frac{a_1\alpha_2}{2\alpha_1}, b_2 = a_1\left(\frac{\alpha_2}{2\alpha_1} - \frac{2\alpha_3}{\alpha_2}\right).
\end{aligned} \tag{22}$$

SET-II:

$$\begin{aligned}
c &= \pm\frac{\alpha_2\sqrt{\alpha_2^2 - 4\alpha_1\alpha_3}}{4\alpha_3}, d = \frac{\sqrt{4\alpha_1\alpha_3 - \alpha_2^2}}{2\sqrt{2}\sqrt{\alpha_3}\sqrt{\eta}}, \\
a_2 &= \mp\frac{a_1\sqrt{\alpha_2^2 - 4\alpha_1\alpha_3}}{\alpha_2}, b_1 = -\frac{a_1\alpha_2}{2\alpha_1}, b_2 = a_1\left(\frac{\alpha_2}{2\alpha_1} - \frac{2\alpha_3}{\alpha_2}\right).
\end{aligned} \tag{23}$$

Corresponding to the above sets of solutions, the only exact solutions for the FitzHugh–Nagumo equation are given by .

$$v_{13,14}(x,t) = -\frac{2\alpha_1\left(\alpha_2 \pm \sqrt{\alpha_2^2 - 4\alpha_1\alpha_3}\coth(ct+dx)\right)}{4\alpha_1\alpha_3\coth^2(ct+dx) - \alpha_2^2\text{csch}^2(ct+dx)}. \tag{24}$$

where the values of c and d are given by Eqs. (22) and (23).

Finally, we derive a third family of solutions by letting $g_1(\xi) = \tanh(\xi)$, $g_2(\xi) = 0$, $h_1(\xi) = \text{sech}^2(\xi)$ and $h_2(\xi) = 0$. Then Eq. (6) becomes

$$v(\xi) = a_0 + \frac{a_1 + a_2\tanh(\xi)}{b_1 + b_2\text{sech}^2(\xi)}. \tag{25}$$

Substituting this value of $v(\xi)$ in the Eq. (5), we get a system of nonlinear algebraic equations given by

$$\begin{aligned}
&a_2(b_1+b_2)^2c = \alpha_3a_1^3 + a_1^2(b_1+b_2)(3a_0\alpha_3+\alpha_2) + a_0(b_1+b_2)^3(a_0(a_0\alpha_3+\alpha_2)+\alpha_1)\\
&+a_1(b_1+b_2)\left(b_1(a_0(3a_0\alpha_3+2\alpha_2)+\alpha_1)+b_2\left(2a_0\alpha_2+3a_0^2\alpha_3+\alpha_1+2d^2\eta\right)\right),\\
&b_2\left(a_0b_2(a_0(a_0\alpha_3+\alpha_2)+\alpha_1)+a_2c-2a_1d^2\eta\right)=0,\\
&b_2\left(a_2^2(-(3a_0\alpha_3+\alpha_2))+3a_0b_2(b_1+b_2)(a_0(a_0\alpha_3+\alpha_2)\right.\\
&\left.+\alpha_1)+a_2b_2c+a_1\left(b_2\left(2a_0\alpha_2+3a_0^2\alpha_3+\alpha_1-2d^2\eta\right)+6b_1d^2\eta\right)\right)=0,\\
&b_2\left(a_1^2(3a_0\alpha_3+\alpha_2)+3a_0(b_1+b_2)^2(a_0(a_0\alpha_3+\alpha_2)+\alpha_1)\right.\\
&\left.+2a_1\left(b_2\left(2a_0\alpha_2+3a_0^2\alpha_3+\alpha_1+d^2\eta\right)+b_1\left(2a_0\alpha_2+3a_0^2\alpha_3+\alpha_1+4d^2\eta\right)\right)\right)\\
&=a_2\left(a_2(3a_1\alpha_3+b_1(3a_0\alpha_3+\alpha_2)+b_2(3a_0\alpha_3+\alpha_2))+(b_1+b_2)^2c\right),
\end{aligned} \tag{26}$$

$$\begin{aligned}
&b_2\left(a_2\left(b_2\left(2a_0\alpha_2+3a_0^2\alpha_3+\alpha_1+4d^2\eta\right)+6b_1d^2\eta\right)-2a_1b_2c\right)=0,\\
&a_2\left(b_2b_1\left(2a_0\alpha_2+3a_0^2\alpha_3+\alpha_1+4d^2\eta\right)+b_2(a_1(3a_0\alpha_3+\alpha_2)\right.\\
&\left.+b_2\left(2a_0\alpha_2+3a_0^2\alpha_3+\alpha_1+4d^2\eta\right))+b_1^2\left(-d^2\right)\eta\right)=\alpha_3a_2^3+2ca_1b_2(b_1+2b_2),\\
&2a_1b_2(b_1+b_2)c+a_2\left(-3\alpha_3a_1^2-2a_1b_2(3a_0\alpha_3+\alpha_2)\right.\\
&-b_2^2\left(2a_0\alpha_2+3a_0^2\alpha_3+\alpha_1+4d^2\eta\right)+b_1^2\left(-2a_0\alpha_2-3a_0^2\alpha_3-\alpha_1+2d^2\eta\right)\\
&\left.-2b_1\left(a_1(3a_0\alpha_3+\alpha_2)+b_2\left(2a_0\alpha_2+3a_0^2\alpha_3+\alpha_1+d^2\eta\right)\right)\right)=0.
\end{aligned} \tag{27}$$

Solving this system of algebraic equations, the following sets of different solutions are obtained.

SET-I:

$$\begin{aligned}
&c = \pm\frac{\alpha_2\sqrt{\alpha_2^2-4\alpha_1\alpha_3}}{4\alpha_3},\ d = -\frac{\sqrt{4\alpha_1\alpha_3-\alpha_2^2}}{2\sqrt{2}\sqrt{\alpha_3}\sqrt{\eta}},\\
&a_1 = -\frac{\alpha_2b_1}{2\alpha_3},\ a_2 = \mp\frac{\sqrt{\alpha_2^2-4\alpha_1\alpha_3b_1}}{2\alpha_3},\ b_2 = \frac{1}{4}\left(\frac{\alpha_2^2}{\alpha_1\alpha_3}-4\right)b_1.
\end{aligned} \tag{28}$$

SET-II:

$$c = \pm - \frac{\alpha_2\sqrt{\alpha_2^2 - 4\alpha_1\alpha_3}}{4\alpha_3}, d = \frac{\sqrt{4\alpha_1\alpha_3 - \alpha_2^2}}{2\sqrt{2}\sqrt{\alpha_3}\sqrt{\eta}},$$
$$a_1 = -\frac{\alpha_2 b_1}{2\alpha_3}, a_2 = \mp\frac{\sqrt{\alpha_2^2 - 4\alpha_1\alpha_3 b_1}}{2\alpha_3}, b_2 = \frac{1}{4}\left(\frac{\alpha_2^2}{\alpha_1\alpha_3} - 4\right) b_1. \tag{29}$$

Corresponding to the above sets of solutions, the only exact solutions for the FitzHugh–Nagumo equation are given by .

$$v_{15,16}(x,t) = -\frac{2\alpha_1 \cosh^2(ct+dx)\left(\alpha_2 \pm \sqrt{\alpha_2^2 - 4\alpha_1\alpha_3}\tanh(ct+dx)\right)}{4\alpha_1\alpha_3 \sinh^2(ct+dx) + \alpha_2^2}. \tag{30}$$

where the values of c and d are given by Eqs. (28) and (29). Graphical representation of the first solution $u_{15}(x,t)$ is given in Fig. 2.

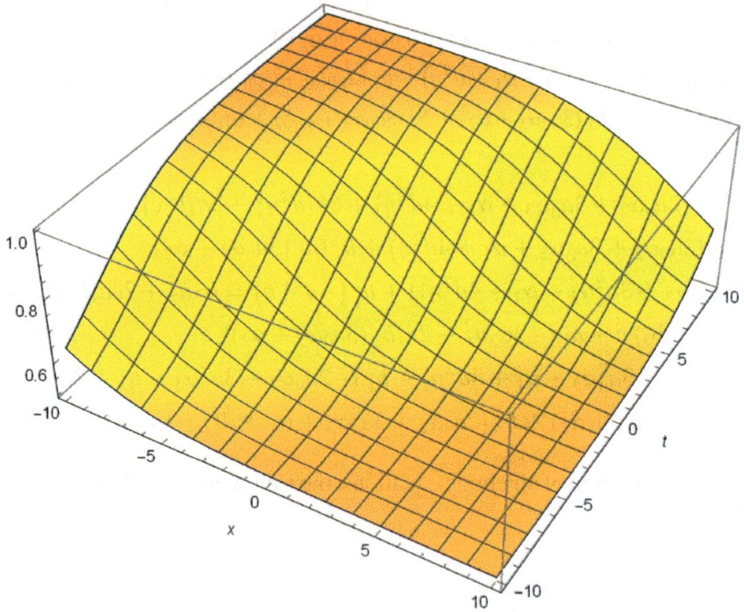

Fig. 2 Graphical representation of the exact solution $v_{15}(x,t)$ given by Eq. (30) of the FitzHugh–Nagumo equation, where $\eta = 1, \alpha_1 = -\frac{1}{2}, \alpha_2 = \frac{3}{2}, \alpha_3 = -1$

4 Discussion

Three families of new exact solutions to the nonlinear evolution Eq. (1) are derived in this paper. This equation represents several physical and biological processes such as transmission of nerve impulses and the evolutionary rescue from biological invasion. The number of existing exact solutions for this equation are very rare. But, exact solutions for such equations are required for a better understanding of the physical problem and for checking accuracy of any numerical algorithm developed for solving the real problems. Since it is difficult to derive the exact solutions using any analytical methods, we employ ansatz method to derive the exact solutions. But, when using this method, a large nonlinear system of equations are obtained corresponding to each ansatz form. It is needed to find out the nontrivial solutions that simultaneously satisfy this system of algebraic equations. This is can be done by using any of the computational algebra system. Other exact solutions for FitzHugh–Nagumo equation can also be derived by assuming similar ansatz forms for exact solutions.

Appendix

Substituting this value of $v(\xi)$ in Eq. (5), we get a system of nonlinear algebraic equations given by

$$\alpha_3 a_2^3 + 2a_2 b_2^2 d^2\eta = 0, \qquad \alpha_3 a_3^3 + 2a_3 b_3^2 d^2\eta = 0,$$

$$a_2 b_2\left(-a_2\left(3a_0\alpha_3+\alpha_2\right)+b_2 c+6b_1 d^2\eta\right)=0,$$

$$a_2\left(-3a_2\left(a_1+a_2-a_3\right)\alpha_3+b_1\left(3b_2 c-a_2\left(3a_0\alpha_3+\alpha_2\right)\right)\right.$$
$$\left.+b_2^2\left(2a_0\alpha_2+3a_0^2\alpha_3+\alpha_1-2d^2\eta\right)+6b_1^2 d^2\eta+6b_2 b_3 d^2\eta\right)=0,$$

$$a_2\left(-a_2 b_3\left(3a_0\alpha_3+\alpha_2\right)+2b_2\left(\left(a_1+a_2-a_3\right)\left(3a_0\alpha_3+\alpha_2\right)\right.\right.$$
$$\left.\left.+2b_3 c\right)+b_1\left(b_2\left(4a_0\alpha_2+6a_0^2\alpha_3+2\alpha_1-4d^2\eta\right)+16b_3 d^2\eta\right)+2b_1^2 c\right)$$
$$=b_2\left(a_0 b_2^2\left(a_0\left(a_0\alpha_3+\alpha_2\right)+\alpha_1\right)-a_1\left(b_2 c+2b_1 d^2\eta\right)+a_3\left(b_2 c+2b_1 d^2\eta\right)\right),$$

$$a_3 b_3\left(a_3\left(3a_0\alpha_3+\alpha_2\right)+b_3 c+6b_1 d^2\eta\right)=0,$$

$$3\alpha_3 a_2^3+a_2^2\left(3\left(2a_1-3a_3\right)\alpha_3+2b_1\left(3a_0\alpha_3+\alpha_2\right)\right)+a_2\left(b_1\left(2\left(a_1-a_3\right)\left(3a_0\alpha_3+\alpha_2\right)-2b_2 c+5b_3 c\right)\right.$$
$$+b_1^2\left(2a_0\alpha_2+3a_0^2\alpha_3+\alpha_1-8d^2\eta\right)+2b_2 b_3\left(2a_0\alpha_2+3a_0^2\alpha_3+\alpha_1+d^2\eta\right)+3\left(\alpha_3\left(a_1-a_3\right)^2+4b_3^2 d^2\eta\right)$$
$$+b_2^2\left(-2a_0\alpha_2-3a_0^2\alpha_3-\alpha_1+2d^2\eta\right)\Big)+b_2\left(-3a_0 b_1 b_2\left(a_0\left(a_0\alpha_3+\alpha_2\right)+\alpha_1\right)\right.$$
$$+a_3\left(b_2\left(2a_0\alpha_2+3a_0^2\alpha_3+\alpha_1-8d^2\eta\right)-b_1 c-6b_3 d^2\eta\right)$$
$$\left.+a_1\left(b_2\left(-2a_0\alpha_2-3a_0^2\alpha_3-\alpha_1+2d^2\eta\right)+b_1 c+6b_3 d^2\eta\right)\right)=0,$$

$$a_2^2\left(b_2-2b_3\right)\left(3a_0\alpha_3+\alpha_2\right)+a_2\left(2b_2\left(\left(a_1-2a_3\right)\left(3a_0\alpha_3+\alpha_2\right)+2b_3 c\right)\right.$$
$$+b_3\left(-2\left(a_1-a_3\right)\left(3a_0\alpha_3+\alpha_2\right)-3b_3 c\right)$$

$$
\begin{aligned}
&+2b_1\left(b_2\left(2a_0\alpha_2+3a_0^2\alpha_3+\alpha_1+d^2\eta\right)+b_3\left(-2a_0\alpha_2-3a_0^2\alpha_3-\alpha_1+11d^2\eta\right)\right)\\
&+2b_1^2c+b_2^2c\big)+b_2\left(a_1^2\left(3a_0\alpha_3+\alpha_2\right)+a_3^2\left(3a_0\alpha_3+\alpha_2\right)\right.\\
&+3a_0\left(b_1^2+b_2b_3\right)\left(a_0\left(a_0\alpha_3+\alpha_2\right)+\alpha_1\right)\\
&+a_1\left(-2a_3\left(3a_0\alpha_3+\alpha_2\right)+2b_1\left(2a_0\alpha_2+3a_0^2\alpha_3+\alpha_1+d^2\eta\right)+b_2c\right)\\
&-2a_3\left(b_1\left(2a_0\alpha_2+3a_0^2\alpha_3+\alpha_1-2d^2\eta\right)+2b_2c\right)\big)=0,
\end{aligned}
$$

$$
\begin{aligned}
&\alpha_3a_3^3+9a_2^2\alpha_3a_3+a_3\left(3\alpha_3a_1^2+b_1\left(2a_1\left(3a_0\alpha_3+\alpha_2\right)+6b_2c+b_3c\right)\right.\\
&+b_1^2\left(2a_0\alpha_2+3a_0^2\alpha_3+\alpha_1-2d^2\eta\right)+2b_2b_3\left(2a_0\alpha_2+3a_0^2\alpha_3+\alpha_1+7d^2\eta\right)\\
&+b_2^2\left(-2a_0\alpha_2-3a_0^2\alpha_3-\alpha_1+20d^2\eta\right)+2b_3^2d^2\eta\big)\\
&+a_2\left(-3\alpha_3a_1^2+12a_3\alpha_3a_1-9a_3^2\alpha_3+2a_0\alpha_2b_3^2+3a_0^2\alpha_3b_3^2\right.\\
&-b_1\left(2\left(a_1-2a_3\right)\left(3a_0\alpha_3+\alpha_2\right)+b_2c+6b_3c\right)-2b_2b_3\left(2a_0\alpha_2+3a_0^2\alpha_3+\alpha_1+7d^2\eta\right)\\
&+b_1^2\left(-2a_0\alpha_2-3a_0^2\alpha_3-\alpha_1+2d^2\eta\right)+\alpha_1b_3^2-2b_2^2d^2\eta-20b_3^2d^2\eta\big)\\
&=a_1^3\alpha_3+a_2^3\alpha_3+\alpha_3a_0^3b_1\left(b_1^2+6b_2b_3\right)+\alpha_2a_0^2b_1\left(b_1^2+6b_2b_3\right)\\
&+a_0b_1\left(3\left(a_2^2+a_3^2\right)\alpha_3+\alpha_1b_1^2+6\alpha_1b_2b_3\right)+a_2^2\alpha_2b_1+a_3^2\alpha_2b_1\\
&+a_1^2b_1\left(3a_0\alpha_3+\alpha_2\right)+a_1\left(3\alpha_3a_2^2+3a_3^2\alpha_3\right.\\
&+b_1^2\left(a_0\left(3a_0\alpha_3+2\alpha_2\right)+\alpha_1\right)+2b_2b_3\left(2a_0\alpha_2+3a_0^2\alpha_3+\alpha_1+6d^2\eta\right)\\
&+b_1\left(b_2+b_3\right)c+2b_2^2d^2\eta+2b_3^2d^2\eta\big),
\end{aligned}
$$

$$
\begin{aligned}
&a_3\left(3a_3^2\alpha_3-3a_1a_3\alpha_3-3a_2a_3\alpha_3-2a_0\alpha_2b_3^2-3a_0^2\alpha_3b_3^2\right.\\
&-b_1\left(a_3\left(3a_0\alpha_3+\alpha_2\right)+3b_3c\right)-\alpha_1b_3^2-6b_1^2d^2\eta+2b_3^2d^2\eta-6b_2b_3d^2\eta\big)=0,
\end{aligned}
$$

$$
\begin{aligned}
&3\alpha_3a_3^3+a_3\left(3\alpha_3a_1^2+6a_2\alpha_3a_1+3a_2^2\alpha_3-2a_0\alpha_2b_3^2-3a_0^2\alpha_3b_3^2\right.\\
&+b_1\left(2\left(a_1+a_2\right)\left(3a_0\alpha_3+\alpha_2\right)+5b_2c-2b_3c\right)
\end{aligned}
$$

$$
\begin{aligned}
&+b_1^2\left(2a_0\alpha_2+3a_0^2\alpha_3+\alpha_1-8d^2\eta\right)\\
&+2b_2b_3\left(2a_0\alpha_2+3a_0^2\alpha_3+\alpha_1+d^2\eta\right)-\alpha_1b_3^2+12b_2^2d^2\eta+2b_3^2d^2\eta\big)\\
&+b_3\left(3a_0b_1b_3\left(a_0\left(a_0\alpha_3+\alpha_2\right)+\alpha_1\right)+a_2\left(b_3\left(2a_0\alpha_2+3a_0^2\alpha_3\right.\right.\right.\\
&\left.+\alpha_1-8d^2\eta\right)-b_1c-6b_2d^2\eta\big)+a_1\left(b_3\left(2a_0\alpha_2+3a_0^2\alpha_3+\alpha_1-2d^2\eta\right)\right.\\
&-b_1c-6b_2d^2\eta\big)\big)=a_3^2\left(3\left(2a_1+3a_2\right)\alpha_3+2b_1\left(3a_0\alpha_3+\alpha_2\right)\right),
\end{aligned}
$$

$$
\begin{aligned}
&a_3\left(b_3\left(\left(2a_1+4a_2-a_3\right)\left(3a_0\alpha_3+\alpha_2\right)+b_3c\right)+2b_2\left(2b_3c-\left(a_1+a_2-a_3\right)\right.\right.\\
&\left.\left(3a_0\alpha_3+\alpha_2\right)\right)+2b_1\left(b_3\left(2a_0\alpha_2+3a_0^2\alpha_3+\alpha_1+d^2\eta\right)\right.\\
&+b_2\left(-2a_0\alpha_2-3a_0^2\alpha_3-\alpha_1+11d^2\eta\right)\big)+2b_1^2c-3b_2^2c\big)\\
&=b_3\left(a_1^2\left(3a_0\alpha_3+\alpha_2\right)+a_2^2\left(3a_0\alpha_3+\alpha_2\right)\right.\\
&+3a_0\left(b_1^2+b_2b_3\right)\left(a_0\left(a_0\alpha_3+\alpha_2\right)+\alpha_1\right)\\
&+a_1\left(2a_2\left(3a_0\alpha_3+\alpha_2\right)+2b_1\left(2a_0\alpha_2+3a_0^2\alpha_3+\alpha_1+d^2\eta\right)+b_3c\right)\\
&+a_2\left(b_1\left(4a_0\alpha_2+6a_0^2\alpha_3+2\alpha_1-4d^2\eta\right)+4b_3c\right)\big),
\end{aligned}
$$

$$
\begin{aligned}
&b_3\left(-a_0b_3^2\left(a_0\left(a_0\alpha_3+\alpha_2\right)+\alpha_1\right)+a_1\left(b_3c+2b_1d^2\eta\right)+a_2\left(b_3c+2b_1d^2\eta\right)\right)\\
&=a_3\left(2\left(a_1+a_2-a_3\right)b_3\left(3a_0\alpha_3+\alpha_2\right)+b_2\left(a_3\left(3a_0\alpha_3+\alpha_2\right)+4b_3c\right)\right.\\
&+2b_1\left(b_3\left(2a_0\alpha_2+3a_0^2\alpha_3+\alpha_1-2d^2\eta\right)+8b_2d^2\eta\right)+2b_1^2c\big).
\end{aligned}
$$

References

1. Banasiak, J., Mokhtar-Kharroubi, M. (Eds), Evolutionary Equations with Applications in Natural Sciences. Springer (2014)
2. Bell, G.: Evolutionary rescue. Annu. Rev. Ecolo. Evolut. Syst. **48**, 605–27 (2017)
3. Carlson, S.M., Cunningham, C.J., Westley, P.A.H.: Evolutionary rescue in a changing world. Trends Ecolo. Evolut. **29**, 521–530 (2014)
4. FitzHugh, R.: Mathematical Models of Excitation and Propagation in Nerve, Biological Engineering. In: Schwann H. (ed.), pp. 1–85. McGraw-Hill, New York (1969)
5. Foroutan, M., Manafian, J., Taghipour-Farshi, H.: Exact solutions for Fitzhugh-Nagumo model of nerve excitation via Kudryashov method. Opt Quant. Electron. **49**, 352 (2017)
6. Gawlik, A., Vladimirov, V., Skurativskyi, S.: Solitary wave dynamics governed by the modified fitzhugh–nagumo equation. J. Comput. Nonlinear Dynam. **15**(6), 061003 (2020)
7. Gawlik, A., Vladimirov, V., Skurativskyi, S.: Existence of the solitary wave solutions supported by the modified FitzHugh-Nagumo system. Nonlinear Anal. Model. Control **25**(3), 482–501 (2020)
8. Gonzalez, A., Ronce, O., Ferriere, R., Hochberg, M.E.: Evolutionary rescue: an emerging focus at the intersection between ecology and evolution. Philos. Trans. Royal Soc. B, 20120404 (2013)
9. Hafez, M.G.: New travelling wave solutions of the (1+1)-dimensional cubic nonlinear Scrodinger equation using novel (G/G)-expansion method. Beni-Suef Univ. J. Basic Appl. Sci. **5**(2), 5109–118 (2016)
10. Khan, Y.: A variational approach for novel solitary solutions of FitzHugh-Nagumo equation arising in the nonlinear reaction-diffusion equation. Int. J. Numer. Methods Heat Fluid Flow (2020). https://doi.org/10.1108/HFF-05-2020-0299
11. McKean, H., Jr.: Nagumo's equation. Adv. Math. **4**(3), 209–223 (1970)
12. Nagumo, J., Arimoto, S., Yoshizawa, S.: An active impulse transmission line simulating Nerve Axon. Proc. IRE **50**(10), 2061–2070 (1962)
13. Nucci, M.C., Clarkson, P.A.: The nonclassical method is more general than the direct method for symmetry reductions: an example of the Fitzhugh-Nagumo equation. Phys. Lett. A **164**(1), 49–56 (1992)
14. Orr, H.A., Unckless, R.L.: The population genetics of evolutionary rescue. PLoS Genetics **10**, e1004551 (2014)
15. Samani, P., Bell, G.: The ghosts of selection past reduces the probability of plastic rescue but increases the likelihood of evolutionary rescue to novel stressors in experimental populations of wild yeast. Ecol. Lett. **19**, 289–298, e1004551 (2016)
16. Liu, S., Fu, Z., Liu, S., Zhao, Q.: Jacobi elliptic function expansion method and periodic wave solutions of nonlinear wave equations **289**(1–2), 69–749 (2001)
17. Shen, Y., He, J.-H.: Variational principle for a generalized KdV equation in a fractal space. Fractals **28**(4), 2050069 (2020). https://doi.org/10.1142/S0218348X20500693
18. Tanaka, H., Stone, A., Nelson, D.R.: Spatial gene drives and pushed genetic waves, PNAS **114** (32) 8452–8457 (2017)
19. Van Dyken, J.D.: Evolutionary rescue from a wave of biological invasion. Amer. Naturalist **195**(1) (2020) https://doi.org/10.1086/706181
20. Wazwaz, A.M.: A sine-cosine method for handling nonlinear wave equations. Math. Comput. Model **40**, 499–508 (2004)
21. Zayed, E.M.E., Abdelaziz, M.A.M.: Exact solutions of the nonlinear Schrodinger equation with variable coefficients using the generalized extended tanh-function method, the sin-cosine and the exp-function methods. Appl. Math. Comput. **218**, 2259–2268, e1004551 (2011)

A Multi-criteria Model of Selection of Students for Project Work Based on the Analysis of Their Performance

Sukarna Dey Mondal, Dipendra Nath Ghosh, and Pabitra Kumar Dey

Abstract Project work does not always imply extensive knowledge, but it does imply the application of such information. Through this project work, students are exposed to educational ideas as well as technical ideas. Therefore, the selection and evaluation of students are crucial parts of any project for any education organization concerning excellence. So, an attempt has been made to draw a mathematical model with the help of several MCDM techniques and Statistics from which it will be very easy to evaluate and select a suitable student for the project. First of all, a payoff matrix has been created with the AHP method. Entropy is used to calculate the total weight. Then utility based, distance based, and out-ranking based MCDM techniques are applied to get several ranking structures. Ultimately, through a voting method, the study offers a ranking of 5 students under student excellence.

Keywords AHP · Entropy · TOPSIS · VIKOR · COPRAS · PROMETHEE-2 · WSM · Voting system

1 Introduction

In recent times, project work is gradually becoming compulsory in schools and colleges. A properly decorated project can carry extra marks in an interview or an exam. Project work is essential to escalating the volume of conception. Thus, command of the project matter, technical skill, ability to communicate, discipline & behavior, experience, leadership and managing power, and stress tolerance are

S. Dey Mondal (✉)
Department of Mathematics, Dr. B.C. Roy Engineering College, MAKAUT, Kolkata, West Bengal, India
e-mail: sukarnadey@gmail.com

D. Nath Ghosh
Controller of Examinations, Kazi Nazrul University, Asansol, West Bengal, India

P. Kumar Dey
Department of Computer Applications, Dr. B.C. Roy Engineering College, MAKAUT, Durgapur, West Bengal, India

S. Banerjee and A. Saha (eds.), *Nonlinear Dynamics and Applications*,
Springer Proceedings in Complexity,
https://doi.org/10.1007/978-3-030-99792-2_73

important requirements towards performance analysis of students. MCDM approach delivers upgraded knowledge to scrutinize the potential of a student in an education organization for ranking a project. Project work means that not only good students are eligible, but also all types of students have to do project work. In many cases due to their lack of technical intelligence, very good students did not get a place in that project. So, finding a skilled student for a project is an important part of the research. A mathematical model has been created by looking into everything from which it will be very easy to rank all these skilled students.

A scientific model related to students' project work has already been created where AHP, PROMETHEE-2, and TOPSIS were used. With the previous model in mind, another new scientific model has been developed and several MCDM approaches like Entropy, VIKOR, COPRAS, and WSM have been applied. Whereas early model only one ranking was found as two MCDM methods were used but now there are several rankings for using several MCDM methods. For this model, the opinions of several project guides have been taken to the performance of the students concerning some pre-assigned criteria. Ultimately, an experiment has been made to explicate in particular how the voting system has been transported from several ranking structures to the single ranking structure to take the final decision.

2 Literature Review

The pair-wise comparison method and the hierarchical model were developed in 1980 by Saaty in the context of the Analytical Hierarchy Process (AHP) [1, 2]. A pis and nis based method (TOPSIS) is sketched [3]. To inspect the quality of performance assessment, a study has been developed on performance management [4]. "The Effects of the Performance Evaluation Process on Academic employees in Higher Education Institutions" is an important research work in today's scenario with the help of the performance estimation [5]. Research work has been carried out with the help of different multi-criteria-decision-making like AHP, Fuzzy-AHP, COPRAS, TOPSIS, Cooperative Game Theory, Compromise Programming, and Group Decision to analyse the performance of a teacher [6]. To examine the performance of supporting staff, a mathematical model is designed using multi-criteria decision making by eight methodologies containing AHP, COPRAS, SAW, TOPSIS, Fuzzy-TOPSIS, PROMETHEE-II, Compromise Programming, Normalized-Weighted-Average and Group Decision-Making method [7]. Another research work is done on how Employee Management can be effectively managed in the future by using performance management, performance appraisal [8]. In recent situations, educational organizations have grown day by day. However, the quality and effective performance among them has not increased proportionally. Already an innovative statistical model has been initiated to explore the NAAC rating of a well-known Engineering College using Multi-Criteria Decision-Making Methods, Statistics, and Group Decision Making [9]. Another research work has been carried out to select the best engineering college by using AHP, TOPSIS, and Fuzzy AHP [10]. College placement is a procedure in which professionals assess each candidate's performance against a set of pre-determined criteria. So, using TOPSIS and AHP in an Interval Types 1 and 2 fuzzy environment (IT1F, IT2F), a mathematical model has been developed to analyse specific criteria [11]. A preference function approach (PROMETHEE-2) is introduced [12]. It is one of the versions, out of 5 of PROMETHEE, i.e., 1–5 [13, 14]. The scholars of V.G.T. University, (Zavadskas et al. 1996) initiated the COPRAS method (Complex Proportional Assessment), applicable for max. and min. criteria values. In 2007, the substitute MCDM method VIKOR was approached by Opricovic et al. [15]. Another method used as a benchmark solution in the situations (Entropy Method) approximates the weights of the criteria. Hwang and Yoon [16] the Entropy Method simplifies disparities between sets of data. Weighted Average Method is a software-kind MCDM approach. Here objective functions are transformed by defining as weighed sums of various objects [17, 18].

3 Proposed Methodology

In the beginning, a literature inspection was conducted. Five different categories of students were observed and arbitrarily nominated for the current study w.r.to 5 scale rating.

It is considered as

Two students are below average, one is normal, and two are above average.

The sample of five different categories must be preserved separately, according to NBA (National Board of Accreditation) or NAAC (National Assessment and Accreditation Council) rules. In the same way, 5 students have been given importance in this matter.

An additional survey was arranged between Experts (E1, E2), for adding/removing criteria. They furnished their valuable judgment w.r.to the following criteria (Fig. 1 and Table 1).

C1—command of the project matter
C2—technical skill
C3—ability to communicate clearly
C4—discipline & behavior
C5—past experience
C6—leadership and managing power
C7—stress tolerance

3.1 Proposed Flowchart

See Fig. 2.

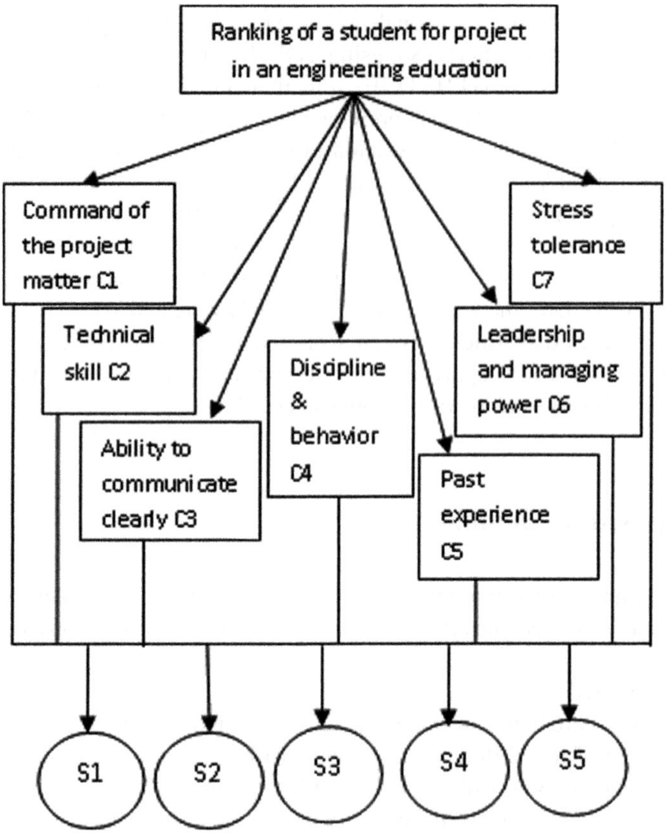

Fig. 1 Hierarchy of ranking students based on the below mentioned criteria

Table 1 Experts' opinion against each student with respect to each criteria

Students/Experts	Criteria													
	C1		C2		C3		C4		C5		C6		C7	
	E1	E2	E1	E2	E1	E2	E1	E2	E1	E2	E1	E2	E1	E2
S1	VG	E	E	VG	E	G	G	G	E	G	B	G	G	G
S2	G	G	G	G	B	A	VG	E	E	G	VG	G	VG	A
S3	E	E	VG	G	E	E	VG	VG	G	E	B	G	B	G
S4	G	VG	B	G	E	VG	G	B	VG	VG	VG	G	G	VG
S5	VG	VG	G	E	G	A	G	VG	VG	VG	G	B	A	B

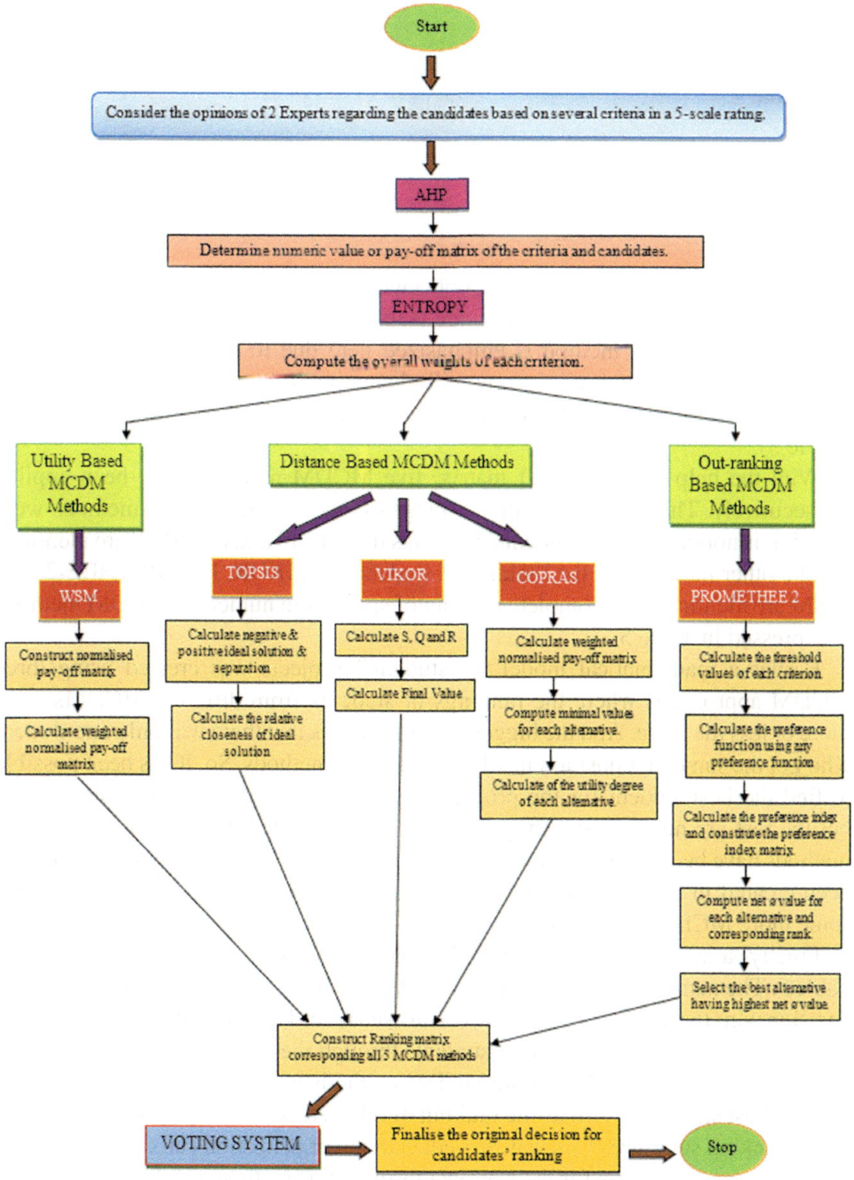

Fig. 2 Proposed flowchart for selection of students for project work

3.2 Proposed Algorithm

See Table 2.

4 Analysis and Discussion

After using the opinions which are collected from the experts, prepare a pay-off matrix and, using the Entropy technique (see Table 4), determine the weight of the criteria. The Entropy method is emphasized here due to the important aspect of entropy, where the result of entropy will change with the change of data. For this characteristic, Entropy is different from the rest of the methods for estimating weights (Table 3).

With the help of the pay-off matrix, five MCDM methods have been applied consecutively. Then from there the rankings of students in different methods were found. It is noticed that the students' rankings in TOPSIS and COPRAS are identical. On the other hand, again the ranking of students in VIKOR, PROMETHEE-2, and WSM are identical. The complete ranking of students in numerous MCDM methods is expressed in Table 5.

Earlier a mathematical model for students' projects was created using three MCDM approaches where the rankings of students from different MCDMs were more or less the same. And the biggest advantage was that there was only one student who got the same first position in all three MCDM methods. So, it has been possible to find the best student for the project very easily from that mathematical model but the rest are not so easy. This work has been extended where frequent MCDM methods have been applied. Along with this model, another mathematical model has been created in this current study whereas previous work has been extended with some more MCDM approaches.

Finally, a system has been applied to the rankings of students obtained from Table 5. The name of the system is Voting System. It is a majority system. "Majority decision wins the match". This fact has been used here. Therefore, it has been possible to get a single ranking structure of students through the voting system based on the majority ranking obtained from the MCDM methods from Table 6. It is very easy to find the best student as well as the ranking of other students for the project from this single ranking structure. This way any educational organization can effortlessly find the best team of students for their project through this mathematical model. In such a manner, the validity of the voting system has been maintained here (Table 6).

Table 2 Proposed algorithm for selection of students for project work

Step 1	Consider the opinions of 2 Experts regarding the candidates
Step 2	Apply pairwise comparison method AHP to find the numeric value or pay-off matrix of the criteria and candidates
Step 3	Using the Entropy approach, measure the weight of each condition
MCDM Approaches	
Utility/Priority-Based MCDM Methods	**WSM**
Step 4.1.1	Normalize the pay-off matrix obtained from AHP
Step 4.1.2	Calculate weighted normalized matrix
Step 4.1.3	Rank the candidates in descending order
Distance-Based MCDM Methods	**TOPSIS**
Step 5.1.1	Compute the pis and nis as for each criterion: $A^* = \{v_1^*, v_2^*, \ldots\ldots, v_n^*\}$ where v_n^* gives the maximum value of nth criteria $A^- = \{v_1^-, v_2^-, \ldots\ldots v_n^-\}$ where v_n^- gives the minimum value of nth criteria
Step 5.1.2	Calculate the distance of individual alternatives from PIS and NIS, as well as their relative proximity to the optimal answer, where there are J alternatives and n criteria $CC_i = \frac{d_i^-}{d_i^* + d_i^-}$ i = 1,2, 3…., J
Step 5.1.3	Rank the candidates in descending order
	VIKOR
Step 5.2.1	Calculate R, S, and Q $Q_i = \vartheta\left[\frac{S_i - S^*}{S^- - S^*}\right] + (1 - \vartheta)\left[\frac{R_i - R^*}{R^- - R^*}\right]$ where Q_i signifies the i-th VIKOR value, $i = 1, 2, 3 \ldots .m$; $S^* = Min(S_i)$; $S^- = Max(S_i)$; $R^* = Min(R_i)$; $R^- = Max(R_i)$ and ϑ is the utmost group utility's value (usually it is to be set to 0.5)
Step 5.2.2	The choice with the lowest VIKOR value is considered to be the right approach
Step 5.2.3	Rank the candidates in descending order
	COPRAS
Step 5.3.1	Calculate weighted normalized decision matrix

(continued)

Table 2 (continued)

Step 1	Consider the opinions of 2 Experts regarding the candidates
Step 5.3.2	Obtain the sums of the threshold values for each choice (bigger values are preferred) (optimization direction is maximization)
Step 5.3.3	Obtain the sums of the threshold values for each choice (smaller values are preferred) (optimization direction is minimization)
Step 5.3.4	Determine the minimal value
Step 5.3.5	Calculate the utility degree of each alternative
Step 5.3.6	Rank the candidates in descending order
Outranking-Based MCDM Methods	**PROMETHEE-2**
Step 6.1.1	Normalize all the values w.r.to any normalization method
Step 6.1.2	Calculate the threshold values of each criterion
Step 6.1.3	Calculate the preference function using any preference function; here Gaussian function is used as a preference function
Step 6.1.4	Construct the preferences factor matrices by estimating the preference score
Step 6.1.5	Analyze net ø value of each option and the associated rank
Step 6.1.6	Choose the best option that has the biggest ø value
Step 6.1.7	Rank the candidates in descending order
	VOTING SYSTEM
Step 7.1.1	Construct Ranking Matrix corresponding to all 5 MCDM methods
Step 7.1.2	Apply voting system to finalize the original decision

Table 3 Saaty's 9-point of pairwise comparison

Scale	Compare factor of i and j
1	Equally important
3	Weakly important
5	Strongly important
7	Very strongly important
9	Extremely important
2,4,6,8	Intermediate value between adjacent scales

Table 4 Weights of criteria by Entropy method

Criteria	Weights
C1	0.1438
C2	0.1407
C3	0.1477
C4	0.1442
C5	0.1386
C6	0.1426
C7	0.1425

Table 5 Ranking of students in different MCDM methods

Students	MCDM methods				
	TOPSIS	VIKOR	COPRAS	PROMETHEE-2	WSM
S1	2	3	2	3	3
S2	4	4	4	4	4
S3	1	1	1	1	1
S4	3	2	5	2	2
S5	5	5	3	5	5

Table 6 Final ranking of students based on the Voting System

Students	MCDM methods					Voting system
	TOPSIS	VIKOR	COPRAS	PROMETHEE-2	WSM	
S1	2	3	2	3	3	3
S2	4	4	4	4	4	4
S3	1	1	1	1	1	1
S4	3	2	5	2	2	2
S5	5	5	3	5	5	5

5 Conclusion

Student selection and evaluation for the project are significant parts of any educational organization. Improving the teaching–learning in school or college mostly depends on the student's project. Through this project, various educational organizations have often received different funds. Therefore, appropriate students are very much desirable to accomplish the projects. So, this study establishes an innovative MCDM model that amalgamates AHP, Entropy, TOPSIS, VIKOR, COPRAS, PROMRTHEE-2, and WSM to keep up with students' performance ranking decisions. Finally, an experiment was conducted to demonstrate how the voting system is transformed from many ranking structures to a single ranking structure to get the conclusion. So, with the help of this model not only the educational organization will benefit but will also work in the same way where performance is evaluated with judgment on different criteria.

References

1. Saaty, T.L.: The Analytic Hierarchy Process. McGraw-Hill, New York (1980)
2. Saaty, T.L.: Priority setting in complex problems. IEEE Trans. Eng. Manage. **30**(3), 140–155 (1983)
3. Hwang, C.L., Yoon, K.: Multiple Attribute Decision Making Methods and Applications. Springer, Berlin Heidelberg (1981)
4. Lohman, L.: Evaluation of university teaching as sound performance appraisal, Elsevier.https://doi.org/10.1016/j.stueduc.2021.101008
5. Dasanayaka, C.H., Abeykoon, C., Ranaweera, R.A.A.S., Koswatte, I.: The Impact of the Performance Appraisal Process on Job Satisfaction of the Academic Staff in Higher Educational Institutions. https://doi.org/10.3390/educsci11100623 (2021)
6. Dey, S., Ghosh, D.N.: An integrated approach of multi-criteria group decision making techniques to evaluate the overall performance of teachers. Int. J. Adv. Res. Comput. Sci. **7**(5) (2016)
7. Dey, S., Ghosh, D.N.: Non-teaching staff performance analysis using multi-criteria group decision making approach. Int. J. Educ. Learn. **4**(2), 35–50 (2015)
8. Brown, T.C., O'Kane, P., Mazumdar, B., McCracken. M.: Performance management: a scoping review of the literature and an agenda for future research **18**(1), 47–82 (2019)
9. Dey, S., Ghosh, D.N., Dey, P. K.: Prediction of NAAC grades for affiliated institute with the help of statistical multi criteria decision analysis: national conference on recent trends in IOT. Mach. Learn. Artif. Intell. Appl. **1**(2), 116–126 (2021)
10. Rana, S., Dey, P.K., Ghosh, D.N.: Best engineering college selection through fuzzy multi-criteria decision making approach: a case study: universal journal of applied computer science and technology **2**(2), 246–256 (2012)
11. Dey, S., Ghosh, D.N.: Comparative evaluation of students' performance in campus recruitment of a technical institution through Fuzzy-MCDM techniques. Int. J. Comput. Sci. Eng. **7**(Special issue), 1
12. Brans, J.P., Vincke, Ph., Mareschal, B.: How to select and how to rank projects: the Promethee method. Europ. J. Oper. Res. **24**, 228–238 (1986)
13. Taleb, M.F.A., Mareschal, B.: Water resources planning in the middle east: application of the Promethee v Multicriteria method. Eur. J. Oper. Res. **81**, 500–511 (1995)
14. Pomerol, J.Ch., Romero, S.B.: Multi-criterion Decision in Management: Principles and Practice, Kluwer Academic, Netherlands (2000)
15. Opricovic, S., Treng, G.H.: Compromise solution by MCDM methods: a comparative analysis of VIKOR and TOPSIS. Eur. J. Oper. Res. **156**, 445–455 (2004)
16. Hwang, C., Yoon, K.: Multiple Attribute Decision-Making. Springer-Verlag, Methods and Application. A State-of-the-Art Survey (1981)
17. Loucks, D.P., Stedinger, J.R., Haith, D.A.: Water Resources Systems Planning and Analysis. Prentice-Hall, Englewood Cliffs, New Jersey (1981)
18. Vedula, S., Mujumdar, P.P.: Water Resources Systems, Modelling Techniques, and Analysis. Tata McGraw-Hill Publishing Company Limited, New Delhi (2005)

Mathematical Modeling of Thermal Error Using Machine Learning

Rohit Ananthan and N. Rino Nelson

Abstract On many types of machine tools, thermal effects produce the most of machining defects, with linear expansion and deformation of structural parts creating undesired movement between the tool and the workpiece. Thermal flaws are difficult to control without some type of compensation because heat inputs that produce temperature rise and gradients occur from a variety of sources both within and outside the machine tool. Moreover, heat generation also cannot be prevented. As a result, the goal of this research is to focus on thermal error modelling and evaluate the various machine learning algorithms to discover the most effective solution.

Keywords CNC machine · Multiple regression · Thermal expansion · Linear regression

1 Introduction

Because of the expanding need in the contemporary industry, thermal compensation has been thoroughly explored. These complex machines generate a lot of heat, which causes the machine tools and the work piece to distort, which is the main source of inaccuracy [1]. According to Bryan's research published in 1990, the thermal error accounts for 40–70% of the total error [2]. In general, there are two kinds of heat sources in machine tools, namely internal and external heat sources, bringing about the temperature rising and thermal errors [3].

- Internal heat sources: heat generated from cutting process; heat generated from frictions in ball screws, spindle, gear box, guides, etc.; heat generated in motor; heating or cooling influences provided by the various cooling systems.
- External heat sources: environmental temperature variation; solar and personal radiations.

R. Ananthan · N. Rino Nelson (✉)
Mechanical Engineering, Indian Institute of Information Technology, Design and Manufacturing, Kancheepuram, Chennai 600127, India
e-mail: rino@iiitdm.ac.in

S. Banerjee and A. Saha (eds.), *Nonlinear Dynamics and Applications*,
Springer Proceedings in Complexity,
https://doi.org/10.1007/978-3-030-99792-2_74

As the core component in machine tool, the spindle would generate large amounts of heat when it is running at a high speed. Among the heat sources listed above, the spindle is considered as an important one [4].

In order to minimize the spindle thermal error, there are namely three methods:

1. Thermal error avoidance
2. Thermal error control
3. Thermal error compensation

Here we have taken up thermal error compensation as our topic of research. Compared with other two types of methods, thermal error compensation is more convenient and cost-efficient [5]. The most obvious method for reducing the implications of thermal error in machine tool is to compensate for the changes. The simplest and most widely used way is to record temperature and thermal displacement, which may then be used to create a model. The measured temperatures will be input to the developed mode for predicting the thermal displacement of the spindle which will be compensated using a controller. Error compensation approaches try to create a artificial error in order to compensate for the real one [6]. The foundation of accurate thermal error identification is advanced detection technology. A laser interferometer or other measuring instruments can easily measure the geometric error as well as thermally induced errors. Yang suggested a new spindle thermal error monitoring method based on a ball bar system rather than a capacitance sensor system.

Thermal errors on CNC machine tools can be well predicted by splitting the machine tool into its constituent elements and modelling just those portions exhibiting the highest thermal movement. They also demonstrated that machine tool structural elements may be accurately simulated using rectangle-based prisms. White shown that in order to predict the two-dimensional deformation of a machine tool structural element over a wide range of machine operating conditions, it was necessary to know both the magnitude and position of temperature gradients.

2 Creation of a Thermal Model

The created thermal model is used to compensate for thermal errors. Temperature and displacement sensors are measured from key spots on the machine tool, as well as other potential locations (Fig. 1).

Heat emanates from a variety of places in precision machine tools.

- The heat generated during the cutting process. The work piece, the chips, and the coolant [7] all exposed to the heat. The bulk of heat is taken away by chips (60–80%) and finally passed to the coolant.
- Heat was generated as a result of mechanical, electrical, and hydraulic losses [7]. Motion (spindle bearing) losses are used to indicate mechanical losses. The heat generated in motors and drives represents the electrical losses. Fluid dynamic

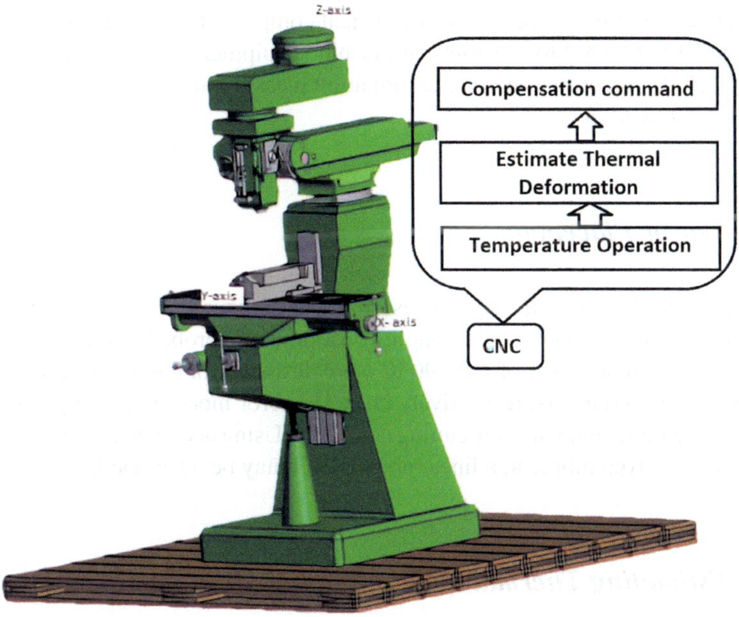

Fig. 1 Concept diagram of thermal compensation system

friction energy losses in the collet closer, hydraulic, and coolant pump constitute hydraulic losses [7].

- Changes in the ambient environment's thermal influence.

Thermal deformation of machine tool components can be caused by both internal and external heat sources, resulting in poor geometric precision in work parts after machining. Temperature and thermal deformation issues account for approximately 40–70% of total machining errors. As a result, developing solutions for reducing and eliminating thermal deformation faults is crucial and attracts a lot of attention in high-precision machining [8]. Temperature-sensitive points, also known as key temperature measurement sites on a machine tool, are locations where temperature changes have a strong relationship with thermal displacement.

3 Thermal Modelling

One of the most important tasks in successful thermal error modelling and correction is to identify the source of the error. The temperature of the primary locations on the machine, as well as other available factors, can be used to determine deformations or displacement changes at specific places of machine tools owing to thermal inaccuracy [9]. This approach makes use of temperature factors, which show a strong link between thermal deformation and critical components. This effective approach has now been adopted for a variety of machine equipment. Thermal error compensation includes recording thermal deformation measurement, create an error model and then compensate it.

3.1 Sensor Placement

This research [10] established a measuring and compensating control system for machine tool spindle thermal expansion. The tool setting probe MP4 and the low-cost yet precise thermal sensor of AD 590 IC were designed for temperature and spindle expansion measurements, respectively [10]. The error model is developed from the cutting state rather than the non-cutting condition. Using accessible temperature data points at any given moment, a linear error model may be developed.

3.2 Estimating Thermal Deformation of Main Spindle

The primary spindle rotation causes the greatest amount of thermal deformation due to internal heating. The spindle speed affects this property of thermal deformation. As a result, precisely estimating thermal deformation proved challenging. To overcome this challenge, we used experimental formulations based on spindle speed parameters to characterize these deformation characteristics, and we added the formula for compensating in a continuously transient condition into the computation for predicting thermal deformation. This has made it possible to precisely predict thermal deformation at all rotation speeds.

3.3 Linear Regression Analysis

Linear-regression models are relatively simple and provide an easy-to-interpret mathematical formula that can generate predictions. The relationship between predictor and responder variables is explicitly described in a data model. Linear regression is used to fit a data model with linear coefficients. A least-squares fit is the most frequent sort of linear regression, and it can fit both lines and polynomials, among other linear models. Linear regression is an extensively used method to compare the correlation between two variables using a linear line. The dependent variable can be predicted using the independent variable as well [11] (Table 1).

$$Y = \beta_0 + \beta_1 * x_1 + \beta_2 * x_2 + \beta_3 * x_3 + \epsilon \tag{1}$$

Table 1 Linear regression results

Training results	
RMSE (Validation)	8.6107
R-Squared (Validation)	1
MSE (Validation)	74.144
MAE (Validation)	6.5657
Prediction speed	13,000 obs/sec
Training time	1.2053s
Test results	
RMSE (Test)	8.3586
R-Squared (Test)	1
MSE (Test)	69.866
MAE (Test)	6.359

$$\varepsilon = \sum_{i=1} \left(y_i - \hat{y}_i\right)^2 \quad (2)$$

x is the independent variable
Y is the dependent variable
β_0 is the constant term
β_1 is the coefficient of x_1
$\in$ is the total error of the actual and predicted value
y_i is the i th input value
$\hat{y}_i$ is the i th predicted value

As shown in the above Fig. 2, the linear graph has been presented with DIA (diameter) versus Temperature recorded in the spindle housing bed using MATLAB regression analyzer.

3.4 ANN (Artificial Neural Networks)

The neural network is a complicated algorithm used for predictive analysis that is physiologically inspired by the structure of the human brain. Time series data may be predicted using neural networks. A neural network may be programmed to recognize patterns in incoming data and provide noise-free output.

Even when the data contains a substantial quantity of noise, neural networks have a high degree of accuracy. This is a significant benefit; if the hidden layer can still uncover correlations in the data despite noise, you may be able to use otherwise worthless data. Multiple variables are employed as input and output in the neural network model, as shown in Fig. 3. In a traditional NN model, there are

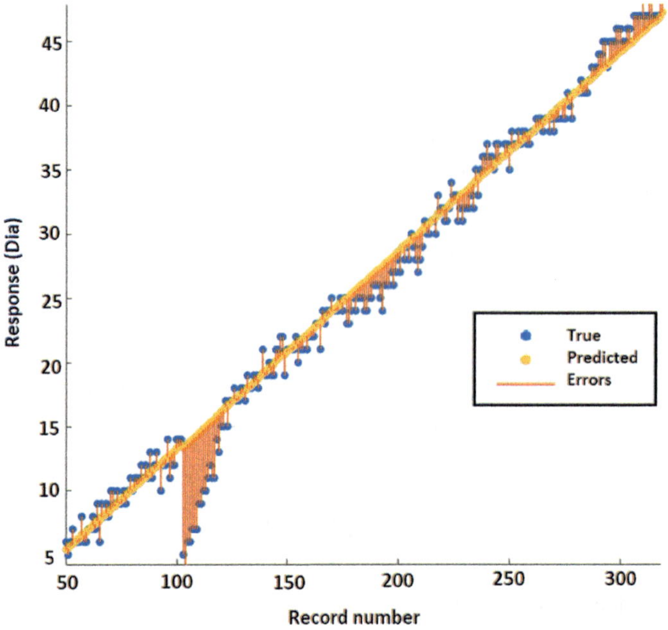

Fig. 2 Linear regression plot

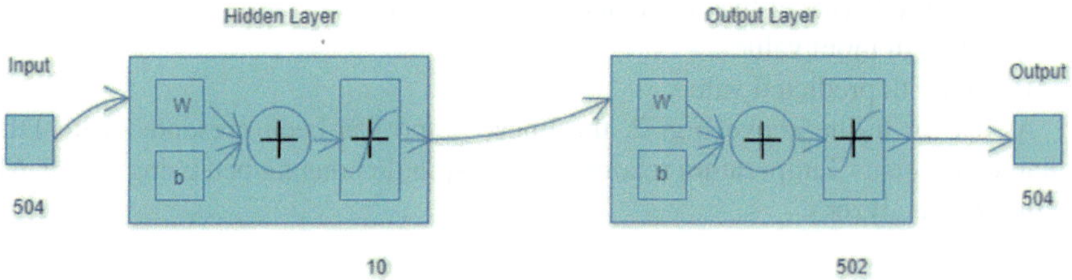

Fig. 3 ANN

three layers (input, output, and hidden). Figure 3 shows how the input layer, hidden layer and output layer minimizes the error during the training phase due to the input being continuously being optimized while moving towards the output layer. External signals and data are accepted by the input layer. The hidden layer is a unit that resides between the input and output layers and cannot be seen from the outside of the system, and it realizes the output of the system processing results.

In the NN training procedure, the Sum-Square Error (SSE) is the object function of the network optimization under the same training epoch (the number of iterations). A high number of training epochs may result in a small training error (SSE), but it does not necessarily lead to a better network. An error goal that is too small will also lead to an over-training problem similar to a surplus of neurones in the hidden layer.

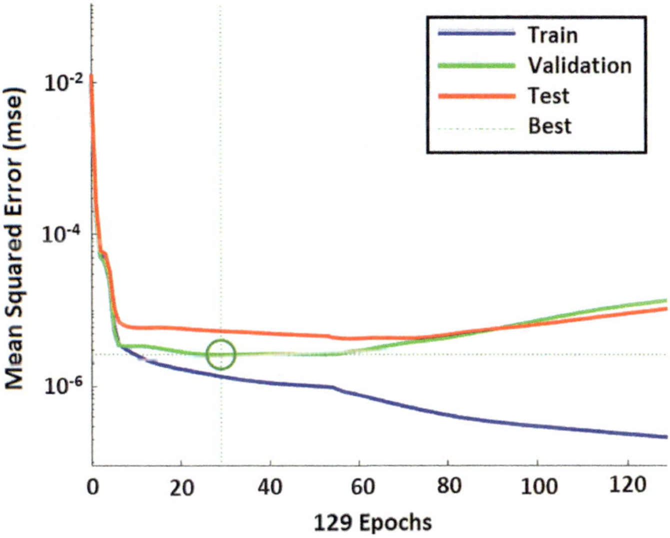

Fig. 4 ANN validation epoch plot

It involves,

1. Processing the dataset
2. Making the artificial neural network
3. Training the ANN
4. Testing the ANN

The following results were acquired after analyzing the input and output layers with MATLAB's neural network toolbox, which were both analyzed and then lowered to smaller values to reduce error. Figure 4 demonstrate how the RMSE value changed over time and at which epoch the performance was at its best, which is 2.6471e-06 at epoch 29.

3.5 SVM

SVM is a pattern recognition classifier derived from the generalized portrait technique. For the leaderway-V450 machining center, Miao et al. developed spindle thermal error models based on SVM and multiple regression in 2013. It was determined that the SVM model employed for thermal error correction not only had high accuracy, but also good robustness, by examining the accuracy of the cross-quarter experimental data [12]. One of the most important aspects of machine learning is data classification. The aim behind a support vector machine is to establish a hyper plane between data sets to show which class they belong to. The goal is to teach the machine to recognize structure in data and map it to the appropriate class label. The hyper plane has the greatest distance to the nearest training data points of any class, resulting in the best outcome.

In kernel SVM, you map your data points onto a Hilbert space H with potentially infinite dimensions. The normal vector to the separating hyperplane (which completely characterizes the decision boundary) turns out to have the form

$$w = \sum_{i=1}^{n} \alpha_i y_i \phi(x_i) \tag{3}$$

We classify a point x via

$$\text{sgn}(b + \langle w, x \rangle) = \text{sgn}\left(b + \sum_{i=1}^{n} \alpha_i y_i K(x_i, x)\right) \tag{4}$$

where K is the kernel function. For K,

$$K(x, x') = \exp\left(-\gamma ||x - x'||^2\right) \tag{5}$$

We have,

$$||x - x'||^2 = ||x||^2 + ||x'||^2 - 2x^T x' \tag{6}$$

Each test point's squared norm is computed by X1, while each training point's squared norm is computed by X2. Then compute the total of these minus two times the data points' inner products, which is the (or at least a) vectorized approach to do all pairwise calculations (Table 2).

$$||x - x'||^2 \tag{7}$$

Table 2 SVM results

Training results	
RMSE (Validation)	10.193
R-Squared (Validation)	1.00
MSE (Validation)	103.9
MAE (Validation)	8.3483
Prediction speed	24,000 0bs/s
Training time	3.1051
Test results	
RMSE (Test)	10.023
R-Squared (Test)	1.00
MSE (Test)	100.46
MAE (Test)	8.3321

3.6 Bagged Trees

The primary idea behind bagged trees is that instead of relying on a single decision tree, you rely on a large number of them, allowing you to combine the insights of multiple models The problem is that any small change in the data might cause significant changes in the model and future forecasts. The rationale for this is that one of the advantages of bagged trees is that it reduces variation while maintaining bias consistency.

Given a set of independent observations,

$$Z_1, \ldots, Z_n$$

Each with variance

$$\sigma^2$$

The variance of Z' is

$$(\sigma^2)/n$$

Calculation using N separate training sets

$$\widehat{f^1}(x), \widehat{f^2}(x), .., \widehat{f^N}(x)$$

And obtain a low variance statistical learning model using

$$\widehat{f_{ave}}(x)\frac{1}{N}\sum_{n=1}^{N}\widehat{f^n}(x) \tag{8}$$

N distinct bootstrapped training data sets are constructed from repeated samples from the taken data set. In order to get

$$\widehat{f^*n}(x)$$

The nth bootstrapped training set is trained.

Finally

$$\widehat{f_{bag}}(x)\frac{1}{N}\sum_{n=1}^{N}\widehat{f^*n}(x) \tag{9}$$

is calculated by averaging all of the forecasts.

Regression trees are not trimmed and have a deep root system. As a result, each tree has a high variation but a low bias. As a result, by averaging these B trees, the

Table 3 Bagged trees results

Training results	
RMSE (Validation)	3.977
R-Squared (Validation)	1.00
MSE (Validation)	15.817
MAE (Validation)	2.4622
Prediction speed	5400 obs/s
Training time	3.5372
Test results	
RMSE (Test)	2.333
R-Squared (Test)	1.00
MSE (Test)	5.443
MAE (Test)	1.4669

variance is reduced. By grouping hundreds or even thousands of trees into a single method, bagging has been shown to provide significant gains in accuracy (Table 3).

4 Conclusion

This work aims to improve the efficiency of compensation system in CNC machine. The location of the sensor is observed to play a significant influence in building a robust thermal model. A total of four key sensing stations were chosen after testing many temperature fields. The following conclusions are derived,

1. The thermal errors are unpredictably variable, resulting in a distinct thermal model each time. The use of linear regression revealed that it is one of the simplest and most straightforward methods.
2. In ANN, the nonlinearity won't be captured if there aren't enough neurons in the hidden layer. In contrast, if we include too many neurons, the ANN suffers from overfitting, resulting in a lack of generalizability. The RMSE value is lower than all other approaches and is beneficial in prediction when a large number of data points are involved.
3. A fully formed decision tree, has a large variance and a low bias. Bagging forest aggregates these high variance models in order to minimize variance and hence improve prediction accuracy. This leads in reduced RMSE values for the bagged trees approach, which has been demonstrated to be more efficient than linear regression.

References

1. Li, Y., Zhao, W., Lan, S., Ni, J., Wu, W., Lu, B.: A review on spindle thermal error compensation in machine tools. Int. J. Mach. Tools Manuf. **95**, 20–38 (2015)
2. Bryan, J.: International status of thermal error research. CIRP Ann. **39**(2), 645–656 (1990)
3. Ramesh, R., Mannan, M., Poo, A.: Error compensation in machine tools—a review: Part ii: thermal errors. Int. J. Mach. Tools Manuf **40**(9), 1257–1284 (2000)
4. Haitao, Z., Jianguo, Y., Jinhua, S.: Simulation of thermal behavior of a cnc machine tool spindle. Int. J. Mach. Tools Manuf **47**(6), 1003–1010 (2007)
5. Aguirre, G., Nanclares, A., Urreta, H.: Thermal error compensation for large heavy-duty milling-boring machines. Proceedings of the 29th Annual Meeting of the American Society for Precision Engineering, pp. 57–62. Zurich, Switzerland (2014)
6. Han, Z.Y., Jin, H.Y., Liu, Y.L., Fu, H.Y.: A review of geometric error modeling and error detection for cnc machine tool. Appl. Mech. Mater. **303**, 627–631 (2013)
7. Kushnir, E.: Thermal compensation algorithm for machine tool. Amer. Soc. Mech. Eng. **493**, 51–60 (2005). New York, N.Y
8. Tsai, P.C., Cheng, C.C., Chen, W.J., Su, S.J.: Sensor placement methodology for spindle thermal compensation of machine tools. Int. J. Adv. Manuf. Technol. **106**(11), 5429–5440 (2020)
9. Chen, T.C., Chang, C.J., Hung, J.P., Lee, R.M., Wang, C.C.: Real-time compensation for thermal errors of the milling machine. Appl. Sci. **6**(4), 4–8 (2016)
10. Fan, K.C.: An intelligent thermal error compensation system for cnc machining centers. J. Chinese Soc. Mech. Eng. **28**(1), 81–90 (2007)
11. Lin, C.J., Su, X.Y., Hu, C.H., Jian, B.L., Wu, L.W., Yau, H.T.: A linear re gression thermal displacement lathe spindle model. Energies **13**(4), 1–12 (2020)
12. Gong, Y.Y., Miao, E.M., Chen, H.D., Cheng, T.J.: Application of support vector regression machine to thermal error modelling of machine tools. Opt. Precis. Eng. **4** (2013)

Establishing the Planting Calendar for Onions (*Allium cepa L*) Using Localized Data on Temperature and Rainfall

Jubert B. Oligo and Julius S. Valderama

Abstract The study aimed to determine a planting calendar for a red variety of onion (*Allium cepa L*) in Aritao, Nueva Vizcaya utilizing the temperature and rainfall data of the locality and matched with the temperature and water requirements of the onion plant. Onion has 9 stages starting from the sowing stage to the fall-down stage that lasted for an average of 126 days. Onions require cooler weather during the early stages of growth while a dry atmosphere with moderately high temperature is necessary for bulb development up to maturation until the harvesting period. Water requirements of the onion plant also vary from every stage; lack of water, as well as excessive water, could be disadvantageous to the plant growth and development. Ten years of data on rainfall and temperature of the locality were sourced-out in the NVSU Agromet station. These data were used to forecast 12 months of data on temperature and rainfall using three forecasting methods of the SPSS, NCSS, and MS Excel. The forecasted rainfall and temperature data on weekly basis were matched to the 9 stages of onions to its temperature and water requirements starting from the sowing stage to the fall-down stage. The study was able to determine the best timing for the plant; it is on the second week of February up to the third week of June. Onions planted at this time interval have a high forecasted percentage of survival as the temperature and water requirements of the plant in its stages were all sustained.

Keywords Climate adaptation · Planting calendar · Time-series analysis

J. B. Oligo
College of Teacher Education, Nueva Vizcaya State University, Bayombong, Philippines

J. S. Valderama (✉)
College of Arts and Sciences, Nueva Vizcaya State University, Bayombong, Philippines
e-mail: valderamajulius@gmail.com

S. Banerjee and A. Saha (eds.), *Nonlinear Dynamics and Applications*,
Springer Proceedings in Complexity,
https://doi.org/10.1007/978-3-030-99792-2_75

1 Introduction

Climate change presents significant risks and opportunities for agriculture [1]. Chmielewski et al. [2], Kalbarczyk [3] confirmed that rapid changes in air temperature in the rest of the world and it is associated to the so-called climate change. According to Altieri et al. [4], climate change impact potentially significant to small farm production is loss of soil organic matter due to soil warming. Higher air temperatures are likely to speed the natural decomposition of organic matter and to increase the rates of other soil processes that affect fertility. Climate changes remote from production areas may also be critical. Irrigated agricultural land comprises less than one-fifth of all cropped area but produces between 40 and 45% of the world's food and water for irrigation is often extracted from rivers that depend upon distant climatic conditions [5]. Climate change produces a basic sense of ethical and existential violation that creates new norms, laws, markets, technologies, understandings of the nation and the state, urban forms, and international cooperation [6].

Air temperature is one of the most important meteorological elements, deciding the rate of a plant's growth and development. The examined onion phenophases were most correlated at $P < 0.01$, with the mean air temperature from the period of 6–9 weeks before the earliest date of their occurrence [3]. Hatfield and Prueger [7] states that changes in short-term temperature extremes can be critical, especially if they coincide with key stages of development. Only a few days of extreme temperature (greater than 32 °C) at the flowering stage of many crops can drastically reduce yield.

The same is true with precipitation—gross production value in agriculture would decrease by 0.24 for every 1 mm increase in precipitation but more number of rain days would increase gross production value by 1.24. An increase in Diurnal range temperature as measured by the difference between the daily maximum and minimum temperature would decrease gross production value by 5.74 [8]. Daymond et al. [9] revealed that Measurements of the ratio of the maximum diameter of the bulb to the minimum diameter of the neck for onions showed that there was little or no influence of CO_2, whereas the effect of temperature was substantial. Bulbing was accelerated by high temperature and was greatly delayed at low temperature.

Onion farmers aimed to produce quality onions. The quantity of their harvest is dependent on the available land area for cultivation, capital for the expenditures, manpower, and some resources which limits the capacity of the farmer to produce a larger quantity. In onion production and on its business side, the quality and quantity of the harvest are two inseparable ideas. Some researchers like Al-jamal et al. [10] have worked on optimizing the yields in onion productions but maintaining the farm inputs practices to produce quality onions. They focused on improving the yield but not sacrificing the quality of the onions. On the other hand, some researchers like Boyhan et al. [11] ventured on producing quality onions, but not sacrificing the yield produced. Moreover, researchers like Piri and Naserin [12], de Santa et al. [13], Zheng et al. [14], Channagoudra et al. [15] who have worked on how to improve the quality of onions produced and the same time the quantity produced.

As a tropical country, the Philippines is acknowledged as an onion-growing state. Because the versatile high-value crop could be grown all over the archipelago, many rice farmers have shifted to onion farming as they tried and proven that they earn more from cultivating the red bulb better known to Filipinos as "sibuyas". In Aritao, Nueva Vizcaya, farmers shifted to growing onions in large-scale production. Onion has become the town's One Town, One Product. The production surplus of onions in the province is also seen as a bright opportunity among farmers as an alternative source of income and to boost the local economy [16]. The study was limited to the red variety of onions as this variety of onions is commonly planted by the farmers in the locality.

Onion farmers used to plant onions (*Allium cepa L*) from December to April when rain is not expected to occur. However, because of climate change, even during these said months, rains already occur. Making the time to plant onions becomes unpredictable and risky, thus resulting in a poor harvest of farmers. It is for this reason that this research was conducted to determine the right timing for farmers to plant onions to obtain a bountiful harvest. In this study, the objectives were: (1) to determine the trend of temperature and rainfall from January, 2008 to March, 2018, (2) to determine the water and temperature requirements of onions, (3) to predict the rainfall and temperature for the next planting seasons of onions, and (4) to establish the planting calendar for onions.

2 Methods

Descriptive—exploratory type of research was used in this study. The descriptive type was used to describe the onion's water and temperature requirements, as well as the trends of the climatic data using 10 years of previous data on temperature and rainfall. The exploratory part of the design was used in the generation of mathematical models for forecasting one-year data for rainfall and temperature. Exploration was also used in establishing the planting calendar.

Physiological characteristics of the red onions, particularly its need requirements for water and temperature was established through literature reviews, guidebooks in onion productions, fact sheets, and internet resources. These onions' characteristics were then presented to experts for their validation and further enhancement or recommendation. Experts in Onion cropping and production, as well as the farmers, were identified with the assistance of the Department of Agriculture—Aritao.

The study was conducted mid-year of 2019, however, the available data on rainfall and temperature used for the analysis includes from January 2008 to March 2019. Temperature and Rainfall data from January 2008 to March 2019 was collected from the Agromet Station of Nueva Vizcaya State University Bayombong (NVSU-PAGASA). The study was conducted at Aritao, Nueva Vizcaya since red-onion production is more abundant in the said municipality.

The growth and production of onions are influenced by several factors like soil moisture and nutrients [17], soil quality or types [18], farm irrigation [19], and others.

Since the study was conducted purposely for Aritao, other parameters were no longer considered as all the red-onion farms in the locality were all paddy, not irrigated, and with soil pH ranging from <4.50 to >6.80.

The data was transformed into weekly temperature and rainfall data by getting the average of the 7 consecutive days. The data was used to determine the climatic condition of Nueva Vizcaya Province. In addition, this was also used to predict the climate condition for the next 48 weeks (1 year) using the three different forecasting models namely SPSS forecast, MS Excel, and NCSS. The average of the obtained weekly forecasted data was then utilized to analyze whether or not possible to plant onion in the week/month of the year 2020 considering the water and temperature requirements of the onions.

3 Results and Discussion

3.1 The Water and Temperature Requirements of Onions

Table 1 reflects the stages of onions starting from sowing to harvesting. The average number of weeks in every stage was also included in the table. Onions cropping is ideal if the plot has provisions for irrigation, however, onions are dependent on rainfall as rainfall could cause excessive water in the plot and later cause the plant to wilt and die.

Land preparation is done one month (4 weeks) prior to transplanting. Transplant seedlings 4–6 weeks after sowing. First side-dressing will be done ten days after transplanting 4,6,8 weeks after transplanting. Depending on soil types, irrigation varies between 4 to7 days. Stop irrigation 2–3 weeks before harvest. Harvest when the tops begin to fold over. Ambient temperature requires the Seedling growth 20–22 °C, before bulbing 15–24 °C and for bulb development 15–24 °C. Bulb Onions grow well in an easily crumbled and well-drained loam soil with good water holding

Table 1 Stages of planting onions

Stages	Allowable Time (in weeks)
Land preparation (Basal)/sowing transplanting and fertilization	4 week
Transplanting	4–6 weeks after sowing
Fertilization (side dress)	4–8 weeks after transplanting
Irrigation	1 week *Note: stop irrigation 2-3 weeks before harvest*
Pest & Disease management	As the need arises
Harvesting	Harvest when the tops begin to fold over

Table 2 Water and temperature requirements of onion on its stages

Growth stage	Average water use rate (mm/day)	Total water use during stage	No. of days
Seed in soil after sowing	33.0	33.0	1.0
Loop stage	4.9	98.0	20.0
First leaf 'flag' stage	3.4	41.0	12.0
Cotyledon senescence	3.6	36.0	10.0
Fourth leaf 'leek' stage	3.6	51.0	14.0
Fall of the first leaf	3.7	52.0	14.0
Start of bulbing	3.6	51.0	14.0
Bulb swelling	2.8	75.0	27.0
Fall-down or soft neck	3.6	50.0	14.0
Total		487.0	126.0

capacity and pH between 6 and 7. The result was supported by Netafim. The Best Management Practices (BMPs) determined the additional proven research becomes available wherein the optimum ambient temperatures for onion are: Seedling growth 20–25 °C, vegetative growth 13–24 °C, before bulbing 15–21 °C and for bulb development 20–25 °C. The soil suitability must be fertile, light, deep friable well-drained fine sandy, loamy and alluvial and the optimum soil pH is 5.8 to 6.5 [20–24].

Table 2 reflects the water and temperature requirements of Onions in all the stages. On average, one-onion cropping lasted for 126 days starting from sowing up to fall-downstage. Included in the table are the data on the average water use rate (mm/day) of the onions, and the total water use for the entire specific growth stage.

The data in Table 2 was supported by the study of Pejic [25] the effect of different irrigation schedules on yield and water use of onion *(Allium cepa L.)*, the values of evapotranspiration of 450–500 mm could be used as a good platform for onion growers in the region in terms of maximum yield and optimum utilization of irrigation water. The average daily water needs of onion during irrigation season grown in a semi-arid climate with a mean temperature of 20 °C needs approximately 6.5 mm of water per day and the indicative values of crop water needs are 350-550 mm. Harvesting starts at 95 DAT or 125 DAS.

3.2 The Monthly Average of the Rainfall and Temperature

Figure 1 reveals the average monthly rainfall for 10 years from January 2008–May 2019 of Nueva Vizcaya particularly. On average, water precipitation from January to April was at the minimum or scarce level, little precipitation occurred from May to July, high volume of water precipitation was down-pour from August to December. These rainfall patterns coincide with the weather seasons of the Philippines.

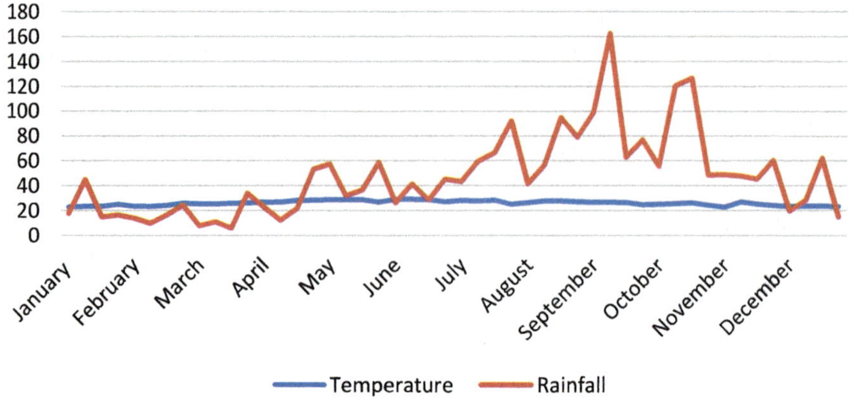

Fig. 1 Average rainfall and temperature from 2008–2019

On the other hand, seemingly there was small variation as to temperature reading from January to December. The coldest months were from December to February, and the hottest months were from March to May.

These patterns or trends of temperature and rainfall occurred repeatedly every year following the weather cycle of the Philippines. Water precipitation increases in the months of June and begins to decrease in the months of November. Temperature also follows a trend.

3.3 Forecasted One-Year Data of Rainfall and Temperature

Table 3 is the weekly forecasted rainfall and temperature of Aritao, Nueva Vizcaya. Model 1 corresponds for time series analysis using SPSS, Model 2 for NCSS, Model 3 for MS Excell, and Ave Model for the average of the values.

Table 3 shows the forecasted weekly rainfall and temperature for January 2020 to March 2021 through the use of SPSS time series (Model 1), NCSS forecasting (Model 2), and MS Excel forecast function (Model 3). The averages of these data were used to determine the planting season onions *(Allium cepa L)* in Aritao to be able to have an effective yield. On the other hand, the average slightest rainfall forecasted is the third week of January 2019 to the second week of April 2019 which is ranging from 10–32 mm while the average forecasted temperature for the whole year round is ranging from 22 °C–28 °C.

Table 3 Forecasted temperature and forecasted rainfall for the January 2020–March 2021

Month/wk	Forecasted temperature				Forecasted rainfall			
	Model 1	Model 2	Model 3	Model Tmp	Model 1	Model 2	Model 3	Model Rnf
January								
w1–w2	23.5	23	24	23.5	84.5	17	34	45.5
w3–w4	24.5	23	25	24.5	31.5	14	17	20.5
February								
w1–w2	25	23.5	24	24	26	11	13	16.5
w3–w4	27	25.5	26	26	25	16	22	21
March								
w1–w2	26.5	23.5	26	25.5	24.5	6	10	13.5
w3–w4	27	23.5	27	25.5	26.5	15.5	21.5	21
April								
w1–w2	29	24	28	27	23	14.5	19	18.5
w3–w4	29.5	24	29	27.5	61	25.5	41	42.5
May								
w1–w2	30	24	30	28	82	24.5	49	52
w3–w4	29.5	24	29	27.5	59.5	32.5	52.5	48
June								
w1–w2	30	24	30	28	66.5	20.5	37	41
w3–w4	29.5	24	29	27.5	56.5	25.5	40	40.5
July								
w1–w2	29	23	28.5	27	93	32	56	60.5
w3–w4	28.5	23	27.5	26	110	49	87	82
August								
w1–w2	28.5	23	27.5	26	66	35.5	54	52
w3–w4	28	22	28.5	26	78	71.5	96	81.5
September								
w1–w2	28.5	22.5	27.5	26	65	107	144	105.5
w3–w4	27	22	26.5	25.5	82	49	77	69.5
October								
w1–w2	25.5	21.5	26	24	49.5	73	97	73
w3–w4	26	21.5	26	24.5	55	71.5	96.5	74
November								
w1–w2	24.5	21	26	24	56	38	53.5	49
w3–w4	24.5	20	25.5	23.5	60	40.5	58	52.5

(continued)

Table 3 (continued)

Month/wk	Forecasted temperature				Forecasted rainfall			
	Model 1	Model 2	Model 3	Model Tmp	Model 1	Model 2	Model 3	Model Rnf
December								
w1–w2	23.5	20	24.5	22.5	18	20	26.5	21.5
w3–w4	23	19	24.5	22	21	38.5	42.5	34
January								
w1–w2	23.5	18.5	24	22	85	13	35	44.5
w3–w4	25	19	25	23	32.5	12.5	17.5	20.5
February								
w1–w2	25	19	24	23	27	10.5	13	16.5
w3–w4	27	19.5	26	24	26	15	22.5	21
March								
w1–w2	27	18.5	26	24	25	5.5	10	14
w3–w4	27	18.5	27	24	27	15	22	21

3.4 The Planting Calendar for Onions

The predicted rainfall data and temperature data in Table 3 were matched to established water and temperature requirements as presented in Table 1. The forecasted rainfall and temperature; and the onion's water and temperature requirements were analyzed to come up with a planting schedule that can sustain the growth and yield production of the plant. This was determined using the identified water and temperature requirements of the plant and matched to the value of the Model average for rainfall and temperature.

$$\text{Wkly prob} = \begin{cases} 1 \text{ if computed Model ave is lower than the plant requirement} \\ 0 \text{ if computed Model ave is higher than the plant requirement} \end{cases}$$

In every stage of the plant, if it is denoted by 1, it means to say that the onion's requirements, both temperature, and rainfall were satisfied. Thus, the plant life can be sustained in that week. On the contrary, if it is denoted by 0, it means that either or both the onion's requirements on water or temperature were not sustained. Thus, the plant life is at risk on that week.

The rate of survival indicates the percentage of the number of weeks the plant sustained over the total number of weeks. Thus, 100% implies that plant life will be sustained throughout the cropping; and 90% could be sustained with 10% risk.

There are six identified planting schedules for onion. These six schedules have a high probability of supporting the water and temperature requirement of onion for 18 weeks from planting up to harvesting. As presented in Table 4, these were the following schedules (planting—harvesting):

Table 4 Schedule of Planting Onions

Month/wk	Forecasted temperature				Forecasted rainfall			
	Model 1	Model 2	Model 3	Model Tmp	Model 1	Model 2	Model 3	Model Rnf
January								
w1–w2	23.5	23	24	23.5	84.5	17	34	45.5
w3–w4	24.5	23	25	24.5	31.5	14	17	20.5
February								
w1–w2	25	23.5	24	24	26	11	13	16.5
w3–w4	27	25.5	26	26	25	16	22	21
March								
w1–w2	26.5	23.5	26	25.5	24.5	6	10	13.5
w3–w4	27	23.5	27	25.5	26.5	15.5	21.5	21
April								
w1–w2	29	24	28	27	23	14.5	19	18.5
w3–w4	29.5	24	29	27.5	61	25.5	41	42.5
May								
w1–w2	30	24	30	28	82	24.5	49	52
w3–w4	29.5	24	29	27.5	59.5	32.5	52.5	48
June								
w1–w2	30	24	30	28	66.5	20.5	37	41
w3–w4	29.5	24	29	27.5	56.5	25.5	40	40.5
July								
w1–w2	29	23	28.5	27	93	32	56	60.5
w3–w4	28.5	23	27.5	26	110	49	87	82
August								
w1–w2	28.5	23	27.5	26	66	35.5	54	52
w3–w4	28	22	28.5	26	78	71.5	96	81.5
September								
w1–w2	28.5	22.5	27.5	26	65	107	144	105.5
w3–w4	27	22	26.5	25.5	82	49	77	69.5
October								
w1–w2	25.5	21.5	26	24	49.5	73	97	73
w3–w4	26	21.5	26	24.5	55	71.5	96.5	74
November								
w1–w2	24.5	21	26	24	56	38	53.5	49
w3–w4	24.5	20	25.5	23.5	60	40.5	58	52.5

(continued)

Table 4 (continued)

Month/wk	Forecasted temperature				Forecasted rainfall			
	Model 1	Model 2	Model 3	Model Tmp	Model 1	Model 2	Model 3	Model Rnf
December								
w1–w2	23.5	20	24.5	22.5	18	20	26.5	21.5
w3–w4	23	19	24.5	22	21	38.5	42.5	34
January								
w1–w2	23.5	18.5	24	22	85	13	35	44.5
w3–w4	25	19	25	23	32.5	12.5	17.5	20.5
February								
w1–w2	25	19	24	23	27	10.5	13	16.5
w3–w4	27	19.5	26	24	26	15	22.5	21
March								
w1–w2	27	18.5	26	24	25	5.5	10	14
w3–w4	27	18.5	27	24	27	15	22	21

1—water and temperature requirements can be sustained
0—water and temperature requirements cannot be sustained
x, y—planting time, harvesting time

Table 4 shows the schedule of planting Onions (*Allium cepa L*) for C.Y. 2020. The requirements for planting Onions as shown in the table using the codes 0 and 1; whereas code 0 means the plant will not survive on that specific stage as the temperature requirement or rainfall requirement or both requirements were not sustained. Code 1 means that both temperature and rainfall were sustained in that specific week. X marks reefers to planting time, and Y represents the harvesting time. The probability of 100% survival indicates that the plant requirement for rainfall and temperature were all sustained.

In the table, the highest probability rate of survival falls under schedule 6 which is the 2nd week of February 2020 to the 3rd week of June 2020 with a probability rate of 94% while the lowest probability rate is schedule 1 which is the 1st week of January 2020 to 2nd week of May 2020 with a probability rate of 72%. For best growth and bulb quality, onion requires cooler weather during the early stages of growth and a dry atmosphere with moderately high temperature for bulb development & maturation until the harvesting period.

4 Conclusions and Recommendations

The study has shown the seasonality of rainfall and temperature of Bayombong Nueva Vizcaya for 10 years. The dry season's starts in the months of January 2020 up to

April 2020 while the wet season starts with the months of May 2020 to November 2020, then another cycle of the dry season in the months of January 2021 up to April 2021. The study suggested the planting seasons of onions for CY 2021 could be scientifically determined using combinations of time-series analysis. Based on the established planting calendar of onions, the highest probability rate of survival is if the plant will be planted on the 2nd week of February 2020 to the 1st week of March 2020 and expected to be harvested on the 3rd week of June to 2nd week of July. The forecasting procedure of the study could be repeated for the year 2022 or in the future years to determine the planting calendar of the red onion, white onion, and yellow onion.

References

1. Cradock-Henry, N.A. et al.: Climate adaptation pathways for agriculture: insights from a participatory process. Environ. Sci. Policy **107**, 66–79 (2020)
2. Chmielewski, F.-M., Rötzer, T.: Annual and spatial variability of the beginning of growing season in Europe in relation to air temperature changes. Climate Res. **19**(3), 257–264 (2002)
3. Kalbarczyk, R.: The effect of climate change in Poland on the phenological phases of onion (Allium cepa L.) between 1966 and 2005. Agric. Conspec. Sci. **74**(4), 297–304 (2009)
4. Altieri, M.A., Koohafkan, P.: Enduring farms: climate change, smallholders and traditional farming communities (2008)
5. Siebert, S., Döll, P.: 2.4 Irrigation water use–a global perspective. Central Asia **14**, 10–2 (2007)
6. Beck, U.: How climate change might save the world. Develop. Soc. **43**(2), 169–183 (2014)
7. Hatfield, J.L., Prueger, J.H.: Temperature extremes: Effect on plant growth and development. Weather Climate Extremes **10**, 4–10 (2015)
8. Dait, J.M.: Effect of climate change on Philippine agriculture. Int. J. Sci. Res. **4**, 1922–1924 (2013)
9. Daymond, A.J., et al.: The growth, development and yield of onion (Allium cepa L.) in response to temperature and CO_2. J. Horticult. Sci. **72**(1), 135–145 (1997)
10. Al-Jamal, M.S., et al.: Computing the crop water production function for onion. Agric. Water Manag. **46**(1), 29–41 (2000)
11. Boyhan, G.E. et al.: Evaluation of poultry litter and organic fertilizer rate and source for production of organic short-day onions. HortTechnology **20**(2), 304–307 (2010)
12. Piri, H., Naserin, A.: Effect of different levels of water, applied nitrogen and irrigation methods on yield, yield components and IWUE of onion. Sci. Hortic. **268**, 109361 (2020)
13. de Santa Olalla, F.M., Domınguez-Padilla, A., Lopez, R.: Production and quality of the onion crop (Allium cepa L.) cultivated under controlled deficit irrigation conditions in a semi-arid climate. Agric. Water Manag. **68**(1), 77–89 (2004)
14. Zheng, J., et al.: Effects of water deficits on growth, yield and water productivity of drip-irrigated onion (Allium cepa L.) in an arid region of Northwest China. Irrigation Science **31**(5), 995–1008 (2013)
15. Channagoudra, R.F., Prabhudeva, A., Kamble, A.S.: Response of onion (Allium cepa L.) to different levels of irrigation and sulphur in alfisols of northern transitional tract of Karnataka. Asian J. Horticult. **4**(1), 152–155 (2009)
16. Perante, Leonardo II. East-West Seed PHL projects sustainable onion industry nationwide. https://businessmirror.com.ph/2018/05/05/east-west-seed-phl-projects-sustainable-onion-industry-nationwide/
17. Kumar, S., Imtiyaz, M., Kumar, A.: Effect of differential soil moisture and nutrient regimes on postharvest attributes of onion (Allium cepa L.). Scientia Horticult. **112**(2), 121–129 (2007)

18. Shock, C.C., Wang, F.-X.: Soil water tension, a powerful measurement for productivity and stewardship. HortScience **46**(2), 178–185 (2011)
19. Mermoud, A., Tamini, T.D., Yacouba, H.: Impacts of different irrigation schedules on the water balance components of an onion crop in a semi-arid zone. Agric. Water Manag. **77**(1–3), 282–295 (2005)
20. López-Urrea, R., et al.: Single and dual crop coefficients and water requirements for onion (Allium cepa L.) under semiarid conditions. Agric. Water Manag. **96**(6), 1031–1036 (2009)
21. De Lis, B.R., et al.: Studies of water requirements of horticultural crops: II. Influence of drought at different growth stages of onion 1. Agron. J. **59**(6), 573–576 (1967)
22. Bekele, S., Tilahun, K.: Regulated deficit irrigation scheduling of onion in a semiarid region of Ethiopia. Agric. Water Manag. **89**(1–2), 148–152 (2007)
23. Ortola, M.P., Knox, J.W.: Water relations and irrigation requirements of onion (Allium cepa L.): a review of yield and quality impacts. Exp. Agric. **51**(2), 210–231 (2015)
24. El Balla, M.M.A.D., Hamid, A.A., Abdelmageed, A.H.A.: Effects of time of water stress on flowering, seed yield and seed quality of common onion (Allium cepa L.) under the arid tropical conditions of Sudan. Agric. Water Manag. **121**, 149–157 (2013)
25. Pejić, B. et al.: Effect of irrigation schedules on yield and water use of onion (Allium cepa L.). African J. Biotechnol. **10**(14), 2644–2652 (2011)

Growth of Single Species Population: A Novel Approach

Suvankar Majee, Soovoojeet Jana, Anupam Khatua, and T. K. Kar

Abstract In this paper, we have proposed a new growth model for a single species population that captures certain features of logistic growth. We have constructed the growth model for a single species population on the basis of the assumptions that the individual reproduction rate is proportional with the available resources and a portion of the population species have no reproduction power. We have shown that this model would give better realistic phenomena than the other existing models, and also, it is capable of making new useful models.

Keywords Ecological problems · Birth-death process · Growth rate · Environmental carrying capacity · Reproduction rate

1 Introduction

In mathematical ecology, the growth rate of a population species is one of the most important aspects. Several growth models have been developed considering different biological organisms. In most cases, the population dynamics are modeled continuously. In their books, Kot [1] and Britton [2] described different growth rates for living creatures. The very basic model to describe the growth of species without

S. Majee (✉) · A. Khatua · T. K. Kar
Department of Mathematics, Indian Institute of Engineering Science and Technology, Shibpur, Howrah 711103, India
e-mail: suvankarmajee2@gmail.com

S. Jana
Department of Mathematics, Ramsaday College, Amta, Howrah 711401, India

S. Banerjee and A. Saha (eds.), *Nonlinear Dynamics and Applications*,
Springer Proceedings in Complexity,
https://doi.org/10.1007/978-3-030-99792-2_76

constraint is exponential growth. In this model, the equation for the single species population whose biomass density is x can be written as follows:

$$\frac{dx}{dt} = rx, \quad x(0) = x_0 \tag{1}$$

Where the single population species has an average birth rate, say, b and, has an average death rate, say, d and then the intrinsic growth rate is defined as $r = b - d$. This type of growth is known as the Malthusian growth. The main drawback of the exponential growth is $x \to \infty$ as $t \to \infty$ if $r > 0$, i.e., the population biomass would go at infinite level along with time. If there are some limitations, the growth must be checked, and then the population would not grow following geometric ratio or exponential ratio. As there are some limitations (like limitations in available food, place for living, etc.) regarding the growth of any population species, therefore the Malthusian growth is modified. The simplest modifications have been done by the logistic type growth where it is considered that environmental carrying capacity is fixed and the growth of the population depends on that capacity. In this model, the growth of a single species population is described by the following equation:

$$\frac{dx}{dt} = rx\left(1 - \frac{x}{K}\right), \quad x(0) = x_0 \tag{2}$$

where the environmental carrying capacity is taken as $K > 0$.

Later on, many researchers have generalized the growth model and also proposed new types of growth functions. Different types of growth functions are available in the literature, for example, generalized logistic growth Nelder [11], Von Bertalanffy's growth [8], Richards's growth [9], Gompertz growth [10]), Weibull function (Rwalings et al. [12]), Allee type growth function (Courchamp et al. [16]), etc. Some other growth functions and the estimation of the system parameters can be found in Tsoularis and Wallace [3], Bhowmick and Bhattacharya [4], Bhowmick et al. [4], Koya and Goshu [7], Crescenzo and Spina [6], Misra and Chaturvedi [13], Misra and Babu [14] and Kumar et al. [15]. The proper growth model is extremely important in studying the biological growth problems.

In the logistic type of growth, we have to consider a non-linear term of the state variable limiting the growth of the population species. This article proposes a new type of growth curve of a single species population. In this model, we consider the growth equation as the Malthusian type, although the intrinsic growth rate varies from time to time, and the intrinsic growth rate depends on the available environmental resource.

1.1 The Growth Model

Obviously, for each species, there are some individuals, who may be identified and differ from others, have no reproduction power. Besides the juvenile and the old individuals, there is a certain number of other individuals who have no reproduction power, and their reproduction power has been destroyed due to some disease or by means of some natural reason. For example, in species like bee, most portions of the individuals have no reproduction power, and thus considering an average per capita birth for each individual is not logical at all. Therefore, to overcome these inconveniences, we can propose two separate ways to consider the birth rate of a species. In the first assumption, we can assume that a certain portion of the species, say, $\rho(0 \leq \rho \leq 1)$ has no reproduction power, and the rest portion $(1 - \rho)$ has that power. On the other hand, the species like bees, etc., we can assume constant biomass, say, m has reproduction power, and the rest of individual biomass $(x - m)$ are not capable of reproducing. However, in both cases, the birth rate of those individuals who have no reproduction power should be zero. In contrast, for the parameter death rate, we don't claim that any particular individual has zero death rates, although it is quite true that some individual has some less death rates and the other has some more death rate. Therefore, without loss of generality, we can assume that the death rate of the single species individual follows a distribution such that the weighted mean of this distribution exists and say it is d, i.e., the per capita death rate of each individual is d. In the rest of the paper, by the phrase 'birth rate' we would mean that the per capita birth rate of those individuals which have reproduction power.

Now we assume that the portion $\rho(0 \leq \rho \leq 1)$ has no reproduction power whether the others are capable of reproducing. We also consider the per capita birth rate is b and the per capita death rate is d. Therefore, the governing equation regarding the change of the biomass of the single species population $x(t)$ with time is given by:

$$\frac{dx}{dt} = (1 - \rho)bx - dx, \text{ with } x(0) = x_0. \tag{3}$$

The system (3) can be written in the following form:

$$\frac{dx}{dt} = (b - d - b\rho)x, \text{ with } x(0) = x_0. \tag{4}$$

From system (4), it may be concluded that, $x(t) \to \infty$ if $\rho < \frac{b-d}{b}$ whereas if $\rho > \frac{b-d}{b}$ then the population species go to extinct and lastly for $\rho = \frac{b-d}{b}$ we have $x(t) = x_0$ for all time t (here we assume that $b > d$). Thus, it can be concluded that if $\rho > \frac{b-d}{b}$, i.e., if the portion of the individual has no reproduction power is greater than some threshold, then the population species may go extinct. Hence for any population species, if the biomass of old age individuals is on the higher side continuously, then that species either goes extinct, or its biomass would be very low. Our present model can draw an extensive light on these types of natural phenomena.

Both the exponential growth and logistic growth models (as presented in Eqs. (1) and (2)), the intrinsic growth rate (r) is considered as an independent model parameter. But according to our explanations, r cannot be an independent parameter, rather, it should depend on mainly three parameters, namely (i) the portion of the population (ρ) which have no reproduction power, (ii) the birth rate (ρ), and (iii) the per capita death rate (d). Thus we can define the intrinsic growth rate as follows:

$$\begin{aligned} r &= b - d, \text{ for those portions which are capable of reproducing offspring,} \\ &= -d, \text{ for those portions which are unable to reproducing offspring.} \end{aligned}$$

Therefore, we can define the intrinsic growth rate r as follows

$$r = (1 - \rho)b - d, \ 0 \leq \rho \leq 1 \tag{5}$$

where ρ, b, d are defined earlier. Here $\rho = \rho(K, t)$ is a function of available resource (i.e., the carrying capacity of the individual K), and it varies with time. Now obviously, ρ will be inversely related to environmental carrying capacity K. Depending upon the numeric value of ρ, the intrinsic growth rate r may be positive or negative. The parameter ρ can be treated as a control parameter, and this parameter has a huge and important role, not only to evaluate the exact value of r but also to determine the exact population biomass level at any time.

With the help of the above definition of r, the growth of the single species population whose biomass is $x(t)$ can be defined as follows:

$$\frac{dx}{dt} = rx, \ x(0) = x_0 \tag{6}$$

where r may be positive or negative or zero. This type of growth is also one type of exponential growth, but the main difference of this type of growth is that here r is not always positive.

2 Explanations of the Model

Now suppose that initially $r > 0$, then after some time, say after $t = t_1$, the biomass of the population will be $x_0 exp(rt_1)$. Now the growth, birth, and death of the population depend on the quantity as well as the quality of the available resource. If the available resource can sufficient to fulfill the requirement of $x = x_1$ number of individuals, then as soon as the population goes beyond the level of x_1, then there will be a competition among them for the resource and due to this competition, some of them unable to avail any resource or a negligible amount of resource. When the population biomass goes beyond this level x_1, then due to the competition, some less amount of biomass than the level x_1 would able to use the resource, and they remain stronger, and the rest portion (which is greater than the $(x - x_1)$ when $x > x_1$) becomes weaker. Now at

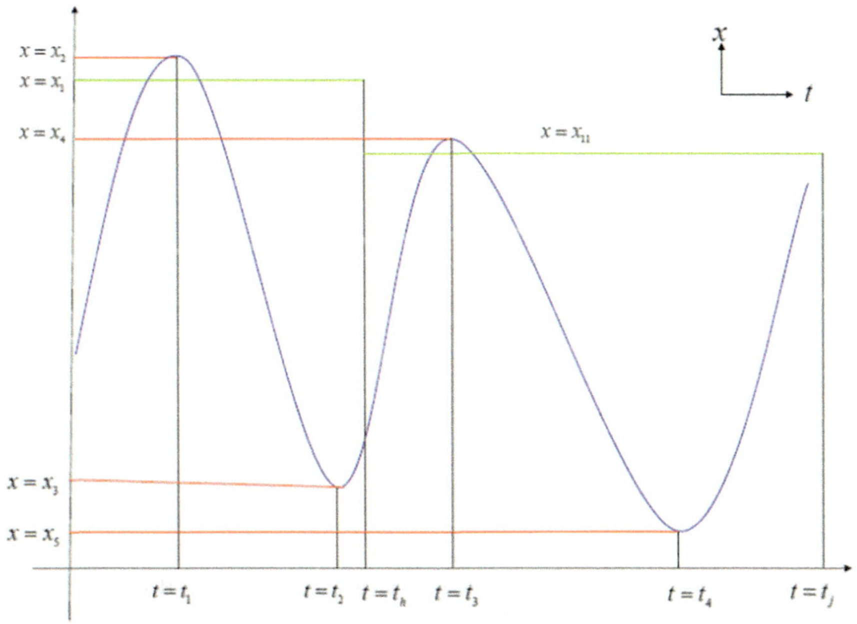

Fig. 1 Growth of population species

$x = x_2$, level, say the individual of biomass $(x_2 - x_1 - \eta x)$, $0 \leq \eta < 1$ don't get any resource to live. Thus we may consider that the individual of biomass $(x_2 - x_1 - \eta x)$ have no birth rate although they have death rate d. Now recalling the definition of r, we can conclude that the intrinsic growth rate r of that portion will be $-d$. Therefore, after a certain amount of time, say, t_1, r will be negative, and then the biomass of the species will decrease instantaneously. Again let after $t = t_2$, the population biomass reaches at x_3, far below the maximum available resource x_1 and then again r changes its sign and r becomes positive, until the population biomass x reaches some x_{11}, (which may be different from x_1 because the available resource may vary time to time) and after that r will again change its sign and will be negative. This process will continue, and if the population follows this type of growth, then the population will remain between some $\underline{x}(\geq 0)$ to some $\bar{x}(< \infty)$ level as the environmental carrying capacity is always finite, however large it may be (Figs. 1 and 2). This type of growth can define the extinction of a population species also. For example, for some time period, r will be less than zero and the available resource also reduces with time and eventually becomes so less that it will make tough to live for individuals of the species, then r will never be positive since the birth rate entirely depends on the resource. In this situation $x(t) \to 0$ after $t > t_1$ and it remains stable there.

The main difference of considering this type growth rate and the logistic type and the classical exponential growth rate are as follows:

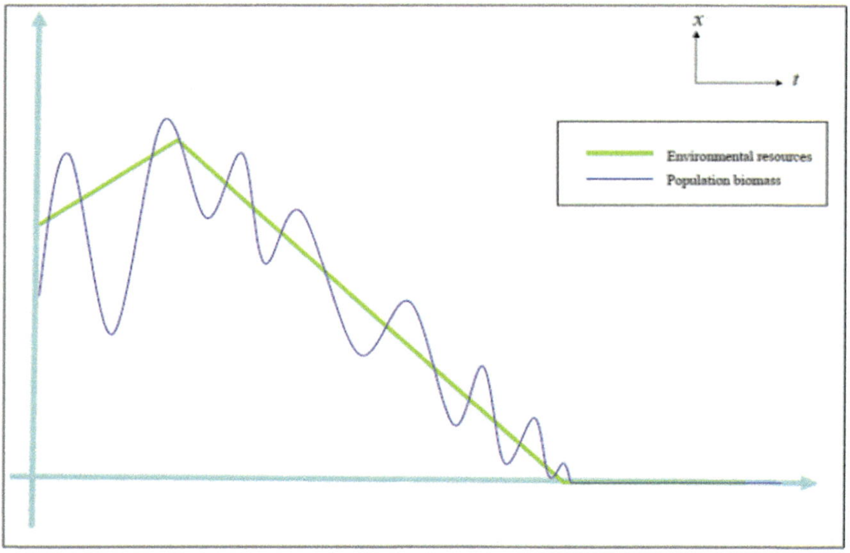

Fig. 2 Extinction of a population species due to unavailability of resource. Here green line represents the carrying capacity of the environment for that population species and deep blue line represents the level of population biomass

(1) Here, we need not consider any intraspecific competition term in the differential equation of the population, i.e., in (6) there is no carrying capacity term. The available resource depends on the growth rate
(2) The carrying capacity is taken as a fixed quantity in logistic type growth model, although in reality, it changes time to time.
(3) The logistic growth model is unable to produce a sufficient explanation of the extinction of a species in a single species model (since for logistic growth, the trivial equilibrium $x(t) = 0$ is an unstable equilibrium).
(4) The logistic type growth explains that $x(t) \to K$ as $t \to \infty$ but here we explain that through our model, $x(t)$ will oscillate within some region unless it goes to extinct, and this phenomenon is a quite more realistic one.
(5) Our model is quite simple than the logistic type growth model (since here (the model (6)) right-hand side is the simple linear function of the single variable x and there is no quadratic term in it) and can describe more broad dynamics. Through our described model (6) can describe the dynamics of any population species.
(6) Through this model (6), we can define the local asymptotic stability of the single species population at its environmental carrying capacity level provided the available resource to live always remains constant, say some k_1. In this situation we have, $\underline{x} = k_1 - \epsilon$ and $\bar{x} = k_1 + \epsilon$ for $\epsilon > 0$ and then the population always remains within $(k_1 - \epsilon,\ k_1 + \epsilon)$ and hence it can be concluded that if the available resource is sufficient for k_1 amount of population biomass, and k_1 will

always remain constant then the population will asymptotically stable at $x = k_1$ (Fig. 3). Thus we can derive the result of the logistic growth model through our model (6).

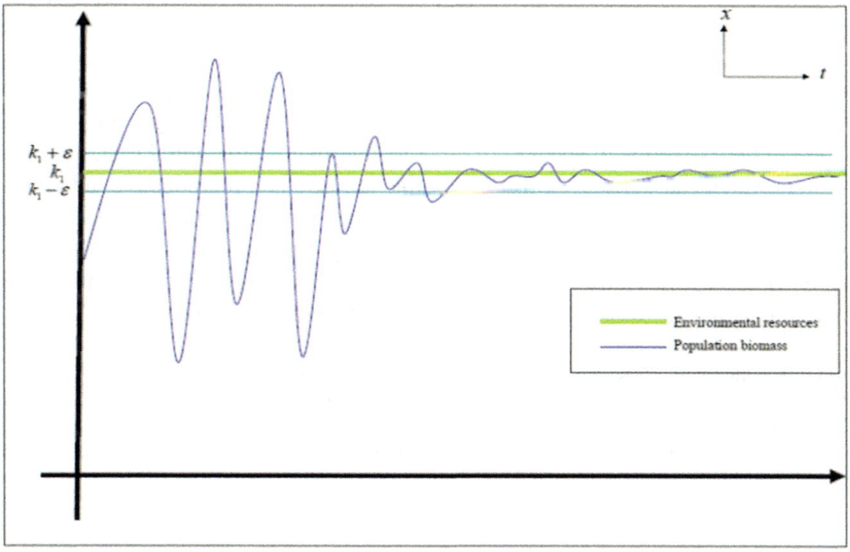

Fig. 3 Asymptotic stability of population biomass near where is the upper limit of the population biomass such that the population can live healthy with the help of available resource which always remains same

3 An Application in Ecology

Our present analysis can be used to solve different ecological problems concerning the growth of population species. For example, when we are concerned about the pest control problem or vector disease problem, to keep our society free from pests or vectors, we have to reduce their corresponding resources. If a living creature can be used to eat same food as those of the larvae of mosquitoes, then the resource for mosquitoes' larvae would be reduced. Thus, the growth of mosquitoes can be checked.

4 Conclusions

This paper introduces a new ordinary differential equation-based mathematical model for a single species model in order to study the dynamics of a single-dimensional population species, where a significant number of individuals have no reproduction power. Our present assumption and the model of type (6) would be very helpful in studying the growth of single species, and it is very close to reality. In spite of being the model the first-order ODE, it is capable of capturing the dynamics of non-linear ODE-based growth like logistic growth. The present study draws light on a possible cause of the extinction of species. In order to control the agricultural pest, notorious vectors which cause diseases, if it is possible to make non-reproductive individuals greater than some threshold for a continuous time period, then those pests and the vector populations can be controlled significantly. On the other hand, if the alive member of a species is very low and the species is threatening to go to extinct, then one significant way to keep that species exist is to increase the birth rate.

Acknowledgements Research work of Suvankar Majee is financially supported by Council of Scientific and Industrial Research (CSIR), India (File No. 08/003(0142)/2020-EMR-I, dated: 18th March 2020). Moreover, the authors are very much thankful to the anonymous reviewers for their constructive comments and helpful suggestions to improve both the quality and presentation of the manuscript significantly.

References

1. Kot, M.: Elements of Mathematical Ecology. Cambridge University Press (2001)
2. Britton, N.F.: Essential Mathematical Biology. Springer (2003)
3. Tsoularis, A., Wallace, J.: Analysis of logistic growth models. Math. Biosci. **179**(1), 21–55 (2002)
4. Bhowmick, A.R., Chattopadhyay, G., Bhattacharya, S.: Simultaneous identification of growth law and estimation of its rate parameter for biological growth data: a new approach. J. Biol. Phys. **40**(1), 71–95 (2014)
5. Bhowmick, A.R., Bhattacharya, S.: A new growth curve model for biological growth: some inferential studies on the growth of Cirrhinus mrigala. Math. Biosci. **254**, 28–41 (2014)
6. Di Crescenzo, A., Spina, S.: Analysis of a growth model inspired by Gompertz and Korf laws, and an analogous birth-death process. Math. Biosci. **282**, 121–134 (2016)
7. Koya, P.R., Goshu, A.T.: Generalized mathematical model for biological growths. Open J. Modell. Simul. (2013)
8. Von Bertalanffy, L.: A quantitative theory of organic growth (inquiries on growth laws. II). Human Biol. **10**(2), 181–213 (1938)
9. Richards, F.J.: A flexible growth function for empirical use. J. Exper. Botany **10**(2), 290–301 (1959)
10. Gompertz, B.: On the nature of the function expressive of the law of human mortality, and on a new mode of determining the value of life contingencies. Philos. Trans. R. Soc. Lond. **155**, 513–583 (1825)
11. Nelder, J.A.: The fitting of a generalization of the logistic curve. Biometrics **17**(1), 89–110 (1961)
12. Rawlings, J.O., Pantula, S.G., Dickey, D.A.: Applied Regression Analysis: A Research Tool. Springer Science & Business Media (2001)

13. Misra, O.P., Chaturvedi, D.: Fate of dissolved oxygen and survival of fish population in aquatic ecosystem with nutrient loading: a model. Model. Earth Syst. Environ. **2**(3), 1–14 (2016)
14. Misra, O.P., Babu, A.R.: Mathematical study of a Leslie-Gower-type tritrophic population model in a polluted environment. Model. Earth Syst. Environ. **2**(1), 29 (2016)
15. Kumar, A., Agrawal, A.K., Hasan, A., Misra, A.K.: Modeling the effect of toxicant on the deformity in a subclass of a biological species. Model. Earth Syst. Environ. **2**(1), 40 (2016)
16. Couchamp, F., Berec, L., Gascoigne, J.: Allee Effects in Ecology and Conservation. OUP Oxford (2008)

A Numerical Approximation of the KdV-Kawahara Equation via the Collocation Method

Seydi Battal Gazi Karakoc and Derya Yıldırım Sucu

Abstract This paper presents a finite element scheme for numerical solution of the Korteweg-de Vries-Kawahara (KdV-K) equation using septic B-spline functions as approximate functions. L_2 and L_∞ error norms are calculated for single solutions to show the practicality and robustness of the proposed scheme. Applying von-Neumann theory, we demonstrate that the scheme is marginally stable. Obtained numerical results have been illustrated with tables and graphics for easy visualization of properties of the problem modelled. Numerical experiment supports the correctness and reliability of the method.

Keywords KdV-Kawahara equation · Finite element method · Collocation

1 Introduction

Almost all physical processes encountered in nature are defined by various types of non-linear partial differential equations (NLPDEs). Understanding the structure of these NLPDEs and seeking their solutions is of prime importance for scientists, as their solutions illuminate the way to understand the behavior of systems and help to predict the development of the process in nature [1]. However, usually it is difficult to find their solutions analytically and sometimes it is almost impossible. Therefore many researchers have been working on to obtain efficient and high accurate numerical algorithms to overcome such problems [2].

In 1895, Dutch mathematician Korteweg and de-Vries defined a PDE (KdV) equation

S. B. G. Karakoc (✉) · D. Y. Sucu
Department of Mathematics, Faculty of Science and Art, Nevsehir Haci Bektas Veli University, Nevsehir 50300, Turkey
e-mail: sbgk44@gmail.com

S. Banerjee and A. Saha (eds.), *Nonlinear Dynamics and Applications*, Springer Proceedings in Complexity,
https://doi.org/10.1007/978-3-030-99792-2_77

$$u_t + uu_x + u_{xxx} = 0, \tag{1}$$

which models the propagation of wide waves in shallow water waves to model Russell's soliton phenomenon. They observed that these solitons treat as particles [3]. Kawahara equation which is known as a fifth-order KdV type equation has the following form

$$u_t + \kappa uu_x + qu_{xxx} - ru_{xxxxx} = 0, \tag{2}$$

where κ, q, r are constants [4, 5]. Another form of the Kawahara equation is following modified Kawahara equation [6, 7]:

$$u_t + \kappa u^2 u_x + qu_{xxx} - ru_{xxxxx} = 0. \tag{3}$$

By coupling the KdV equation and the modified Kawahara equation, one can obtain the following generalized Korteweg-de Vries-Kawahara (GKdV-K) equation [8, 9]

$$u_t + \alpha u_x + \kappa u^p u_x + qu_{xxx} - ru_{xxxxx} = 0, \tag{4}$$

where $p \geqslant 1$ is a positive integer, $\alpha \geq 0, \kappa > 0, q > 0$ and $r > 0$ [10]. In recent years, many authors have been interested in the solution of the equation [11–16].

The main form of the paper can be outlined in brief as follows: In Sect. 2, septic B-spline approximation is introduced and finite element solution of KdV-K (for $p = 1$) equation is proposed. Stability analysis of method has been done in Sect. 4. Sect. 4.1 exhibited numerical application and it's results with table and graphs to see the performance and accuracy of the method. Finally, in Sect. 5, a brief conclusion about the presented method is given.

2 Application of the Numerical Method

In this section, our goal is to find numerical solution of the KdV-K Eq. (4) with following initial and homogeneous boundary conditions below:

$$\begin{array}{ll} u(x,0) = f(x), & a \leq x \leq b, \\ u(a,t) = 0, & u(b,t) = 0, \\ u_x(a,t) = 0, & u_x(b,t) = 0, \\ u_{xx}(a,t) = 0, & u_{xx}(b,t) = 0, \end{array} \tag{5}$$

where $f(x)$ is a detected function. To obtain the solution on the interval $a \leq x \leq b$ division $a = x_0 < x_1 < ... < x_N = b$ of the space domain is imagined scattered uniformly with $h = \frac{b-a}{N} = (x_{m+1} - x_m)$ for $m = 1(1)N$. The septic B-spline functions at the nodes x_m are described on the solution region $[a,\ b]$ by Prenter [17]:

$$\phi_m(x) = \frac{1}{h^7}\begin{cases} l, & [x_{m-4}, x_{m-3}], \\ l - 8m, & [x_{m-3}, x_{m-2}], \\ l - 8m + 28n, & [x_{m-2}, x_{m-1}], \\ l - 8m + 28n - 56p, & [x_{m-1}, x_m], \\ r - 8s + 28y - 56z, & [x_m, x_{m+1}], \\ r - 8s + 28y, & [x_{m+1}, x_{m+2}], \\ r - 8s, & [x_{m+2}, x_{m+3}], \\ r, & [x_{m+3}, x_{m+4}], \\ 0, & otherwise \end{cases} \tag{6}$$

where $l = (x - x_{m-4})^7, m = (x - x_{m-3})^7, n = (x - x_{m-2})^7, p = (x - x_{m-1})^7, r = (x_{m+4} - x)^7$, $s = (x_{m+3} - x)^7$, $y = (x_{m+2} - x)^7$, $z = (x_{m+1} - x)^7$. Now, we continue the numerical treatment, which we will apply using the septic B-spline collocation finite element method, by generating an approximate solution for the equation. Approximate solution $u_N(x, t)$ for analytical solution $u(x, t)$ are sought in the following equality,

$$u_N(x, t) = \sum_{m=-3}^{N+3} \phi_m(x)\sigma_m(t) \tag{7}$$

where $\sigma_m(t)$ are time dependent unknown coefficients specified from the boundary conditions [18]. In each element, when we use the following equality,

$$h\sigma = x - x_m, \quad 0 \le \xi \le 1 \tag{8}$$

Equation (6) is defined in terms of σ on interval [0, 1] as [19]:

$$\begin{aligned}
\phi_{m-3} &= 1 - 7\sigma + 21\sigma^2 - 35\sigma^3 + 35\sigma^4 - 21\sigma^5 + 7\sigma^6 - \sigma^7, \\
\phi_{m-2} &= 120 - 392\sigma + 504\sigma^2 - 280\sigma^3 + 84\sigma^5 - 42\sigma^6 + 7\sigma^7, \\
\phi_{m-1} &= 1191 - 1715\sigma + 315\sigma^2 + 665\sigma^3 - 315\sigma^4 - 105\sigma^5 + 105\sigma^6 - 21\sigma^7, \\
\phi_m &= 2416 - 1680\sigma + 560\sigma^4 - 140\sigma^6 + 35\sigma^7, \\
\phi_{m+1} &= 1191 + 1715\sigma + 315\sigma^2 - 665\sigma^3 - 315\sigma^4 + 105\sigma^5 + 105\sigma^6 - 35\sigma^7, \\
\phi_{m+2} &= 120 + 392\sigma + 504\sigma^2 + 280\sigma^3 - 84\sigma^5 - 42\sigma^6 + 21\sigma^7, \\
\phi_{m+3} &= 1 + 7\sigma + 21\sigma^2 + 35\sigma^3 + 35\sigma^4 + 21\sigma^5 + 7\sigma^6 - \sigma^7, \\
\phi_{m+4} &= \sigma^7.
\end{aligned} \tag{9}$$

The values of u_m and its derivatives at the knots are calculated as:

$$\begin{aligned}
u_N(x_m,t) &= \sigma_{m-3}+120\sigma_{m-2}+1191\sigma_{m-1}+2416\sigma_m+1191\sigma_{m+1}+120\sigma_{m+2}+\sigma_{m+3},\\
u'_m &= \tfrac{7}{h}(-\sigma_{m-3}-56\sigma_{m-2}-245\sigma_{m-1}+245\sigma_{m+1}+56\sigma_{m+2}+\sigma_{m+3}),\\
u''_m &= \tfrac{42}{h^2}(\sigma_{m-3}+24\sigma_{m-2}+15\sigma_{m-1}-80\sigma_m+15\sigma_{m+1}+24\sigma_{m+2}+\sigma_{m+3}),\\
u'''_m &= \tfrac{210}{h^3}(-\sigma_{m-3}-8\sigma_{m-2}+19\sigma_{m-1}-19\sigma_{m+1}+8\sigma_{m+2}+\sigma_{m+3}),\\
u^{iv}_m &= \tfrac{840}{h^4}(\sigma_{m-3}-9\sigma_{m-1}+16\sigma_m-9\sigma_{m+1}+\sigma_{m+3}),\\
u^{v}_m &= \tfrac{2520}{h^5}(-\sigma_{m-3}+4\sigma_{m-2}-5\sigma_{m-1}+5\sigma_{m+1}-4\sigma_{m+2}+\sigma_{m+3}).
\end{aligned} \tag{10}$$

Now, using (7) and (10) into Eq. (4), the following general form of equation is reached:

$$\begin{aligned}
&\dot{\sigma}_{m-3}+120\dot{\sigma}_{m-2}+1191\dot{\sigma}_{m-1}+2416\dot{\sigma}_m+1191\dot{\sigma}_{m+1}+120\dot{\sigma}_{m+2}+\dot{\sigma}_{m+3}\\
&+(\alpha+\kappa Z_m)\tfrac{7}{h}(-\sigma_{m-3}-56\sigma_{m-2}-245\sigma_{m-1}+245\sigma_{m+1}+56\sigma_{m+2}+\sigma_{m+3})\\
&+q\tfrac{210}{h^3}(-\sigma_{m-3}-8\sigma_{m-2}+19\sigma_{m-1}-19\sigma_{m+1}+8\sigma_{m+2}+\sigma_{m+3})\\
&-r\tfrac{2520}{h^5}(-\sigma_{m-3}+4\sigma_{m-2}-5\sigma_{m-1}+5\sigma_{m+1}-4\sigma_{m+2}+\sigma_{m+3})=0,
\end{aligned} \tag{11}$$

where $\dot{\sigma}=\frac{d\sigma}{dt}$ and

$$Z_m=u=(\sigma_{m-3}+120\sigma_{m-2}+1191\sigma_{m-1}+2416\sigma_m+1191\sigma_{m+1}+120\sigma_{m+2}+\sigma_{m+3}).$$

Both the finite difference approach and the Crank-Nicolson diagrams described below can be applied to the Eq. (11):

$$\sigma_i=\frac{\sigma_i^{n+1}+\sigma_i^n}{2},\ \dot{\sigma}_i=\frac{\sigma_i^{n+1}-\sigma_i^n}{\Delta t}. \tag{12}$$

So, the above operation allows us to derive a recursion relationship between two time levels based on the parameters σ_i^{n+1} and σ_i^n as:

$$\begin{aligned}
&\lambda_1\sigma_{m-3}^{n+1}+\lambda_2\sigma_{m-2}^{n+1}+\lambda_3\sigma_{m-1}^{n+1}+\lambda_4\sigma_m^{n+1}+\lambda_5\sigma_{m+1}^{n+1}+\lambda_6\sigma_{m+2}^{n+1}+\lambda_7\sigma_{m+3}^{n+1}\\
&=\lambda_7\sigma_{m-3}^{n}+\lambda_6\sigma_{m-2}^{n}+\lambda_5\sigma_{m-1}^{n}+\lambda_4\sigma_m^{n}+\lambda_3\sigma_{m+1}^{n}+\lambda_2\sigma_{m+2}^{n}+\lambda_1\sigma_{m+3}^{n},
\end{aligned} \tag{13}$$

where

$$\begin{aligned}
&\lambda_1=[1-EZ_m-M+K], && \lambda_2=[120-56EZ_m-8M-4K],\\
&\lambda_3=[1191-245EZ_m+19M+5K], && \lambda_4=[2416],\\
&\lambda_5=[1191+245EZ_m-19M-5K], && \lambda_6=[120+56EZ_m+8M+4K],\\
&\lambda_7=[1+EZ_m+M-K], &&\\
&E=\tfrac{7}{2h}\omega\Delta t,\quad M=\tfrac{105}{h^3}q\Delta t,\ K=\tfrac{2520}{h^5}r\Delta t,\ \omega=(\alpha+\kappa Z_m). &&
\end{aligned} \tag{14}$$

In this algebraic system (13), the number of linear equations are less than the number of unknown coefficients, that is, the system contains the $(N+1)$ equation and $(N+7)$ unknown time-dependent parameters [20]. The best way to obtain a unique solution is to remove the six unknowns $\sigma_{-3}, \sigma_{-2}, \sigma_{-1}, \ldots, \sigma_{N+1}, \sigma_{N+2}$, and σ_{N+3} from the system. This procedure is implemented using the boundary conditions with the values of u and after removing the unknowns, using a matrix system consisting of linear equations $(N+1)$ unknown parameters $d^n = (\sigma_0, \sigma_1, \ldots, \sigma_N)^T$ is obtained the following matrix vector form:

$$R\mathbf{d}^{\mathbf{n}+1} = S\mathbf{d}^{\mathbf{n}}. \tag{15}$$

3 Stability Analysis

In this section, Von-Neumann theory was used for the stability of the algorithm. To display stability analysis, KdV-Kawahara equation was linearized by supposing that quantities u^p in nonlinear term $u^p u_x$ is locally constant. Growth factor ξ of a characteristic Fourier mode is identified as:

$$\sigma_m^n = \xi^n e^{imkh}, \tag{16}$$

here $i = \sqrt{-1}$. Putting the equality (16) into the iterative systems (13), gives the following growth factor

$$\xi = \frac{\rho_1 - i\rho_2}{\rho_1 + i\rho_2}, \tag{17}$$

where

$$\begin{aligned} \rho_1 &= 2\cos(3kh) + 240\cos(2kh) + 2382\cos(kh) + 2416, \\ \rho_2 &= (2M + 2T + 2E)\sin(3kh). \end{aligned} \tag{18}$$

$|\xi| = 1$ is obtained when we take the modulus of Eq. (17). In this way, we demonstrate that scheme (13) is unconditionally stable under the present conditions.

4 Numerical Experiment and Discussion

In this part, we illustrate our method, improved in Sect. 2, to the KdV-K equation for single solitary wave. Effectiveness of the suggested method will be checked with the L_2 and L_∞ error norms given as [21]

$$L_2 = \left\| u^{exact} - u_N \right\|_2 \simeq \sqrt{h \sum_{j=1}^{N} \left| u_j^{exact} - (u_N)_j \right|^2}, \tag{19}$$

and

$$L_\infty = \left\| u^{exact} - u_N \right\|_\infty \simeq \max_j \left| u_j^{exact} - (u_N)_j \right|, \quad j = 1, 2, ..., N. \tag{20}$$

KdV-K equation (4) possesses three conservation constants given by

$$\begin{aligned} I_1 &= \int_a^b U dx \simeq h \sum_{j=1}^N U_j^n, \\ I_2 &= \int_a^b U^2 dx \simeq h \sum_{j=1}^N (U_j^n)^2, \\ I_3 &= \int_{-\infty}^{\infty} U^{p+2}(x,t) dx. \end{aligned} \tag{21}$$

which correspond to conversation of mass, momentum and energy, respectively.

4.1 Case 1

KdV-K equation has the following exact solution:

$$u(x,t) = \frac{105}{169} sech^4 [\frac{1}{2\sqrt{13}} \left(x - \frac{205}{169} t - x_0 \right)], \tag{22}$$

x_0 is the center of the solitary wave. For the equation following initial condition is chosen

$$u(x,0) = \frac{105}{169} sech^4 [\frac{1}{2\sqrt{13}} (x - x_0)]. \tag{23}$$

To illustrate the validity of our numerical scheme, the algorithm was run up to time $t = 30$. The solitary wave has amplitude $A = 0.62130$ at $x = 0$. In simulation calculations, typical values $\Delta t = 0.01$ with $h = 0.1$ were used. In Table 1, the conserved quantities and error norms L_2 and L_∞ for different time levels and different step sizes were presented. The table shows that the three conserved amounts remained nearly constant over time and the changes in the amounts are in good agreement with their analytical values. Thus, the effects of the amount of sorting points on the numerical method can be seen more easily. The calculated L_2 and L_∞ errors were found to be satisfactorily small. These errors hardly change as time progresses. If we examine Fig. 1, two and three dimensional state of the bell shaped solitary wave solutions produced from $t = 0$ to $t = 30$ can be clearly seen. From the figures, it can be noticed that studied method executes the propagation movement of a single wave, it conserves the amplitude and shape.

Table 1 Invariants and error norms for single solitary wave

Time	I_1	I_2	I_3	L_2	L_∞
0	5.9734098	1.2724981	0.6839459	0.00000000	0.00000000
5	5.9734128	1.2724980	0.6839449	0.00012014	0.00014431
10	5.9733848	1.2724980	0.6839444	0.00012188	0.00015202
15	5.9733818	1.2724980	0.6839422	0.00012665	0.00015689
20	5.9733887	1.2724980	0.6839417	0.00013701	0.00016202
25	5.9733943	1.2724980	0.6839418	0.00015282	0.00016746
30	5.9733890	1.2724980	0.6839406	0.00017237	0.00017310

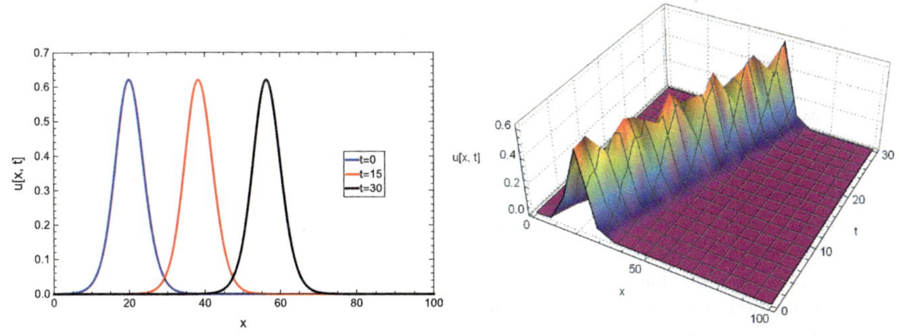

Fig. 1 Numerical solutions of KdV-K equation at different time stages with $\Delta t = 0.01$ and $h = 0.1$

5 Conclusion

In this paper, numerical solution of KdV-K equation has been investigated by considering some fixed selection initial and boundary conditions. Our numerical algorithm is shown to be unconditionally stable. The algorithm has been tested for single solitary wave in which the exact solution is known. L_2 and L_∞ error norms and the invariants are calculated to show the reliability and accuracy of the method. From these calculations, we can say that the proposed method yield good enough results. For our method, we portray some graphical illustrations of the obtained solutions of the equation. It may be concluded that the method used here is a powerful, efficient and powerful technique for solving a wide class of nonlinear evolution equations.

References

1. Sari, M., Gurarslan, M.: A sixth-order compact nite difference scheme to the numerical solutions of Burgers' equation. Appl. Math. Comput. **208**, 475–483 (2009)
2. Saleem, S., Hussain, M.Z.: Numerical solution of nonlinear fifth-order KdV-type partial differential equations via Haar Wavelet. Int. J. Appl. Comput. Math. **6**(164), 1–16 (2020)

3. Goswami, A., Singh, J., Kumar, D.: Numerical simulation of fth order KdV equations occurring in magneto-acoustic waves. Ain Shams Eng. J. **9**, 2265–2273 (2018)
4. Chen, M.: Internal controllability of the Kawahara equation on a bounded domain. Nonlinear Anal. **185**, 356–373 (2019)
5. Kawahara, T.: Oscillatory solitary waves in dispersive media. J. Phys. Soc. Japan **33**, 260–264 (1972)
6. Biswas, A.: Solitary wave solution for the generalized Kawahara equation. Appl. Math. Lett. **22**, 208–210 (2009)
7. Ak, T., Karakoc, S.B.G.: A numerical technique based on collocation method for solving modified Kawahara equation. J. Ocean Eng. Sci. **3**, 67–75 (2018)
8. Ceballos, J.C., Sepúlveda, J.C., Villagrán, O.P.V.: The Korteweg-de Vries-Kawahara equation in a bounded domain and some numerical results. Appl. Math. Comput. **190**, 912–936 (2007)
9. Sepúlveda, M., Villagrán, O.P.V.: Numerical method for a transport equation perturbed by dispersive terms of 3rd and 5th order. Sci. Ser. A Math. Sci. **13**, 13–21 (2006)
10. Wang, X., Cheng, H.: Solitary wave solution and a linear mass-conservative difference scheme for the generalized Korteweg-de Vries-Kawahara equation. Comput. Appl. Math. **40**(273), 1–26 (2021)
11. Assas, L.M.B.: New exact solutions for the Kawahara equation using Exp-function method. J. Comput. Appl. Math. **233**, 97–102 (2009)
12. Ye, Y.H., Mo, L.F.: He's variational method for the Benjamin-Bona-Mahony equation and the Kawahara equation. Comput. Math. Appl. **58**, 2420–2422 (2009)
13. Wazwaz, A.M.: New solitary wave solutions to the modified Kawahara equation. Phys. Lett. A **360**, 588–592 (2007)
14. Kaya, D.: An explicit and numerical solutions of some fifth-order KdV equation by decomposition method. Appl. Math. Comput. **144**, 353–363 (2003)
15. He, D.: Exact solitary solution and a three-level linearly implicit conservative finite difference method for the generalized Rosenau-Kawahara-RLW equation with generalized Novikov type perturbation. Nonlinear Dyn. **85**, 479–498 (2016)
16. Nanta, S., Yimnet, S., Poochinapan, K., Wongsaijai, B.: On the identification of nonlinear terms in the generalized Camassa-Holm equation involving dual-power law nonlinearities. Appl. Numer. Math. **160**, 386–421 (2021)
17. Prenter, P.M.: Splines and Variational Methods. Wiley, New York (1975)
18. Karakoc, S.B.G., Zeybek, H.: A septic B spline collocation method for solving the generalized equal width wave equation. Kuwait J. Sci. **43**, 20–31 (2016)
19. Karakoc, S.B.G., Saha, A., Sucu, D.: A novel implementation of Petrov-Galerkin method to shallow water solitary wave pattern and super peiodic traveling wave and its multistability: Generalized Korteweg-de Vries equation. Chinese J. Phys. **68**, 605–617 (2020)
20. Karakoc, S.B.G., Zeybek, H.: Solitary wave solutions of the GRLW equation using septic B spline collocation method. Appl. Math. Comput. **289**, 159–171 (2016)
21. Karakoc, S.B.G.: A new numerical application of the generalized Rosenau-RLW equation. Scientia Iranica B **27**(2), 772–783 (2020)

Approximate Solutions to Pseudo-Parabolic Equation with Initial and Boundary Conditions

Nishi Gupta and Md. Maqbul

Abstract This manuscript concerned with the pseudo-parabolic equation along with initial and boundary conditions. We prove the existence and uniqueness of a solution with the aid of Rothe's time-discretization technique. We have exemplified the main result.

Keywords Pseudo-parabolic equation · Semidiscretization method · Boundary conditions · Strong solution

1 Introduction

Rothe's time-discretization method has been opted by many researcher for solving differential equations. In this paper, approximate solutions, as well as the existence of unique strong solution, are established by applying Rothe's method for the following pseudo-parabolic equation

$$\frac{\partial g}{\partial t} - \mu \frac{\partial^3 g}{\partial t \partial y^2} - \mu \frac{\partial^2 g}{\partial y^2} = p(t, g(t, y)), \quad t \in (0, \top], \;\; y \in (0, 1) \tag{1}$$

subject to the following initial and the boundary conditions

$$g(0, y) = G_0(y) \quad \text{and} \quad \frac{\partial^2 g}{\partial y^2}(0, y) = G_1(y) \text{ for all } y \in (0, 1), \tag{2}$$

$$g(t, 0) = g(t, 1) \text{ for all } t \in [0, \top], \tag{3}$$

where, $\top > 0$, $\mu > 0$, and p, G_0, and G_1 are some given suitable functions.

N. Gupta (✉) · Md. Maqbul
National Institute of Technology Silchar, Silchar, Assam 788010, India
e-mail: nishigupta4792@gmail.com

S. Banerjee and A. Saha (eds.), *Nonlinear Dynamics and Applications*,
Springer Proceedings in Complexity,
https://doi.org/10.1007/978-3-030-99792-2_78

Type (1) equations have various applications in many physical situations, namely, in the study of homogeneous fluid seepage in fissured rocks [1], heat conduction and newtonian fluids theory [2, 3]. Maqbul and Raheem [5] established unique solution to semilinear pseudo parabolic equation with integral conditions by using Rothe's method.

2 Abstract Formulation and Preliminaries

Defining $s(t, y)$ as

$$s(t, y) = g(t, y) - \mu \frac{\partial^2 g}{\partial y^2}. \tag{4}$$

Then, (1)–(3) becomes

$$\frac{\partial s}{\partial t} + s(t, y) = g(t, y) + p(t, g(t, y)), \tag{5}$$

$$g(0, y) = G_0(y) \quad \text{and} \quad s(0, y) = S_0(y), \tag{6}$$

$$g(t, 0) = g(t, 1), \tag{7}$$

where, $S_0(y) = G_0(y) - \mu G_1(y)$.

Let $L^2(0, 1) = \mathcal{H}$ be the Hilbert space. Defining two functions $g : [0, \mathrm{T}] \to \mathcal{H}$, $s : [0, \mathrm{T}] \to \mathcal{H}$, and the nonlinear map $p : [0, \mathrm{T}] \times \mathcal{H} \to \mathcal{H}$ as

$$g(t)(y) = g(t, y), \quad s(t)(y) = s(t, y), \quad p(t, g(t))(y) = p(t, g(t, y)),$$

respectively. Consider the following presumptions:

(C1) T and μ are positive real numbers.
(C2) $G_0, \ G_1 \in \mathcal{H}$.
(C3) $p : [0, \mathrm{T}] \times \mathcal{H} \to \mathcal{H}$ holds the Lipschitz inequality that $\exists\ \zeta > 0$ as

$$\|p(t, g) - p(\delta, s)\| \le \zeta(|t - \delta| + \|g - s\|) \quad \text{for all } t, \delta \in [0, \mathrm{T}], \quad \text{for all } g, s \in \mathcal{H}.$$

$\mathcal{D}(\cdot)$ and $\mathcal{R}(\cdot)$, domain and range of an operator respectively. Consider the linear operator $\mathcal{A}$ defined as

$$\mathcal{D}(\mathcal{A}) := \Big\{\ell \in \mathcal{H} : \ell'' \in \mathcal{H}, \ell(0) = \ell(1)\Big\}, \quad \mathcal{A}\ell = -\ell''.$$

Then, $-\mathcal{A}$ is an infinitesimal generator of C_0-semigroups of contractions in Hilbert space. Therefore, (4)–(7) become

$$g(t) + \mu \mathcal{A} g(t) = s(t), \tag{8}$$

$$\frac{ds}{dt} + s(t) = g(t) + p(t, g(t)), \tag{9}$$

$$g(0) = G_0, \quad s(0) = S_0. \tag{10}$$

Hence, both (1)–(3) and (8)–(10) represent the same problem.

3 Discretization Scheme and a Priori Estimates

Assume that $(C1) - (C3)$ are true for the rest of this article. Dividing $[0, \top]$ into m subintervals $[t_{k-1}^m, t_k^m]$ with $\hbar_m = \frac{\top}{m}$ for each $m \in \mathbb{N}$, where $t_k^m = k\hbar_m$, for $k = 1, 2, \ldots, m$. Let $g_0^m = G_0$ and $s_0^m = S_0$ for all $m \in \mathbb{N}$. Successively, $\{g_k^m\}$ and $\{s_k^m\}$ are unique solution of the equations.

$$g_k^m + \mu \mathcal{A} g_k^m = s_k^m \tag{11}$$

and

$$\frac{s_k^m - s_{k-1}^m}{\hbar_m} + s_k^m = g_{k-1}^m + p(t_k^m, g_{k-1}^m). \tag{12}$$

Lemma 1 *There exists* $\kappa > 0$ *such that*

$$\|g_k^m\| + \|s_k^m\| \le \kappa, \ k = 1, 2, \ldots, m, \quad m \ge 1. \tag{13}$$

Proof In view of (11) and Theorem 1.4.3 of [6]

$$\|g_k^m\| \le \|s_k^m\|. \tag{14}$$

By (12),

$$\left\langle \frac{1}{\hbar_m} s_k^m, s_k^m \right\rangle + \langle s_k^m, s_k^m \rangle = \left\langle \frac{1}{\hbar_m} s_{k-1}^m + g_{k-1}^m + p(t_k^m, g_{k-1}^m), s_k^m \right\rangle. \tag{15}$$

Therefore,

$$\|s_k^m\| \le \|s_{k-1}^m\| + \hbar_m \|g_{k-1}^m\| + \hbar_m \|p(t_k^m, g_{k-1}^m)\|. \tag{16}$$

Thus,

$$\|s_k^m\| \leq \|s_0^m\| + \hbar_m \sum_{i=1}^{k} (\|g_{i-1}^m\| + \|p(t_i^m, g_{i-1}^m)\|)$$
$$\leq \|S_0\| + (\zeta + 1)\hbar_m \sum_{i=1}^{k} \|s_{i-1}^m\| + \zeta \top^2 + \top\|p(0,0)\|. \quad (17)$$

Therefore,

$$\|s_k^m\| \leq \alpha_1 + \beta_1 \hbar_m \sum_{i=1}^{k} \|s_i^m\|, \quad (18)$$

where, $\alpha_1 = (\zeta\top + \top + 1)\|S_0\| + \zeta\top^2 + \top\|p(0,0)\|$ and $\beta_1 = \zeta + 1$.
By Gronwall's inequality,

$$\|s_k^m\| \leq \frac{\alpha_1}{1 - \beta_1 \hbar_m} \exp\left(\frac{\beta_1 (j-1)\hbar_m}{1 - \beta_1 \hbar_m}\right) \leq \frac{\alpha_1}{1 - \beta_1 \hbar_m} \exp\left(\frac{\top \beta_1}{1 - \beta_1 \hbar_m}\right).$$

Therefore,
$\|s_k^m\| \leq \kappa$, where $\kappa = \sup_{m \in \mathbb{N}} \left[\frac{\alpha_1}{1-\beta_1\hbar_m} \exp\left(\frac{\top\beta_1}{1-\beta_1\hbar_m}\right)\right]$.
By (14), $\|g_k^m\| \leq \kappa$.

Lemma 2 *There exists* $\kappa > 0$ *such that*

$$\|g_k^m - g_{k-1}^m\| + \|s_k^m - s_{k-1}^m\| \leq \hbar_m \kappa, \; k = 1, 2, \ldots, m, \quad m \geq 1. \quad (19)$$

Proof In the view of (11), for $k = 2, 3 \ldots, m$, and Theorem 1.4.3 of [6]

$$\|g_k^m - g_{k-1}^m\| \leq \|s_k^m - s_{k-1}^m\|. \quad (20)$$

By (12), for $k = 2, 3 \ldots, m$,

$$\frac{1}{\hbar_m}\|s_k^m - s_{k-1}^m\| \leq \|g_{k-1}^m - g_{k-2}^m\| + \|p(t_k^m, g_{k-1}^m) - p(t_{k-1}^m, g_{k-2}^m)\|. \quad (21)$$

Therefore, for $k = 2, 3 \ldots, m$,

$$\frac{1}{\hbar_m}\|s_k^m - s_{k-1}^m\| \leq \sum_{i=2}^{k} \Big(\|g_{i-1}^m - g_{i-2}^m\| + \|p(t_i^m, g_{i-1}^m) - p(t_{i-1}^m, g_{i-2}^m)\|\Big)$$
$$\leq \zeta(k-1)h_m + (\zeta + 1)\sum_{i=2}^{k} \|g_{i-1}^m - g_{i-2}^m\|$$
$$(22)$$

Thus,

$$\frac{1}{\hbar_m}\|s_k^m - s_{k-1}^m\| \leq \alpha_2 + \beta_2 \sum_{i=1}^{k} \|s_i^m - s_{i-1}^m\|, \tag{23}$$

where, $\alpha_2 = \top\zeta$ and $\beta_2 = \zeta + 1$.
By Gronwall's inequality,

$$\frac{1}{\hbar_m}\|s_k^m - s_{k-1}^m\| \leq \frac{\alpha_2}{1-\beta_2\hbar_m}\exp\left(\frac{\beta_2(k-1)\hbar_m}{1-\beta_2\hbar_m}\right) \leq \frac{\alpha_2}{1-\beta_2\hbar_m}\exp\left(\frac{\top\beta_2}{1-\beta_2\hbar_m}\right).$$

Therefore,
$\frac{1}{\hbar_m}\|s_k^m - s_{k-1}^m\| \leq \kappa$, where $\kappa = \sup_{m\in\mathbb{N}}\left[\frac{\alpha_2}{1-\beta_2\hbar_m}\exp\left(\frac{\top\beta_2}{1-\beta_2\hbar_m}\right)\right]$.
By (20), $\frac{1}{\hbar_m}\|g_k^m - g_{k-1}^m\| \leq \kappa$.

4 Convergence

Let $m \in \mathbb{N}$. Defining $\mathcal{U}^m(t)$, $\mathcal{W}^m(t)$, $\mathcal{X}^m(t)$ and $\mathcal{Y}^m(t)$ as

$$\mathcal{U}^m(t) = \begin{cases} g_0^m & \text{if} \quad t = t_0^m, \\ g_k^m & \text{if} \quad t \in (t_{k-1}^m, t_k^m], \end{cases} \tag{24}$$

$$\mathcal{W}^m(t) = \begin{cases} s_0^m & \text{if} \quad t = t_0^m, \\ s_k^m & \text{if} \quad t \in (t_{k-1}^m, t_k^m], \end{cases} \tag{25}$$

$$\mathcal{X}^m(t) = \begin{cases} g_0^m & \text{if} \quad t = t_0^m, \\ g_{k-1}^m + (t - t_{k-1}^m)g_k^m & \text{if} \quad t \in (t_{k-1}^m, t_k^m], \end{cases} \tag{26}$$

and

$$\mathcal{Y}^m(t) = \begin{cases} s_0^m & \text{if} \quad t = t_0^m, \\ s_{k-1}^m + \frac{1}{\hbar_m}(t - t_{k-1}^m)(s_k^m - s_{k-1}^m) & \text{if} \quad t \in (t_{k-1}^m, t_k^m], \end{cases} \tag{27}$$

where, $k = 1, 2, \ldots, m$. For $t \in (t_{k-1}^m, t_k^m]$, $1 \leq k \leq m$, define $\mathcal{P}^m(t)$ by

$$\mathcal{P}^m(t) = g_{k-1}^m + p(t_k^m, g_{k-1}^m), \tag{28}$$

then (11) and (12) comes out to be

$$\frac{d^-\mathcal{X}^m}{dt}(t) + \mu\mathcal{A}\mathcal{U}^m(t) = \mathcal{W}^m(t), \quad t \in (0, \top], \tag{29}$$

$$\frac{d^-\mathcal{Y}^m}{dt}(t) + \mathcal{W}^m(t) = \mathcal{P}^m(t), \quad t \in (0, \top], \tag{30}$$

Integrating (30),

$$\mathcal{Y}^m(t) + \int_0^t \mathcal{W}^m(\delta)d\delta = S_0 + \int_0^t \mathcal{P}^m(\delta)d\delta. \tag{31}$$

Lemma 3 *There exists* $g,\ s \in C([0, \top]; L^2(0, 1))$ *such as* $\mathcal{X}^m(t) \to g(t)$ *and* $\mathcal{Y}^m(t) \to s(t)$ *uniformly on* $[0, \top]$. $g(t)$ *and* $s(t)$ *are differentiable a.e. on* $[0, \top]$ *as well.*

Proof Consider $t \in (t^m_{k-1}, t^m_k]$ and $t \in (t^q_{l-1}, t^q_l],\ t^m_k \le t^q_l,\ 1 \le k \le m,\ 1 \le l \le q$. Then, by (29) and Theorem 1.4.3 of [6],

$$\frac{1}{2}\frac{d}{dt}\|\mathcal{X}^m(t) - \mathcal{X}^q(t)\|^2 \le \|\mathcal{W}^m(t) - \mathcal{W}^q(t)\|\|\mathcal{X}^m(t) - \mathcal{X}^q(t)\|. \tag{32}$$

By (30),

$$\begin{aligned}&\left\langle\frac{d^-}{dt}(\mathcal{Y}^m(t) - \mathcal{Y}^q(t)), \mathcal{Y}^m(t) - \mathcal{Y}^q(t)\right\rangle + \langle\mathcal{W}^m(t) - \mathcal{W}^q(t), \mathcal{Y}^m(t) - \mathcal{Y}^q(t)\rangle\\ &= \langle\mathcal{P}^m(t) - \mathcal{P}^q(t), \mathcal{Y}^m(t) - \mathcal{Y}^q(t)\rangle.\end{aligned} \tag{33}$$

By (33) and using Lemmas 1 and 2,

$$\frac{1}{2}\frac{d}{dt}\|\mathcal{Y}^m(t) - \mathcal{Y}^q(t)\|^2 \le 4\kappa^2(\hbar_m + \hbar_q) + \|\mathcal{P}^m(t) - \mathcal{P}^q(t)\|\|\mathcal{Y}^m(t) - \mathcal{Y}^q(t)\|. \tag{34}$$

Consider,

$$\|\mathcal{W}^m(t) - \mathcal{W}^q(t)\| \le 2\kappa(\hbar_m + \hbar_q) + \|\mathcal{Y}^m(t) - \mathcal{Y}^q(t)\|,$$

and

$$\|\mathcal{P}^m(t) - \mathcal{P}^q(t)\| \le (\zeta + 2\kappa + 2\zeta\kappa)(\hbar_m + \hbar_q) + (\zeta + 1)\|\mathcal{X}^m(t) - \mathcal{X}^q(t)\|.$$

Combining (32) and (34),

$$\begin{aligned}&\frac{1}{2}\frac{d}{dt}\left(\|\mathcal{X}^m(t) - \mathcal{X}^q(t)\|^2 + \|\mathcal{Y}^m(t) - \mathcal{Y}^q(t)\|^2\right)\\ &\le \kappa(\hbar_m + \hbar_q)[2\zeta + 10\kappa + 4\zeta\kappa + (\zeta + 3\kappa + 2\zeta\kappa)(\hbar_m + \hbar_q)]\\ &\qquad + \left(1 + \frac{\zeta}{2}\right)(\|\mathcal{X}^m(t) - \mathcal{X}^q(t)\|^2 + \|\mathcal{Y}^m(t) - \mathcal{Y}^q(t)\|^2).\end{aligned}$$

Therefore,

$$\frac{d}{dt}\Big(\|\mathcal{X}^m(t)-\mathcal{X}^q(t)\|^2+\|\mathcal{Y}^m(t)-\mathcal{Y}^q(t)\|^2\Big)$$
$$\leq \varrho_{mq}+(\zeta+2)(\|\mathcal{X}^m(t)-\mathcal{X}^q(t)\|^2+\|\mathcal{Y}^m(t)-\mathcal{Y}^q(t)\|^2),$$

where,

$$\varrho_{mq}=2\kappa(\hbar_m+\hbar_q)[2\zeta+10\kappa+4\zeta\kappa+(\zeta+3\kappa+2\zeta\kappa)(\hbar_m+\hbar_q)].$$

Since $\mathcal{X}^m(0)=G_0$ and $\mathcal{Y}^m(0)=S_0$ for all $m\in\mathbb{N}$, for $t\in[0,\top]$,

$$\|\mathcal{X}^m(t)-\mathcal{X}^q(t)\|^2+\|\mathcal{Y}^m(t)-\mathcal{Y}^q(t)\|^2$$
$$\leq \top\varrho_{mq}+(\zeta+2)\int_0^t(\|\mathcal{X}^m(\delta)-\mathcal{X}^q(\delta)\|^2+\|\mathcal{Y}^m(\delta)-\mathcal{Y}^q(\delta)\|^2)d\delta.$$

Applying Gronwall's inequality,

$$\|\mathcal{X}^m(t)-\mathcal{X}^q(t)\|^2+\|\mathcal{Y}^m(t)-\mathcal{Y}^q(t)\|^2\leq \top\varrho_{mq}e^{(\zeta+2)t}\leq \top\varrho_{mq}e^{(\zeta+2)\top}\ \forall t\in[0,\top]. \tag{35}$$

Thus, $\{\mathcal{X}^m(t)\}$ and $\{\mathcal{Y}^m(t)\}$ are Cauchy sequences in $C([0,\top];L^2(0,1))$. Therefore we get $g,s\in C([0,\top];L^2(0,1))$ such that $\mathcal{X}^m(t)\to g(t)$ and $\mathcal{Y}^m(t)\to s(t)$ uniformly on $[0,\top]$. As $\|\mathcal{X}^m(t)-\mathcal{X}^m(\delta)\|\leq\kappa|t-\delta|$ and $\|\mathcal{Y}^m(t)-\mathcal{Y}^m(\delta)\|\leq\kappa|t-\delta|$ hence $g(t)$ and $s(t)$ are Lipschitz continuous on $[0,\top]$. Therefore, $\frac{dg}{dt},\ \frac{ds}{dt}\in L^\infty([0,\top];L^2(0,1))$.

5 Main Results

Theorem 1 *If the presumptions (C1)–(C3) hold good, then (8)–(10) has a unique strong solution* $(g(t),s(t))$ *on* $[0,\top]$. *Furthermore, the inequality* $\|g(t)\|\leq\|s(t)\|$ *holds for all* $t\in(0,\top]$.

Proof Let $g(t)$ and $s(t)$ functions obtained from Lemma 3 and $t\in(t_{k-1}^m,t_k^m]$, $k=1,2,\ldots,m$, $m\in\mathbb{N}$. Since,

$$\|\mathcal{U}^m(t)-g(t)\|\leq\kappa\hbar_m+\|\mathcal{X}^m(t)-g(t)\|,$$

and

$$\|\mathcal{W}^m(t)-s(t)\|\leq 2\kappa\hbar_m+\|\mathcal{Y}^m(t)-s(t)\|,$$

hence, $\mathcal{U}^m(t) \to g(t)$ and $\mathcal{W}^m(t) \to s(t)$ uniformly on $[0, \top]$. Therefore, $\|\mathcal{U}^m(t)\|$ and $\|\mathcal{W}^m(t)\|$ are bounded uniformly on $[0, \top]$.
Consider,

$$\begin{aligned}\|\mathcal{P}^m(t) - g(t) - p(t, g(t))\| &\leq \zeta \hbar_m + (\zeta + 1)(\|g_{k-1}^m - \mathcal{X}^m(t)\| + \|\mathcal{X}^m(t) - g(t)\|) \\ &\leq (\zeta + \kappa + \zeta\kappa)\hbar_m + (\zeta + 1)\|\mathcal{X}^m(t) - g(t)\|.\end{aligned}$$

Therefore, $\mathcal{P}^m(t) \to g(t) + p(t, g(t))$ uniformly on $[0, \top]$. Thus, $\|\mathcal{P}^m(t)\|$ is bounded uniformly on $[0, \top]$. By Lemmas 1 and 2, $\|\frac{d\mathcal{X}^m}{dt}\|$ and $\|\frac{d\mathcal{Y}^m}{dt}\|$ are bounded uniformly on $[0, \top]$. Hence, $\|\mathcal{AU}^m(t)\|$ is bounded uniformly on $[0, \top]$. From Lemma 2.5 of [4], $\mathcal{AU}^m(t) \rightharpoonup \mathcal{A}g(t)$ on $[0, \top]$. Clearly, (11) becomes

$$\mathcal{U}^m(t) + \mu\mathcal{AU}^m(t) = \mathcal{W}^m(t), \quad t \in (0, \top]. \tag{36}$$

Therefore, $\mathcal{AU}^m(t) = \frac{1}{\mu}(\mathcal{W}^m(t) - \mathcal{U}^m(t)) \to \frac{1}{\mu}(s(t) - g(t))$ uniformly on $[0, \top]$. Hence, $\mathcal{AU}^m(t) \to \mathcal{A}y(t)$ uniformly on $[0, \top]$. Therefore,

$$g(t) + \mu\mathcal{A}g(t) = s(t), \quad t \in (0, \top]. \tag{37}$$

Since $\mathcal{X}^m(t_0^m) = \mathcal{X}^m(0) = g_0^m = G_0$ for all $m \in \mathbb{N}$, $g(0) = G_0$. By (31), for every $\hbar \in \mathcal{H}$,

$$\int_0^t \langle \mathcal{W}^m(\delta), \hbar\rangle d\delta = \langle S_0, \hbar\rangle - \langle \mathcal{Y}^m(t), \hbar\rangle + \int_0^t \langle \mathcal{P}^m(\delta), \hbar\rangle d\delta.$$

By the bounded convergence theorem,

$$\int_0^t \langle s(\delta), \hbar\rangle d\delta = \langle S_0, \hbar\rangle - \langle s(t), \hbar\rangle + \int_0^t \langle g(\delta) + p(\delta, g(\delta)), \hbar\rangle d\delta. \tag{38}$$

Since $s(t)$ is Bochner integrable function on $[0, \top]$, by (38),

$$\begin{cases} \dfrac{ds}{dt} + s(t) = g(t) + p(t, g(t)) \text{ a.e. } t \in [0, \top], \\ s(0) = S_0. \end{cases} \tag{39}$$

Thus, by (37) and (39), doublet of functions g and s is a strong solution of (8)–(10) on $[0, \top]$. By (37) and by Theorem 1.4.3 of [6], $\|g(t)\| \leq \|s(t)\|$ holds for all $t \in (0, \top]$.

Suppose that (g_1, s_1) and (g_2, s_2) are two strong solutions of (8)–(10). Let $g = g_1 - g_2$ and $s = s_1 - s_2$. Then,

$$g(t) + \mu \mathcal{A} g(t) = s(t),$$
$$\frac{ds}{dt} + s(t) = g(t) + p(t, g_1(t)) - p(t, g_2(t)),$$
$$g(0) = s(0) = 0.$$

Applying Theorem 1.4.3 of [6],

$$\|g(t)\| \leq \|s(t)\|, \tag{40}$$

and

$$\frac{d}{dt}(\|s(t)\|^2) \leq 2\|g(t) + p(t, g_1(t)) - p(t, g_2(t))\| \|s(t)\|$$
$$\leq 2(\zeta + 1)\|g(t)\|\|s(t)\| \leq 2(\zeta + 1)\|s(t)\|^2.$$

By Gronwall's inequality, $\|s(t)\|^2 = 0 \ \forall \ t \in [0, \top]$. By (40), $\|g(t)\| = 0$ for all $t \in [0, \top]$. Thus, $g_1(t) = g_2(t)$ and $s_1(t) = s_2(t) \ \forall \ t \in [0, \top]$.

6 Applications

Example 1 Consider the following pseudo-parabolic equation

$$\frac{\partial l}{\partial t} - \frac{\partial^3 l}{\partial t \partial y^2} - \frac{\partial^2 l}{\partial y^2} = \cos t + \varrho_1 l(t, y), \quad t \in (0, \top], \ y \in (0, 1) \tag{41}$$

with initial conditions

$$l(0, y) = \sin y \quad \text{and} \quad \frac{\partial^2 l}{\partial y^2}(0, y) = \cos y \text{ for all } y \in (0, 1), \tag{42}$$

and boundary condition (3), where $\top > 0$ and $\varrho_1 > 0$, and the unknown function $l : [0, \top] \to L^2(0, 1)$. Here, $\mu = 1$, $L_0(y) = \sin y$, $L_1 = \cos y$, and $p(t, l(t, y)) = \cos t + \varrho_1 l(t, y)$. Clearly, both functions L_0 and L_1 belong to $L^2(0, 1)$. Then, we have

$$\|p(t, l) - p(\delta, s)\| \leq \zeta(|t - \delta| + \|l - s\|),$$

where $\zeta^2 = 2\max\{1, \varrho_1^2\}$. Therefore, from Theorem 1, the problem (41)–(42) with the boundary condition (3) has a unique strong solution $l(t, y)$.

7 Conclusions

Here, we considered a pseudo-parabolic equation (1) that is subjected to initial conditions (2) and a boundary condition (3). First, the problem (1)–(3) is reduced in the form of coupled Eqs. (8)–(10). To prove the existence and uniqueness of the solution of (1)–(3), Rothe's time-discretization technique is adopted. A pair of sequence of function is fabricated, and then established their uniform convergence to the unique pair of strong solutions. Eventually, we provided an example in support of the results.

References

1. Barenblbatt, G., Zheltov, I.P., Kochina, I.N.: Basic concepts in the theory of seepage of homogeneous liquids in fissured rocks. J. Appl. Math. Mech. **24**, 1286–1303 (1960)
2. Chen, P.J., Gurtin, M.E.: On a theory of heat conduction involving two temperatures. Z. Angew. Math. Phys. **19**, 614–627 (1968)
3. Coleman, B.D., Noll, W.: An approximation theorem for functionals, with applications in continuum mechanics. Arch. Rational Mech. Anal. **6**, 355–370 (1960)
4. Kato, T.: Nonlinear semigroup and evolution equations. Math. Soc. Jpn. **19**, 508–520 (1967)
5. Maqbul, Md., Raheem, A.: Time-discretization schema for a semilinear pseudo-parabolic equation with integral conditions. Appl. Num. Math. **148**, 18–27 (2020)
6. Pazy, A.: Semigroup of Linear Operators and Application to Partial Differential Equations. Springer-Verlag, New York (1983)

Mathematical Model for Tumor-Immune Interaction in Imprecise Environment with Stability Analysis

Subrata Paul, Animesh Mahata, Supriya Mukherjee, Prakash Chandra Mali, and Banamali Roy

Abstract We introduce a tumor model with a tri-trophic level of prey, intermediate predator and top predator in an imprecise environment. The model consists of tumor cells, hunting predator cells, and resting predator cells in a three-dimensional predictable system. We investigated the non-negativity and boundedness of the system's solutions and identified all equilibrium points of the model along with their existence conditions. In the imprecise environment, stability analysis was performed and presented at all of the model system's equilibrium points. We also explain the global simulation study of such equilibrium position using an appropriate Lyapunov function. Detailed numerical simulations to investigate the dynamical behavior of the model are performed.

Keywords Tumor model · Stability analysis · Numerical simulation

1 Introduction

Over the last three decades, mathematical modeling has focused mostly on tumor development and immune system dynamics. Tumors are cancerous growths of aberrant cells that can infect tissues and cause death. Cancer cells can spread throughout

S. Paul (✉)
Department of Mathematics, Arambagh Govt. Polytechnic, Arambagh, West Bengal, India
e-mail: paulsubrata564@gmail.com

A. Mahata
Mahadevnagar High School, Maheshtala, Kolkata, West Bengal 700141, India

S. Mukherjee
Department of Mathematics, Gurudas College, Kolkata, West Bengal 700054, India

P. C. Mali
Department of Mathematics, Jadavpur University, Kolkata 700032, India
e-mail: pcmali@math.jdvu.ac.in

B. Roy
Department of Mathematics, Bangabasi Evening College, Kolkata, West Bengal 700009, India

S. Banerjee and A. Saha (eds.), *Nonlinear Dynamics and Applications*,
Springer Proceedings in Complexity,
https://doi.org/10.1007/978-3-030-99792-2_79

the body via the bloodstream and lymphatic system. A tumor is a mass or lumps of tissue generated by an accumulation of aberrant cells, and it is the most serious sickness in medical science [1–3]. Evidence is accumulating in recent years demonstrating that the immune system may detect and eradicate malignant tumors. Before a tumor grows to the point that it kills the patient or permanently impairs their quality of life, it goes through numerous phases. Cancer is a group of illnesses distinguished by uncontrolled cell proliferation. Tumors can develop and cause problems with the digestive, neurological, and circulatory systems, as well as releasing hormones that cause changes in bodily processes. Our immune response serves a critical role in preventing the spread of malignant cells. Chemotherapy, immunotherapy, radiation therapy, surgery, and other treatments are available to patients with cancer. The type of treatment depends on the position and severity of the tumor, the phase of the disease, the patients' medical condition, and their age. The objective of treatment is to completely remove the cancer without causing harm to the rest of the body. However, the majority of cancer therapies have a harmful effect on normal bodily cells. The natural control techniques that exist for tumors must be considered while studying the growth and regulation of cancer. The use of mathematical models in the theoretical research of cancer is a particularly important strategy for shaping our understanding of tumor resistance dynamics [4–7]. Mathematical tools have been used in a range of studies in this field [8–10].

There are several studies of mathematical models of tumor development and interactions between the tumor and the immune system. The early research on tumor formation was aimed at figuring out how "normal" cells may transform become cancer cells. Cancer research, with a focus on theoretically and experimentally immunotherapy, receives a significant amount of human and financial resources, with both successful and unsuccessful outcomes. Despite the fact that cancer is one of the world's leading causes of death, it is common in the medical field for patients to present with advanced cancer that cannot be cured. This awe-inspiring phenomenon of unexpected cancer remission lives on in medical history, completely unexplainable but genuine. A significant amount of effort has gone into developing dynamical models that may be used to explain and predict tumor progression during the last few decades. There are a few more studies on tumor-immune interactions [11–15].

Researchers make various assumptions in mathematical modeling in order to replicate facts in a truncated but adequately meaningful way. Interaction between biotic and abiotic entities has been represented throughout the field of mathematical biology and ecology by many functions including many parameters, with the greatest of these values being regarded constant. However, due to incorporation of various human and ecological elements, it is widely accepted that uncertainty and imprecision about these characteristics cannot be disregarded. Despite the fact that a lot of work has previously been done in this topic involving uncertainty theory [16–18], there is still a lot of opportunities to improve and expand this area. The parameters of a tumor model with a tri-trophic level of prey-predator system are represented as fuzzy interval numbers in this paper. The findings have confirmed the imprecise solution's viewpoint, which has been visually and quantitatively connected.

2 Preliminaries

Definition 1 The interval $I = [\alpha, \beta]$ can be represented as $f(\mu) = (\alpha)^{\mu-1}(\beta)^{\mu}$ for $\mu \in [0, 1]$, where $f(\mu)$: interval valued function.

3 Model Formulation

To formulate the model framework, the following assumption is made:

$x(t)$	the number of tumor cells present,
$y(t)$	the number of intermediate (hunting) predator cells present,
$z(t)$	the number of top (resting) predator cells present,
r	the development rate of tumor cells,
q	the conversion of normal cells to malignant ones (fixed input),
k_1	the maximum carrying or packing capacity of tumor cells,
k_2	the maximum carrying capacity of resting cells (also, $k_1 > k_2$),
a	rate of predation of tumor cells by the hunting cells,
b	resting cell to hunting cell conversion rate,
c	natural death rate of hunting cell,
d	development rate of resting predator cell,
e	natural death rate of resting cell.

$$\begin{aligned}
\frac{dx}{dt} &= q + rx\left(1 - \frac{x}{k_1}\right) - axy \\
\frac{dy}{dt} &= byz - cy \\
\frac{dz}{dt} &= dz\left(1 - \frac{z}{k_2}\right) - byz - ez
\end{aligned} \tag{1}$$

In the imprecise environment where all the coefficients are interval numbers, the above-mentioned tumor model (1) can be altered as follows:

$$\begin{aligned}
\frac{dx}{dt} &= q + \tilde{r}x\left(1 - \frac{x}{k_1}\right) - \tilde{a}xy \\
\frac{dy}{dt} &= \tilde{b}yz - \tilde{c}y \\
\frac{dz}{dt} &= \tilde{d}z\left(1 - \frac{z}{k_2}\right) - \tilde{b}yz - \tilde{e}z
\end{aligned} \tag{2}$$

where, $\tilde{r} = [r_m, r_n]$, $\tilde{a} = [a_m, a_n]$, $\tilde{b} = [b_m, b_n]$, $\tilde{c} = [c_m, c_n]$, $\tilde{d} = [d_m, d_n]$, $\tilde{e} = [e_m, e_n]$ and $r_m, r_n, a_m, a_n, b_m, b_n, c_m, c_n, d_m, d_n, e_m, e_n$ all are positive.

Using Definition 1, the system (2) can be written as

$$\begin{aligned}
\frac{dx}{dt} &= q + r_m^{1-p} r_n^p x\left(1 - \frac{x}{k_1}\right) - a_n^{1-p} a_m^p xy \\
\frac{dy}{dt} &= b_m^{1-p} b_n^p yz - c_n^{1-p} c_m^p y \\
\frac{dz}{dt} &= d_m^{1-p} d_n^p z\left(1 - \frac{z}{k_2}\right) - b_n^{1-p} b_m^p yz - e_n^{1-p} e_m^p z
\end{aligned} \tag{3}$$

where $p \in [0, 1]$.

3.1 Non-negativity and Boundedness

The non-negativity and boundedness of solutions of the system will be discussed in this section.

Theorem 1 All the solutions of the system (3) are positive.

Proof From the system (3) we have.

$$\frac{dx}{dt} = q + r_m^{1-p} r_n^p x\left(1 - \frac{x}{k_1}\right) - a_n^{1-p} a_m^p xy.$$

Now, integrating in $[0, t]$ we get,

$$x(t) = x(0)e^{\int_0^t \Psi(x,y,z)dt} \geq 0, \forall t$$

$x(t) = x(0)e^{\int_0^t \Psi(x,y,z)dt} \geq 0, \forall t$ as, $x(0) \geq 0$ where $\Psi(x, y, z) = \frac{q}{x} + r_m^{1-p} r_n^p\left(1 - \frac{x}{k_1}\right) - a_n^p a_m^{1-p} y$.

Again, from second equation we have

$$\frac{dy}{dt} = b_m^{1-p} b_n^p yz - c_n^{1-p} c_m^p y$$

Integrating the above equation we get, $y(t) = y(0)e^{\int_0^t \mu(x,y,z)dt} \geq 0 \forall t$ as, $y(0) \geq 0$ where $\mu(x, yz) = b_m^{1-p} b_n^p z - c_n^p c_m^{1-p}$. $\frac{dz}{dt} = d_m^{1-p} d_n^p z\left(1 - \frac{z}{k_2}\right) - b_n^{1-p} b_m^p yz - e_n^{1-p} e_m^p z$

At last, integrating the above equation, we get, $z(t) = z(0)e^{\int_0^t \omega(x,y,z)dt} \geq 0, \forall t$ as, $z(0) \geq 0$ where $\omega(x, y, z) = d_m^{1-p} d_n^p \left(1 - \frac{z}{k_2}\right) - b_n^{1-p} b_m^p y - e_n^{1-p} e_m^p$.

Thus, all solutions of the system (3) are non-negative.

4 Stability Analysis

All the equilibrium points of system (3) as well as their existence requirements are presented in this section.

4.1 Equilibrium Points and Existence Criteria

The equilibrium points of the system (3) are as follows:

(i) Boundary equilibrium point $E_1(x_1, 0, 0)$

where $x_1 = \frac{k_1}{2}\left(1 + \sqrt{1 + \frac{4q}{k_1 r_m^{1-p} r_n^p}}\right)$.

(ii) Planar equilibrium point $E_2(x_2, 0, y_2)$

where $x_2 = \frac{k_1}{2}\left(1 + \sqrt{1 + \frac{4q}{k_1 r_m^{1-p} r_n^p}}\right)$ and $y_2 = k_2\left(1 - \frac{e_n^{1-p} e_m^p}{d_m^{1-p} d_n^p}\right)$.

The existence of the planar equilibrium point depends on the criteria: $\frac{e_n^{1-p} e_m^p}{d_m^{1-p} d_n^p} < 1$.

(iii) Interior equilibrium point $E_3(x^*, y^*, z^*)$

where, x^* is the solution of the equation,

$$\frac{r_m^{1-p} r_n^p}{k_1} x^{*2} + \left[\frac{a_n^{1-p} a_m^p d_m^{1-p} d_n^p}{b_n^{1-p} b_m^p}\left(1 - \frac{c_n^{1-p} c_m^p}{k_2 b_m^{1-p} b_n^p}\right) - \frac{a_n^{1-p} a_m^p e_n^{1-p} e_m^p}{b_n^{1-p} b_m^p} - r_m^{1-p} r_n^p\right] x^* - q = 0,$$

$y^* = \frac{d_m^{1-p} d_n^p}{b_m^{1-p} b_n^p}\left(1 - \frac{c_n^{1-p} c_m^p}{k_2 b_m^{1-p} b_n^p}\right) - \frac{e_n^{1-p} e_m^p}{b_n^{1-p} b_m^p}$ and $z^* = \frac{c_n^{1-p} c_m^p}{b_m^{1-p} b_n^p}$. $E_3(x^*, y^*, z^*)$ will exist if $b_m^{1-p} b_n^p k_2 (d_m^{(1-p)} d_n^p - e_n^{(1-p)} e_m^p) > c_n^{(1-p)} c_m^p d_m^{(1-p)} d_n^p$.

4.2 Local Stability Analysis

The Jacobian matrix at (x, y, z) of the system (3) can be written as

$$J(x, y, z) = \begin{pmatrix} r_m^{1-p} r_n^p\left(1-\frac{2x}{k_1}\right) - a_n^{1-p}a_m^p y & -a_n^{1-p}a_m^p x & 0 \\ 0 & b_m^{1-p}b_n^p z - c_n^{1-p}c_m^p & b_m^{1-p}b_n^p y \\ 0 & -b_n^{1-p}b_m^p z & d_m^{1-p}d_n^p\left(1-\frac{2z}{k_1}\right) - b_n^{1-p}b_m^p y - e_n^{1-p}e_m^p \end{pmatrix}$$

Theorem 2 The system (3) displays unstable behavior at the boundary equilibrium point $E_1(x_1, 0, 0)$.

Proof The Jacobian matrix at the boundary equilibrium point $E_1(x_1, 0, 0)$ is.

$$\begin{pmatrix} -r_m^{1-p} r_n^p\sqrt{1+\frac{4q}{k_1 r_m^{1-p} r_n^p}} & -\frac{a_n^{1-p}a_m^p k_1}{2}\left(1+\sqrt{1+\frac{4q}{k_1 r_m^{1-p} r_n^p}}\right) & 0 \\ 0 & -c_n^{1-p}c_m^p & 0 \\ 0 & 0 & d_m^{1-p}d_n^p - e_n^{1-p}e_m^p \end{pmatrix}$$

The eigen values of this matrix are $\lambda_1 = -r_m^{1-p} r_n^p\sqrt{1+\frac{4q}{k_1 r_m^{1-p} r_n^p}} < 0$, $\lambda_2 = -c_n^{1-p}c_m^p < 0$ and $\lambda_3 = d_m^{1-p}d_n^p - e_n^{1-p}e_m^p > 0$

(if the planar equilibrium point E_2 exist, then $\lambda_3 > 0$).

Therefore, the system is saddle at the boundary equilibrium point and we can say that, it is unstable at this point.

Theorem 3 The model (3) is unstable at $E_2(x_2, 0, y_2)$.

Proof Now the Jacobian matrix at the planer equilibrium point $E_2(x_2, 0, y_2)$ is.

$$\begin{pmatrix} -r_m^{1-p} r_n^p\sqrt{1+\frac{4q}{k_1 r_m^{1-p} r_n^p}} & -\frac{a_n^{1-p}a_m^p k_1}{2}\left(1+\sqrt{1+\frac{4q}{k_1 r_m^{1-p} r_n^p}}\right) & 0 \\ 0 & b_m^{1-p}b_n^p k_2\left(1-\frac{e_n^{1-p}e_m^p}{d_m^{1-p}d_n^p}\right) - c_n^{1-p}c_m^p & 0 \\ 0 & -b_n^{1-p}b_m^p k_2\left(1-\frac{e_n^{1-p}e_m^p}{d_m^{1-p}d_n^p}\right) & -\left(d_m^{1-p}d_n^p - e_n^{1-p}e_m^p\right) \end{pmatrix}$$

The eigen values of this Jacobian matrix are $\lambda_1^/ = -r_m^{1-p} r_n^p\sqrt{1+\frac{4q}{k_1 r_m^{1-p} r_n^p}} < 0$, $\lambda_2^/ = b_m^{1-p}b_n^p k_2\left(1-\frac{e_n^{1-p}e_m^p}{d_m^{1-p}d_n^p}\right) - c_n^{1-p}c_m^p > 0$ (from the existence criteria of E_3) and $\lambda_3^/ = -\left(d_m^{1-p}d_n^p - e_n^{1-p}e_m^p\right) < 0$. Hence the model (3) is unstable at $E_2(x_2, 0, y_2)$ if E_3 exist.

Theorem 4 The model (3) is asymptotically stable at $E_3(x^*, y^*, z^*)$.

Proof Finally, the Jacobian matrix at the interior equilibrium point E_3 is.

$$\begin{pmatrix} M & -a_n^{1-p}a_m^p x^* & 0 \\ 0 & 0 & b_m^{1-p}b_n^p\left[\frac{d_m^{1-p}d_n^p}{b_n^{1-p}b_m^p}\left(1-\frac{c_n^{1-p}c_m^p}{k_2 b_m^{1-p}b_n^p}\right)-\frac{e_n^{1-p}e_m^p}{b_n^{1-p}b_m^p}\right] \\ 0 & -\frac{b_n^{1-p}b_m^p c_n^{1-p}c_m^p}{b_m^{1-p}b_n^p} & b_m^{1-p}b_n^p\left[\frac{d_m^{1-p}d_n^p}{b_n^{1-p}b_m^p}\left(1-\frac{c_n^{1-p}c_m^p}{k_2 b_m^{1-p}b_n^p}\right)-\frac{e_n^{1-p}e_m^p}{b_n^{1-p}b_m^p}\right] \end{pmatrix}$$

where,

$$M = -\sqrt{\left[\frac{a_n^{1-p}a_m^p d_m^{1-p}d_n^p}{b_n^{1-p}b_m^p}\left(1-\frac{c_n^{1-p}c_m^p}{k_2 b_m^{1-p}b_n^p}\right)-\frac{a_n^{1-p}a_m^p e_n^{1-p}e_m^p}{b_n^{1-p}b_m^p}-r_m^{1-p}r_n^p\right]^2+\frac{4r_m^{1-p}r_n^p q}{k_1}}$$

The eigen values of the Jacobian matrix are

$$\lambda_1^* = -\sqrt{\left[\frac{a_n^{1-p}a_m^p d_m^{1-p}d_n^p}{b_n^{1-p}b_m^p}\left(1-\frac{c_n^{1-p}c_m^p}{k_2 b_m^{1-p}b_n^p}\right)-\frac{a_n^{1-p}a_m^p e_n^{1-p}e_m^p}{b_n^{1-p}b_m^p}-r_m^{1-p}r_n^p\right]^2+\frac{4r_m^{1-p}r_n^p q}{k_1}},$$

$$\lambda_2^* = \frac{-h+\sqrt{h^2-4g}}{2} \text{ and } \lambda_3^* = \frac{-h-\sqrt{h^2-4g}}{2}$$

where,

$$h = \left[\frac{2d_m^{1-p}d_n^p c_n^{1-p}c_m^p}{k_2 b_m^{1-p}b_n^p}-\frac{b_n^{1-p}b_m^p d_m^{1-p}d_n^p c_n^{1-p}c_m^p}{k_2 b_m^{1-p}b_n^p}\right] > 0,$$

$$g = d_m^{1-p}d_n^p c_n^{1-p}c_m^p\left(1-\frac{c_n^{1-p}c_m^p}{k_2 b_m^{1-p}b_n^p}\right)-c_n^{1-p}c_m^p e_n^{1-p}e_m^p > 0$$

(from the existence condition of E_3).

Since, $\lambda_1^* < 0$ and λ_2^*, λ_3^* have negative real part. Thus, the system is asymptotically stable around E_3.

4.3 *Global Stability Analysis*

The coexistence critical point's global stability indicates that all solution trajectories connected with the system (3) approach the point's trajectories.

Theorem 5 The system (3) is globally asymptotically stable at $E_3(x^*, y^*, z^*)$ in the region $D = \left\{(x, y, z) \in \mathbb{R}^3 : 0 < \frac{x}{x^*} = \frac{y}{y^*} = \frac{z}{z^*} < 1\right\}$.

Proof Taking the appropriate Lyapunov function into consideration.

$$L = A\left[(x - x^*) - x^* \log\left(\frac{x}{x^*}\right)\right] + B\left[(y - y^*) - y^* \log\left(\frac{y}{y^*}\right)\right]$$
$$+ C\left[(z - z^*) - z^* \log\left(\frac{z}{z^*}\right)\right]$$

Taking time derivative, we get

$$\begin{aligned}
\frac{dL(t)}{dt} &= A\frac{x - x^*}{x}\frac{dx(t)}{dt} + B\frac{y - y^*}{y}\frac{dy(t)}{dt} + C\frac{z - z^*}{z}\frac{dz(t)}{dt} \\
&= A\frac{x - x^*}{x}\left[q + r_m^{1-p} r_n^p x\left(1 - \frac{x}{k_1}\right) - a_n^{1-p} a_m^p xy\right] + B\frac{y - y^*}{y}\left[b_m^{1-p} b_n^p yz - c_n^{1-p} c_m^p y\right] \\
&\quad + C\frac{z - z^*}{z}\left[d_m^{1-p} d_n^p z\left(1 - \frac{z}{k_2}\right) - b_n^{1-p} b_m^p yz - e_n^{1-p} e_m^p z\right] \\
&= A(x - x^*)\left[\frac{q}{x} + r_m^{1-p} r_n^p\left(1 - \frac{x}{k_1}\right) - a_n^{1-p} a_m^p y - \frac{q}{x^*} - r_m^{1-p} r_n^p\left(1 - \frac{x^*}{k_1}\right) + a_n^{1-p} a_m^p y^*\right] \\
&\quad + B(y - y^*)\left[b_m^{1-p} b_n^p z - c_n^{1-p} c_m^p - b_m^{1-p} b_n^p z^* + c_n^{1-p} c_m^p\right] \\
&\quad + C(z - z^*)\left[d_m^{1-p} d_n^p\left(1 - \frac{z}{k_2}\right) - b_n^{1-p} b_m^p y - e_n^{1-p} e_m^p - d_m^{1-p} d_n^p\left(1 - \frac{z^*}{k_2}\right) + b_n^{1-p} b_m^p y^* + e_n^{1-p} e_m^p\right] \\
&= A\frac{q(x - x^*)^2}{(xx^*)} - A\left(\frac{r_m^{1-p} r_n^p}{k_1}\right)(x - x^*)^2 - C\frac{d_m^{1-p} d_n^p}{k_2}(z - z^*)^2 - a_n^{1-p} a_m^p (x - x^*)(y - y^*) \\
&\quad - Bb_m^{1-p} b_n^p (y - y^*)\frac{(x^* z^* y - xzy^*)}{yy^*} - Cb_m^{1-p} b_n^p (y - y^*)(z - z^*).
\end{aligned}$$

So $\frac{dL(t)}{dt} < 0$, since $D = \left\{(x, y, z) \in \mathbb{R}^3 : 0 < \frac{x}{x^*} = \frac{y}{y^*} = \frac{z}{z^*} < 1\right\}$.
Hence the system (3) is globally asymptotically stable at $E_3(x^*, y^*, z^*)$.

5 Numerical Simulation

We use rigorous numerical study to assess and confirm the analytical conclusions of our model system in this section. To graphically forecast the model's solution, we utilize MATLAB (2018).

In this scenario, we simulate the system (3) using the model parameter values provided in Table 1 and pick the value of parameter 'p' into three levels that satisfy the requirement specified in Theorem 4. At this equilibrium position, we ran the model system and computed the eigen values of the Jacobi matrix. Table 2 displays the eigen values. As a result all the eigen values of Jacobi matrix are negative, we may deduce that the interior equilibrium point $E_3(x^*, y^*, z^*)$ is stable in nature when model parameters satisfy the criteria given in Theorem 4 (Figs. 1 and 2).

Table 1 Parametric values for interior equilibrium point

Parameters	Values (for interior)
$\tilde{r}$	[0.9, 0.95]
$\tilde{a}$	[0.3, 0.35]
$\tilde{b}$	[0.1, 0.2]
$\tilde{c}$	[0.02, 0.03]
$\tilde{d}$	[0.8, 0.9]
$\tilde{e}$	[0.03, 0.04]
k_1	0.85
k_2	0.7
q	12

Table 2 Eigen values and nature of equilibrium point $E_3(x^*, y^*, z^*)$ for different levels of p

p	Equilibrium point	Eigen value	Nature
0	(3.524, 1.638, 0.2912)	(−7.293, −0.63, −0.02)	Stable
0.5	(3.354, 2.628, 0.1813)	(−7.26, −0.048, −0.314)	Stable
1	(3.198, 3.768, 0.112)	(−7.33, −0.0305, −0.1705)	Stable

6 Conclusion

We have presented a tumor model for a tri-trophic level with prey, intermediate predator, and top predator in an imprecise environment. We have formulated the model system in an imprecise situation and realized all its stable state points. The feasibility criteria of the system equilibrium points in the supporting environment are studied along with their stability analysis. To justify and confirm the model's analytical conclusions, careful numerical simulations were performed. In nature, foraging behavior is frequent, and the best foraging technique plays an essential role in prey-predator engagements. The tumor growth mathematical models help us comprehend the nature of tumor–immune interactions. The major objective of health administrators, policymakers, and researchers is to create diverse cancer medications and identify the most effective therapy against tumor cell spread. Our model research is a little step in the right direction.

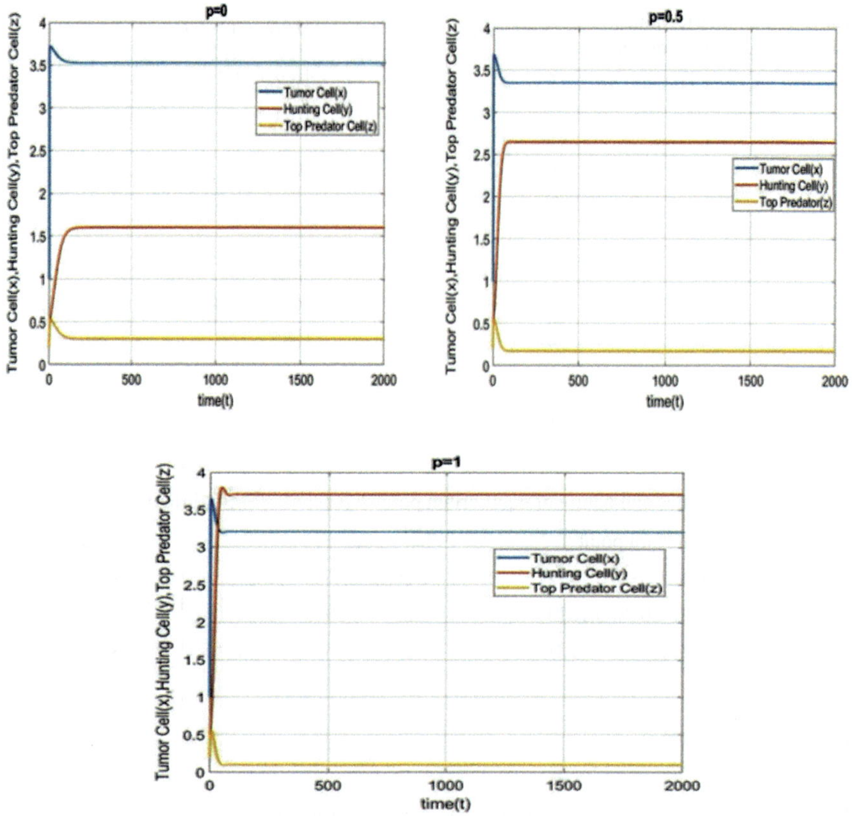

Fig. 1 Time series plot of system (3) in [0, 2000] for various values of parameter 'p'. This figure shows that the interior equilibrium point $E_3(x^*, y^*, z^*)$ is stable

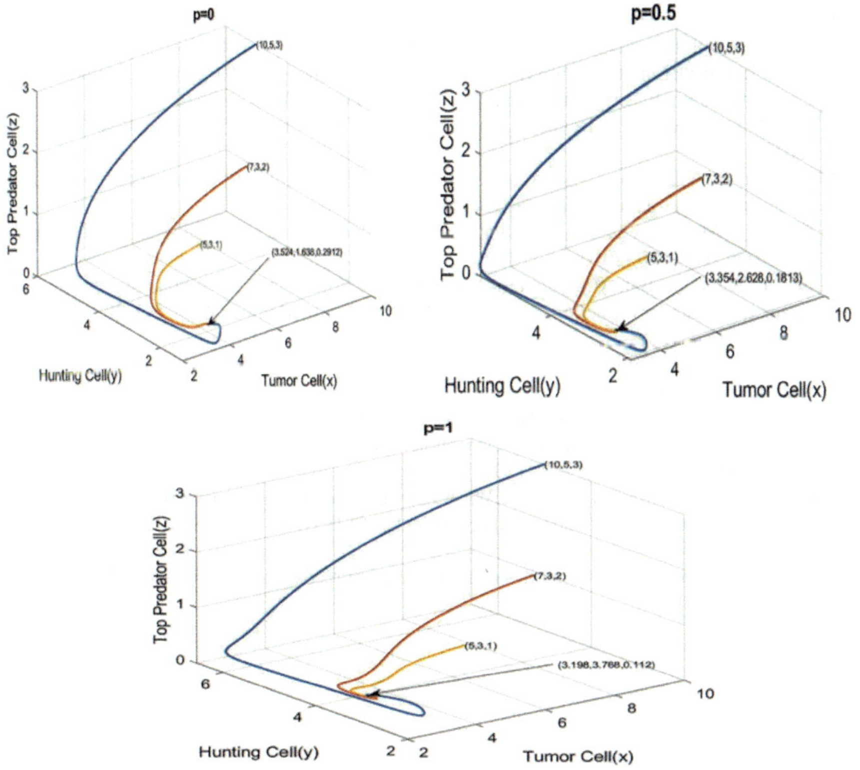

Fig. 2 When numerical simulation is started from different beginning points under the option of model parameter values as indicated in Table 1, the system (3) approaches a stable interior equilibrium point $E_3(x^*, y^*, z^*)$ for different values of the parameter p

References

1. Sarkar, R.R., Chattopadhyay, J.: Occurrence of planktonic blooms under environmental fluctuations and its possible control mechanism—mathematical models and experimental observations. J. Theor. Biol. **224**, 501 (2003)
2. Rockne, R.C., Scott, J.G.: Introduction to mathematical oncology. JCO Clin. Cancer Inform. 1–4 (2019).
3. Jackson, T., Komarova, N., Swanson, K.: Mathematical oncology: using mathematics to enable cancer discoveries. Amer. Math. Monthly **121**, 840–856 (2014)
4. Abernathy, K., Abernathy, Z., Baxter, A., Stevens, M.: Global dynamics of a breast cancer competition model. Differ. Equ. Dyn. Syst. (2017)
5. Sarkar, R.R., Chattopadhyay, J.: The role of environmental stochasticity in a toxic phytoplankton—non-toxic phytoplankton—zooplankton system. Environmetrics **14**, 775 (2003)
6. Sharma, S., Samanta, G.P.: Dynamical behaviour of a tumor-immune system with chemotherapy and optimal control. J. Nonlinear Dyn. **608598**, 13 (2013)
7. Crespi, B., Summers, K.: Evolutionary biology of cancer. Trends Ecol. Evol. **20**(10), 545–552 (2005)
8. Pacheco, J.M., Santos, F.C., Dingli, D.: The ecology of cancer from an evolutionary game theory perspective. Interface Focus **4**(4), 20140019 (2014)

9. Xu, S., Feng, Z.: Analysis of a mathematical model for tumor growth under indirect effect of inhibitors with time delay in proliferation. J. Math. Anal. Appl. **374**, 178–186 (2011)
10. Li, F., Liu, B.: Bifurcation for a free boundary problem modeling the growth of tumors with a drug induced nonlinear proliferation rate. J. Differ. Equ. **263**, 7627–7646 (2017)
11. Fouad, Y.A., Aanei, C.: Revisiting the hallmarks of cancer. Am. J. Cancer Res. **7**(5), 1016 (2017)
12. Basanta, D., Anderson, A.R.: Exploiting ecological principles to better understand cancer progression and treatment. Interface Focus **3**(4), 20130020 (2013)
13. Escher, J., Matioc, A.V.: Radially symmetric growth of nonnecrotic tumors. Nonlinear Differ. Equ. Appl. **17**, 1–20 (2010)
14. Xu, S., Zhou, Q., Bai, M.: Qualitative analysis of a time-delayed free boundary problem for tumor growth under the action of external inhibitors. Math. Methods Appl. Sci. **38**, 4187–4198 (2015)
15. Pan, H., Xing, R.: Bifurcation for a free boundary problem modeling tumor growth with ECM and MDE interactions. Nonlinear Anal. RWA **43**, 362–377 (2018)
16. Salahshour, S., Ahmadian, A., Mahata, A., Mondal, S.P., Alam, S.: The behavior of logistic equation with alley effect in fuzzy environment: fuzzy differential equation approach. Int. J. Appl. Comput. Math. **4**(2), 1–20 (2018)
17. Obajemu, O., Mahfouf, M., Catto, J.W.: A new fuzzy modeling framework for integrated risk prognosis and therapy of bladder cancer patients. IEEE Trans. Fuzzy Syst. **26**(3), 1565–1577 (2017)
18. Tudu, S., Alam, S.: Tumour model with different imprecise coefficients. Int. J. Hybrid Intell. **2**(1), 15–25 (2021)

Dromion Lattice Structure for Coupled Nonlinear Maccari's Equation

J. Thilakavathy, K. Subramanian, R. Amrutha, and M. S. Mani Rajan

Abstract In this article, we are interested to obtain analytic solutions of coupled nonlinear Maccari's equation. We employ Truncated Painlevé Approach to construct dromion lattice structure. The solution of the Maccari's equation is expressed in the form of arbitrary functions. Further, we have constructed dromion lattice structure graphically by considering suitable arbitrary functions. It is seen that the amplitude of the dromion lattice is stable and does not move during the time evolution. The coupled nonlinear Maccari's equation have wide applications in ocean wave theory.

Keywords Coupled nonlinear Maccari's equation · Trunctated Painlevé approach · Dromion lattice

1 Introduction

Dromions [1–3] are the localized solutions that travel with a constant speed without dispersion or dissipation. Dromions have their origin at the intersection of two line-solitons. They have exponentially decaying tails in all the directions and in contrast to lumps that decay only algebraically. Earlier, dromion solution was constructed for

J. Thilakavathy (✉)
Department of Science and Humanities, Jerusalem College of Engineering, Chennai, India
e-mail: cthilakay@gmail.com

K. Subramanian
Department of Physics, SRM Institute of Science and Technology, Ramapuram Campus, Chennai, India
e-mail: subramak2@srmist.edu.in

R. Amrutha
Department of Physics, KCG College of Technology, Chennai, India
e-mail: amrutha@kcgcollege.com

M. S. Mani Rajan
Department of Physics, University College of Engineering, Anna University, Ramanathapuram, India

S. Banerjee and A. Saha (eds.), *Nonlinear Dynamics and Applications*, Springer Proceedings in Complexity,
https://doi.org/10.1007/978-3-030-99792-2_80

the Davey–Stewartson I equation [4, 5] which is a (2 + 1) dimensional generalization of the nonlinear Schrödinger equation [6].

In this work, Truncated Painlevé Approach [7–9] is applied to obtain the localized solution such as dromion lattice structure [17] for the coupled nonlinear Maccari's equation [10–12]. The Maccari's equation was derived by Maccari from the Kadomtsev–Petviashvili equation, by means of a reduction method based on Fourier decomposition and space–time rescaling. Noteworthy developments have been established for examining the closed form solutions of Maccari's equation in recent years. Plentiful effective tools have been utilized to handle Maccari's equation, such as, Exp-function method [13], generalized Riccati relation [14], extended Fan sub-equation method [15], bilinear method [16] etc. and closed form solutions with arbitrary parameters are successfully obtained.

This paper is systematized as follows: In Sect. 2, the solution of coupled nonlinear Maccari's equation has shown by using the tool Trunctated Painlevé Approach. The Sect. 3 is devoted for the discussion on the dromion lattice structure. Finally, we have concluded with notes and comments.

2 Solution of Coupled Nonlinear Maccari's Equation by Truncated Painlevé Approach

We consider the Maccari's equation

$$iS_t + S_{xx} + LS = 0, \tag{1}$$

$$iK_t + K_{xx} + LK = 0, \tag{2}$$

$$L_y = (SS^* + KK^*)_x. \tag{3}$$

By considering, $S = \alpha$, $S^* = \beta$, $K = \gamma$, $K^* = \delta$, then Eqs. (1)–(3) can be given as

$$i\alpha_t + \alpha_{xx} + L\alpha = 0, \tag{4}$$

$$-i\beta_t + \beta_{xx} + L\beta = 0, \tag{5}$$

$$i\gamma_t + \gamma_{xx} + L\gamma = 0, \tag{6}$$

$$-i\delta_t + \delta_{xx} + L\delta = 0, \tag{7}$$

$$L_y = (\alpha\beta)_x + (\gamma\delta)_x. \tag{8}$$

The truncated Laurent series of the solutions of Eqs. (4)–(8) results the following Bäcklund transformation

$$\alpha = \frac{\alpha_0}{\phi} + \alpha_1,\ \beta = \frac{\beta_0}{\phi} + \beta_1,\ \gamma = \frac{\gamma_0}{\phi} + \gamma_1,\ \delta = \frac{\delta_0}{\phi} + \delta_1 \ and\ L = \frac{L_0}{\phi^2} + \frac{L_1}{\phi} + L_2. \tag{9}$$

We assume the vacuum solutions as $\alpha_1 = \beta_1 = \gamma_1 = \delta_1 = 0$ and

$$L_2 = L_2(x, t). \tag{10}$$

By substituting Eq. (9) with the above vacuum solutions into Eqs. (4)–(8) and equate the like coefficients of ϕ^{-3} to zero, one gets

$$2\alpha_0\phi_x^2 + L_0\alpha_0 = 0, \tag{11}$$

$$2\beta_0\phi_x^2 + L_0\beta_0 = 0, \tag{12}$$

$$2\gamma_0\phi_x^2 + L_0\gamma_0 = 0, \tag{13}$$

$$2\delta_0\phi_x^2 + L_0\delta_0 = 0, \tag{14}$$

$$2(\alpha_0\beta_0\phi_x + \gamma_0\delta_0\phi_x - L_0\phi_y) = 0. \tag{15}$$

From Eqs. (11)–(14), the value of L_0 is determined by

$$L_0 = -2\phi_x^2 \tag{16}$$

nd

$$\alpha_0\beta_0 + \gamma_0\delta_0 = -2\phi_x\phi_y. \tag{17}$$

Equating the like coefficients of ϕ^{-2} to zero yields

$$-i\alpha_0\phi_t - 2\alpha_{0x}\phi_x - \alpha_0\phi_{xx} + L_1\alpha_0 = 0, \tag{18}$$

$$i\beta_0\phi_t - 2\beta_{0x}\phi_x - \beta_0\phi_{xx} + L_1\beta_0 = 0, \tag{19}$$

$$-i\gamma_0\phi_t - 2\gamma_{0x}\phi_{xx} - \gamma_0\phi_{xx} + L_1\gamma_0 = 0, \tag{20}$$

$$i\delta_0\phi_y - 2\delta_{0x}\phi_x - \delta_0\phi_{xx} + L_1\delta_0 = 0, \tag{21}$$

$$L_{0y} - L_1\phi_y = (\alpha_0\beta_0 + \gamma_0\delta_0)_x. \tag{22}$$

Using Eqs. (16) and (17) in Eq. (22), we get

$$L_1 = 2\frac{(\phi_{xx}\phi_y - \phi_{xy}\phi_x)}{\phi_y}. \tag{23}$$

By substituting Eq. (23) into Eqs. (18)–(21), the field variables α_0, β_0, γ_0 and δ_0 can be determined as

$$\alpha_0 = U(y)\exp\left[\frac{1}{2}\int \frac{-i\phi_t + \phi_{xx} - \frac{2\phi_x\phi_{xy}}{\phi_y}}{\phi_x}dx\right], \tag{24}$$

$$\beta_0 = U(y)\exp\left[\frac{1}{2}\int \frac{i\phi_t + \phi_{xx} - \frac{2\phi_x\phi_{xy}}{\phi_y}}{\phi_x}dx\right], \tag{25}$$

$$\gamma_0 = V(y)\exp\left[\frac{1}{2}\int \frac{-i\phi_t + \phi_{xx} - \frac{2\phi_x\phi_{xy}}{\phi_y}}{\phi_x}dx\right], \tag{26}$$

$$\delta_0 = V(y)\exp\left[\frac{1}{2}\int \frac{i\phi_t + \phi_{xx} - \frac{2\phi_x\phi_{xy}}{\phi_y}}{\phi_x}dx\right], \tag{27}$$

where $U(y)$ and $V(y)$ are arbitrary functions.

When the like coefficients of ϕ^{-1} is equated to zero, it gives

$$i\alpha_{0t} + \alpha_{0xx} + L_2\alpha_0 = 0, \tag{28}$$

$$-i\beta_{0t} + \beta_{0xx} + L_2\beta_0 = 0, \tag{29}$$

$$i\gamma_{0t} + \gamma_{0xx} + L_2\gamma_0 = 0, \tag{30}$$

$$-i\delta_{0t} + \delta_{0xx} + L_2\delta_0 = 0, \tag{31}$$

$$L_{1y} = 0. \tag{32}$$

The Eq. (32) can be expressed as

$$\phi_{xxy}\phi_y^2 + \phi_x\phi_{xy}\phi_{yy} - \phi_{xy}^2\phi_y - \phi_x\phi_{xyy}\phi_y = 0. \tag{33}$$

The Eq. (33) can be solved as

$$\phi = \phi_1(x) + \phi_2(t), \tag{34}$$

where $\phi_1(x)$and $\phi_2(t)$ are arbitrary functions.

When Eqs. (24)–(27) and (34) are plugged into Eq. (17), one obtains

$$V(y) = iU(y). \tag{35}$$

When the like coefficients of ϕ^0 is equated to zero, it gives

$$L_{2y} = 0. \tag{36}$$

The Eq. (36) is solved into

$$L_2 - \frac{1}{2}\int \frac{-\phi_{2tt}}{\phi_{1x}}dx - \frac{\phi_{1xxx}}{2\phi_{1x}} + \frac{\phi_{2t}^2 + \phi_{1xx}^2}{4\phi_{1x}^2}. \tag{37}$$

With the help of Eqs. (16), (23), (24)–(27), (34), (35) and (37) into Eq. (9), the solution of Eqs. (1)–(3) is obtained as

$$S = \frac{U(y)\exp\left[\frac{1}{2}\int \frac{-i\phi_{2t}+\phi_{1xx}}{\phi_{1x}}dx\right]}{\phi_1(x)+\phi_2(t)}, \tag{38}$$

$$K = \frac{iU(y)\exp\left[\frac{1}{2}\int \frac{-i\phi_{2t}+\phi_{1xx}}{\phi_{1x}}dx\right]}{\phi_1(x)+\phi_2(t)}, \tag{39}$$

$$L = \frac{-2\phi_{1x}^2}{(\phi_1(x)+\phi_2(t))^2} + \frac{2\phi_{1xx}}{(\phi_1(x)+\phi_2(t))} - \frac{1}{2}\int \frac{\phi_{2tt}}{\phi_{1x}}dx - \frac{\phi_{1xxx}}{2\phi_{1x}} + \frac{\phi_{2t}^2 + \phi_{1xx}^2}{4\phi_{1x}^2}. \tag{40}$$

The magnitudes of Eqs. (38) and (39) are represented as

$$|S|^2 = \frac{U^2(y)\phi_{1x}}{(\phi_1(x)+\phi_2(t))^2}, \tag{41}$$

$$|K|^2 = \frac{-U^2(y)\phi_{1x}}{(\phi_1(x)+\phi_2(t))^2}. \tag{42}$$

3 Dromion Lattice Solution

The dromion lattice solution is constructed by setting the $\phi_1(x) = \exp(d_1x); \phi_2(t) = 1 + \exp(d_2t)$; $F_1(y) = \exp(cos(y) - d_3)$. The snapshot of dromion lattice structure is shown in Fig. 1. It is noted that the peaks of dromion lattice is constant and it does not move during the time evolution.

Notes and Comments. In this paper, the solution of coupled integrable Maccari's equation is obtained by the tool Truncated Painlevé Approach. The solution contains

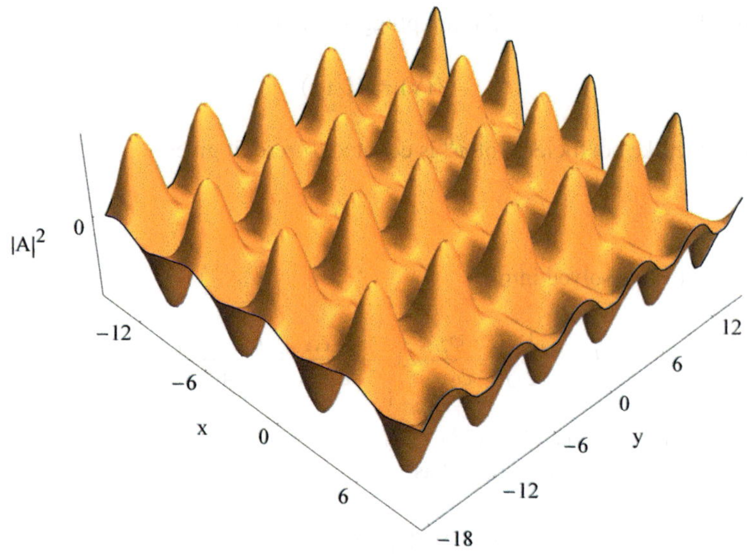

Fig. 1 Dromion lattice pattern for arbitrary constants $d_1 = 1; d_2 = 1; d_3 = 1;$ at t = -10

three arbitrary functions of space and time. By setting suitable arbitrary functions, the localized solution for dromion lattice has constructed. It is noted that the arbitrary manifold is independent of only one space variable y. To understand the physical behavior of the dromion lattice structure, it is also illustrated graphically. It would be interesting to note that the amplitude of dromion lattice is constant and the wave pattern is static during the time evolution. This eccentric nature occurs due the complete separation of arbitrary functions in space (x, y) and time (t).

References

1. Subramanian, K., Senthil Kumar, C., Radha, R., Alagesan, T.: Elusive noninteracting localized solutions of (2+1) dimensional Maccari equation. Romanian. Rep. Phys. **69**, 1–16 (2017)
2. Thilakavathy, R., Amrutha, K., Subramanian, M., Mani Rajan, S.: Different wave patterns for (2+1) dimensional Maccari's equation. Nonlinear Dyn. **107** (2022)
3. Boiti, M., Leon, J.J.P., Martina, L., Pempinelli, F.: Scattering of localized solitons in the plane. Phys. Lett. A. **32**, 432–439 (1988)
4. Fokas, A.S., Santini, P.M.: Dromions and a boundary value problem for the Davey-Stewartson I equation. Physica D. **44**, 99–130 (1990)
5. Radha, R., Vijayalakshmi, S., Lakshmanan, M.: Explode-decay dromions in the non-isospectral Davey-Stewartson (DSI) equation. J. Nonlinear Math. Phys. **6** (1999)
6. Radha, R., Vijayalakshmi, S., Lakshmanan ,M.: Explode-decay dromions in the non-isospectral Davey-Stewartson (DSI) equation. J. Nonlinear Math. Phys. **6** (1996)
7. Radha, R., Lakshmanan, M.: A new class of induced localized coherent structures in the (2+1) dimensional nonlinear Schrodinger equation. J. Phys. A: Math. Gen. **30** (1997)

8. Radha, R., Tang, X.Y., Lou, S.Y.: Truncated Painlevé Expansion—A unified approach to exact solutions and dromion interactions of (2+1)-dimensional nonlinear systems. Z. Naturforsch. **62a**, 107 –116 (2007)
9. Radha, R., Senthil Kumar, C., Subramanian, K.: Drone like dynamics of dromion pairs in the (2+1) AKNS equation. Comput. Math. Appl. **75**, 2356–2364 (2018)
10. Maccari, A.: Universal and integrable nonlinear evolution systems of equations in 2+1 dimensions. J. Math. Phys. **38**, 4151–4164 (1997)
11. Maccari, A.: The Kadomtsev-Petviashvili equation as a source of integrable model equations. J. Math. Phys. **37**, 6207–6212 (1996)
12. Yuan, F., Rao, J., Porsezian, K., Mihalache, D., He., J.: Various exact rational solutions of the two-dimensional Maccari's system. Rom. Journ. Phys. **61**, 378–399 (2016)
13. Zhang, S.: Exp-function method for solving Maccari system. Phys. Lett. A. **371**, 65–71 (2007)
14. Ting, P.J., Xun, G.L.: Exact solutions to Maccari system. Commun. Theor. Phys. **48**, 12–23 (2007)
15. Chemaa, N., Younis, M.: New and more exact traveling wave solutions to integrable (2+1)-dimensional Maccari system. Nonlinear Dyn. **83**, 1395 (2016)
16. Hua, W.G., Hong, W.L., Guang, R.J., Song, H.J.: New patterns of the two-dimensional rogue waves: (2 + 1)-dimensional Maccari system. Commun. Theor. Phys. **67** (2017)
17. Lai, D.W.C., Chow, K.W.: Coalescence of ripplons, breathers, dromions and dark solitons. J. Phys. Soc. Jpn. **70**, 666–677 (2001)

Solving Non-linear Partial Differential Equations Using Homotopy Analysis Method (HAM)

Ajay Kumar and Ramakanta Meher

Abstract In this research paper, a semi-analytical method,i.e.,homotopy analysis method, is implemented for finding the solution of non-linear partial differential equations. Homotopy analysis method is very effective, and easy to evaluate as compared to other numerical methods. The results through HAM on illustrative examples are compared with two well-known methods, namely, Variational Iteration Method(VIM) and Adomian Decomposition Method(ADM). The comparisons of obtained solutions results in the high accuracy through HAM when compared to the other competing methods. Hence, the solution obtained using HAM has much faster convergence to the exact solution.

Keywords Non-linear partial differential equation · Numerical results · Homotopy analysis method

1 Introduction

Solving a partial differential equation (PDE) involves lot of computations and when the PDE is non-linear it become really tough for solving and getting solutions. For solving non-linear PDE we have many numerical methods which provide numerical solutions. Also we solve non-linear PDE using analytic methods. The main difference between analytical and numerical approach is that numerical solution provides solution at discrete points while analytical technique provides continuous graph for the solution. Non-linear equations are not easy to solve, it needs better computational software and systems.

Homotopy analytical technique was firstly proposed by Liao [1–3] in 1992. Homotopy analytical method is based on the concept of homotopy of topology. The method has been widely implemented for solving several nonlinear problems in physical sci-

A. Kumar (✉) · R. Meher
Department of Mathematics and Humanities, Sardar Vallabhbhai National Institute of Technology, Surat 395007, Gujarat, India
e-mail: ajaykhator123@gmail.com

S. Banerjee and A. Saha (eds.), *Nonlinear Dynamics and Applications*,
Springer Proceedings in Complexity,
https://doi.org/10.1007/978-3-030-99792-2_81

ence & engineering [4–12], after Liao published a book[13] in 2004. For solving non-linear problems, HAM [14–17] is a powerful mathematical tool. Adjusting and controlling the convergence region and rate of convergence of the series solution with the help of certain auxiliary parameter h is the main advantage of HAM.

The objective of the current paper is to solve initial value problems of non-linear partial differential equations by HAM and to make comparison of obtained approximate solutions by HAM with that already obtained by Variational Iteration Method [18–20] and Adomian Decomposition Method [21–23].

2 Analysis of Homotopy Analysis Method

Let's consider a non-linear differential equation, for the better understanding of the method,

$$N[f(\zeta,\eta)] = 0, \tag{1}$$

where N is representing as operator for non-linearity, ζ denotes spatial variables, η stands for time-related independent variables, & $f\ (\zeta, \eta)$ is taken as an unknown function.

For quick understanding and simplification, each initial conditions and boundary conditions are intentionally ignored, which can be explained with similar way. According to the concept of homotopy, the so called 0th-order deformation equation can be constructed as follow,

$$(1-q)\mathcal{L}[\psi(\zeta,\eta;q) - f_0(\zeta,\eta)] = hqH(\zeta,\eta)\mathcal{N}[\psi(\zeta,\eta;q)], \tag{2}$$

where $q\ (0 \le q \le 1)$ is known as embedding parameter, h is known as a non-zero auxiliary parameter, $H\ (\zeta, \eta)$ is known as a non-zero auxiliary function, $\mathcal{L}$ is known as an auxiliary operator for linearity, $f_0\ (\zeta, \eta)$ is an initial assumption of the solution $f\ (\zeta, \eta)$, and $\psi\ (\zeta, \eta; q)$ is an unspecified function. It must be noted that there is no restriction in choosing the initial assumption, $\mathcal{L}$ the auxiliary operator for linearity , h the auxiliary parameter, and $H\ (\zeta, \eta)$ the auxiliary function.

For $q = 0$ & $q = 1$, $\psi\ (\zeta, \eta; q)$ can be expressed as follows

$$\psi(\zeta,\eta;0) = f_0(\zeta,\eta), \quad \psi(\zeta,\eta;1) = f(\zeta,\eta) \tag{3}$$

So, $\psi\ (\zeta, \eta; q)$ reaches from the initial assumption $f_0\ (\zeta, \eta)$ to $f\ (\zeta, \eta)$ as a solution, when the value of q increases from 0 to 1. Expansion of $\psi\ (\zeta, \eta; q)$ in the Taylor series concerning the parameter q, we obtain

$$\psi(\zeta,\eta;q) = f_0(\zeta,\eta) + \sum_{n=1}^{\infty} f_n(\zeta,\eta)q^n, \tag{4}$$

Here $f_n(\zeta, \eta)$ stands for

$$f_n(\zeta, \eta) = \frac{1}{n!} \frac{\partial^n \psi(\zeta, \eta; q)}{\partial q^n}\bigg|_{q=0} \tag{5}$$

At $q = 1$, the choice of auxiliary operator for linearity $\mathcal{L}$, the initial assumption $f_0(\zeta, \eta)$, the auxiliary parameter h and the auxiliary function $H(\zeta, \eta)$ is proper then above series converges, and we obtain

$$f(\zeta, \eta) = f_0(\zeta, \eta) + \sum_{k=1}^{\infty} f_k(\zeta, \eta), \tag{6}$$

Equation (6) is original nonlinear equation's one of the solution. As shown in Eq. (6), the controlling equation can be obtained from 0th-order deformation equation. Define vector,

$$\bar{f}(\zeta, \eta) = \{f_0(\zeta, \eta), f_1(\zeta, \eta), f_2(\zeta, \eta), ..., f_n(\zeta, \eta)\} \tag{7}$$

Taking n-times differentiation of 0th-order deformation equation concerning embedding parameter q, putting $q = 0$ and after dividing with $n!$, the deformation equation of nth-order can be presented as follows

$$\mathcal{L}[f_n(\zeta, \eta) - \chi_n f_{n-1}(\zeta, \eta)] = hH(\zeta, \eta)R_n(\bar{f}_{n-1}) \tag{8}$$

where $R_n(\bar{f}_{n-1})$ stands for

$$R_n(\bar{f}_{n-1}) = \frac{1}{(n-1)!} \frac{\partial^{n-1}\mathcal{N}[\psi(\zeta, \eta; q)]}{\partial q^{n-1}}\bigg|_{q=0} \tag{9}$$

and

$$\chi_n = \begin{cases} 0, & n \leq 1 \\ 1, & n > 1. \end{cases} \tag{10}$$

Here, It is surely noted that when $n \geq 1$, $f_n(\zeta, \eta)$ is controlled by the linear deformation equation of nth-order .

3 Numerical Application

Example 1 Consider the 1st-order quasi-linear homogeneous partial differential equation[24]

$$\frac{\partial u(\zeta, \eta)}{\partial \eta} + (1 + u(\zeta, \eta)) \frac{\partial u(\zeta, \eta)}{\partial \zeta} = 0, \quad 0 \leq \eta \tag{11}$$

with the initial condition

$$u(\zeta, 0) = \frac{\zeta - 1}{2}, \tag{12}$$

Equation (11) can be simply expressed as

$$\frac{\partial u}{\partial \eta} + \left(\frac{\partial u}{\partial \zeta}\right) + u\left(\frac{\partial u}{\partial \zeta}\right) = 0 \tag{13}$$

Exact solution for this problem is $u(\zeta, \eta) = \frac{\zeta-\eta-1}{\eta+2}$, which can also be verified.

For solving problem in Eq. (11) using HAM we need to consider initial approximation, letting the initial approximation to be

$$u_0(\zeta, \eta) = \frac{\zeta - 1}{2} \tag{14}$$

Here, nonlinear operator for Eq. (11) can be defined as

$$\mathcal{N}[\psi(\zeta, \eta; q)] = \frac{\partial \psi(\zeta, \eta; q)}{\partial \eta} - \left(\frac{\partial \psi(\zeta, \eta; q)}{\partial \zeta}\right) - \psi(\zeta, \eta; q)\left(\frac{\partial \psi(\zeta, \eta; q)}{\partial \zeta}\right) \tag{15}$$

Now, we have to construct the 0th-order deformation equation as

$$(1 - q)\mathcal{L}[\psi(\zeta, \eta; q) - u_0(\zeta, \eta)] = hq\mathcal{N}[\psi(\zeta, \eta; q)]. \tag{16}$$

where an auxiliary linear operator $\mathcal{L}$ is expressed as

$$\mathcal{L}[\psi(\zeta, \eta; q)] = \frac{\partial \psi(\zeta, \eta; q)}{\partial \eta} \tag{17}$$

Also, at $q = 0$ & 1, we have $\psi(\zeta, \eta; q)$ as

$$\psi(\zeta, \eta; 0) = u_0(\zeta, 0), \quad \psi(\zeta, \eta; 1) = u(\zeta, \eta). \tag{18}$$

Here, $\psi(\zeta, \eta; q)$ varies from initial guess $u_0(\zeta, \eta) = \frac{\zeta-1}{2}$ to the solution $u(\zeta, \eta)$, as q varies from 0 to 1.

Now, we will have the nth-order deformation equation as

$$(1 - q)\mathcal{L}[u_n - \chi_n u_{n-1}] = h\mathcal{R}_n(\vec{u}_{n-1}), \tag{19}$$

where

$$\mathcal{R}_n(\vec{u}_{n-1}) = \frac{\partial u_{n-1}}{\partial \eta} - \frac{\partial u_{n-1}}{\partial \zeta} - \left(\sum_{i=0}^{n-1} u_i \frac{\partial u_{n-1-i}}{\partial \zeta}\right) \tag{20}$$

Therefore,

$$u_n = \chi_n u_{n-1} + h\mathcal{L}^{-1}[\mathcal{R}_n(\overrightarrow{u}_{n-1})], \tag{21}$$

where

$$\mathcal{L}^{-1} = \int (\cdot) d\eta \tag{22}$$

So, Eq. (19) becomes

$$u_n = \chi_n u_{n-1} + h \int \mathcal{R}_n(\overrightarrow{u}_{n-1}) d\eta, \tag{23}$$

The first few terms of the solution, with the help of $u_0(\zeta, \eta)$ from Eq. (14) can be expressed as follows:

$$u_1(\zeta, \eta) = h\left(\frac{\eta}{4} + \frac{1}{4}\zeta\eta\right)$$

$$u_2(\zeta, \eta) = \frac{1}{4}h\eta + \frac{1}{4}h\zeta\eta + \frac{1}{8}\eta^2h^2 + \frac{1}{8}\eta^2h^2\zeta + \frac{1}{4}h^2\eta + \frac{1}{4}h^2\zeta\eta$$

$$u_3(\zeta, \eta) = \frac{1}{4}h\eta + \frac{1}{4}h\zeta\eta + \frac{1}{4}\eta^2h^2 + \frac{1}{4}\eta^2h^2\zeta + \frac{1}{2}h^2\eta + \frac{1}{2}h^2\zeta\eta + \frac{1}{16}\eta^3h^3 + \frac{1}{16}\eta^3h^3\zeta + \frac{1}{4}\eta^2h^3 + \frac{1}{4}\eta^2h^3\zeta + \frac{1}{4}h^3\eta + \frac{1}{4}h^3\zeta\eta$$

Thus, the approximated series solution

$$\begin{aligned} u(\zeta, \eta) &= u_0(\zeta, \eta) + u_1(\zeta, \eta) + u_2(\zeta, \eta) + u_3(\zeta, \eta) + \ldots \\ &= \frac{\zeta - 1}{2} + h\left(\frac{\eta}{4} + \frac{1}{4}\zeta\eta\right) + \frac{1}{4}h\eta + \frac{1}{4}h\zeta\eta + \frac{1}{8}\eta^2h^2 + \frac{1}{8}\eta^2h^2\zeta + \frac{1}{4}h^2\eta \\ &+ \frac{1}{4}h^2\zeta\eta + \frac{1}{4}h\eta + \frac{1}{4}h\zeta\eta + \frac{1}{4}\eta^2h^2 + \frac{1}{4}\eta^2h^2\zeta + \frac{1}{2}h^2\eta \\ &+ \frac{1}{2}h^2\zeta\eta + \frac{1}{16}\eta^3h^3 + \frac{1}{16}\eta^3h^3\zeta + \frac{1}{4}\eta^2h^3 + \frac{1}{4}\eta^2h^3\zeta + \frac{1}{4}h^3\eta + \frac{1}{4}h^3\zeta\eta + \cdots \end{aligned}$$

Example 2 Consider the second order nonlinear hyperbolic equation [25]

$$\frac{\partial^2 u(\zeta, \eta)}{\partial \eta^2} = \frac{\partial}{\partial \zeta}\left(u(\zeta, \eta)\frac{\partial u(\zeta, \eta)}{\partial \zeta}\right), \quad 0 \leq \eta \tag{24}$$

with the initial conditions

$$u(\zeta, 0) = \zeta^2, \quad u_\eta(\zeta, 0) = -2\zeta^2 \tag{25}$$

Equation (24) can be simply expressed as

$$\frac{\partial^2 u}{\partial \eta^2} = \left(\frac{\partial u}{\partial \zeta}\right)^2 + u\left(\frac{\partial^2 u}{\partial \zeta^2}\right) \tag{26}$$

Exact solution for this problem Eq. (24) is $u(\zeta,\eta) = \left(\frac{\zeta}{1+\eta}\right)^2$, which can be verified also.

For solving Eq. (24) using HAM, let the initial approximation be

$$u_0(\zeta,\eta) = \zeta^2\,(1 - 2\,\eta) \tag{27}$$

Here, nonlinear operator for Eq. (24) can be defined as

$$\mathcal{N}[\psi(\zeta,\eta;q)] = \frac{\partial^2 \psi(\zeta,\eta;q)}{\partial \eta^2} - \left(\frac{\partial \psi(\zeta,\eta;q)}{\partial \zeta}\right)^2 - \psi(\zeta,\eta;q)\left(\frac{\partial^2 \psi(\zeta,\eta;q)}{\partial \zeta^2}\right) \tag{28}$$

Now, we have to construct the 0th-order deformation equation as

$$(1-q)\mathcal{L}[\psi(\zeta,\eta;q) - u_0(\zeta,\eta)] = hq\mathcal{N}[\psi(\zeta,\eta;q)]. \tag{29}$$

where an auxiliary linear operator $\mathcal{L}$ is expressed as

$$\mathcal{L}[\psi(\zeta,\eta;q)] = \frac{\partial^2 \psi(\zeta,\eta;q)}{\partial \eta^2} \tag{30}$$

Also, at $q = 0$ & 1, we have $\psi(\zeta,\eta;q)$ as

$$\psi(\zeta,\eta;0) = u_0(\zeta,0),\quad \psi(\zeta,\eta;1) = u(\zeta,\eta). \tag{31}$$

Here, $\psi(\zeta,\eta;q)$ varies from initial guess $u_0(\zeta,\eta) = \zeta^2\,(1-2\,\eta)$ to the solution $u(\zeta,\eta)$, as q varies from 0 to 1.

Now, we will have the nth-order deformation equation as

$$(1-q)\mathcal{L}[u_n - \chi_n u_{n-1}] = h\mathcal{R}_n(\vec{u}_{n-1}), \tag{32}$$

where

$$\mathcal{R}_n(\vec{u}_{n-1}) = \frac{\partial^2 u_{n-1}}{\partial \eta^2} - \left(\sum_{i=0}^{n-1} \frac{\partial u_i}{\partial \zeta}\frac{\partial u_{n-1-i}}{\partial \zeta}\right) - \left(\sum_{i=0}^{n-1} u_i \frac{\partial^2 u_{n-1-i}}{\partial \zeta^2}\right) \tag{33}$$

Therefore,

$$u_n = \chi_n u_{n-1} + h\mathcal{L}^{-1}[\mathcal{R}_n(\vec{u}_{n-1})], \tag{34}$$

where

$$\mathcal{L}^{-1} = \int\int (\cdot) d\eta d\eta \tag{35}$$

So, Eq. (32) becomes

$$u_n = \chi_n u_{n-1} + h \int\int \mathcal{R}_n(\overrightarrow{u}_{n-1}) d\eta d\eta, \tag{36}$$

The first few terms of the solution, with the help of $u_0(\zeta, \eta)$ from Eq. (27) can be expressed as follows :

$$u_1(\zeta, \eta) = h\left(-2\eta^4\zeta^2 + 4\eta^3\zeta^2 - 3\eta^2\zeta^2\right)$$

$$\begin{aligned} u_2(\zeta, \eta) = &-2\eta^4\zeta^2 h + 4\eta^3\zeta^2 h - 3\eta^2\zeta^2 h - \frac{4\eta^7 h^2\zeta^3}{21} - \frac{8\eta^7 h^2\zeta^2}{7} + \frac{2}{3}\eta^6\zeta^3 h^2 \\ &+ 4\eta^6\zeta^2 h^2 - \eta^5 h^2\zeta^3 - 6\eta^5 h^2\zeta^2 + 1/2\,\eta^4\zeta^3 h^2 + \eta^4\zeta^2 h^2 + 4\eta^3\zeta^2 h^2 - 3\eta^2\zeta^2 h^2 \end{aligned}$$

$$\begin{aligned} u_3(\zeta, \eta) = &\frac{20\eta^9\zeta^2 h^3}{7} - 12\eta^5 h^2\zeta^2 - 3\eta^2\zeta^2 h^3 - 6\eta^2\zeta^2 h^2 - 3\eta^2\zeta^2 h \\ &+ 4\eta^4\zeta^2 h^3 + \frac{32\eta^7 h^3\zeta^2}{7} - \eta^5 h^3\zeta^3 - 2\eta^4\zeta^2 h + \frac{80\eta^9\zeta^3 h^3}{189} \\ &+ 2\eta^4\zeta^2 h^2 + \frac{16\eta^7 h^3\zeta^3}{21} - \frac{16\eta^{10}\zeta^3 h^3}{189} + \frac{1}{3}\eta^6\zeta^3 h^3 \\ &+ \frac{2}{3}\eta^6\zeta^3 h^2 + 4\eta^3\zeta^2 h^3 - 4/7\,\eta^{10}\zeta^2 h^3 + 5\eta^6\zeta^2 h^3 + 4\eta^3\zeta^2 h + 8\eta^3\zeta^2 h^2 \\ &- 12\eta^5 h^3\zeta^2 + \frac{1}{2}\eta^4\zeta^3 h^2 - \frac{4\eta^7 h^2\zeta^3}{21} - \frac{20\eta^8\zeta^3 h^3}{21} \\ &- \frac{16\eta^7 h^2\zeta^2}{7} - \frac{45\eta^8\zeta^2 h^3}{7} + \frac{1}{2}\eta^4\zeta^3 h^3 + 8\eta^6\zeta^2 h^2 - \eta^5 h^2\zeta^3 \end{aligned}$$

Thus, the approximated series solution

$$u(\zeta, \eta) = u_0(\zeta, \eta) + u_1(\zeta, \eta) + u_2(\zeta, \eta) + u_3(\zeta, \eta) + \ldots \tag{37}$$

From calculated $u_i(\zeta, \eta)$'s, $i = 0, 1, \ldots$ the above series solution when $h = -1$ becomes

$$\begin{aligned} u(\zeta, \eta) = &-8\eta^7\zeta^2 - 6\eta^5\zeta^2 + 1/2\,\eta^4\zeta^3 + 7\eta^6\zeta^2 - 2\zeta^2\eta + \zeta^2 - \frac{8\eta^7\zeta^3}{7} + \eta^6\zeta^3 - \eta^5\zeta^3 \\ &- \frac{20\eta^9\zeta^2}{7} + \frac{20\eta^8\zeta^3}{21} + 4/7\,\eta^{10}\zeta^2 - \frac{80\eta^9\zeta^3}{189} + \frac{45\eta^8\zeta^2}{7} + \frac{16\eta^{10}\zeta^3}{189} + 5\eta^4\zeta^2 \\ &- 4\eta^3\zeta^2 + 3\eta^2\zeta^2 + \ldots \end{aligned}$$

4 Results and Discussion

This section discusses the numerical values of the solution of some non-linear partial differential equations that has been solved using HAM. Table 1, shows the error of solution obtained by HAM that is compared with the errors of ADM [24] and VIM [24]. Table 2 discusses the error of solution obtained by HAM and is compared with the errors of ADM [25]. It is apparent from the tables that the error is lower with Homotopy Analysis Method as compared to Adomian Decomposition Method and Variational Iteration Method.

Table 1 Comparison of the error in Example 1 of the solutions by u_{ADM}[24], u_{VIM}[24] and $u_{HAM}(h = -0.77)$

ζ	η	$\lvert u_{Exact} - u_{ADM}\rvert$	$\lvert u_{Exact} - u_{VIM}\rvert$	$\lvert u_{Exact} - u_{HAM}\rvert$
−1	0	0	0	0
−0.8	0.1	6×10^{-7}	7.7×10^{-6}	3.3×10^{-5}
−0.6	0.2	1.8×10^{-5}	1.2×10^{-4}	6.5×10^{-5}
−0.4	0.3	1.3×10^{-4}	5.4×10^{-4}	5.8×10^{-5}
−0.2	0.4	5.3×10^{-4}	1.6×10^{-4}	2.9×10^{-5}
0	0.5	1.6×10^{-3}	3.6×10^{-3}	5.2×10^{-6}
0.2	0.6	3.7×10^{-3}	7.1×10^{-3}	1×10^{-10}
0.4	0.7	7.8×10^{-3}	1.2×10^{-2}	1.1×10^{-5}
0.6	0.8	1.5×10^{-2}	2×10^{-2}	1.0×10^{-4}
0.8	0.9	2.5×10^{-2}	2.9×10^{-2}	4.4×10^{-6}
1	1	4.2×10^{-2}	4.2×10^{-2}	1.2×10^{-3}

Table 2 Comparison of the error in Example 2 of the solutions by u_{ADM}[25] and u_{HAM} $(h = -1)$

ζ	η	$\lvert u_{Exact} - u_{ADM}\rvert$	$\lvert u_{Exact} - u_{HAM}\rvert$
0	0	0	0
0.1	0.1	5.4×10^{-7}	4.0×10^{-8}
0.2	0.2	6.2×10^{-5}	4.1×10^{-6}
0.3	0.3	9.7×10^{-4}	5.1×10^{-5}
0.4	0.4	6.7×10^{-3}	2.5×10^{-4}
0.5	0.5	3.0×10^{-2}	5.8×10^{-4}
0.6	0.6	9.8×10^{-2}	1.6×10^{-4}
0.7	0.7	2.7×10^{-1}	4.3×10^{-3}
0.8	0.8	6.5×10^{-1}	2.0×10^{-2}
0.9	0.9	1.4×10^{-0}	5.9×10^{-2}
1	1	2.8×10^{-0}	1.3×10^{-1}

5 Conclusion

In this paper, two PDEs' analytical and approximate solutions are obtained using HAM. It is found that the obtained solutions using the HAM approach are very close to exact solutions as compared to the solutions of other available methods, as mentioned in the section. It can be seen from the comparison results that HAM is efficient and robust in finding analytical solutions for a broader class of problems. Furthermore, HAM gives us a free hand on adjusting and controlling the series solution's convergence by taking appropriate values of homotopy & auxiliary parameters accordingly. In conclusion, our method HAM also provides exact solutions accurately for various problems.

References

1. Liao, S.J.: The proposed homotopy analysis techniques for the solution of nonlinear problems. Ph.D. dissertation, Shanghai Jiao Tong University, Shanghai (1992)
2. Liao, S.J.: A kind of linear invariance under homotopy and some simple applications of it in mechanics, Bericht Nr. 520. Institute fuer Sciffbau der Universitaet Hamburg (1992)
3. Liao, S.J.: Notes on the homotopy analysis method: some definitions and theorems. Commun. Nonlinear Sci. Numer. Simul. **14**(4), 983–997 (2009)
4. Kumar, S., Kumar, A., Odibat, Z.: A nonlinear fractional model to describe the population dynamics of two interacting species. Math. Methods Appl. Sci. **40** (11), 4134–4148 (2017)
5. Massa, F., Lallemand, B., Tison, T.: Multi-level homotopy perturbation and projection techniques for the reanalysis of quadratic eigenvalue problems: the application of stability analysis. Mech. Syst. Signal Process. **52**, 88–104 (2015)
6. Kumar, S., Singh, J., Kumar, D., Kapoor, S.: New homotopy analysis transform algorithm to solve Volterra integral equation. Ain Shams Eng. J. **5**(1), 243–246 (2014)
7. Martin, O.: On the homotopy analysis method for solving a particle transport equation. Appl. Math. Model. **37**(6), 3959–3967 (2013)
8. Nave, O., Gol'dshtein, V., Ajadi, S.: Singularly perturbed homotopy analysis method applied to the pressure driven flame in porous media. Combust. Flame **162**(3), 864–873 (2015)
9. Sardanyés, J., Rodrigues, C., Januário, C., Martins, N., Gil-Gómez, G., Duarte, J.: Activation of effector immune cells promotes tumor stochastic extinction: a homotopy analysis approach. Appl. Math. Comput. **252**(1), 484–495 (2015)
10. Molabahrami, A., Khani, F.: The homotopy analysis method to solve the Burgers-Huxley equation. Nonlinear Anal. Real World Appl. **10**(2), 589–600 (2009)
11. Shivanian, E., Abbasbandy, S.: Predictor homotopy analysis method: two points second order boundary value problems,. Nonlinear Anal. Real World Appl. **15**, 89–99 (2014)
12. Yang, Z., Liao, S.: A HAM-based wavelet approach for nonlinear partial differential equations: two dimensional Bratu problem as an application. Commun. Nonlinear Sci. Numer. Simul. **53**, 249–262 (2017)
13. Liao, S.J.: Beyond Perturbation: Introduction to Homotopy Analysis Method. Chapman & Hall/CRC (2004)
14. Liao, S.J.: Homotopy analysis method and its applications in mathematics. J. Basic Sci. Eng. **5**(2), 111–125 (1997)
15. Gohil, V.P., Meher, R.: Effect of viscous fluid on the counter-current imbibition phenomenon in two-phase fluid flow through heterogeneous porous media with magnetic field. Iran. J. Sci. Technol. Trans. A Sci. **43**(4), 1799–810 (2019)

16. Kesarwani, J., Meher, R.: Modeling of an imbibition phenomenon in a heterogeneous cracked porous medium on small inclination. Spec. Top. I Rev. Porous Media Int. J. **12**(1) (2021)
17. Gohil, V.P., Meher, R.: Homotopy analysis method for solving counter current imbibition phenomena of the time positive fractional type arising in heterogeneous porous media. Int. J. Math. I Comput. **28**(2), 77–85 (2017)
18. Momani, S., Abuasad, S.: Application of He's variational iteration method to Helmholtz equation. Chaos, Solitons Fractals **27**(5), 1119–1123 (2006)
19. He, J.H.: Variational iteration method for delay differential equations. Commun. Nonlinear Sci. Numer. Simul. **2**(4), 235–236 (1997)
20. He, J.H.: Variational principle for some nonlinear partial differential equations with variable coefficients. Chaos, Solitons Fractals **19**(4), 847–851 (2004)
21. Adomian, G.: A review of the decomposition method in applied mathematics. J. Math. Anal. Appl. **135**, 501–544 (1988)
22. Wazwaz, A., El-Sayed, S.: A new modification of the Adomian decomposition method for linear and nonlinear operators. Appl. Math. Comput. **122**, 393–405 (2001)
23. Ray, S.S., Bera, R.K.: Solution of an extraordinary differential equation by adomian decomposition method. J. Appl. Math. **4**, 331–338 (2004)
24. Bildik, N., Konuralp, A.: Two-dimensional differential transform method, Adomian decomposition method and variational iteration method for partial differential equations. Int. J. Comput. Math. **83**, 973–987 (2006)
25. Odibat, Z., Momani, S.: Numerical methods for nonlinear partial differential equations of fractional order. Appl. Math. Model. **32**, 28–39 (2008)

Nonlinear Modelling and Analysis of Longitudinal Dynamics of Hybrid Airship

Abhishek Kumar and Om Prakash

Abstract The objective of the paper is to develop a Nonlinear mathematical model for analyzing the longitudinal dynamics of a hybrid small sized airship that includes wings and elevators as a control surface. In this model will propose a single body longitudinal dynamics model for the hybrid airship and two body dynamics model for the hybrid airship with tethered suspended payload. Apparent mass matrix is taken care in the modelling of hybrid airship. Analysis of the system is done on MATLAB with simulated results.

Keywords Hybrid airship · Nonlinear longitudinal dynamics · Multi body dynamics

1 Introduction

This paper deals with the area of dynamics and modelling of lighter than air vehicles. More specifically modelling of hybrid airship is dealt. Modelling is usually done using some subset of the inputs and outputs, taken from flight data according to the objective. Due to huge cost, these plants are rarely accessible to do experiment design on it. Modelling work is carried out with the help of flight data, geometry data and aerodynamic data by understanding the mathematics and physics of the proposed model. Airships are being proposed for heavy freight aerial transport and also for surveillance tasks. The Hybrid airship system consists of semi rigid airship hull mass (including apparent mass and included air mass) and rigid wing attached below the hull and a payload is attached through number of riser lines. The suspension lines

A. Kumar (✉)
Department of Electrical Engineering, Manipal University Jaipur, Jaipur 303007, Rajasthan, India
e-mail: abhishekkumr@gmail.com

A. Kumar · O. Prakash
Department of Aerospace Engineering, University of Petroleum and Energy Studies, Dehradun 248007, India
e-mail: omprakash@ddn.upes.ac.in

S. Banerjee and A. Saha (eds.), *Nonlinear Dynamics and Applications*,
Springer Proceedings in Complexity,
https://doi.org/10.1007/978-3-030-99792-2_82

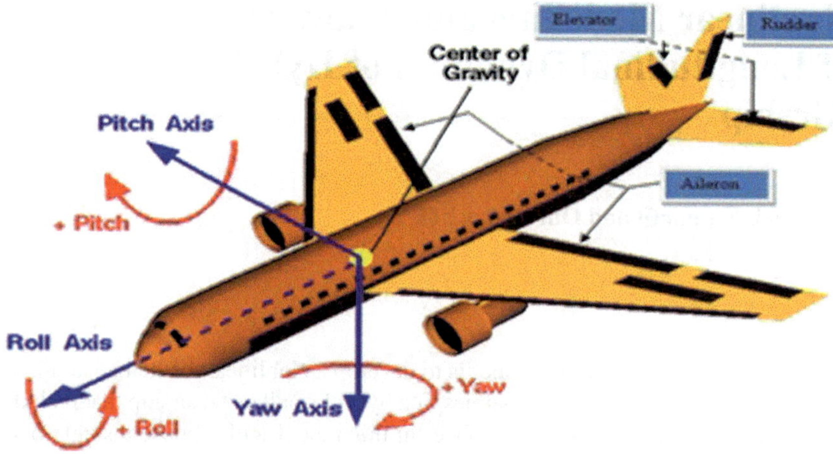

Fig. 1 Body axes system of an aircraft with primary control surfaces

coming down from the hull are attached to the payload riser through a connection point allowing independent rotational motion of hull and payload.

The idea of the hybrid airship model has been taken from [1, 2] and Autonomous parafoil payload delivery systems. Here in Fig. 1. the positions of the three primary control surfaces (i.e. aileron δa or ξ, elevator δe or η and rudder δr or τ) are shown properly. Centre of gravity is shown as a yellow dot near the join position of two wings and is also known as the origin of the body axes system. X-axis is in the forward direction of the pilot also known as Roll axis, Y-axis is towards the direction of right wing known as Pitch axis and Z-axis is in downward direction towards the gravity known as Yaw axis. Angular velocities or applied moments about the x, y and z body axes are described with the adjectives roll, pitch and yaw respectively. These moments are positive if they follow the directions as shown in the Fig. 1. The rigging angle is a critical design parameter as we have the provision to control the roll moment and pitch moment with deflecting the control lines attached to airship as shown in the Fig. 5. Improper choice of rigging angle can lead to adverse dynamics during deployment. The type of uncertainty expected for the trajectory tracking is primarily due to the inability to accurately predict the aerodynamics and control coefficients for the hybrid airship.

However, controller designed for airship must guarantee stability for the system and provide a satisfied control performance. The final control objective based on obtained design parameter such as rigging angle for two body dynamics is the robustness of stability and tracking performance to the presence of modelling uncertainty. Moreover, a control strategy is desired for which stability and robustness to uncertainty, collision avoidance and trajectory tracking can be guaranteed, and a solution can be obtained in real time. The parafoil/payload system described in reference [3] can be utilised for safe guided delivery of payload to a specific target region or to numerous targets from a single launch. In reference [2–4], a full 9 DOF or 4 DOF two-body dynamic model for parafoil payload delivery system is used to capture these distinct motions along with pendulum stability effect and the hybrid airship 4 DOF model has been proposed with this concept. Unlike an aeroplane, where the wings are the most deformable, an airship's hull is the most flexible component. The hull shape of both semi-rigid and non-rigid airships is maintained by a pressure level greater than the surrounding air pressure. apparent mass matrix is taken care in modelling of hybrid airship. Mathematical modelling of hybrid airship with payload is more complicated compared with either Aircraft or with Parafoil delivery system.

2 Aircraft Dynamics and Parameter Definition

Flight test data is required for aerodynamic parameter estimation of a postulated mathematical model of an aircraft. The mathematical model consists of both the aircraft equations of motion and the equations for aerodynamic forces and moments, known as the aerodynamic model equations. State variables and control inputs of an aircraft are defined as shown in Fig. 2. An inertial coordinate system is fixed by earth's surface. The origin of the body-fixed coordinate system coincides with the centre of mass of the aircraft.

Here some description of the notations used in Fig. 2. and they are as follows:
u, v, w = body-axis components of aircraft velocity relative to Earth axes.
p, q, r = body-axis components of aircraft angular velocity.
Xb, Yb, Zb = body-axis components of aerodynamic force acting on the aircraft.
L, M, N = body-axis components of aerodynamic moment acting on the aircraft.

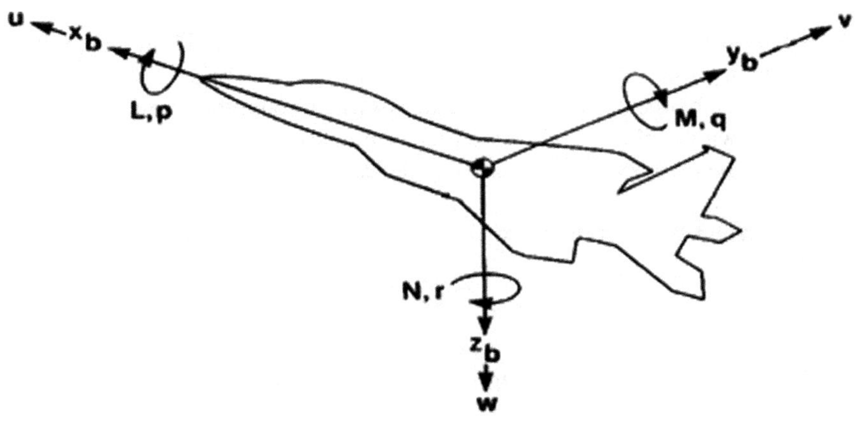

Fig. 2 Airplane notation and sign convention

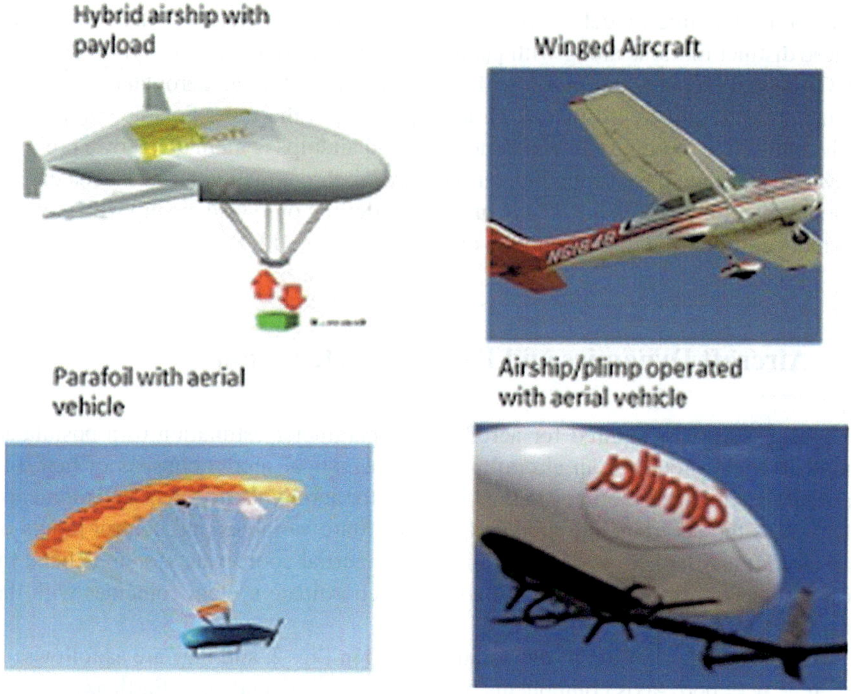

Fig. 3 Pictorial Idea for winged hybrid airship flight vehicle with suspended payload

3 Modelling of Hybrid Airship

As shown in Fig. 3, the proposed model named as "winged hybrid airship flight vehicle with suspended payload", located at 1st row 1st column, having developed with combining the modelling features of winged aircraft, airship with aircraft feature known as 'plimp' and parafoil payload delivery flight system. The hybrid airship geometric, apparent mass and Inertia, aerodynamics data and payload data are taken from A.F.A. Gaffar [1] for dynamic simulation of Hybrid-Airship with payload system. The Hybrid Airship is assumed to carry attached payload (Tables 1 and 2).

4 Mathematical Modelling of Hybrid Airship

4.1 Single Body Dynamics for Hybrid Airship

The complexity in mathematical derivation of hybrid airship is because of the centre of mass or centre of gravity and centre of volume is positioned at two distinct locations

Table 1 Hull geometry

Hull Parameter	Value
Hull mass m_k	10 kg
Overall length,L	3.75 m
Maximum Diameter, D	1.6 m
Volume,V	5.6 m^3
Hull Reference Area, S_k	3.16 m^2
Hull Reference length, c	1.78 m
Ellipsoid semi-minor axis,b	0.8 m
Wing span$b$	3.06 m
Wing area,S_t	1.72 m^2
tail spanb_t	3.06 m
tail area,S_t	0.916 m^2
Rck	0.8 m

Table 2 Payload geometry

Payload Parameter	Value
mb	2 kg
Sb	0.1 × 0.2 m^2
Rcb	0.25 m
$C_D{}^B$	1.05

as shown in the Fig. 4. Moment equation and force equation is obtained easily on centre of mass or centre of gravity. In Hybrid airship, the centre of volume is almost

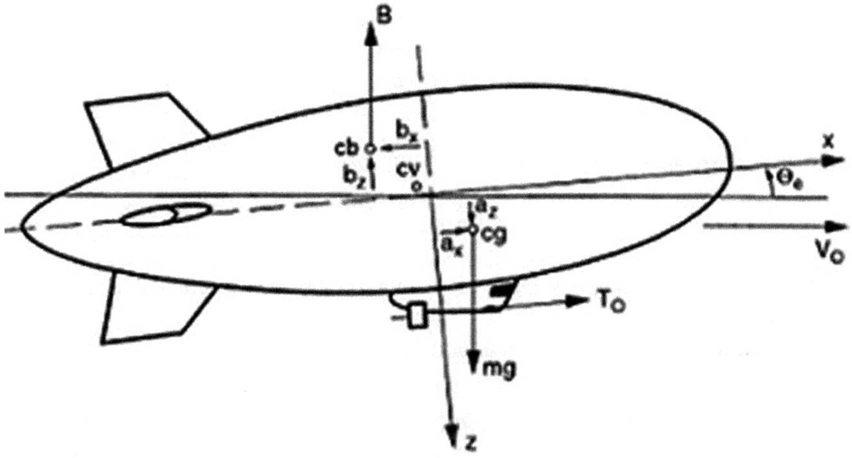

Fig. 4 Coordinate location of CG and CV

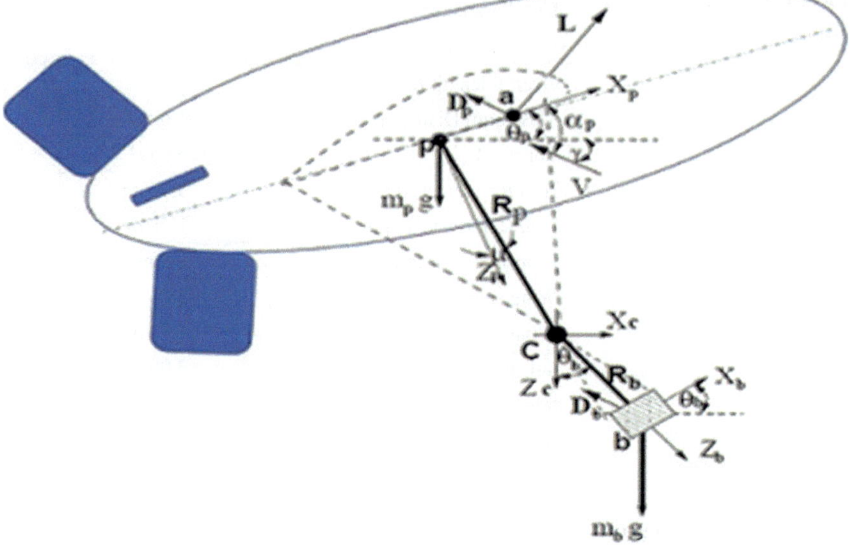

Fig. 5 4 DOF Longitudinal Two body dynamics for Hybrid Airship-payload system

located at the centre position of HULL that is filled with Helium or Hydrogen gas. Centre of mass or centre of gravity in hybrid airship is located near gondola where payload is attached, its mass is larger compared to HULL mass. Airship and hybrid airship is lighter than air vehicle, so we must take care of apparent mass of the system in the mathematical derivation which is not required in the final equation of motion of the aircraft as shown in Eq. (3).

For calculating the lift in airship design, Archimedes principle required volume for hydrostatic lift

$$L_{hst} = vol(\rho_a - \rho_g) \quad (1)$$

where $\rho_a and \rho_g$ are densities of air and lifting gas.

$$w_{net} = (m_{GTw} \times g - L_{hst}) \quad (2)$$

Net weight w_{net} is gross take-off weight minus the weight balanced by the hydrostatic lift.

Hybrid airship 6 DOF Non-linear equation of motion:

$$\begin{bmatrix} mI_3 + M' & -mr_G^X \\ mr_G^X & I_0 + I_0' \end{bmatrix} \begin{bmatrix} \dot{V}_0 \\ \dot{\omega}_0 \end{bmatrix} + \begin{bmatrix} m(\omega_0 \times V_0 + \omega_0 \times (\omega_0 \times r_G)) \\ \omega_0 \times (I_0\omega_0) + mr_G \times (\omega_0 \times V_0) \end{bmatrix} = \begin{bmatrix} F \\ \zeta \end{bmatrix} \quad (3)$$

$$r_G^X = \begin{bmatrix} 0 & -a_z & a_y \\ a_z & 0 & -a_x \\ -a_y & a_x & 0 \end{bmatrix} \tag{4}$$

$$\left(r_G^X\right)^T = -\left(r_G^X\right) \tag{5}$$

Aircraft 6 DOF Non-linear equation of motion:

$$\begin{bmatrix} mI_3 & 0 \\ 0 & I_0 \end{bmatrix}\begin{bmatrix} \dot{V}_0 \\ \dot{\omega}_0 \end{bmatrix} + \begin{bmatrix} m(\omega_0 \times V_0) \\ \omega_0 \times (I_0\omega_0) \end{bmatrix} = \begin{bmatrix} F \\ \zeta \end{bmatrix} \tag{6}$$

Airship and hybrid airship is lighter than air vehicle, so we have to take of apparent mass of the system in the mathematical derivation which is not required in the final equation of motion of the aircraft as shown in Eq. (3) and (6).

The simplified version of the equation for propulsion system of hybrid airship is given in Eq. (7).

$$M_P = M_T = Td_z \tag{7}$$

where M_T is the moment due to propulsion system or thrust, T is total thrust generated by the engine and d_z is the distance of propulsion system from CV. For simplifying the complexity in the 6DOF non-linear dynamic equations of motion for hybrid airship (3), the dynamics can be analysed in two parts as longitudinal dynamics consists of 3DOF equation of motion and lateral dynamics consists of another 3DOF equation of motion. This can be done because, the shape of the hybrid airship is also symmetrical along x-axis just as an aircraft. The longitudinal dynamics consists of axial force equation ($\dot{U}$), normal force equation ($\dot{W}$), pitching moment equation ($\dot{q}$) and kinematic equation ($\dot{\theta}$). All the 6 DOF non-linear equation of motion is given in body axis frame of reference. The longitudinal 3 DOF non-linear equation of motion for single body dynamics of hybrid airship by considering the associated terms used for longitudinal dynamics and dropping out the remaining terms in the given 6 DOF nonlinear equation of motion for airship is proposed here. The simulation is done in wind axis coordinate system, so the terms used for simulation is $\dot{V}$ (resultant of U and W velocity in wind axis), $\dot{\gamma}(gamma)$, $\dot{q}$ and $\dot{\theta}$ for longitudinal dynamics of the system.

$$\dot{v} = \frac{1}{mx} \times (T \times \cos(\alpha) + \overline{q} \times s \times (-1) \times C_d - (m \times g - B) \times \sin(\gamma)) \tag{8}$$

$$\dot{\gamma} = \frac{1}{(mz \times v)} \times (T \times \sin(\alpha) + FAY - (m \times g - B) \times \cos(\gamma)) \tag{9}$$

$$\dot{q} = \frac{(((M_q - m \times a_x \times v \times \cos(\alpha) - m \times a_z \times v \times \sin(\alpha)) \times q) + M_a + M_s + M_T)}{I_H} \tag{10}$$

$$\dot{\theta} = q \tag{11}$$

$$FAY = \overline{q} \times s \times C_l + (-1) \times 0.5 \times \rho \times \frac{v}{2} \times S \times MAC \times c_{l_q} \times 0.5 \tag{12}$$

$$\text{Mq} = (\rho \text{X } v) \text{ X S X MAC}^2 \text{X Cmq}) \tag{13}$$

4.2 Two Body Dynamics for Hybrid Airship with Suspended Payload

Mathematical model of the hybrid airship for getting the desired objective is developed by applying physical laws which describes aerodynamics and non-linear dynamics of the hybrid airship. In reference [3], The aerodynamic forces and moments acting at parafoil canopy mass center are modelled in terms of aerodynamic force and moment coefficients as:

$$C_L = C_L(\alpha_P, \delta s) + c_{L\delta a}\delta_a \tag{14}$$

$$C_D^P = C_D^P(\alpha_P, \delta s) + C_{D_{\delta a}}\delta_a + c_{L\delta e}\delta_e \tag{15}$$

$$C_Y = C_Y\beta + C_{Y_r} r_P \frac{b}{2V_P} + c_{Y_{\delta a}}\delta_a \tag{16}$$

In terms of hybrid airship-fixed axis coefficients:

$$C_X = \left(-C_D^P u_p + C_L w_P\right)/V_P \tag{17}$$

$$C_Y = C_Y \tag{18}$$

$$C_z = \left(-C_D^P w_P - C_L u_P\right)/V_P \tag{19}$$

$$C_l = C_{l_\beta} + C_{lp} P_P \frac{b}{2V_P} + Cl_{\delta_a}\delta_a \tag{20}$$

$$C_m = \left\{C_{m_{C/4}}(\alpha_P, \delta_s) + x_{pa} C_z\right\} + C_{mq} q_P \frac{c}{2V_P} + C_{m_{\delta a}}\delta_a + c_{L\delta e}\delta_e \tag{21}$$

$$C_n = C_{n\beta}\beta + Cn_P P_P \frac{b}{2V_P} + Cn_{\delta a}\delta_a \tag{22}$$

$$M_s = -(mga_z + Bb_z)\sin\theta - (mga_x + Bb_x)\cos\phi\cos\theta \tag{23}$$

where symmetric brake deflection δs corresponds both right and left brakes equally down, while asymmetric brake deflection δ_a is defined as $\delta_a = \mathrm{l/c}$, where l is length of control lines pulled down. Positive δ_a represents right brake down and negative δ_a represents left brake down. Positive δ_e is considered in downward deflection. As shown in Fig. 5, the rigging angle μ is the angle between the line joining mid-baseline point of the hull to joint C and the line parallel to Zh axis passing through the mid-baseline point. Therefore,

$$z_{C_P} = R_{ch}\cos\mu \tag{24}$$

$$x_{cp} = R_{ch}\sin\mu \tag{25}$$

$$y_{cp} = 0 \tag{26}$$

The 4DOF model of hybrid airship with suspended payload has been proposed as shown in Fig. 5 by combing the 3DOF longitudinal dynamics of hybrid airship with wing and the 3DOF equation of motions of suspended payload. The idea to develop the 4 DOF model of hybrid airship come from 9 DOF model of hybrid airship [2], the 9DOF and 4DOF model of parafoil payload system [4, 5]. 4-DOF model of winged hybrid airship flight vehicle -payload system is formed with B vector in Eq. (27a to f) and A system matrix in Eq. (28) and by deriving dynamic equations for winged hybrid airship flight vehicle. The 4DOF model of hybrid airship will consist of a Submodel of winged hybrid airship flight vehicle canopy, airship wing link and joint C and payload submodel consisting of payload, payload link and joint C, by separating the winged hybrid airship flight vehicle -payload system at the link joint C and considering components of internal joint forces Fx, Fz, and their moment about airship CG and payload CG respectively. It is expected that the outcome of hybrid airship with wing and suspended payload is a combined effect of the dynamics of parafoil payload delivery system and dynamics of aircraft. For Level Flight Thrust is equal to drag force for airship dynamic model with velocity assumed to be constant at 20 m/s. In the moment equation the effect of elevator at wing and moment at fin is also considered. Apparent mass is also taken care in the modelling.

$$b_1 = -m_b g\sin\theta_b - \mathbb{Q}_b S_b C_{D_b}\cos\alpha_b \tag{27a}$$

$$b_2 = m_b g\cos\theta_b - Q_b S_b C_{D_b}\sin\alpha_b + m_b R_b q_b^2 \tag{27b}$$

$$b_3 = 0 \tag{27c}$$

$$b_4 = -m_p g \sin\theta_p + \mathbb{Q}_p S_p C_x + m_{pc} R_p \sin\mu q_p^2 - (C - A)(u_C \sin\theta_P + w_C \cos\theta_p) q_p \tag{27d}$$

$$b_5 = m_P g \cos\theta p + Q_p S_p C_z + m_{pA} R_p \cos^{\mu} q_P^2 - (C - A)(U_C \cos\theta_P - w_C \sin\theta_p) q_p \tag{27e}$$

$$b_6 = -x_{pa} Q_p S_p C_z + Q_p S_{PC} C_M \tag{27f}$$

The two body dynamic model as 4DOF is developed for Parafoil payload delivery system and wing payload delivery system. For hybrid airship with payload delivery system, the two body dynamic 4DOF model has been proposed here and simulated on MATLAB and verified the result with single body 3DOF result and the result of Ghaffar [1].

$$\begin{bmatrix} m_b \cos\theta_b & -m_b \sin\theta_b & m_b R_b & 0 & \cos\theta_b & -\sin\theta_b \\ m_b \sin\theta_b & m_b \cos\theta_b & 0 & 0 & \sin\theta_b & \cos\theta_b \\ 0 & 0 & I_b & 0 & -R_b \cos\theta_b & R_b \sin\theta_b \\ m_{pA} \cos\theta p & -m_{pA} \sin\theta p & 0 & m_{PA} Rp\cos\mu & -\cos\theta_p & \sin\theta_p \\ m_{PC} \sin\theta_P & m_{PC} \cos\theta_P & 0 & -m_{pc} R_P \sin\mu & -\sin\theta_p & -\cos\theta_p \\ 0 & 0 & 0 & I_{PF} & z_{cp} & -x_{Cp} \end{bmatrix} \begin{bmatrix} \dot{u}_c \\ \dot{w}_C \\ \dot{q}_b \\ \dot{q}_p \\ F_x \\ F_z \end{bmatrix} = \begin{Bmatrix} b_1 \\ b_2 \\ b_3 \\ b_4 \\ b_5 \\ b_6 \end{Bmatrix} \tag{28}$$

5 Simulation Result

As in Fig. 6, By changing the rigging angle μ, the payload attached with airship moves backward and forward. If the payload movement is in backward side, then the airship noses up and vice-versa. So from changing the rigging angle also we

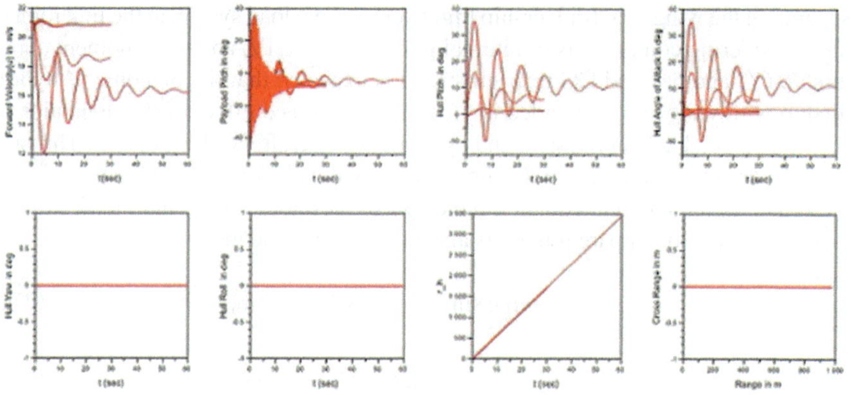

Fig. 6 Simulation result of 4DOF longitudinal dynamics of hybrid airship with suspended payload

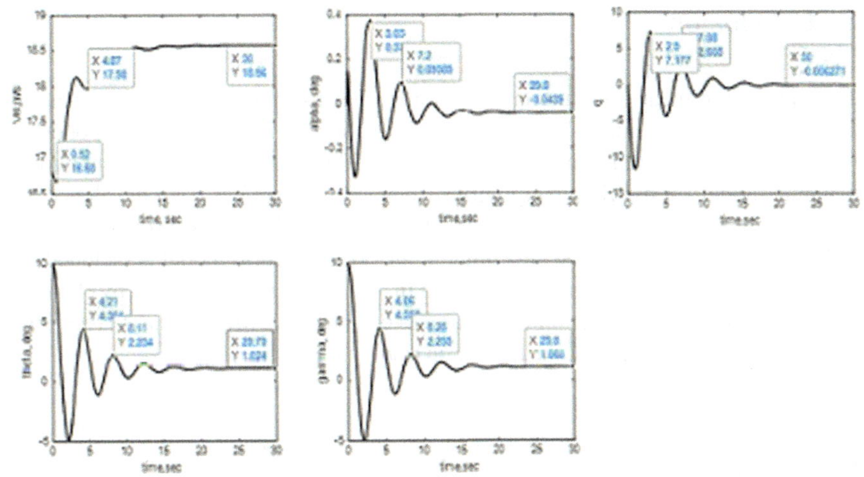

Fig. 7 Simulation result of 3DOF longitudinal dynamics of hybrid airship at delfin 0 degree deflection

can control the pitching moment of the airship. In Fig. 7, the values of alpha, q, theta and gamma trim at almost 0 degree which match with trim values of A.F.A. Gaffar [1]. The velocity trims at 18.5 m/s with initial value provided with 16.8 m/s. All the output settles the trim condition within 20 s. For q, time period is 4.1 s and oscillation frequency as 1.5 rad/sec. Time period of theta is 3.9 s with angular frequency as 1.6 rad/sec. The system is stable as it settles within 20 s. By deflecting the delfin also the trim value of the longitudinal dynamics of hybrid airship is noted down and it also settles down within 20 s. The trajectory variation is observed in longitudinal dynamics by looking the trajectory of flight path angle and theta angle, when a deflection happens in delfin control surface as shown in Fig. 8.

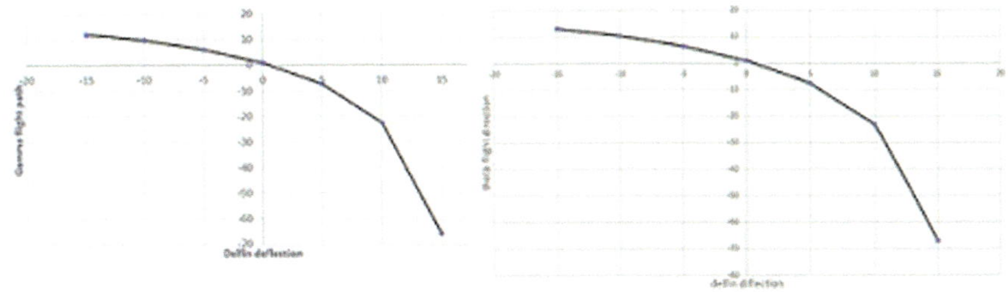

Fig. 8 Flight path trajectory with delfin deflection

6 Conclusion

The 3DOF nonlinear longitudinal dynamic model for hybrid airship is developed and the simulation result is verified with trim values of A.F.A. Gaffar [1] obtained for linearized longitudinal model for the same. The 4 DOF nonlinear longitudinal dynamic model for hybrid airship with suspended payload has been developed and the simulated result is almost behaving like the result obtained from single body 3 DOF longitudinal model. The new thing in the 3 DOF hybrid model is included and that is fin deflection control maneuvers and in 4DOF, the new finding is by changing the rigging angle also pitch control of airship can be done. More accurate result will come if wind tunnel data for the system is available.

References

1. Ghaffar, A.F.A.: The development of a mathematical model of a hybrid airship. Master of Science Thesis Aerospace Engineering, University of Southern California (2012)
2. Prakash, O., Purohit, S.: Multibody dynamics of winged hybrid airship payload delivery system. AIAA Aviation 2020 Forum (2020). https://doi.org/10.2514/6.2020-3200
3. Prakash, O., Anathkrishnan, N.: NDI based generic heading tracking control law for parafoil/payload system. AIAA Aviation 2020 Forum (2020). https://doi.org/10.2514/6.2020-3195
4. Prakash, O., Ananthkrishnan, N.: Modeling and simulation of 9-DOF parafoil-payload system flight dynamics. In: AIAA Atmospheric Flight Mechanics Conference and Exhibit, AIAA, Keystone, Colorado USA (2006). https://doi.org/10.2514/6.2006-6130
5. Prakash, O., Daftary, A., Ananthkrishnan, N.: Bifurcation analysis of parafoil-payload system flight dynamics. In: AIAA Atmospheric flight mechanics conference and exhibit, AIAA, San Francisco, California, USA (2005). https://doi.org/10.2514/6.2005-5806
6. Tiwari, A., Vora, A., Sinha, N.K.: Airship trim and stability analysis using bifurcation techniques. In: 7th International Conference on Mechanical and Aerospace Engineering, IEEE Xplore (2016). 978-1-4673-8829-0/16

A New Two-Parameter Odds Generalized Lindley-Exponential Model

Sukanta Pramanik

Abstract A new two-parameter lifetime model, called Odds Generalized Lindley-Exponential distribution (OGLED), is proposed for modelling life time data. A detailed structural and reliability properties of the new distribution is derived. The mle method has been derived for estimating the model parameters. A real life time data set has been analysed to illustrates as application.

Keywords T-X family of distributions · Exponential distribution · Lindley distribution

1 Introduction

Fitting real-life dataset with a new probability distribution and synthesising information becomes a more challenging work for researchers. Classical probability distributions e.g. binomial, gamma, exponential, hypergeometric, Poisson, normal are insufficient to get information properly from the dataset. In nature now a day it is very complex to generate data from the day-to-day work environment. Statistical distributions are very important for parametric inferences and the applications to fit real-world phenomena. Various methods have been developed to generate statistical distributions in the literature to fit and analyse complex data. Some methods were developed prior to 1980s s like the Pearsonian system by Pearson [11], Johnson [6], and Tukey [14]. Azzalini [3], McDonald [10], Marshall and Olkin [9] are also developed since 1980s. In this current century, Eugene et al. [5] developed the beta generated family of distributions, Jones [7] and Cordeiro and de Castro [4] developed the beta generated family of distributions by using Kumaraswamy distribution in the place of beta distribution.

S. Pramanik (✉)
Department of Statistics, Siliguri College, North Bengal University,
Siliguri 734 001, West Bengal, India
e-mail: skantapramanik@gmail.com
URL: http://siliguricollege.org.in/

S. Banerjee and A. Saha (eds.), *Nonlinear Dynamics and Applications*,
Springer Proceedings in Complexity,
https://doi.org/10.1007/978-3-030-99792-2_83

Alzaatreh et al. [2] has proposed a new generalized T-X (Transformed-Transformer) family of distributions. Using the T-X family of distributions, I have defined positive support of a generalized class of any distribution. Taking, the odds function $W(F_\theta(x)) = \frac{F_\theta(x)}{1-F_\theta(x)}$, the cdf of my proposed distribution is given by

$$F(x;\lambda,\theta) = \int_0^{\frac{F_\theta(x)}{1-F_\theta(x)}} f_\lambda(t)dt. \quad (1)$$

Here, $\frac{F_\theta(x)}{1-F_\theta(x)} = \frac{F_\theta(x)}{\bar{F}_\theta(x)} = \infty$ as $x \to \infty$ (assuming $\frac{1}{0} = \infty$). The resulting distribution is not only generalized but also added with some parameter(s) to this base distribution. I called this kind of class of distributions as Odds Generalized family of distributions (OGFD).

In this present article, I choose the Lindley distribution as base distribution with pdf $f_\lambda(x) = \frac{\lambda^2(1+x)}{1+\lambda}e^{-\lambda x}$ and the transformer distribution is exponential distribution in (1) i.e. $F_\theta(x) = 1 - e^{-\theta x}$. So, I call this distribution as a new two-parameter odds generalized Lindley-exponential distribution (OGLED).

In this article in Sect. 2, I developed the new modelling of the distribution. The mathematical, structural and reliability properties of the new model is derived in Sect. 3. In Sect. 4, I estimate the parameters. In Sect. 5, the applicability i.e. the application of the distribution is provided. At last in Sect. 6 conclusions.

2 The Pdf and Cdf of the OGLED

The c.d.f of the OGLED is given by

$$F(x;\lambda,\theta) = \int_0^{\frac{F(x)}{1-F(x)}} f_\lambda(x)dx = 1 - \frac{1+\lambda e^{\theta x}}{1+\lambda}e^{-\lambda(e^{\theta x}-1)} \quad (2)$$

Also the p.d.f of the OGLED is given by

$$f(x;\lambda,\theta) = \frac{\lambda^2\theta e^{2\theta x}}{(1+\lambda)}e^{-\lambda(e^{\theta x}-1)} \quad (3)$$

with range$(0,\infty)$. Figure 1 is the pdf plot for λ =2 and 3 with different values of θ. Figure 2 is the pdf plot for fixed θ with different values of λ.

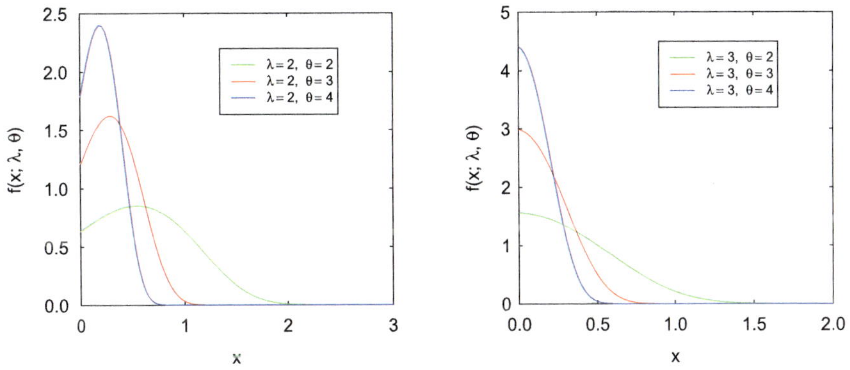

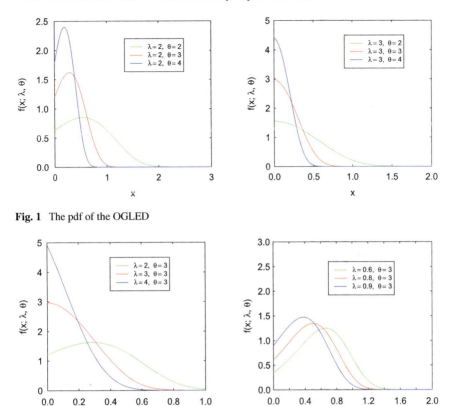

Fig. 1 The pdf of the OGLED

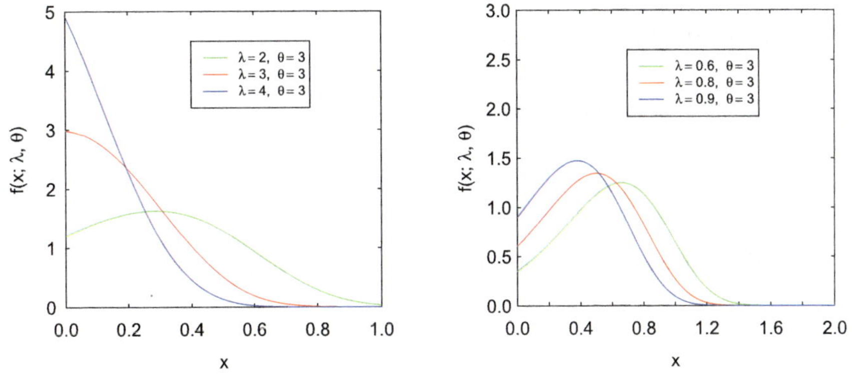

Fig. 2 The pdf of the OGLED

3 Properties of the Distribution

3.1 The Limits of the Distribution

Since the cdf of this distribution is $F(x; \lambda, \theta) = 1 - \frac{1+\lambda e^{\theta x}}{1+\lambda} e^{-\lambda(e^{\theta x}-1)}$
so, $\lim_{x \to 0} F(x; \lambda, \theta) = \lim_{x \to 0}(1 - \frac{1+\lambda e^{\theta x}}{1+\lambda} e^{-\lambda(e^{\theta x}-1)}) = 0$
i.e. $F(0) = 0$

Now $\lim_{x \to \infty} F(x; \lambda, \theta) = \lim_{x \to \infty}(1 - \frac{1+\lambda e^{\theta x}}{1+\lambda} e^{-\lambda(e^{\theta x}-1)}) = 1$ i.e. $F(\infty) = 1$.

Figure 3 is the plot of the cdf for different values of λ and θ.

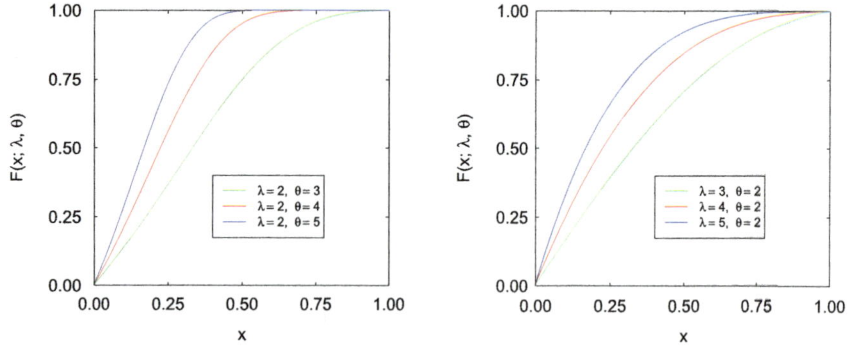

Fig. 3 The cdf of the OGLED

3.2 Structural Properties of the OGLED

The **mean** value of this OGLED is:

$$\mu_1^{'} = \frac{e^{\lambda}}{(1+\lambda)\theta}\left[\Gamma^{(1)}(2,\lambda) - \ln\lambda.\Gamma(2,\lambda)\right]$$

The **median** value of our distribution is solving the following equation using numerical method

$$1 - \frac{1+\lambda e^{\theta M}}{1+\lambda} e^{-\lambda(e^{\theta M}-1)} = \frac{1}{2}$$

The **mode** value of our distribution: mode= arg max(f(x))
So that $\log f_{l.e}(x) = 2\log\lambda + \log\theta + 2\theta x - log(1+\lambda) - \lambda(e^{\theta x} - 1)$
Now $\frac{d}{dx}\log f_{l.e}(x) = 2\theta - \lambda\theta e^{\theta x} = 0$, i.e. $x = \frac{1}{\theta}\log\left(\frac{2}{\lambda}\right)$,
Thus the mode is $\frac{1}{\theta}\log\left(\frac{2}{\lambda}\right)$.
The rth raw moment of the OGLED is as follows:

$$E(X^r) = \frac{e^{\lambda}}{(1+\lambda)\theta^r}\sum_{j=0}^{r}(-1)^{(r-j)}\binom{r}{j}(\ln\lambda)^{-(j-r)}\,\Gamma^{(j)}(2,\lambda) \tag{4}$$

Now putting suitable values of r in the above equation, I can have the Variance (μ_2), Skewness (γ_1), and Kurtosis (γ_2) of the OGLED.

Moment Generating Function(MGF):

$$M_X(t) = \sum_{r=0}^{\infty}\sum_{j=0}^{r}\frac{t^r}{r!}\frac{e^{\lambda}}{(1+\lambda)\theta^r}(-1)^{-(j-r)}\binom{r}{j}(\ln\lambda)^{-(j-r)}\,\Gamma^{(j)}(2,\lambda) \tag{5}$$

Characteristic Function(CF):

$$\Psi_X(t) = \sum_{r=0}^{\infty} \frac{(it)^r}{r!} \frac{e^{\lambda}}{(1+\lambda)\theta^r} \sum_{j=0}^{r} (-1)^{-(j-r)} \binom{r}{j} (\ln \lambda)^{-(j-r)} \Gamma^{(j)}(2, \lambda) \tag{6}$$

Cumulant Generating Function(CGF):

$$K_X(t) = \ln_e \left[\sum_{r=0}^{\infty} \frac{t^r}{r!} \frac{e^{\lambda}}{(1+\lambda)\theta^r} \sum_{j=0}^{r} (-1)^{-(j-r)} \binom{r}{j} (\ln \lambda)^{-(j-r)} \Gamma^{(j)}(2, \lambda) \right] \tag{7}$$

Mean Deviation:
The mean deviation about the mean is

$$MD_{\mu} = \frac{2e^{\lambda}}{(1+\lambda)} \left[\frac{1}{\theta} \left\{ \Gamma^{(1)}(2, \lambda e^{\theta\mu}) - \ln \lambda . \Gamma(2, \lambda e^{\theta\mu}) \right\} - \mu(1+\lambda e^{\theta\mu}) e^{-\lambda e^{\theta\mu}} \right] \tag{8}$$

also mean deviation about the median is

$$MD_M = -\mu + \frac{2e^{\lambda}}{(1+\lambda)\theta} \left[\Gamma^{(1)}(2, \lambda e^{\theta M}) - \ln \lambda . \Gamma(2, \lambda e^{\theta M}) \right] \tag{9}$$

respectively, where $\mu = E(X)$ and $M = Median(X)$.

Conditional Moments: The rth raw moment of the residual life is

$$m_r(t) = \frac{\sum_{j=0}^{r} \sum_{k=0}^{j} \frac{(-1)^j}{\theta^j} \binom{r}{j} t^{r-j} (-1)^{(j-k)} \binom{j}{k} (\ln \lambda)^{(j-k)} \Gamma^{(k)}(2, \lambda e^{\theta t})}{(1+\lambda e^{\theta t}) e^{-\lambda e^{\theta t}}}$$

The rth raw moment of the reversed residual life is

$$\bar{m}_r(t) = \frac{e^{\lambda} \sum_{j=0}^{r} \sum_{k=0}^{j} \frac{(-1)^j}{\theta^j} \binom{r}{j} t^{(r-j)} \binom{j}{k} (-\ln \lambda)^{j-k} \left[\gamma^{(k)}(2, \lambda e^{\theta t}) - \gamma^{(k)}(2, \lambda) \right]}{\left\{ 1 + \lambda - (1+\lambda e^{\theta x}) e^{-\lambda(e^{\theta x}-1)} \right\}}$$

Quantile function($Q(p)$): Let X denote a r.v. with the pdf in (3).

$$1 - \frac{1+\lambda e^{\theta(Q(p))}}{1+\lambda} e^{-\lambda[e^{\theta(Q(p))}-1]} = p, \tag{10}$$

where, $F(Q(p)) = p$

3.3 Incomplete Moment, Bonferroni and Lorenz Curves

The rth order incomplete moment of the OGLED is

$$m_r^I(t) = \frac{e^{\lambda}\lambda^2}{\theta^r(1+\lambda)}\left[\Gamma^{(r)}(2,\lambda) - \Gamma^{(2)}(2,\lambda e^{\theta t})\right] \tag{11}$$

The Bonferroni and Lorenz curves are defined by

$$B(p) = \frac{m_1^I(x_p)}{p\mu} \tag{12}$$

and

$$L(p) = \frac{m_1^I(x_p)}{\mu} \tag{13}$$

respectively, where, $\mu = E(X)$ and $x_p = F^{-1}(p)$ which is to be calculated numerically using (10) for given p.

3.4 Entropies

The Rényi entropy for a new two-parameter model is

$$H_R(\beta) = -\log\theta + \frac{\lambda\beta}{1-\beta} - \frac{\beta}{1-\beta}\log(1+\lambda) - \frac{2\beta}{1-\beta}\log\beta + \frac{\log\Gamma(2\beta,\lambda\beta)}{1-\beta}$$

Shannon measure of entropy for a new two parameter model is

$$H(f) = E[-\log f(X)] = -2\log\lambda - \log\theta - \lambda + \log(1+\lambda) + \frac{e^{\lambda}}{1+\lambda}\Gamma(3,\lambda) - \frac{2e^{\lambda}}{(1+\lambda)}\left[\Gamma^{(1)}(2,\lambda) - \log\lambda.\Gamma(2,\lambda)\right]$$

3.5 Reliability Properties

The survival or reliability function is given by

$$R(x) = \frac{1+\lambda e^{\theta x}}{1+\lambda}e^{-\lambda(e^{\theta x}-1)} \tag{14}$$

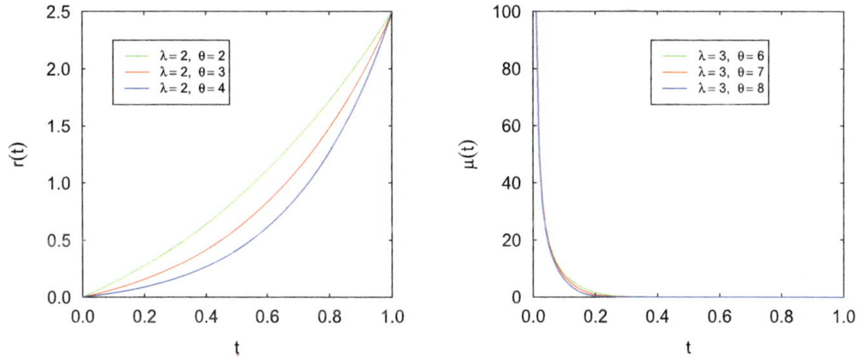

Fig. 4 Hazard rate and Reversed Hazard Rate of the OGLED

also the hazard rate function is given by

$$r(t) = \frac{\lambda^2 \theta e^{2\theta t}}{1 + \lambda e^{\theta t}} \tag{15}$$

Now, $\log f(x) = 2\log\lambda + \log\theta + \lambda - \log(1+\lambda) + 2\theta x - \lambda e^{\theta x}$
So, $\frac{d}{dx}\log f(x) = 2\theta - \lambda\theta e^{\theta x}$ and $\frac{d^2}{dx^2}\log f(x) = -\lambda\theta^2 e^{\theta x}$
Since $\lambda > 0, \theta > 0$ and $x > 0$, $\frac{d^2}{dx^2}\log f(x) < 0$. Thus, the it is log-concave distribution. So, the distribution possesses an Increasing failure rate (IFR) and Decreasing Mean Residual Life (DMRL) property. Figure 4 is the Hazard rate and Reversed Hazard Rate of the OGLED plot for $\lambda = 1.1$ and 2.1 with different values of θ.

The expression of **mean residual life (MRL)** function is

$$m_1(t) = \frac{e^{\lambda e^{\theta t}}}{1 + \lambda e^{\theta t}} \left[\frac{1}{\theta} \Gamma^{(1)}(2, \lambda e^{\theta t}) - \left(t + \frac{\ln\lambda}{\theta} \right) \Gamma(2, \lambda e^{\theta t}) \right]. \tag{16}$$

Reversed Hazard rate:

$$\mu(x) = \frac{f(x)}{F(x)} = \frac{\frac{\lambda^2\theta e^{2\theta x}}{(1+\lambda)} e^{-\lambda(e^{\theta x}-1)}}{1 - \frac{1+\lambda e^{\theta x}}{1+\lambda} e^{-\lambda(e^{\theta x}-1)}} \tag{17}$$

The expression of **mean reversed residual life (MRRL)** function is

$$\bar{m}_1(t) = \frac{e^{\lambda}\left[\left(t + \frac{\ln\lambda}{\theta}\right) \{\gamma(2,\lambda) - \gamma(3, \lambda e^{\theta t})\} - \frac{1}{\theta}\{\gamma^{(2)}(2,\lambda) - \gamma^{(2)}(3, \lambda e^{\theta t})\}\right]}{1 - (1 + \lambda e^{\theta t}) e^{-\lambda(e^{\theta t}-1)} + \lambda}$$

Figure 5 is the MRL and MRRL of the OGLED plot.

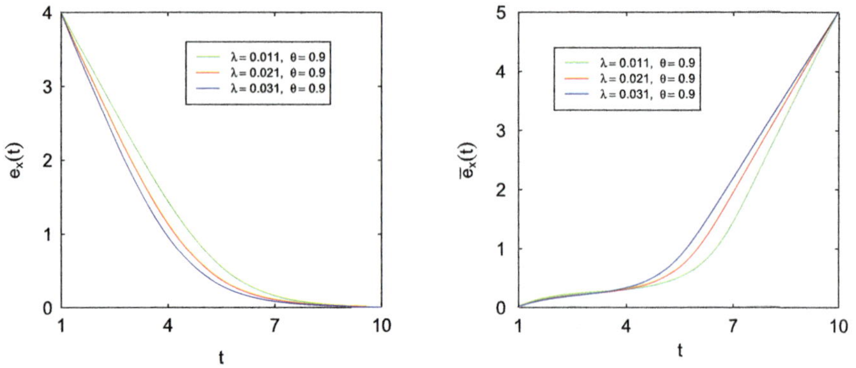

Fig. 5 Plots of the MRL and MRRL of the OGLED

3.6 *Stress-Strength Reliability*

The Stress-Strength reliability model describes the probability of surviving the life of a component with a random strength(X) subjected to a random stress(Y). Let $X \sim OGLED(\lambda_1, \theta_1)$ and $Y \sim OGLED(\lambda_2, \theta_2)$ be iid's. Thus,

$$R = P(X > Y) = 1 - \frac{e^{\lambda_1+\lambda_2}\lambda_1{}^2\theta_1}{(1+\lambda_1)(1+\lambda_2)} \int_0^\infty [1+\lambda_2 e^{\theta_2 x}]e^{2\theta_1 x} e^{-\lambda_1 e^{\theta_1 x}-\lambda_2 e^{\theta_2 x}} dx$$

When both θ_1 and θ_2 are equal to θ, then

$$R = 1 - \frac{\lambda_1{}^2}{(1+\lambda_1)(1+\lambda_2)}\left[\frac{(1+\lambda_2)}{(\lambda_1+\lambda_2)} + \frac{(1+2\lambda_2)}{(\lambda_1+\lambda_2)^2} + \frac{2\lambda_2}{(\lambda_1+\lambda_2)^3}\right]$$

4 Estimation of the Parameters

Here, I estimate the parameters of the OGLED by using the method of Maximum Likelihood Estimation (MLE).

The MLEs of λ and θ are the roots of

$$\frac{\partial \ln L(x; \lambda, \theta)}{\partial \lambda} = 0 \text{ and } \frac{\partial \ln L(x; \lambda, \theta)}{\partial \theta} = 0$$

Here using numerical methods, the two parameters λ and θ are to be estimated.

Table 1 Summarized table for fitting the data set

Distributions	Estimate of the parameter	AIC
Gamma-Half Normal	$\hat{\theta} = 0.3934$, $\hat{\alpha} = 2.8794$, $\hat{\beta} = 3.1725$	105.3572
OGLED	$\hat{\lambda} = 0.2202$, $\hat{\theta} = 1.3773$	102.3232

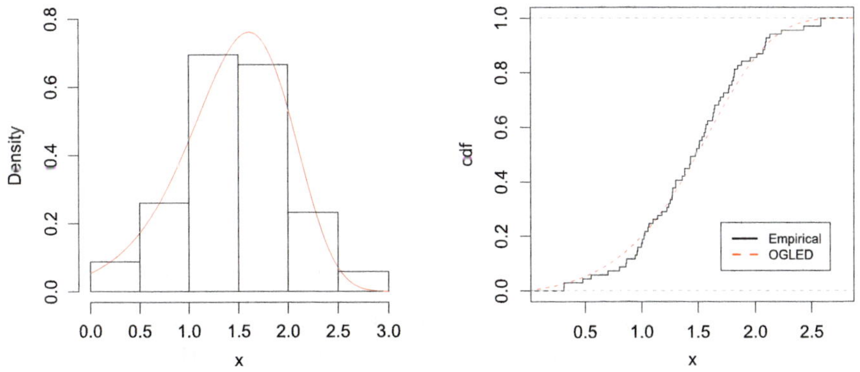

Fig. 6 Plot of the OGLED model(pdf and cdf) for the above data set

5 Application

Here, I fit the new two-parameter OGLED model to a real data set which was obtained from Raqab et al. [12] in this section and the data set is the tensile strength, measured in GPa for single carbon fibers which were tested at 20 mm. Gamma-Half Normal distribution has been fitted by Alzaatreh and Knight [1]. $\hat{\lambda} = 0.2202$ and $\hat{\theta} = 1.3773$ are the estimated parameter values of the of OGLED and AIC = 102.3232. In the OGLED fitting, only two parameters are estimated to minimize estimation error (Table 1 and Fig. 6).

6 Conclusion

The above article, studied a new two-parameter Odds generalized Lindley-exponential distribution. The distribution is a particular form of the Transformed-Transformer (or T-X) family of distributions. The different properties like structural and reliability properties have been studied. The parameters are estimated by the maximum likelihood method of estimation. The appropriateness of fitting the new two-parameter Odds generalized Lindley-exponential model has been established by fitting a real-life data set.

Acknowledgements The author thank the anonymous referee for constructive suggestions, which immensely helped to improve the paper.

References

1. Alzaatreh, A., Knight, K.: On the gamma-half normal distribution and its applications. J. Mod. Appl. Stat. Methods **12**(1), 103–119 (2013)
2. Alzaatreh, A., Lee, C., Famoye, F.: A new method for generating families of continuous distributions. Metron **71**, 63–79 (2013)
3. Azzalini, A.: A class of distributions which includes the normal ones. Scand. J. Stat. **12**, 171–178 (1985)
4. Cordeiro, G.M., de Castro, M.: A new family of generalized distributions. J. Stat. Comput. Simul. **81**(7), 883–898 (2011)
5. Eugene, N., Lee, C., Famoye, F.: Beta-normal distribution and its applications. Comm. Statist. Theory Methods **31**(4), 497–512 (2002)
6. Johnson, N.L.: Systems of frequency curves generated by methods of translation. Biometrika **36**, 149–176 (1949)
7. Jones, M.C.: Kumaraswamy's distribution: a beta-type distribution with tractability advantages. Stat. Methodol **6**, 70–81 (2009)
8. Lindley, D.V.: Fiducial distributions and Bayes' theorem. J. Roy. Stat. Soc. B **20**, 102–107 (1958)
9. Marshall, A.W., Olkin, I.: A new method for adding a parameter to a family of distributions with applications to the exponential and weibull families. Biometrika **84**, 641–652 (1997)
10. McDonald, J.B.: Some generalized functions for the size distribution of income. Econometrica **52**, 647–663 (1984)
11. Pearson, K.: Contributions to the mathematical theory of evolution to skew variation in homogeneous material. Philos. Trans. R. Soc. Lond. A **186**, 343–414 (1895)
12. Raqab, M., Madi, M., Kundu, D.: Estimation of $P(Y<X)$ for the 3-parameter generalized exponential distribution. Commun. Stat. Theory Methods **37**(18), 2854–2864 (2008)
13. Rényi, A.: On measures of entropy and information, In: Proceedings of the Fourth Berkeley Symposium on Mathematical Statistics and Probability I, pp. 547–561. University of California Press, Berkeley, CA (1961)
14. Tukey, J.W.: The Practical Relationship between the Common Transformations of Percentages of Counts and Amounts, Technical Report 36, Princeton. Princeton University, Statistical Techniques Research Group, NJ (1960)

Stability Switching in a Cooperative Prey-Predator Model with Transcritical and Hopf-bifurcations

Sajan, Ankit Kumar, and Balram Dubey

Abstract In nature organisms attempt to adopt new techniques to diminish the possibilities of being falling prey. Interspecies cooperation is one of these approaches which two different types of prey can use against a common predator. Inspired by this, we purpose a prey-predator model having two prey who cooperate with each other while interacting with a predator. For making the model more general and realistic, the interactions between prey and predator are handled through general Holling type-IV and Crowley-Martin functional responses. For well-posedness of the proposed model, firstly, its boundedness is investigated which is followed by the vigorous proofs for the existence of equilibrium points, their stability analysis, evaluation of conditions for occurrence of transcritical and Hopf-bifurcations. Numerically, we observe that as the inverse measure of predator's immunity from first prey and coefficient of cooperation from first prey to second prey crosses some respective critical values, there is occurrence of Hopf-bifurcation.Transcritical bifurcation is also depicted numerically for the intrinsic growth rate of first prey and the death rate of predator species. Several phase portraits, bifurcation diagrams are drawn to support our analytical findings. We also endorse the attribute of bistability, and basins of attraction for both stable equilibrium points are also drawn.

Keywords Prey-predator · Cooperation · Bifurcation · Bistability

Sajan acknowledges the Ph.D. fellowship [File No. 09/719(0104)/2019-EMR-I] received from the Council of Scientific & Industrial Research, New Delhi, India.

Sajan (✉) · A. Kumar · B. Dubey
Department of Mathematics, BITS Pilani, Pilani Campus, Pilani 333031, Rajasthan, India
e-mail: sajanmaths7@gmail.com

A. Kumar
e-mail: ankitmaths2738@gmail.com

B. Dubey
e-mail: bdubey@pilani.bits-pilani.ac.in

S. Banerjee and A. Saha (eds.), *Nonlinear Dynamics and Applications*,
Springer Proceedings in Complexity,
https://doi.org/10.1007/978-3-030-99792-2_84

1 Introduction

The study of prey-predator interrelationships is the foundation for research in the field of ecology. Exploring this relationship helps us to disclose various phenomena happening in ecology. Prey can use various techniques to reduce their predation like; camouflage, imitation of some dangerous breed, escape instead of direct encounter, use of harmful chemicals in the form of chemical defence, changing body color, formation of a team, etc. [1] There can be a reasonable possibility of cooperation among two types of prey to fight against a predator in an ecological sector. A nice introduction about teaming technique can be seen in [2]. Tripathi et al. [3] examined a three-dimensional model with two prey who help each other against a predator, although these two prey are competing internally. They discussed the persistence, permanence, and global stability of the model about an interior stationary point. Mishra and Raw [4] studied a prey-predator system having two prey and a predator involving teaming of prey. They used modified Holling type-IV and Holling type-II functional response to study the interaction against the predator.

Recently, Mondal and Samanta [5] worked on a three species population model having two prey and a predator. They assumed that prey help each other by forming team against the predator, and neglected the inter-species competition between these two types of prey. They used functional response of Holling type II, I and a single discrete delay in the model. Alsakaji et al. [6] extended the cooperation model by Tripathi et al. [3], by replacing the linear functional responses by Monod-Haldane functional responses with inclusion of two gestation delays in each response. They evaluated the critical values of delays for occurrence of Hopf-bifurcation. Ferrara et al. [7] extended the work of Mishra and Raw [4] with incorporation of two discrete delays. They did the Hopf-bifurcation and local stability analysis of the proposed model.

To the best knowledge of authors, a tri-trophic model with two prey and one predator model with cooperation between prey, where first prey is consumed by general Holling type-IV response while the interaction of predator with second one is dealt by Crowley-Martin functional response has not been studied yet.

2 Construction of Mathematical Model

Tripathi et al. [8] again investigated a two-prey one predator competitive system with help. But this time, they assumed that the predator consumes both the prey species via Beddington-DeAngelis response (model (1)). In this, the terms a_4xyz and b_4xyz correspond to the help provided by both prey species x and y to each other against the predator z, whereas the terms a_3xy and b_3xy signify the interspecies competition among both the prey species. The rest of the parameters in model (1) have their conventional meanings.

$$
\begin{aligned}
\frac{dx}{dt} &= a_1 x\left(1-\frac{x}{k_1}\right)-\frac{a_2 x z}{m+m_1 x+m_2 z}-a_3 x y+a_4 x y z,\\
\frac{dy}{dt} &= b_1 y\left(1-\frac{y}{k_1}\right)-\frac{b_2 y z}{n+n_1 y+n_2 z}-b_3 x y+b_4 x y z,\\
\frac{dz}{dt} &= -c_1 z-c_2 z^2+\frac{c_3 x z}{m+m_1 x+m_2 z}+\frac{c_4 x z}{n+n_1 x+n_2 z}.
\end{aligned}
\tag{1}
$$

Inspired from studies [3, 4, 8], we aim to study a 3-dimensional autonomous system of ordinary differential equations reflecting an interaction between a predator and a team of two prey species, in which these two prey species help each other against the predator to save themselves. A realistic example for this can be seen in forest when Antelope and Zebras cooperate in fighting against the predator like cheetah, lion, etc. Let $x_1(t)$ and $x_2(t)$ are the densities of two prey species where $y(t)$ is the density of predator population. The purposed model can be written in the form of a system of three non-linear differential equations and which is based upon the following assumptions:

1. Both the prey, x_1 and x_2 grows logistically in the absence of predator y, which is reflected by the terms $r_1x_1\left(1-\frac{x_1}{K_1}\right)$ and $r_2x_2\left(1-\frac{x_2}{K_2}\right)$, respectively. Here, r_i is the intrinsic growth rate and K_i is the carrying capacity of habitat, for prey x_i, $i=1,2$.
2. Prey x_1 is assumed to show the characteristic of group defence, hence the interaction between x_1 and y is addressed by generalised Holling type-IV functional response. This is given by the function $f(x_1)=\frac{x_1}{a{x_1}^2+bx_1+c}$, where a is the inverse measure of predator's immunity from prey, b is the handling time for each prey and c is the half saturation constant.
3. Prey x_2 is assumed to be consumed by Crowley-Martin response, given by function, $g(x_2, y)=\frac{x_2}{(1+\alpha x_2)(1+\beta y)}$. In this, α is the handling time for each prey and β is coefficient of interference among predator species for this prey population.
4. The terms $\sigma_1x_1x_2y$ and $\sigma_2x_1x_2y$ represent cooperation among both the prey species against the predator y. Here, σ_i is the coefficient of cooperation when x_j helps x_i, $i\neq j=1,2$.
5. Predator y is assumed to be a specific predator that is dependent only upon x_1 and x_2. Hence, in the absence of these prey species it dies, which is shown by the term $\delta_0 y$ and the intraspecific competition among predator species is indicated by the term $\delta_1 y^2$. The remaining parameters have their standard meanings.
   With all above assumptions, the purposed model is:

$$
\begin{aligned}
\frac{dx_1}{dt} &= r_1x_1\left(1-\frac{x_1}{K_1}\right)-\frac{m_1x_1y}{ax_1{}^2+bx_1+c}+\sigma_1x_1x_2y = F_1(x_1,x_2,y),\\
\frac{dx_2}{dt} &= r_2x_2\left(1-\frac{x_2}{K_2}\right)-\frac{m_2x_2y}{(1+\alpha x_2)(1+\beta y)}+\sigma_2x_1x_2y = F_2(x_1,x_2,y),\\
\frac{dy}{dt} &= \frac{c_1m_1x_1y}{ax_1{}^2+bx_1+c}+\frac{c_2m_2x_2y}{(1+\alpha x_2)(1+\beta y)}-\delta_0y-\delta_1y^2 = F_3(x_1,x_2,y),
\end{aligned}
\tag{2}
$$

where $r_i,\ K_i,\ m_i,\ a,\ b,\ c,\ \alpha,\ \beta,\ \sigma_i,\ c_i,\ \delta_0,\ \delta_1 \in (0,\infty)$ for $i = 1, 2$. The relevant biological initial condition for model (2) are $x_1(0) \geq 0,\ x_2(0) \geq 0,\ y(0) \geq 0$.

3 Kinetics of the Model

In this section, firstly we show that the solution of our system is bounded, which restricts the population explosion in finite time. Then we find all the possible equilibrium points and check their local and global behavior.

3.1 Boundedness

Theorem 1 *The set* $\Omega = \Big\{(x_1, x_2, y) : 0 \leq x_1 \leq K_1,\ 0 \leq x_2 \leq K_2,\ 0 \leq c_1x_1 + c_2x_2 + y \leq \frac{K}{\delta^*}\Big\}$, *is a positive invariant set for all the solutions starting inside of the positive octant, where* $K = 2c_1r_1K_1 + 2c_2r_2K_2 + \frac{c_1^2\sigma_1^2K_1^2K_2^2}{2\delta_1} + \frac{c_2^2\sigma_2^2K_1^2K_2^2}{2\delta_1}$ *and* $\delta^* = min\{r_1, r_2, \delta_0\}$.

Proof From the first and second equation of model (2), it is easy to get

$$\limsup_{t\to\infty} x_1(t) \leq K_1 \text{ and } \limsup_{t\to\infty} x_2(t) \leq K_2.$$

Now, let $M = c_1x_1 + c_2x_2 + y$, which implies $\frac{dM}{dt} = c_1\frac{dx_1}{dt} + c_2\frac{dx_2}{dt} + \frac{dy}{dt}$, as mentioned in [3, 8], to avoid population explosion in upcoming time, we also assume that the cooperation terms in model (2) are dominated by the prey-predator interaction terms, thus using this fact we get

$$
\begin{aligned}
\frac{dM}{dt} &\leq c_1r_1\left(1-\frac{x_1}{K_1}\right)+c_1\sigma_1x_1x_2y+c_2r_2\left(1-\frac{x_2}{K_2}\right)+c_2\sigma_2x_1x_2y-\delta_0y-\delta_1y^2\\
&\leq 2c_1r_1K_1+2c_2r_2K_2-\frac{\delta_1}{2}\left(y-\frac{c_1\sigma_1K_2x_1}{\delta_1}\right)^2+\frac{c_1^2\sigma_1^2K_2^2x_1^2}{2\delta_1}-\frac{\delta_1}{2}\left(y-\frac{c_2\sigma_2K_1x_2}{\delta_1}\right)^2+\\
&\quad \frac{c_2^2\sigma_2^2K_1^2x_2^2}{2\delta_1}-c_1r_1x_1-c_2r_2x_2-\delta_0y
\end{aligned}
$$

$\leq 2c_1r_1K_1 + 2c_2r_2K_2 + \frac{c_1^2\sigma_1^2K_2^2x_1^2}{2\delta_1} + \frac{c_2^2\sigma_2^2K_1^2x_2^2}{2\delta_1} - c_1r_1x_1 - c_2r_2x_2 - \delta_0 y.$

So, $\frac{dM}{dt} = K - \delta^*(c_1x_1 + c_2x_2 + y)$.

Thus $\limsup_{t\to\infty} M(t) \leq \frac{K}{\delta^*}$, where $K = 2c_1r_1K_1 + 2c_2r_2K_2 + \frac{c_1^2\sigma_1^2K_1^2K_2^2}{2\delta_1} + \frac{c_2^2\sigma_2^2K_1^2K_2^2}{2\delta_1}$ and $\delta^* = min\{r_1, r_2, \delta_0\}$.

Thereby, the solutions of the purposed model are bounded.

3.2 Existence of Biomass Stationary Points

Model (2) has seven equilibrium points, namely, $E_0(0,0,0)$, $E_1(K_1,0,0)$, $E_2(0,K_2,0)$, $E_3(K_1,K_2,0)$, $E_4(0,\bar{x}_2,\bar{y})$, $E_5(\tilde{x}_1,0,\tilde{y})$ and $E^*(x_1^*,x_2^*,y^*)$. The equilibrium points $E_0(0,0,0)$, $E_1(K_1,0,0)$, $E_2(0,K_2,0)$ and $E_3(K_1,K_2,0)$ exist trivially. Now we drive the conditions under which remaining equilibrium points exist.

- Existence of $E_4(0,\bar{x}_2,\bar{y})$: $x_2 = \bar{x}_2,\ y = \bar{y}$ are the positive solutions of a system of equations given as

$$r_2\left(1 - \frac{\bar{x}_2}{K_2}\right) = \frac{m_2\bar{y}}{(1+\alpha\bar{x}_2)(1+\beta\bar{y})} = 0,\quad \frac{c_2m_2\bar{x}_2}{(1+\alpha\bar{x}_2)(1+\beta\bar{y})} - \delta_0 - \delta_1\bar{y} = 0. \tag{3}$$

Solving above both the equations, we obtain a 5-degree algebraic polynomial in terms of $\bar{x}_2$, given as:

$$A_5\bar{x}_2^5 + A_4\bar{x}_2^4 + A_3\bar{x}_2^3 + A_2\bar{x}_2^2 + A_1\bar{x}_2 + A_0 = 0, \tag{4}$$

where, $A_5 = \left(\frac{c_2{r_2}^3\beta^2\alpha^2}{{K_2}^2}\right)$, $A_4 = \left(-\frac{2c_2{r_2}^2\beta^2\alpha^2}{K_2} + \frac{2c_2{r_2}^2\beta^2\alpha}{{K_2}^2}\right)$, $A_3 = \left(-\frac{4c_2{r_2}^2\beta^2\alpha}{K_2} + c_2{r_2}^2\beta^2\alpha^2 + \frac{c_2{r_2}^2\beta^2}{{K_2}^2} + \frac{2c_2m_2r_2\beta\alpha}{K_2} - \frac{\delta_0r_2\beta\alpha^2}{K_2} + \frac{\delta_1r_2\alpha^2}{K_2}\right)$, $A_2 = \left(2c_2{r_2}^2\beta^2\alpha - \frac{2c_2{r_2}^2\beta^2}{K_2} - 2c_2m_2r_2\beta\alpha + \frac{2c_2m_2r_2\beta}{K_2} + \delta_0\alpha^2r_2\beta - \frac{2\delta_0r_2\beta\alpha}{K_2} - \delta_1r_2\alpha^2 + \frac{2\delta_1r_2\alpha}{K_2}\right)$, $A_1 = \left(c_2{r_2}^2\beta^2 - 2c_2m_2\beta r_2 + \delta_0r_2\beta(2\alpha - \frac{1}{K_2}) - \delta_1r_2(2\alpha - \frac{1}{K_2}) + c_2{m_2}^2 - \delta_0m_2\alpha\right)$, $A_0 = -\delta_0(m_2 - r_2\beta) - \delta_1r_2$.

By using Descarte's rule of sign, (4) has at least one positive $\bar{x}_2$ if

$$m_2 > r_2\beta.$$

Using this $\bar{x}_2$ in first equation of (3), we get $y = \frac{r_2(1+\alpha\bar{x}_2)\left(1-\frac{\bar{x}_2}{K_2}\right)}{m_2 - r_2\beta(1+\alpha\bar{x}_2)\left(1-\frac{\bar{x}_2}{K_2}\right)}$, which is positive under the condition $m_2 > r_2\beta(1+\alpha\bar{x}_2)\left(1-\frac{\bar{x}_2}{K_2}\right)$. Therefore, $E_4(0, \bar{x}_2, \bar{y})$ exists if

$$m_2 > max\left\{1, (1+\alpha\bar{x}_2)\left(1-\frac{\bar{x}_2}{K_2}\right)\right\}.$$

- Existence of $E_5(\tilde{x}_1, 0, \tilde{y})$: $x_1 = \tilde{x}_1,\ y = \tilde{y}$ are the positive solutions of a system of equations given by

$$r_1\left(1-\frac{\tilde{x}_1}{K_1}\right) = \frac{m_1\tilde{y}}{a\tilde{x}_1^2 + b\tilde{x}_1 + c}, \ \frac{c_1 m_1\tilde{x}_1}{a\tilde{x}_1^2 + b\tilde{x}_1 + c} - \delta_0 - \delta_1\tilde{y} = 0. \tag{5}$$

By solving above both equations, again we get a 5-degree polynomial as in the previous one, and it is in terms of $\tilde{x}_1$, given as

$$B_5\bar{x}_1^5 + B_4\bar{x}_1^4 + B_3\bar{x}_1^3 + B_2\bar{x}_1^2 + B_1\bar{x}_1 + B_0 = 0, \tag{6}$$

where, $B_5 = \left(\frac{a^2\delta_1 r_1}{m_1 K_1}\right)$, $B_4 = \left(-\frac{\delta_1 r_1 a^2}{m_1} + \frac{2ab\delta_1 r_1}{m_1 K_1}\right)$, $B_3 = \left(-\frac{2ab\delta_1 r_1}{m_1} + \frac{2ac\delta_1 r_1}{m_1 K_1} + \frac{\delta_1 r_1 b^2}{m_1 K_1}\right)$, $B_2 = \left(-\frac{\delta_1 r_1 b^2}{m_1} - \frac{2ac\delta_1 r_1}{m_1} + \frac{2bc\delta_1 r_1}{K_1 m_1} - \delta_0 a\right)$, $B_1 = \left(-\frac{2\delta_1 r_1 bc}{m_1} + \frac{c^2\delta_1 r_1}{m_1 K_1} - \delta_0 b + c_1 m_1\right)$, $B_0 = -\frac{\delta_1 r_1 c^2}{m_1} - \delta_0 c$. Again from Descarte's rule of sign, (6) has at least one positive root $\tilde{x}_1$ and using this in first equation of (5) we get

$$\tilde{y} = \frac{r_1}{m_1}\left(1-\frac{\tilde{x}_1}{K_1}\right)(a\tilde{x}_1^2 + b\tilde{x}_1 + c),$$

and hence we obtain $E_5(\tilde{x}_1, 0, \tilde{y})$.

- Existence of $E^*(x_1^*, x_2^*, y^*)$:
Interior equilibrium E^* is the positive solution of the system of equations given as:

$$r_1\left(1-\frac{x_1^*}{K_1}\right) - \frac{m_1 y^*}{a x_1^{*2} + b x_1^* + c} + \sigma_1 x_2^* y^* = 0, \tag{7}$$

$$r_2\left(1-\frac{x_2^*}{K_2}\right) - \frac{m_2 y^*}{(1+\alpha x_2^*)(1+\beta y^*)} + \sigma_2 x_1^* y^* = 0, \tag{8}$$

$$\frac{c_1 m_1 x_1^*}{a x_1^* + b x_1^* + c} + \frac{c_2 m_2 x_2^*}{(1+\alpha x_2^*)(1+\beta y^*)} - \delta_0 - \delta_1 y^* = 0. \tag{9}$$

Since Eqs. (7–9) are complex, and hence it is not easy to prove the existence of interior equilibrium analytically. Thus we will show its existence numerically.

3.3 Stability Analysis

For local stability analysis of model (2), we have following findings:

- It is easy to see that the equilibrium point $E_0(0, 0, 0)$ is a saddle point with unstable manifold along the x_1x_2 plane and stable manifold along the y-axis.
- $E_1(K_1, 0, 0)$ is a saddle point with one dimensional stable manifold and two dimensional unstable manifold if $\frac{c_1m_1K_1}{aK_1{}^2+bK_1+c} > \delta_0$ or with two dimensional stable manifold and one dimensional unstable manifold if $\frac{c_1m_1K_1}{aK_1{}^2+bK_1+c} < \delta_0$.
- $E_2(0, K_2, 0)$ is a saddle point with similar interpretation as of E_1.
- $E_3(K_1, K_2, 0)$ is locally asymptotically stable point, provided $\frac{c_1m_1K_1}{aK_1{}^2+bK_1+c} + \frac{c_2m_2K_2}{(1+\alpha K_2)} < \delta_0$, otherwise a saddle point.
- One of the eigenvalue of variational matrix for $E_4(0, \bar{x}_2, \bar{y})$ is $r_1 - \frac{m_1\bar{y}}{c} + \sigma_1\bar{x}_2\bar{y}$ and rest two of eigenvalues are given by the roots of equation given below.

$$\xi^2 + C_1\xi + C_2 = 0, \tag{10}$$

where, $C_1 = \left(\frac{r_2\bar{x}_2}{K_2} - \frac{m_2\alpha\bar{x}_2\bar{y}}{(1+\alpha\bar{x}_2)^2(1+\beta\bar{y})} + \frac{c_2m_2\beta\bar{x}_2\bar{y}}{(1+\alpha\bar{x}_2)(1+\beta\bar{y})^2} + \delta_1\bar{y}\right)$ and $C_2 = \left(\frac{r_2\bar{x}_2}{K_2} - \frac{m_2\alpha\bar{x}_2\bar{y}}{(1+\alpha\bar{x}_2)^2(1+\beta\bar{y})}\right)\left(\frac{c_2m_2\beta\bar{x}_2\bar{y}}{(1+\alpha\bar{x}_2)(1+\beta\bar{y})^2} + \delta_1\bar{y}\right) + \left(\frac{m_2\bar{x}_2}{(1+\alpha\bar{x}_2)(1+\beta\bar{y})^2}\right)\left(\frac{c_2m_2\bar{y}}{(1+\alpha\bar{x}_2)^2(1+\beta\bar{y})}\right)$. From Routh–Hurwitz criteria the real part of all roots of (10) are negative iff $C_1 > 0$ and $C_2 > 0$. Thus we can state the following theorem

Theorem 2 *Model (2) is asymptotically stable in neighbourhood of stationary point E_4 with three dimensional stable manifold, if $r_1 < \frac{m_1\bar{y}}{c} - \sigma_1\bar{x}_2\bar{y}$ and $r_2 > \frac{m_2K_2\alpha\bar{y}}{(1+\alpha\bar{x}_2)^2(1+\beta\bar{y})}$.*

- Similarly one of the eigenvalue of Jacobian matrix corresponding to $E_5(\tilde{x}_1, 0, \tilde{y})$ is $r_2 - \frac{m_2\tilde{y}}{1+\beta\tilde{y}} + \sigma_2\tilde{x}_1\tilde{y}$ and rest of two eigenvalues are the roots of equation given by

$$\Pi^2 + D_1\Pi + D_2 = 0, \tag{11}$$

where, $D_1 = (\frac{r_1\tilde{x}_1}{K_1} - \frac{m_1\tilde{y}(2a\tilde{x}_1^2+b\tilde{x}_1)}{(a\tilde{x}_1^2+b\tilde{x}_1+c)^2} + \delta\tilde{y})$ and $D_2 = \left(\frac{r_1\tilde{x}_1}{K_1} - \frac{m_1\tilde{y}(2a\tilde{x}_1^2+b\tilde{x}_1)}{(a\tilde{x}_1^2+b\tilde{x}_1+c)^2}\right)\delta\tilde{y} + \left(\frac{m_1\tilde{x}_1}{a\tilde{x}_1^2+b\tilde{x}_1+c}\right)\left(\frac{c_1m_1\tilde{y}(-a\tilde{x}_1^2+c)}{(a\tilde{x}_1^2+b\tilde{x}_1+c)^2}\right)$. Again from Routh–Hurwitz criteria, we have the following theorem.

Theorem 3 *The system (2) is asymptotically stable around the equilibrium E_5 if $r_2 < \frac{m_2\tilde{y}}{1+\beta\tilde{y}} - \sigma_2\tilde{x}_1\tilde{y}$ and $r_1 > \frac{m_1K_1\tilde{y}(2a\tilde{x}_1^2+b\tilde{x}_1)}{\tilde{x}_1(a\tilde{x}_1^2+b\tilde{x}_1+c)^2}$.*

– The variational matrix about $E^*(x_1^*, x_2^*, y^*)$ is

$$J|_{E^*} = \begin{bmatrix} -\frac{r_1 x_1^*}{K_1} + \frac{m_1 y^*(2ax_1^{*2}+bx_1^*)}{(ax_1^{*2}+bx_1^*+c)^2} & \sigma_1 x_1^* y^* & -\frac{m_1 x_1^*}{ax_1^{*2}+bx_1^*+c} + \sigma_1 x_1^* x_2^* \\ \sigma_2 x_2^* y^* & -\frac{r_2 x_2^*}{K_2} + \frac{m_2 \alpha x_2^* y^*}{(1+\alpha x_2^*)^2(1+\beta y^*)} & -\frac{m_2 x_2^*}{(1+\alpha x_2^*)(1+\beta y^*)^2} + \sigma_2 x_1^* x_2^* \\ \frac{c_1 m_1 y^*(-ax_1^{*2}+c)}{(ax_1^{*2}+bx_1^*+c)^2} & \frac{c_2 m_2 y^*}{(1+\alpha x_2^*)^2(1+\beta y^*)} & -\frac{c_2 m_2 \beta x_2^* y^*}{(1+\alpha x_2^*)(1+\beta y^*)^2} - \delta_1 y^* \end{bmatrix}.$$

Characteristic equation corresponding to above variational matrix is

$$\lambda^3 + E_1\lambda^2 + E_2\lambda + E_3 = 0. \tag{12}$$

In equation (12), the multipliers $E_1,\ E_2$ and E_3 are given as $E_1 = -(k_{11} + k_{22} + k_{33}),\ E_2 = (k_{22}k_{33} - k_{23}k_{32}) + (k_{11}k_{33} - k_{13}k_{31}) + (k_{11}k_{22} - k_{12}k_{21})$, and $E_3 = -(k_{11}(k_{22}k_{33} - k_{23}k_{32}) - k_{12}(k_{21}k_{33} - k_{23}k_{31}) + k_{13}(k_{21}k_{32} - k_{22}k_{31}))$, where, k_{mn} for $m, n = 1, 2, 3$ represent an entry in $J|_{E^*}$, in m^{th} row and n^{th} column. Again all eigenvalues of $J|_{E^*}$ have negative real part iff $E_1 > 0,\ E_3 > 0$ and $E_1E_2 > E_3$. Thus, we can state the following theorem.

Theorem 4 *The interior equilibrium $E^*(x_1^*, x_2^*, y^*)$ is asymptotically stable iff $E_1 > 0,\ E_3 > 0$ and $E_1E_2 - E_3 > 0$.*

Theorem 5 *The system (2) is globally asymptotically stable about the interior equilibrium E^* under the following conditions:* $\frac{r_1}{K_1} > \frac{m_1 y^*(a(K_1+x_1^*)+b_1)}{(ax_1^{*2}+bx_1^*+c)c}$, $\frac{r_2}{K_2} > \frac{\alpha m_2 y^*}{(1+\alpha x_2^*)(1+\beta y^*)}$, $(\sigma_1 + \sigma_2 l_1)^2 y_M^2 < \left(\frac{r_1}{K_1} - \frac{m_1 y^*(a(K_1+x_1^*)+b_1)}{(ax_1^{*2}+bx_1^*+c)c}\right)\left(\frac{r_2}{K_2} - \frac{\alpha m_2 y^*}{(1+\alpha x_2^*)(1+\beta y^*)}\right)$, $\sigma_2{}^2 l_1{}^2 x_1^{*2} < \left(\frac{r_2}{K_2} - \frac{\alpha m_2 y^*}{(1+\alpha x_2^*)(1+\beta y^*)}\right)\left(\delta_1 + \frac{c_2 m_2 x_2^* \beta}{(1+\beta y_M)(1+\alpha x_2)(1+\beta y^*)}\right)$, $\left(\sigma_1 x_2^* + \frac{l_2 c_1 m_1 a K_1 x_1^*}{(ax_1^{*2}+bx_1^*+c)c}\right)^2 < \left(\frac{r_1}{K_1} - \frac{m_1 y^*(a(K_1+x_1^*)+b_1)}{(ax_1^{*2}+bx_1^*+c)c}\right)\left(\delta_1 + \frac{c_2 m_2 x_2^* \beta}{(1+\beta y_M)(1+\alpha x_2)(1+\beta y^*)}\right)$.

Proof Consider a positive definite function V as an appropriate Lyapunov function around E^* given as $V(x_1, x_2, y) = \left(x_1 - x_1^* - x_1^* \ln \frac{x_1}{x_1^*}\right) + l_1\left(x_2 - x_2^* - x_2^* \ln \frac{x_2}{x_2^*}\right) + l_2\left(y - y^* - y^* \ln \frac{y}{y^*}\right)$, where $l_1 = \frac{l_2 c_2(1+\beta y^*)}{(1+\alpha x_2^*)}$ and $l_2 = \frac{(ax_1^{*2}+bx_1^*+c)}{c_1 c}$. Now, after differentiating V with respect to time along with solution of (2) and with some algebraic manipulations, we get $\frac{dV}{dt} = -\frac{1}{2}A_{11}(x_1 - x_1^*)^2 + A_{12}(x_1 - x_1^*)(x_2 - x_2^*) - \frac{1}{2}A_{22}(x_2 - x_2^*)^2 - \frac{1}{2}A_{22}(x_2 - x_2^*)^2 + A_{23}(x_2 - x_2^*)(y - y^*) - \frac{1}{2}A_{33}(y - y^*)^2 - \frac{1}{2}A_{11}(x_1 - x_1^*)^2 + A_{13}(x_1 - x_1^*)(y - y^*) - \frac{1}{2}A_{33}(y - y^*)^2$,
where, $A_{11} = \frac{r_1}{K_1} - \frac{m_1 y^*(a(x_1+x_1^*)+b_1)}{(ax_1^{*2}+bx_1^*+c)(ax_1{}^2+bx_1+c)}$, $A_{22} = \frac{r_2}{K_2} - \frac{\alpha m_2 y^*}{(1+\alpha x_2)(1+\alpha x_2^*)(1+\beta y^*)}$, $A_{33} = \delta_1 + \frac{c_2 m_2 x_2^* \beta}{(1+\beta y)(1+\alpha x_2^*)(1+\beta y^*)}$, $A_{12} = \sigma_1 y + \sigma_2 l_1 y$, $A_{13} = \sigma_1 x_2^* - \frac{l_2 c_1 m_1 a x_1 x_1^*}{(ax_1^{*2}+bx_1^*+c)(ax_1{}^2+bx_1+c)}$, $A_{23} = \sigma_2 l_1 x_1^*$.

So, using Sylvester's criterion, $\frac{dV}{dt}$ is negative definite under the hypothesis given in statement of the theorem. Therefore, E^* is globally asymptotically stable with the assumed conditions.

3.4 Bifurcation Assessment

Theorem 6 *Coexistence equilibrium E^* is bifurcated from the planner equilibrium E_4, through a transcrtical bifurcation, when parameter r_1, crosses its transcritical threshold value $r_1 = r_1^{tc} = \frac{m_1\bar{y}}{c} - \sigma_1\bar{x}_2\bar{y}$.*

Proof To prove this theorem, we have to satisfy the conditions given in Sotomayor Theorem [9]. Consider, $V = (v_1, v_2, v_3)^T$, $W = (w_1, w_2, w_3)^T$ as the eigenvectors corresponding to zero eigenvalue of $J|_{E_4}$ and $J^T|_{E_4}$, respectively. Here, $v_1 = \frac{a_{22}a_{33}-a_{23}a_{32}}{a_{32}a_{21}-a_{22}a_{31}}$, $v_2 = \frac{a_{23}a_{31}-a_{33}a_{21}}{a_{32}a_{21}-a_{22}a_{31}}$ and $v_3 = 1$, where a_{ij}, for $i, j = 1, 2, 3$ is an entry in $J|_{E_4}$ in i^{th} row and j^{th} column. Also $W = (w_1, w_2, w_3)^T = (1, 0, 0)^T$. Define, $U(x_1, x_2, y) = (F_1, F_2, F_3)^T$. Therefore, $U_{r_1} = \left[\frac{\partial F_1}{\partial r_1}, \frac{\partial F_2}{\partial r_1}, \frac{\partial F_3}{\partial r_1}\right]^T = \left[x_1\left(1 - \frac{x_1}{K_1}\right), 0, 0\right]^T$. One can easily verify the transversality conditions: $W^T U_{r_1}(E_4, r_1^{tc}) = 0$, $W^T[DU_{r_1}(E_4, r_1^{tc})]V = v_1 \neq 0$, $W^T[D^2U(E_4, r_1^{tc})(V, V)] = (-\frac{2r_1^{tc}}{K_1} + \frac{2m_1b\bar{y}}{c^2})v_1v_1 + 2\sigma_1\bar{y}v_1v_2 - 2\left(\frac{2m_1b}{c^2} + \sigma_1\bar{x_2}\right)v_1v_3$. Hence, if $(-\frac{2r_1^{tc}}{K_1} + \frac{2m_1b\bar{y}}{c^2})v_1v_1 + 2\sigma_1\bar{y}v_1v_2 - 2\left(\frac{2m_1b}{c^2} + \sigma_1\bar{x_2}\right)v_1v_3 \neq 0$ then can say that at $r_1 = r_1^{tc}$, E^* is bifurcated from E_4 by the transcritical bifurcation.

Note: A similar result can be established for transcritical bifurcation involving interior equilibrium E^* and predator free equilibrium E_3 (depicted in numerical simulation). We also provide conditions under which our 3D-model (2) undergoes Hopf-bifurcation around the positive equilibrium E^* when the parameter α passes its some critical value α^{hf}.

Theorem 7 *The necessary and sufficient prerequisites for occurrence of Hopf-bifurcation of system (2) around E^* at $\alpha = \alpha^{hf}$ are*

1. $E_1 > 0,\ E_3 > 0$.
2. $E_1E_2 - E_3 = 0$.
3. $\frac{d}{d\alpha}(Re(\lambda))_{\alpha=\alpha^{hf}} \neq 0$. This condition holds iff $\frac{dR}{d\alpha} \neq 0$, where $R = E_1E_2 - E_3$.

Proof Proof of this theorem readily follows as in [8].

4 Numerical Simulation

$$r_1 = 7,\ K_1 = 10,\ m_1 = 0.45,\ a = 0.025,\ b = 0.1,\ c = 7,\ \sigma_1 = 0.001,\ r_2 = 5,\ K_2 = 15,\ m_2 = 0.1,$$
$$\alpha = 0.07,\ \beta = 0.0025,\ \sigma_2 = 0.004,\ c_1 = 0.2222,\ c_2 = 0.7,\ \delta_0 = 0.01,\ \delta_1 = 0.001. \quad (13)$$

For the set of parameters give in (13), $E_0(0, 0, 0)$, $E_1(10, 0, 0)$, $E_2(0, 15, 0)$, $E_3(10, 15, 0)$, $E_4(0,\ 1.183603,\ 56.973631)$, $E_5(5.600137, 0, 57.108613)$ are saddle points whereas $E^*(3.982544,\ 0.573808,\ 73.695332)$ is a stable focus with $(-0.033869 \pm$

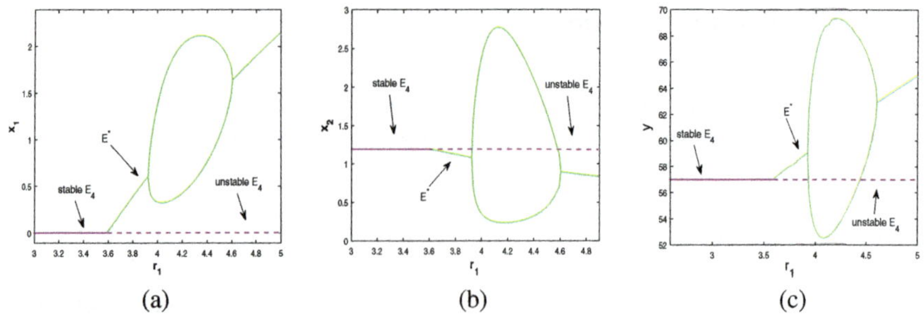

Fig. 1 E_4 is going for transcritical bifurcation with loosing its stability which leads to generation of E^* and E^* changes its stability twice via Hopf-bifurcation on further increase of r_1 with parameters same as in (13)

0.436450i, −2.108830) as eigenvalues. The system goes through transcritical as well as Hopf-bifurcation for the parameter r_1 (Fig. 1). As long as r_1 lies in the interval $[0.1, r_1^{tc})$, E_4 is a stable focus and in this interval E^* is not biologically feasible. At $r_1 = r_1^{tc} = 3.595156$, E_4 undergoes transcritical bifurcation to become unstable with, $\dim(W^u(E_4)) = 1$ and $\dim(W^s(E_4)) = 2$. The stable and unstable branches of E_4 are shown by solid and dashed curves in purple colour, respectively. At this critical value, r_1^{tc}, E^* is generated as a spiral sink. From Fig. 1, we can observe that, on further increase of r_1, E^* passes through Hopf-bifurcation at $r_1 = r_1^{hf_1} = 3.930962$, and on keep increasing r_1, E^* again endures Hopf-bifurcation at $r_1 = r_1^{hf_2} = 4.599987$ to become stable and it maintains this behavior on further increment of r_1. So, we observed that E^* is bifurcated from E_4 via a transcritical bifurcation and E^* switches its stability two times by means of Hopf-bifurcation with respect to parameter r_1. In this figure, E^* is indicated by the green colour.

Now, Keeping all the parameters fixed given in (13), when we use inverse measure of predator's immunity a, as a control parameter then system (2) experiences Hopf-bifurcation in the vicinity of E^*. For $a < a^{hf} = 0.072198$, E^* remains stable and at $a = a^{hf}$, E^* becomes non-hyperbolic and then turn into a spiral source on continue increment of a. For an example, concerning with a as a bifurcation parameter, we simulate model (2) for $a = 0.04 < a^{hf}$ and $a = 0.08 > a^{hf}$, showing the stable and unstable nature of the system, respectively. From Fig. 2(a), we can see that as long as $a < a^{hf}$, system remains stable. As a crosses its critical value a^{hf}, system go through Hopf-bifurcation to become unstable. The unstable nature of E^* is depicted in Fig. 2(b).

Now, we talk about the coefficient of help σ_2. When help provided by x_1 to x_2 is less than a threshold value, system remains stable but as it becomes $\sigma_2 = \sigma_2^{hf} = 0.005389$, E^* suffers Hopf-bifurcation. So, the model (2) turns to be an unstable system. We have presented two illustrations of 3D-phase portraits for this case in Fig. 3. Firstly for $\sigma_2 = 0.0045$ (Fig. 3(a)), we see that, E^* remains continue to be a stable focus until σ_2 reaches its critical value. Secondly from $\sigma_2 = 0.0065$, (Fig. 3(b)), it is clear that system becomes unstable for $\sigma_2 > \sigma_2^{hf}$. So, the excess amount of help

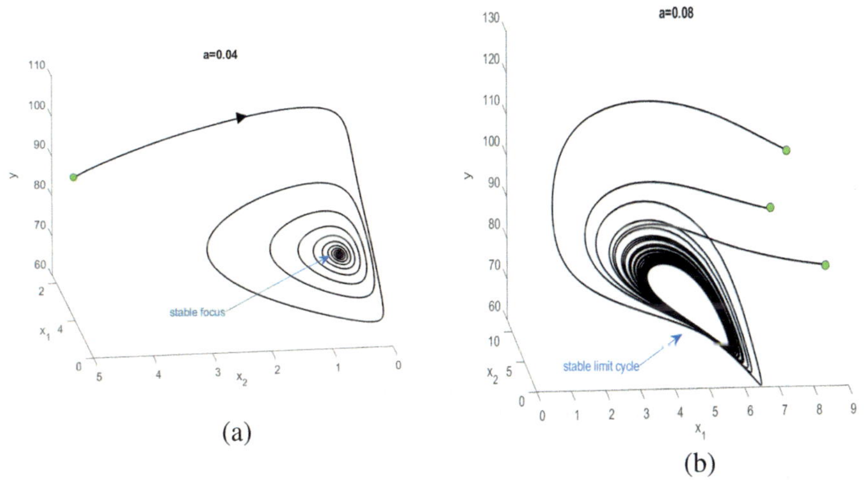

Fig. 2 Stable nature of system for $a = 0.04 < a^{hf}$ (Fig. a) and unstable nature for $a = 0.08 > a^{hf}$ (Fig. b) with all parameters same as in (13)

from x_1 to x_2 makes the system unstable.

For δ_0, the system experiences the change of stability twice by Hopf-bifurcation and then a transcritical bifurcation on its further increment. This involves the predator free equilibrium E_3 and the coexistence equilibrium E^*. For $\delta_0 < \delta_0^{hf_1} = 0.049805$, E^* remains a stable focus and at $\delta_0^{hf_1}$, the system suffers Hopf-bifurcation around E^*. Then it again become stable around E^* via Hopf-bifurcation at $\delta_0^{hf_2} = 0.102547$. The interval $(\delta_0^{hf_1}, \delta_0^{hf_2})$ corresponds to a stable limit cycle. For $\delta_0 < \delta_0^{tc} = 0.607423$, E_3 continue to be a proper unstable node. At δ_0^{tc}, E_3 becomes a stable node, and E^* disappears through the transcritical bifurcation. These two bifurcations are shown simultaneously in Fig. 4 through three corresponding bifurcation diagrams. In this figure, E^* is indicated with green colour where unstable and stable components of E_3 are shown by dashed and solid curves in purple colour, respectively.

Moreover, we have drawn Fig. 5 to depict the distribution of regions according to the feasibility of E_3, E_4 and E^* in the $r_1\delta_0$-plane. In region R_1, E_3 and E_4 are the feasible equilibria. Region R_1 and R_2 are separated by the transcritical curve marked with red colour. When we moves from R_1 to R_2, the interior equilibrium E^* is generated via transcritical bifurcation. Region R_2 and R_3 are separated by a blue coloured line $\delta_0 = 0.512195$. When δ_0 becomes greater than 0.512195, y component of E_4 remains negative with all values of r_1, so E_4 becomes non-feasible in R_3. Next, when $\delta_0 > 0.607423$, E^* disappears beyond the transcritical curve $\delta_0 = 0.607423$ marked with pink colour. So in region R_4, E_3 is the only feasible equilibrium point.

To show the bistablity phenomenon in the system, we choose the following set of parameters:

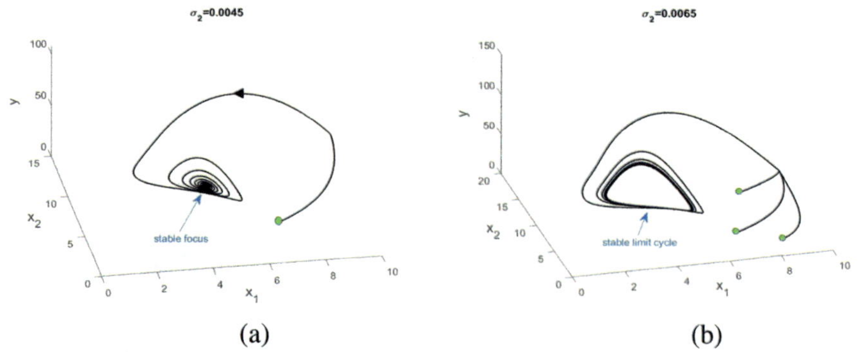

Fig. 3 Using σ_2 as the control parameter, phase portrait (Fig. a) shows the stable focus for $\sigma_2 = 0.0045$ and (Fig. b) exhibits a stable limit cycle for $\sigma_2 = 0.0065$ with parameters fixed in (13)

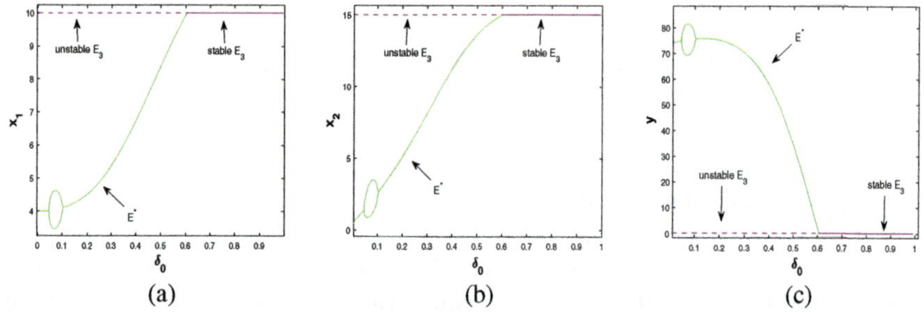

Fig. 4 E^* changes its stability twice by Hopf-bifurcation and on further increase of δ_0, E^* disappears and E_3 becomes stable via transcritical bifurcation, with parameters fixed in (13)

$$r_1 = 3,\ K_1 = 10,\ m_1 = 0.4,\ a = 0.1,\ b = 0.08,\ c = 0.11,\ \sigma_1 = 0.01,\ r_2 = 4,\ K_2 = 7,\ m_2 = 0.62,$$
$$\alpha = 0.01,\ \beta = 0.03,\ \sigma_2 = 0.02,\ c_1 = 0.4,\ c_2 = 0.63,\ \delta_0 = 0.1,\ \delta_1 = 0.02. \quad (14)$$

For set of parameters (14), we have two locally asymptotically stable equilibrium points; one is coexistence equilibrium $E^*(7.792154, 0.593490, 12.534084)$, and the other one is planner equilibrium $E_4(0, 0.751129, 7.025585)$. We have illustrated this characteristic of the system through a phase portrait diagram given in Fig. 6(a) in which two solutions started from two different initial conditions converge to these two different equilibrium points. We have also plotted their basin of attraction in Fig. 6(b).

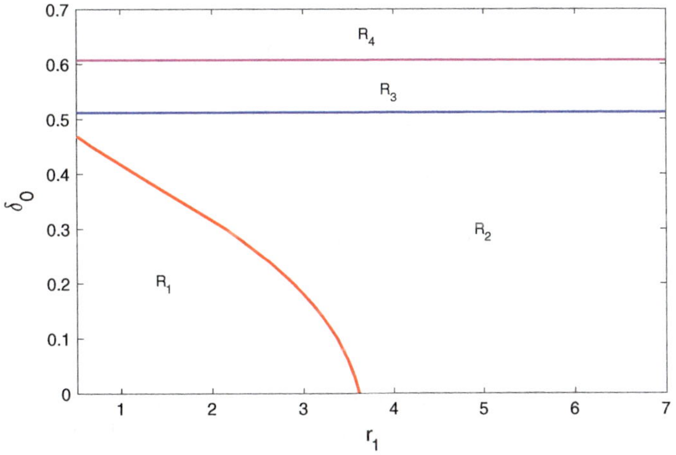

Fig. 5 Distribution of regions for existence of E_3, E_4 and E^* in $r_1\delta_0$-plane with parameters same as in (13)

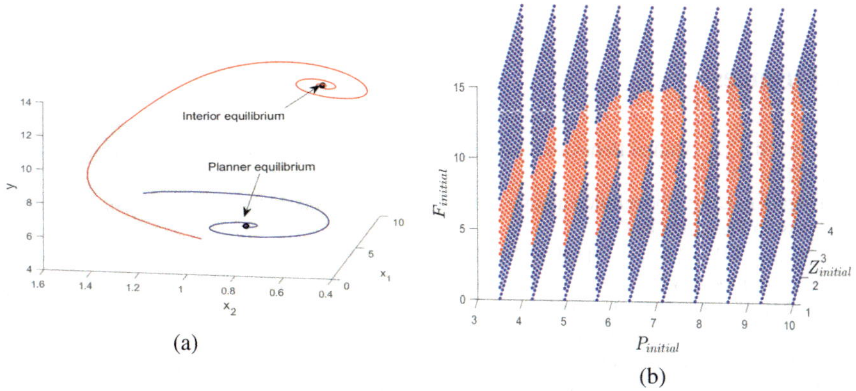

Fig. 6 In Fig. (a) solutions in red and blue color are tending to E^* and E_4, respectively. Figure (b) represent the basin of attractions (red dots for interior and blue dots for planner equilibrium point) where $P(0) \in [3.5, 10]$, $Z(0) \in [1, 4]$ and $F(0) \in [0.1, 15]$, with parameters fixed in (14)

5 Conclusion

In the present work, we have formulated a prey-predator model with cooperation between two types of prey against a predator. Both the prey species are assumed to grow logistically whereas predator consume the one prey via Holling type-IV type response and another prey via Crowley–Martin type response. Firstly, we have proved that the model is biologically well-posed having the quality of being bounded in a compact domain Ω of R^3_+. In the dynamics of the system, we study the existence of seven equilibrium points. During the stability analysis, we notice that E_0, E_1 and

E_2 always behave like saddle points, while E_3, E_4 and E_5 are stable under some conditions. We also studied the global stability of E^* by picking an appropriate Lyapunov function. Further, we have analyzed the transcritical as well as Hopf-bifurcations for the system involving E_4, E^* and E_3, E^*. From the bifurcation diagrams for r_1, it can be noticed that on increase of r_1, x_1 increases which increase the efficiency of their group defence and so predator shifts its attention to prey x_2 which decreases the density of x_2 population. When we use a as the bifurcation parameter, x_1 increases with a partial decrement in x_2. Now, as when we increase a more than a threshold value $a = a^{hf}$, there is a rise of periodic oscillations due to Hopf-bifurcation. Most importantly, when the help (σ_2) provided by x_1 to x_2 is increased, there is a growth in the population density of x_2 but this increased help imparts a negative effect on x_1 population due to its sacrifice in helping x_2. Raising this help against a certain limit becomes a cause of instability in the system. The system also shows the property of bistability. From numerical simulation, we can see that our system shows much more rich dynamical behavior than the existing studies. Thus we hope that our work can be a useful source for ecologists to explore different segments of experimental and theoretical ecology in a better way.

References

1. Barnard, C.J.: Animal Behaviour: Ecology and Evolution. Springer Science & Business Media (2012)
2. Dugatkin, L.A.: Cooperation among Animals: An Evolutionary Perspective. Oxford University Press on Demand (1997)
3. Tripathi, J.P., Abbas, S., Thakur, M.: Local and global stability analysis of a two prey one predator model with help. Commun. Nonlinear Sci. Numer. Simul. **19**(9), 3284–3297 (2014)
4. Mishra, P., Raw, S.N.: Dynamical complexities in a predator-prey system involving teams of two prey and one predator. J. Appl. Math. Comput. **61**(1–2), 1–24 (2019)
5. Mondal, S., Samanta, G.P.: Dynamical behaviour of a two-prey and one-predator system with help and time delay. Energy Ecology Environ. **5**(1), 12–33 (2020)
6. Alsakaji, H.J., Kundu, S., Rihan, F.A.: Delay differential model of one-predator two-prey system with Monod-Haldane and Holling type II functional responses. Appl. Math. Comput. **397**, 125919 (2021)
7. Ferrara, M., Gangemi, M., Pansera, B.A.: Dynamics of a delayed mathematical model for one predator sharing teams of two preys. Appl. Sci. **23**, 52–61 (2021)
8. Tripathi, J.P., Jana, D., Tiwari, V.: A Beddington-DeAngelis type one-predator two-prey competitive system with help. Nonlinear Dyn. **94**(1), 553–573 (2018)
9. Perko, L.: Differential Equations and Dynamical Systems, vol. 7. Springer Science & Business Media (2013)

Mathematical Model of Solute Transport in a Permeable Tube with Variable Viscosity

M. Varunkumar

Abstract The purpose of this paper is to investigate the influence of variable viscosity on solute transfer in fluid flow through a permeable tube with possible applications to the blood flow in glomerular capillaries. The difference in transcapillary hydrostatic pressure and the equivalent difference in colloid osmotic pressure regulates solute transport through the glomerular capillary wall(Starling's law). Fluid flow in a capillary is assumed to be viscous, incompressible, and Newtonian with variable viscosity. The nonlinear and coupled equations regulating fluid flow and solute transport are solved analytically and numerically. Graphs have been used to discuss the impacts of varying viscosity and flow parameters on hydrostatic and osmotic pressures, and solute concentrations using a set of physiological data. It is observed that increasing the viscosity coefficient raises the hydrostatic pressure while decreasing the osmotic pressure at the capillary's end. As the viscosity coefficient increases, the solute concentration at the exit falls and the solute clearance increases through the capillary wall.

Keywords Starling's law · Ultrafiltration · Variable viscosity · Permeable wall · Finite difference method

1 Introduction

The kidneys have two primary functions. They perform two functions: first, they excrete the large majority of waste products produced during metabolism, and second, they regulate the concentrations of the vast majority of body fluids. Insight into the mechanisms of glomerular ultrafiltration and blood waste removal, which

M. Varunkumar (✉)
Department of Basic Sciences and Humanities, GMR Institute of Technology, Rajam 532127, Andhra Pradesh, India
e-mail: varun.nitw@gmail.com

Department of Mathematics, School of Advanced Sciences, VIT-AP University, Amaravati 522237, Andhra Pradesh, India

S. Banerjee and A. Saha (eds.), *Nonlinear Dynamics and Applications*, Springer Proceedings in Complexity,
https://doi.org/10.1007/978-3-030-99792-2_85

create urine. Normally, the capillaries walls pass roughly one-fifth to one-third of the amount of blood plasma entering each glomerulus. This fluid's composition is close to that of an ideal ultrafiltrate, with solute concentrations similar to plasma water. The development of this ultrafiltrate is controlled by the driving factors that dictate fluid flow through the capillaries, specifically the imbalance between transcapillary hydrostatic and colloid osmotic pressures. Ultrafiltrate plasma travels into Bowman's Space as a result of these pushing factors. Several theoretical and mathematical models developed to describe the dynamics of ultrafiltration assumed that the glomerulus's local driving factors for fluid transfer were evenly distributed throughout the entire glomerular capillary network [1–4].

A study by Brenner et al. [5, 6] focused on glomerular pressure measurements at the afferent and efferent ends of the glomerulus, as well as the glomerular filtering rate of single kidney nephrons, in addition to extravascular pressures in Bowman's space, solute concentrations, and colloid osmotic pressure. Deen et al. [7] examined the relationship between ultrafiltration and the rise in plasma protein concentration caused by ultrafiltration, and their findings led them to the conclusion that the overall filtration rate rises as blood volume flow rises. The models for glomerular ultrafiltration proposed by Marshal and Trowbridge [8] and Huss et al. [9] eliminated the assumption of a constant axial pressure gradient and regarded the intraluminal pressure gradient to be dependent on the axial distance. As seen in Papenfuss and Gross [10], a model developed by them examined the effects of intraluminal pressure drop and wall permeability on glomerular ultrafiltration. Papenfuss and Gross [11] and Salathe [12] investigated fluid exchange and solute transport in capillary tissue under the assumption that concentrations were identical at each cross-section. Axially, the concentration profiles were determined by assuming the capillary wall as impermeable and applying a constant hydrostatic pressure to the solute transfer in capillary, as done by Deen et al. [13].

Ross [14] proposed a mathematical model for mass transfer in fluid flow through a capillary membrane with a tiny radial fluid flux (ultrafiltration) assumed to be zero by zero osmotic pressure. In addition, it should be mentioned that the ultrafiltration process is regulated by hydrostatic and oncotic pressure differences (Starling's theory) as well as the solute transported across the permeable wall by diffusion and convection as investigated by Chaturani and Ranganatha [15]. In a porous tube with varying permeability, Varun and Muthu [16] obtained a solution for the transport of solutes.

The relationship between fluid viscosity and solute concentration is significant in the filtering process. The viscosity is treated as a constant in the previous research, but in reality, it is dependent on a wide range of fluid characteristics ([17, 18]). A steady-state boundary layer with changing diffusivity and viscosity, as explored by Davis and Leight [20], was considered by Davis and Sherwood [19]. For concentration-dependent viscosity and diffusion coefficient, Bowen and Williams [21] introduced cross flow ultrafiltration. As a result of this inspiration, the impact of varying viscosity is examined in this article.

2 Mathematical Model

Consider a fluid flow through a rigid cylindrical permeable tube with a radius R and a length L (refer Fig. 1). The set-up is considered to be axisymmetric, with axial and radial directions z and r, and corresponding velocity components $v(r, z)$(radial) and $u(r, z)$(axial). The following are the equations that govern viscous, incompressible Newtonian fluid flow and solute transport ([15, 16]):

$$\frac{\partial u}{\partial z} + \frac{\partial v}{\partial r} + \frac{v}{r} = 0 \tag{1}$$

$$u\frac{\partial u}{\partial z} + v\frac{\partial u}{\partial r} + \frac{1}{\rho}\frac{\partial P}{\partial z} - 2\mu(r)\frac{\partial^2 u}{\partial z^2} - \frac{1}{\rho r}\frac{\partial}{\partial r}\left(r\mu(r)\left(\frac{\partial v}{\partial z} + \frac{\partial u}{\partial r}\right)\right) = 0 \tag{2}$$

$$u\frac{\partial v}{\partial z} + v\frac{\partial v}{\partial r} + \frac{1}{\rho}\frac{\partial P}{\partial r} - 2\frac{\partial}{\partial r}\left(\mu(r)\frac{\partial v}{\partial r}\right) - \frac{2}{r}\mu(r)\left(\frac{\partial v}{\partial r} - \frac{v}{r}\right) - \frac{\partial}{\partial z}\left(\mu(r)\left(\frac{\partial v}{\partial z} + \frac{\partial u}{\partial r}\right)\right) = 0 \tag{3}$$

$$v\frac{\partial c}{\partial r} + u\frac{\partial c}{\partial z} - D\left(\frac{\partial^2 c}{\partial r^2} + \frac{\partial^2 c}{\partial z^2} + \frac{1}{r}\frac{\partial c}{\partial r}\right) = 0 \tag{4}$$

where P, $\mu(r)$ and ρ are the fluid's pressure, variable viscosity, and density, respectively. The solute concentration is denoted by c, while the diffusion coefficient is denoted by D.

The boundary conditions are expressed as follows ([15, 22, 23]):
At $z = 0$,

$$c = c_0, \quad \Delta P = \Delta P_a, \quad \int_0^R 2\pi r u(r, 0)dr = Q_0 \tag{5}$$

At $r = 0$,

$$v = 0, \quad \frac{\partial u}{\partial r} = 0, \quad \frac{\partial c}{\partial r} = 0 \tag{6}$$

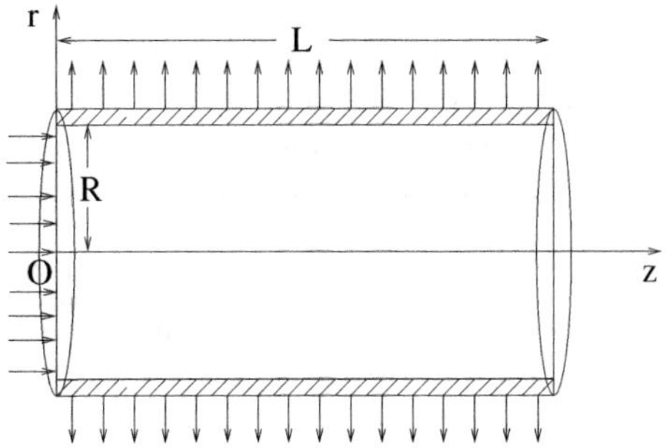

Fig. 1 Geometric model of the glomerular capillary

At $r = R$,

$$v = k(\Delta P - \sigma \Delta\pi) = V_R(z), u = 0, -D\frac{\partial c}{\partial r} = (T_R - 1)V_R\Phi + h(c - c_T) \quad (7)$$

The second boundary condition in (7) is stated according to Starling's hypothesis, which states that the flow velocity through the capillary wall is proportional to the difference between the hydrostatic and osmotic pressure differences. According to the third boundary condition in (7), the solute flow through membrane equals the solute flux via the interface. The product of solute permeability (h) and concentration difference ($c(r = R) - c_T$) gives the solute transported by ordinary diffusion via the membrane interface. ($(T_R - 1)V_R$) is the solute flux through membrane pores. $\Phi = \begin{cases} c; & V_R > 0 \\ c_T; & V_R < 0 \end{cases}$, $\Delta P = P - P_T$, $\Delta\pi = \pi - \pi_T$, the hydrostatic pressures within and outside the capillary wall are P and P_T, respectively, while the corresponding osmotic pressures are π and π_T. The volume flow rate and hydrostatic pressure at channel's entry are Q_0 and P_a. The wall's hydraulic permeability is k, and the reflection coefficient is σ([24]). P_T and π_T are assumed to be constants. Here, c_0 and c_T are represent the solute concentration at the tube's entry, and concentration outside the channel, respectively. h and T_R are represent the solute permeability at the wall, and transmittance coefficient, respectively.

In the present study, we considered the connection between osmotic pressure π and solute concentration $c(r, z)$ as:

$$\pi(c) = 0.009\, c^3 + 0.16\, c^2 + 2.1\, c \quad (8)$$

2.1 Non-dimensionalization

The non-dimensional quantities listed below are introduced in Eqs. (1)–(8): $\hat{z} = z/R, \hat{r} = r/R, \hat{u} = u/U_0, \hat{v} = v/U_0, \hat{V_R} = V_R/U_0, \hat{c} = c/c_0, \hat{c_T} = c_T/c_0$, $\Delta\hat{\pi} = \Delta\pi/\Delta P_a, \Delta\hat{P} = \Delta P/\Delta P_a, \hat{b}_1 = b_1/(\Delta P_a/c_0), \hat{b}_2 = b_2/(\Delta P_a/c_0^2), \hat{b}_3 = b_3/(\Delta P_a/c_0^3)$, $\hat{J_S} = J_S/(c_0 D/R), \hat{J_C} = J_C/Q_0 c_0, \hat{\mu} = \mu/\mu_0, \hat{Q} = Q/Q_0, Q_0 = \pi R^2 U_0$. In glomerular capillaries, axial diffusion is modest in comparison to radial diffusion [7]. There are no end effects since the tube length to radius ratio is expected to be so enormous. In addition, net radial flow is likewise minimal in comparison to average axial flow and inertial effects can be ignored (The Reynolds number is of 10^{-3}). The governing equations (1)–(8), which are based on non-dimensional quantities and assumptions, become (after dropping caps)

$$\frac{\partial u}{\partial z} + \frac{1}{r}\frac{\partial}{\partial r}(rv) = 0 \quad (9)$$

$$\frac{\partial(\Delta P)}{\partial z} = \frac{1}{R_p}\frac{1}{r}\left[\frac{\partial}{\partial r}\left(r\mu(r)\frac{\partial u}{\partial r}\right)\right] \tag{10}$$

$$\frac{\partial(\Delta P)}{\partial r} = 0 \tag{11}$$

$$v\frac{\partial c}{\partial r} + u\frac{\partial c}{\partial z} = \frac{1}{Pe}\left[\frac{\partial^2 c}{\partial r^2} + \frac{1}{r}\frac{\partial c}{\partial r}\right] \tag{12}$$

The non-dimensional boundary conditions leading to,
At $z = 0$,

$$c = 1, \quad \Delta P = 1, \quad \int_0^1 2ru(r,0)dr = 1 \tag{13}$$

At $r = 0$,

$$v = 0, \quad \frac{\partial u}{\partial r} = 0, \quad \frac{\partial c}{\partial r} = 0 \tag{14}$$

At $r = 1$,

$$v = \epsilon R_p(\Delta P - \sigma\Delta\pi) = V_R(z), u = 0, \frac{\partial c}{\partial r} = Pe(1 - T_R)V_R\Phi + Sh(c_T - c) \tag{15}$$

where, $Pe = U_0R/D, Sh = hR/D, R_p = (R\Delta P_a)/(\mu_0 U_0)$, and $\varepsilon = k\mu_0/R$ denote the Peclet number, the Sherwood number, the non-dimensional parameter and the filtration coefficient, respectively.

3 Method of Solution

The axial and radial velocities are calculated by solving Eqs. (9) and (10) using the conditions (14) and (15) as follows:

$$u(r,z) = -\frac{R_p}{2}\frac{d(\Delta P)}{dz}\int_r^1 \frac{r}{\mu(r)}dr \tag{16}$$

$$v(r,z) = \frac{R_p}{4}\frac{d^2(\Delta P)}{dz^2}\left(r\int_r^1 \frac{r}{\mu(r)}dr + \frac{1}{r}\int_0^r \frac{r^3}{\mu(r)}dr\right). \tag{17}$$

Equation (17) and the second boundary condition of (15) produce the equation for hydrostatic pressure,

$$\frac{d^2(\Delta P)}{dz^2} + \frac{4\epsilon}{I_1}(\sigma \Delta \pi - \Delta P) = 0. \tag{18}$$

The second and third conditions of (13) are written in the following form,

$$\Delta P = 1, \frac{d\Delta P}{dz} = -\frac{1}{R_p I_2} \quad \text{at} \quad z = 0 \tag{19}$$

$$\Delta \pi = 0.009[c^3 - c_T^3] + 0.16[c^2 - c_T^2] + 2.1[c - c_T], \tag{20}$$

$$I_1 = \int_0^1 \frac{r^3}{\mu(r)} dr, I_2 = \int_0^1 r \left(\int_r^1 \frac{r}{\mu(r)} dr \right) dr \tag{21}$$

The viscosity affects the integrations in Eqs. (16)–(19). This study used the exponential viscosity model, which is provided as [17].

$$\mu(r) = e^{-\alpha r} \tag{22}$$

where α denotes viscosity. The exponential type of viscosity was used in all of the calculations. Due to their dependency, the solutions of Eqs. (12) and (18), hydrostatic pressure (ΔP), and solute concentration (c) are difficult to derive analytically in a closed form expression. Using the relevant boundary conditions of (13), (14) and (15), get numerical solutions for (12) and (18) correspondingly. For solute concentration, Eq. (12) is solved, which is linked with Eqs. (18), (16), (17) and (20), along with u, v and ΔP quantities. The solution technique was explained in detail by Varun and Muthu [16].

4 Results and Discussions

To understand the influence of variable viscosity during filtering, a mathematical model of solute transfer through a permeable tube was developed. Computations were performed utilizing physiological data from rat glomerular capillaries (Chaturani and Ranganatha [15]). The observed findings for $\alpha = 0$(zero viscosity) are in excellent accord with earlier research [15, 16]. The viscosity is modelled as an exponentially decreasing function of radial distance, with decreasing fluid viscosity as the viscosity parameter is increased. Viscosity parameter (α) and emerging parameters have substantial influence on flow quantities such as hydrostatic and osmotic pressures, and solute concentration investigated.

Hydrostatic Pressure and Osmotic Pressure: Figures 2 and 3 illustrate the distribution of hydrostatic pressure (ΔP) and osmotic pressure ($\Delta \pi$) profiles for distinct values of viscosity parameter α and other emerging parameters with regard to axial distance z. In accordance with the experimental results, the hydrostatic and

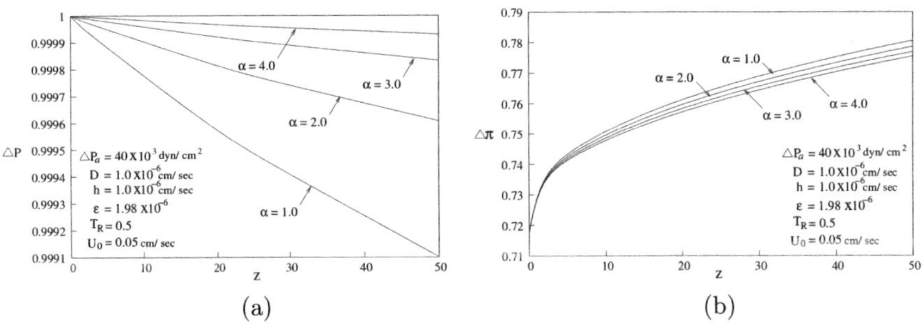

Fig. 2 Effect of viscosity parameter (α) on hydrostatic and osmotic pressure distributions

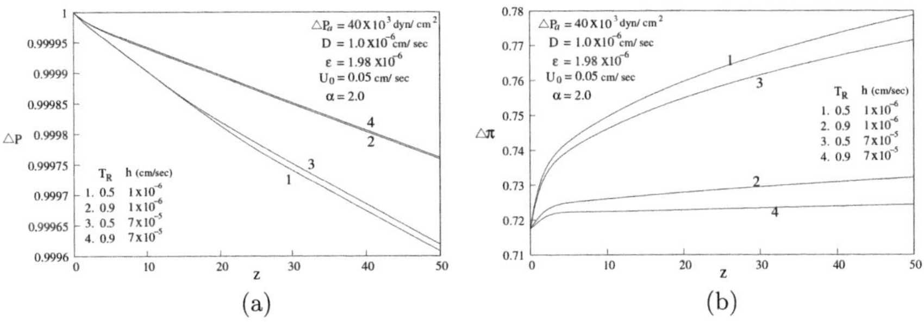

Fig. 3 Effects of T_R and h on hydrostatic and osmotic pressure distributions

osmotic pressure curves may be seen in [23]. Osmotic pressure $\Delta\pi$ rises nonlinearly as the axial length of the capillary decreases because of the linear reduction in ΔP.

Figure 2 shows the effect of the viscosity parameter (α) on hydrostatic pressure and osmotic pressure. It is noticed that as the parameter α is increased, ΔP increases along the axial length. This indicates that driving fluids with greater viscosities requires a higher pressure (Fig. 2a). When predicted, the osmotic pressure ($\Delta\pi$) values drop as the viscosity parameter (α) is increased in the $z-$ direction (Fig. 2b). Figure 3 shows that solute wall permeability h and transmittance coefficient T_R have only a little impact on the ΔP profiles. The osmotic pressure falls throughout the axial length as T_R and h rise, indicating that the solute has crossed the channel wall. The increase in T_R values indicates greater solute transfer across the capillaty wall and as a result, $\Delta\pi$ falls in the axial direction.

Concentration Profiles c: Figure 4 depicts the impact of the viscosity parameter (α) on concentration patterns at two distinct positions. The concentration at the centerline grows insignificantly while the concentration near the wall drops as (α) increases. The solute concentration falls as you get closer to the wall and reaches its maximum on the wall, as seen in this diagram. For various T_R and h, Fig. 5 depicts the distribution of solute concentration at two distinct positions along the tube's length. Because of the delicate balance between convective protein transport and diffuse protein transport, at any given axial location the concentration of solutes at wall is

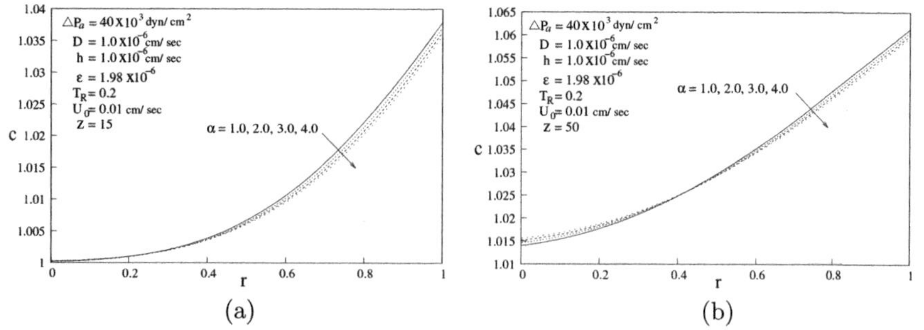

Fig. 4 Effect of viscosity parameter (α) on concentration (c) with r at different locations **a** $z = 15$ and **b** $z = 50$

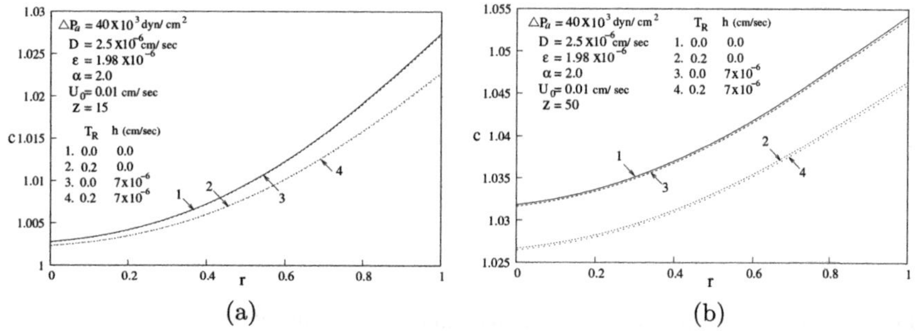

Fig. 5 Effects of T_R and h on solute concentration (c) with r at $z = 15$ and $z = 50$

larger than at centerline. The solute concentration increases with radial distance at any constant cross-sectional tube because solute particles remove via ultrafiltration.

The effect of ε and D on concentration distribution is seen in Fig. 6. The concentration of solute increases as ε increases, indicating that the solute volume per unit volume at the wall has risen. With a rise in D, the solute concentration decreases along the axis. It is true that in the situation of $\varepsilon = 0$(zero ultrafiltration), it is a fact that no solute is transferred through the wall. That is, from the initial constant value ($c_0 = 1$) to zero at cross-section $z = 15$, the radial concentration profile falls considerably.

Concentration at the wall c_w : The impact of the viscosity parameter (α) on wall concentration is seen in Fig. 7, along axial direction. It is noted that variation in the concentration exists only near the permeable wall. It is also worth noting that as (α), the concentration near the wall drops. The wall concentration with axial distance for two cases $\Delta\pi \neq 0$ and $\Delta\pi = 0$ with various values of the ultrafiltration parameter ε are shown in Fig. 8. To enforce the scenario where there is no osmotic pressure across the permeable border explicitly, $\Delta\pi = 0$ is used in Eq. (18). In both situations of $\Delta\pi$, as ε rises, the wall concentration rises. In the case of $\Delta\pi = 0$, the numerical value of the wall concentration is higher than in the case of $\Delta\pi \neq 0$.

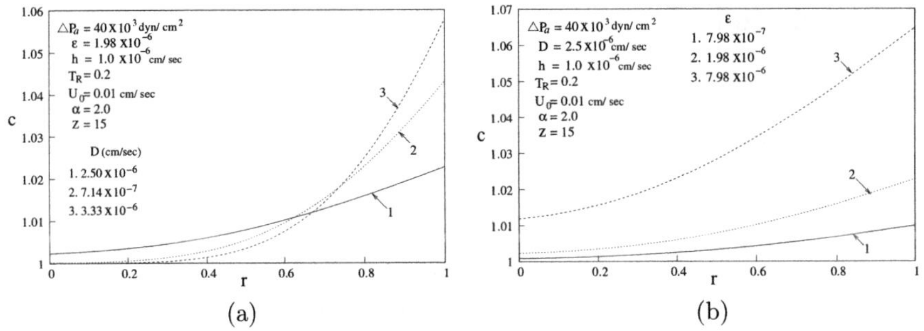

Fig. 6 Effects of D and ε on concentration of solute (c) with r

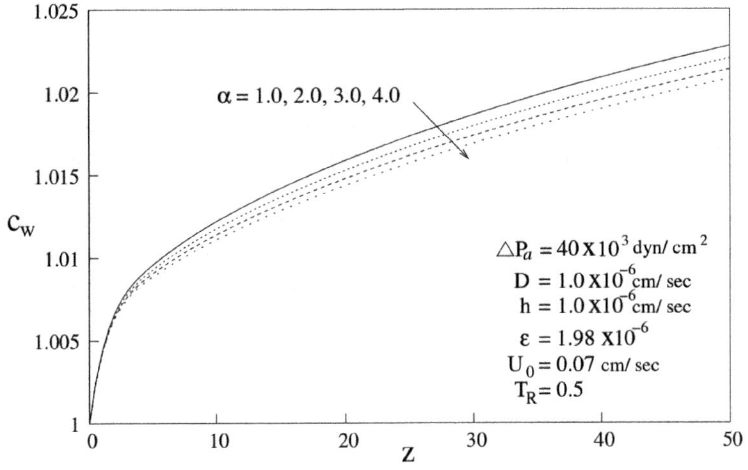

Fig. 7 Effect of viscosity parameter (α) on wall solute concentration (c_w)

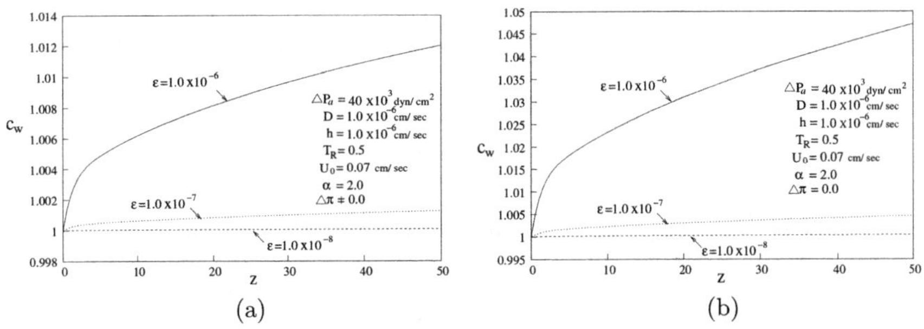

Fig. 8 Effects of ε on wall concentration c_w for $\Delta\pi = 0$ and $\Delta\pi \neq 0$

5 Conclusions

This study concentrated on the quantitative evaluation of fluid and solute transfer in a tube with a permeable wall under the influence of varying viscosity. The controlling fluid flow and solute transport equations have been solved analytically and numerically. It is noteworthy to note that the hydrostatic pressure loss is quite minimal in the current results, whereas it is considered constant in some of the earlier models. The consideration of solute transport via the permeable tube in conjunction with exponentially varying viscosity, specifically in the flow through glomerular capillaries, is one of the study's unique aspects. The results are obtained for hydrostatic and osmotic pressures and, concentration profiles, demonstrating the impact of various viscosity parameter values and other factors. The following are the study's key findings:

(i) The hydrostatic pressure drops linearly throughout the capillary's length. The ΔP decreases as the viscosity parameter (α) rises. The osmotic pressure rises in a nonlinear fashion throughout the capillary's length. As (α) increase, the $\Delta\pi$ lowers.
(ii) As the viscosity parameter (α) is increased, the concentration increases along the axis and drops along the wall.
(iii) As the viscosity parameter decreases, so does the concentration near the wall. For small values of ε, the findings for two specific instances $\Delta\pi = 0$ and $\Delta\pi \neq 0$ are close to each other. The discrepancy between the two outcomes increases as ε rises.

References

1. Berman, A.S.: Laminar flow in channels with porous walls. J. Appl. Phys. **24**, 1232–1235 (1953)
2. Berman, A.S.: Laminar flow in an annulus with porous walls. J. Appl. Phys. **29**, 71–75 (1958)
3. Apelblat, A., Katchasky, A.K., Silberberg, A.: A mathematical analysis of capillary tissue fluid exchange. Biorheology **11**, 1–49 (1974)
4. Guyton, A.C.: Text Book of Medical Physiology, 7th edn. W. B, Saunders Company (1986)
5. Brenner, B.M., Troy, J.L., Daugharty, T.M., Deen, W.M.: Dynamics of glomerular ultrafiltration in the rat II plasma flow dependence of GFR. Am. J. Physiol. **223**, 1184–1190 (1972)
6. Brenner, B.M., Baylis, C., Deen, W.M.: Transport of molecules across renal glomerular capillaries. Physiol. Rev. **56**, 502–534 (1978)
7. Deen, W.M., Robertson, C.R., Brenner, B.M.: A model of glomerular ultrafiltration in the rat. Am. J. Physiol. **223**, 1178–1183 (1972)
8. Marshall, E.A., Trowbridge, E.A.: A mathematical model of the ultrafiltration process in a single glomerular capillary. J. Theor. Biol. **48**, 389–412 (1974)
9. Huss, R.E., Marsh, D.J., Kalaba, R.E.: Two models of glomerular filtration rate and renal blood flow in the rat. Ann. Biomed. Eng. **3**, 72–99 (1975)
10. Papenfuss, H.D., Gross, J.F.: Analytical study of the influence of capillary pressure drop and permeability on glomerular ultrafiltration. Microvasc. Res. **16**, 59–72 (1978)
11. Papenfuss, H.D., Gross, J.F.: Transcapillary exchange of fluid and plasma protiens. Biorheology **24**, 319–335 (1987)
12. Salathe, E.P.: Mathematical studies of capillary tissue exchange. Bull. Math. Biol. **50–3**, 289–311 (1988)

13. Deen, W.M., Robertson, C.R., Brenner, B.M.: Concentration polarization in an ultrafiltering capillary. BioPhys. J. **14**, 412–431 (1974)
14. Ross, M.S.: A mathematical model of mass transport in a long permeable tube with radial convection. J. Fluid Mech. **63**, 157–175 (1974)
15. Chaturani, P., Ranganatha, T.R.: Solute transfer in fluid in permeable tubes with application to flow in glomerular capillaries. Acta Mechanica **96**, 139–154 (1993)
16. Varunkumar, M., Muthu, P.: Fluid flow and solute transfer in a tube with variable wall permeability. Zeitschrift für Naturforschung A—A J. Phys. Sci. **74–12**, 1057–1067 (2019)
17. Umavathi, J.C.: Combined effect of variable viscosity and variable thermal conductivity on double-diffusive convection flow of a permeable fluid in a vertical channel. Transp. Porous. Med. **108**, 659–678 (2015)
18. Herterich, J.G., Griffiths, I.M., Vella, D., Field, R.W.: The effect of a concentration-dependent viscosity on particle transport in a channel flow with porous walls. AIChE J. **60**, 1891–1904 (2014)
19. Davis, R.H., Sherwood, J.D.: A similarity solution for steady-state crossflow microfiltration. Chem. Eng. Sci. **45**, 3203–3209 (1990)
20. Davis, R.H., Leighton, D.T.: Shear-induced transport of a particle layer along a porous wall. Chem. Eng. Sci. **42**, 275–281 (1987)
21. Bowen, W.R., Williams, P.M.: Prediction of the rate of cross-flow ultrafiltration of colloids with concentration-dependent diffusion coefficient and viscosity-theory and experiment. Chem. Eng. Sci. **56**, 3083–3099 (2001)
22. Moustafa, E.: Blood flow in capillary under starling hypothesis. Appl. Math. Comput. **149**, 431–439 (2004)
23. Pollak, M.R., Susan, E.Q., Melanie, P.H., Lance, D.D.: The glomerulus: the sphere of influence. Clin. J. Am. Soc. Nephrol. **9**, 1461–1469 (2014)
24. Regirer, S.A.: Quasi one dimensional model of transcapillary filtration. J. Fluid Dyn. **10**, 442–446 (1975)

Dynamical Systems: Chaos, Complexity and Fractals

Impact of Cooperative Hunting and Fear-Induced in a Prey-Predator System with Crowley-Martin Functional Response

Anshu, Sourav Kumar Sasmal, and Balram Dubey

Abstract Cooperative hunting among predators and the fear-induced growth rate reduction in prey populations is an ecologically significant phenomenon. Many researchers have studied the effects of hunting cooperation and fear independently, but there has not been much research on the combined effect. This study analyzed a classical predator-prey system incorporating hunting cooperation and fear effect with Crowley-Martin functional response. We have done the basic analysis, including positivity, boundedness of solutions, existence and stability analysis of equilibria, Hopf-bifurcation, saddle-node bifurcation. We analyzed that incorporating cooperative hunting among predators may destabilize the system dynamics by producing limit cycles via Hopf-bifurcation. Furthermore, we noticed that the system shows bi-stability behavior between predator-free equilibrium and the coexistence equilibrium. Also, analysis shows that the system becomes unstable for a fixed hunting cooperation parameter on increasing the strength of fear. To validate the analytical conclusions, numerical simulations are conducted.

Keywords Prey-predator dynamics · Fear effect · Hunting cooperation · Stability analysis · Bifurcation.

1 Introduction

For many species, social interactions between individuals constitute an important aspect of their life histories. Cooperative behavior among animals is a common and essential phenomenon from a biological perspective. Cosner et al. [3] derived

Anshu (✉) · S. K. Sasmal · B. Dubey
Department of Mathematics, BITS Pilani, Pilani Campus, Pilani 333031, Rajasthan, India
e-mail: anshumor028@gmail.com

S. K. Sasmal
e-mail: sourav.kumar@pilani.bits-pilani.ac.in

B. Dubey
e-mail: bdubey@pilani.bits-pilani.ac.in

S. Banerjee and A. Saha (eds.), *Nonlinear Dynamics and Applications*,
Springer Proceedings in Complexity,
https://doi.org/10.1007/978-3-030-99792-2_86

a functional response depending on the spatial distribution of predators when the predators aggregate for capturing prey. Berec [2] studied a prey-predator model and discovered that hunting cooperation destabilizes the system by affecting the encounter rate between prey and predator. Alves and Hilker [1] discovered that cooperative hunting might benefit the predator population by increasing the encounter rate, but it may also lead to a sudden collapse of the predator population.

Due to predation fear, preys exhibit a wide range of anti-predator behaviors, including habitat change, reduced foraging activities, reducing prey's per capita growth rate. Zanette et al. [11] conducted an experiment on song sparrows and discovered that only predation fear could reduce the reproduction rate of song sparrows by 40%. Wang et al. [10] studied the dynamics of a predator-prey model incorporating the fear effect and analyzed that relatively high values of the fear parameter may stabilize the system by excluding the existence of limit cycles. Sasmal and Takeuchi [9] investigated the effects of fear in a predator-prey model and analyzed that fear can greatly affect the system dynamics. In literature, many researchers have studied the effects of cooperative hunting and fear on the dynamics of the prey-predator system, but not much work has been done on the combined effects. Pal et al. [7] studied the dynamics of a prey-predator model incorporating hunting cooperation and fear and discovered that fear-induced due to cooperative hunting might destabilize the system by producing periodic oscillations.

Crowley-Martin type functional response shows that higher predator density reduces predator feeding rate due to interference among themselves for a limited resource [4]. This assumption makes the functional response more pragmatic from the ecological point of view. Kumar and Dubey [5] studied a predator-prey model with Crowley-Martin functional response and analyzed that preserving the prey population below a certain threshold level is beneficial to both the species. Maiti et al. [6] studied the dynamics of a stage-structured predator-prey system with Crowley-Martin type functional response. To the best of the authors' knowledge, the combined effect of hunting cooperation and fear induced by a predator on prey in a predator-prey system with Crowley-Martin type functional response has not been studied. Thus, we propose a mathematical model to study the dynamics of the prey-predator system incorporating the above aspects.

2 Mathematical Model

We consider a habitat in which prey of density $x(t)$ and specialist predator of density $y(t)$ live together at any time t. We assume that the prey species is growing logistically with intrinsic growth rate r and carrying capacity k, thus its dynamics leads to the following ODE:

$$\frac{dx}{dt} = rx\left(1 - \frac{x}{k}\right). \tag{1}$$

Since the predator population may decline at high predator density due to interference among them for common limited resources, hence to capture this aspect Crowley-Martin [4–6] type functional response is more realistic. This functional response is given by

$$f(x, y) = \frac{\beta x}{(1 + ax)(1 + by)},$$

where β is attack rate, a is handling time required per prey and b is the magnitude of interference among predator individuals. It may be pointed out here that when $a > 0$ and $b = 0$, then $f(x, y)$ becomes Holling type-II functional response; when $a = 0$ and $b > 0$, then $f(x, y)$ describes a saturation response with respect to predator; when $a = 0$ and $b = 0$, $f(x, y)$ denotes linear mass-action response.

Now, we assume that predators cooperate among themselves to encounter a strong prey, and in such a case the attack rate β is given by $\beta = \alpha_0 + \alpha y$, where α_0 is the capture rate without considering the cooperative hunting among predators and α is hunting cooperation parameter among predators. These predators induce fear among prey which causes a decrease in the growth rate of prey. This fear function is described by [5, 9, 10]

$$g(e, \alpha, y) = \frac{1}{1 + e\alpha y},$$

where e is the cost of fear. Keeping all the above aspects in view, the dynamics of our proposed system can be governed by the following system of ODEs:

$$\begin{aligned} \frac{dx}{dt} &= \frac{rx}{(1 + e\alpha y)}(1 - \frac{x}{k}) - \frac{(\alpha_0 + \alpha y)}{(1 + ax)(1 + by)}xy, \\ \frac{dy}{dt} &= \frac{c(\alpha_0 + \alpha y)}{(1 + ax)(1 + by)}xy - \delta_0 y - \delta_1 y^2, \end{aligned} \tag{2}$$

$$x(0) \geq 0, \quad y(0) \geq 0.$$

In the above model, $c(0 < c < 1)$ is the conversion rate from prey to predator density, δ_0 is the predators' natural mortality rate, and δ_1 is the intraspecific interference coefficient among predators. It may be pointed out here that in case of strong prey, α may be large and b negligible; and in case of weak prey α may be negligible and b large.

3 Mathematical Analysis

Now, we will do the basic mathematical analysis of the model (2). All the parameters involved in our model are positive.

3.1 Basic Analysis

The proofs of the following two theorems are similar to [5] and hence, omitted.

Theorem 1 *All the solutions* $\phi(t) = (x(t), y(t))$ *with initial conditions* $\phi_0(t) = (x_0, y_0) \in \mathbb{R}^2_+$ *remains positive in the first quadrant.*

Theorem 2 $\Omega_1 = \{(x, y) : 0 \leq x \leq k, 0 \leq x + \frac{1}{c}y \leq \frac{2rk}{\delta_{min}}\}$ *is a positively invariant set for all the solutions originating from the first quadrant, where* $\delta_{min} = min\{r, \delta_0\}$.

3.2 Equilibria Analysis

The system (2) can have the following non-negative equilibria:

The trivial equilibrium $E_0 = (0, 0)$, and the predator-free equilibrium $E_1 = (k, 0)$ always exist. We can find the interior equilibrium by solving the following set of equations:

$$\begin{aligned} \frac{r}{1+e\alpha y}(1-\frac{x}{k}) - \frac{(\alpha_0+\alpha y)y}{(1+ax)(1+by)} &=: f(x,y) \\ \frac{c(\alpha_0+\alpha y)x}{(1+ax)(1+by)} - \delta_0 - \delta_1 y &=: g(x,y) \end{aligned} \tag{3}$$

From $f(x, y) = 0$, it follows that:
When $y = 0$, then $x_* = k$. When $x = 0$, we get a cubic equation in y i.e. $e\alpha^2 y^3 + \alpha(1 + e\alpha_0)y^2 + (\alpha_0 - rb)y - r = 0$ which has a positive root (using Descarte's rule of sign). Now

$$\frac{dy}{dx} = -\frac{\left(\frac{r}{k(1+e\alpha y)} - \frac{ay(\alpha_0+\alpha y)}{(1+ax)^2(1+by)}\right)}{\left(\frac{re\alpha}{(1+e\alpha y)^2}(1-\frac{x}{k}) + \frac{(\alpha_0+2\alpha y+b\alpha y^2)}{(1+ax)(1+by)^2}\right)}.$$

From above analysis, we notice that $f(x, y) = 0$ passes through the points $(k, 0)$ and $(0, y_1)$ and it may increase or decrease depending upon the sign of $\frac{dy}{dx}$.
From $g(x, y) = 0$, it follows that:

When $y = 0$, then $x_1 = \frac{\delta_0}{c\alpha_0 - a\delta_0}$. When $x = 0$, then $y_2 = -\frac{\delta_0}{\delta_1}$.

$$\frac{dy}{dx} = \frac{c(\alpha_0+\alpha y)}{(1+ax)^2(1+by)\left(\frac{c(b\alpha_0-\alpha)x}{(1+ax)(1+by)^2} + \delta_1\right)}.$$

Therefore, the above analysis shows that $g(x, y) = 0$ passes through $(x_1, 0)$ and $(0, y_2)$ and it may increase or decrease depending on the sign of $\frac{dy}{dx}$.
Now, based on the above analysis, we state the following theorems.

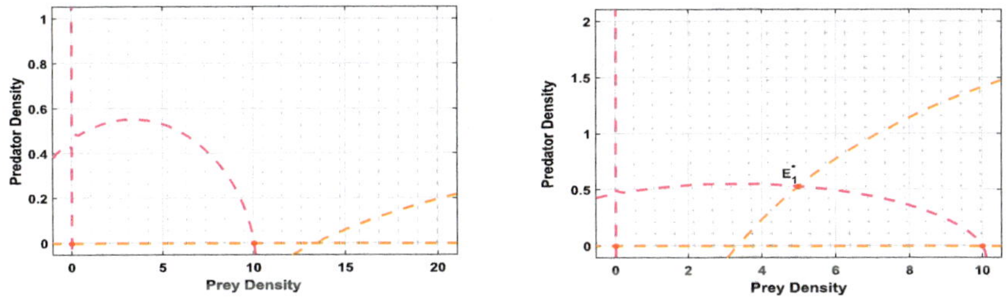

(a) No interior equilibrium point for $\delta_0 = 0.8$.

(b) Unique interior equilibrium point for $\delta_0 = 0.5$.

Fig. 1 Existence of equilibria for model (2) with varying δ_0. Remaining parameters are fixed as $r = 4, k = 10, e = 10, c = 0.1, \alpha_0 = 3, \alpha = 0.5, a = 0.3, b = 0.5$ and $\delta_1 = 0.03$. Here, the slope of predator isocline is positive

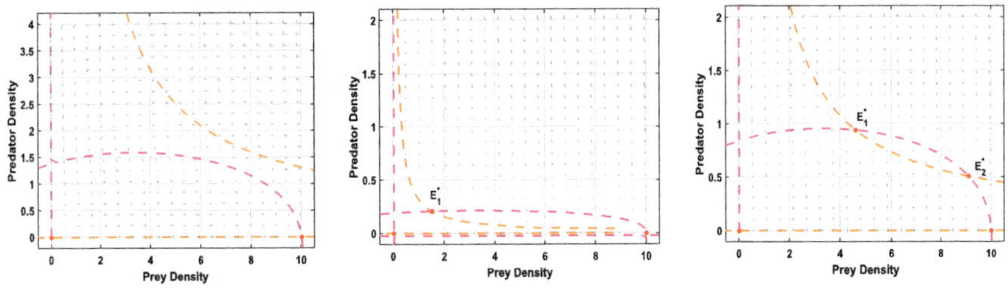

(a) No interior equilibrium point for $\alpha = 0.2$.

(b) Unique interior equilibrium point for $\alpha = 6$.

(c) Two interior equilibrium points for $\alpha = 0.5$.

Fig. 2 Existence of equilibrium points for the model (2) with varying α. Remaining parameters are fixed as $r = 4, k = 10, e = 10, c = 0.5, \alpha_0 = 0.5, a = 0.3, b = 0.01, \delta_0 = 0.9$ and $\delta_1 = 0.03$. Here, the slope of predator isocline is negative

Theorem 3 *The system will have atmost one interior equilibrium if* $\left(\frac{c(b\alpha_0-\alpha)x}{(1+ax)(1+by)^2} + \delta_1\right) > 0$ *and* $k(c\alpha_0 - a\delta_0) > \delta_0$ *hold (see Fig. 1).*

Theorem 4 *The system will have atmost two interior equilibrium if* $\left(\frac{c(b\alpha_0-\alpha)x}{(1+ax)(1+by)^2} + \delta_1\right) < 0$ *holds (see Fig. 2).*

Now we have the following theorems for the stability analysis of different equilibria corresponding to system (2).

Theorem 5 *The trivial equilibrium* E_0 *is always a saddle point.*

Proof The eigenvalues of the Jacobian matrix at the trivial equilibrium E_0 are given by $\lambda_1 = r(> 0)$ and $\lambda_2 = -\delta_0(< 0)$.

Theorem 6 *The axial equilibrium $E_1 = (k, 0)$ always exists and is locally asymptotically stable if $ck\alpha_0 < \delta_0(1 + ak)$.*

Proof The eigenvalues of the Jacobian matrix at the predator-free equilibrium E_1 are given by $\lambda_1 = -r(< 0)$ and $\lambda_2 = \frac{ck\alpha_0}{1+ak} - \delta_0$. Thus, E_1 is locally asymptotically stable if $ck\alpha_0 < \delta_0(1 + ak)$.

Theorem 7 *The interior equilibrium $E^* = (x^*, y^*)$ is locally asymptotically stable if $tr(J_{E^*}) < 0$ and $det(J_{E^*}) > 0$, where J_{E^*} is the Jacobian matrix evaluated at E^*.*

Proof The Jacobian matrix J_{E^*} at the interior equilibrium $E^* = (x^*, y^*)$ is given by:

$$J_{E^*} = \begin{bmatrix} -\frac{rx^*}{k(1+e\alpha y^*)} + \frac{a(\alpha_0+\alpha y^*)x^*y^*}{(1+ax^*)(1+by^*)} & \frac{-e\alpha rx^*}{(1+e\alpha y^*)^2}(1-\frac{x^*}{k}) - \frac{(\alpha_0+2\alpha y^*+b\alpha(y^*)^2)x^*}{(1+ax^*)(1+by^*)^2} \\ \frac{c(\alpha_0+\alpha y^*)y^*}{(1+ax^*)^2(1+by^*)} & -(\frac{c(b\alpha_0-\alpha)x^*y^*}{(1+ax^*)(1+by^*)^2} + \delta_1 y^*) \end{bmatrix}.$$

From above matrix, we have the characteristic equation as:

$$\begin{aligned} \lambda^2 - tr(J_{E^*})\lambda + det(J_{E^*}) &= 0, \\ \lambda^2 + \left[\frac{rx^*}{k(1+e\alpha y^*)} - \frac{a(\alpha_0+\alpha y^*)x^*y^*}{(1+ax^*)(1+by^*)} + (\frac{c(b\alpha_0-\alpha)x^*y^*}{(1+ax^*)(1+by^*)^2} + \delta_1 y^*)\right]\lambda + \\ \left[\left(\frac{rx^*}{k(1+e\alpha y^*)} - \frac{a(\alpha_0+\alpha y^*)x^*y^*}{(1+ax^*)(1+by^*)}\right)\left(\frac{c(b\alpha_0-\alpha)x^*y^*}{(1+ax^*)(1+by^*)^2} + \delta_1 y^*\right)\right] + \\ \left[\left(\frac{e\alpha rx^*}{(1+e\alpha y^*)^2}(1-\frac{x^*}{k}) + \frac{(\alpha_0+2\alpha y^*+b\alpha(y^*)^2)x^*}{(1+ax^*)(1+by^*)^2}\right)\left(\frac{c(\alpha_0+\alpha y^*)y^*}{(1+ax^*)^2(1+by^*)}\right)\right] &= 0 \end{aligned} \quad (4)$$

Thus, $E^* = (x^*, y^*)$ is locally asymptotically stable if $tr(J_{E^*}) < 0$ and $det(J_{E^*}) > 0$ (using Routh-Hurwitz criterion).

Remark: It may be noted that if:

$$\frac{r}{k(1+e\alpha y^*)} > \frac{a(\alpha_0+\alpha y^*)y^*}{(1+ax^*)(1+by^*)}$$

holds, then $tr(J_{E^*}) < 0$ and $det(J_{E^*}) > 0$, and hence $E^* = (x^*, y^*)$ is locally asymptotically stable.

4 Bifurcation Analysis

Next, we investigate the possibility of existence of limit cycle via Hopf-bifurcation near the interior equilibrium E_1^* with respect to the parameter α.

The characteristic equation evaluated from the Jacobian matrix at E_1^* is:

$$\lambda^2 - tr(J_{E_1^*})\lambda + det(J_{E_1^*}) = 0.$$

For the Hopf- bifurcation to occur, we need $tr(J_{E_1^*}) = 0$, and $det(J_{E_1^*}) > 0$. From the above two conditions, we calculate the critical value $\alpha = \alpha_*$ of hunting cooperation parameter.

Then we check the transversality condition:

$$\frac{\partial}{\partial\alpha}|tr(J_{E_1^*})|_{\alpha=\alpha_*} = -\left[\frac{rex^*y^*}{k(1+e\alpha y^*)^2} + \frac{ax^*(y^*)^2}{(1+ax^*)(1+by^*)} + \frac{cx^*y^*}{(1+ax^*)(1+by^*)}\right] \neq 0$$

Hence, the system experiences Hopf-bifurcation at the equilibrium point E_1^* when $\alpha = \alpha_*$.

Theorem 8 *The system (2) goes through a saddle-node bifurcation around the equilibrium point $E(\bar{x}, \bar{y})$ as the cooperation parameter α crosses the bifurcation value $\alpha = \alpha_c$ if and only if*

$$-\left(\frac{-re\bar{x}\bar{y}}{(1+e\alpha_c\bar{y})^2}\left(1-\frac{\bar{x}}{k}\right) - \frac{\bar{x}\bar{y}^2}{(1+a\bar{x})(1+b\bar{y})} - \frac{f_{\bar{x}}}{g_{\bar{x}}(1+a\bar{x})(1+b\bar{y})}\right) \neq 0,$$

$$\left[\left(f_{\bar{x}\bar{x}} - \frac{f_{\bar{x}}}{f_{\bar{y}}}(f_{\bar{x}\bar{y}} + f_{\bar{y}\bar{x}}) + \frac{f_{\bar{x}}^2}{f_{\bar{y}}^2} f_{\bar{y}\bar{y}}\right) - \frac{f_{\bar{x}}}{g_{\bar{x}}}\left(g_{\bar{x}\bar{x}} - \frac{f_{\bar{x}}}{f_{\bar{y}}}(g_{\bar{x}\bar{y}} + g_{\bar{y}\bar{x}}) + \frac{f_{\bar{x}}^2}{f_{\bar{y}}^2} g_{\bar{y}\bar{y}}\right)\right] \neq 0,$$

where

$$f_{\bar{x}} = -\frac{r_1\bar{x}}{k(1+e\alpha_c\bar{y})} + \frac{a(\alpha_0+\alpha_c\bar{y})\bar{x}\bar{y}}{(1+a\bar{x})(1+b\bar{y})},$$

$$f_{\bar{y}} = \frac{-e\alpha_c r x^*}{(1+e\alpha_c y^*)^2}\left(1-\frac{\bar{x}}{k}\right) - \frac{(\alpha_0 + 2\alpha_c\bar{y} + b\alpha_c(\bar{y})^2)\bar{x}}{(1+a\bar{x})(1+b\bar{y})^2},$$

$$g_{\bar{x}} = \frac{c(\alpha_0+\alpha_c\bar{y})\bar{y}}{(1+a\bar{x})^2(1+b\bar{y})},$$

$$g_{\bar{y}} = -\left(\frac{c(b\alpha_0-\alpha_c)\bar{x}\bar{y}}{(1+a\bar{x})(1+b\bar{y})^2} + \delta_1\bar{y}\right).$$

Proof The Jacobian matrix calculated at the interior equilibrium $E(\bar{x}, \bar{y})$ is :

$$J_{E(\bar{x},\bar{y})} = \begin{bmatrix} -\frac{r\bar{x}}{k(1+e\alpha\bar{y})} + \frac{a(\alpha_0+\alpha\bar{y})\bar{x}\bar{y}}{(1+a\bar{x})(1+b\bar{y})} & \frac{-e\alpha r\bar{x}}{(1+e\alpha\bar{y})^2}\left(1-\frac{\bar{x}}{k}\right) - \frac{(\alpha_0+2\alpha\bar{y}+b\alpha(\bar{y})^2)\bar{x}}{(1+a\bar{x})(1+b\bar{y})^2} \\ \frac{c(\alpha_0+\alpha\bar{y})\bar{y}}{(1+a\bar{x})^2(1+b\bar{y})} & -\left(\frac{c(b\alpha_0-\alpha)\bar{x}\bar{y}}{(1+a\bar{x})(1+b\bar{y})^2} + \delta_1\bar{y}\right) \end{bmatrix}.$$

Now differentiating the given system w.r.t. α, we get

$$F_\alpha(\bar{x}\bar{y}) = \begin{bmatrix} \frac{-re\bar{x}\bar{y}}{(1+e\alpha y)^2}\left(1-\frac{\bar{x}}{k}\right) - \frac{\bar{x}\bar{y}^2}{(1+a\bar{x})(1+b\bar{y})} \\ \frac{c\bar{x}\bar{y}^2}{(1+a\bar{x})(1+b\bar{y})} \end{bmatrix}$$

$$B = Df(E(\bar{x}, \bar{y}), \alpha_c) = \begin{bmatrix} f_x & f_y \\ g_x & g_y \end{bmatrix}$$

One can see that the eigenvector corresponding to zero eigenvalue of matrix $B = Df(E(\bar{x}, \bar{y}), \alpha_c)$ is $v' = \begin{bmatrix} 1 & \frac{-f_x}{f_y} \end{bmatrix}^T$, and the eigenvector corresponding to zero eigenvalue of matrix $B^T = [Df(E(\bar{x}, \bar{y}), \alpha_c)]^T$ is: $w' = \begin{bmatrix} 1 & \frac{-f_x}{g_x} \end{bmatrix}^T$.
Using Sotomayor theorem [8] for saddle-node bifurcation, we get the conditions for saddle-node bifurcation by doing simple calculations.

5 Numerical Simulations

In this section, we perform some numerical simulations to show the population insights from our analytical findings.

In Fig. 3, we fixed the parameters as $r = 4$, $k = 10$, $e = 10$, $c = 0.1$, $\alpha_0 = 3$, $\delta_0 = 0.5$, $a = 0.3$, $b = 0.5$ and $\delta_1 = 0.03$. For $\alpha = 0.2$, the system (2) has a stable unique interior equilibrium point i.e. $E_1^*(5, 0.53)$ as a spiral sink (Fig. 3a) and for $\alpha = 2$, the system (2) has a stable limit cycle around the interior equilibrium point $E_1^*(3.2, 0.271)$ i.e. E_1^* is a spiral source (Fig. 3b).

In Fig. 4, we fixed the parameters values as $r = 4, k = 10, e = 10, c = 0.5, \alpha_0 = 0.5, \delta_0 = 0.8, a = 0.3, b = 0.01$ and $\delta_1 = 0.03$ and here, the system has two interior equilibrium points E_1^* and E_2^*. For $\alpha = 0.2$, the system (2) has two possible attractors $E_1(10, 0)$ and $E_1^*(6.25, 1.45)$ (Fig. 4a). Moreover for $\alpha = 0.25$, the system (2) has a stable equilibrium point $E_1(10, 0)$ and also, a spiral source equilibrium $E_1^*(5.2, 1.3)$ (4b). In Fig. 4a, we notice that the system exhibits bi-stability between the interior equilibrium $E_1^*(6.25, 1.45)$ and the axial equilibrium $E_1(10, 0)$. Moreover, in Fig. 5,

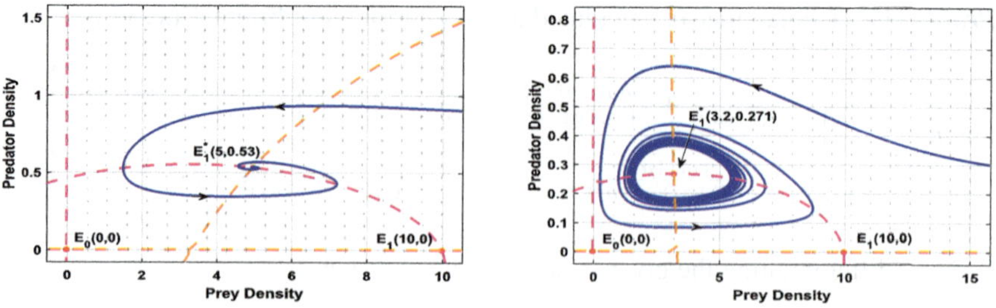

(a) $\alpha = 0.2$, the unique interior equilibrium $E_1^* = (5, 0.53)$ is locally asymptotically stable.

(b) $\alpha = 2$, $E_1^* = (3.2, 0.271)$ is a spiral source.

Fig. 3 All other parameters are fixed as $r = 4, k = 10, e = 10, c = 0.1, \alpha_0 = 3, \delta_0 = 0.5, a = 0.3$, $b = 0.5$ and $\delta_1 = 0.03$. Here, $E_(0, 0)$ and $E_1(10, 0)$ are always saddle points

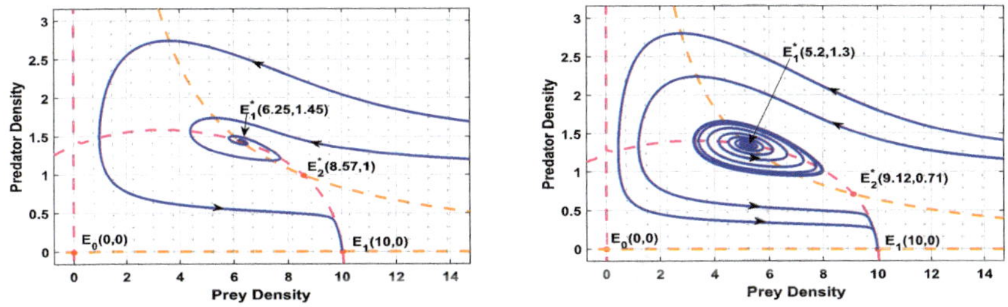

(a) $\alpha = 0.2$, $E_1(10,0)$ and $E_1^*(6.25, 1.45)$ are locally asymptotically stable.

(b) $\alpha = 0.25$, $E_1(10,0)$ is locally asymptotically stable and $E_1^*(5.2, 1.3)$ is a spiral source.

Fig. 4 All other parameters are fixed as $r = 4$, $k = 10$, $e = 10$, $c = 0.5$, $\alpha_0 = 0.5$, $\delta_0 = 0.8$, $a = 0.3$, $b = 0.01$ and $\delta_1 = 0.03$. Here, E_0 and E_2^* are always saddle points

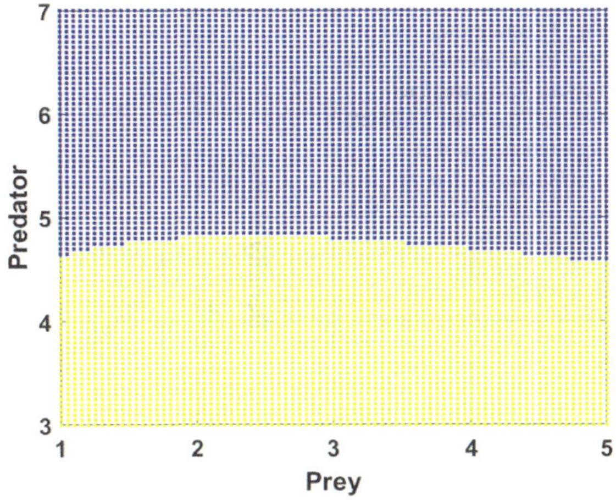

Fig. 5 Basin of attraction for the prey-only equilibrium $E_1(10, 0)$ and the interior equilibrium point E_1^*. Here, all the other parameters are fixed as $r = 4$, $k = 10$, $e = 10$, $c = 0.5$, $\alpha_0 = 0.5$, $\alpha = 0.22$, $a = 0.3$, $b = 0.01$, $\delta_0 = 0.8$ and $\delta_1 = 0.03$. Blue and yellow represents the convergence region of E_1 and E_1^*, respectively

we show the basin of attraction for the predator-free equilibrium and the interior equilibrium. Basin of attraction for an equilibrium point is a set of initial points for which the solutions will converge to the same equilibrium point.

Next, we plot the bifurcation diagram of prey with respect to α in Fig. 6. We notice that for relatively small cooperation parameter α values, both prey and predator species have stable coexistence. As the cooperation parameter α increases, Hopf-bifurcation occurs, and the stable coexistence equilibrium loses its stability and produces periodic oscillations. In Fig. 7, we show the bifurcation plot of prey with respect to the fear parameter e. We also observe that for a fixed value of cooperation parameter α, the system tends to become unstable with an increase in the fear parameter e. Here, we have only shown bifurcation plot for prey population because prey and predator has the same bifurcation behavior.

Figure 8 depicts saddle-node bifurcation with respect to the parameter α. We have fixed all the parameters as $r = 4$, $k = 10$, $e = 10$, $c = 0.5$, $\alpha_0 = 0.5$, $a = 0.3$,

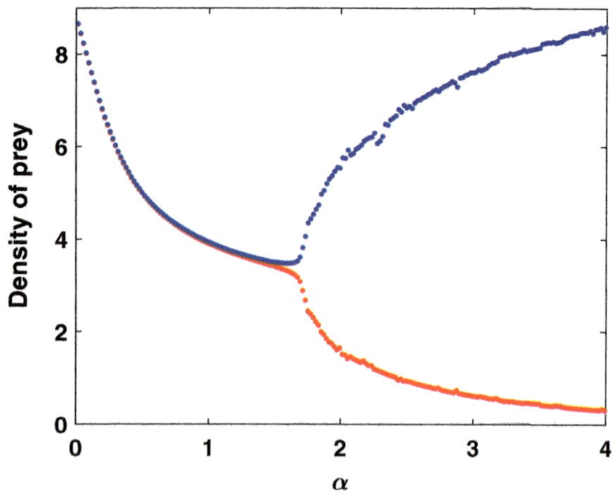

Fig. 6 Bifurcation plot of prey for the model (2) with varying hunting cooperation parameter α. Here, all the other parameters are fixed as $r = 4$, $k = 10$, $e = 10$, $c = 0.1$, $\alpha_0 = 3$, $a = 0.3$, $b = 0.5$, $\delta_0 = 0.5$ and, $\delta_1 = 0.03$

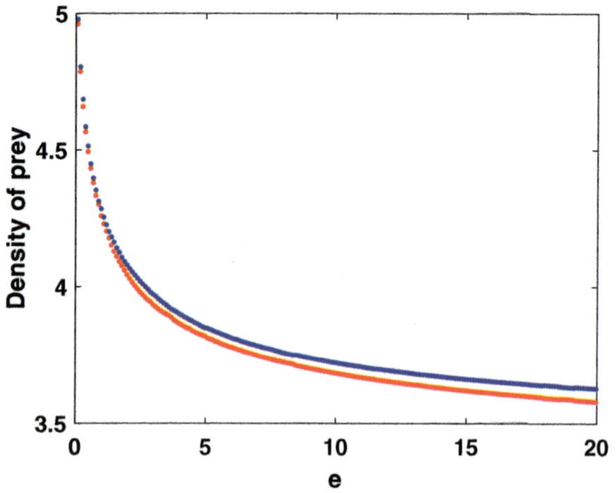

Fig. 7 Bifurcation plot of prey for the model (2) with varying fear parameter e. Here, all the other parameters are fixed as $r = 4$, $k = 10$, $c = 0.1$, $\alpha_0 = 3$, $\alpha = 1.2$, $a = 0.3$, $b = 0.5$, $\delta_0 = 0.5$ and, $\delta_1 = 0.03$

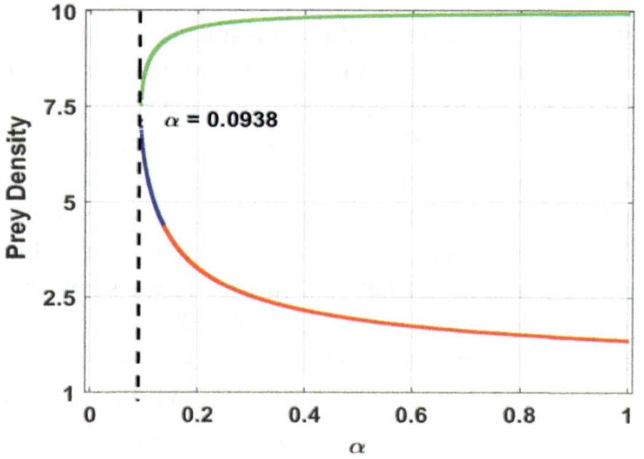

Fig. 8 Plot of saddle-node bifurcation with respect to the parameter α. Remaining parameters are fixed as $r = 4$, $k = 10$, $e = 10$, $c = 0.5$, $\alpha_0 = 0.5$, $a = 0.3$, $b = 0.01$, $\delta_0 = 0.8$ and $\delta_1 = 0.03$. Here, blue colour indicates stable equilibrium point; red colour indicates source; green colour indicates saddle point

$b = 0.01$, $\delta_0 = 0.8$ and $\delta_1 = 0.03$. Here, we observe that the system goes through saddle node bifurcation at $\alpha = 0.0938$ and the set of parameters satisfy the analytical conditions.

6 Discussion and Concluding Remarks

Social interactions within a population are a common and essential phenomenon from an ecological perspective. In particular, predators cooperate during hunting to increase the success rate of catching prey. Due to predation fear, prey shows a variety of anti-predator behavior, decreasing prey's per capita growth rate. The present manuscript considered a model incorporating both hunting cooperation and fear effect with Crowley-Martin functional response. First, we observed that incorporating hunting cooperation may produce limit cycles via Hopf-bifurcation and, hence, destabilizing the system. Also, the system undergoes saddle-node bifurcation with respect to the parameter α under certain conditions. In addition, we noticed that the system shows bi-stability behavior in which the solution tends to prey-only equilibrium or coexisting equilibrium state. Furthermore, we have seen that for a fixed value of hunting cooperation parameter, increasing the strength of fear makes the system unstable. Hence, hunting cooperation and fear effect is of great significance from an ecological aspect.

Acknowledgements The authors are grateful to the anonymous referees for the critical review and suggestions that improved the quality of the paper. The first author, Anshu, acknowledges the Junior Research Fellowship received from UGC, New Delhi, India.

References

1. Alves, M.T., Hilker, F.M.: Hunting cooperation and allee effects in predators. J. Theor. Biol. **419**, 13–22 (2017)
2. Berec, L.: Impacts of foraging facilitation among predators on predator-prey dynamics. Bull. Mathem. Biol. **72**(1), 94–121 (2010)
3. Cosner, C., DeAngelis, D.L., Ault, J.S., Olson, D.B.: Effects of spatial grouping on the functional response of predators. Theor.. Popul. Biol. **56**(1), 65–75 (1999)
4. Crowley, P.H., Martin, E.K.: Functional responses and interference within and between year classes of a dragonfly population. J. North Am. Benthol. Soc. **8**(3), 211–221 (1989)
5. Kumar, A., Dubey, B.: Modeling the effect of fear in a prey-predator system with prey refuge and gestation delay. Int. J. Bifurc. Chaos **29**(14), 1950195 (2019)
6. Maiti, A.P., Dubey, B., Chakraborty, A.: Global analysis of a delayed stage structure prey-predator model with crowley-martin type functional response. Mathem. Comput. Simul. **162**, 58–84 (2019)
7. Pal, S., Pal, N., Chattopadhyay, J.: Hunting cooperation in a discrete-time predator-prey system. Int. J. Bifurc. Chaos **28**(07), 1850083 (2018)

8. Perko, L.: Differential equations and dynamical systems. Springer-Verlag New York **5** (2001)
9. Sasmal, S.K., Takeuchi, Y.: Dynamics of a predator-prey system with fear and group defense. J. Mathem. Anal. Appl. **481**(1), 123471 (2020)
10. Wang, X., Zanette, L., Zou, X.: Modelling the fear effect in predator-prey interactions. J. Mathem. Biol. **73**(5), 1179–1204 (2016)
11. Zanette, L.Y., White, A.F., Allen, M.C., Clinchy, M.: Perceived predation risk reduces the number of offspring songbirds produce per year. Science **334**(6061), 1398–1401 (2011)

Chaotic Dynamics of Third Order Wien Bridge Oscillator with Memristor Under External Generalized Sinusoidal Stimulus

Aniruddha Palit

Abstract The qualitative behaviour of the signals generated by a third order Wien bridge oscillator with memristor under external generalized sinusoidal stimulus is studied. The bifurcation of the nature of the solution for different range of the parameters of the system reveal that the external stimulus generates an added layer of security which can be used to build a secure communication channel using the synchronization of chaos. Some specific regimes of the parameters of the external stimulus are identified over which such secure channel can be established. The 0-1 test of chaos has been employed to verify the chaotic nature of the output signal.

Keywords Wien bridge oscillator · Memristor · Bifurcation · Chaos

1 Introduction

The study of dynamical behaviour of the signals in the circuit theory has always been a topic of interest because of the complex dynamics and noise arising in the output and the analysis of their characteristic properties is a matter of great importance in the transmission of signals to make a consistent communication system. In the year 1971 Leon Chua [4] first observed that four fundamental variables, namely charge (q), current (i), flux (ϕ) and voltage (v) arise in mathematical formulation of a circuit. Determination of these four variables require four relations involving them out of which three relations can be generated by the axioms of classical two terminal circuit elements, namely inductor (relation involving i and ϕ), resistor (relation involving i and v) and capacitor (relation involving v and q). However, one relation between ϕ and q remains undefined. Chua postulated this missing element, named as memristor which was realized recently [20] by Stan Williams group of HP Labs in 2008. The character of memristor is nonlinear and unique in the sense that no combination of nonlinear resistive, capacitive and inductive components can duplicate its excellent

A. Palit (✉)
Department of Mathematics, Surya Sen Mahavidyalaya, Siliguri 734004, India
e-mail: mail2apalit@gmail.com

S. Banerjee and A. Saha (eds.), *Nonlinear Dynamics and Applications*,
Springer Proceedings in Complexity,
https://doi.org/10.1007/978-3-030-99792-2_87

feature of memory and neuromorphic property. As a consequence the application of memristor has drawn the attention of many researchers in the construction of non-volatile memory [16], neural network [12], nonlinear circuits [23] and various other fields.

In the circuit theory wien bridge oscillator is used to generate sinusoidal signals and is composed of resistors and capacitors. Recently memristors are used in such circuit exhibiting complex dynamic phenomena such has chaotic [22], hyperchaotic [25] behaviours as well as periodic and quasiperiodic behaviours for different regimes of the parameters present in the system. Wu et al. [22] designed a fourth order chaotic oscillator by the construction of a generalized memristor. Ye et al. [25] manufactured a fifth-order Wien-bridge hyperchaotic circuit. However, these higher order circuits are difficult to analyze due to their complex dynamics. Bao et al. [1] presented a third-order RLCM-four-elements-based autonomous memristive chaotic circuit by an active oscillator and a memristor. Rajagopal et al. [17] studied a third order Wien bridge oscillator (WBO) with fractional order memristor. Xu et al. [24] introduced external sinusoidal voltage stimulus in WBO and studied different complex behaviour. This kind of systems possessing complex dynamic behaviour can be used in information engineering such as generation of pseudorandom sequences in various information encryption, chaotic communication systems and synchronization etc.

Various electrical circuits such as Chua circuit, WBO etc. produce chaotic output which are categorized into self-excited and hidden attractors. The basin of attraction of a self-excited attractor is connected with neighbourhoods of unstable equilibrium point. Therefore, such attractors can be identified numerically following standard computational procedures in which starting from a sufficiently small neighbourhood of an unstable equilibrium point and after a transient process a trajectory is attracted to a state of oscillation and then traces it. On the other hand the basin of a hidden attractor is not connected with equilibria and hence it is very much challenging to visualize. Leonov et al. [11] classified hidden and self-excited attractors which captured the attention of the scientific community. Burkin and Khien [2] introduced an analytical-numerical method for localization of hidden attractors. Dynamics of self-excited and hidden attractors have been studied by Chen et al. [3], Stankevich et al. [19] and many more. Synchronization of such attractors in Chua circuit have been investigated by Kiseleva et al. [10]. Therefore, identification of chaotic attractors and realization of its nature is a challenging and active research of interest and is expected to contribute significantly in the transmission of confidential information through a secure communication channel.

Secure communication with chaos is an important field in engineering. A continuously changing chaotic signal is used as a carrier signal in the chaotic communication in contrast to a fixed carrier signal as used in classical communication. As a result it becomes very much difficult to predict the signal transmitted through chaotic communication thereby increases the security level and consequently it has very important place in secure communication system. Oppenheim et al. [14] first made the study of chaotic communication in 1992 and this field has been further explored by different researchers studying 5D hyperchaotic systems [26], chaotic systems with no equilibrium [27] or hidden attractor [15]. Even chaotic system of fractional order [15], jerk chaotic system using sliding mode control [6] and several other interesting studies have been reported in literature. Therefore, identification of the regimes of the parameters of an WBO has significant applications in chaotic communication. One must apply proper verification method to identify the chaotic nature of a signal, without which definite decisions cannot be made. The computation of the maximal Lyapunov exponent is one such technique to determine chaos for quite a long time in the literature. However, we have used another method proposed by Gottwald and Melbourne [8] for detecting chaotic dynamics, known as 0-1 Test, which is based on time series data and has shown its potential in the determination of chaotic as well as non-chaotic nature of a signal in the recent past. The test can be applied in higher order systems without any practical difficulties.

In this article we have proposed a model of WBO under externally driven generalized sinusoidal voltage stimulus and studied the nature of the output signal. The objective of this paper is to check the potential of the driven stimulus on the output and to identify regimes of the parameters of these external forces for which chaotic signals can be generated by this kind of circuits. Such external stimulus enables us to make extra layer of security over the system parameters making the transmission of the signals in chaotic communication system more unpredictable and the system becomes more reliable.

The article is arranged as follows. A stability analysis of the equilibrium points of the WBO is presented in Sect. 2. The bifurcation of the output signals of driven WBO have been studied in Sect. 3 and the subsection therein and regimes of the parameters of the external stimulus are identified for which chaotic output is produced. A specimen of the parameter values are chosen and the 0-1 test is employed on the corresponding output signal to ensure its chaotic nature in Sect. 4. Conclusions and future aspects of this model are discussed in Sect. 5.

2 Stability Analysis of Wien Bridge Oscillator

The idea of memristor was first introduced by Chua [4] and its mathematical model was first proposed in [5]. A third order Wien bridge oscillator can be written in dimensionless form [17] as

$$\dot{x} = x\left(a - 1 - cz^2\right) - by \tag{1a}$$

$$\dot{y} = ax - by \tag{1b}$$

$$\dot{z} = -x - z\left(d - x^2\right) \tag{1c}$$

where a, b, c and d are parameters of the oscillator. In an electrical circuit the resistors and capacitors are responsible for the loss of electrical energy and so the external stimulus should be introduced in the circuit to restore this loss in order to execute the operation of the circuit for a long time. The external stimulus produces electromotive force (EMF) which is responsible to maintain voltage difference between the nodes of a circuit. Various external energy sources such as battery, generators etc. can be introduced at different nodes to boost the voltage differences in order to produce chaotic output of the circuit making it difficult to predict in advance unless one precisely knows the values of the parameters involved in the system and such output can be made to build a secure chaotic communication system. In order to enhance the security of the chaotic communication system more than one sources for production of EMF can be introduced in the above circuit having various magnitude, phase and frequency. We consider three different external sinusoidal stimulus having different magnitudes in the above system and check if the parameters of these external stimuli make any additional layer of security over the system parameters by producing chaotic output signal. Since, an autonomous system representation of a differential system is not unique, for the sake of simplicity, we can study the effect of these external stimuli separately in each of the above equations so that it can be written in generalized form as

$$\dot{x} = x\left(a - 1 - cz^2\right) - by + f_1 \cos(\omega t) \tag{2a}$$

$$\dot{y} = ax - by + f_2 \cos(\omega t) \tag{2b}$$

$$\dot{z} = -x - z\left(d - x^2\right) + f_3 \cos(\omega t) \tag{2c}$$

and study its qualitative behaviour under the influence of the parameters of these external stimuli.

The unperturbed system (1) has the fixed points at

$$P_0 = (0, 0, 0),\ P_1 = \left(x_{P_1}, \frac{a}{b}x_{P_1}, \frac{x_{P_1}}{\left(x_{P_1}\right)^2 - d}\right),\ P_2 = \left(x_{P_2}, \frac{a}{b}x_{P_2}, \frac{x_{P2}}{\left(x_{P_2}\right)^2 - d}\right)$$

$$P_3 = \left(-x_{P_1}, -\frac{a}{b}x_{P_1}, -\frac{x_{P_1}}{\left(x_{P_1}\right)^2 - d}\right) \text{ and } P_4 = \left(-x_{P_2}, -\frac{a}{b}x_{P_2}, -\frac{x_{P_2}}{\left(x_{P_2}\right)^2 - d}\right)$$

where,

$$x_{P_1} = \frac{\sqrt{-c + \sqrt{c}\sqrt{c - 4d} + 2d}}{\sqrt{2}},\ x_{P_2} = -\frac{\sqrt{-c - \sqrt{c}\sqrt{c - 4d} + 2d}}{\sqrt{2}}.$$

It is notable that x_{P_1} and x_{P_2} are not real for any real value of c and d so that the only effective real fixed point of the system (1) is the point P_0. By linearizing the system near the point P_0 the Jacobian Matrix can be expressed as

$$J = \begin{bmatrix} -1 + a & -b & 0 \\ a & -b & 0 \\ -1 & 0 & -d \end{bmatrix}$$

the characteristic equation of which is

$$\lambda^3 - (-1 + a - b - d)\lambda^2 - (-b - d + ad - bd)\lambda + bd = 0.$$

The eigen values of J at the fixed point P_0 are the roots of the above characteristic equation given by

$$\lambda_1 = -d,\ \lambda_{2,3} = \frac{1}{2}\left(-1 + a - b \pm \sqrt{-4b + (1 - a + b)^2}\right).$$

It is clear that the eigen values λ_2 and λ_3 will be complex conjugates when the discriminant is negative i.e., if

$$-4b + (1 - a + b)^2 < 0 \text{ i.e. if } 1 - 2\sqrt{b} + b \leq a \leq 1 + 2\sqrt{b} + b.$$

In this interval if particularly $a < b + 1$ along with $d > 0$, the real parts of λ_2 and λ_3 will be negative and in that case the equilibrium point will be asymptotically stable. However, if $a > b + 1$ with $d > 0$, we have $\lambda_1 < 0$, Re $(\lambda_2) > 0$ and Re $(\lambda_3) > 0$ so that the points P_0 becomes an unstable equilibrium point.

The oscillator (1) is generally known to generate sinusoidal waves in a large range of frequencies. However, we observe that the system has chaotic solution for different values of the parameters. Rajagopal et al. [17] studied the Wien Bridge Oscillator with fractional order memristor considering the values

$$b = 1,\ c = 0.5,\ d = 2 \tag{3}$$

as the parameter a increases. In this section we highlight similar nature of the solution of the system (1) for the above mentioned set of values of the parameters. The solution orbit for some discrete values of a are shown in Fig. 1.

We may observe different kind of orbits in these figures. Notably, a small change in a from 5.96 in the Fig. 1c to 5.97 in the Fig. 1d produces a bifurcation from an irregular orbit to a perfectly periodic orbit. The nature of the irregularity cannot be explicitly identified unless we perform certain test for confirmation of chaotic behaviour. One

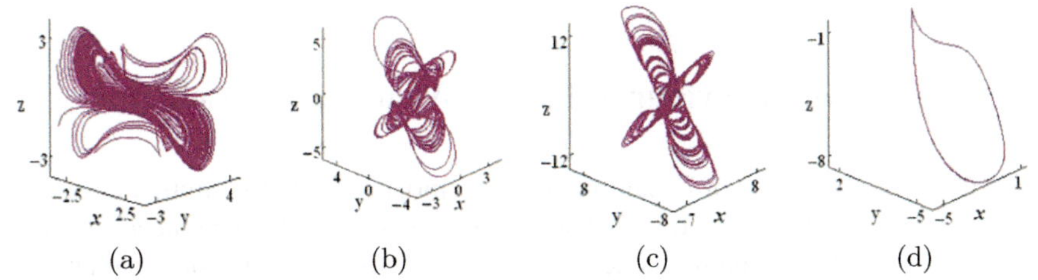

Fig. 1 Solution orbit of the WBO system (1) with parameter values given by (3) for $a =$ 2.55, 3, 5.96, 5.97 respectively in the subfigures $(a) - (d)$

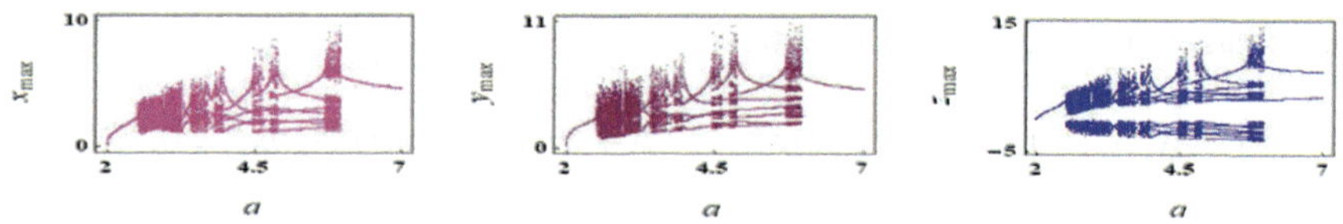

(a) (b) (c)

Fig. 2 Bifurcation diagram of $x_{\max}$, $y_{\max}$ and $z_{\max}$ of the solution orbit of the WBO system (1) for $700 \leq t \leq 1000$ against the parameter a when $b = 1,\ c = 0.5,\ d = 2$

can only state that the parameters present in the system have significant effect not only on the nature of the equilibrium point P_0, but also on the stability of the periodic solution. In order to have a proper idea regarding the effect of the parameter a on the periodic or chaotic nature of the solution orbit we draw the bifurcation diagram of $x_{\max}$, $y_{\max}$ and $z_{\max}$ for $700 \leq t \leq 1000$ against a as shown in Fig. 2.

It is clear from these figures that the system (1) exhibits sensitive dependence on the values the parameter a. The periodic and chaotic regimes occur alternately in the spectrum of the parameter a. Extensive study has been performed [1, 17, 25] for this classical third order WBO. A transmitter made by WBO generates sinusoidal waves which can be easily captured by a receiver thereby making a consistent communication system. However, such periodic transmission can be captured by any receiver synchronizing the parameter values without facing much difficulties, which makes the communication system open to all. Confidential information cannot be transmitted through such a system. A confidential information requires the data to be encrypted in such a manner that cannot be decrypted by anyone without the detailed specification of the communication system. Such a secure communication channel can be established by the synchronization of chaos, through which information can be transmitted and received and finally decrypted only when the chaotic data can be generated identically in the receiver end. Sensitive dependence of the system on more than one parameters like a may produce an additional layer of security and in order to introduce such extra layer we apply external generalized sinusoidal stimuli to the system (1) and construct the generalized forced system (2) and check if the externally excited system produce chaotic output signal controlled by the parameters $f_1,\ f_2,\ f_3$ and ω.

3 Qualitative Behaviour of Forced Wien Bridge Oscillator

The dynamic behaviour of nonautonomous memristor oscillator circuits have been investigated by several authors [24] showing complex dynamics induced by variations of the amplitude and frequency of the external stimulus and system parameters. Several experimental, simulative and theoretical aspects of the memristor have been studied by many authors [1, 17, 22, 25] in the last decade exhibiting different higher and lower order complex dynamics. Such complex dynamical behaviour of the system are very much useful in the construction of chaotic communication system and remarkable results have been achieved by many researchers [6, 15, 26, 27]. The characters of chaotic signals, such as nonlinearity, unstability, aperiodicity etc. have made it attractive for the use in the chaotic communication to enhance the level of security of the transmission of the encrypted data. In this section we shall investigate how the external periodic forces applied on the traditional wien bridge oscillator impact on the generation of chaotic solution and construct an enhanced layer of security over the system parameters.

3.1 *Effect of Amplitude of the Driven Forces*

We first study the forced Wien bridge oscillator (FWBO) given by (2) for $a = 2$, $b = 1$, $c = 0.5$, $d = 2$ and check the behaviour of its solution for the special case when $f_1 = f_2 = f_3 = f$ (say) and f increases in the range $0 \leq f \leq 4$, where we have chosen $\omega = 1$. We present the bifurcation diagram of the x, y coordinates of the points on the Poincare section of the solution by the half plane $z = 0,\ x \geq 0$ along with the distance $r = \sqrt{x^2 + y^2}$ from the fixed point $P_0\ (0, 0, 0)$ in Fig. 3a-c as f increases in the range $0 \leq f \leq 4$. We observe that the deformation of the periodic orbit starts at $f = 3.54$ and finally lead to chaotic orbit as f increases. We choose a specimen value $f = 3.65$ from the chaotic regime and draw the solution orbit as shown in the Fig. 3d. The Poincare section of this orbit by the half plane $z = 0, x \geq 0$ is presented in the Fig. 3e. The corresponding plots of $x(t)$, $y(t)$, $z(t)$ are shown in Fig. 3f for $700 \leq t \leq 1000$.

The bifurcation diagrams shown in Fig. 3a-c clearly show that the FWBO produces periodic oscillation for a wide range of values of f. However, this periodic nature of the solution breaks and it deforms into unstable orbit which ultimately produces chaotic behaviour for little increase in the value of the amplitude f. In order to get a clear idea the solution orbit for $f = 3.65$ is displayed in Fig. 3d which evidently shows that the orbit is not periodic at all for $700 \leq t \leq 1000$. It is clear that the orbit is chaotic in nature and becomes evident from its Poincare section in Fig. 3e, though some decisive tests such as 0-1 test is required to be performed as discussed in Sect. 4. The solutions x, y, z plotted against t in the Fig. 3f do not have any periodic nature. If the value of the parameter f is increased further the chaotic solution again deforms to period one. We clearly obtain periodic orbit when f increases beyond 3.765. Thus, we identify a small window of f for which the FWBO produces chaotic solution. Identification of such small window is necessary for the enhancement of the security of the chaotic communication system.

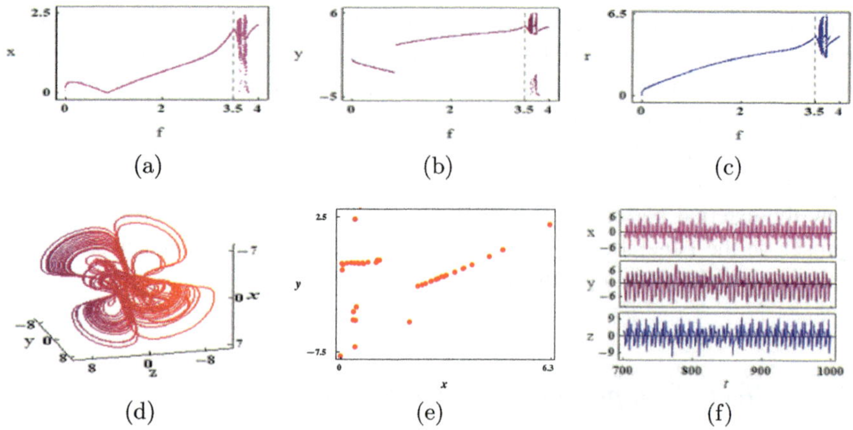

Fig. 3 Bifurcation diagram of **a** the x coordinate, **b** the y coordinate, **c** the distance $r = \sqrt{x^2 + y^2}$ from $P_0\,(0, 0, 0)$ for the points of the Poincare section of the solution of the FWBO system (2) by the half plane $z = 0,\ x \geq 0$ against the parameter f. **d** The solution orbit of the system (2) for $f = 3.65$, **e** corresponding Poincare section by the half plane $z = 0$, $x \geq 0$ and **f** plots of $x(t)$, $y(t)$, $z(t)$ for $700 \leq t \leq 1000$. Here, $a = 2, b = 1, c = 0.5, d = 2, \omega = 1$ and $f_1 = f_2 = f_3 = f$ in all the subfigures

3.2 *Effect of Frequency of the Driven Forces*

We next study the effect of the parameter ω on the system (2). The vertical gridlines in Fig. 3a-c at $f = 3.5$ show that the solution of this system is periodic for $\omega = 1$. We now see how the behaviour of the solution changes when the frequency ω of the externally applied force vary. In order to study elaborately we may inspect the bifurcation diagram of the x, y coordinates of the points on the Poincare section of the solution along with the distance $r = \sqrt{x^2 + y^2}$ when $700 \leq t \leq 1000$ in Fig. 4a-c as ω increases from 0.95 to 1. We find that as ω decreases, the nature of the solution of the system deforms from periodic to chaotic through the route of periodic doubling bifurcation, starting from $\omega = 0.9872$. The chaotic orbit is observed when ω decreases further and becomes less than 0.9829. Similar bifurcation diagrams are also plotted in Fig. 4d-f when ω increases from 1.3 to 1.44. One may observe bifurcation of the periodic solution starting from $\omega = 1.379$ and generates chaotic solution, but in this case it does not follow the route of periodic doubling. Thus, we identify two ranges of ω, viz. $0.95 \leq \omega \leq 0.9829$ and $1.379 \leq \omega \leq 1.44$ in which the system (2) exhibits chaotic behaviour. Therefore, signals generated by FWBO in this kind of unusual windows of the frequency ω may be securely transmitted exploiting its chaotic nature.

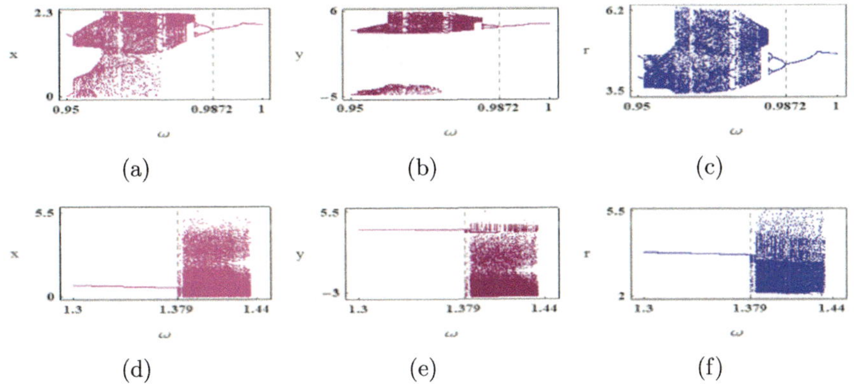

Fig. 4 Bifurcation diagram of **a** the x coordinate, **b** the y coordinate, **c** the distance $r = \sqrt{x^2 + y^2}$ from P_0 (0, 0, 0) for the points of the Poincare section of the solution of the FWBO system (2) by the half plane $z = 0,\ x \geq 0$ against the parameter ω when $0.95 \leq \omega \leq 1$. Similar bifurcation diagram of **d** the x coordinate, **e** the y coordinate, **f** the distance $r = \sqrt{x^2 + y^2}$ against the parameter ω when $1.3 \leq \omega \leq 1.44$. Here, $a = 2, b = 1, c = 0.5, d = 2$ and $f_1 = f_2 = f_3 = f = 3.5$ in all the subfigures

3.3 *Effect of Variable Amplitude of the Driven Forces*

We have observed the effect of the external forces on the FWBO and identified some ranges of the amplitude f and frequency ω in which the solution of the system has chaotic behaviour. So far we have imposed the restriction that $f_1 = f_2 = f_3 = f$ (say). Here, we investigate the system under little general criteria assuming $f_3 \neq f = f_1 = f_2$. Similar to the Fig. 3d we fix $f_3 = 3.65$, but decrease the value of f and investigate the behaviour of the system. The Fig. 5 show the bifurcation of the x, y coordinates of the points on the Poincare section of the solution along with the distance $r = \sqrt{x^2 + y^2}$ by the half plane $z = 0,\ x \geq 0$ for the interval $700 \leq t \leq 1000$. It is interesting to see that the solution does not intersect the half plane for $1.82725 \leq f \leq 2.16375$. We, therefore, check the section of the solution orbit by the half plane $z = -2,\ x \geq 0$ and find that it is periodic in this range and so no such diagram is presented here. Thus, we find that the system produces chaotic oscillation for quite a large value of the amplitude f of the driven forces.

4 The 0-1 Test for Chaos Applied on the FWBO

One standard technique for determination of the chaotic nature of a solution is to compute the maximal Lyapunov exponent [9]. Recently Gottwald and Melbourne proposed a new method [8] for detecting chaotic dynamics, known as 0-1 Test, which can be applied to ordinary and partial differential equations as well as on

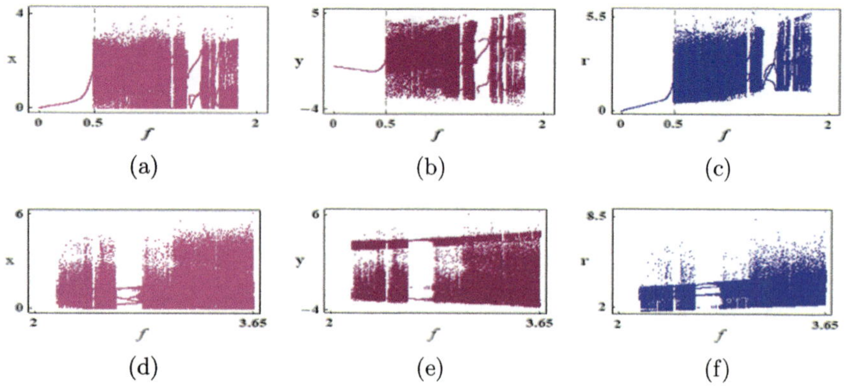

Fig. 5 Bifurcation diagram of **a** the x coordinate, **b** the y coordinate, **c** the distance $r = \sqrt{x^2 + y^2}$ from P_0 (0, 0, 0) for the points of the Poincare section of the solution of the FWBO system (2) by the half plane $z = 0,\ x \geq 0$ against the parameter f when $0 \leq f \leq 2$. Similar bifurcation diagram of **d** the x coordinate, **e** the y coordinate, **f** the distance $r = \sqrt{x^2 + y^2}$ against the parameter f when $2 \leq f \leq 3.65$. Here, $a = 2$, $b = 1$, $c = 0.5$, $d = 2$, $f_3 = 3.65$ and $f_1 = f_2 = f$ in all the subfigures

maps. Even if the deterministic time series data is provided, one can bypass the phase space reconstruction technique using this 0-1 test and determine whether a solution is chaotic or non-chaotic. The simplicity of the technique has drawn the attention of many researchers in the recent past [7, 13, 18]. The main advantages of this test are (i) it is binary i.e., the output of this test can be 0 or 1, (ii) the nature of the vector field and the dimension of the system do not impose any practical limitations, (iii) the difficulty of the phase space reconstruction process can be avoided for time series data. In this section we apply the 0-1 test on FWBO for specific values of the parameters, as a specimen, to check if the signal so generated is chaotic in nature. We choose the parameter values given by (3) and

$$a = 2,\ f_1 = f_2 = 1,\ f_3 = 3.65,\ \omega = 1.4 \tag{4}$$

and verify whether the corresponding solutions $x(t)$, $y(t)$ and $z(t)$ of FWBO are chaotic.

We briefly review this test in the context of the solution $x(t)$. The test has undergone through different equivalent modifications [8]. We are following the version, known as correlation method, discussed in [21] where a pseudo code is provided to determine the output K of the test. The values of $x(t)$ are discretized to $x_j = x\left(t_j\right)$ for $j = 1, 2, \ldots, N$ and generate the translation variables

$$p_n(c) = \sum_{j=1}^{n} x_j \cos(jc), q_n(c) = \sum_{j=1}^{n} x_j \sin(jc) \tag{5}$$

for $n = 1, 2, \ldots, N$, where $c \in (0, \pi)$, with little abuse of notation of the parameter c in the FWBO (2). One important property of the test is that the method is independent of the discrete values of the solution and almost any choice of x_j will serve the requirement. The only precaution one should take is that discrete values must be chosen after sufficient long time so that the trajectories lie on the attractor or remain close to the attractor. Keeping this in mind we choose t_j in the domain [700, 10000] by considering $t_1 = 700$ and choosing the subsequent values of t at a distance $\Delta t = 2$ so that $t_j = t_1 + (j - 1)\,\Delta t$. The second important necessary requirement for the scheme is that the time series should be long enough to allow for asymptotic behaviour of $p_n(c)$ and $q_n(c)$, which means that the value of N should be taken sufficiently large.

It is notable that the output of the test is independent of the choice of the parameter c. It can be rigorously shown that $p_n(c)$ and $q_n(c)$ are bounded if the underlying dynamics is periodic or quasiperiodic, whereas they behave like Brownian motion for large class of chaotic dynamical systems. These behaviours can be investigated by analyzing the mean square displacement defined by

$$M_n(c) = \lim_{n\to\infty} \frac{1}{N} \sum_{j=1}^{N} \left[p_{n+j}(c) - p_j(c)\right]^2 + \left[q_{n+j}(c) - q_j(c)\right]^2 \tag{6}$$

for $n = 1, 2, \ldots, N_c \ll N$. In practice we have the approximate formula

$$M_n(c) \simeq \frac{1}{N} \sum_{j=1}^{N} \left[p_{n+j}(c) - p_j(c)\right]^2 + \left[q_{n+j}(c) - q_j(c)\right]^2 \tag{7}$$

where N is taken as a sufficiently large positive integer. The mean square displacement is a bounded function if the underlying signal is periodic or quasiperiodic, whereas it becomes unbounded for chaotic data. The asymptotic growth rate $K(c)$ is defined by

$$K(c) = \lim_{n\to\infty} \rho(\mathbf{t}_n, \mathbf{M}_n(c)) \tag{8}$$

where $\mathbf{t}_n = (t_1, t_2, \ldots, t_n)$ and $\mathbf{M}_n(c) = (M_1(c), M_2(c), \ldots, M_n(c))$ and ρ designates the correlation coefficient between $\mathbf{t}_n$ and $\mathbf{M}_n(c)$. This $K(c)$ is practically computed by the approximation $K_n(c) = \rho(\mathbf{t}_n, \mathbf{M}_n(c))$ for $n = N_c$. The process involves different approximations of $M_n(c)$ and $K(c)$ which may produce little variation in the asymptotic growth rate. The value of $K(c)$ will be 1 for chaotic signal and 0 for periodic or quasiperiodic signal. However, the approximation process involved in the computation scheme will produce the value of $K(c)$ close to either 1 or 0. In our computation we have taken $N = 3200$, $N_c = 400$, $c = 1$ and the plot of p_n versus q_n is shown in Fig. 6a along with the graph of $M_n(c)$ in Fig. 6b. The graph of $K_n(c)$ is plotted in Fig. 6c for $n \le N_c$ showing that $K_n(c) \simeq 1$ at $n = N_c$. Although the quantity $K(c)$ is expected to produce a value which does not depend on c, the approximation process involved in this scheme may produce unexpected variation

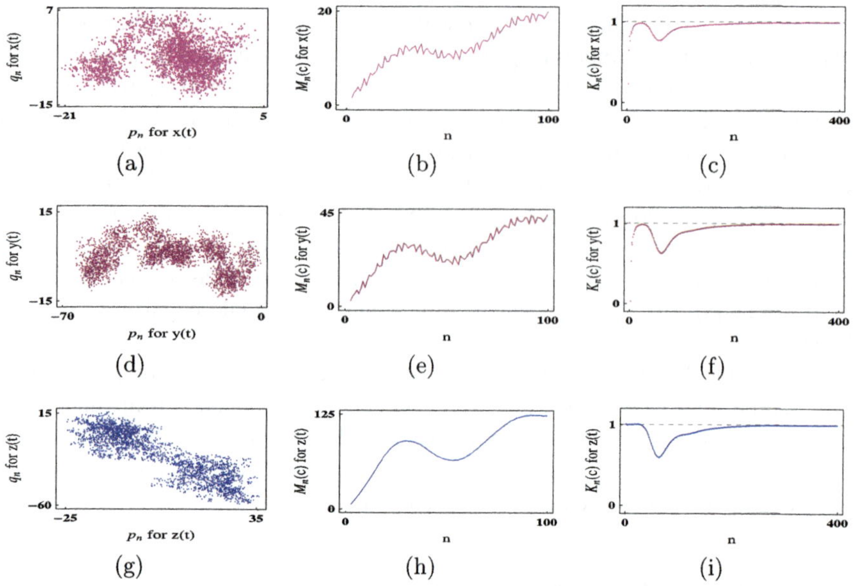

Fig. 6 **a** The p_n versus q_n plot, **b** the graphs of M_n, **c** the graph of K_n computed for the solution $x(t)$ of the FWBO system (2). Analogously, **d** The p_n versus q_n plot, **e** the graphs of M_n, **f** the graph of K_n computed for the solution $y(t)$ of the system (2). Finally, **g** The p_n versus q_n plot, **h** the graphs of M_n, **i** the graph of K_n computed for the solution $z(t)$ of the system (2). Here, the values of the parameters are given by (3) and (4) in all the subfigures

in the approximation of $K(c)$ for some isolated value of c, such as resonant points. This problem is treated by considering $K = \text{median}\{K(c_1), K(c_2), \ldots, K(c_m)\}$, where $c_1, c_2, \ldots, c_m$ are chosen randomly from the domain $(0, \pi)$. The median is taken in place of mean because of the fact that median is robust and less sensitive against outliers associated to resonances. We have restricted, for simplicity, the domain of c to $[0.5, 1.5]$ and taken equidistant values of c as

$$c_1 = 0.5,\ c_2 = 0.75,\ c_3 = 1, c_4 = 1.25, c_5 = 1.5$$

and found the approximations $K(c_1) = 0.990825$, $K(c_2) = 0.988733$, $K(c_3) = 0.994847$, $K(c_4) = 0.995746$, $K(c_5) = 0.995214$ and get $K = 0.994847 \simeq 1$ showing that the solution $x(t)$ is chaotic.

Analogous computations are performed for $y(t)$ and $z(t)$. The p_n versus q_n plot, the graphs of $M_n(c)$ and $K_n(c)$ are shown in the Fig. 6d-f with $c = 1$ for $y(t)$. Similar plots for $z(t)$ are shown in Fig. 6g-i. The signals represented by $y(t)$ produce the approximations $K(c_1) = 0.997263$, $K(c_2) = 0.991210$, $K(c_3) = 0.995412$, $K(c_4) = 0.997774$, $K(c_5) = 0.978373$ so that $K = 0.995412 \simeq 1$. The signals represented by $z(t)$ produce the approximations $K(c_1) = 0.998093$,

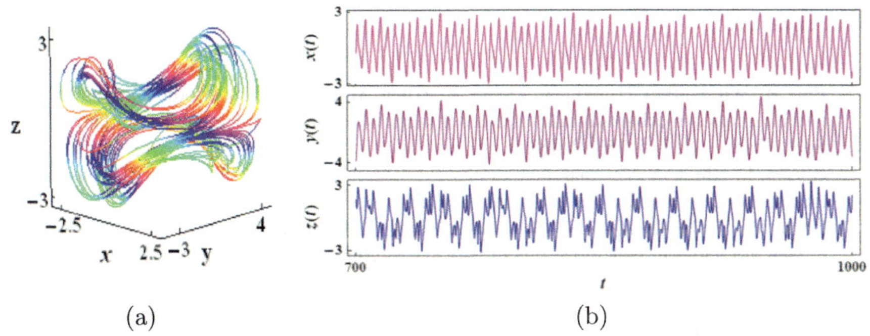

Fig. 7 **a** The solution orbit of the FWBO system (2) and **b** the corresponding $x(t)$, $y(t)$, $z(t)$. Here, the values of the parameters are given by (3) and (4) in all the subfigures

$K(c_2) = 0.995041,\ K(c_3) = 0.991666,\ K(c_4) = 0.995254,\ K(c_5) = 0.994567$ so that $K = 0.995041 \simeq 1$. Thus, we observe that the solutions $y(t)$ and $z(t)$ produce chaotic signals which can be used in the chaotic transmission. Therefore, the 0-1 test confirms the chaotic nature of the output signals generated by the FWBO where the parameter values of the system are given by (3) and (4). The solution orbit of this system is displayed in the Fig. 7a and the corresponding $x(t)$, $y(t)$, $z(t)$ are presented in Fig. 7b.

5 Conclusion

A model of Wien bridge oscillator driven by external sinusoidal stimulus have been studied and regimes of the parameters of these external forces have been identified over which the signal so generated becomes very much unpredictable enhancing the security in chaotic communication system. A theoretical analysis of the system dynamics has been performed using phase portraits, bifurcation diagrams. It is found that the external periodic stimuli are responsible to generate chaotic carrier signals. The signals generated by the circuit have been tested by 0-1 test confirming its chaotic nature. The behaviour of the output generated under other kind of external stimulus such as digital, trapezoidal, triangular, sawtooth signals remains an open problem and will be studied elsewhere.

References

1. Bao, B., Wu, P., Bao, H., Wu, H., Zhang, X., Chen, M.: Symmetric periodic bursting behavior and bifurcation mechanism in a third-order memristive diode bridge-based oscillator. Chaos, Solit. Fract. **109**, 146–153 (2018)
2. Burkin, I., Khien, N.N.: Analytical-numerical methods of finding hidden oscillations in multidimensional dynamical systems. Diff. Equ. **50**(13), 1695–1717 (2014)
3. Chen, M., Li, M., Yu, Q., Bao, B., Xu, Q., Wang, J.: Dynamics of self-excited attractors and hidden attractors in generalized memristor-based chua's circuit. Nonl. Dyn. **81**(1), 215–226 (2015)
4. Chua, L.: Memristor-the missing circuit element. IEEE Transactions on circuit theory **18**(5), 507–519 (1971), publisher: IEEE
5. Chua, L.O., Kang, S.M.: Memristive devices and systems. Proceedings of the IEEE **64**(2), 209–223 (1976), publisher: IEEE
6. Çiçek, S., Kocamaz, U.E., Uyaroğlu, Y.: Secure chaotic communication with jerk chaotic system using sliding mode control method and its real circuit implementation. Iranian J. Sci. Technol. Trans. Elect. Eng. **43**(3), 687–698 (2019)
7. Dawes, J., Freeland, M.: The '0–1 test for chaos' and strange nonchaotic attractors. preprint (2008)
8. Gottwald, G.A., Melbourne, I.: On the implementation of the 0–1 test for chaos. SIAM J. Appl. Dyn. Syst. **8**(1), 129–145 (2009)
9. Kantz, H., Schreiber, T.: Nonlinear time series analysis, vol. 7. Cambridge university press (2004)
10. Kiseleva, M., Kudryashova, E.V., Kuznetsov, N.V., Kuznetsova, O.A., Leonov, G.A., Yuldashev, M.V., Yuldashev, R.V.: Hidden and self-excited attractors in chua circuit: synchronization and spice simulation. Int. J. Paral. Em. Distr. Syst. **33**(5), 513–523 (2018)
11. Leonov, G., Kuznetsov, N.: Localization of hidden oscillations in dynamical systems (plenary lecture). In: 4th International Scientific Conference on Physics and Control (2009)
12. Lv, M., Ma, J.: Multiple modes of electrical activities in a new neuron model under electromagnetic radiation. Neurocomputing **205**, 375–381 (2016)
13. Martinsen-Burrell, N., Julien, K., Petersen, M.R., Weiss, J.B.: Merger and alignment in a reduced model for three-dimensional quasigeostrophic ellipsoidal vortices. Phys. Fluids **18**(5), 057101 (2006)
14. Oppenheim, A.V., Wornell, G.W., Isabelle, S.H., Cuomo, K.M.: Signal processing in the context of chaotic signals. In: icassp. vol. 4, pp. 117–120 (1992)
15. Pham, V.T., Volos, C., Jafari, S., Vaidyanathan, S., Kapitaniak, T., Wang, X.: A chaotic system with different families of hidden attractors. Int. J. Bifurc. Chaos **26**(08), 1650139 (2016)
16. Rabbani, P., Dehghani, R., Shahpari, N.: A multilevel memristor-cmos memory cell as a reram. Microelect. J. **46**(12), 1283–1290 (2015)
17. Rajagopal, K., Li, C., Nazarimehr, F., Karthikeyan, A., Duraisamy, P., Jafari, S.: Chaotic dynamics of modified wien bridge oscillator with fractional order memristor. Radioengineering **28**(1), 165–174 (2019)
18. Skokos, C., Antonopoulos, C., Bountis, T., Vrahatis, M.: Detecting order and chaos in hamiltonian systems by the sali method. J. Phy. A: Mathem. General **37**(24), 6269 (2004)
19. Stankevich, N.V., Kuznetsov, N.V., Leonov, G.A., Chua, L.O.: Scenario of the birth of hidden attractors in the chua circuit. Int. J. Bifurc. Chaos **27**(12), 1730038 (2017)
20. Strukov, D.B., Snider, G.S., Stewart, D.R., Williams, R.S.: The missing memristor found. Nature **453**(7191), 80–83 (2008)
21. Tosin, M., Issa, M.V., Matos, D., Do Nascimento, A., Cunha Jr, A.: Employing 0-1 test for chaos to characterize the chaotic dynamics of a generalized gauss iterated map. In: XIV Conferência Brasileira de Dinâmica, Controle e Aplicações (DINCON 2019) (2019)
22. Wu, H., Bao, B., Liu, Z., Xu, Q., Jiang, P.: Chaotic and periodic bursting phenomena in a memristive wien-bridge oscillator. Nonl. Dyn. **83**(1), 893–903 (2016)

23. Xia, X., Zeng, Y., Li, Z.: Coexisting multiscroll hyperchaotic attractors generated from a novel memristive jerk system. Pramana **91**(6), 1–14 (2018)
24. Xu, Q., Zhang, Q., Jiang, T., Bao, B., Chen, M.: Chaos in a second-order non-autonomous wien-bridge oscillator without extra nonlinearity. Circuit world (2018)
25. Ye, X., Mou, J., Luo, C., Wang, Z.: Dynamics analysis of wien-bridge hyperchaotic memristive circuit system. Nonl. Dyn. **92**(3), 923–933 (2018)
26. Zarei, A.: Complex dynamics in a 5-d hyper-chaotic attractor with four-wing, one equilibrium and multiple chaotic attractors. Nonl. dyn. **81**(1), 585–605 (2015)
27. Zhao, Q., Wang, C., Zhang, X.: A universal emulator for memristor, memcapacitor, and meminductor and its chaotic circuit. Chaos: An Interdisciplinary Journal of Nonlinear Science **29**(1), 013141 (2019)

The Electrodynamic Origin of the Wave-Particle Duality

Álvaro García López

Abstract A derivation of pilot waves from electrodynamic self-interactions is presented. For this purpose, we abandon the current paradigm that describes electrodynamic bodies as point masses. Beginning with the Liénard-Wiechert potentials, and assuming that inertia has an electromagnetic origin, the equation of motion of a nonlinear time-delayed oscillator is obtained. We analyze the response of the uniform motion of the electromagnetic charged extended particle to small perturbations, showing that very violent oscillations are unleashed as a result. The frequency of these oscillations is intimately related to the *zitterbewegung* frequency appearing in Dirac's relativistic wave equation. Finally, we compute the self-energy of the particle. Apart from the rest and the kinetic energy, we uncover a new contribution presenting the same fundamental physical constants that appear in the quantum potential.

Keywords Nonlinear dynamics · Chaos · Delay differential equations · Electrodynamics · Retarded potentials · Pilot waves · Quantum mechanics

1 Introduction

Recently developed models of silicon droplets have shown deep connections between quantum mechanical systems and classic hydrodynamics, allowing nonlinear dynamicists to grasp how the complex motion of a quantum particle can be [1, 2]. More specifically, these macroscopic systems describe the unpredictable dynamics of walking droplets as the result of a feedback interaction between the bouncing particle and the waves that it produces when it strikes the surface of a fluctuating medium underneath, close to and even beyond the Faraday threshold. Fortunately, and contrary to quantum mechanical models, these hydrodynamic analogs are investigated in terms of understandable and firmly established principles of chaos theory and nonlinear dynamical systems.

Á. García López (✉)
Universidad Rey Juan Carlos, Móstoles s/n 28933, Madrid, Spain
e-mail: alvaro.lopez@urjc.es

S. Banerjee and A. Saha (eds.), *Nonlinear Dynamics and Applications*,
Springer Proceedings in Complexity,
https://doi.org/10.1007/978-3-030-99792-2_88

Although the pilot wave dynamics of silicon droplets has been proposed as a candidate to comparatively investigate quantum systems, a specific homologous mechanism that can give rise to the wave-particle duality in the microscopic realm has not been rigorously developed until very recently [3]. In the present paper we provide strong evidence suggesting that the wave-particle duality has its basis in the theory of classical electromagnetism. For this purpose, we show that extended charged bodies can self-interact when they are accelerated. A certain region of the particle can emit radiation, which later on affects a different region of the same particle. This phenomenon introduces a time-delay in the self-force of the extended body.

Consequently, the description of the dynamics of charged bodies must be posed by means of retarded differential equations. As it is well-known, the solutions to these differential equations frequently present limit cycle behaviour as a consequence of the Andronov-Hopf bifurcation [4, 5]. The feedback interaction of radiative and Coulombian fields among different charged parts of the particle can trigger a fast oscillation, destabilizig its uniform motion. These fields produce dissipation and antidamping as a consequence of radiation reaction. In the thermodynamic context of open systems, such a periodic motion has been recently referred as a self-oscillation [6]. In this manner, we show that the wave-particle duality is just an immediate consequence of the self-oscillation of extended electrodynamic bodies, which can be regarded as dissipative structures.

2 Electrodynamics of an Extended Body

We model the electrodynamics of an extended charged body by using the Lagrangian density of Maxwell's theory of electrodynamics with sources. This density is written as

$$\mathcal{L} = -\frac{1}{4\mu_0} F_{\mu\nu} F^{\mu\nu} - A_\mu J^\mu, \tag{1}$$

where J^μ denotes the four-current density representing the sources, and $F^{\mu\nu}$ is the Faraday tensor. Then, Maxwell's equations can be derived in covariant form from the previous action by differentiation, yielding

$$\partial_\mu F^{\mu\nu} = \mu_0 J^\nu. \tag{2}$$

To describe the dynamics of the source of charge, we express the four-density as $J^\mu = \rho_0 U^\mu$, where the density of charge ρ_0 in the proper frame and the four-velocity U^μ have been introduced. The charge density ρ in some inertial reference frame can be related to the density in the proper frame by using the Lorentz factor γ, through the relation $\rho = \gamma \rho_0$. Then, the four-current is simply written as $J^\mu = (\rho c, J)$, with $J = \rho v_s$ the euclidean current density. Here we are considering a rigid distribution of charge. This assumption allows us to write the charge density at time t as $\rho(x, t) = \rho(x - x_s(t))$, where the vector $x_s(t)$ represents the position of the particle's centre of mass at time t. Under these assumptions, Maxwell's equations can be solved in terms of retarded potentials, which allow to generally derive the fields by means of Jefimenko's equations. Using the fact that $F_{\mu\nu} = \partial_\mu A_\nu - \partial_\nu A_\mu$, the four-potential in the Lorenz gauge can then be written as

$$A^\mu(x,t) = \frac{1}{4\pi\epsilon_0 c^2}\int \frac{J^\mu(x,t_r)}{|x-x'|}d^3x', \tag{3}$$

where we have introduced the retarded time $t_r = t - |x - x'|/c$. The analysis presented ahead considers a very simple rigid charged distribution comprised of two point particles at a fixed distance. Therefore, in order to compute the self-force, we can simply use the *Liénard-Wiechert potential*. This potential is the solution to a charged point particle, whose charge density can be represented by means of the Dirac delta distribution, in the form $\rho(x,t) = q\delta(x - x_s(t))$. In the Lorenz gauge we have the solution

$$A^\mu(x,t) = \frac{q}{4\pi\epsilon_0 c}\left(\frac{v^\mu}{(1 - n_s\cdot\beta_s)|x - x_s|}\right)_{t=t_r}, \tag{4}$$

where the relative speed of the source $\beta_s(t) = v_s(t)/c$ has been defined, together with the time-like four-vector $v_s^\mu = (c, v_s(t))$. Finally, the unit vector $n_s(x,t) = (x - x_s(t))/|x - x_s(t)|$ has also been introduced. It points to the spatial point x, where we want to compute the value of the fields. Its application point is located at the position $x_s(t)$, where the charge was placed at the retarded time $t = t_r$.

2.1 A Model of an Electron

In the preset work we consider the charge density $\rho(x,t) = -(e/2)\delta(x - x_s(t))\delta(z)$ $(\delta(y + d/2) + \delta(y - d/2))$. It is the most simple model of an electron, represented as an "extended" electrodynamic body, as shown in Fig. 1. This charge density represents a body formed by two points with charge $-e/2$ placed at a fixed distance d in the y-axis, which moves along the x-axis. We restrict the motion to transversal displacements to simplify our study, since the non-conservative character of electrodynamics with sources makes the computations very entangled, due to the fact that the Liénard-Wiechert potential is retarded in time. By restricting ourselves to a one-dimensional translational motion, we avoid the more complicated three-dimensional problem, including self-torques. This fudamental model has been designed in previous works to underpin the use of the *Abraham-Lorentz* force and also to study the contribution of electromagnetic mass to inertia [7]. We use this elementary model hereafter, which suffices to explain the physical mechanism leading to the wave-particle duality.

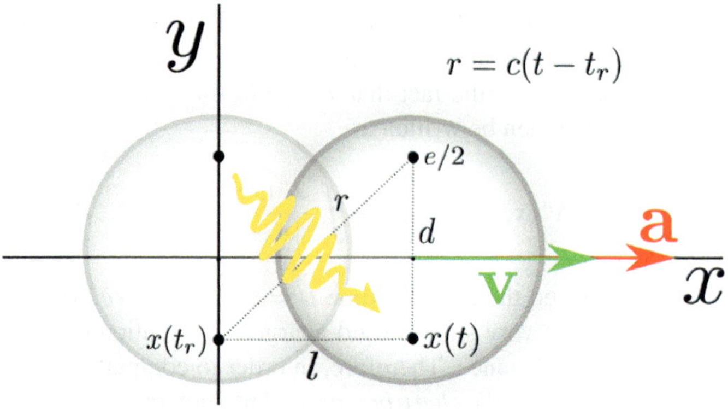

Fig. 1 An electron is shown at the retarded time t_r and at the present time t. It consists of two point charges placed along the y-axis at a constant distance d. From t_r to t, the particle accelerates and advances a distance l along the x-axis. A field perturbation is shown emerging from the upper point at the retarded time (yellow photon). Later on, this perturbation exerts a force on the second point at the lower side of the body. In this manner, an extended corpuscle can feel itself in the past. The speed v and the acceleration a of the particle are represented in green and red, for clearness

2.2 *The Self-Interaction*

As shown in Fig. 1, an extended electrodynamic body can interact with itself. This kind of interaction is frequently called a *self-interaction* [7]. The upper charged point exerts an electromagnetic force on the lower point a short time later. The self-force appears because electromagnetic waves can travel between the two points of the electrodynamic body. Apart from a force of inertia, a term of damping and a restoring elastic force, it can induce the self-exicted motion of the particle due to a radiation reaction force. We now compute the electric field produced by the upper point at the lower point of the body. If we utilize the Liénard-Wiechert potential of a point particle, we obtain

$$\boldsymbol{E} = \frac{q}{8\pi\epsilon_0}\frac{r}{(\boldsymbol{r}\cdot\boldsymbol{u})^3}\left(\boldsymbol{u}(1-\beta^2) + \frac{1}{c^2}\boldsymbol{r}\times(\boldsymbol{u}\times\boldsymbol{a})\right), \tag{5}$$

where the relative position between the two points at different times is $\boldsymbol{r}(t_r)$, the normalized velocity is $\boldsymbol{\beta}(t_r) = \boldsymbol{v}(t_r)/c$, the acceleration $\boldsymbol{a}(t_r)$ and the vector $\boldsymbol{u} = \hat{\boldsymbol{r}} - \boldsymbol{\beta}$ has been introduced. We highlight that these kinematic variables depend on the retarded time $t_r = t - r/c$. This time-delay appears because electromagnetic field perturbations travel with limited velocity in spacetime, according to the principle of *causality*. This limitation puts a constraint $r = c(t - t_r)$ on the self-interaction, assigning a specific event in the past light cone from which the signals coming from one point of the particle can affect the remaining point.

According to Fig. 1, we can write the position as $\boldsymbol{r} = l\hat{\boldsymbol{x}} + d\hat{\boldsymbol{y}}$, the velocity as $\boldsymbol{\beta} = v/c\hat{\boldsymbol{x}}$ and the acceleration vector as $\boldsymbol{a} = a\hat{\boldsymbol{x}}$. These relations allow to compute the vector $\boldsymbol{u}$ as $\boldsymbol{u} = (l - r\beta)/r\hat{\boldsymbol{x}} + d/r\hat{\boldsymbol{y}}$. Then, making use of the identity $r^2 = (x(t) - x(t_r))^2 + d^2$, the following inner product $\boldsymbol{r} \cdot \boldsymbol{u} = r - l\beta$ results. Regarding the radiative component of the field, we need to expand the double cross-product in the form $\boldsymbol{r} \times (r\boldsymbol{u} \times \boldsymbol{a}) = -d^2 a\hat{\boldsymbol{x}} + dal\hat{\boldsymbol{y}}$. The total self-force on the center of mass of the particle can be written as $\boldsymbol{F}_{\text{self}} = -e\boldsymbol{E}$, where the symmetry of the arrangement has been taken into consideration. Because of the rigidity of the charge density, we recall that the the magnetic attractive forces and the electric repulsive forces all cancel each other along the y-axis. The resulting force that the particle exerts on itself is

$$\boldsymbol{F}_{\text{self}} = \frac{e^2}{8\pi\epsilon_0} \frac{1}{(r - l\beta)^3} \left((l - r\beta)(1 - \beta^2) - \frac{d^2}{c^2} a \right) \hat{\boldsymbol{x}}. \tag{6}$$

2.3 *Time-Delayed Equation of Motion*

Following the tradition, we could now invoke Newton's second law of mechanics. For a non-relativistic particle it is written as $\boldsymbol{F}_{\text{self}} = m\boldsymbol{a}$, with m the electron's bare mechanical mass. However, it has been shown in recent works that the self-force can be expanded by using a Taylor series of the time-delay r/c [3]. Among the infinite linear and nonlinear terms that contribute to the self-force, the most well-known are the Lorentz-Abraham force, which consists of a linear term proportional to the jerk ($\dot{a}$) of the particle, and the term of inertia, which is proportional to the acceleration. This term dominates over all other terms in the limit of very small d/c, what allows to approximate the self-force as $\boldsymbol{F}_{\text{self}} = -m_e\boldsymbol{a}$ for non-relativistic velocities. To this end, we simply define the electromagnetic mass as $m_e c^2 = e^2/16\pi\epsilon_0 d$, where Einstein's mass-energy relation has been obviously used.

In the present work we are assuming that the rest mass of the electron comes entirely from its electrostatic energy, so that its bare mass can be made equal to zero. For if mass is not a fundamental property of particles, but just energy, all the mass in our model must come from the electrostatic energy of the two point charges. Then, if we use Sommerfeld's relation for the fine structure constant, the rest mass of the electron is

$$m_e = \frac{\hbar\alpha}{4dc}. \tag{7}$$

Using this relation, we can approximate a electron radius of $r_e = d/2 = 0.35$ fm. Naturally, this value is closely related to the the electron's classical radius.

Therefore, this approach does not artificially introduce bare mechanical inertia in the theory of classical electromagnetism. Instead, we use the *principle of D'Alembert* that, in the approximation of macroscopic objects, leads to Newton's second law. Thus classical mechanics should be considered an emergent theory resulting from averaging magnitudes over large numbers of electrodynamic bodies. As a corollary, we predict that the gravitational force between two particles have an electrodynamic origin as well. Consequently, we propose to replace Newton's second law by the static problem

$$\boldsymbol{F}_{\text{ext}} + \boldsymbol{F}_{\text{self}} = 0. \tag{8}$$

As long as we can approximate $\boldsymbol{F}_{\text{self}} = -m_e \boldsymbol{a}$ in the macroscopic limit, we see that Newton's second law naturally arises from Maxwell's dynamical theory of the electromagnetic field. The force of inertia reveals in this way as an electromagnetic force of *self-induction*, coming from the interior of the body as a consequence of Faraday's law. This statement opposes to Mach's principle, which tries to justify the origin of inertial forces on external distant masses.

To study the "free" particle, we can settle the external forces to zero, thus we have the simple law of motion $\boldsymbol{F}_{\text{self}} = 0$. Its solution describes the geodesic motion of the electrodynamic body in the same way as in the theory of general relativity, for example. The differential equation of motion is

$$\frac{d^2}{c^2} a(t_r) + \frac{r}{c}\left(1 - \frac{v^2(t_r)}{c^2}\right) v(t_r) + \left(1 - \frac{v^2(t_r)}{c^2}\right)(x(t_r) - x(t)) = 0. \tag{9}$$

The difficulty with this state-dependent *delayed differential equation* [8] is that most kinematic variables are specified at the retarded time $t_r = t - r/c$. If we translate them to the present time $t \to t + r/c$, we obtain

$$a(t) + \frac{r}{d}\frac{c}{d}\left(1 - \frac{v^2(t)}{c^2}\right) v(t) + \left(\frac{c}{d}\right)^2 \left(1 - \frac{v^2(t)}{c^2}\right)\left(x(t) - x\left(t + \frac{r}{c}\right)\right) = 0. \tag{10}$$

This differential equation clearly evokes a nonlinear oscillator [6]. We can identify a term of Newtonian inertia and a typical linear oscillating term representing an elastic restoring force. But we can see two more nonlinear contributions, as well. Firstly, the contribution appearing in the second term acts as a nonlinear damping force, producing the system's dissipation. Secondly, the advanced potential produces a non-conservative force of antidamping.

The frequency of oscillation can be approximated as $\omega_0 = c/d$, what yields a value of the period $T_0 = 4\pi r_e/c = 1.18 \times 10^{-22}$s, if we use the electron's classical radius. Thus the electron oscillates very fast describing a deterministic motion. However, this motion resembles a stochastic motion at large enough time scales. This jittery dynamics and the specific value of its period are very familiar to quantum mechanical theorists. They are closely related to the trembling motion appearing in Dirac's wave equation equation for relativistic particles, commonly known as *zitterbewegung*.

Importantly, we notice that the time-delay and the damping term involve an *arrow of time*. Irreversibility is inherent to non-conservative dynamical systems presenting limit cycle behavior. It is also conventional in time-delayed systems, whose trajectories are not specified by some initial conditions, but rely on the complete knowledge of functions describing part of their previous history. Of course, this non-conservative dynamics only appears when we try to describe the motion of the particle solely, without reference to the dynamical fields.

In summary, fundamental particles can be considered as open dissipative structures. They are locally active and operate far from equilibrium by taking and releasing electromagnetic energy to their surroundings. The dissipative nature of classical electrodynamics with sources becomes manifest by the fact that a Lagrangian density for the motion of the particle cannot be written by using the traditional minimal coupling, as it is frequently done in quantum particle physics.

3 Stability Analysis

Now we prove that transversal motion at constant speed is not stable. Consequently, self-oscillatory motion is the only possibility, irrespective of the periodicity of this nonlinear oscillation. For this purpose, we consider the differential Eq. (10) and pose it in the phase space canonical variables. We obtain

$$\begin{aligned}\dot{x} &= v,\\ \dot{v} &= -\frac{c}{d}\frac{r}{d}\left(1-\frac{v^2}{c^2}\right)v-\left(\frac{c}{d}\right)^2\left(1-\frac{v^2}{c^2}\right)(x-x_\tau)\,.\end{aligned} \tag{11}$$

The variable x_τ has been introduced. It represents the electron's position evaluated at time $t+\tau$, where we recall that $\tau = r/c$. Consider that motion at constant speed βc is feasible. Since $x(t) = vt$, we also have $x(t+r/c) = vt + vr/c$, what yields $x - x_\tau = -vr/c$. If we now substitute in Eq. (11) we get

$$\begin{aligned}\dot{x} &= v,\\ \dot{v} &= -\frac{c}{d}\frac{r}{d}\left(1-\frac{v^2}{c^2}\right)v+\frac{c}{d}\frac{r}{d}\left(1-\frac{v^2}{c^2}\right)v = 0.\end{aligned} \tag{12}$$

Therefore, every uniform motion is an invariant solution of our delayed dynamical system. We now prove that these solutions are unstable as well. For this purpose, we compute the variational equations

$$\begin{aligned}\delta\dot{x} &= \delta v,\\ \delta\dot{v} &= -\frac{c}{d}\frac{\delta r}{d}\left(1-\frac{v^2}{c^2}\right)v-\frac{c}{d}\frac{r}{d}\left(1-\frac{v^2}{c^2}\right)\delta v+\frac{c}{d}\frac{r}{d}\frac{2v^2}{c^2}\delta v-\\ &\quad -\frac{c}{d}\frac{r}{d}\frac{2v^2}{c^2}\delta v-\left(\frac{c}{d}\right)^2\left(1-\frac{v^2}{c^2}\right)(\delta x-\delta x_\tau)\,.\end{aligned} \tag{13}$$

We follow by computing δr when $\dot{v} = 0$, with $v = \beta c$. For this purpose, we use the relation $r^2 = (x(t) - x(t_r))^2 + d^2$ and Eq. (9). If we combine these two equations we can derive polynomial of degree two in r. If we also introduce the Lorentz factor $\gamma = (1 - \beta^2)^{-1/2}$, the solution to this polynomial yields

$$r = \gamma d\sqrt{1 + \gamma^6 \dot{\beta}^2 \left(\frac{d}{c}\right)^2} + \gamma^4 c\beta\dot{\beta}\left(\frac{d}{c}\right)^2. \tag{14}$$

Recall, the speed and the acceleration appearing in Eq. 14 are evaluated at time t_r. Interestingly, the *time-delay* becomes a function of the kinematic variables. When the particle increases its speed, the self-force is originated at an earlier past time, because the light cone of the particle evolves with its motion. Evaluating the Eq. (14) at time t, the computation of the variations of r can be done immediately, yielding

$$\delta r(t) = \gamma^4 \beta \left(\frac{d}{c}\right)^2 \delta\dot{v}(t) + d\delta\gamma(t). \tag{15}$$

Grouping terms and using the equation $r = \gamma d$ for $\dot{v} = 0$, we obtain the variational equation

$$\begin{aligned} \delta\dot{x} &= \delta v, \\ \delta\dot{v}\gamma^2 &= -\frac{c}{d}\gamma\delta v - \left(\frac{c}{d}\right)^2 \left(1 - \beta^2\right)\left(\delta x - \delta x_\tau\right). \end{aligned} \tag{16}$$

If we consider exponential solutions $\delta x = Ae^{\lambda t}$, the characteristic equation of the dynamical system (16) can be found. It reads

$$\mu^2 + \mu + (1 - \beta^2)(1 - e^{\mu}) = 0, \tag{17}$$

where the variable $\mu = \lambda\gamma d/c$ has been defined. With the exception of one eigenvalue, the solutions to this equation always present a positive real part, what guarantees their unstability for all values of β. As depicted in Fig. 2, this analytical statement is confirmed by numerical simulation. An infinite spectrum of *eigenvalues* of the frequency is obtained, with the following quantization rule

$$\omega_n = \eta_n \frac{c}{\gamma d}. \tag{18}$$

The appearence of the γ factor is due to the Lorentz boost, which produces a time dilation. The form factor η_n is characteristic of the geometry of the body. In summary, we have proved mathematically the existence of oscillatory motion in our dynamical system for any value of the relative velocity β. Importantly, we would like to mention

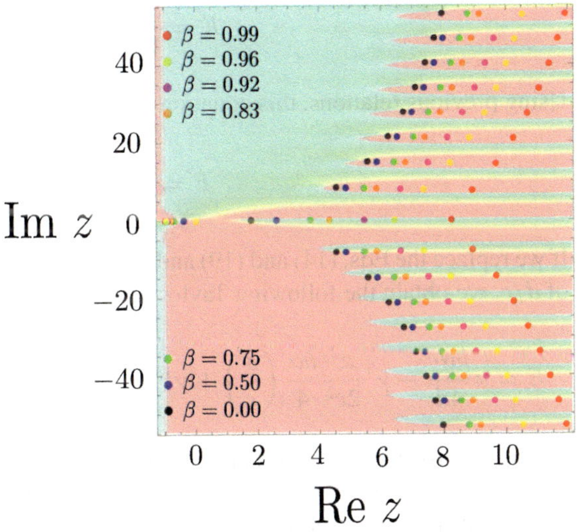

Fig. 2 The complex function $f(z) = z^2 + z + (1 - \beta^2)(1 - e^z)$ and its roots are computed for seven different values of the velocity, with the aid of Newton-Raphson method. Since $z = \gamma d/c$, we get the spectrum of eigenfrequencies of the self-oscillation, which can be approximated as $\omega_n \propto nc/\gamma d$. The background corresponds to a representation of the function for $\beta = 0$, using a domain coloring technique

that this instability depends on the shape of the particle. If the geometry of the electrodynamic body is switched from oblate to prolate, a Hopf bifurcation occurs [5].

4 The Quantum Potential

We conclude the present paper by deriving the relativistic kinetic energy and the quantum potential from the Liénard-Wiechert potential. The insertion of Eq. (14) into the equation $r^2 = l^2 + d^2$ allows to exactly obtain l as a function of the relative speed β and the relative acceleration $\dot{\beta}$. The result yields the equation

$$l = \sqrt{\gamma^2 c^2 \beta^2 \left(\frac{d}{c}\right)^2 + \gamma^8 c^2 \dot{\beta}^2 (1 + \beta^2) \left(\frac{d}{c}\right)^4 + 2c^2 \gamma^5 \beta \dot{\beta} \left(\frac{d}{c}\right)^3 \sqrt{1 + \gamma^6 \dot{\beta}^2 \left(\frac{d}{c}\right)^2}}. \tag{19}$$

Now we denote the self-energy of the particle as E, which we define as the energy of non-dissipative origin needed to build the charge and achieve its particular dynamical state of motion. As it is well-known, the magnetic fields perform no work, and because dissipated energy is being disregarded, we focus on the curl-free part of the electric field. Bearing in mind these considerations, the potential energy E of the electrodynamic particle can be defined by means of the Liénard-Wiechert potential as $E = -ecA^0/2$, what yields

$$E = \frac{e^2}{16\pi\epsilon_0}\frac{1}{\boldsymbol{r}\cdot\boldsymbol{u}}. \tag{20}$$

Using previous relations, this equation is written as

$$E = \frac{\hbar\alpha c}{4(r - l\beta)}. \tag{21}$$

If we replace the Eqs. (14) and (19) and expand the resulting self-potential in powers of d/c, we obtain the following Taylor series expansion

$$E = \gamma\frac{\hbar\alpha c}{4d} - \gamma^7\frac{a^2}{2c^2}\frac{\hbar\alpha}{4}\left(\frac{d}{c}\right) + \gamma^{13}\frac{3a^4}{8c^4}\frac{\hbar\alpha}{4}\left(\frac{d}{c}\right)^3 - \gamma^{19}\frac{5a^6}{16c^6}\frac{\hbar\alpha}{4}\left(\frac{d}{c}\right)^5 + \ldots \tag{22}$$

Again, we assume the idea that mass and inertia have a total electromagnetic origin. Therefore, the size of the particle can be written using Eq. (7) as $d = \hbar\alpha/4m_e c$. Noticeably, mass m_e is proportional to $\hbar$. This relation implies that any kind of energy or canonical momentum can be written as proportional to Planck's constant. Furthermore, if the speed of the particle is related to the group velocity of the pilot wave, then it seems reasonable to consider that the relation $p = \hbar k$ holds. This introduces De Broglie's relation connecting the velocity of the particle and the wavelength of the electromagnetic pilot wave. Substitution of Eq. (7) in Eq. (22) yields

$$E = \gamma m_e c^2 + Q, \tag{23}$$

where we have introduced the potential

$$Q = \frac{\hbar^2}{2m_e}\frac{\alpha^2}{8d^2}\gamma\sum_{n=1}^{\infty}\frac{(-1)^n(2n-1)!!}{2^n n!}\gamma^{6n}\frac{a^{2n}}{c^{2n}}\left(\frac{d}{c}\right)^{2n}. \tag{24}$$

We detect two well differentiated terms in Eq. (23). The former corresponds to the famous relativistic equation representing the energy of a particle in the theory of special relativity. It contains the rest energy of the particle together with its kinetic energy. Importantly, we stress that these two magnitudes are not fundamental and correspond to plain electrodynamic energy. Additionaly, apart from the rest mass and the kinetic energy, the new potential energy Q has appeared. By using a quadrature related to the coefficients in Eq. (24), the series can be computed and one last integration yields [3] the expression

$$Q = -\frac{\hbar^2}{2m_e}\frac{\alpha^2}{8d^2}\gamma\left(1 - 1/\sqrt{1 + \gamma^6\dot{\beta}^2\,(d/c)^2}\right). \tag{25}$$

This new contribution vanishes for uniform motion. The Lorentz factor prevents the particle from traveling at velocities equal or above the speed of light. The constant term $\hbar^2/2m_e$ preceding this potential is identical to the quantum potential appearing in Bohmian mechanics [9], which can be written as $Q = -(\hbar^2/2m_e)\nabla^2 R/R$. This potential can not be derived from a Hamiltonian including an external source of potential, and involves a self-organising process produced by the internal electromagnetic field [3]. The quantum potential entails an interpretation of classical electrodynamic phenomena in terms of an emergent hydrodynamic theory [10, 11], which overcomes the representation of complicated internal self-interactions by using the concept of quantum pressure.

In theory, once the dynamical system approaches its asymptotic limit set, a functional relation between the position of the particle in the configuration space and its acceleration can be provided. This relation can be replaced in $Q(x, t)$, allowing to compute the function $R(x, t)$. Then, we can pose the Hamilton-Jacobi equation. If the particle is also subjected to the influence of a newtonian external potential $V(x, t)$ and its average velocity is not relativistic, such an equation reads

$$\frac{\partial S}{\partial t} + \frac{1}{2m_e}(\nabla S)^2 + Q + V = 0. \tag{26}$$

After solving the previous equations, and making use of the information about the particle's trajectory, the wave function can be built using the polar expression $\psi(x, t) = R(x, t) \exp(iS(x, t)/\hbar)$. Importantly, we deduce from these relations that the wave function is not an ordinary probabilistic entity, but a *real* physical field [9] related to external and internal electrodynamic fields. These fields describe the pilot wave of the particle, which can produce well-known physical phenomena, as for example interference and diffraction.

A conservative approximation of the quantum potential has been derived in recent works, connecting it to typical potentials that break fundamental symmetries [3]. In particular, it has been claimed that the quantum potential can produce the symmetry breaking of the Lorentz group. Importantly, we recall that *symmetry breaking* is an essential feature in the study of nonlinear dynamics [12].

5 Conclusions and Discussion

The dynamics of an extended electrodynamic body has major similarities with the motion of silicon droplets found in many experiments during the past decade. In the present model, quantum waves have their origin in the self-oscillation of the electron, produced by the feedback interaction of the particle with its own electromagnetic field. This self-interaction enforces a pilot wave travelling with the corpuscle. Therefore, the wave-particle duality is immediately explained, since quantum waves appear naturally as perturbations of the dynamic electromagnetic fields. These self-interactions and their concomitant forces of recoil, perhaps together with external *zero-point fluctuations* [13], can prevent the collapse of the hydrogen atom [14].

The fact that electromagnetic mass allows to derive the exact relativistic kinetic energy analytically from the electrodynamic potentials strongly suggests that inertial mass is not a fundamental concept in physics, but an emergent one. This conclusion points towards the fact that gravitational mass is also a redundant idea in fundamental physics, and that the force of gravity has an electromagnetic origin as well. In this perspective, an electrodynamic theory of the gravitational force would also explain in simple terms the principle of equivalence. The equality of gravitational mass and inertial mass would then be explained because of their common origin in the electromagnetic force. Importantly, our finding that Newton's second law can be deduced from classical electrodynamics shows that classical mechanics is an emergent theory based on classical electromagnetism. Just in the same way as thermodynamics laws result from averaging mechanical properties over large number of ensembles. Consequently, equations in which the concept of mass appears as an elementary parameter, as it occurs with the Schödinger or the Dirac equations, should not be considered fundamental in physics.

Finally, the model presented in this work is not very realistic because it considers a rigid charge density, which is structurally unstable. More reasonably, it is expected that fundamental particles can arise from self-confined fields as a consequence of the rotation of the fields and their electromagnetic stress, stabilizing the electron. This idea suggests that particles are *electromagnetic solitons* [15, 16], perhaps arising in the context of the Einstein-Maxwell equations. Then, *zitterbewegung* could be regarded from the point of view of a field theory as a quasi-breather solution.

Theories including more complex lagrangian densities with other general relativistic invariants or even nonlinear electrodynamic fields [17] might also allow to understand fundamental particles as compact field configurations. Then, the electric charge could be explained as the *topological charge* of the solitary wave, and not as a fundamental parameter. Anyway, no theory of elementary particles can be considered a fundamental theory as long as it does not conform to the principle of general covariance. Only this principle allows adopting any reference frame to describe the motion of dynamical fields, dispensing with the metaphysical concept of absolute spacetime.

References

1. Protière, S., Boudaoud, A., Couder, Y.: Particle-wave association on a fluid interface. J. Fluid Mech. **544**, 85–108 (2006)
2. Fort, E., Eddi, A., Boudaoud, A., Moukhtar, J., Couder, Y.: Path-memory induced quantization of classical orbits. Proc. Natl. Acad. Sci. **107**, 17515–17520 (2010)
3. López, A.G.: On an electrodynamic origin of quantum fluctuations. Nonl. Dyn. **102**, 621–634 (2020)
4. Ghaffari, A., Tomizuka, M., Soltan, R.A.: The stability of limit cycles in nonlinear systems. Nonl. Dyn. **56**, 269–275 (2009)
5. López, A.G.: Stability analyisis of the uniform motion of electrodynamic bodies. Phys. Scr. **96**, 015506 (2021)
6. Jenkins, A.: Self-oscillation. Phys. Rep. **525**, 167–222 (2013)

7. Griffiths, D.J.: Introduction to Electrodynamics. Prentice Hall, New Jersey (1989)
8. Keane, A., Krauskopf, B., Dijkstra, H.A.: The effect of state dependence in a delay differential equation model for the El Niño Southern Oscillation. Phil. Trans. R. Soc. A **377**, 20180121 (2019)
9. Bohm, D.: A suggested interpretation of the quantum theory in terms of "hidden" variables. I. Phys. Rev. **85**, 166–179 (1952)
10. Schönberg, M.: On the hydrodynamical model of the quantum mechanics. Il Nuovo Cimento. **12**, 103–133 (1954)
11. López, A.G., Ali, R., Mandi, L., Chatterjee, P.: Average conservative chaos in quantum dusty plasmas. Chaos **31**, 013104 (2021)
12. López, A.G., Benito, F., Sabuco, J., Delgado-Bonal, A.: The thermodynamic efficiency of the Lorenz system. Available at arXiv:2202.07653 [cond-mat.stat-mech] (2022)
13. de la Peña, L., Cetto, A.M., Valdés-Hernandes, A.: The zero-point field and the emergence of the quantum. Int. J. Mod. Phys. E **23**, 1450049 (2014)
14. Raju, C.K.: The electrodynamic 2-body problem and the origin of quantum mechanics. Found. Phys. **34**, 937–963 (2004)
15. Faber, M.: Particles as stable topological solitons. In Journal of Physics: Conference Series **361**, 012022. IOP Publishing (2012)
16. Rañada, A.F., Trueba, J.: Electromagnetic knots. Phys. Lett. A **202**, 337–342 (1995)
17. Gullu, I., Mazharimousavi, S.H.: Black holes in double-Logarithmic nonlinear electrodynamics. Phys. Scr. **96**, 095213 (2021)

Randomness and Fractal Functions on the Sierpinski Triangle

A. Gowrisankar and M. K. Hassan

Abstract The notion of fractal and its inherent characters are, in general, explained through the classical examples Cantor set and Sierpinski triangle. This article incorporates the probability and randomness on the dyadic Sierpinski triangle as follows. The construction process of dyadic Sierpinski triangle starts with an equilateral triangle as an initiator. Then, the generator divides the initiator into four equal triangles, by connecting the midpoints of three sides and removing the middle interior triangle with probability $(1 - p)$, here the probability gears the randomness. Further, the homogeneous relation between the fractal dimension of the dyadic Sierpinski triangle and its randomness is investigated. Finally, the fractal interpolation function with variable scaling is implemented on the Sierpinski triangle by defining its Laplacian.

1 Introduction

Fractal analysis is introduced to describe the irregular objects which are traditionally observed as too complex to describe using classical Euclidean geometry. Although the phenomena of a fractal does have a long history in mathematics, it is precisely defined as a particular class of processes called iterated function systems in which the so-called selfsimilar or fractal sets can be produced as follows. Let (X, d) be a complete metric space and $(\mathcal{H}(X), H_d)$ denotes the corresponding hyperspace of nonempty compact subsets of X where H_d is Hausdorff metric. For $n \in \mathbb{N}$, let $\mathbb{N}_n := \{1, 2, \ldots, n\}$. A complete metric space X consisting of a finite family of contractions f_k with the ratios α_k, for $k \in \mathbb{N}_n$, constitutes an iterated function system (IFS) and it is symbolized by $\{X; f_k : k \in \mathbb{N}_n\}$. Define a self map $\mathcal{F}$ on the complete metric space $\mathcal{H}(X)$ by

A. Gowrisankar (✉)
Department of Mathematics, School of Advanced Sciences, Vellore Institute of Technology, Vellore 632 014, Tamil Nadu, India
e-mail: gowrisankargri@gmail.com

M. K. Hassan
Department of Physics, University of Dhaka, Dhaka 1000, Bangladesh

S. Banerjee and A. Saha (eds.), *Nonlinear Dynamics and Applications*, Springer Proceedings in Complexity,
https://doi.org/10.1007/978-3-030-99792-2_89

$$\mathcal{F}(B) = \bigcup_{k \in \mathbb{N}_n} f_k(B) \tag{1}$$

which is a contraction with the ratio $\alpha = \max\{\alpha_k : k \in \mathbb{N}_n\}$, and thus it has a unique invariant point B^* in $\mathcal{H}(X)$. This invariant point B^* is referred as a deterministic fractal constructed by the IFS $\{X; f_1, f_2, \ldots, f_n\}$. The fixed point B^* satisfies the Eq. (1), thus B^* can be written as the finite copies of itself. The fractal B^* generated by the above process obviously have exact self-similarity and hence B^* is called the deterministic fractal generated by the finite family of contraction mappings. For a detailed exposition of the iterated function system reader may refer to [1–4]. The powerful method to construct the deterministic fractals is iterated function system [2]. With the help of iterated function system, Barnsley [2] constructed the fractal interpolation function (FIF) as follows which is development over the interpolation techniques. A data set $\{(x_k, y_k) \in \mathbb{R}^2 : k \in \mathbb{N}_n\}$ with $x_1 < x_2 < \cdots < x_n$ is given and x_i's are not necessarily equidistant. Let I and I_k denote the closed intervals $[x_1, x_n]$ and $[x_k, x_{k+1}]$, respectively, for $k \in \mathbb{N}_{n-1}$ and $L_k : I \to I_k, k \in \mathbb{N}_{n-1}$ be $(n-1)$ contraction homeomorphisms such that

$$L_k(x_1) = x_k, L_k(x_n) = x_{k+1}. \tag{2}$$

For $r_k \in [0, 1), k \in \mathbb{N}_{n-1}$ and $X := I \times \mathbb{R}$. Set $R_k : I \times \mathbb{R} \to \mathbb{R}$ be the $n-1$ continuous mappings satisfying

$$\begin{aligned} R_k(x_1, y_1) = y_k, R_k(x_n, y_n) = y_{k+1} \\ |R_k(x, y_1) - R_k(x, y_2)| \leq r_k|y_1 - y_2|, x \in I, y, y^* \in \mathbb{R}. \end{aligned} \tag{3}$$

That is, R_k is contraction mapping with respect to second variable. Define functions $f_k : X \to I_k \times \mathbb{R}$ by $f_k(x, y) = (L_k(x), R_k(x, y))$, for $k \in \mathbb{N}_{n-1}$. Associated with the IFS $\{X; f_k : k \in \mathbb{N}_{n-1}\}$, a set valued mapping F is defined on $\mathcal{K}(X)$ by

$$F(A) = \bigcup_{k \in \mathbb{N}_{n-1}} f_k(A),$$

for any $A \in \mathcal{K}(X)$. $\mathcal{K}(X)$ together with the Hausdorff metric h is a complete metric space, since X is complete. Moreover, by the theory of IFS the contraction mapping F on $\mathcal{K}(X)$ has a unique invariant set G such that $G = F(G)$ and G is the graph of a continuous function $g : I \to \mathbb{R}$ satisfying $g(x_k) = y_k$ for $k \in \mathbb{N}_n$. For the data set $\{(x_k, y_k) \in \mathbb{R}^2 : k \in \mathbb{N}_n\}$, the function g whose graph is the attractor of an IFS $\{X; f_k : k \in \mathbb{N}_{n-1}\}$ is called a fractal interpolation function.

Let a set of interpolation data $\{(x_k, y_k) \in [x_1, x_k] \times \mathbb{R} : k \in \mathbb{N}_n\}$ be given. Then, the following process explains the construction of an iterated function system in $\mathbb{R}^2$ such that its attractor is the graph of the interpolation function of given data. The interpolation function f^* is generated as the fixed point of the self mapping $T : \mathbb{G} \to \mathbb{G}$ defined by

$$(Tf)(u) = R_k(L_k^{-1}(u), f \circ L_k^{-1}(u)),\ u \in I_k, k \in \mathbb{N}_{n-1},$$

where $\mathbb{G} = \{h : I \to \mathbb{R} | h$ is continuous on $I, h(x_1) = y_1, h(x_n) = y_n\}$ is a compete metric space equipped with the uniform metric $\delta(f, g) = \max\{|f(u) - g(u)| : u \in I\}$. Then T is a contraction map on $(\mathbb{G}, \delta)$ with contractivity factor $r = \max\{r_k : k \in \mathbb{N}_{n-1}\} < 1$. The invariant point f^* of T is the fractal interpolation function obeying the fixed point equation

$$f^*(u) = R_k(L_k^{-1}(u), f^* \circ L_k^{-1}(u)),\ u \in I_k,\ k \in \mathbb{N}_{n-1}. \tag{4}$$

As widened applications of FIF for approximating the naturally occurring functions, there have been sequel studies reported in the literature [10–20]. Mostly, the FIFs are generated from the IFS of the form

$$L_k(x) = a_k x + b_k,\ R_k(x, y) = \alpha_k(x) y + q_k(x),\ k \in \mathbb{N}_{n-1}, \tag{5}$$

where $\{\alpha_k : \alpha_k \in (-1, 1), k \in \mathbb{N}_{n-1}\}$ is a family of parameters named as vertical scaling factors. The shape and fractal dimension of the fractal interpolation function are heavily influenced by the set of vertical scaling factors. The FIF generated by the constant vertical scaling parameter have self-similar character which could lead to loss of flexibility and may cause approximation error. Since, the natural functions are not necessarily self-similar and nature always like to enjoy the freedom of choice not determinism. To overcome this problem, a class of FIF with variable scaling parameter is introduced and analyzed based on the affine iterated function system by considering $R_k(x, y) = \alpha_k(x) y + q_k(x), k \in \mathbb{N}_{n-1}$, where $\alpha_k \in C(I)$ satisfying $\|\alpha_k\|_\infty = \sup\{|\alpha_k| : x \in I\} < 1$ (for more details, refer [7]). The fractal interpolation functions (FIFs) on the Sierpinski gasket is defined in [8] and some basic properties like finite energy, min-max property of uniform fractal interpolation functions explored in [9]. These study motivates to investigate the FIF with variable scaling parameter on the Sierpinski triangle by defining its Laplacian. Further, this paper explores the analogous facts about randomness on the Sierpinski triangle and generalized Cantor set, which should be regarded as the classical examples of a fractal supporting a concept of random fractals.

This paper is organized in the following manner. Section 2 starts with the construction of deterministic Sierpinski triangle and its fractal dimension is discussed through the scaling law. Additionally, the fractal dimension is estimated by defining the qth moment of the remaining smaller equilateral triangle as sum of qth power of its size. The dyadic Sierpinski triangle is explored by introducing the probability in the process of removing its middle part and its fractal dimension is measured through the moments. Further, the generalized Cantor set is presented in term of the partition value of the closed interval [0, 1] and the effect of partition value in the fractal dimension is investigated in Sect. 2. In Section 3, a new family of FIF with variable scaling parameter is described on the Sierpinski triangle which gives the degrees of freedom in the choice of vertical scaling of the fractal function.

2 Randomness on the Sierpinski Triangle

The pedagogical importance and impact of the Sierpinski triangle (ST) motivated us to investigate interesting variants of the Sierpinski triangle in which both probability and randomness are included in a logical progression in this section.

2.1 *Sierpinski Triangle*

Generalization of the Cantor set into higher dimension is Sierpinski triangle in this case the initiator is an equilateral triangle S_0 and the generator divides it into four equal triangles, by connecting the midpoints of the sides and remove the interior triangle whose vertices are the midpoints of each side of the initiator leaving the boundary of the triangle. The resultant is $S_1 \subseteq S_0$. Next each of the remaining three triangles are split up into four smaller triangles with side length $1/4$, and three middle triangles are taken away. The resultant is $S_2 \subseteq S_1$. The generator is then applied over and over again to all the available triangles, this process gives sequence S_n of sets such that $S_0 \supseteq S_1 \supseteq S_2 \supseteq \ldots$. The limit of S_n is called the Sierpinski triangle, thus $S = \bigcap_{n\in\mathbb{N}} S_n$.

It is easy to find out that at the nth step, there are $N = 3^n$ triangles of side $\delta = 2^{-n}$. So the total area of S_n is $3^n.(1/2^n)^2.\sqrt{3}/4$. As $n \to \infty$, S_n approaches to 0. The total area of the Sierpinski triangle is 0. Further, all the subsequent approximations $S_n, n \geq k$ will have the line segments that constitute the boundary of one of the triangles of S_n. Hence, the set S includes at least all boundaries. In S_n, there are 3^n triangles, individually having 3 edges of length 2^{-n}. Therefore, the length of entire S is at least $3^n.3.2^{-n}$. This approaches ∞ as $n \to \infty$. It gives that the total length of Sierpinski triangle is infinite. Moreover, eliminating n from the scaling law $N(\delta) \sim \delta^{-d_f}$, here $N(\delta)$ is the number of boxes with size δ required to cover the given object X and d_f is the fractal dimension of X. That is, $3^n \sim (1/2)^{-nd_f}$ provides $d_f = \frac{\ln 3}{\ln 2}$ which is the fractal dimension of the Sierpinski triangle.

Suppose $x_1, x_2, \ldots, x_N$ are size of each equilateral triangle. Define the qth moment M_q of the remaining equilateral triangle by

$$M_q = \sum_i^N x_i^q, \tag{6}$$

where $x_1 = x_2 = \cdots = x_N = 2^{-n}$ and $N = 3^n$ at the nth step of the construction process. Observe, M_0, M_1 are increasing quantity with n whereas M_2 is decreasing quantity with n. Hence, there exists $q \in (1, 2)$ such that $M_q = \sum_i^N x_i^q = 1$,

$$M_q = e^{n\ln 3 - qn\ln 2}. \tag{7}$$

If $q = \frac{\ln 3}{\ln 2}$, then

$$M_{\frac{\ln 3}{\ln 2}} = \sum_{i=1}^{3^n} \left(\frac{1}{2^n}\right)^{\frac{\ln 3}{\ln 2}} = 1, \tag{8}$$

independent of n and $d_f = \frac{\ln 3}{\ln 2}$ moment equal to size of S_0. At the nth step there are $N = 3^n$ triangles of side $\delta = 2^{-n}$. Eliminating n in favor of δ yields $N \sim \delta^{-\frac{\ln 3}{\ln 2}}$. One can also show that $M \sim L^{\frac{\ln 3}{\ln 2}}$ and $M_{-\frac{\ln 3}{\ln 2}} = 1$. The conservation law

$$M_{-\frac{\ln 3}{\ln 2}} = 1 \tag{9}$$

is obeyed here as well regardless of n when the object is Sierpinski triangle.

2.2 *Dyadic Sierpinski Triangle*

Starts with a equilateral triangle as an initiator. The generator divides the initiator into four equal triangles, by connecting the midpoints of three sides and removing the middle interior triangle with probability $(1 - p)$. Now system have $(3 + p)$ average number of smaller triangle each of size $1/2$ because after step one the middle interior triangle is present with probability p and it has probability $(1 - p)$ for absent. In second step, the generator is employed to each of the leftover $(3 + p)$ smaller triangles to separate them into four equal parts. Thus, each of 3 smaller triangles in step one will have 3 triangles with absolute certainty and one triangle with probability p. Then, the middle interior triangle in step one will have 3 triangles with probability p and one triangle with probability p^2. Thus, after step two system will have $(3 + p)^2$ average number of smaller triangles of size $\frac{1}{2^2}$ and $(1 - p)(3 + p)$ average number of smaller triangles are removed. First and second iteration are showed in Fig. 1. Continue this process recursively to all the remaining smaller triangles at each step, then the resultant system is dyadic Sierpinski triangle provided number of iteration approaches to infinity.

In nth step of the construction process of dyadic Sierpinski triangle there are $(3 + p)^n$ average number of triangles with size $\frac{1}{2^n}$. By Eq. (6)

$$\begin{aligned} M_q &= \sum_{i=1}^{(3+p)^n} \left(\frac{1}{2^n}\right)^q \\ &= e^{n \ln(3+p) - nq \ln 2}, \end{aligned} \tag{10}$$

for all $p \in (0, 1)$. Thus, $d_f = \frac{\ln(3+p)}{\ln 2}$ provides

$$M_{d_f} = \sum_{i=1}^{(3+p)^n} 2^{-nd_f} = 1. \tag{11}$$

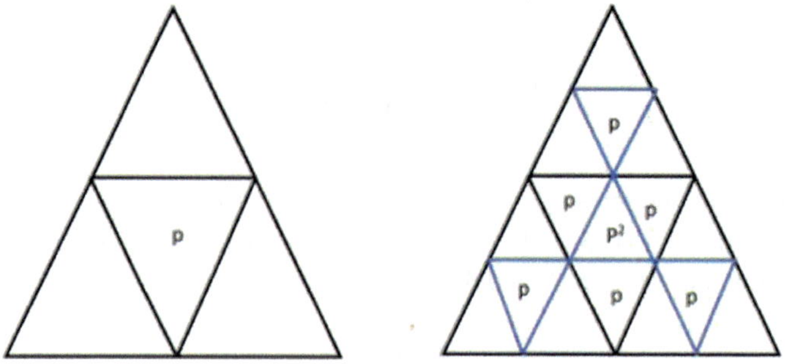

Fig. 1 First and second iteration of the dyadic Sierpinski triangle

It is independent of n. Therefore, d_f^{th} moment M_{d_f} of the $(3+p)^n$ average number of remaining triangles of size $1/2^n$ is a conserved quantity. Meantime, adding all removed triangles at each step gives followings. After step 1 there are 3 triangles each of area $1/4$. One has area $1/4$ with probability p. The one which is thrown out has area $1/4$ with probability $1-p$. If we sum them all we get

$$\frac{3}{4}+\frac{p}{4}+\frac{1-p}{4}=1, \tag{12}$$

which is the area of the initiator. After step 2, there are 9 triangles each of area $1/4^2$, 6 of them have area $1/4^2$ with probability p, one of them have area $1/4^2$ with probability p^2. The ones are thrown are as follows. Three of them have area $1/4^2$ with probability $(1-p)$ and one of them have area $1/4^2$ with probability $p(1-p)$. If we now sum all the surviving triangles and the ones in the bin we get

$$\frac{9}{4^2}+\frac{6p}{4^2}+\frac{p^2}{4^2}+\frac{3(1-p)}{4^2}+\frac{p(1-p)}{4^2}+\frac{1-p}{4}=1, \tag{13}$$

the size of the initiator. Observe that the last term on the L.H.S. is the size of the thrown out quantity in the bin.

We thus see that the amount which is thrown in the bin at step 1 is $(1-p)/4$. In step 2 we throw

$$\frac{3(1-p)}{4^2}+\frac{p(1-p)}{4^2}=\frac{1-p}{4}\frac{(3+p)}{4}. \tag{14}$$

Similarly, we can find that after step 3 the amount we throw to the bin is $\frac{1-p}{4\left(\frac{(3+p)}{4}\right)^2}$. We can now add all the quantities in the bin and find the following series

$$\frac{1-p}{4}\left[1+\frac{(3+p)}{4}+\left(\frac{(3+p)}{4}\right)^2+\left(\frac{(3+p)}{4}\right)^3+\ldots+\ldots\right]=\frac{1-p}{4}\frac{1}{1-\frac{(3+p)}{4}}=1. \tag{15}$$

However, the d_f^{th} moment equals the size of the initiator providing that there are infinitely many smaller triangles in the resultant set.

Example 1 Let X be an equilateral triangle with the vertices $v_1 = (0, 0)$, $v_2 = (1, 0)$ and $v_3 = (1/2, \sqrt{3}/2)$ and consider the IFS on X consists of the following three contractions, $f_1(u, v) = (\frac{1}{2}u, \frac{1}{2}v)$; $f_2(u, v) = (\frac{1}{2}u + \frac{1}{2}, \frac{1}{2}v)$; $f_3(u, v) = (\frac{1}{2}u + \frac{1}{4}, \frac{1}{2}v + \frac{\sqrt{3}}{4})$. All these three contractions have contraction factor 1/2. The resulting fractal is called Sierpinski triangle. If the contraction factor is chosen as an arbitrary number in (0, 1), say α, then it provides different fractal as function of α. Thus, $G : (0, 1) \to \mathcal{H}(X)$ defined by $G(\alpha) = A^*_\alpha$, where A^*_α is a fixed point of the IFS including three contractions

$$\begin{aligned} f_1(u, v) &= (\alpha u, \alpha v); \\ f_2(u, v) &= (\alpha(u + 1), \alpha v); \\ f_3(u, v) &= \left(\alpha(u + \frac{1}{2}), \alpha(v + \frac{\sqrt{3}}{2})\right). \end{aligned}$$

2.3 Generalized Triadic Cantor Set

The generalized triadic Cantor set is constructed as follows. Consider the unit length line segment as closed interval $C_0 = [0, 1]$ which is called as an initiator. Initially to generate the set, divide the initiator into $k(3 \le k < \infty)$ equal pieces and remove the centre piece among k subintervals. It remains two closed intervals $[0, \frac{1}{k}]$ and $[\frac{k-1}{k}, 1]$ in step 1 each of length $\frac{1}{k}$. Thus, $C_1 = [0, \frac{1}{k}]\bigcup[\frac{k-1}{k}, 1]$. Now remove the middle parts from the leftover intervals thus, obtaining four equal pieces of closed interval of length $\frac{1}{k^2}$ and they are $[0, \frac{1}{k^2}]$, $[\frac{k-1}{k^2}, \frac{1}{k}]$, $[\frac{k-1}{k}, \frac{k^2-k-1}{k^2}]$, $[\frac{k^2-1}{k^2}, 1]$. This gives

$$C_2 = [0, \frac{1}{k^2}]\bigcup[\frac{k-1}{k^2}, \frac{1}{k}]\bigcup[\frac{k-1}{k}, \frac{k^2-k-1}{k^2}]\bigcup[\frac{k^2-1}{k^2}, 1].$$

Now remove the middle parts from each of the remaining four intervals to create eight smaller closed intervals. This continuous process gives the generalized triadic Cantor set as a limit C of the decreasing sequence $(C_n)_{n\in\mathbb{N}}$ of sets. Hence, the limit is the intersection of the sets, $C = \cap_{n\in\mathbb{N}} C_n$. If one can continue the above construction process through infinitely many steps then the following questions naturally arise : (i) What is the cardinality of C? (ii) How much of the initiator removed in [0, 1]?

Start to generate the set by splitting the initiator into k equal parts and hence there are k number of smaller intervals which has size $1/k$. Removing the middle parts from the k subintervals means $k - 2$ of them are removed in step one so that two closed intervals $[0, 1/k]$ and $[\frac{k-1}{k}, 1]$ is remained in the system. The sum of the sizes of all the intervals removed at this stage is $\frac{k-2}{k}$. In step two, each of the two subsets is divided into k intervals and remove the middle $k - 2$ number of smaller intervals so there will be four intervals $[0, 1/k^2]$ and $[k - 1/k^2, 1/k]$. The size of the intervals being removed are $\frac{k-2}{k}$, $\frac{2(k-2)}{k^2}$, $\frac{4(k-2)}{k^3}$ etc. If we sum them up, we get

$$\frac{k-2}{k}\sum_{n}\left(\frac{2}{k}\right)^{n} = \frac{k-2}{k}\frac{1}{1-2/k} = 1.$$

Moreover, the set C_n contains 2^n disjoint closed intervals, each of length $(\frac{1}{k})^n$ in step n. So the total length of C_n, (i.e.) the sum of the lengths, is $(2/k)^n$ and its limit is $\lim_{n\to\infty}(\frac{2}{k})^n = 0$. Therefore, the total length of the Cantor set is zero. $k = 3$ gives the total length of all the thrown intervals is equal to the size of the initiator and length of C is zero. If k approaches to infinity, then the total length of all intervals begin thrown away is zero as well as length of C is zero. This is quite surprising as it implies that there is hardly anything left in the Cantor set. However, there are infinite number of points in the Cantor set. Since construction of Cantor set started with the initiator [0, 1] and the endpoints 0 and 1 belong to all of the succeeding sets C_{n_0}, $n_0 \geq n$, and hence belong to the intersection C. Thus, C is nonempty and taking every endpoints of all the intervals in every approximations C_n provides an infinite number of endpoints which are belonging to C.

Define the qth moment M_q of the leftover intervals at the nth step of the construction process as

$$M_q = \sum_{i}^{2^n} x_i^q. \tag{16}$$

Note that each of the leftover intervals at the nth step are of equal size $x_i = k^{-n}$ and hence can write

$$M_q = e^{n\ln 2 - qn\ln k}. \tag{17}$$

It means that if $q = \frac{\ln 2}{\ln k}$ then

$$M_{\frac{\ln 2}{\ln k}} = 1, \tag{18}$$

independent of n. That is, this result is true even in the limit $n \to \infty$. Thus, it concludes that the set is nonempty which is another surprising fact of the Cantor set. In the construction process of Cantor set, the nth step starts with deleting the middle parts from each of the leftover 2^{n-1} intervals in the $(n-1)^{th}$ step to produce 2^n closed intervals of size $1/k^n$ for each n. If one can take yardstick as $\delta = k^{-n}$ then the number of yardsticks $N(\delta)$ required to cover the set C is equal to $N = 2^n$. Removing n from $N = 2^n$ with the help of δ provides $N(\delta)$ is direct proportion to δ^{-d_f} that is $N(\delta) \sim \delta^{-d_f}$ where $d_f = \frac{\ln 2}{\ln k}$ is the box-counting or fractal dimension of the Cantor set.

Remark 1 The fractal dimension of generalized Cantor set is $d_f = \frac{\ln 2}{\ln k}$ which is decreasing quantity whenever k is increasing. Thus, if the partition of the initiator increased then it will give the more irregular Cantor type set.

If the partition of the initiator $k = 3$, then the generalized triadic Cantor set is known as the classical Cantor set and its fractal dimension d_f is $\frac{\ln 2}{\ln 3}$. Further, by applying probability on the construction of the Cantor set, one can get the dyadic Cantor set as follows: The generator divides the initiator [0, 1] into two equal parts. In which delete one subinterval with probability $(1 - p)$, remains another subinterval with probability p. Now, the system will have an average $(1 + p)$ number of sub-intervals each of size $1/2$. By continuing the process over and over again on all the available intervals at each step recursively, the dyadic Cantor set is obtained . This result is investigated in [5]. If we find the d_fth moment M_{d_f} of the $(1 + p)^n$ available intervals of size 2^{-n} by using the Eq. (6), then it provides that M_{d_f} is conserved quantity. Hence, by changing the partition value k and applying above procedure one can get different types of the Cantor set with fractal dimension $d_f = \frac{\ln 2}{\ln k}$.

Example 2 Let $X = [0, 1]$ and consider the IFS on X with the following two mappings, $f_1(x) = \frac{x}{3}$, $f_2(x) = \frac{(x+2)}{3}$. The self mappings f_1, f_2 are contractions with contraction factor $1/3$ and the resulting fractal of the $IFS\ \{X; f_1, f_2\}$ is Cantor set. If the contraction factor α is chosen as an arbitrary number in $(0, 1)$, then one can get different fractal as function of α. Thus, $G : (0, 1) \to \mathcal{H}(X)$ defined by $G(\alpha) = A^*_\alpha$, where A^*_α is a fixed point of the IFS including two contractions $f_1(x) = \alpha x$, $f_2(x) = \alpha(x + 2)$. As per the remark 1, if $\alpha_1 \leq \alpha_2$, then $d_f(A^*_{\alpha_1}) \leq d_f(A^*\alpha_2)$. The relation between the contraction factor α and the number of partition k in the construction of generalized Cantor set is $\alpha = 1/k$.

The influence of contraction factor in the shape of the resultant fractals is explained in Example 1 and Example 2.

3 Fractal Function on the Sierpinski Triangle

This section demonstrates the fractal interpolation function with variable scaling on the Sierpinski triangle by defining the Laplacian on the Sierpinski triangle (ST). That is the class of FIF with vertical scaling parameters $\alpha_{w_k} : ST \to (-1, 1)$, here α_{w_k} is continuous and satisfies $\|\alpha_{w_k}\|_\infty = \sup\{|\alpha_{w_k}| : \forall x \in ST\} < 1$, are considered in this study.

3.1 *Laplacian on the Sierpinski Triangle*

Let $V_0 = \{p_1, p_2, p_3\}$ be a set of vertices of an equilateral triangle, say X, in $\mathbb{R}^2$. Define $f_k(x) = \frac{1}{2}(x + p_k)$ on X. Then, the attractor of the iterated function system $\{X; f_k : k = 1, 2, 3\}$ is a Sierpinski triangle. Define a sequence of finite sets $(V_m)_{m\geq 0}$ by $V_{m+1} = \cup_{k=1}^3 f_k(V_m)$. We write $f_w = f_{w_1} \circ f_{w_2} \circ \dots f_{w_m}$ for any finite sequence $w = (w_1, w_2, \dots, w_m)$ of length $|w| = m$, $k \in \{1, 2, 3\}$ i.e. $w \in \{1, 2, 3\}^m$.

The union of the image of V_0 under the iteration of f_w constitutes the set of m vertices V_m. For any $p \in V_m$, let $N_{m,p}$ be the collection of the direct neighborhood of p in V_m.

Observe that, the cardinality of $N_{m,p}$ is

$$|N_{m,p}| = \begin{cases} 4 \text{ if } p \notin V_0 \\ 2 \text{ if } p \in V_0. \end{cases}$$

Let $C(V_m) = \{h : h \text{ is continuous on } V_m \text{ to } \mathbb{R}\}$ and define the linear operator $\Delta_m : C(V_m) \to C(V_m)$ by

$$5^m(\Delta_m h)(p) = \sum_{q \in N_{m,p}} (h(q) - h(p))$$

for all $h \in C(V_m)$ and all $p \in V_m$. Then the Laplacian on the Sierpinski triangle is defined by

$$(\Delta_m h)(p) \to (\Delta h)(p)$$

as $m \to \infty$. A function $f \in C(V_m)$ is called as harmonic on V_m if $(\Delta_m f)(p) = 0$ for all $p \in V_m / V_0$. A continuous function $f : ST \to \mathbb{R}$ is said to be harmonic, if its restriction to V_m is harmonic for all m.

In [8], the first degree polynomials are constructed as classical harmonic functions on an interval and by substituting them on the Sierpinski triangle using the standard Laplacian on SG of fractal analysis, the theorem of interpolation for ST is obtained. Whereas, this study extends the class of FIF by considering vertical scaling parameters as $\alpha_{w_k} : ST \to (-1, 1)$, here α_{w_k} is continuous and satisfies $\|\alpha_{w_k}\|_\infty = \sup\{|\alpha_{w_k}| : \forall x \in ST\} < 1$ in place of constant scaling factors $\alpha_{w_k} \in (-1, 1)$, for all $k \in \{1, 2, 3\}$, in [6, 7]. Note that $\alpha_w = \alpha_{w_1} \circ \alpha_{w_2} \circ \ldots \alpha_{w_m}$.

Theorem 1 *Let $h \in C(V_m)$ be given and any $\alpha_w(x)$ for $w \in \{1, 2, 3\}^m$ with $\|\alpha_w\| < 1$. Then there is a unique continuous function $g : ST \to \mathbb{R}$ satisfying $g|V_m = h$ and*

$$g(f_w(x)) = \alpha_w(x) g(x) + q_w(x)$$

where q_w is harmonic function on ST for all $w \in \{1, 2, 3\}^m$.

Let $\mathcal{T} = \{t : ST \to \mathbb{R} \text{ continuous function with } t(p_k) = h(p_k), k = 1, 2, 3\}$. Then $\mathcal{T}$ is complete with respect to uniform metric. Define $T : \mathcal{T} \to \mathcal{T}$ by

$$(Tt)(x) = \alpha_w(x) t(f_w^{-1}(x)) + q_w(f_w^{-1}(x))$$

where q_w is the harmonic function on ST with $q_w(p_k) = h(f_w(p_k)) - \alpha_w(x) h(p_k)$ for $k = 1, 2, 3$. It is observed that, T is well defined, contractive with ratio $\|\alpha_w\|$ and $(Tt)(p_k) = h(p_k)$. Then, the unique fixed point of T obeys

$$g(x) = \alpha_w(x)g(f_w^{-1}(x)) + q_w(f_w^{-1}(x))$$

for $x \in f_w(ST)$ and $w \in \{1, 2, 3\}^m$,

$$g(f_w(x)) = \alpha_w(x)g(x) + q_w(x).$$

Remark 2 If $\alpha_{wk}(x) = \alpha_{wk}$, then Theorem 1 provides the results in [8] and [9]. In, [8, 9], authors have considered the scaling factors α_{wk} as constant value in between 1 and -1 whereas in this paper scaling factors are considered as a function scaling.

4 Conclusion

In the present study, the dyadic Sierpinski triangle is introduced which is a classical example of the fractal supporting a concept of random fractal. Further, the fractal dimension of the dyadic Sierpinski triangle estimated by moments of available triangle in each iteration narrates that there is a conservative law on it since $M_{-\frac{\ln 3}{\ln 2}} = 1$ which is independent of its iteration n. Consequently, the generalized Cantor set presented in term of the partition value of the closed interval in [0, 1] and the influence of partition value in the fractal dimension is investigated which generalizes the dyadic Cantor set considered in [5, 6]. Finally, a new set of fractal functions with variable scaling parameter on Sierpinski triangle is demonstrated which gives the more flexibilities and degrees of freedom in the choice of vertical scaling of the fractal function.

References

1. Barnsley, M.F.: Fractals Everywhere. Academic Press, Dublin (1988)
2. Barnsley, M.F.: Fractal functions and interpolation. Constr. Approx. **2**(1), 303–329 (1986)
3. Santo Banerjee, Hassan, M.K., Sayan Mukherjee, Gowrisankar, A.: Fractal patterns in nonlinear dynamics and applications. CRC Press, Baco Raton (2019)
4. Santo Banerjee, Easwaramoorthy, D., Gowrisankar, A.: Fractal Functions, Dimensions and Signal Analysis. 1st ed. Springer, Cham (2021)
5. Hassan, M.K., Pavel, N.I., Pandit, R.K., Kurths, J.: Dyadic Cantor set and its kinetic and stochastic counterpart. Chaos Solit. Fract. **60**, 31–39 (2014)
6. Hassan, M.K.: Is there always a conservation law behind the emergence of fractal and multifractal? Eur. Phys. J.: Spec. Top. **228**(1), 209–232 (2019)
7. Wang, H.Y., Yu, J.S.: Fractal interpolation functions with variable parameters and their analytical properties. J. Approx. Theory **175**, 1–8 (2013)
8. Celik, D., Kocak, S., Ozdemir, Y.: Fractal interpolation on the Sierpinski gasket. J. Math. Anal. Appl. **337**, 343–347 (2008)
9. Ri, Song-Gyong., Ruan, Huo-Jun.: Some properties of fractal interpolation functions on Sierpinski gasket. J. Math. Anal. Appl. **380**, 313–322 (2011)

10. Barnsley, M.F., Elton, J., Hardin, D., Massopust, P.: Hidden variable fractal interpolation functions. SIAM J. Math. Anal. **20**(5), 1218–1242 (1989)
11. Navascués, M. A.: Fractal polynomial interpolation. Z. Anal. Anwend. **25**(2), 401–418 (2005)
12. Gowrisankar, A., Uthayakumar, R.: Fractional calculus on fractal interpolation function for a sequence of data with countable iterated function system. Mediterr. J. Math. **13**(6), 3887–3906 (2016)
13. Secelean, N.A.: The existence of the attractor of countable iterated function systems. Mediterr. J. Math. **9**, 61–79 (2012)
14. Priyanka, T.M.C., Gowrisankar, A.: Riemann-Liouville fractional integral of non-affine fractal interpolation function and its fractional operator. Eur. Phys. J.: Spec. Top. **230**, 3789–3805 (2021)
15. Priyanka, T.M.C., Gowrisankar, A.: Analysis on Weyl-Marchaud fractional derivative for types of fractal interpolation function with fractal dimension. Fractals **29**(7), 2150215 (2021)
16. Chand, A.K.B., Katiyar, S.K., Viswanathan, P.: Approximation using hidden variable fractal interpolation functions. J. Fractal Geom. **2**(1), 81–114 (2015)
17. Fataf, N.A.A., Gowrisankar, A.: Santo Banerjee: in search of self-similar chaotic attractors based on fractal function with variable scaling. Phys. Scr. **95**, 075206 (2020)
18. Min, Wu.: The Hausdorff measure of some Sierpinski carpets. Chaos Solit. Fract. **24**(3), 717–731 (2005)
19. Rani, M., Goel, S.: Categorization of new fractal carpets. Chaos Solit. Fract. **41**, 1020–1026 (2009)
20. Aslan, N., Saltan, M., Demir, B.: The intrinsic metric formula and a chaotic dynamical system on the code set of the Sierpinski tetrahedron. Chaos Solit. Fract. **123**, 422–428 (2019)

Bifurcation Analysis of a Leslie-Gower Prey-Predator Model with Fear and Cooperative Hunting

Ashvini Gupta and Balram Dubey

Abstract The current work examines the dynamical features of a Leslie-Gower prey-predator model. The effects of fear and group defense among prey with the mechanism of cooperative hunting by predators are incorporated. The existence and uniqueness of the interior equilibrium are explained. We obtained sufficient conditions for the local and global stability behavior. With regard to the fear parameter and cooperation strength parameter, the proposed system undergoes Hopf-bifurcation, transcritical bifurcation, and saddle-node bifurcation. Moreover, the system exhibits the property of bi-stability between two interior equilibrium points. The basin of attraction of these points is also plotted. All theoretical results are verified numerically by MATLAB R2021a.

Keywords Prey-predator · Fear · Cooperative hunting · Bifurcation

1 Introduction

Understanding of prey-predator interactions via differential equations is a classical application of mathematics in ecology. The dynamics of such systems are often altered due to various ecological factors. Employing these factors makes the system more consistent with the real world. Introducing fear may cause the interacting populations to oscillate or stabilitate about their steady-state [12]. These oscillations are most commonly due to the occurrence of Hopf-bifurcation [4]. In population dynamics, group defense is a common concept that describes an instance in which

The author (Ashvini Gupta) is grateful to the University Grants Commission, New Delhi, India, for the Junior Research Fellowship.

A. Gupta (✉) · B. Dubey
Department of Mathematics, BITS Pilani, Pilani Campus, Pilani 333031, Rajasthan, India
e-mail: aashvinig28@gmail.com

B. Dubey
e-mail: bdubey@pilani.bits-pilani.ac.in

S. Banerjee and A. Saha (eds.), *Nonlinear Dynamics and Applications*,
Springer Proceedings in Complexity,
https://doi.org/10.1007/978-3-030-99792-2_90

prey form groups to defend against the predator, which can cut off the predation rate. Considering Holling type IV or Monod-Haldane type functional response is the most frequent and accessible technique to implement group defense [6, 11]. It is evident from the research that dominance of defense could lead to predator's extinction [5]. The cooperation among predators to hunt down the target significantly increase the chances of their survival [1]. Saha and Samanta [10] extensively studied a 3-D prey-predator model involving cooperative hunting strategy and group defense mechanism. They observed transcritical bifurcation, saddle-node bifurcation, Hopf-bifurcation, and many other type of bifurcations. Pal et al. [8] studied the combined effect of fear and cooperative hunting and they observed various bifurcations and bi-stability in their model. The predator often switches to a different food to prevent extinction and becomes a generalist. The standard way to incorporate this feature is to use the modified Leslie-Gower scheme. Many authors [2, 3, 7] remarked the persistence of species in the modified Leslie-Gower prey-predator model. As per our knowledge, there is no work done comprising fear, group hunting, and group defense in a Leslie-Gower prey-predator model. Hence our main purpose is to study the effects of group defense in prey, group hunting in predator and fear induced by predator in prey on the dynamical behavior of prey-predator system.

2 The Mathematical Model

The survival of species is one of the most fundamental and significant issues in ecology. The modified Leslie-Gower prey-predator model formulation is an interesting approach in species conservation. According to this scheme, the predator acts as a generalist, which increases their chances of survival. Inspiring from aforementioned facts and pioneering literature as cited in the introduction, we consider an ecosystem where prey-predator species are interacting with each other through modified Leslie-Gower scheme. Prey species defend themselves against predators for their survival. Predator species hunt prey in groups and induce fear in the prey. With all these assumptions, we propose the following model:

$$\begin{aligned} \frac{dx}{dt} &= \frac{rx}{1+Ky} - r_0 x - r_1 x^2 - \frac{(\alpha+\lambda y)xy}{a+x^2} := f(x,y), \\ \frac{dy}{dt} &= sy - \frac{\omega y^2}{b+x} := g(x,y), \\ x(0) &\geq 0,\ y(0) \geq 0. \end{aligned} \tag{1}$$

The variables and parameters involved in the model are listed in Table 1 wih their biological meaning and dimensions.

Table 1 Biological explanation and dimension of variables/parameters employed in model (1)

Variables/Parameters	Biological explanation	Dimensions
x	Prey density	Biomass
y	Predator density	Biomass
r	Birth rate of prey	Time^{-1}
K	Cost of fear	Biomass^{-1}
r_0	Prey mortality rate	Time^{-1}
r_1	Death rate of prey due to competition among them	Biomass^{-1} Time^{-1}
α	Predation rate	Biomass Time^{-1}
λ	Cooperation strength of predators	Time^{-1}
a	Half saturation constant of prey	Biomass^2
b	Half saturation constant of predator	Biomass
s	Intrinsic growth rate of predator	Time^{-1}
ω	The highest rate of predator eradication per capita	Time^{-1}

3 Dynamics of the System

The model (1) can be re-written as

$$\frac{dx}{dt} = x\phi_1(x, y), \quad \frac{dy}{dt} = y\phi_2(x, y),$$

where

$$\phi_1(x, y) = \frac{r}{1 + Ky} - r_0 - r_1 x - \frac{(\alpha + \lambda y)y}{a + x^2}, \quad \phi_2(x, y) = s - \frac{\omega y}{b + x}.$$

It follows that

$$x(t) = x(0)e^{\int_0^t \phi_1(x(\theta), y(\theta))d\theta} \geq 0, \quad y(t) = y(0)e^{\int_0^t \phi_2(x(\theta), y(\theta))d\theta} \geq 0.$$

Hence, in R_+^2, all $(x(t), y(t))$ solutions with the positive starting point stay positive.

Nature does not enable any species to spread rapidly due to a lack of resources. As a result, it is critical to ensure that the defined model is bounded.

Theorem 1 *All solutions initiating in R_+^2 are contained in the domain $\Omega = \{(x, y) \in R_+^2 : 0 \leq x \leq K_1,\ 0 \leq y \leq \frac{s(b+K_1)}{\omega}\}$, where $K_1 = \frac{r-r_0}{r_1}$.*

Proof We may write the first equation of the model as

$$\dot{x} \leq rx - r_0 x - r_1 x^2.$$

This implies

$$\limsup_{t\to\infty} x(t) \leq \frac{r-r_0}{r_1} := K_1.$$

To show the boundedness of $y(t)$, we can write

$$\dot{y} \leq sy - \frac{\omega y^2}{b+K_1}.$$

This entails

$$\limsup_{t\to\infty} y(t) \leq \frac{(b+K_1)s}{\omega} := K_2,$$

which completes the proof.

3.1 Equilibrium Points

The proposed system has four feasible equilibrium points: extinction equilibrium; $E_0(0,0)$, predator-free equilibrium; $E_1(K_1,0)$, prey-free equilibrium; $E_2(0,\frac{bs}{\omega})$ and interior equilibrium; $E^*(x^*,y^*)$. Here x^* is a positive root of the following quartic equation:

$$A_1x^4 + A_2x^3 + A_3x^2 + A_4x + A_5 = 0, \tag{2}$$

where $A_1 = r_1\omega^2 sK$, $A_2 = \omega^2(r_0Ks + r_1\omega + r_1Ksb) + s^3\lambda K$, $A_3 = \omega^2Ks(r_0b + r_1a) + 3b\lambda Ks^3 + s^2\omega(\lambda + \alpha K) - (r - r_0)\omega^3$, $A_4 = (r_0Ksa + r_1a(\omega + Ksb))\omega^2 + 3\lambda Kb^2s^3 + sa\omega^2 + 2bs^2\omega(\lambda + \alpha K)$, $A_5 = \lambda Kb^3s^3 + b^2s^2\omega(\lambda + \alpha K) + b\omega^2s(\alpha + r_0aK) - (r - r_0)a\omega^3$.

Since A_1, A_2 and A_4 are positive. Therefore, according to the Descartes' rule of signs, Eq. (2) will have unique, two, three or no positive root based on the sign of A_3 and A_5 (refer to Table 2 and Fig. 1). It is worthy to note here that when $A_3 > 0$ and

Table 2 Existence of positive root of (2)

A_3	A_5	Number of positive roots	Color
-	+	2 or 0	yellow
-	-	3 or 1	red
+	-	1	green
+	+	0	blue

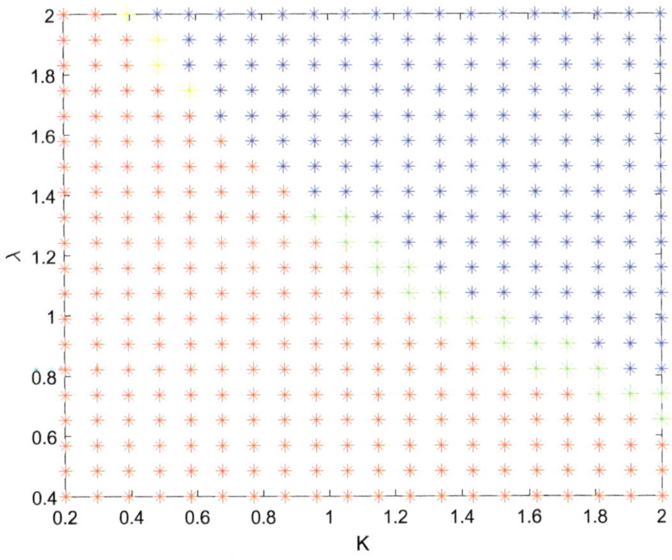

Fig. 1 Different colors showing all cases of Table 2 in $K\lambda$-plane, where $r = 0.6,\ a = 1,\ \alpha = 0.4,\ \omega = 1,\ b = 1,\ s = 0.4,\ r_0 = 0.05,\ r_1 = 0.05$

$A_5 < 0$, Eq. (2) has a unique positive solution x^*. On obtaining x^* from Eq. (2), we can easily determine y^* from the relation

$$y^* = \frac{s(b+\omega)}{x^*}.$$

3.2 *Stability Analysis*

The local stability feature of any equilibrium can be established using eigenvalue theory. The boundary equilibria with their local stability feature are described in Table 3.

Table 3 The local stability characteristics of system (1)'s boundary equilibria

Equilibrium points	Stability characteristics
$E_0(0, 0)$	Unconditionally unstable
$E_1(K_1, 0)$	Always saddle point
$E_2(0, \frac{bs}{\omega})$	Asymptotically stable if $r < (1 + \frac{bKs}{\omega})(r_0 + \frac{(\alpha\omega+\lambda bs)bs}{a\omega^2})$; saddle if $r > (1 + \frac{bKs}{\omega})(r_0 + \frac{(\alpha\omega+\lambda bs)bs}{a\omega^2})$

Biological significance: As per the concept of the modified Leslie-Gower prey-predator model, the predator can switch to other food when prey is absent. Moreover, predator performs cooperative hunting and induce fear in prey. Due to all these factors, predators may not become extinct. Therefore, when the prey's birth rate is less than a critical value, they might become extinct, nevertheless, predator always persists due to their generalist nature. Hence the prey-free equilibrium $E_2(0, \frac{bs}{\omega})$ can be stable, but the extinction state $E_0(0, 0)$ and predator-free state $E_1(K_1, 0)$ can never be stable.

Theorem 2 *$E^*(x^*, y^*)$ is locally asymptotically stable if and only if $B_1 > 0$ and $B_2 > 0$, where B_1 and B_2 are stated in the proof.*

Proof The Jacobian matrix, computed at positive equilibrium $E^*(x^*, y^*)$ is given by

$$J|_{E^*} = \begin{pmatrix} -r_1 x^* + \frac{2(\alpha+\lambda y^*)x^{*2}y^*}{(a+x^{*2})^2} - \frac{rKx^*}{(1+Ky^*)^2} & -\frac{(\alpha+2\lambda y^*)x^*}{a+x^{*2}} \\ \frac{\omega y^{*2}}{(b+x^*)^2} & -\frac{\omega y^*}{b+x^*} \end{pmatrix}.$$

The characteristic equation for the aforementioned matrix is as follows:

$$\xi^2 + B_1\xi + B_2 = 0, \tag{3}$$

where $B_1 = -tr(J|_{E^*})$ and $B_2 = det(J|_{E^*})$.
As per the Routh-Hurwitz criterion, the interior equilibrium $E^*(x^*, y^*)$ is locally asymptotically stable if and only if $B_1 > 0$ and $B_2 > 0$.

Remark. If $r_1 > \frac{2(\alpha+\lambda y^*)x^*y^*}{(a+x^{*2})^2}$, then $E^*(x^*, y^*)$ is locally asymptotically stable.
In a two-dimensional system, the possible attractors inside the positive invariant set could be equilibrium points and periodic solutions. If we are able to show that no periodic solution exists, and all boundary equilibrium points are unstable, then, in that case, all trajectories starting in the positive invariant region will eventually converge to the interior equilibrium E^* if it exists uniquely.

Theorem 3 *Let the positive equilibrium E^* exists uniquely. Then it is globally asymptotically stable under the following conditions:*

(i) $r > (1 + \frac{bKs}{\omega})(r_0 + \frac{(\alpha\omega+\lambda bs)bs}{a\omega^2})$,
(ii) $\frac{3\sqrt{3}(\alpha+\lambda K_2)}{8a\sqrt{a}} < \frac{r_1}{K_2} + \frac{\omega}{K_1(b+K_1)}$.

Proof If (i) holds, it directly implies $E_2(0, \frac{bs}{\omega})$ is a saddle point. Now, to show the non-existence of periodic solution, consider a function that is continuously differentiable in R_+^2, $H = \frac{1}{xy}$ and we define

$$\nabla = \frac{\partial}{\partial x}(fH) + \frac{\partial}{\partial y}(gH).$$

Simple calculation yields

$$\nabla = -\frac{r_1}{y} + \frac{2x(\alpha + \lambda y)}{(a + x^2)^2} - \frac{\omega}{x(b + x)}.$$

∇ remains negative if $\frac{3\sqrt{3}(\alpha+\lambda K_2)}{8a\sqrt{a}} < \frac{r_1}{K_2} + \frac{\omega}{K_1(b+K_1)}$.
Hence, system (1) cannot have a closed trajectory in the interior of the positive xy-plane, according to the Bendixson-Dulac criteria. In such a case, all solutions starting in Ω will converge to the interior equilibrium E^*, if it exists uniquely.

3.3 Bifurcation Analysis

Theorem 4 *System (1) experiences a transcritical bifurcation between the axial equilibrium $E_2(0, \frac{bs}{\omega})$ and interior equilibrium $E^*(x^*, y^*)$ with respect to the parameter K at $K^{[tc]} = \frac{\omega}{bs}\left(\frac{(r-r_0)a\omega^2-(\alpha\omega+\lambda bs)bs}{r_0a\omega^2+(\alpha\omega+\lambda bs)bs}\right)$ if $(r - r_0)a\omega^2 > (\alpha\omega + \lambda bs)bs$ and $\delta_3 \neq 0$, where δ_3 is defined in the proof.*

Proof At $K = K^{[tc]}$,

$$A = J|_{E_2} = \begin{pmatrix} 0 & 0 \\ \frac{s^2}{\omega} & -s \end{pmatrix}.$$

$v = (1, \frac{s}{\omega})$ and $w = (1, 0)$ are the eigenvectors of matrix A and A^T for the zero eigenvalue, respectively. Let $F = (f, g)^T$, where f and g are the RHS functions of model (1). Now, we define
$\delta_1 = w^T F_K(E_2, K^{[tc]})$, $\delta_2 = w^T[DF_K(E_2, K^{[tc]})v]$, and $\delta_3 = w^T[D^2F(E_2, K^{[tc]})(v, v)]$.

Simple computation yields

$$\delta_1 = 0,\ \delta_2 = -\frac{rbs\omega}{(\omega + bKs)^2} < 0$$

and

$$\delta_3 = -2r_1 - \frac{2}{ra^2\omega^4 b}(r_0a\omega^2 + (\alpha\omega + \lambda bs)bs)((r - r_0)a\omega^2 - (\alpha\omega + \lambda bs)bs).$$

If $\delta_3 \neq 0$, then all the conditions of the Sotomayor's Theorem [9] are satisfied. Hence, the system experiences a transcritical bifurcation at $K = K^{[tc]} = \frac{\omega}{bs}\left(\frac{(r-r_0)a\omega^2-(\alpha\omega+\lambda bs)bs}{r_0a\omega^2+(\alpha\omega+\lambda bs)bs}\right)$ between prey-free equilibrium E_2 and coexistence equilibrium E^*.

Theorem 5 *Let us assume that B_2 is positive. Then system (1) experiences a Hopf-bifurcation with respect to the cooperation strength λ at $\lambda = \lambda^{[hf]}$ around the coexistence equilibrium E^*.*

Proof It can be noted that

(i) When $B_1 > 0$ and $B_2 > 0$, E^* is locally asymptotically stable for $\lambda < \lambda^{[hf]}$.
(ii) When $B_1 < 0$ and $B_2 > 0$, E^* is unstable for $\lambda > \lambda^{[hf]}$.

Here B_1 and B_2 are defined in Eq. (3). This indicates that there is a switching of stability when cooperative strength λ crosses the critical value $\lambda = \lambda^{[hf]}$. At this point, $B_1 = 0$ and $B_2 > 0$, which implies that the eigenvalues are purely imaginary. Furthermore, we check the transversality condition *viz.*,

$$\left.\frac{dB_1}{d\lambda}\right|_{\lambda=\lambda^{[hf]}} = -\frac{2x^{*2}y^{*2}}{(a+x^{*2})^2} < 0.$$

Therefore, by the Andronov-Hopf bifurcation theorem, the system undergoes Hopf-bifurcation at $\lambda = \lambda^{[hf]}$ near the equilibrium point E^*.

4 Numerical Simulation

We use MATLAB R2021a to run numerical simulations to validate our analytical results of the model. The dataset we have picked is as follows:

$$r = 0.6,\ \lambda = 0.7,\ K = 0.1,\ a = 1,\ \alpha = 0.0005,\ \omega = 1,\ b = 1,\ s = 0.4,\ r_0 = 0.05,\ r_1 = 0.05 \tag{4}$$

For $\lambda = 0.005$ and other parameters from (4), the predator-only state $(0, 0.4)$ is a saddle-point. As per the Theorem 3, we obtain $\frac{3\sqrt{3}(\alpha+\lambda K_2)}{8a\sqrt{a}} - \frac{r_1}{K_2} + \frac{\omega}{K_1(b+K_1)} = -0.002$, implying that the system cannot have a closed trajectory in R^2_+. The interior equilibrium $E^*(7.8437, 3.5374)$ exists uniquely, and is a globally stable focus with eigenvalues $-0.395 \pm 0.2044i$. This phenomenon can be seen in Fig. 2.

In the model, fear parameter K plays a vital role. As per Theorem 4, we obtain $K^{[tc]} = 6.7478$ and $\delta_3 = -0.3367 \neq 0$. All conditions of the theorem are satisfied, hence the system undergoes a transcritical bifurcation at $K = K^{[tc]}$. The phenomenon of transcritical bifurcation is easy to understand with the help of a bifurcation diagram. It can be depicted from Fig. 3, E^* is stable and E_2 is unstable when $K < K^{[tc]}$. In this range, the value of ∇ remains negative. Therefore, E^* is globally stable. After crossing the threshold value of the fear parameter, the stability of E^* is transferred to E_2 via a transcritical bifurcation.

The traits of the system (1) are not limited to transcritical bifurcation. It has been observed that there are three positive equilibrium points, out of which two are stable, and the other is a saddle-point for the parameters given in (4) with $\alpha = 0.7$. The stable point $E^{*(1)}$ and saddle-point $E^{*(2)}$ approach towards each other with the variation in cooperation strength. At $\lambda = \lambda^{[sn]}$, they annihilate one another by means of a saddle-node bifurcation (see Fig. 4).

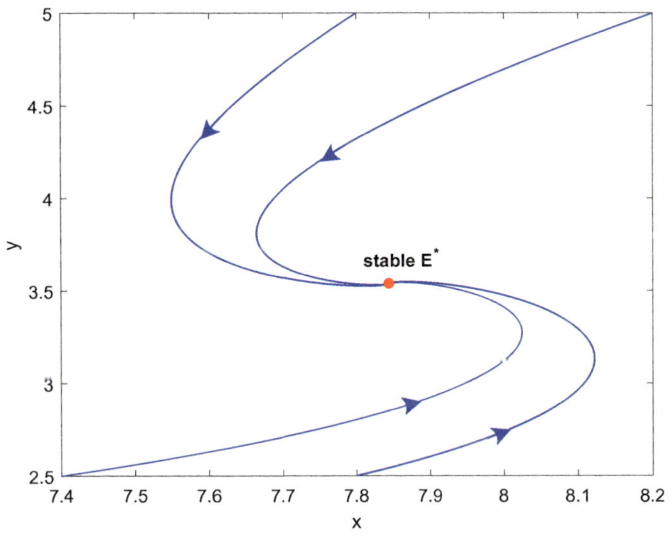

Fig. 2 Phase portrait showing global stable node E^*

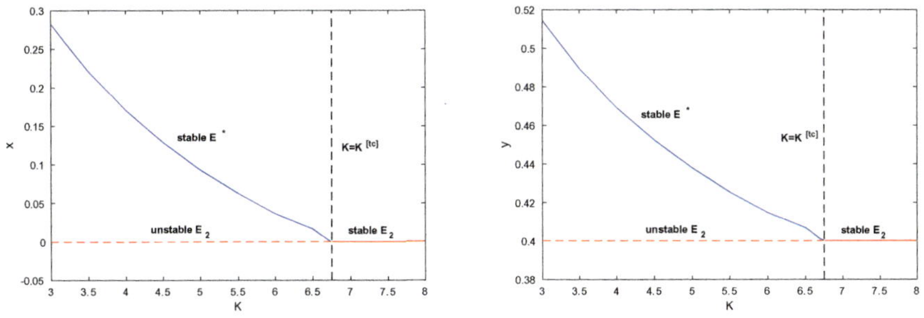

Fig. 3 Transcritical bifurcation diagram with respect to fear parameter K. The other parameters are from (4)

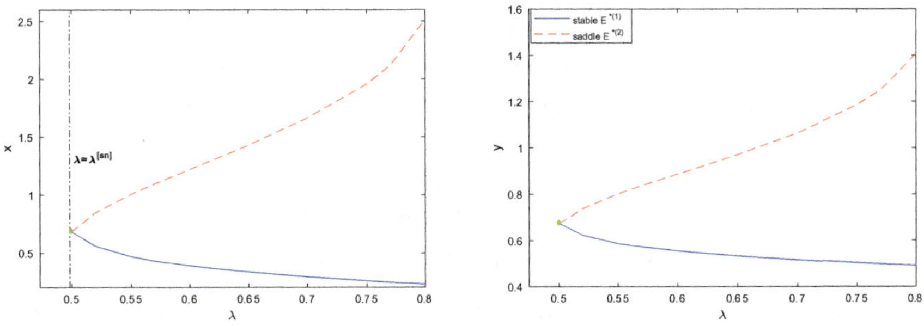

Fig. 4 Saddle-node bifurcation diagram with respect to cooperation strength λ, where $\alpha = 0.7$ and other parameters are same as in (4)

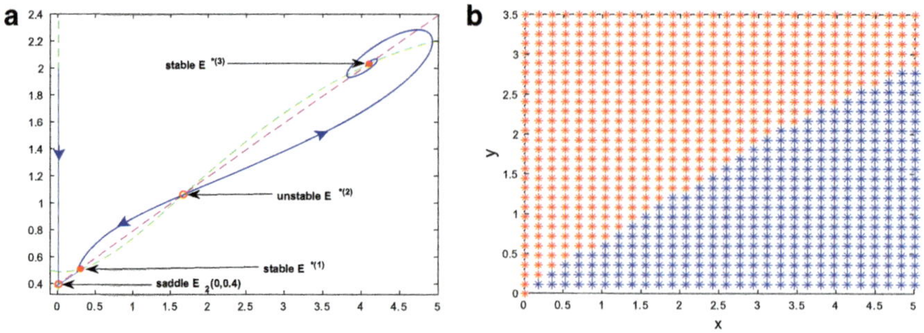

Fig. 5 **a** Bi-stability between two interior equilibrium points at $\alpha = 0.7$ and other parameters are from (4). Here green and magenta color dashed curves represent the prey and predator null-clines, respectively. **b** The basin of attraction for two stable points is shown by blue color for $E^{*(3)}(4.088, 2.0352)$ and red color for $E^{*(1)}(0.2929, 0.5171)$

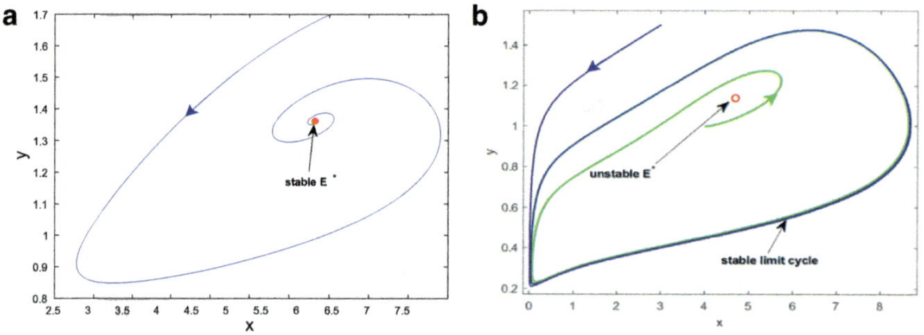

Fig. 6 **a** Phase portrait showing E^* as stable focus at $\lambda = 3$. **b** After $\lambda > \lambda^{[hf]}$, stable limit cycle surrounding unstable E^* at $\lambda = 3.9$

The phase portrait diagram illustrating bi-stability between two interior equilibrium points ($E^{*(1)}$, $E^{*(3)}$) along with one saddle interior equilibrium $E^{*(2)}$ and saddle prey-free equilibrium E_2 is shown in Fig. 5a. In such a case, the initial condition of the solution decides its convergence. Here, the solutions from red color $*$ will eventually go to the attractor $E^{*(1)}(0.2929, 0.5171)$. On the other hand, if the solution begins from blue color $*$, it will approach $E^{*(3)}(4.088, 2.0352)$ in the future (refer to Fig. 5b).

For $s = 0.2$, $\alpha = 0.7$ and keeping other parameters same as in (4), we compute the value of Hopf-bifurcation point $\lambda^{[hf]}$ by equating B_1 to zero, and we obtained $\lambda^{[hf]} = 3.6567$. At this value, $B_2 = 0.0424 > 0$, and $\left.\frac{dB_1}{d\lambda}\right|_{\lambda=\lambda^{[hf]}} = -0.1064 < 0$. Hence, according to the above theorem, the system experiences Hopf-bifurcation at $\lambda^{[hf]} = 3.6567$ around $E^*(5.0208, 1.2041)$.

For the lower value of cooperation strength λ, both species fluctuate for a finite time around their steady-state. They eventually reach the positive equilibrium E^* (see the phase portrait in Fig 6a at $\lambda = 3 < \lambda^{[hf]}$). When the value of λ is increased, E^*

loses its stability with the formation of a stable limit cycle through Hopf-bifurcation at $\lambda = \lambda^{[hf]} = 3.6567$. The phase portrait after Hopf-bifurcation is depicted in Fig. 6b at $\lambda = 3.9 > \lambda^{[hf]}$.

5 Conclusion

In the present manuscript, we proposed a modified Leslie-Gower predator-prey model employing ordinary differential equations. While formulation, we considered that the birth rate of the prey population is reduced due to the fear induced by predators. Therefore we multiply the birth rate of the prey population with the decreasing function of the predator population size, $\phi(K, y) = \frac{1}{1+Ky}$. Moreover, we assumed that predators cooperate for hunting a common target. This mechanism affects the predation rate significantly. Therefore, the group hunting term $\alpha + \lambda y$ is incorporated in the predation term. Prey species perform group defense for their survival in this situation, which is shown in the model through simplified Holling type IV functional response.

To ensure the biological validity of the system, we proved that all solutions are positive and bounded in R^2_+. We determined all feasible equilibrium points and analyzed their stability. The extinction state $E_0(0, 0)$ and predator-free state $E_1(K_1, 0)$ are always unstable. When the prey's birth rate is less than a critical value, the prey-free equilibrium $E_2(0, \frac{bs}{\omega})$ is stable. All cases of the existence of positive equilibrium E^* are discussed. We obtained sufficient conditions for the local and global stability of E^*.

It is noticed that the fear parameter K and the cooperation strength parameter λ play a crucial role in the system's dynamics. The system experiences transcritical bifurcation for the fear parameter. Moreover, we remarked that a high level of fear might cause the prey species to be extinct. The system shows a feature of bi-stability between two interior points, and it undergoes a saddle-node bifurcation with respect to λ. We noticed that both species start to fluctuate about their co-existence state when the cooperation strength λ is more than a critical value $\lambda^{[hf]}$. This change in dynamics is due to the Hopf-bifurcation at $\lambda = \lambda^{[hf]}$.

References

1. Mickaël Teixeira Alves, Frank M Hilker.: Hunting cooperation and Allee effects in predators. J. Theor. Biol. **419**:13–22 (2017)
2. Aziz-Alaoui, M.A., Daher Okiye, M.: Boundedness and global stability for a predator-prey model with modified Leslie-Gower and Holling-type II schemes. Appl. Mathem. Lett. **16**(7):1069–1075 (2003)
3. Chen, Fengde, Chen, Liujuan, Xie, Xiangdong: On a Leslie-Gower predator-prey model incorporating a prey refuge. Nonl. Anal. Real World Appl. **10**(5), 2905–2908 (2009)

4. Dubey, B., Sajan Sajan, Ankit Kumar.: Stability switching and chaos in a multiple delayed prey-predator model with fear effect and anti-predator behavior. Mathem. Comput. Simul. **188**:164–192 (2021)
5. Freedman, H.I., Wolkowicz, G.S.K.: Predator-prey systems with group defence: The paradox of enrichment revisited. Bull. Mathem. Biol. **48**, 493–508 (1986)
6. Gupta, Ashvini, Dubey, Balram: Bifurcations and multi-stability in an eco-epidemic model with additional food. European Phys. J. Plus **137**(118), 1–20 (2022)
7. Nindjin, A.F., Aziz-Alaoui, M.A.: Persistence and global stability in a delayed Leslie-Gower type three species food chain. J. Mathem. Analy. Applic. **340**(1), 340–357 (2008)
8. Pal, Saheb, Pal, Nikhil, Samanta, Sudip, Chattopadhyay, Joydev: Effect of hunting cooperation and fear in a predator-prey model. Ecolog. Compl. **39**, 100770 (2019)
9. Lawrence Perko.: *Differential Equations and Dynamical Systems*, vol. **7** (2000)
10. Sangeeta Saha, Samanta, G.P.: A prey–predator system with disease in prey and cooperative hunting strategy in predator. J. Phy. A: Mathem. Theor. **53**(48):485601 (2020)
11. Shang, Zuchong, Qiao, Yuanhua: Bifurcation analysis of a Leslie-type predator-prey system with simplified Holling type IV functional response and strong Allee effect on prey. Nonl. Anal. Real World Appl. **64**, 103453 (2022)
12. Wang, Xiaoying, Zanette, Liana, Zou, Xingfu: Modelling the fear effect in predator-prey interactions. J. Mathem. Biol. **73**(5), 1179–1204 (2016)

Chaotic Behavior in a Novel Fractional Order System with No Equilibria

Santanu Biswas, Humaira Aslam, Satyajit Das, and Aditya Ghosh

Abstract This article takes into consideration a novel chaotic system of four dimensional fractional order having no equilibria. We cannot use mathematical methods such as Melnikov's and Shilnikov's method to prove that the given system is chaotic. We shall analyse the dynamical features of the fractional order system by using predictor-corrector algorithm. This method reports chaotic dynamics. We shall apply the basic ideas of non linear dynamical analysis such as bifurcation diagrams and Lyapunov exponents to recognise the chaotic behavior for the given system. One interesting phenomena for the system is that it has cascade of period doubling bifurcations and chaotic attractors without having any equilibrium points.

Keywords Fractional calculus · Lyapunov exponents · Chaotic dynamics · Predictor-corrector algorithm · No equilibrium point

1 Introduction

Fractional calculus is an ongoing topic which is being used since the last 300 years, however its implementation has been inflated in recent years. The mathematical phenomenon describe real objects more precisely than the classical integer methods. The concept of fractional calculus for example has been implemented for modelling circuit theory [1], control systems [13] etc. Li et al. [10] fractional order Chua's circuit, [22] fractional order Rossler system, [16], which describes that fractional order systems can also behave chaotically. Moreover all these fractional order systems deal with either one or maybe more than one equilibrium points; except for only few systems that have already been discovered exhibiting chaos without any equilibrium points until now.

S. Biswas (✉) · H. Aslam · S. Das · A. Ghosh
Department of Mathematics, Adamas University, Kolkata 700126, India
e-mail: santanubiswas1988@gmail.com

S. Biswas
Department of Mathematics, Jadavpur University, Kolkata 700032, India

S. Banerjee and A. Saha (eds.), *Nonlinear Dynamics and Applications*,
Springer Proceedings in Complexity,
https://doi.org/10.1007/978-3-030-99792-2_91

Sprott [14] found 19 chaotic models with five terms and six terms, including linear and quadratic terms by exhaustive computer searching with no more than three equilibrium points. Inspired by his pioneering work, the chaotic dynamics in an integer order model has been examined in [2, 9, 18, 20]. References [17, 19] introduced and analyzed new chaotic systems having no equilibrium points. The presence of chaos cannot be verified by Shilnikov method as, they can not have homoclinic or heteroclinic orbits. Referring to the fractional order system, as far as our knowledge goes only two systems [3, 11, 12] describe a system having chaotic dynamics without any equilibria.

In this article, we have described a novel 4D fractional jerk system with hidden attractors. The dynamics of the non commensurate order fractional model & the commensurate order fractional model, has been explained individually. The article can be helpful to solve the problem of sudden chaotic oscillation caused by hidden attractors, thus providing a good reference and inspiration for solving similar engineering oscillation problems.

Based on the given chaotic attractors it is exciting to develop the chaos theory in order to create new systems. From this point of view a novel fractional order system with chaotic dynamics is described in the given article. The presence of chaos is illustrated by using various bifurcation diagrams and maximum Lyapunov exponents. Using Adams–Bashforth Moulton algorithm we have solved the given fractional order system. One interesting phenomena for the above said system is that it has cascade of period doubling bifurcations and chaotic attractors without having any equilibrium points.

The remaining article is organized in the following manner: Sect. 2 focuses on the development of the integer order model with its basic dynamics; In Sect. 3, we deal with the fractional order model having no equilibrium. Elaborate discussion of the dynamical analysis of the fractional model is done in Sect. 4. The predictor corrector algorithm is described in Sect. 4 as well. We conclude the article with a discussion.

2 Proposed System

In the search for chaotic flows, we were inspired by [14] case K system,

$$\begin{aligned} \tfrac{dx}{dt} &= xy - z \\ \tfrac{dy}{dt} &= x - y \\ \tfrac{dz}{dt} &= x + 0.3z. \end{aligned} \tag{1}$$

Equation 1 has two equilibria.

We performed a search for additional chaotic system with no equilibria. We added a constant to each of the derivatives in Eq. 1 with the simplest four dimensional extension of the system using linear feedback control. We consider a general parametric form as

$$
\begin{aligned}
\frac{dx}{dt} &= ay^2 + \gamma_1 xy + \gamma_2 x^2 + \gamma_3 z^2 + \gamma_4 xz + \gamma_5 yz - z + a_1 \\
\frac{dy}{dt} &= bx - y + a_2 \\
\frac{dz}{dt} &= cx + dz + a_3 + k_1 w \\
\frac{dw}{dt} &= -k_2 z
\end{aligned} \tag{2}
$$

We search for the cases where we can show algebraically that the equilibrium points are imaginary. An extensive search for the chaotic system with no equilibrium points found the example

$$
\begin{aligned}
\frac{dx}{dt} &= ay^2 - z + a_1 \\
\frac{dy}{dt} &= bx - y \\
\frac{dz}{dt} &= cx + dz + k_1 w \\
\frac{dw}{dt} &= -k_2 z
\end{aligned} \tag{3}
$$

where x, y, z, w are state variables and $a, a_1, b, c, d, k_1, k_2$ are real constant parameters. Eq. 3 is dissipative if $d < 1$.

The system (3) can not have chaotic solution in few cases. So, we proved the following Theorem.

Theorem 1 *Suppose that the following conditions hold:*

1. $k_1 = 0, c = 0, a > 0$ *and* $a_1 > 0$
2. $k_1 = 0, c = 0, a < 0$ *and* $a_1 < 0$

then system (3) does not have bounded chaotic attractors.

Proof From system (3) we get,

$$\ddot{y} - ab\dot{y}^2 + (1+d)\dot{y} + dy = \int_0^t [bd(ay^2 + a_1) - bk_1 w - bcx]dy + C. \tag{4}$$

Under the above said assumptions Eq. 4 has a monotone left hand side. Arguing as [8] we can say that the system can not be chaotic.

2.1 Dynamics of the System (3)

The model (3) has no point of equilibria for $a_1 > 0$. Phase plot for the system (3) is drawn in Fig. 1 with parameter values as $a = a_1 = 0.1$; $b = 0.1$; $c = 0.4$; $k_1 = k_2 = 0.01$; $d = 0.25$ and the initial value was assumed as $[1, 1, 0, 1]$. Figure 1 depicted chaotic dynamics for the system (3). Due to the chaotic behaviour of the system, the trajectories diverges from $[1, 0.1, 0, 40]$ to the higher values. Small changes in the initial value may results dramatically different trajectories.

The divergence and convergence exponential rates of the trajectories close by in the phase plane of the given chaotic system is measured by the Lyapunov exponents.

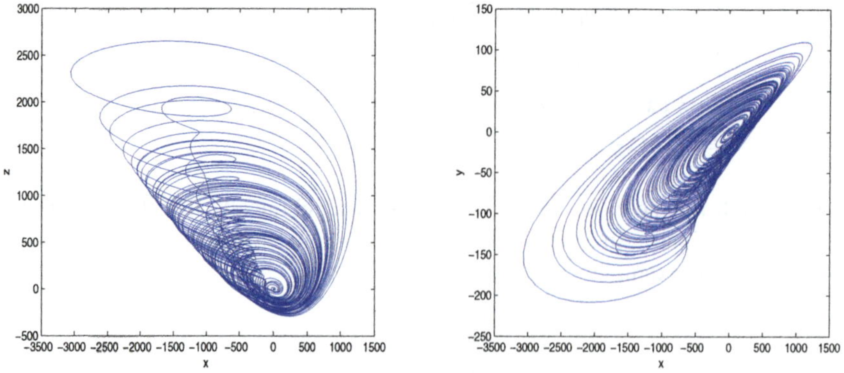

(a) Phase plot for the system 3 in x-z plane

(b) Phase plot for the system 3 in x-y plane

Fig. 1 Phase plot for the system equation 3 in x-z plane and x-y plane

Hence we have constructed these Lyapunov exponents for the given system (3) by making use of the algorithm in [21] to verify that the given system is chaotic or not. We calculated the Lyapunov exponents and plot in the Fig. 2b for the system (3). We see that the two Lyapunov exponents is negative (cyan, red), one is zero (green) and another is positive (blue).

We define the Lyapunov dimension by

$$D_{ky} = j + \frac{1}{L_{j+1}} \Sigma_{i=1}^{j} L_i,$$

where j is the largest integer satisfying $\Sigma_{i=1}^{j} L_i \geq 0$ and $\Sigma_{i=1}^{j+1} L_i < 0$. For the system equation 3, $D_{ky} = 3.1456 > 3$. Hence it is an indication of a strange attractor.

The bifurcation diagram for the given system (3) is depicted in Fig. 2a with reference to the parameter d in the range of $d \in [-0.02, 0.25]$. The values of the given parameters are fixed as $a = a_1 = 0.1;\ b = 0.1;\ c = 0.4;\ k_1 = k_2 = 0.01$. Although the system (3) has no equilibrium but still the system (3) has period doubling bifurcation route to chaos. After d crosses the critical value 0, the system equation 3 losses it's stability and undergoes a period-doubling bifurcation. Gradual increase of d makes the system chaotic. We can observe that the bifurcation diagram and the Lyapunov exponents spectrum very well coincide.

3 Fractional Order Model

By fractional-order systems, we refer to dynamical systems which can be modeled using a fractional differential equation along with a non-integral derivative. Fractional

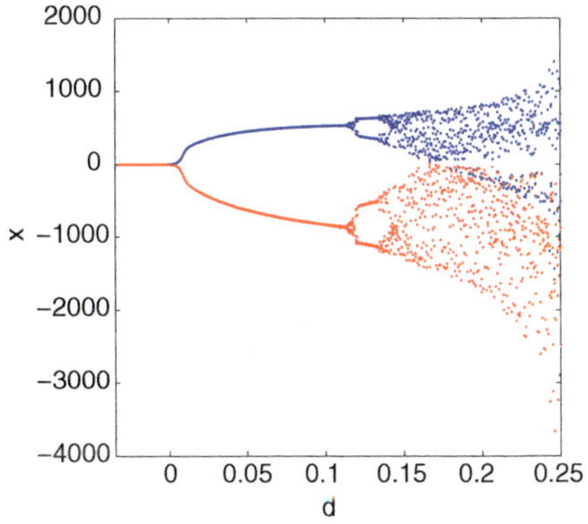

(a) Bifurcation diagram for the system 3 with respect to parameter d

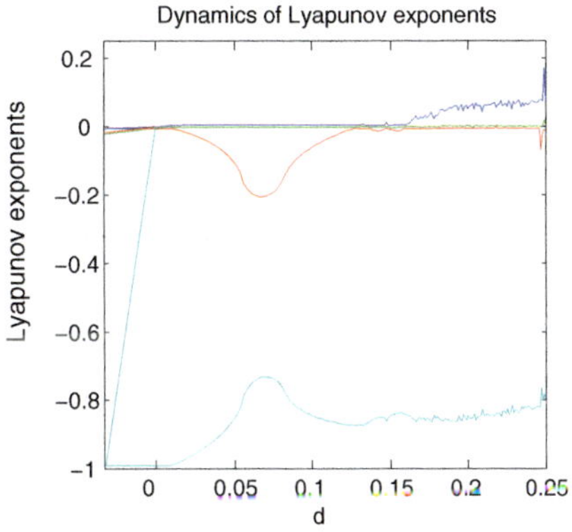

(b) Lyapunov exponent diagram for the system 3 with respect to parameter d

Fig. 2 Chaotic dynamics for the system equation 3 with respect to parameter d

order systems are quite useful in understanding the characteristics of the dynamical system in various fields of electrochemistry, biology, physics, and chaotic systems. Recently fractional-order systems have captured a lot of interest and recognition because of their usefulness in providing an exact description of various nonlinear phenomena. Control theory has a lot of applications in fractional-order systems since several physical systems cannot be effectively modeled using differential equations of integer order, hence fractional-order systems have an important role to play here. The presence of non-integral derivatives is the primary reason for selecting fractional-order systems in mathematical modeling problems. The fractional-order controller gives comparatively more adjustable time and frequency response for any given control system. Being a modified form of the integer order systems fractional order systems adjust the controllers more accurately according to the system requirements. Hence fractional-order systems are preferred over integer order systems for modeling various dynamical systems.

In this section we consider the fractional-order system denoted as Eq. 5. The standard derivative is replaced by a fractional derivative as follows:

$$\begin{aligned} D^{q_1}x &= ay^2 - z + a_1 \\ D^{q_2}y &= bx - y \\ D^{q_3}z &= cx + dz + k_1 w \\ D^{q_4}w &= -k_2 z \end{aligned} \tag{5}$$

where $0 < q_1, q_2, q_3, q_4 \leq 1$; D^{q_i} denote the Caputo fractional operator with initial time $t_0 = 0$. When $q_1 = q_2 = q_3 = q_4 = 1$, the above system becomes Eq. 3.

Next, we consider two methods to solve the system (5). We discuss the methods in next two sections.

4 Adams–Bashforth–Moulton Method

Diethelm and Ford [6] already discussed about converting to Volterra integral equations from fractional differential equations with initial conditions. Now, by applying the predictor-corrector algorithm [7], the solution of the system (5) can be written as:

$$\begin{aligned}
x_{n+1} &= x_0 + \frac{h_1^q}{\Gamma(q_1+2)}\{ay_{n+1}^p y_{n+1}^p - z_{n+1}^p + a_1 + \Sigma_{j=0}^n a_{1,j,n+1}(ay_j^2 - z_j + a_1)\}\\
y_{n+1} &= y_0 + \frac{h_2^q}{\Gamma(q_2+2)}\{bx_{n+1}^p - y_{n+1}^p + \Sigma_{j=0}^n a_{2,j,n+1}(bx_j - y_j)\}\\
z_{n+1} &= z_0 + \frac{h_3^q}{\Gamma(q_3+2)}\{cx_{n+1}^p + dz_{n+1}^p + k_1 w_{n+1}^p + \Sigma_{j=0}^n a_{3,j,n+1}(cx_j + dz_j + k_1 w_j)\}\\
w_{n+1} &= w_0 + \frac{h_4^q}{\Gamma(q_4+2)}\{-k_2 z_{n+1}^p + \Sigma_{j=0}^n a_{4,j,n+1}(-k_2 z_j)\}
\end{aligned} \tag{6}$$

in which

$$\begin{aligned}
x_{n+1}^p &= x_0 + \frac{1}{\Gamma(q_1)}\{\Sigma_{j=0}^n b_{1,j,n+1}(ay_j^2 - z_j + a_1)\}\\
y_{n+1}^p &= y_0 + \frac{1}{\Gamma(q_2)}\{\Sigma_{j=0s}^n b_{2,j,n+1}(bx_j - y_j)\}\\
z_{n+1}^p &= z_0 + \frac{1}{\Gamma(q_3)}\{\Sigma_{j=0}^n b_{3,j,n+1}(cx_j + dz_j + k_1 w_j)\}\\
w_{n+1}^p &= w_0 + \frac{1}{\Gamma(q_4)}\{\Sigma_{j=0}^n b_{4,j,n+1}(-k_2 z_j)\}
\end{aligned} \tag{7}$$

4.1 Chaos and Bifurcations with $q = q_1 = q_2 = q_3 = q_4$ for the System (5)

Assuming the parameter values as $a = 0.1$; $b = 0.1$; $c = 0.4$; $k_1 = k_2 = 0.01$; $a_1 = .001$; $d = 0.245$; $q = q_1 = q_2 = q_3 = q_4 = 0.98$ and initial conditions as = $(1, 0.1, 0, 40)$ we get the following phase space trajectory in Fig. 3. The chaotic motion which was identified in Fig. 3 is confirmed by the maximum Lyapunov exponents following by [15] and plotted in Fig. 4.

In order to make it direct, a bifurcation diagram shall be drawn with respect to $q = q_1 = q_2 = q_3 = q_4$ for $0.86 \le q \le 0.99$ and the rest of the parameter values are fixed like in Fig. 3. The bifurcation diagram with respect to q plotted in Fig. 5, depicts the complex dynamical features in our presented model equation 5 from the limit cycle to chaos and on gradually increasing q we observe that the system switches its stability such that from being stable focus to limit cycle oscillation it becomes limit cycle oscillation to chaotic oscillation. We can observe that in Fig. 5

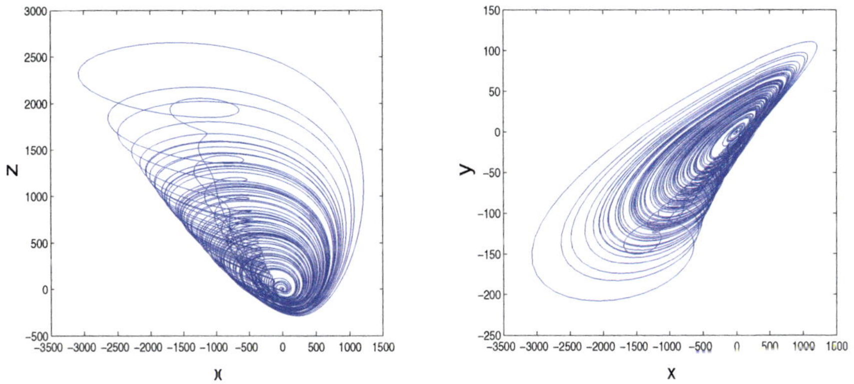

(a) Phase plot for the system 5 in x-y plane

(b) Phase plot for the system 5 in x-z plane

Fig. 3 Phase plot for the system (5) with commensurate fractional order

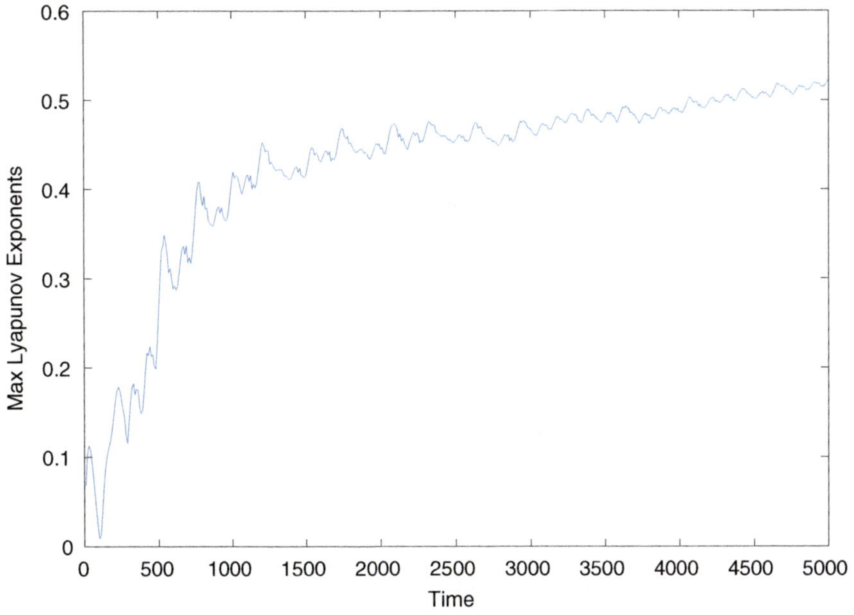

Fig. 4 Maximum Lyapunov exponents for the system (5)

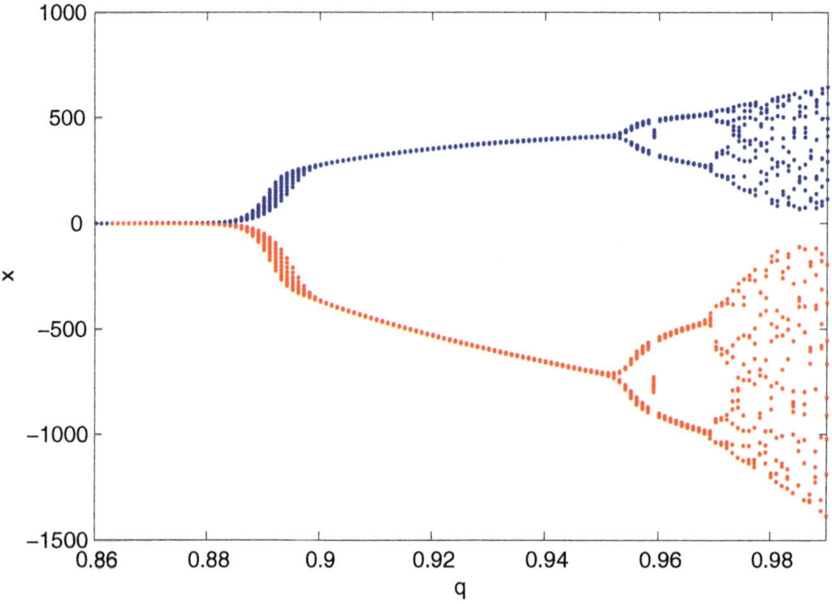

Fig. 5 Bifurcation diagram for the system (5) with respect to the parameter q

for $q \in [0.86, 0.885)$ the system is stable, for $q \in [0.885, 0.968)$ it exhibits limit cycle oscillations and for $q \in [0.968, 0.99]$ it shows a comparatively higher periodic as well as chaotic oscillations.

4.2 *Chaos and Bifurcations with Different q_i for the System (5)*

For the incommensurate order case we find dynamics present in the system (5) for $q_1 = 0.98$; $q_2 = 0.85$; $q_3 = 0.98$; $q_4 = 0.4$ as well. The corresponding phase plots and maximum Lyapunov exponents [4, 5] for the system equation 5 has been drawn in Fig. 6 and Fig. 7 respectively. We did not try to find the lowest order chaos for the incommensurate order case.

In order to understand the dynamics of the given system (5) with different q_i We shall consider the following three cases:

1. First with respect to q_1 for $0.7 < q_1 \leq 0.99$, we shall draw a bifurcation diagram where the remaining parameters are fixed as in Fig. 3. The bifurcation diagram with reference to q_1 shown in Fig. 8a, identifies the complex dynamical behaviour in the model (5) from the limit cycle to chaos. As we increase the values of q_1 the system (5) converts from a stable focus to limit cycle oscillation and then limit cycle oscillation to chaotic oscillation. Figure 8a depicts that for $q_1 \in (0.7, 0.78)$

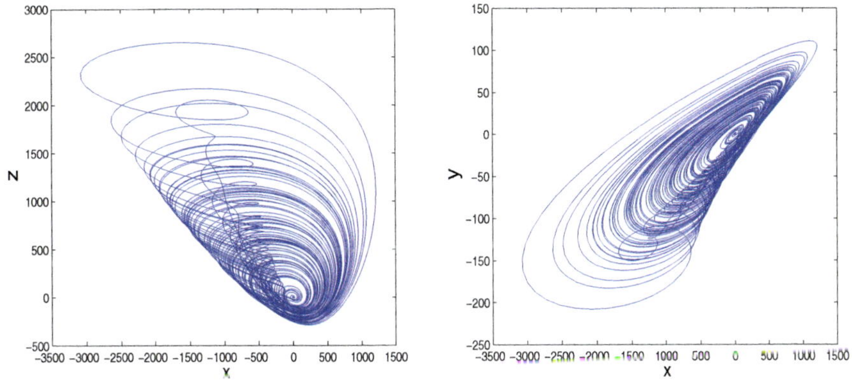

(a) Phase plot for the system 5 in x-y plane

(b) Phase plot for the system 5 in x-z plane

Fig. 6 Phase plot for the system (5) with non-commensurate fractional order

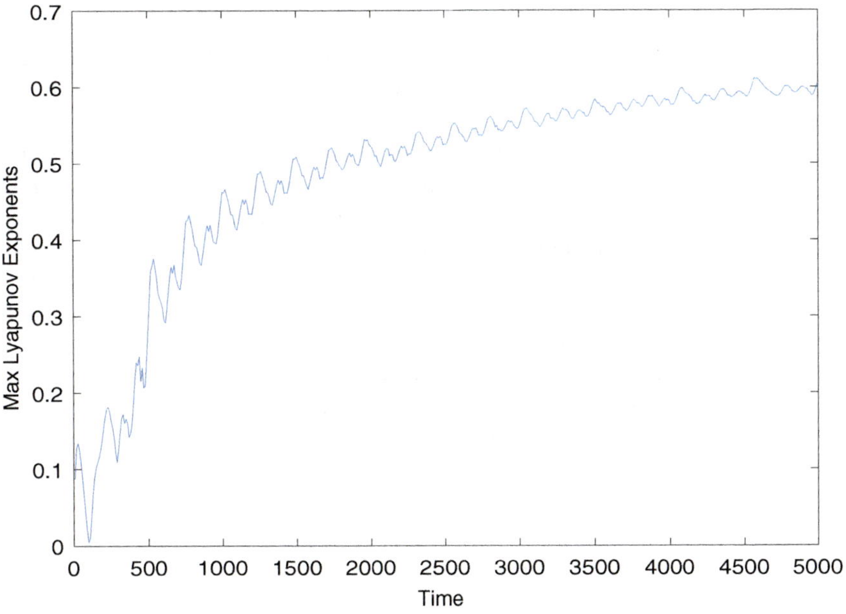

Fig. 7 Maximum Lyapunov exponents for the system (5)

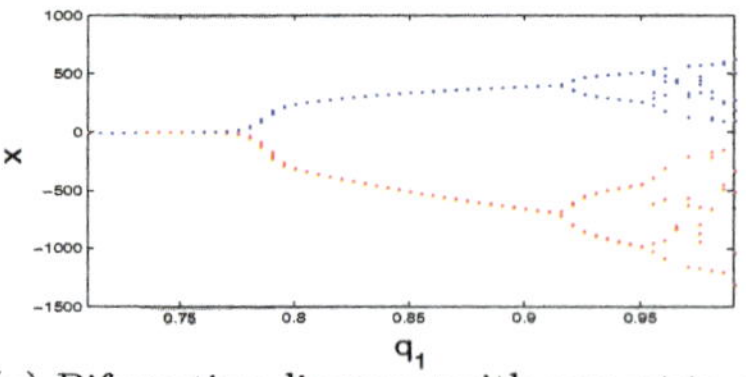

(a) Bifurcation diagrams with respect to q_1.

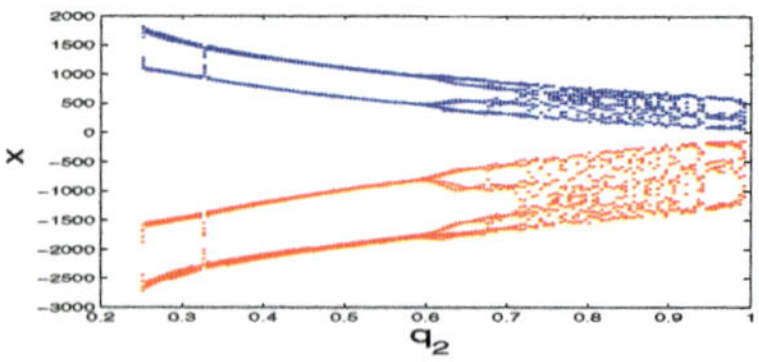

(b) Bifurcation diagrams with respect to q_2 .

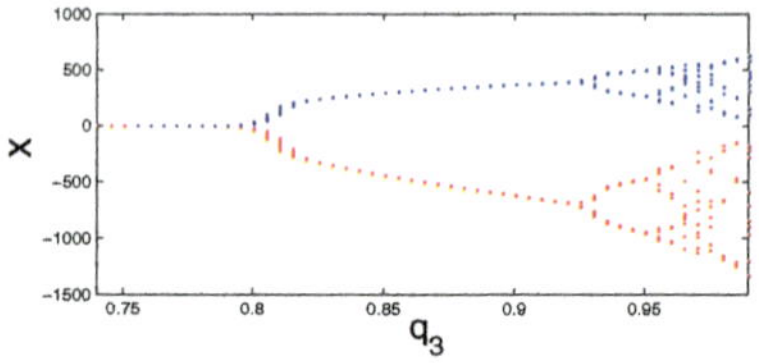

(c) Bifurcation diagrams with respect to q_3.

Fig. 8 Bifurcation diagrams with respect to different q_i for the system (5)

the system equation 5 is stable, for $q_1 \in [0.78, 0.95)$ it exhibits limit cycle oscillations, and for $q_1 \in [0.95, 0.99]$ it shows a higher periodic as well as chaotic oscillation.

2. In the next step we shall draw the bifurcation diagram with reference to the parameter q_2 for $q_2 \in [0.25, 0.99]$ where q_1 is kept constant at 0.98 and $q_3 = q_4 = 0.98$. From the Fig. 8b the complex dynamical behaviour of the system (5) with reference to q_2 including chaos is clearly evident. We can also notice that for $q_2 \in [0.25, 0.60)$ the system exhibits 4 - periodic solution, a 8 - periodic solution can be seen for $0.60 < q_2 < 0.70$ and for $q_2 \in [0.70, 0.99]$ the system exhibits higher periodic and chaotic oscillations.
3. On gradually increasing the parameter q_3 the system equation 5 exhibits chaotic dynamics. Figure 8c shows that q_3 behaves exactly in the same way as q_1. We omit the details. Here, all other parameters are fixed as in Fig. 3.

4.3 Chaos and Bifurcations with Different d for the System (5)

Let us take the parameter values as $a = 0.1$; $b = 0.1$; $c = 0.4$; $k_1 = k_2 = 0.01$; $a_1 = 0.001$; $q_1 = q_2 = q_3 = q_4 = 0.98$ and vary d from -0.02 to 0.25 with the initial condition $(1, 0.1, 0, 40)$. The bifurcation diagram is shown in Fig. 9. On comparing the fractional order system (5) with the integer order system (3) we see that they both have the same kind of tendency with the parameter d. For $d \in [-0.02, 0.06)$ the system shows stable behaviour, for $d \in [0.06, 0.18)$ the system exhibits limit cycle behaviour and higher periodic, chaotic oscillations can be seen for $d \in [-0.18, 0.25]$.

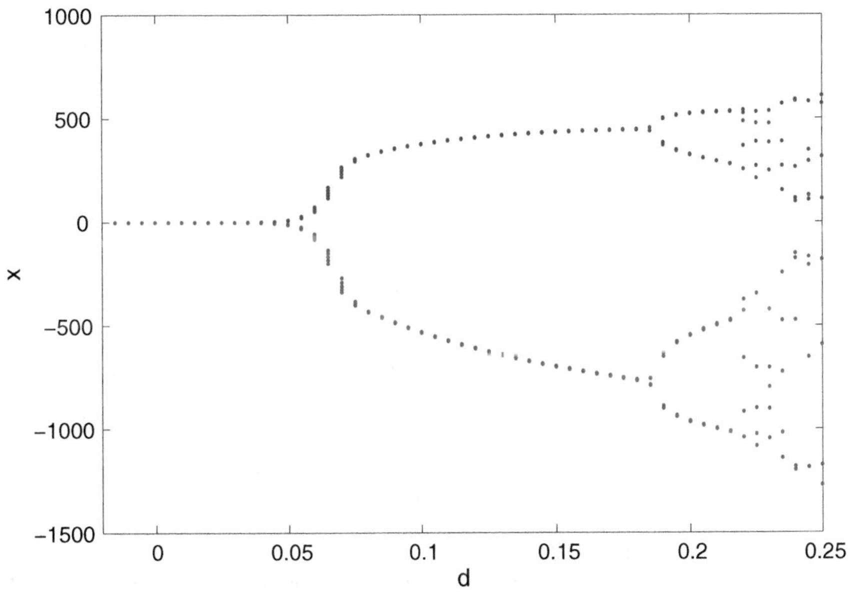

Fig. 9 Bifurcation diagram for the system (5) with respect to the parameter d

5 Conclusion

The dynamics for the proposed system Eqs. 3 (integer order) and 5 (fractional order) has been studied in this article. We solve the system (5) using Adams-Bashforth-Moulton method. The method describes chaotic dynamics in the system (5) though the system has no equilibria. Several bifurcation diagrams and maximum Lyapunov exponents are drawn to establish our results by taking different values of the parameter d and the fractional order q_i. The existence of the attractors of the same kind in the integer order system and fractional order system both are a new field to explore, also it represents a new exciting phenomenon, which may serve helpful in the forthcoming research regarding the relations between integer order system and fractional order system. We leave synchronization control to the considered fractional model for future work.

Acknowledgements Research of Santanu Biswas is supported by Dr. D. S. Kothari Postdoctoral Fellowship under University Grants Commission scheme (Ref. No. F.4-2/2006 (BSR)/MA/19-20/0057). Special thanks to Prof. Sudeshna Banerjee for her valuable suggestions to improve the quality of the article.

References

1. Arena, P., Caponetto, R., Fortuna, L., Porto, D.: Nonlinear Non Integer Order Circuits and Systems-an Introduction. World Scientific, Singapore (2000)
2. Bayani, A., Rajagopal, K., Khalaf, A., Jafari, S., Leutcho, G., Kengne, J.: Dynamical analysis of a new multistable chaotic system with hidden attractor: antimonotonicity, coexisting multiple attractors, and offset boosting. Phys. Lett. A **383**(13), 1450–1456 (2019)
3. Cafagna, D., Grassi, G.: Chaos in a fractional-order Rossler system. Commun. Nonlinear Sci. Numer. Simul. **19**, 2919–2927 (2014)
4. Danca, M.: Matlab code for Lyapunov exponents of fractional-order systems, part ii: the non-commensurate case. Int. J. Bifurc. Chaos **31**(12), 2150187 (2021)
5. Danca, M., Kuznetsov, N.: Matlab code for Lyapunov exponents of fractional-order systems. Int. J. Bifurc. Chaos **28**(05), 1850067 (2018)
6. Diethelm, K., Ford, N.: Analysis of fractional differential equations. J. Math. Anal. Appl. **265**, 229–248 (2002)
7. Diethelm, K., Ford, N., Freed, A.: A predictor-corrector approach for the numerical solution of fractional differential equations. Nonlinear Dyn. **29**, 3–22 (2002)
8. Fu, Z., heidel, J.: Non chaotic behaviour in three-dimensional quadratic systems. Nonlinearity 1289–1303 (1997)
9. Jafari, S., Sprott, J., Hashemi Golpayegani, S.: Elementary quadratic chaotic flows with no equilibria. Phys. Lett. A **377**, 699–702 (2013)
10. Li, C., Deng, W., Xu, D.: Chaos synchronization of the Chua system with a fractional order. Phys. Lett. A **360**, 171–85 (2006)
11. Li, H., Liao, X., Luo, M.: A novel non-equilibrium fractional-order chaotic system and its complete synchronization by circuit implementation. Nonlinear Dyn. **68**, 137–49 (2012)
12. Liu, T., Yan, H., Banerjee, S., Mou, J.: A fractional-order chaotic system with hidden attractor and self-excited attractor and its DSP implementation. Chaos Solitons & Fractals **145**, 110791 (2021)
13. Podlubny, I.: Fractional Differential Equations. Academic, New York (1999)
14. Sprott, J.: Some simple chaotic flows. Phys. Rev. E **50**(2) (1994)
15. Sprott, J.: Chaos and Time Series Analysis (Chap. 5). Oxford University Press, Oxford (2003)
16. Sun, K., Wang, X., Sprott, J.: Bifurcations and chaos in fractional-order simplified Lorentz system. Int. J. Bifurc. Chaos **20**(4), 1209–1219 (2010)
17. Tahir, F.R., Jafari, S., Pham, V., Volos, C., Wang, X.: A novel no-equilibrium chaotic system with multiwing butterfly attractors. Int. J. Bifurc. Chaos **25**(4), 1550056 (2015)
18. Tian, H., Wang, Z., Zhang, P., Chen, M., Wang, Y.: Dynamic analysis and robust control of a chaotic system with hidden attractor. Complexity (2021)
19. Wang, Z., Cang, S., Ochola, E., Sun, Y.: A hyperchaotic system without equilibrium. Nonlinear Dyn. **69**, 531–7 (2012)
20. Wei, Z.: Dynamical behaviors of a chaotic system with no equilibria. Phys. Lett. A **376**, 102–8 (2011)
21. Wolf, A., Swift, J., Swinney, H., Vastano, J.: Determining Lyapunov exponents from a time series. Phys. D **16**, 285–317 (1985)
22. Zhang, W., Zhao, S., Li, H., Zhu, H.: Chaos in a fractional-order Rossler system. Chaos Solitons Fractals **42**, 1684–1691 (2009)

Soliton Dynamics in a Weak Helimagnet

Geo Sunny, L. Kavitha, and A. Prabhu

Abstract We considered a Helimagnetic nanowire, with an antisymmetric spin interaction known as the Dzyaloshinskii-Moriya (DM) interaction in analogy with Cholesteric Liquid Crystal (CLC) model. We derive the nonlinear dynamical equation after boronizing the nanowire with the Holstein–Primakoff (HP) transformation aided with Glauber's coherent-state representation. The governing equation of motion is the celebrated Discrete Non-Linear Schrodinger (DNLS) equation for the Helimagnetic nanowire. We attempt to solve the DNLS equation, using Jacobian elliptical function (JEF) technique, and analyzed the competency of the helicity and the weak DM interaction on the dynamics of helimagnetic nanowire.

Keywords Soliton · Helimagnet · Dzyaloshinskii–Moriya (DM) interaction

1 Introduction

Recently, helimagnetic systems have been gaining lots of attention due to its contribution in the field of data storage technology [1, 2] as it can be easily fabricated into different structures like arrays of wires, dots, rings and sheets which have varied possibilities in the development of magnetic storage devices. Amid this, helimagnetic nanowire exhibits uniqueness due to its tunable magnetization properties which arises from its inherent shape anisotropy [3]. The current research in the field of helimagnetic nano-wire is the development of ultra-high density magnetic recordings incorporating the DM interaction. The DM interaction plays a crucial role in the for-

G. Sunny · L. Kavitha (✉)
Department of Physics, School of Basic and Applied Sciences, Central University of Tamil Nadu, Thiruvarur 610005, India
e-mail: lkavitha@cutn.ac.in

L. Kavitha
The Abdus Salam International Centre for Theoretical Physics, Trieste, Italy

A. Prabhu
Department of Physics, Periyar University, Salem 636011, India

S. Banerjee and A. Saha (eds.), *Nonlinear Dynamics and Applications*,
Springer Proceedings in Complexity,
https://doi.org/10.1007/978-3-030-99792-2_92

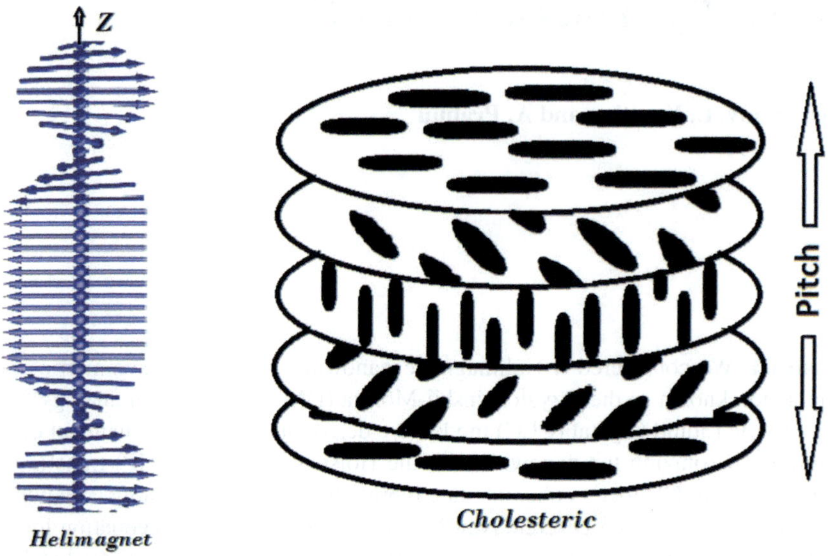

Fig. 1 Schematic representation of a helimagnet

mation of chiral spin texture. It is generated by the strong spin-orbit coupling (SOC) between the atomic spins due to the chiral interaction between them. Recent study by Sampaio et al. demonstrated the influence of DM interaction in the magnetic memory cell performance [4]. Various experimental researches are also being conducted in this field, these provide validation to some of the theoretical works [5–7].

Various models have been proposed to investigate the dynamics of helimagnetic systems [8]. Chandra et al. showed that the large quantum fluctuations induce an anisotropy in the helimagnet [9]. Beula et al. investigated the influence of constant magnetic field on the dynamics of an anisotropic helimagnet and found that the spin configurations are unstable when the applied field is normal to the anisotropic axis [10]. Martin et al. studied the ground state of MnGe cubic alloy and observed proliferation in the long wavelength with gapless spin fluctuations which is associated with evolution of the helical correlation length [11]. Daniel et al. considered a Helimagnetic model in analogous to cholesteric liquid crystal and found that the soliton excitations govern the nonlinear dynamics of the helimagnet. They found that the helicity does not alter the nature of the soliton during propagation, however it suffers with some fluctuations in the localized region [12] (Fig. 1). The dynamics of Discrete Breather (DB) in an antiferromagnetic system under the influence of DM interaction has already been explored [13, 14]. The discrete Breathers are spatially localized nonlinear excitations which appear in classical discrete lattice systems [13, 15–17].

It has been a concern for the scientists to find an exact solution for the nonlinear dynamic equation. These solutions may provide more insight towards physical phenomenons happening in biology, chemistry, physics and various other fields. To obtain an exact solution to these dynamical equations various methods have been implemented such as the inverse scattering method, the tangent hyperbolic function method, the Jacobi elliptical function method, the sine-cosine function method, the trail function method, the nonlinear transformation method and so on. In this paper, we attempt to derive an exact solution for the helimagnetic system under the influence of DM interaction using jacobi elliptical method.

2 Governing Dynamical Model and Equation

We consider an one dimensional ferromagnetic system in analogous to the CLC model in order to incorporate the helicity. As the play role of the anisotropy is much significant for the intrinsic localisation of nonlinear spin waves, we consider an anisotropic weak ferromagnetic spin chain with appreciable helicity as represented by the Hamiltonian,

$$\tilde{H} = -\sum_{n}[J(\vec{S}_n \cdot \vec{S}_{n+1}) + \vec{D} \cdot (\vec{S}_n \times \vec{S}_{n+1}) + \tau \left\{[\hat{k} \cdot (\vec{S}_n \times \vec{S}_{n+1})]^2 - q_1^2\right\} - A(S_n^z)^2 - A'(S_n^z)^4] \tag{1}$$

where $\vec{S}_n = (S_n^x, S_n^y, S_n^z)$ represents the local spin vector at the lattice site 'n'. $J > 0$ characterize the nearest-neighbor exchange interaction, which is a short ranged strong spin-spin exchange coupling. The second term designates the presence of DM interaction, and lead to the canting of spins which depends on the direction of the monoaxial vector $\vec{D} = D\hat{e}_z$, restricted to z axis. The cross product $\vec{D}_n \cdot (\vec{S}_n \times \vec{S}_{n+1})$ characterize the spin-flop hopping arises due to the presence of DM interaction which occurs in the systems lacking inversion symmetry. The term represents the helical spin interaction similar to that of the molecular interaction in a CLC. We adopt a similar kind of helical twisting designated as $\left\{[\hat{k} \cdot (S_n \times S_{n+1})]^2 - q_1^2\right\}^2$, where $k = (0, 0, 1)$, q_1 is the helical pitch which controls the long-range ordering of spins. The last two terms designate the lower and higher order anisotropy mainly arised due to the combined effect of crystal field effect and spin-orbit interaction. A and A' respectively are the lower and higher order single-ion unizxial anisotropy parameters, when $A(> 0)$, we assume that all spins align along the z axis, being the easy axis of magnetization in the ground state. We map the spin operators of our one dimensional helimagnetic systems of spin-s moments on a discrete lattice to bosonic creation $a_i^\dagger$ and annihilation operator a_i using the Holstein–Primakoff (H–P) representation [18]

$$\hat{S}_i^+ = (2S)^{1/2}\Big[1 - \frac{a_i^\dagger a_i}{2S}\Big]^{1/2} a_i,$$

$$\hat{S}_i^- = (2S)^{1/2} a_i^\dagger\Big[1 - \frac{a_i^\dagger a_i}{2S}\Big]^{1/2},$$

$$\hat{S}_i^z = \Big[S - a_i^\dagger a_i\Big].$$

where $a_i^\dagger(a_i)$ is a bosonic creation(annihilation) operator at site $'i'$ satisfies the bosonic commutation relations in the second quantisation formulation of the helimagnetic spin lattice as $[a_j, a_i^\dagger] = \delta_{ij}$, $[a_j, a_i] = [a_j^\dagger, a_i^\dagger] = 0$. and $n_i = a_i a_i^\dagger$ represents the number operator. In this mapping, each Holstein–Primakoff bosonic operator represents a spin-1 moment in the $-z$ direction and the vaccum state of the bosons, i.e. $|n = 0>$ has a spin of $+S$ in the z direction. In this mapping, the vaccum state is not always the ground state, thereby representing a perturbation from the classical ferromagnetic ground state. Conceived by this physical picture, it is manifested that the factor $\sqrt{2S - n_i}$, limits the number of HP bosons to $2S$ on a given site i, since the z-projection of the spin moment at a given site 'i' must be between $-S$ and $+S$. Since low temperatures, the number of perturbations about the classical ground state is very small $n_i << S$, we invoke power series expansion of the HP transformation for the spin operators as,

$$\hat{S}_n^+ = \sqrt{2}\Big[1 - \frac{\epsilon^2}{4} a_n^\dagger a_n - \frac{\epsilon^4}{32} a_n^\dagger a_n a_n^\dagger a_n - O\epsilon^6\Big]\epsilon a_n,$$

$$\hat{S}_n^- = \sqrt{2}\epsilon a_n^\dagger\Big[1 - \frac{\epsilon^2}{4} a_n^\dagger a_n - \frac{\epsilon^4}{32} a_n^\dagger a_n a_n^\dagger a_n - O\epsilon^6\Big],$$

$$\hat{S}_n^z = [1 - \epsilon^2 a_n^\dagger a_n]. \tag{2}$$

where $\epsilon = 1/\sqrt{S}$

$$i\hbar\frac{\partial a_n}{\partial t} = [a_n, H] = F(a_n^\dagger, a_n, a_{n+1}^\dagger, a_{n+1}).$$

We then introduce the Glauber's coherent—state representation [19] defined by the product of the multimode coherent states $|u>= \prod_n |u_n>$ with $<u|u>= 1$. Here $|u(n)>$ is an eigenstate vector of the annihilation operator a_n i.e., $a_n|u>= u_n|u>$, and u_n is the coherent amplitude. The p-representation of nonlinear equation leads,

$$
\begin{aligned}
i\frac{du_n}{dt} &= \epsilon^2\Big[2(\tau - A - 2A')u_n - J(u_{n-1} + u_{n+1}) - iD^z(u_{n+1} - u_{n-1})\Big] \\
&+ \frac{\epsilon^4}{4}\Big[J[| u_n |^2 (u_{n+1} + u_{n-1}) + u_n^2(u_{n+1}^* + u_{n-1}^*) + | u_{n+1} |^2 u_{n+1} + | u_{n-1} |^2 u_{n-1}] \\
&- 4(J + 2\tau)[| u_{n+1} |^2 + | u_{n-1} |^2]u_n + 8\tau[u_{n+1}^2 + u_{n-1}^2]u_n^* + 8(A + A') | u_n |^2 u_n \\
&+ iD^z[2(u_{n+1} - u_{n-1}) | u_n |^2 - (u_{n+1}^* - u_{n-1}^*)u_n^2 + | u_{n+1} |^2 u_{n+1} - | u_{n-1} |^2 u_{n-1}]\Big] \\
&+ iD^z\frac{\epsilon^5}{32}\Big[3 | u_n |^4 (u_{n+1} - u_{n-1}) - 2 | u_n |^2 u_n^2(u_{n+1}^* - u_{n-1}^*) + | u_{n+1} |^4 u_{n+1} - | u_{n-1} |^4 u_{n-1}\Big] \\
&+ \frac{\epsilon^6}{32}\Big[J\Big(2u_n^2(| u_{n+1} |^2 u_{n+1}^* + | u_{n-1} |^2 u_{n-1}^*) + 4 | u_n |^2 (| u_{n+1} |^2 u_{n+1} + | u_{n-1} |^2 u_{n-1}) \\
&- 3 | u_n |^4 (u_{n+1} + u_{n-1}) - 2 | u_n |^2 u_n^2(u_{n+1}^* + u_{n-1}^*) - (| u_{n+1} |^4 u_{n+1} - | u_{n-1} |^4 u_{n-1})\Big) \\
&+ 16\tau\Big(2u_n^*(| u_{n+1} |^2 u_{n+1}^2 + | u_{n-1} |^2 u_{n-1}^2) + 3 | u_n |^2 u_n^*(u_{n+1}^2 + u_{n-1}^2) - 2(| u_{n+1} |^4 + | u_{n-1} |^4)u_n \\
&- 4 | u_n |^2 (| u_{n+1} |^2 + | u_{n-1} |^2)u_n + u_n^3(u_{n+1}^{*2} + u_{n-1}^{*2})\Big) + 384A' | u_n |^4 u_n \\
&- iD^z\Big(2 | u_n |^2 (| u_{n+1} |^2 u_{n+1} - | u_{n-1} |^2 u_{n-1}) - u_n^2(| u_{n+1} |^2 u_{n+1}^* + | u_{n-1} |^2 u_{n-1}^*)\Big)\Big].
\end{aligned}
\tag{3}
$$

Equation (3) represents the spin dynamics of an anisotropic weak helimagnet. This discrete equation leads to several nonlinear excitations. The combination of nonlinearity and discreteness gives rise to new types of nonlinear excitations which are not present in the continuum models. These complexities make it difficult to solve the equation directly.

3 Kink Solitonic Profile

The exact solution to the Eq. (3) can be obtained by the use of Jacobi elliptic function method [20]. We introduce the transformations [21]

$$u_n = e^{i\theta_n}\phi_n(\xi_n), \tag{4}$$

where,

$$\theta_n = pn + \omega t + \theta_0, \quad \xi_n = kn + ct + \chi_0,$$

using the trignometric relation $e^{\pm ip}$=$\cos(p) \pm i\sin(p)$, separate the real and imaginary parts, we obtain following set of equations:

$$\omega\phi_n + \epsilon^2\big[2(J - A - 2A')\phi_n - J(\phi_{n+1} + \phi_{n-1})(\cos p + D^z \sin(p))\big] +$$
$$(1/4)\epsilon^4\Big[J\big[(3\phi_n^2(\phi_{n+1} + \phi_{n-1})\cos(p)) + (\phi_{n+1}^3 + \phi_{n-1}^3)\cos(p)\big] - 4(J + 2\tau)$$
$$(\phi_{n+1}^2 + \phi_{n-1}^2)\phi_n + 8\tau\phi_n(\phi_{n+1}^2 + \phi_{n-1}^2)\cos(2p) + 8(A + 6A')\phi_n^3 + D^z \sin(p)$$
$$(2\phi_n^2(\phi_{n+1} - \phi_{n-1}) - \phi_{n+1} + \phi_{n-1} + \phi_{n+1}^3 - \phi_{n-1}^3)) - (D^z\epsilon^5 \sin(p)/32)(5\phi_n^4$$
$$(\phi_{n+1} + \phi_{n-1}) + \phi_{n+1}^5 + \phi_{n-1}^5) - (\epsilon^6/32)(J\cos(p)(\phi_n^2(\phi_{n+1}^3 + \phi_{n-1}^3)) - 5\phi_n^4$$
$$(\phi_{n+1} + \phi_{n-1}) - \phi_{n+1}^5 - \phi_{n-1}^5) + 16\tau(\phi_n(\cos(2p) - 1)2(\phi_{n+1}^4 + \phi_{n-1}^4) + 4$$
$$(\cos(2p) - 1) \times \phi_n^3(\phi_{n+1}^2 + \phi_{n-1}^2)) + 384A'\phi_n^5 + 3D^z \sin(p)\phi_n^2(\phi_{n+1}^3 + \phi_{n-1}^3) = 0, \quad (5)$$

$$-c\phi_n' - \epsilon^2(J(\phi_{n+1} - \phi_{n-1})\sin(p) + D^z\cos(p)(\phi_{n+1} - \phi_{n-1})) + (1/4)\epsilon^4(J$$
$$(\phi_{n+1}^3 + \phi_{n-1}^3)\sin(p) + 8\tau\phi_n(\phi_{n-1}^2 - \phi_{n+1}^2)\sin(2p) + D^z\cos(p)(\phi_n^2(\phi_{n+1}$$
$$-\phi_{n-1}) + \phi_{n+1}^3 - \phi_{n-1}^3)) + (1/32)D^z\epsilon^5\cos(p) \times (\phi_n^4(\phi_{n+1} - \phi_{n-1}) + \phi_{n+1}^5$$
$$-\phi_{n-1}^5) - (1/32)\epsilon^6(J\sin(p) \times (2\phi_n^2(\phi_{n+1}^3 - \phi_{n-1}^3) - \phi_n^4(\phi_{n+1} - \phi_{n-1})$$
$$-\phi_{n+1}^5 + \phi_{n-1}^5) + 32\tau\sin(2p)(\phi_n(\phi_{n+1}^4 - \phi_{n-1}^4) + \phi_n^3(\phi_{n+1}^2 - \phi_{n-1}^2))$$
$$-D^z\cos(p)\phi_n^2(\phi_{n+1}^3 - \phi_{n-1}^3)) = 0. \quad (6)$$

We use the following series of expression [22] as a solution:

$$\phi_n(\xi_n) = a_0 + a_1 sn(\xi_n),$$
$$\phi_{n+1}(\xi_n) = a_0 + a_1\frac{sn(\xi_n)cn(k,m)dn(k,m) + sn(k,m)cn(\xi_n)dn(\xi_n)}{1 - m^2sn^2(\xi_n)sn^2(k)},$$
$$\phi_{n-1}(\xi_n) = a_0 + a_1\frac{sn(\xi_n)cn(k,m)dn(k,m) - sn(k,m)cn(\xi_n)dn(\xi_n)}{1 - m^2sn^2(\xi_n)sn^2(k)}. \quad (7)$$

Further substituting Eq. (7) into Eqs. (5) and (6) and after equating the coefficients of all power to zero, we get a series of algebraic equations. After solving these equations using symbolic computation, we obtain the following solution

$$\phi_n(\xi_n) = a_0 + \left[\frac{2D^Z\sin(p) - 15360\epsilon\tau a_0 A' + 3\epsilon J\cos(p)a_0}{-3072\epsilon A'}\right]\tanh(kn + ct + \chi_0), \quad (8)$$

Upon substituting Eq. (8) in Eq. (4), we write the exact travelling solitary solution as,

$$u(n,t) = \left[a_0 + \left[\frac{2D^Z\sin(p) - 15360\epsilon\tau a_0 A' + 3\epsilon J\cos(p)a_0}{-3072\epsilon A'}\right]\tanh(kn + ct + \chi_0)\right]e^{i(pn+\omega t+\theta_0)} \quad (9)$$

The solution of Eq. (9) is plotted with set of parameters ($a_0 = 0.2$, $J = 0.2$ and $A' = 0.1$) and the Fig. 3 shows that the magnetic soliton assumes Kink soliton profile and propagates along the chain.

4 Results and Discussion

We have investigated the soliton profile of the Helimagnetic nanowire by varying the strength of helicity and DM interaction. It can be observed from the Fig. 2 that there is a significant change in the soliton profile when the value of helicity (τ) is changed. When the value of τ is increased from 0.1 to 0.9, nature of soliton profile

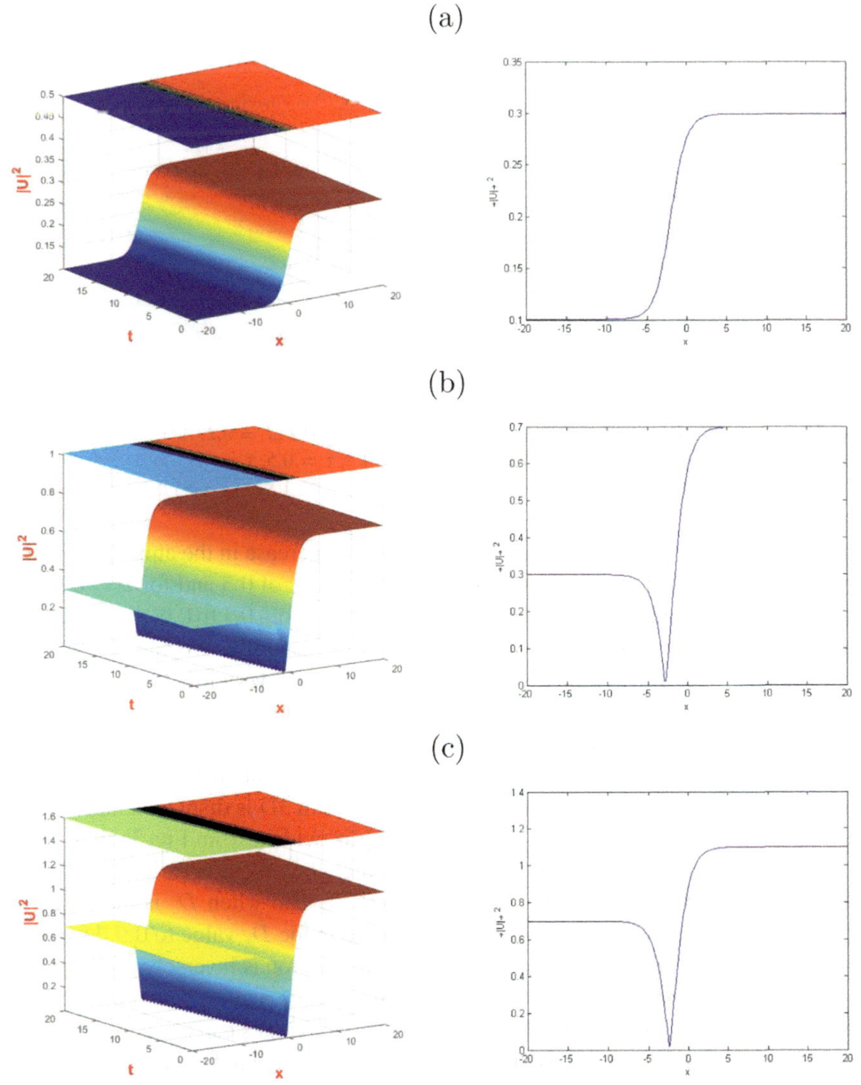

Fig. 2 Soliton profile for $D^z = 0.1, a_0 = 0.2, J = 0.2, A' = 0.1$ and **a** $\tau = 0.1$ **b** $\tau = 0.5$ **c** $\tau = 0.9$

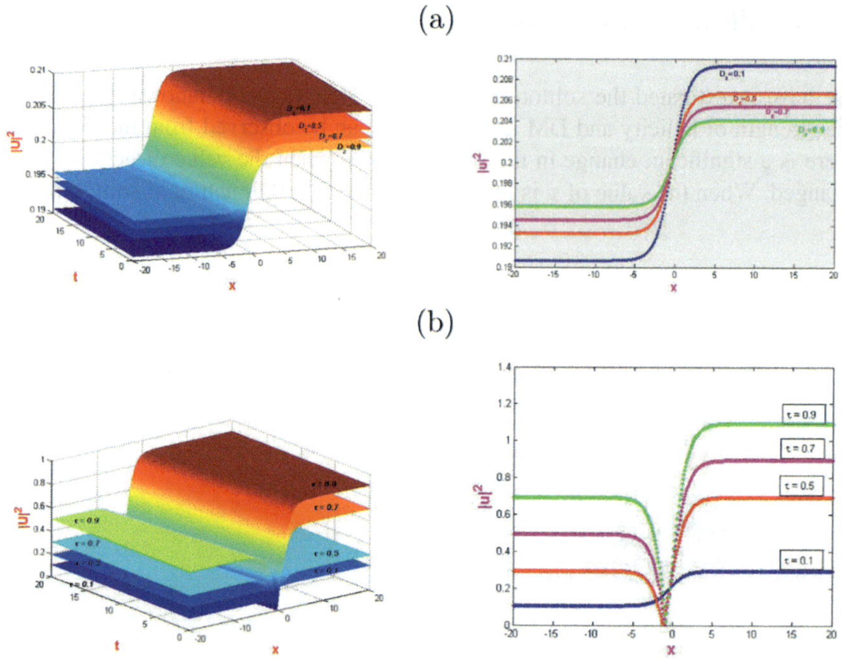

Fig. 3 Cumulative plots for the soliton profile for values $a_0 = 0.2$,$J = 0.2$, $A' = 0.1$ and **a** $D^z = 0.1$, $D^z = 0.3$, $D^z = 0.5$ and $D^z = 0.9$ **b** $\tau = 0.1$, $\tau = 0.3$, $\tau = 0.5$, $\tau = 0.9$

changes from kink to antisoliton and there is also an increase in the amplitude of the profile. When τ is 0.1, the kink soliton has an amplitude of 0.3 and for $\tau = 0.9$, the antisoliton has an amplitude of 1.1. So as τ is increased from 0.1 to 0.9, the amplitude has increased at the blistering speed and grows exponentially. We can see that the difference between the height of the two maxima's is also very less for $\tau = 0.1$ and is significantly larger when τ reaches 0.9. So it can be inferred that the structure of helimagnet in analogy with CLC stabilizes the soliton and settle in a more robust and coherent profile fashion. In Fig. 3 we have plotted the cumulative plots for various values of τ and D_z. On the left side, cumulative plot in 3D is displayed and the 2D plot for the same is displayed on the right side. It is evident from Fig. 3a that, there is a decrease in the amplitude of the soliton profile when there is an increase in the DM interaction. When D_z is 0.1, amplitude is 0.209 and when D_z is increased to 0.5, amplitude decreases to 0.207. Further increase in the D_z value to 0.9 decreases the amplitude to 0.204. Hence, the amplitude of the soliton is inversely proportional to the strength of the DM interaction in the Helimagnetic nanowire. Additionally, a marginal difference between the heights of the two maxima's can be seen, when the strength of the DM interaction is increased. It can be noted that there is no significant changes in the nature of soliton profile.

5 Conclusions

In this paper, we have investigated the effect of DM interaction along with the influence of helicity on a weak Helimagnetic nanowire. We have obtained a DNLS evolution equation using the HP transformation aided with the Glauber's coherent-state representation. The DNLS equation thus obtained is solved for exact solution using the JEF approach. Analyzing the soliton profile obtained, it is observed that the value of helicity plays a major role in the profile of the antisoliton. Even though the changes in the Kink soliton profile due to the varying DM interaction is significantly low, amplitude is influenced by the strength of the DM interaction. These results emphasize the significance of helicity and DM interaction in materials used for the memory storage devices.

Acknowledgements G.S gratefully acknowledges DST, India for the INSPIRE fellowship (IF160862). L.K acknowledges the financial support from UGC-DAE (Ref.No:CSR-KN/CRS-102/2019–20), India, CSIR (Ref.No:03(1414)/17/EMR-II), India and DST-SERB (Ref.No.:MTR /2017/000314/MS), India in the form of a major research project and ICTP, Italy in the form of a Regular Associateship.

References

1. Beg, M., Carey, R., Wang, W., Cortés-Ortuño, D., Vousden, M., Bisotti, M.-A., Albert, M., Chernyshenko, D., Hovorka, O., Stamps, R.L., et al.: Ground state search, hysteretic behaviour and reversal mechanism of skyrmionic textures in confined helimagnetic nanostructures. Sci. Rep. **5**(1), 1–14 (2015)
2. Rybakov, F.N., Borisov, A.B., Blügel, S., Kiselev, N.S.: New type of stable particlelike states in chiral magnets. Phys. Rev. Lett. **115**(11), 117201 (2015)
3. Chui, C., Ma, F., Zhou, Y.: Geometrical and physical conditions for skyrmion stability in a nanowire. AIP Adv. **5**(4), 047141 (2015)
4. Sampaio, J., Khvalkovskiy, A., Kuteifan, M., Cubukcu, M., Apalkov, D., Lomakin, V., Cros, V., Reyren, N.: Disruptive effect of Dzyaloshinskii-Moriya interaction on the magnetic memory cell performance. Appl. Phys. Lett. **108**(11), 112403 (2016)
5. Osorio, S., Laliena, V., Campo, J., Bustingorry, S.: Creation of single chiral soliton states in monoaxial helimagnets. Appl. Phys. Lett. **119**(22), 222405 (2021)
6. Laliena, V., Bustingorry, S., Campo, J.: Dynamics of chiral solitons driven by polarized currents in monoaxial helimagnets. Sci. Rep. **10**(1), 1–10 (2020)
7. Saravanan, M.: Electromagnetic soliton propagation in an anisotropic Heisenberg helimagnet. Phys. Lett. A **378**(41), 3021–3027 (2014)
8. Sunny, G., Kavitha, L.: Modulational instability induced generation of solitary wave profile of an anisotropic-ferromagnetic nanowire with asymmetric Dzyaloshinskii-Moriya interaction. Mater. Today: Proc. (2020)
9. Chandra, P., Coleman, P., Larkin, A.: A quantum fluids approach to frustrated Heisenberg models. J. Phys.: Condens. Matter **2**(39), 7933 (1990)
10. Beula, J., Daniel, M.: Nonlinear spin excitations in a classical Heisenberg anisotropic helimagnet. Phys. D **239**(8), 397–406 (2010)
11. Martin, N., Mirebeau, I., Franz, C., Chaboussant, G., Fomicheva, L., Tsvyashchenko, A.: Partial ordering and phase elasticity in the MnGe short-period helimagnet. Phys. Rev. B **99**(10), 100402 (2019)

12. Daniel, M., Beula, J.: Soliton spin excitations in a Heisenberg helimagnet. Chaos Solitons & Fractals **41**(4), 1842–1848 (2009)
13. Kavitha, L., Parasuraman, E., Gopi, D., Prabhu, A., Vicencio, R.A.: Nonlinear nano-scale localized breather modes in a discrete weak ferromagnetic spin lattice. J. Magn. Magn. Mater. **401**, 394–405 (2016)
14. Kavitha, L., Sathishkumar, P., Saravanan, M., Gopi, D.: Soliton switching in an anisotropic Heisenberg ferromagnetic spin chain with octupole-dipole interaction. Phys. Scr. **83**(5), 055701 (2011)
15. Kavitha, L., Srividya, B., Dhamayanthi, S., Kumar, V.S., Gopi, D.: Collision and propagation of electromagnetic solitons in an antiferromagnetic spin ladder medium. Appl. Math. Comput. **251**, 643–668 (2015)
16. Kavitha, L., Mohamadou, A., Parasuraman, E., Gopi, D., Akila, N., Prabhu, A.: Modulational instability and nano-scale energy localization in ferromagnetic spin chain with higher order dispersive interactions. J. Magn. Magn. Mater. **404**, 91–118 (2016)
17. Darvishi, M., Louis, K., Najafi, M., Senthil Kumar, V.: Elastic collision of mobile solitons of a (3 + 1)-dimensional soliton equation. Nonlinear Dyn. **86** (2016). https://doi.org/10.1007/s11071-016-2920-0
18. Holstein, T., Primakoff, H.: Field dependence of the intrinsic domain magnetization of a ferromagnet. Phys. Rev. **58**(12), 1098 (1940)
19. Glauber, R.J.: Coherent and incoherent states of the radiation field. Phys. Rev. **131**(6), 2766 (1963)
20. Jacobi, C.: New foundations of the theory of elliptic functions. Konigsberg, Borntraeger 1829 (2012)
21. Kovacic, I., Cveticanin, L., Zukovic, M., Rakaric, Z.: Jacobi elliptic functions: a review of nonlinear oscillatory application problems. J. Sound Vib. **380**, 1–36 (2016)
22. Wazwaz, A.-M.: A sine-cosine method for handling nonlinear wave equations. Math. Comput. Model. **40**(5–6), 499–508 (2004)

Delay-Resilient Dynamics of a Landslide Mechanical Model

Srđan Kostić and Nebojša Vasović

Abstract In present paper we analyze dynamics of a simple landslide mechanical model induced by the co-action of included time delay between the motion of the neighboring blocks and their coupling strength. Analyzed mechanical model represents an idealized interaction between accumulation and feeder slope at the accumulation coast. Dynamics of the proposed dynamical system is examined by applying standard bifurcation analysis: we derive explicit relations between time delay, spring stiffness and control parameters, while bifurcation curves are derived numerically. The results of the presented research indicate the onset of Hopf bifurcation, i.e. occurrence of instability for rather high values of the assumed time delay and spring stiffness, which indicates that slope instability occurs only in case when feeder and accumulation slope are observed as strongly coupled, but with a significant delay in interaction. Moreover, we showed that the increase of friction force along the sliding surface suppresses the effect of time delay, indicating that sliding surfaces with low friction parameters are prone to onset of instability.

Keywords Landslide · Time delay · Spring stiffness · Friction · Bifurcation

1 Introduction

Mechanical models are commonly used to successfully simulate dynamics of many natural phenomena, since their dynamics could be reliably described by a set of ordinary/partial differential equations, whose solutions for different initial conditions and values of control parameters could also be confirmed by real physical simulation of the process under study. The most remarkable example of this approach is certainly Burridge–Knopoff model of the earthquake nucleation process, composed of series

S. Kostić (✉)
Geology Department, Jaroslav Černi Water Institute, Jaroslava Černog 80, 11226 Belgrade, Serbia
e-mail: srdjan.kostic@jcerni.rs

N. Vasović
Department of Applied Mathematics and Informatics, Faculty of Mining and Geology, University of Belgrade, Đušina 7, 11000 Belgrade, Serbia

S. Banerjee and A. Saha (eds.), *Nonlinear Dynamics and Applications*,
Springer Proceedings in Complexity,
https://doi.org/10.1007/978-3-030-99792-2_93

of blocks, interconnected only to neighboring blocks, while sliding over the rough surface [1]. In their original paper, Burridge and Knopoff showed that sudden triggering events ("spikes") which occur during the sliding of the blocks, follow macroseismological Gutenberg–Richter and Omori–Utsu laws. This original model has been in succeeding years used as a common model of seismogenic fault movement, whose dynamics has been described by different equations capturing different effect of controlling parameter and interactions among the blocks, including the variable friction laws [2–5]. Apart from the use of this model for description of earthquake nucleation process, it was also used for description of other processes, such as landslide dynamics. In a dynamical sense, both earthquakes and landslides act similarly: a period of no movement or with only small displacements is followed by the period of sudden large displacements, when accumulated energy is being dissipated. The first attempt of using spring-block model for simulation of landslide dynamics was made by Davis [6], who formulated simple mechanical model of two interconnected blocks (accumulation and feeder slope) sliding along the accumulation coast. In his paper, Davis recorded a certain time delay between the movement of the feeder and accumulation slope, which was not included in further analysis, but certainly have effect on the landslide dynamics. This model of Davis [6] was further examined by Morales et al. [7], who studied effect of different friction laws on the landslide dynamics.

In present paper, we start from the model proposed by Davis [6] and Morales et al. [7], but also explicitly include time delay effect between the feeder and accumulation slope, in order to analyze its effect on the onset of instability. Influence of time delay is examined in co-action with coupling strength between the neighboring blocks and frictional parameters along the sliding surface.

2 Description of the Proposed Model

We start from the model proposed by Davis [6]:

$$m_1\dot{V}_1 = W_1 sin\beta_1 - S_1 - F$$

$$m_2\dot{V}_2 = W_2 sin\beta_2 - S_2 + F$$

$$\dot{F} = k(V_1 - V_2) + c\left(\dot{V}_1 - \dot{V}_2\right) \quad (1)$$

where the superposed dot denotes differentiation with respect to time, and: $W = m_i g$—block weight, g—acceleration of gravity; S_i—sliding resistance on failure surface along each block, F—combined elastic and viscous forces, k—spring constant, c—dash pot constant, and β_i = slope angle. Sliding resistance along the failure surface is defined using conventional effective stress model for frictional strength: $S_i = W_i(1 - \alpha_i)cos\beta_i f(V_i)$, where α represents the effect of piezometric elevation on the effective stress which acts on the failure surface: $\alpha_i = \gamma_w h_{wi}/\gamma_{hi}$, where γ and γ_w are the unit weights of the slide material and the pore water respectively, and h_i and h_{wi} are layer thickness and groundwater depth within the layer. The function $f(V_i)$ represents the mobilized strength on the failure surface:

$$f(V_i)\begin{cases} = tan\varphi \;\; for V_i > 0 \\ \leq tan\varphi \;\; for V_i = 0 \end{cases} \tag{2}$$

where φ represents the effective stress friction angle appropriate to the slide material.

In contrast to Davis [6] and model (1), we consider the following:

- two blocks (upper feeder and accumulation slide) on an inclined plane are only connected by elastic springs (Fig. 1a), without the dash pot, which also reduces the effect of F only to elastic force;
- sliding resistance on failure surface is assumed to have the following general form:

$$S = aV_i^3 + bV_i^2 + cV_i \tag{3}$$

  which is the nonmonotonic friction law according to Morales et al. [7], and it describes a smooth spinodal friction law similar to the one introduced in Cartwright et al. [3], see Fig. 1b.
- values of frictional parameters a, b, and c are chosen according to Morales et al. [7], where cubic friction force is given for $a = 1$, where a is the location of the local minimum, i.e. the transition point from the velocity weakening ($b < v < a$) to the velocity strengthening regime ($v > a$).

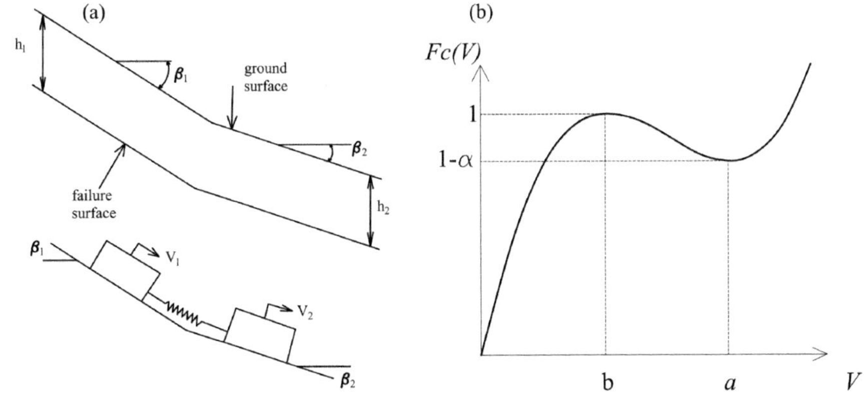

Fig. 1 **a** (top) Typical accumulation slide profile; (bottom) idealized model for accumulation slide. **b** Assumed friction law along the sliding surface

Given this, we propose the model for landslide dynamics in the following general form:

$$\frac{dU_n}{dt} = V_n$$

$$m\frac{dV_n}{dt} = k(U_{n+1} - 2U_n + U_{n-1}) - F(V_0 + V_n) + G \tag{4}$$

where G is the tangential component of the gravity force, and $F(V)$ is a velocity-dependent friction force. A steady state of (4) exists when the block achieves a constant velocity motion $dU/dt = V_0$, and then $F(V_0) = G$, so Eq. (4) represents a dynamical system moving at velocity V_0. Hence, equilibrium state of the examined model is considered as a creep regime, with initial conditions set to $(U_i, V_i) = (0.01, 0.02)$. Such setup of the examined model corresponds to the old existing landslide where creep along the sliding surface is permanently observed and considered as equilibrium state, e.g. landslide "Plavinac" in Smederevo (Serbia).

Model (4) actually represents an infinite chain of identical blocks linearly coupled though Hookean springs of stiffness k that slips at the constant velocity V0 over an inclined surface.

For two coupled blocks, model (4) becomes:

$$\frac{dU_1}{dt} = V_1$$

$$m_1\frac{dV_1}{dt} = k(U_2 - U_1) - F(V_0 + V_1) + F(V_0)$$

$$\frac{dU_2}{dt} = V_2$$

$$m_2\frac{dV_2}{dt} = k(U_1 - U_2) - F(V_0 + V_2) + F(V_0) \tag{5}$$

Model (5) with the included delayed interaction becomes:

$$\frac{dU_1(t)}{dt} = V_1(t)$$

$$\frac{dV_1(t)}{dt} = \frac{1}{m}[k(U_2(t-\tau) - U_1(t)) - F(V_0 + V_1(t)) + F(V_0)]$$

$$\frac{dU_2(t)}{dt} = V_2(t)$$

$$\frac{dV_2(t)}{dt} = [k(U_1(t-\tau) - U_2(t)) - F(V_0 + V_2(t)) + F(V_0)] \tag{6}$$

3 Results

Linearization of the system (6) and substitution $U_1 = A_1 e^{\lambda t}$, $U_2 = A_2 e^{\lambda t}$, $U_1(t-\tau) = A_1 e^{\lambda(t-\tau)}$, $U_2(t-\tau) = A_2 e^{\lambda(t-\tau)}$, $V_1 = B_1 e^{\lambda t}$ and $V_2 = B_2 e^{\lambda t}$ results in a system of algebraic equations for the constants A_1, A_2, B_1 and B_2. This system has a nontrivial solution if the following is satisfied:

$$\left[\lambda\left(\lambda + \frac{1}{m}\frac{dF(V_0+v)}{dv}|v \equiv 0\right) + \frac{k}{m}\right]^2 - \left[\frac{k}{m}e^{-\lambda\tau}\right]^2 = 0 \tag{7}$$

Equation (7) is the characteristic equation of the system (6) Infinite dimensionality of the system (6) is reflected in the transcendental character of (7). By substituting $\lambda = i\omega$ in Eq. (7) we obtain:

$$\left[\frac{1}{m}\frac{dF(V_0+v)}{dv}|v \equiv 0\right] \bullet \omega = \pm\frac{k}{m} sin\omega t \tag{8}$$

In this way, one obtains parametric representations of the relations between τ and the parameters, which correspond to the bifurcation values $\lambda = i\omega$. The general form of such relations is illustrated by the following formula for k as a function of ω:

$$k = \frac{m}{2}\omega^2 + \frac{1}{2m}\left(\frac{dF(V_0+v)}{dv}|v \equiv 0\right)^2 \tag{9}$$

and for τ as a function of ω:

$$\tau = \frac{1}{\omega} arctan\left(\frac{\frac{1}{m}\frac{dF(V_0+v)}{dv}|v \equiv 0}{\omega^2 - \frac{k}{m}}\right) + (2n+1)\pi, n = 0, 1, 2, \ldots \tag{10}$$

One should know that although the solution of characteristic equation is indicative of Hopf bifurcations, proof of this claim is rather lengthy to convey [8]. Instead, it could be numerically shown that the above parametric equations for k and τ coincide with the Hopf bifurcation curve illustrated in Fig. 2, which shows the transition from constant slow creep of the landslide (which we consider as equilibrium state) to oscillatory periodic motion (which we consider as instable region).

As one can see a supercritical Hopf bifurcation occurs for rather strong spring stiffness k (>6) and high values of time delay τ (>1). In present paper, we consider that onset of regular periodic oscillations indicates instability along the slope. Onset of instability for high spring stiffness indicates that system under study (conditionally stable slope) needs to be observed as a system of strongly coupled accumulation and feeder slope, in order to exhibit the instability. On the other hand, occurrence of instability for high values of time delay shows the high resilience of the conditionally

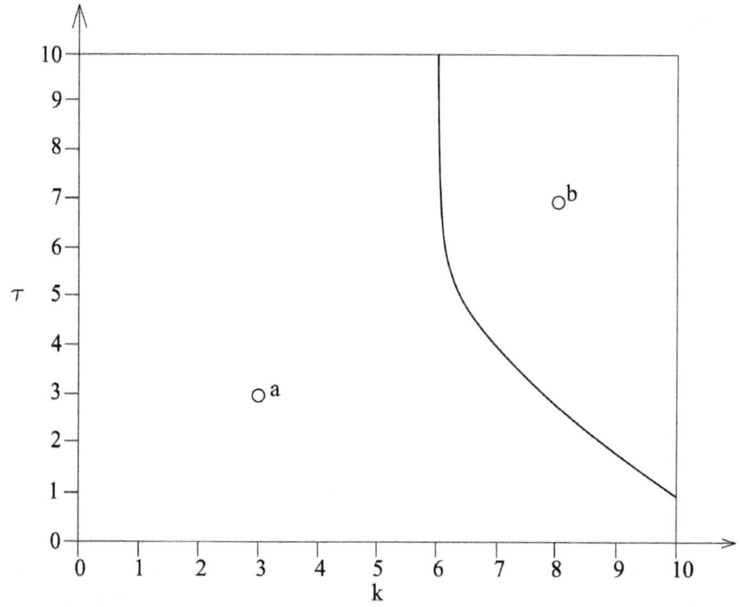

Fig. 2 Hopf bifurcation curve k(τ), for the fixed values of parameters $V_0 = 0.1$, $a = 3.2$, $b = -7.2$ and $c = 4.8$. Initial conditions are set to $(Ui, Vi) = (0.01, 0.02)$. The appropriate time series which correspond to points *a* and *b* are shown in Fig. 3

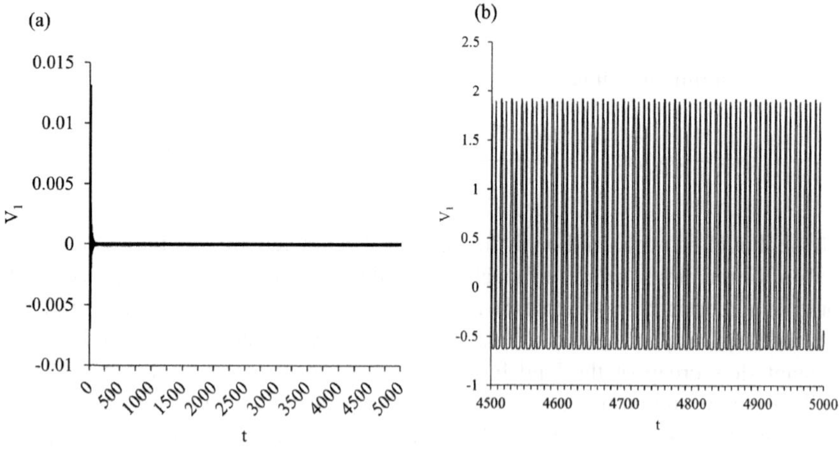

Fig. 3 Temporal evolution of variable V_1 for **a** $\tau = 3$, $k = 3$, **b** $\tau = 7$, $k = 8$. Time series are constructed for the fixed values of parameters $V_0 = 0.1$, $a = 3.2$, $b = -7.2$ and $c = 4.8$

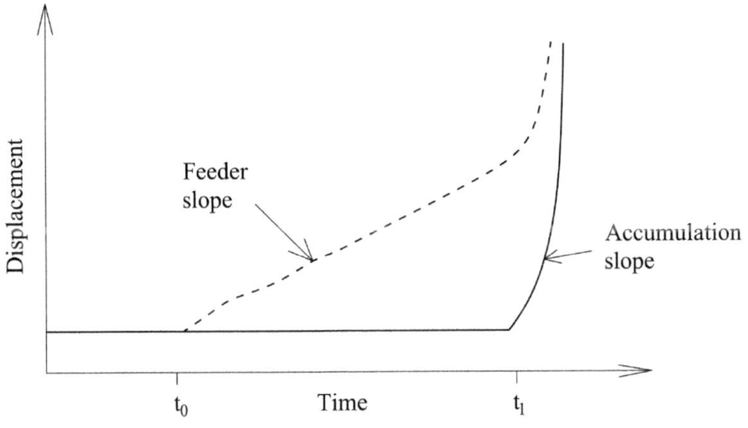

Fig. 4 Time delay between the onset of feeder and accumulation slope, as suggested by Davis [6]

stable slope to occurrence of time delay between the motion of feeder and accumulation slope. This may indicate that the time delay indicated originally by Davis [6], as shown in Fig. 4, maybe does not have significant influence on the system dynamics, for the chosen values of friction parameters.

If we want further to examine the effect of time delay, let us analyze the influence of the frictional parameters on the effect of τ. If one holds value of time delay and spring constant above the bifurcation curve, increase of parameters *a, b* and *c* suppress the effect of the introduced time delay (Fig. 5). In particular, for lower values of parameter a (Fig. 5a), b (Fig. 5b) and c (Fig. 5c), observed dynamical system is in unstable regime (periodic motion), while further increase of the friction effect induces the transition to equilibrium state. This indicates that sliding surfaces with low friction parameters are more susceptible to the onset of instability.

4 Conclusions

In present paper we analyze the sensitivity of the landslide mechanical model to the effect of time delay between the displacement of the accumulation and feeder slope. The research was performed using standard bifurcation analysis, while bifurcation curves were constructed numerically. Assumed friction law along the existing sliding surface is assumed to have cubic expression. The case analyzed represents the case of the landslides with slow permanent displacement. Results obtained indicate that for the analyzed friction parameter values, instability occurs for high values of time delay and coupling strength, which indicates high resilience of the model under study to the effect of delayed interaction between the accumulation and feeder slope. Moreover, we showed that sliding surface with low frictional parameters is more prone to the onset of instability.

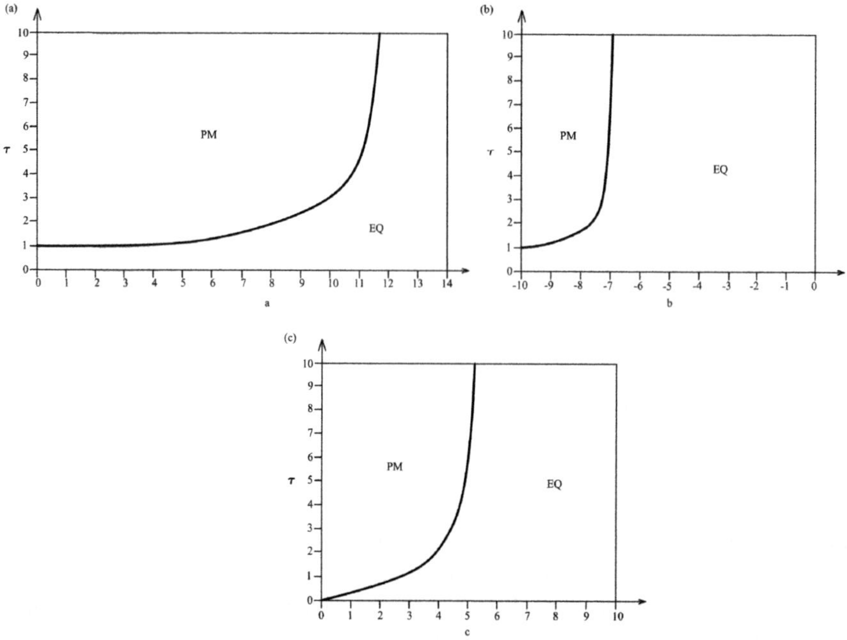

Fig. 5 Bifurcation diagrams regarding the effect of frictional parameters on the onset of instability: **a** $\tau = f(a)$, **b** $\tau = f(b)$, **c**, $\tau = f(c)$, for fixed parameters values: $V_0 = 0.1$, $K = 7$, $a = 3.2$, $b = -7.2$ and $c = 4.8$. EQ stands for the equilibrium state, while PM denote periodic (oscillatory) regime

Further research on this topic should include additional inquiries on the types of instabilities which are formed with the increase of time delay, with the emphasis of irregular or stick–slip regime, which could be treated as an adequate representative of the real landslide motion. Moreover, one could analyze the effect of the choice of various friction laws on the onset of instability.

References

1. Burridge, R., Knopoff, L.: Model and theoretical seismicity. Bull. Seismol. Soc. Am. **57**, 341–371 (1967)
2. Carlson, J.M., Langer, J.S.: Mechanical model of an earthquake fault. Phys. Rev. A **40**(11), 6470–6484 (1989)
3. Cartwright, J.H.E., Hernández-García, E., Piro, O.: Burridge-Knopoff models as elastic excitable media. Phys. Rev. Lett. **79**(3), 527–530 (1997)
4. Clancy, I., Corcoran, D.: State-variable friction for the Burridge-Knopoff model. Phys. Rev. E Stat. Nonlinear Soft Matter Phys. **80**(1), 016113 (2009)

5. Nkomom, T.N., Ndzana, F.I., Okaly, J.B., Mvogo, A.: Dynamics of nonlinear waves in a Burridge and Knopoff model for earthquake with long-range interactions, velocity-dependent and hydrodynamics friction forces. Chaos Solitons Fractals **150**, 111196 (2021)
6. Davis, R.O.: Modelling stability and surging in accumulation slides. Eng. Geol. **33**, 1–9 (1992)
7. Morales, J.E.M., James, G., Tonnelier, A.: Travelling waves in a spring-block chain sliding down a slope. Phys. Rev. E **96**, 012227 (2017)
8. Belair, J., Campbell, S.A.: Stability and bifurcations of equilibria in a multiple delayed differential equation. SIAM Appl. Math. **54**, 1402–1424 (1994)

The Fifth Order Caudrey–Dodd–Gibbon Equation for Exact Traveling Wave Solutions Using the $(G'/G,\ 1/G)$-Expansion Method

M. Mamun Miah

Abstract In this investigation, I am trying to extract abundant exact traveling wave solutions for the nonlinear partial fifth order Caudrey–Dodd–Gibbon (CDG) differential equation via the $(G'/G,\ 1/G)$-expansion method. Here I accomplish varieties types of wave solutions as like, trigonometric, hyperbolic, and rational function solution. Since new solutions provided us new physical explanation of the mathematical model for engineering applications and nonlinear sciences. So this article is very effective to extract abundant new analytic traveling wave solitons. Graphical representations of the obtained solutions are also portrayed and the shapes of the new solutions are bright soliton, dark soliton, periodic soliton etc. This eminent method is more applicable and easier to analysis nonlinear partial differential models.

Keywords Nonlinear partial differential equation · The fifth order Caudrey-Dodd-Gibbon equation · Traveling wave solutions · The $(G'/G,\ 1/G)$-expansion method

1 Introduction

Most of the physical conditions exist in all branches of engineering applications and scientific fields such as plasma physics, optical fibers, fluid mechanics, elastic media, solid state physics etc. may be expressed as in terms of mathematical models i.e. nonlinear partial differential equations (NLPDEs). So the studies of NLPDEs are most interesting topics in modern researcher. For the consequences of modern researcher there are many methods invented to investigate the nonlinear system, such as the homogeneous balance method [1, 2], the Jacobi elliptic function method [3], the tanh-coth method [4], the first integral method [5], the Kudryashov method [6, 7], the (G'/G)-expansion method [8], the unified method [9] etc. Recently a new technique is discovered for investigating nonlinear evolution equations (NLEEs) and the name

M. Mamun Miah (✉)
Department of Mathematics, Khulna University of Engineering & Technology, Khulna, Bangladesh
e-mail: mamun0954@math.kuet.ac.bd

S. Banerjee and A. Saha (eds.), *Nonlinear Dynamics and Applications*, Springer Proceedings in Complexity,
https://doi.org/10.1007/978-3-030-99792-2_94

of this method is the $(G'/G, 1/G)$-expansion method [10–12]. Many researcher are used this method and get outstanding performance for studying NLEEs. At first Li et al. [13] invented this eminent method and investigated the Zakharov equations for extracting abundant new traveling wave solutions. Recently, Chowdhury et al. [14] investigated the integro-differential equations make use of this eminent method. Very recently, Iqbal et al. [15] applied this method on Date–Jimbo–Kashiwara–Miwa equation with conformable derivative and obtained abundant exact traveling wave solutions. I have seen that, by using this renowned method there extract huge closed form wave solutions. Since no one scrutinized the fifth order CDG equation by means of the indicated method, so I used this method.

Our article is scheduled as following instruction: In Sect. 2, the applied method explanation. In Sect. 3, exact solutions of the fifth order CDG equation is scrutinized. In Sect. 4, graphical representations are delivered and finally, Sect. 5, the conclusions are given.

2 Explanation of the $(G'/G, 1/G)$-Expansion Method

Here, I designate the brief explanation of the applied method for extracting wave solutions of denoted NLEE. At first I suppose that the auxiliary equation as,

$$G''(\xi) + \lambda\, G(\xi) = \mu \tag{1}$$

where both two of μ and λ arbitrary constants and setup the expression as follows,

$$\varphi = G'/G,\ \psi = 1/G \tag{2}$$

Thus,

$$\varphi' = -\varphi^2 + \mu\,\psi - \lambda,\ \psi' = -\varphi\,\psi \tag{3}$$

Depends on λ, I have discussed three cases:
Case 1: For $\lambda < 0$, the general exact solution of Eq. (1),

$$G(\xi) = A_2\ \cos(\sqrt{-\lambda}\,\xi) + A_1\ \sinh(\sqrt{-\lambda}\,\xi) + \frac{\mu}{\lambda}, \tag{4}$$

where above two constants A_1 and A_2 are arbitrary. Consequently,

$$\psi^2 = \frac{-\lambda\left(\varphi^2 - 2\mu\psi + \lambda\right)}{\lambda^2\sigma + \mu^2}, \tag{5}$$

wherein, $\sigma = A_1^2 - A_2^2$.

Case 2: For $\lambda > 0$,

$$G(\xi) = A_2\ \cos(\sqrt{\lambda}\ \xi) + A_1 \sin(\sqrt{\lambda}\ \xi) + \frac{\mu}{\lambda}, \tag{6}$$

and hence

$$\psi^2 = \frac{\lambda(\varphi^2 - 2\mu\psi + \lambda)}{\lambda^2\rho - \mu^2}, \tag{7}$$

wherein $\rho = A_1^2 + A_2^2$.

Case 3: For $\lambda = 0$,

$$G(\xi) = \frac{\mu}{2}\xi^2 + A_1\ \xi + A_2, \tag{8}$$

and I obtain,

$$\psi^2 = \frac{(\varphi^2 - 2\mu\psi)}{A_1^2 - 2\ \mu\ A_2}. \tag{9}$$

Assume the following NLEE is in two variables x and t which are independent,

$$T(u, u_x, u_{xt}, u_t, u_{tt}, u_{xx}, u_{xxt} \ldots) = 0, \tag{10}$$

here T is a function of nonlinear polynomial of u and its derivatives partially. Now to apply our desired method I consider the following steps:

Step 1: For transferring to ordinary from partial differential equation, I consider the wave variable as,

$$u(x,\ t) = u(\xi) \text{ and } \xi = x - v\ t \tag{11}$$

where v is a constant which takes arbitrary values.

From Eqs. (11) to (10),

$$M(u, u', -vu',\ v^2u'', u'' \ldots) = 0, \tag{12}$$

wherein M is a function of nonlinear polynomial of u and its derivatives ordinary.

Step 2: Let us consider the solution of Eq. (12),

$$u(\xi) = \sum_{i=0}^{N} a_i \varphi(\xi)^i + \sum_{i=1}^{N} b_i \varphi(\xi)^{i-1} \psi(\xi), \tag{13}$$

where above two constants a_i and b_i are arbitrary and for both $i = 1, 2, 3, 4 \ldots$.

Step 3: By the use of balance principal, the value of N and after setting the value of N into Eq. (13) and inserting this modified Eqs. (13) into (12), using (3) and (5) (for case 1), after this performances the left-hand side of Eq. (12) moves into a polynomial of φ and ψ, in which the degree of φ and ψ are zero to any positive integer and less than one respectively. Equating coefficient of the same powers to zero, gives a set of equations in arbitrary constants and solving these system yield the values of required arbitrary constants for $\lambda < 0$.

Step 4: Utilizing these obtained arbitrary values in step 3 and back substituting in Eqs. (10), (11) and (12), I obtain our desired wave solutions of the NLEEs.

Step 5: Again applying step 3 and step 4, back substituting Eqs. (13), (12), (3) and (7) for $\lambda > 0$ (or from Eqs. (3) to (9) for $\lambda = 0$), the exact solutions of Eq. (12) i.e. Eq. (10) demonstrated by trigonometric function solutions (or by the rational function solutions) respectively. The details of our applied method are given in Ref. [10–15].

3 Exact Solutions of the CDG Equation

First I introduce the fifth order CDG equation [16, 17],

$$u_t + u_{xxxxx} + 30\, u\, u_{xxx} + 30\, u_x\, u_{xx} + 180\, u^2 u_x = 0. \tag{14}$$

Equation (14) moves to ordinary differential equation by applying the wave transformation Eq. (11) as,

$$-v\, u' + u^{(5)} + 30\, u\, u' + 30\, u'\, u'' + 180\, u^2 u' = 0. \tag{15}$$

Integrating Eq. (15) and using c as integrating constant,

$$c - v\, u + u^{(4)} + 30\, u\, u'' + 60\, u^3 = 0. \tag{16}$$

Now applying the idea of homogeneous balance number between $u^{(4)}$ and u^3, assume the solution of Eq. (16),

$$u(\xi) = a_0 + a_1 \varphi(\xi) + a_2 \varphi(\xi)^2 + b_1 \psi(\xi) + b_2 \varphi(\xi)\, \psi(\xi), \tag{17}$$

where b_1, b_2, a_0, a_1 and a_2 all are arbitrary constant which are determine below. Here I discuss three cases for solving the NLEE Eq. (14).

Case 1. When $\lambda < 0$, the final exact solution in terms of hyperbolic function are given as,

$$
\begin{aligned}
u(\xi) = & -\frac{5\lambda}{6} + \frac{\lambda\left\{A_2 \sinh(\sqrt{-\lambda}\,\xi) + A_1 \cosh(\sqrt{-\lambda}\,\xi)\right\}^2}{\left\{A_2 \cosh(\sqrt{-\lambda}\xi) + A_1 \sinh(\sqrt{-\lambda}\,\xi) + \frac{\mu}{\lambda}\right\}^2} \\
& + \frac{\mu}{A_2 \cosh(\sqrt{-\lambda}\xi) + A_1 \sinh(\sqrt{-\lambda}\,\xi) + \frac{\mu}{\lambda}} \\
& \pm \frac{\sqrt{\mu^2 + \lambda^2\sigma}\left\{A_2 \sinh(\sqrt{-\lambda}\,\xi) + A_1 \cosh(\sqrt{-\lambda}\,\xi)\right\}}{\left\{A_2 \cosh(\sqrt{-\lambda}\,\xi) + A_1 \sinh(\sqrt{-\lambda}\xi) + \frac{\mu}{\lambda}\right\}^2},
\end{aligned}
\tag{18}
$$

wherein $\xi = x + \frac{\lambda^3}{9}\,t$, $v = \lambda^2$, $\sigma = A_1^2 - A_2^2$. For special case if $A_2 = 0$, $A_1 \neq 0$ and $\mu = 0$ in Eq. (18), the traveling exact wave solution set as,

$$
\begin{aligned}
u(x,\ t) = & -\frac{5\lambda}{6} + \lambda \coth^2\left(\sqrt{-\lambda}\ \left(x + \frac{\lambda^3}{9}\ t\right)\right) \\
& \pm \lambda \coth\left(\sqrt{-\lambda}\ \left(x + \frac{\lambda^3}{9}\ t\right)\right) \operatorname{cos}ech\left(\sqrt{-\lambda}\ \left(x + \frac{\lambda^3}{9}\ t\right)\right)
\end{aligned}
\tag{19}
$$

Case 2. When $\lambda > 0$, the exact solution in terms of trigonometric function are given as follows,

$$
\begin{aligned}
u(\xi) = & -\frac{5\lambda}{6} - \frac{\lambda\left\{-A_2 \sin(\sqrt{\lambda}\,\xi) + A_1 \cos(\sqrt{\lambda}\,\xi)\right\}^2}{\left\{A_2 \cos(\sqrt{\lambda}\xi) + A_1 \sin(\sqrt{\lambda}\,\xi) + \frac{\mu}{\lambda}\right\}^2} \\
& + \frac{\mu}{A_2 \cos(\sqrt{\lambda}\xi) + A_1 \sin(\sqrt{\lambda}\,\xi) + \frac{\mu}{\lambda}} \\
& \pm \frac{\sqrt{\lambda^2\rho - \mu^2}\left\{-A_2 \sin(\sqrt{\lambda}\,\xi) + A_1 \cos(\sqrt{\lambda}\,\xi)\right\}}{\left\{A_2 \cos(\sqrt{\lambda}\,\xi) + A_1 \sin(\sqrt{\lambda}\xi) + \frac{\mu}{\lambda}\right\}^2},
\end{aligned}
\tag{20}
$$

wherein $\xi = x + \frac{\lambda^3}{9}\,t$, $v = \lambda^2$, $\rho = A_2^2 + A_1^2$. Again I consider $A_1 = 0$, $A_2 \neq 0$, $\mu = 0$ in (20),

$$
\begin{aligned}
u(x,\ t) = & -\frac{5\lambda}{6} - \lambda \tan^2\left(\sqrt{\lambda}\ \left(x + \frac{\lambda^3}{9}\ t\right)\right) \\
& \pm \lambda \tan\left(\sqrt{\lambda}\ \left(x + \frac{\lambda^3}{9}\ t\right)\right) \sec\left(\sqrt{\lambda}\ \left(x + \frac{\lambda^3}{9}\ t\right)\right)
\end{aligned}
\tag{21}
$$

For $A_2 = 0$, $A_1 \neq 0$ and $\mu = 0$ in (20),

$$u(x,\ t) = -\frac{5\lambda}{6} - \lambda\cot^2\left(\sqrt{\lambda}\left(x + \frac{\lambda^3}{9}t\right)\right)$$
$$\pm\lambda\cot\left(\sqrt{\lambda}\left(x + \frac{\lambda^3}{9}t\right)\right)\cos ec\left(\sqrt{\lambda}\left(x + \frac{\lambda^3}{9}t\right)\right). \quad (22)$$

Case 3. When $\lambda = 0$, the exact solution in terms of rational function solution are given as follows,

$$u(\xi) = a_0 - \frac{(\mu\xi + A_1)^2}{2(\frac{\mu}{2}\xi^2 + A_1\xi + A_2)^2} + \frac{\mu}{2(\frac{\mu}{2}\xi^2 + A_1\xi + A_2)}$$
$$\pm\frac{\sqrt{A_1^2 - 2\mu A_2}(\mu\xi + A_1)}{2\left(\frac{\mu}{2}\xi^2 + A_1\xi + A_2\right)^2}. \quad (23)$$

where $\xi = x + 120a_0^3t$, $v = 180a_0^2$, a_0, A_1 and A_2 are arbitrary constants. If I choose arbitrary constants to zero, further traveling exact wave solutions to the fifth order CDG equation can be extracted, but limitation of the article pages have not been sketched. The above solutions of Eq. (14) gives in our article are correct and new other than the solutions remaining in the research fields.

4 Graphical and Physical Explanation

In this section, I discussed about the graphical representation and physical explanation of some soliton solutions. The extracted solutions of our desire equation are different kind such as rational, hyperbolic and trigonometric function. Here, I plotted and discussed about three types of solutions. Figures 1 and 2 shows that the bright solitary wave of solution Eq. (18) for 3D plot and contour plot for the values of, $\lambda = -2,\ \mu = 1,\ A_1 = 2,\ A_2 = 1$ within $x \in [-5,\ 5]$ and $t \in [-5,\ 5]$. Figures 3 and 4 shows the periodic solitary wave of solution Eq. (20) for 3D plot and contour plot for the values of, $\lambda = 2,\ \mu = 1,\ A_1 = 1,\ A_2 = 1$ within $x \in [-5,\ 5]$ and $t \in [-5,\ 5]$. Figures 5 and 6 shows the dark solitary wave of solution Eq. (23) for 3D and contour plot for the values of, $a_0 = 1,\ \mu = 2,\ A_1 = 3,\ A_2 = 1$ within $x \in [-5,\ 5]$ and $t \in [-5,\ 5]$. If I plotted the solution in Eq. (19), it's give same figure of Eq. (18) and similarly same shaped are obtained for the Eqs. (20) and (21). The following obtained figures have been plotted with the help of computation package program like Maple.

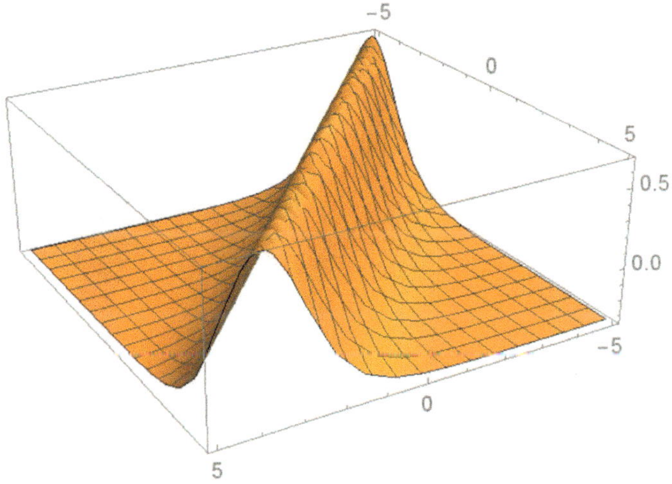

Fig. 1 Bright solitary wave of solution Eq. (18), figure is 3D plot and right one is Contour Plot for the values $\lambda = -2$, $\mu = 1$, $A_1 = 2$ and $A_2 = 1$

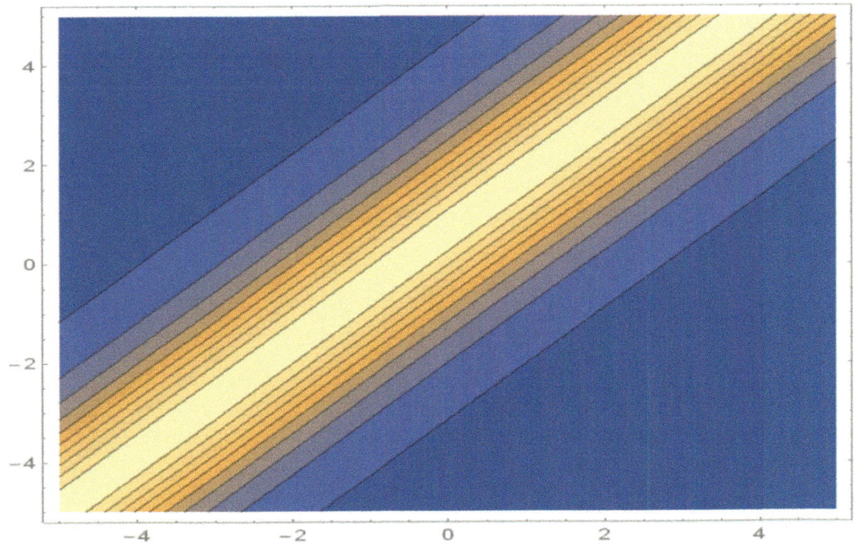

Fig. 2 Bright solitary wave of solution Eq. (18), the figure is Contour Plot for the values, $\lambda = -2$, $\mu = 1$, $A_1 = 2$ and $A_2 = 1$

5 Conclusion

In our article, I extract numerous new exact solutions for the fifth order CDG equation and trace out the graphical representations of these results. There are special types of solutions are founded such as, bright soliton, dark soliton, periodic soliton etc.

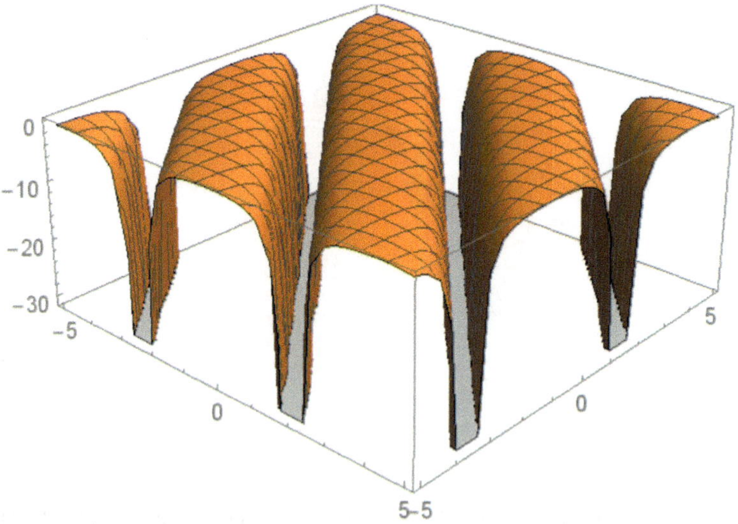

Fig. 3 Periodic solitary wave of solution Eq. (20), the figure is 3D plot and right one is Contour Plot for the values $\lambda = 2$, $\mu = 1$, $A_1 = 1$ and $A_2 = 1$

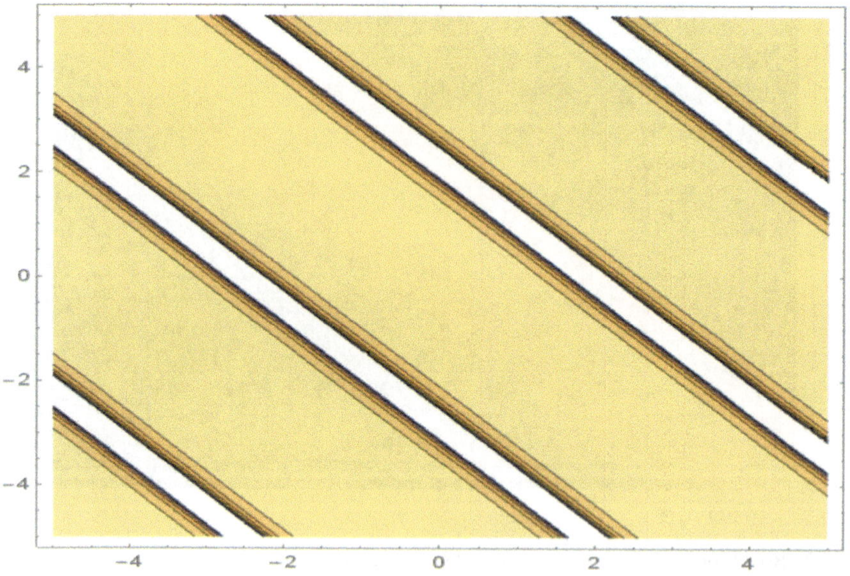

Fig. 4 Periodic solitary wave of solution Eq. (20), the figure is Contour Plot for the values, $\lambda = 2$, $\mu = 1$, $A_1 = 1$ and $A_2 = 1$

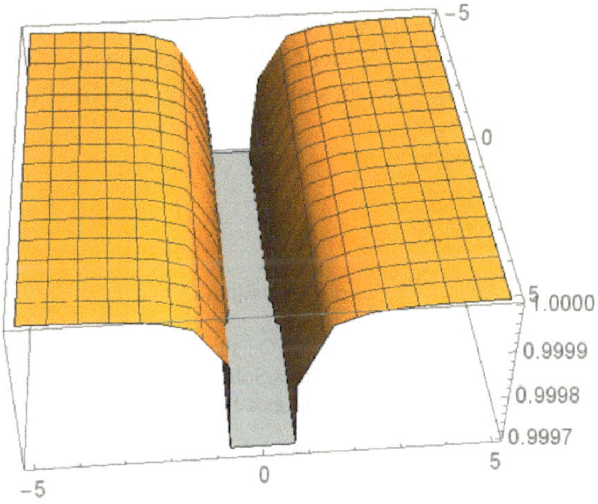

Fig. 5 Dark solitary wave of solution (23), the figure is 3D plot for the values of $a_0 = 1$, $\mu = 2$, $A_1 = 3$ and $A_2 = 1$

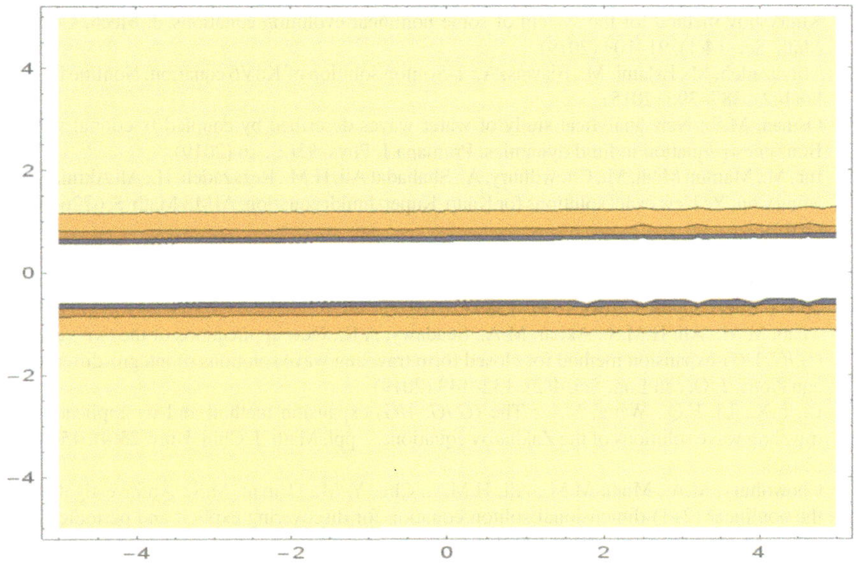

Fig. 6 Dark solitary wave of solution Eq. (23), the figure is Contour Plot for the values of $a_0 = 1$, $\mu = 2$, $A_1 = 3$ and $A_2 = 1$

and such solution pattern are important in nonlinear sciences. The obtained traveling wave solutions might have significant impact for further investigation of the fifth order CDG equation. For the performance of the $(G'/G, 1/G)$-expansion method, I conclude that the method is easier and faster compare to other method by means of computational package program like Mathematica or Maple. Finally I conclude that, our investigation can be extended to other NLPDEs which arise in nonlinear physics, applied mathematics and other branches of engineering and nonlinear science.

References

1. Wang, M.: Solitary wave solutions for variant Boussinesq equations. Phys. Lett. A **199**, 169–172 (1995)
2. Zayed, E.M.E., Zedan, H.A., Gepreel, K.A.: On the solitary wave solutions for nonlinear Hirota-Satsuma coupled KdV equations. Chaos Solitons Fractals **22**, 285–303 (2004)
3. Liu, G.T., Fan, T.Y.: New applications of developed Jacobi elliptic function expansion methods. Phys. Lett. A **345**, 161–166 (2005)
4. Wazwas, A.M.: The tanh-coth method for solitons and kink solutions for nonlinear parabolic equations. Appl. Math. Comput. **188**, 1467–1475 (2007)
5. Bekir, A., Unsal, O.: Periodic and solitary wave solutions of coupled nonlinear wave equations using the first integral method. Phys. Scr. **85**, 065003 (2012)
6. Eslami, M.: Exact traveling wave solutions to the fractional coupled nonlinear Schrodinger equations. Appl. Math. Comput. **285**, 141–148 (2016)
7. Ali, H.M.S., Habib, M.A., Miah, M.M., Akbar, M.A.: A modification of the generalized Kudryshov method for the system of some nonlinear evolution equations. J. Mech. Cont. & Math. Sci. **14**(1), 91–109 (2019)
8. Mirzazadeh, M., Eslami, M., Biswas, A.: 1-Soliton solution of KdV6 equation. Nonlinear Dyn. **80**(1–2), 387–396 (2015)
9. Osman, M.S.: New analytical study of water waves described by coupled fractional variant Boussinesq equation in fluid dynamics. Pramana J. Phys. **93**(2), 26 (2019)
10. Inc, M., Mamun Miah, M., Chowdhury, A., Shahadat Ali, H.M., Rezazadeh, H., Ali Akinlar, M., Ming Chu, Y.: New exact solutions for Kaup-Kupershmidt equation. AIMS Math. **5**, 6726–6738 (2020)
11. Zayed, E.M.E., Alurrfi, K.A.E.: The $(G'/G, 1/G)$-expansion method and its applications to two nonlinear Schrodinger equations describing the propagation of femtosecond pulses in nonlinear optical fibers. Optik **127**(4), 1581–1589 (2016)
12. Miah, M.M., Ali, H.M.S., Akbar, M.A., Seadawy, A.R.: New applications of the two variable $(G'/G, 1/G)$-expansion method for closed form traveling wave solutions of integro-differential equations. J. Ocean Eng. Sci. **4**(2), 132–143 (2019)
13. Li, L.X., Li, E.Q., Wang, M.L.: The $(G'/G, 1/G)$-expansion method and its application to traveling wave solutions of the Zakharov equations. Appl. Math. J. Chin. Univ. **25**(4), 454–462 (2010)
14. Chowdhury, M.A., Miah, M.M., Ali, H.M.S., Chu, Y.-M., Osman, M.S.: An investigation to the nonlinear (2+1)-dimensional soliton equation for discovering explicit and periodic wave solutions. Res. Phys. **23**, 10401 (2021)
15. Iqbal, M.A., Wang, Y., Miah, M.M., Osman, M.S.: Study on Date-Jimbo-Kashiwara-Miwa equation with conformable derivative dependent on time parameter to find the exact dynamical wave solutions, Fractal Fract. **6**(4) (2022)
16. Xu, Y.G., Zhou, X.W., Yao, L.: Solving the fifth order Caudrey-Dodd-Gibbon (CDG) equation using the exp-function method. Appl. Math. Comput. **206**(1), 70–73 (2008)
17. Jaradat, H.M., Syam, M., Jaradat, M., Mustafa, Z., Momani, S.: New solitary wave and multiple soliton solutions for fifth order nonlinear evolution equation with time variable coefficients. Res.Phys. **8**, 977–980 (2018)

Results on Fractal Dimensions for a Multivariate Function

T. M. C. Priyanka and A. Gowrisankar

Abstract In the present work the fractal dimensions for the fractional integral on the multivariate function is explored. In particular, the upper bound of box dimension and Hausdorff dimension for the mixed Riemann–Liouville fractional integral of a multivariate function, which belongs to the class of Hölder continuous functions, is investigated. Further, if the multivariate function satisfies Lipschitz condition, the box dimension and the Hausdorff dimension of its mixed Riemann–Liouville fractional integral are estimated.

Keywords Multivariate function · Fractional integral · Hausdorff dimension · Box dimension

1 Introduction

Fractal dimension is the fundamental feature which distinguishes the naturally occurring functions into smooth and coarse functions. The most surprising aspect of fractal dimension is that it is not always an integer but can also be a fractional number. Box dimension and Hausdorff dimension are the widely discussed quantifiers in the fractal approximation theory. For the basic definitions of fractal dimension and fractal functions, the reader is encouraged to refer the textbooks [1–3].

The interesting connection between the fractal geometry and fractional calculus is the evaluation of fractal dimension for the graphs of fractional derivatives (integrals) of various types of fractal curves. Fractional calculus for different types of fractal functions has been discussed in [4–6]. Many elegant and simple results have been proved concerning the fractal dimension of univariate and bivariate continuous functions. For the Riemann–Liouville (RL) fractional integral of 1D continuous function of bounded variation, the box dimension is proved to be 1, in [7]. Liang has

T. M. C. Priyanka · A. Gowrisankar (✉)
Department of Mathematics, School of Advanced Sciences, Vellore Institute of Technology, Vellore 632014, Tamil Nadu, India
e-mail: gowrisankargri@gmail.com

S. Banerjee and A. Saha (eds.), *Nonlinear Dynamics and Applications*,
Springer Proceedings in Complexity,
https://doi.org/10.1007/978-3-030-99792-2_95

investigated the box dimension of 1D continuous functions of unbounded variation in [8] and its RL fractional integral in [9]. Fractal dimension of bivariate continuous functions of bounded variation and its fractional integral has been explored in [10]. Subhash and Syed have examined the box dimension of Katugampola fractional integral of 2D continuous functions in [11]. For more details on fractional calculus and fractal dimension, refer [13–16]. Several researchers have also discussed the fractal dimension of fractal functions, see for instance [17–19]. Recently, the bounds of Hausdorff dimension and box dimension of mixed RL fractional integral of bivariate continuous function is estimated in [12]. The aforementioned result stimulates an important question of whether an analogous result exists for a multivariate function. This paper estimates the bounds of Hausdorff dimension and box dimension of mixed RL fractional integral for a multivariate continuous function.

The rest of the paper is structured as follows: Preliminaries such as definition of mixed RL fractional integral of a multivariate function, box dimension, Hausdorff dimension and other basic terminologies are presented in Sect. 2. Estimation of bounds for the fractal dimensions of the mixed RL fractional integral of a multivariate function, which belongs to the class of Holder continuous functions, is discussed in Sect. 3. Conclusion of the paper is presented in Sect. 4.

2 Preliminaries

The basic terminologies that are necessary for the current study are precisely overviewed in this section.

Definition 1 Suppose the function f is defined on $[a_1, b_1] \times [a_2, b_2] \times \cdots [a_n, b_n]$ with $a_1, a_2, \ldots, a_n \geq 0$, then its mixed RL integral is given by

$$\mathcal{I}^{\delta} f(z_1, z_2, \ldots, z_n) = \frac{1}{\Gamma(\delta_1)\Gamma(\delta_2)\ldots\Gamma(\delta_n)} \int_{a_1}^{z_1}\int_{a_2}^{z_2}\cdots\int_{a_n}^{z_n} (z_1 - w_1)^{\delta_1 - 1}(z_2 - w_2)^{\delta_2 - 1} \ldots (z_n - w_n)^{\delta_n - 1} f(w_1, w_2, \ldots, w_n) dw_1 dw_2 \ldots dw_n,$$

where $\delta = (\delta_1, \delta_2, \ldots, \delta_n)$ with $\delta_1, \delta_2, \ldots, \delta_n > 0$.

Definition 2 Consider a non-empty subset V of $\mathbb{R}^n$, the diameter of V is given by

$$|V| = \sup\{|u - t|;\ u, t \in V\}.$$

Consider the countable collection of sets $\{V_k\}$ of diameter at most ϵ which can cover A. That is,

$$A \subset \bigcup_{k=1}^{\infty} V_k,\ 0 < |V_k| \leq \epsilon,$$

here $\{V_k\}$ is a ϵ-cover of A for each k. If A is a subset of $\mathbb{R}^n$ and for any $\epsilon > 0$, define

$$\mathcal{H}^d_\epsilon(A) = \inf\left\{\sum_{k=1}^{\infty} |V_k|^d : \{V_k\}\text{is a } \epsilon\text{-cover of A}\right\}$$

where d is any non-negative real number. Then, for the set A, the d-dimensional Hausdorff measure is given by

$$\mathcal{H}^d(A) = \lim_{\epsilon \to 0} \mathcal{H}^d_\epsilon(A).$$

Definition 3 For the subset A of $\mathbb{R}^n$ and $d > 0$,

$$\begin{aligned} dim_{\mathcal{H}}(A) &= \sup\{d : \mathcal{H}^d(A) = \infty\} \\ &= \inf\{d : \mathcal{H}^d(A) = 0\} \end{aligned}$$

is known as the Hausdorff dimension of A. Suppose $d = dim_{\mathcal{H}}(A)$, $\mathcal{H}^d(A)$ may be 0 or ∞, or may obeys

$$0 < \mathcal{H}^d(A) < \infty \tag{1}$$

If a Borel set obeys (1), it is referred as a d-set.

Definition 4 Consider a non-empty bounded subset A of $\mathbb{R}^n$. The least number of sets required to cover A is denoted by $N_\epsilon(A)$ with diameter at most ϵ. Then, the lower and upper box dimensions are, respectively, given by

$$\begin{aligned} \underline{\dim}_B(A) &= \underline{\lim}_{\epsilon \to 0} \frac{\log N_\epsilon(A)}{-\log \epsilon}, \\ \overline{\dim}_B(A) &= \overline{\lim}_{\epsilon \to 0} \frac{\log N_\epsilon(A)}{-\log \epsilon}. \end{aligned}$$

If $\underline{\dim}_B(A) = \overline{\dim}_B(A)$, the common value is termed as the box dimension for A. (i.e.,)

$$\dim_B(A) = \lim_{\epsilon \to 0} \frac{\log N_\epsilon(A)}{-\log \epsilon}.$$

Let $Gr(f, J_1 \times J_2 \times \cdots \times J_n)$ denote the graph of the function f defined on $J_1 \times J_2 \times \cdots \times J_n$ where $J_i \subset \mathbb{R}$, for $i = 1, 2, \ldots, n$ and $\dim_{\mathcal{H}} Gr(f, J_1 \times J_2 \times \cdots \times J_n)$,$\dim_B Gr(f, J_1 \times J_2 \times \cdots \times J_n)$ and $\overline{\dim}_B Gr(f, J_1 \times J_2 \times \cdots \times J_n)$ represent the Hausdorff dimension, box dimension and upper box dimension of f defined on $J_1 \times J_2 \times \cdots \times J_n$, respectively, in this entire paper.

Lemma 1 ([11]) *If* $f \in \mathcal{C}(J_1 \times J_2 \times \cdots \times J_n)$ *and for some* $C > 0$, $0 \leq v \leq 1$,

$$|f(p_1, p_2, \ldots, p_n) - f(q_1, q_2, \ldots, q_n)| \leq C\|(p_1, p_2, \ldots, p_n) - (q_1, q_2, \ldots, q_n)\|_2^v, \quad (2)$$

$\forall(p_1, p_2, \ldots, p_n),\ (q_1, q_2, \ldots, q_n) \in J_1 \times J_2 \times \cdots \times J_n$. *Then the inequality*

$$n \leq \dim_{\mathcal{H}} Gr(f, J_1 \times J_2 \times \cdots \times J_n) \leq \overline{\dim}_B Gr(f, J_1 \times J_2 \times \cdots \times J_n) \leq n + 1 - v$$

remains true if (2) holds with $\|(p_1, p_2, \ldots, p_n) - (q_1, q_2, \ldots, q_n)\|_2 < \epsilon$ *for some* $\epsilon > 0$. *Suppose* $v = 1$, *the function* f *becomes Lipschitz continuous.*

Lemma 2 [11] *Let*

$$\begin{aligned} &H^v(J_1 \times J_2 \times \cdots \times J_n) \\ &\quad = \{f(z_1, z_2, \ldots, z_n) : |f(z_1 + l_1, \ldots, z_n + l_n) - f(z_1, \ldots, z_n)| \leq C\|(l_1, l_2, \ldots, l_n)\|_2^v\}, \end{aligned}$$

$\forall(z_1 + l_1, \ldots, z_n + l_n),\ (z_1, \ldots, z_n) \in J_1 \times J_2 \times \cdots \times J_n$, *be the set of all Hölder continuous functions with holder exponent* v. *If* $f \in H^v(J_1 \times J_2 \times \cdots \times J_n)$, *then*

$$n \leq \dim_{\mathcal{H}} Gr(f, J_1 \times J_2 \times \cdots \times J_n) \leq \overline{\dim}_B Gr(f, J_1 \times J_2 \times \cdots \times J_n) \leq n + 1 - v.$$

3 Fractal Dimensions of the Mixed RL Fractional Integral on Multivariate Function

This section establishes the relation between the two fractal dimensions namely box dimension and Hausdorff dimension of mixed RL fractional integral of a Hölder continuous multivariate function.

Theorem 1 *Let* $f(z_1, z_2, \ldots, z_n) \in H^v(J_1 \times J_2 \times \cdots \times J_n)$ *on* $[a_1, b_1] \times [a_2, b_2] \times \cdots \times [a_n, b_n]$ *such that* $f(0, 0, \ldots, 0) = (0, 0, \ldots, 0)$ *and suppose its mixed RL type integral exists, then*

$$\dim_{\mathcal{H}} Gr(\mathcal{I}^\delta f, J_1 \times J_2 \times \cdots \times J_n) \leq \overline{\dim}_B Gr(\mathcal{I}^\delta f, J_1 \times J_2 \times \cdots \times J_n) \leq n + 1 - v,$$

$0 < \delta_1, \delta_2, \ldots, \delta_n < 1$.

Proof Let $0 < \delta_1, \delta_2, \ldots, \delta_n < 1$ and $0 \leq a_1 \leq z_1 \leq z_1 + l_1 \leq b_1, 0 \leq a_2 \leq z_2 \leq z_2 + l_2 \leq b_2, \ldots, 0 \leq a_n \leq z_n \leq z_n + l_n \leq b_n$. Then

$$(\mathcal{I}^{\delta} f)(z_1+l_1, z_2+l_2, \ldots, z_n+l_n) - (\mathcal{I}^{\delta} f)(z_1, z_2, \ldots, z_n)$$

$$= \frac{1}{\Gamma(\delta_1)\Gamma(\delta_2)\ldots\Gamma(\delta_n)} \int_{a_1}^{z_1+l_1} \int_{a_2}^{z_2+l_2} \cdots \int_{a_n}^{z_n+l_n} (z_1+l_1-w_1)^{\delta_1-1}$$
$$(z_2+l_2-w_2)^{\delta_2-1} \ldots (z_n+l_n-w_n)^{\delta_n-1} f(w_1, w_2, \ldots, w_n) dw_1 dw_2 \ldots dw_n$$
$$- \frac{1}{\Gamma(\delta_1)\Gamma(\delta_2)\ldots\Gamma(\delta_n)} \int_{a_1}^{z_1} \int_{a_2}^{z_2} \cdots \int_{a_n}^{z_n} (z_1-w_1)^{\delta_1-1}$$
$$(z_2-w_2)^{\delta_2-1} \ldots (z_n-w_n)^{\delta_n-1} f(w_1, w_2, \ldots, w_n) dw_1 dw_2 \ldots dw_n$$
$$= \mathcal{I}_1 + \mathcal{I}_2 + \mathcal{I}_3 + \cdots + \mathcal{I}_{n+1} + \mathcal{I}_{n+2} + \mathcal{I}_{n+3},$$

where

$$\mathcal{I}_1 = \frac{1}{\Gamma(\delta_1)\Gamma(\delta_2)\ldots\Gamma(\delta_n)} \int_{a_1}^{a_1+l_1} \int_{a_2}^{a_2+l_2} \cdots \int_{a_n}^{a_n+l_n} (z_1+l_1-w_1)^{\delta_1-1}$$
$$(z_2+l_2-w_2)^{\delta_2-1} \ldots (z_n+l_n-w_n)^{\delta_n-1} f(w_1, w_2, \ldots, w_n) dw_1 dw_2 \ldots dw_n,$$
$$\mathcal{I}_2 = \frac{1}{\Gamma(\delta_1)\Gamma(\delta_2)\ldots\Gamma(\delta_n)} \int_{a_1}^{a_1+l_1} \int_{a_2+l_2}^{z_2+l_2} \cdots \int_{a_n}^{a_n+l_n} (z_1+l_1-w_1)^{\delta_1-1}$$
$$(z_2+l_2-w_2)^{\delta_2-1} \ldots (z_n+l_n-w_n)^{\delta_n-1} f(w_1, w_2, \ldots, w_n) dw_1 dw_2 \ldots dw_n,$$

$$\mathcal{I}_3 = \frac{1}{\Gamma(\delta_1)\Gamma(\delta_2)\ldots\Gamma(\delta_n)} \int_{a_1+l_1}^{z_1+l_1} \int_{a_2}^{a_2+l_2} \cdots \int_{a_n}^{a_n+l_n} (z_1+l_1-w_1)^{\delta_1-1}$$
$$(z_2+l_2-w_2)^{\delta_2-1} \ldots (z_n+l_n-w_n)^{\delta_n-1} f(w_1, w_2, \ldots, w_n) dw_1 dw_2 \ldots dw_n,$$
$$\vdots$$
$$\mathcal{I}_{n+1} = \frac{1}{\Gamma(\delta_1)\Gamma(\delta_2)\ldots\Gamma(\delta_n)} \int_{a_1}^{a_1+l_1} \int_{a_2}^{a_2+l_2} \cdots \int_{a_n+l_n}^{z_n+l_n} (z_1+l_1-w_1)^{\delta_1-1}$$
$$(z_2+l_2-w_2)^{\delta_2-1} \ldots (z_n+l_n-w_n)^{\delta_n-1} f(w_1, w_2, \ldots, w_n) dw_1 dw_2 \ldots dw_n,$$
$$\mathcal{I}_{n+2} = \frac{1}{\Gamma(\delta_1)\Gamma(\delta_2)\ldots\Gamma(\delta_n)} \int_{a_1+l_1}^{z_1+l_1} \int_{a_2+l_2}^{z_2+l_2} \cdots \int_{a_n+l_2}^{z_n+l_n} (z_1+l_1-w_1)^{\delta_1-1}$$
$$(z_2+l_2-w_2)^{\delta_2-1} \ldots (z_n+l_n-w_n)^{\delta_n-1} f(w_1, w_2, \ldots, w_n) dw_1 dw_2 \ldots dw_n,$$
$$\mathcal{I}_{n+3} = \frac{1}{\Gamma(\delta_1)\Gamma(\delta_2)\ldots\Gamma(\delta_n)} \int_{a_1}^{z_1} \int_{a_2}^{z_2} \cdots \int_{a_n}^{z_n} (z_1-w_1)^{\delta_1-1}$$
$$(z_2-w_2)^{\delta_2-1} \ldots (z_n-w_n)^{\delta_n-1} f(w_1, w_2, \ldots, w_n) dw_1 dw_2 \ldots dw_n.$$

By applying variable transformation in $\mathcal{I}_{n+3}$,

$$\mathcal{I}^*_{n+3} = \frac{1}{\Gamma(\delta_1)\Gamma(\delta_2)\ldots\Gamma(\delta_n)} \int_{a_1+l_1}^{z_1+l_1}\int_{a_2+l_2}^{z_2+l_2}\cdots\int_{a_n+l_n}^{z_n+l_n} (z_1+l_1-w_1)^{\delta_1-1}$$
$$(z_2+l_2-w_2)^{\delta_2-1}\ldots(z_n+l_n-w_n)^{\delta_n-1}$$
$$\times f(w_1-l_1, w_2-l_2, \ldots, w_n-l_n)dw_1dw_2\ldots dw_n$$

$$\mathcal{I}_{n+2}-\mathcal{I}^*_{n+3} = \mathcal{I}_{n+4} = \frac{1}{\Gamma(\delta_1)\Gamma(\delta_2)\ldots\Gamma(\delta_n)} \int_{a_1+l_1}^{z_1+l_1}\int_{a_2+l_2}^{z_2+l_2}\cdots\int_{a_n+l_2}^{z_n+l_n} (z_1+l_1-w_1)^{\delta_1-1}$$
$$(z_2+l_2-w_2)^{\delta_2-1}\ldots(z_n+l_n-w_n)^{\delta_n-1}$$
$$[f(w_1-l_1, w_2-l_2, \ldots, w_n-l_n) - f(w_1, w_2, \ldots, w_n)]dw_1dw_2\ldots dw_n$$

$$|\mathcal{I}_{n+4}| \leq \frac{1}{\Gamma(\delta_1)\Gamma(\delta_2)\ldots\Gamma(\delta_n)} \int_{a_1+l_1}^{z_1+l_1}\int_{a_2+l_2}^{z_2+l_2}\cdots\int_{a_n+l_2}^{z_n+l_n} |(z_1+l_1-w_1)^{\delta_1-1}$$
$$(z_2+l_2-w_2)^{\delta_2-1}\ldots(z_n+l_n-w_n)^{\delta_n-1}$$
$$\times [f(w_1-l_1, w_2-l_2, \ldots, w_n-l_n) - f(w_1, w_2, \ldots, w_n)]|dw_1dw_2\ldots dw_n$$

As the function $f(z_1, z_2, \ldots, z_n) \in H^{\upsilon}(J_1 \times J_2 \times \cdots \times J_n)$ on $[a_1, b_1] \times [a_2, b_2] \times \cdots \times [a_n, b_n]$,

$$|J_{n+4}| \leq \frac{C\|l_1, l_2, \ldots, l_n\|_2^{\upsilon}}{\Gamma(\delta_1)\Gamma(\delta_2)\ldots\Gamma(\delta_n)} \int_{a_1+l_1}^{z_1+l_1}\int_{a_2+l_2}^{z_2+l_2}\cdots\int_{a_n+l_n}^{z_n+l_n} |(z_1+l_1-w_1)^{\delta_1-1}$$
$$(z_2+l_2-w_2)^{\delta_2-1}\ldots(z_n+l_n-w_n)^{\delta_n-1}|dw_1dw_2\ldots dw_n$$
$$= \frac{C\|l_1, l_2, \ldots, l_n\|_2^{\upsilon}}{\Gamma(\delta_1+1)\Gamma(\delta_2+1)\ldots\Gamma(\delta_n+1)}(z_1-a_1)^{\delta_1}(z_2-a_2)^{\delta_2}\ldots(z_n-a_n)^{\delta_n}$$

For $(z_1, z_2, \ldots, z_n) \in [a_1, b_1] \times [a_2, b_2] \times \cdots \times [a_n, b_n]$, one can get

$$|\mathcal{I}_{n+4}| \leq \frac{C\|l_1, l_2, \ldots, l_n\|_2^{\upsilon}}{\Gamma(\delta_1+1)\Gamma(\delta_2+1)\ldots\Gamma(\delta_n+1)}(b_1-a_1)^{\delta_1}(b_2-a_2)^{\delta_2}\ldots(b_n-a_n)^{\delta_n}$$
$$|\mathcal{I}_{n+4}| \leq C\|l_1, l_2, \ldots, l_n\|_2^{\upsilon}$$

where

$$C = \frac{(b_1-a_1)^{\delta_1}(b_2-a_2)^{\delta_2}\ldots(b_n-a_n)^{\delta_n}}{\Gamma(\delta_1+1)\Gamma(\delta_2+1)\ldots\Gamma(\delta_n+1)}.$$

Now, similarly applying the above steps, the bound of J_1 is found to be

$$
\begin{aligned}
|\mathcal{I}_1| \le& \frac{1}{\Gamma(\delta_1)\Gamma(\delta_2)\ldots\Gamma(\delta_n)} \int_{a_1}^{a_1+l_1}\int_{a_2}^{a_2+l_2}\cdots\int_{a_n}^{a_n+l_n} |(z_1+l_1-w_1)^{\delta_1-1} \\
&(z_2+l_2-w_2)^{\delta_2-1}\ldots(z_n+l_n-w_n)^{\delta_n-1}| \\
&\times |f(w_1,w_2,\ldots,w_n)-f(0,0,\ldots,0)|dw_1dw_2\ldots dw_n \\
\le& \frac{C\|l_1,l_2,\ldots,l_n\|_2^{\upsilon}}{\Gamma(\delta_1)\Gamma(\delta_2)\ldots\Gamma(\delta_n)} \int_{a_1}^{a_1+l_1}\int_{a_2}^{a_2+l_2}\cdots\int_{a_n}^{a_n+l_n} |(z_1+l_1-w_1)^{\delta_1-1} \\
&(z_2+l_2-w_2)^{\delta_2-1}\ldots(z_n+l_n-w_n)^{\delta_n-1}|dw_1dw_2\ldots dw_n \\
\le& \frac{C\|l_1,l_2,\ldots,l_n\|_2^{\upsilon}}{\Gamma(\delta_1)\Gamma(\delta_2)\ldots\Gamma(\delta_n)} \int_{a_1}^{a_1+l_1}\int_{a_2}^{a_2+l_2}\cdots \\
&\int_{a_n}^{a_n+l_n} |(a_1+l_1-w_1)^{\delta_1-1}(a_2+l_2-w_2)^{\delta_2-1}\ldots(a_n+l_n-w_n)^{\delta_n-1}|dw_1dw_2\ldots dw_n \\
=& \frac{C\|l_1,l_2,\ldots,l_n\|_2^{\upsilon}}{\Gamma(\delta_1+1)\Gamma(\delta_2+1)\ldots\Gamma(\delta_n+1)} l_1^{\delta_1}l_2^{\delta_2}\ldots l_n^{\delta_n} \\
\le& C\|l_1,l_2,\ldots,l_n\|_2^{\upsilon}.
\end{aligned}
$$

where

$$
C = \frac{l_1^{\delta_1}l_2^{\delta_2}\ldots l_n^{\delta_n}}{\Gamma(\delta_1+1)\Gamma(\delta_2+1)\ldots\Gamma(\delta_n+1)}.
$$

Similarly, the bounds of remaining integrals are obtained as

$$
\begin{aligned}
|\mathcal{I}_2| &\le C\|l_1,l_2,\ldots,l_n\|_2^{\upsilon}, \text{ where } C = \frac{l_1^{\delta_1}(b_2-a_2)^{\delta_2}\ldots l_n^{\delta_n}}{\Gamma(\delta_1+1)\Gamma(\delta_2+1)\ldots\Gamma(\delta_n+1)} \\
|\mathcal{I}_3| &\le C\|l_1,l_2,\ldots,l_n\|_2^{\upsilon}, \text{ where } C = \frac{(b_1-a_1)^{\delta_1}l_2^{\delta_2}\ldots l_n^{\delta_n}}{\Gamma(\delta_1+1)\Gamma(\delta_2+1)\ldots\Gamma(\delta_n+1)} \\
&\vdots \\
|\mathcal{I}_{n+1}| &\le C\|l_1,l_2,\ldots,l_n\|_2^{\upsilon}, \text{ where } C = \frac{l_1^{\delta_1}l_2^{\delta_2}\ldots(b_n-a_n)^{\delta_n}}{\Gamma(\delta_1+1)\Gamma(\delta_2+1)\ldots\Gamma(\delta_n+1)}.
\end{aligned}
$$

As a consequence, for a suitable constant C,

$$
\begin{aligned}
|(\mathcal{I}^{\delta}f)(z_1+l_1,z_2+l_2,\ldots,z_n+l_n)-&(\mathcal{I}^{\delta}f)(z_1,z_2,\ldots,z_n)| \\
&\le |\mathcal{I}_1|+|\mathcal{I}_2|+|\mathcal{I}_3|+\cdots+|\mathcal{I}_{n+1}|+|\mathcal{I}_{n+2}|+|\mathcal{I}_{n+3}| \\
&\le C\|l_1,l_2,\ldots,l_n\|_2^{\upsilon}.
\end{aligned}
$$

The proof follows from Lemma 2.

Theorem 2 *If the continuous function* $f(z_1,z_2,\ldots,z_n)$ *defined on* $[a_1,b_1]\times[a_2,b_2]\times\cdots\times[a_n,b_n]$ *with* $f(0,0,\ldots,0)=(0,0,\ldots,0)$ *obeys Lipschitz condition, then for* $0<\delta_1,\delta_2,\ldots,\delta_n<1$,

$$\dim_{\mathcal{H}} Gr(\mathcal{I}^{\delta} f, J_1 \times J_2 \times \cdots \times J_n) = \dim_B Gr(\mathcal{I}^{\delta} f, J_1 \times J_2 \times \cdots \times J_n) = n.$$

The proof follows from Lemma 1 and Theorem 1.

Remark 1 For any fractal function $f(z_1, z_2, \ldots, z_n)$ having box dimension $n + 1 - v$, the upper box dimension of its RL type fractional integral is not more than $n + 1 - v$.(i.e.,) If

$$\dim_B G(f, J_1 \times J_2 \times \cdots \times J_n) = n + 1 - v,$$

then

$$\overline{\dim}_B G(\mathcal{I}^{\delta} f, J_1 \times J_2 \times \cdots \times J_n) \leq n + 1 - v.$$

Hence,

$$\overline{\dim}_B Gr(\mathcal{I}^{\delta} f, J_1 \times J_2 \times \cdots \times J_n) \leq \dim_B Gr(f, J_1 \times J_2 \times \cdots \times J_n) = n + 1 - v.$$

4 Conclusion

In this article, the bounds for both the Hausdorff dimension and upper box dimension of the mixed RL fractional integral of a multivariate function are found to be $n + 1 - v$, when it satisfies Hölder condition. On the other hand, when it obeys Lipchitz condition, it is illustrated that the box dimension and Hausdorff dimension of a multivariate function are n.

References

1. Falconer, K.: Fractal Geometry: Mathematical Foundations and Applications. John Wiley & Sons Ltd, England (1990)
2. Barnsley, M.F.: Fractals Everywhere, 2nd edn. Academic, USA (1993)
3. Banerjee, S., Hassan, M.K., Mukherjee, S., Gowrisankar, A.: Fractal Patterns in Nonlinear Dynamics and Applications. CRC Press, Baco Raton (2019)
4. Gowrisankar, A., Uthayakumar, R.: Fractional calculus on fractal interpolation function for a sequence of data with countable iterated function system. Mediterr. J. Math. **13**(6), 3887–3906 (2016)
5. Priyanka, T.M.C., Gowrisankar, A.: Riemann-Liouville fractional integral of non-affine fractal interpolation function and its fractional operator. Eur. Phys. J.: Spec. Top. **230**, 3789–3805 (2021)
6. Gowrisankar, A., Prasad, M.G.P.: Riemann-Liouville calculus on quadratic fractal interpolation function with variable scaling factors. J. Anal. **27**(2), 347–363 (2019)
7. Liang, Y.S.: Box dimension of Riemann-Liouville fractional integrals of continuous functions of bounded variation. Nonlinear Anal. **72**(11), 4304–4306 (2010)
8. Liang, Y.S., Su, W.Y.: Fractal dimension of certain continuous functions of unbounded variation. Fractals **25**(01), 1750009 (2017)

9. Liang, Y.S., Su, W.Y.: Riemann-Liouville fractional calculus of one dimensional continuous functions. Sci. Sin. Math. **46**(4), 423–438 (2016)
10. Verma, S., Viswanathan, P.: Bivariate functions of bounded variation: fractal dimension and fractional integral. Indag. Math. **31**(2), 294–309 (2020)
11. Chandra, S., Abbas, S.: Box dimension of mixed Katuhgampola fractional integral of two dimensional continuous functions. arXiv:2105.01885
12. Chandra, S., Abbas, S.: Analysis of fractal dimension of mixed Riemann-Liouville fractional integral. arXiv:2105.06648
13. Banerjee, S., Easwaramoorthy, D., Gowrisankar, A.: Fractal Functions, Dimensions and Signal Analysis, 1st edn. Springer, Cham (2021)
14. Verma, S., Viswanathan, P.: A note on Katugampola fractional calculus and fractal dimensions. Appl. Math. Comput. **339**, 220–230 (2018)
15. Verma, S., Viswanathan, P.: Katugampola fractional integral and fractal dimension of bivariate functions. Results Math. (2021). arXiv:2101.06093
16. Frank, B.: Tatom: the relationship between fractional calculus and fractal. Fractals **3**(1), 217–229 (1995)
17. Chandra, S., Abbas, S.: Analysis of mixed Weyl-Marchaud fractional derivative and Box dimensions. Fractals (2021). https://doi.org/10.1142/S0218348X21501450
18. Priyanka, T.M.C., Gowrisankar, A.: Analysis on Weyl-Marchaud fractional derivative for types of fractal interpolation function with fractal dimension. Fractals **29**(7), 2150215 (2021)
19. Peng, W.L., Yao, K., Zhang, X., Yao, J.: Box dimension of Weyl-Marchaud fractional derivative of linear fractal interpolation functions. Fractals **27**(4), 1950058 (2019)

Stochastic Predator-Prey Model with Disease in Prey and Hybrid Impulses for Integrated Pest Management

Shivani Khare, Kunwer Singh Mathur, and Rajkumar Gangele

Abstract The stochastic effect is sometimes crucial in the case of integrated pest management due to fluctuating exotic environmental and climate conditions, and it also affects the resources required for pest extinction. Hence, we extend the classical predator-prey model into an impulsive control system by incorporating disease in the prey along with a stochastic element, which helps in controlling optimal pesticide level more accurately in most economic means for pest eradication. Further in analysis, it is obtained that the solution of the proposed model is positively bounded and globally attractive. The long-term behavior of the model is examined, and the condition for pest eradication is driven. The analysis shows that pest control becomes more complex due to the high amplitude of impulsive period and higher intensities of external interference. Finally, we perform some numerical simulations to support our analytical findings and their interpretation.

Keywords Predator-prey model · Integrated pest management · Impulsive control · Pest eradication · Permanence · Stochastic differential equations

1 Introduction

Integrated pest management (IPM) is one of the most effective methods to reduce the pest level. It is also an environmentally friendly approach that relies on the combination of common-sense practices. The IPM approach mainly focuses on reducing pest damage by the most economical means and the least possible hazard to people, property, and the environment. Biological control, chemical control, mechanical control, microbial control, remote sensing are the most valuable and significant methods to

The research work of the first author is supported by the DST-INSPIRE Fellowship (No. DST/INSPIRE Fellowship/2019/IF190224).

S. Khare · K. S. Mathur (✉) · R. Gangele
Department of Mathematics and Statistics, Dr. Harisingh Gour Vishwavidyalaya, Sagar 470003, Madhya Pradesh, India
e-mail: ksmathur1709@gmail.com

S. Banerjee and A. Saha (eds.), *Nonlinear Dynamics and Applications*, Springer Proceedings in Complexity,
https://doi.org/10.1007/978-3-030-99792-2_96

suppress the pest population. Mechanical control is a traditional method requiring more human efforts, while the chemical control is used to eradicate the pest by using chemical pesticides. But, heavy use of chemical pesticides creates more problems to the environment, and it also affects human health, i.e., overuse of DDT damage bird eggs. On the other hand, the biological control involves controlling populations of pests by releasing other living organisms that are commonly called natural predators. Microbial control with pathogens is also a part of biological control. Microbial control with the pathogen can suppress pests by releasing infected pests into the region of consideration, which is a natural process in controlling pests without harming the crop. In this paper, the hybrid approach including biological and chemical controls will be used, which is more effective in controlling pests.

Further, the mathematical models play a vital role in describing the impact of various pest control strategies in IPM systems. Mathematical modeling is recently catching a lot of attention in the field of Integrated pest management systems [14, 19]. Mathematical models have also been proposed to study the hybrid effect of biological and chemical pesticides together in the eradication of pest population [4, 8]. Moreover, some models involving chemical pesticides and disease incorporation (microbial control) in the pest population are presented in [9, 13]. Further, the impulsive differential equations serve as a tool to study the dynamic processes that are subject to a sudden change in the state of the population [10, 18]. Many researchers have developed a state-dependent impulsive differential equations system in order to study the pest control system by releasing pest controlling agents when the pest population exceeds the economic threshold level [15, 16]. Many authors considered time-dependent impulsive differential systems because sometimes pest control agents are released in a periodic manner [12, 21]. They considered either biological or hybrid approaches, e.g., biological (including microbial) and chemical control.

Besides, the real world is full of randomness. Random fluctuations occur due to drought, earthquake, flooding, harvesting fire, sudden rainfall, temperature rise. These random events affect the dynamics of the ecosystem, and these fluctuations can not be neglected. Generally, in the deterministic system, it is taken that involved parameters are constant, but due to random fluctuations, these parameters fluctuate around some average value, and this fluctuation in parameters affects the system's dynamics. Therefore, to capture the effect of random fluctuations, some authors presented a stochastic prey-predator system for integrated pest management [1, 2, 7, 11]. But generally, these models incorporate single impulsive control strategy i.e. biological control or chemical control [3, 6, 17, 20]. Hence stochastic prey-predator model, which incorporates natural enemies, disease, and chemical pesticides in an impulsive manner, is a new challenging problem to the researchers and has not been investigated yet.

Motivated from the literature, this paper considers two species predator-prey model with stochastic effect for integrated pest management. Also, we apply microbial control with the pathogen, natural enemies, and chemical control methods in an impulsive manner to suppress the pest.

The main aim of our paper is to analyze the dynamics of the proposed Impulsive Eco-epidemic prey-predator system by considering white noise and finding out the condition for pest extinction and permanence of populations. The presented paper is organized as follows: In Sect. 2, we construct the stochastic Eco-epidemic model. Also, some assumptions of the model and essential lemmas are formulated. In Sect. 3, we investigate the susceptible pest extinction solution and the condition for the permanence of the system. Some numerical simulations are carried out to validate the theoretical results in Sect. 4. Finally, we discuss some conclusions of our proposed model and the scope of the work in the last section.

2 Model Development

A predator-prey model with disease in prey and two impulses for integrated pest management is proposed analyzed by Shi et al. in [13]. In this model, two compartments are considered for the pest population: Susceptible pest S(t) and Infected pest I(t). It is considered that the pest can be suppressed with the help of a biological approach using microbial control with the pathogen and releasing of natural enemies simultaneously. However, the growth of pests and natural enemies is affected due to several random fluctuations, like, varying environmental and climate conditions which are not considered in [13] and hence the model proposed in [13] is not able to describe biological phenomena realistically. In this case, only stochastic differential equations can be used to model this biological situation. Keeping in mind the facts, we will propose a stochastic prey-predator model with disease in prey with hybrid impulsive pest management strategies with the following assumptions:

(A1) Due to continuous environment fluctuations, the intrinsic growth rate of susceptible pest population r, natural death rate of infected pest d_1 and death rate of natural enemies d_2 fluctuates around some average value. We construct stochastic model by perturbing $r \to r + \sigma_1 dB_1(t)$, $d_1 \to d_1 - \sigma_2 dB_2(t)$ and $d_2 \to d_2 - \sigma_3 dB_3(t)$.

(A2) Both chemical and biological control are used to suppress the pest population. It is assumed that a portion δ_n of susceptible pest is reduced impulsively by spraying chemical pesticides while a portion p_n of infected pest and q_n of natural enemies are released impulsively at every moment of time $t = nT$.

(A3) Moreover, it is considered that natural enemies consume the susceptible pests with the predation rate β_2 with conversion efficiency rate k.

Therefore, with the above assumptions, we formulate the following stochastic prey-predator model for integrated pest management:

$$\begin{cases} \left.\begin{aligned} dS(t) &= \left[rS(t)\left(1-\tfrac{S(t)}{K}\right)-\beta_1 S(t)I(t)-\beta_2 S(t)N(t)\right]dt \\ &\quad +\sigma_1 S(t)dB_1(t), \\ dI(t) &= [\beta_1 S(t)I(t)-d_1 I(t)]dt+\sigma_2 I(t)dB_2(t), \\ dN(t) &= [k\beta_2 S(t)N(t)-d_2 N(t)]dt+\sigma_3 N(t)dB_3(t), \end{aligned}\right\} \quad t \neq nT, \\ \left.\begin{aligned} S(t^+) &= (1-\delta_n)S(t), \\ I(t^+) &= (1+p_n)I(t), \\ N(t^+) &= (1+q_n)N(t), \end{aligned}\right\}, \quad t=nT,\ n\in z_+. \end{cases} \tag{2.1}$$

Here $S(t), I(t)$ and $N(t)$ denotes the population densities of the susceptible pest, infected pest and natural enemies at time t. σ_1, σ_2, and σ_3 are the coefficient effect of environmental stochastic perturbation on susceptible pest, infected pest and natural enemies, B_i $i = 1, 2, 3$ are independent from other standard Wiener process.

3 Model Analysis

This section is devoted to stating and proving our main results.

Theorem 1 *For all $t \in [0, \infty)$, the model (2.1) has a unique solution (S(t),I(t),N(t)) for any initial condition $(S_0, I_0, N_0) \in R^3_+$.*

Proof First we consider the model (2.1) without impulses:

$$\begin{cases} dS_1(t) = \left[rS_1(t)\left(1-\prod_{0<nT<t}(1-\delta_n)\frac{S_1(t)}{K}\right)-\beta_1\prod_{0<nT<t}(1+p_n)I_1(t)S_1(t)\right. \\ \qquad\left.-\beta_2\prod_{0<nT<t}(1+q_n)N_1(t)S_1(t)\right]dt+\sigma_1 S_1(t)dB_1(t), \\ dI_1(t) = \left[\beta_1\prod_{0<nT<t}(1-\delta_n)S_1(t)I_1(t)-d_1 I_1(t)\right]dt+\sigma_2 I_1(t)dB_2(t), \\ dN_1(t) = \left[k\beta_2\prod_{0<nT<t}(1-\delta_n)S_1(t)N_1(t)-d_2 N_1(t)\right]dt+\sigma_3 N_1(t)dB_3(t), \end{cases} \tag{3.1}$$

with the initial value $(S_{10}, I_{10}, N_{10}) = (S_0, I_0, N_0)$. The classical theory of stochastic differential equations without impulses suggests that there is a unique global positive solution of the system (3.1).

Let

$$S(t) = \prod_{0<nT<t}(1-\delta_n)S_1(t),\ I(t) = \prod_{0<nT<t}(1+p_n)I_1(t),\ N(t) = \prod_{0<nT<t}(1+q_n)N_1(t),$$

with the initial condition $(S_{10}, I_{10}, N_{10}) = (S_0, I_0, N_0)$. $S(t), I(t)$, and $N(t)$ are continuous on every interval $t \in (nT, (n+1)T)$, $n \in Z^+ = 0, 1, 2, \ldots$.

For $S(t)$, we have

$$dS(t) = d\left[\prod_{0<nT<t}(1-\delta_n)S_1(t)\right] = \prod_{0<nT<t}(1-\delta_n)dS_1(t),$$
$$= \left[rS(t)\left(1-\frac{S(t)}{K}\right) - \beta_1 S(t)I(t) - \beta_2 S(t)N(t)\right]dt + \sigma_1 S(t)dB_1(t),$$

for every $n \in N$ and $t \neq nT$.

$$S(nT^+) = \lim_{t\to nT^+} S(t) = \lim_{t\to nT^+} \prod_{0<iT<t}(1-\delta_i)S_1(t) = (1-\delta_n)S(nT),$$

for every $n \in N$. And

$$S(nT^-) = \lim_{t\to nT^-} S(t) = \lim_{t\to nT^-} \prod_{0<iT<t}(1-\delta_i)S_1(t) = (1-\delta_n)S(nT).$$

For I(t), we have,

$$dI(t) = d\left[\prod_{0<nT<t}(1+p_n)I_1(t)\right] = \prod_{0<nT<t}(1+p_n)dI_1(t)$$
$$= [\beta_1 S(t)I(t) - d_1 I(t)]\,dt + \sigma_2 I(t)dB_2(t),$$

for every $n \in N$ and $t \neq nT$.

$$I(nT^+) = \lim_{t\to nT^+} I(t) = \lim_{t\to nT^+} \prod_{0<iT<t}(1+p_i)I_1(t) = (1+p_n)I(nT),$$

for every $n \in N$. Also,

$$I(nT^-) = \lim_{t\to nT^-} I(t) = \lim_{t\to nT^-} \prod_{0<iT<t}(1+p_i)I_1(t) = (1+p_n)I(nT),$$

for every $n \in N$. Similarly, for N(t), we have

$$dN(t) = d\left[\prod_{0<nT<t}(1+q_n)N_1(t)\right] = \prod_{0<nT<t}(1+q_n)dN_1(t)$$
$$= [k\beta_2 S(t)N(t) - d_2 N(t)]\,dt + \sigma_3 N(t)dB_3(t), \text{ for every } n \in N \text{ and } t \neq nT.$$

$$N(nT^+) = \lim_{t\to nT^+} N(t) = \lim_{t\to nT^+} \prod_{0<iT<t}(1+q_i)N_1(t) = (1+q_n)N(nT),$$

for every $n \in N$. Further more,

$$N(nT^-) = \lim_{t \to nT^-} N(t) = \lim_{t \to nT^-} \prod_{0<iT<t} (1 + q_i)N_1(t) = (1 + q_n)N(nT),$$

for every $n \in N$. Hence, the theorem is proved.

Theorem 2 *For any initial value $(S_0, I_0, N_0) \in R_+^3$, there exists functions u(t), U(t), v(t), V(t), w(t), W(t) such that the positive solution of the model (2.1) satisfies the following inequalities.*

$$u(t) \le S(t) \le U(t), \quad v(t) \le I(t) \le V(t), \quad w(t) \le N(t) \le W(t), \quad t \ge 0 \ a.s. \tag{3.2}$$

Proof The model (2.1) has a positive solution, therefore,

$$dS(t) \le \left[rS(t) \left(1 - \frac{S(t)}{K} \right) \right] dt + \sigma_1 S(t) dB_1(t).$$

We formulate the following equations:

$$\begin{cases} dU(t) = \left[rU(t) \left(1 - \frac{U(t)}{K} \right) \right] dt + \sigma_1 U(t) dB_1(t), & t \ne nT, \\ U(t^+) = (1 - \delta_n) U(t), \quad t = nT, \\ U(0) = S_0 \end{cases} \tag{3.3}$$

Obviously, there is a global continuous positive solution of the system (3.3) with initial value S_0.

$$U(t) = \frac{\prod\limits_{0<nT<t} (1 - \delta_n) \exp\left[\int_0^t (r - 0.5\sigma_1^2) ds + \sigma_1 \int_0^t dB_1(s) \right]}{\frac{1}{S_0} + \int_0^t \prod\limits_{0<nT<s} (1 - \delta_n) \frac{r}{K} \exp\left[\int_0^s (r - 0.5\sigma_1^2) d\tau + \sigma_1 \int_0^s dB_1(\tau) \right] ds}.$$

According to the comparison theorem of stochastic differential equations, we get $S(t) \le U(t), t \in [0, t^*), a.s.$ Now from the second equation of the model (2.1),

$$dI(t) \ge -d_1 I(t) dt + \sigma_2 I(t) dB_2(t),$$

we formulate the following equation:

$$\begin{cases} dv(t) = -d_1 v(t) dt + \sigma_2 v(t) dB_2(t), & t \ne nT, \\ v(t^+) = (1 + p_n) v(t), \quad t = nT, \\ v(0) = I_0. \end{cases} \tag{3.4}$$

Obviously, $v(t) = \dfrac{\prod\limits_{0<nT<t}(1+p_n)\exp\left[-0.5\sigma_2^2 t + \sigma_2 B_2(t)\right]}{\frac{1}{I_0} + d_1\int_0^t \prod\limits_{0<nT<t}(1+p_n)\exp\left[-0.5\sigma_2^2 s + \sigma_2 B_2(s)\right]ds}$.

And we get $I(t) \geq v(t), t \in [0, t^*), a.s.$ Moreover, from the third equation of the model (2.1), we can get the inequalities, $dN(t) \geq -d_2 N(t)dt + \sigma_3 N(t)dB_3(t)$, we construct the following equations:

$$\begin{cases} dw(t) = -d_2 w(t)dt + \sigma_3 w(t)dB_3(t), & t \neq nT, \\ w(t^+) = (1+q_n)w(t), \quad t = nT, \\ w(0) = N_0. \end{cases} \tag{3.5}$$

Obviously, $w(t) = \dfrac{\prod\limits_{0<nT<t}(1+q_n)\exp\left[-0.5\sigma_3^2 t + \sigma_3 B_3(t)\right]}{\frac{1}{N_0} + d_2\int_0^t \prod\limits_{0<nT<t}(1+q_n)\exp\left[-0.5\sigma_3^2 s + \sigma_3 B_3(s)\right]ds}$.

And we have $N(t) \geq w(t), t \in [0, t^*), a.s.$

From second equation of the model (2.1), we have

$$dI(t) \leq \beta_1 U(t)I(t)dt - d_1 I(t)dt + \sigma_2 I(t)dB_2(t).$$

In the similar way, we get, $I(t) \leq V(t), t \in [0, t^*), a.s.$

Here, $V(t) = \dfrac{\prod\limits_{0<nT<t}(1+p(nT))\exp\left[-0.5\sigma_2^2 t + \sigma_2 B_2(t)\right]}{\frac{1}{I_0} + \beta_1\int_0^t \prod\limits_{0<nT<t}(1+p(nT))\exp\left[-0.5\sigma_2^2 s + \sigma_2 B_2(s)\right]U(s)ds}$.

Besides, from the third equation of the model (2.1), we have

$$dN(t) \leq k\beta_2 U(t)N(t)dt - d_2 N(t)dt + \sigma_3 N(t)dB_3(t).$$

Obviously, $W(t) = \dfrac{\prod\limits_{0<nT<t}(1+q(nT))\exp\left[-0.5\sigma_3^2 t + \sigma_3 B_3(t)\right]}{\frac{1}{N_0} + k\beta_2\int_0^t \prod\limits_{0<nT<t}(1+q(nT))\exp\left[-0.5\sigma_3^2 s + \sigma_3 B_3(s)\right]U(s)ds}$

represents the solution of the system of equations,

$$\begin{cases} dW(t) = k\beta_2 U(t)W(t)dt - d_2 W(t)dt + \sigma_3 W(t)dB_3(t), & t \neq nT, \\ W(t^+) = (1+q_n)W(t), \quad t = nT, \\ W(0) = N_0. \end{cases} \tag{3.6}$$

and $N(t) \leq W(t), t \in [0, t^*), a.s.$

The first equation of the model (2.1) follows that

$$dS(t) \geq \left[rS(t)\left(1 - \frac{S(t)}{K}\right) - \beta_1 S(t)V(t) - \beta_2 S(t)W(t)\right]dt + \sigma_1 S(t)dB_1(t).$$

According to the comparison system of the SDE, we have $S(t) \geq u(t), t \in [0, t^*), a.s.$ Here,

$$u(t) = \frac{\prod\limits_{0<nT<t} (1-\delta(t)) \exp\left[(r - 0.5\sigma_1^2)t - \beta \int_0^t (V(s) + W(s))ds + \sigma_1 B_1(t)\right]}{\frac{1}{S_0} + \frac{r}{K}\int_0^t \prod\limits_{0<nT<t} (1-\delta(t)) \exp\left[(r - 0.5\sigma_1^2)s - \beta \int_0^s (V(\tau) + W(\tau))d\tau + \sigma_1 B_1(s)\right] ds}.$$

and $\beta = max(\beta_1, \beta_2)$. By the following inequalities, we get that,

$$u(t) \leq S(t) \leq U(t), \quad v(t) \leq I(t) \leq V(t), \quad w(t) \leq N(t) \leq W(t), \quad t \in [0, t^*), a.s.$$

Hence, the theorem is verified.

Theorem 3 *If* $\lim_{t\to\infty} \frac{\sum_{0<nT<t} ln(1-\delta_n)}{t} < 0.5{\sigma_1}^2 - r$, *then the susceptible pests of SDE model (2.1) with any positive initial value tend to extinction according to probability* 1.

Proof Consider the transformation, $S(t) = \prod\limits_{0<nT<t} (1-\delta_n)\xi(t)$, and the lyapunov function, $V(t) = \ln \xi(t), \quad t \geq 0$.

Applying *Itô* formula to the first equation of the model (2.1), we have

$$d \ln \xi(t) = \frac{d\xi(t)}{\xi(t)} - \frac{(d\xi(t))^2}{2\xi^2(t)} \leq \left[r - 0.5\sigma_1^2 - \frac{\prod\limits_{0<nT<t} (1-\delta_n)\xi(t)}{K} \right] dt + \sigma_1 dB_1(t), \quad (3.7)$$

With the help of integration between 0 and t, we obtain that

$$\ln \xi(t) = \ln(S_0) + \int_0^t \left[r - 0.5\sigma_1^2 - \frac{S(s)}{K} \right] ds + M_1(t), \quad (3.8)$$

where $M_1(t)$is a martingale and $M_1(t) = \sigma_1 \int_0^t dB(s)$, the quadratic variation of the martingale is $\langle M_1(t), M_1(t)\rangle = \sigma_1^2 t$. The theory of large numbers for local martingales implies that $\lim_{t\to\infty} \frac{M_1(t)}{t} = 0$.

Therefore, from (3.8),

$$\sum_{0<nT<t} \ln(1-\delta_n) + \ln \xi(t) - \ln S_0 = \sum_{0<nT<t} \ln(1-\delta_n) + \int_0^t \left[r - 0.5\sigma_1^2 - \frac{S(s)}{K} \right] ds + M_1(t).$$

We have, $\prod\limits_{0<nT<t} \ln(1-\delta_n)\xi(t) - \ln S_0 = \sum\limits_{0<nT<t} \ln(1-\delta_n) + \int_0^t \left[r - 0.5\sigma_1^2 - \frac{S(s)}{K} \right] ds$ $M_1(t)$.

Therefore, $\ln S(t) - \ln S_0 \leq \sum_{0<nT<t} \ln(1-\delta_n) + \int_0^t \left[r - 0.5\sigma_1^2\right] ds + M_1(t)$.

According to the hypothesis, $\lim_{t\to\infty} S(t) = 0$.

4 Numerical Simulation and Discussion

This section is devoted to performing some numerical simulations to verify our theoretical findings. Here, we consider the numerical values of some parameters from realistic sources, and some are assumed in a realistic sense (see Table 1). To find the strong solution of the stochastic system (2.1), Higham introduced the Milstein method in [5]. By using this method, we transform the system (2.1) in discrete form and demonstrate our main results. Here the main aim is to investigate the effects of the impulsive time period, pulse releasing amount of chemical pesticides to harvest the susceptible pest, pulse releasing amount of infected pest, pulse releasing amount of natural enemies, and the intensity of stochastic perturbations on the integrated pest management system. The threshold which governs the permanence of all the species and pest extinction is obtained in this section. We choose the initial values $S(0) = 5$, $I(0) = 3$, $N(0) = 3$, controlling parameters $\delta_n = 0.1$, $p_n = 0.4208$, $q_n = 0.4208$ and other parameters values from Table 1. Then we analyze the stochastic system numerically to validate our analytic results and to see the effect of external interference in the system. Theorem 1 shows that there exists a positive solution of the system (2.1). In order to derive numerical simulation to validate the above results, first, we choose $\sigma_1 = \sigma_2 = \sigma_3 = 0$ in Fig. 1 and draw the time series plot of the model (2.1). In Fig. 2 we take $\sigma_1 = \sigma_2 = \sigma_3 = 0.1$. From Fig. 3 it can be seen that as we increase the value of noise $\sigma_1 = \sigma_2 = \sigma_3 = 0.3$, we found that the solution of the system (2.1) fluctuates with high amplitude. Chaos may occur, and pulse phenomena are covered, so pulse can not be produced clearly. Further, consider the system (2.1) again. By computations we get, for the parameters values $r(t) = 0.2$, $\delta_n = 0.1$, $\sigma_1 = 0.6$, $T = 5$ and $\lim_{t\to\infty} \frac{\sum_{0<nT<t} ln(1-\delta_n)}{t} = -0.0211 < 0.5{\sigma_1}^2 - r = -0.020$. Therefore, from the result of Theorem 3 we can determine that the pest population will extinct for these values of parameters. We can illustrate this result in Fig. 4. To see the significant effect of chemical pesticides in the suppression of susceptible pest populations, we take parameters values $r(t) = 0.3$, $\delta_n = 0.1$, $\sigma_1 = 0.6$, $T = 5$. From Fig. 5, we can analyze that the susceptible pest population is not suppressed completely. In order to eradicate the susceptible pest population, we have to increase the pulse releasing the amount of chemical pesticides $\delta_n = 0.49$ (see in Fig. 5). Moreover, we can analyze the effect of impulsive releasing amount of infected pest and natural enemies on the system (2.1) with different intensities of stochastic perturbations. We choose parameters $r(t) = 3$, $\delta_n = 0.1$, $p_n = 0.4208$, $q_n = 0.4208$, $\sigma_1 = \sigma_2 = \sigma_3 = 0$, $T = 5$. From Fig. 1 we can see that all the species coexists. If we increase pulse releasing amount of infected pest $p_n = 1.719$ and natural enemies $q_n = 1.719$ then suscep-

Table 1 Description of parameters values selection for model (2.1)

Parameter	Description	Values	Unit	Source
r	Intrinsic growth Rate	3	day^{-1}	[6]
K	Carrying capacity	12	ind	[6]
β_1	Number of susceptible pest become infected at per unit time due to direct contact with infected pest	0.3	$\text{ind}^{-1}\ \text{day}^{-1}$	Assumed
β_2	Attack rate of natural enemies to catch susceptible pest	0.6	$\text{ind}^{-1}\ \text{day}^{-1}$	[6]
k	Conversion efficiency of natural enemies	0.5	–	[6]
d_1	Death rate of infected prey	0.2	day^{-1}	Assumed
d_2	Death rate of natural enemies	0.2	day^{-1}	[6]

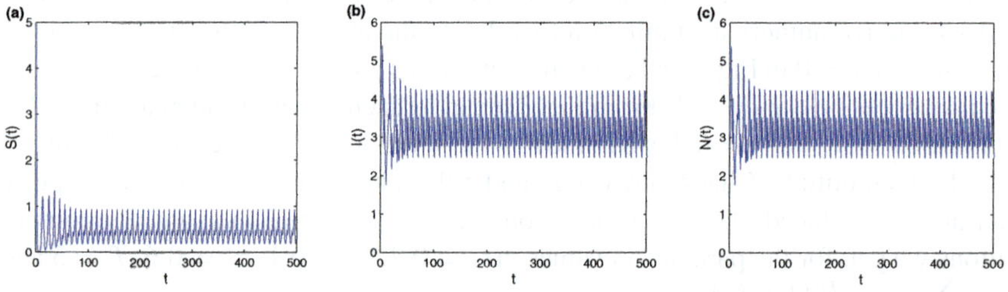

Fig. 1 Stability analysis of system (2.1) with $\delta_n = 0.1$, $p_n = 0.4208$, $q_n = 0.4208$, $T = 5$, $\sigma_1 = \sigma_2 = \sigma_3 = 0$, **a–c** Population dynamics in $t - S(t)$, $t - I(t)$, $t - N(t)$ planes, respectively

tible pest population will die out (see in (a–c) of Fig. 6). Also from (d–f) of Fig. 6 we can observe that if there is small stochastic perturbation in the system (2.1), then for the suppression of susceptible pest population we increase the value of $p_n = 1.8$ and $q_n = 1.8$ (see in (a-c) of Fig. 7). Similarly, if stochastic perturbation is large (i.e. $\sigma_1 = \sigma_2 = \sigma_3 = 0.3$), then we increase the pulse releasing amount of infected pest $p_n = 2.5$ and natural enemies $q_n = 2.5$ to eradicate susceptible pest population completely (see in (d–f) of Fig. 7).

Further, impulsive time period and intensity of noise also affect the stochastic persistence and extinction of susceptible pest populations. In order to eradi-

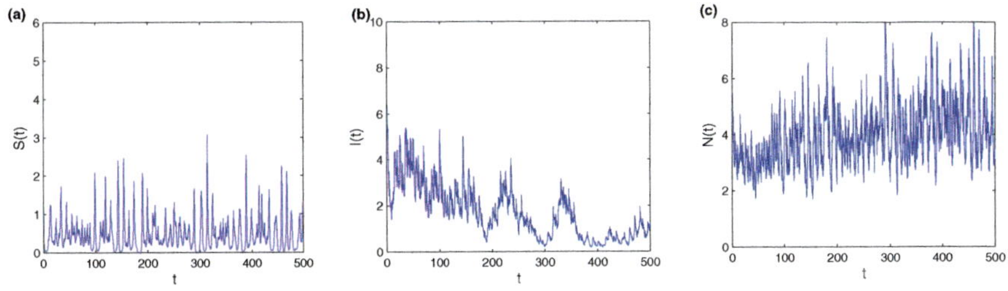

Fig. 2 Stability analysis of system (2.1) with $\delta_n = 0.1$, $p_n = 0.4208$, $q_n = 0.4208$, $T = 5$, $\sigma_1 = \sigma_2 = \sigma_3 = 0.1$, **a–c** Population dynamics in $t - S(t)$, $t - I(t)$, $t - N(t)$ planes, respectively

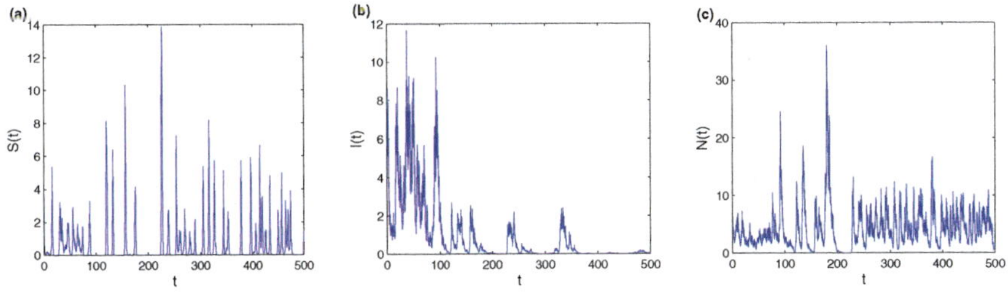

Fig. 3 Stability analysis of system (2.1) with $\delta_n = 0.1$, $p_n = 0.4208$, $q_n = 0.4208$, $T = 5$, $\sigma_1 = \sigma_2 = \sigma_3 = 0.3$, **a–c** Population dynamics in $t - S(t)$, $t - I(t)$, $t - N(t)$ planes, respectively

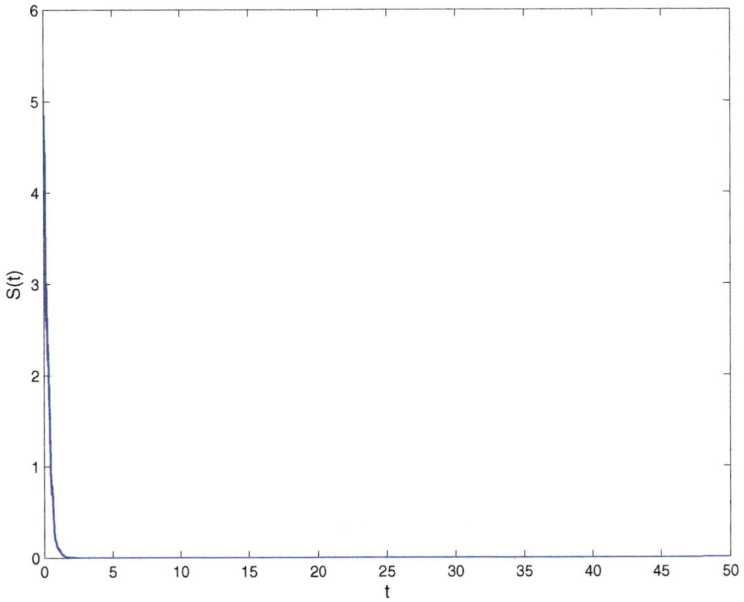

Fig. 4 Pest population of system (2.1) with $S(0) = 5$, $r = 0.2$, $\delta_n = 0.1$, $\sigma_1 = 0.6$, $T = 5$

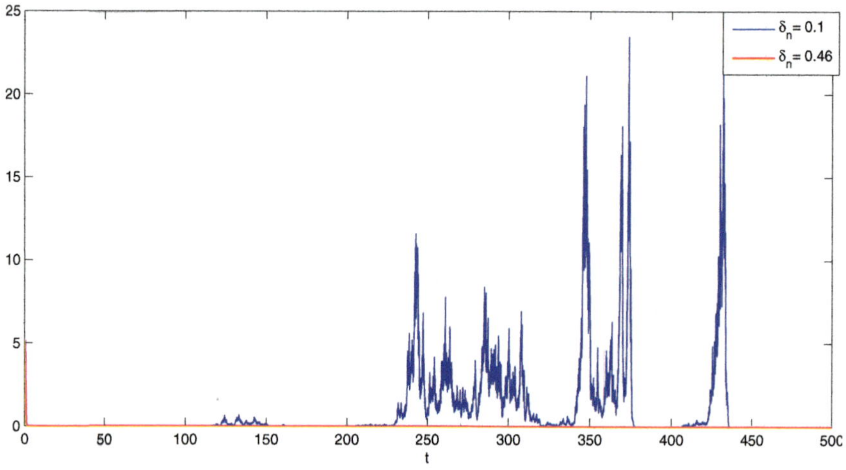

Fig. 5 Pest population of the system (2.1) with $S(0) = 5$, $r = 0.3$, $T = 5$, $\sigma_1 = \sigma_2 = \sigma_3 = 0.6$

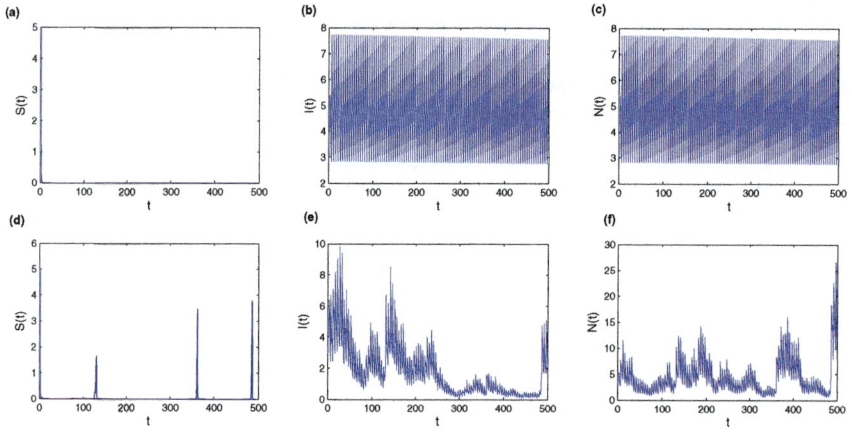

Fig. 6 Stability analysis of system (2.1), **a–c** with $\delta = 0.1$, $p_n = 1.719$, $q_n = 1.719$, $T = 5$, $\sigma_1 = \sigma_2 = \sigma_3 = 0$, Population dynamics in $t - S(t)$, $t - I(t)$, $t - N(t)$ planes, respectively and (d-f) with $\sigma_1 = \sigma_2 = \sigma_3 = 0.1$ Population dynamics in $t - S(t)$, $t - I(t)$, $t - N(t)$ planes, respectively

cate the susceptible pest population we take controlling parameters $\delta_n = 0.1$, $p_n = 0.4208$, $q_n = 0.4208$ and the rest of the parameters are the same. If the intensity of noise is $\sigma_1 = \sigma_2 = \sigma_3 = 0$, then we get a threshold value for impulsive period $T^* = 1.8285$. From Fig. 8 we can see that for $T = 1.7557 < T^* = 1.8285$ pest population become extinct and other populations will survives. If the intensity of noise is $\sigma_1 = \sigma_2 = \sigma_3 = 0.1$, then we have $T^* = 1.7511$. Similarly for $\sigma_1 = \sigma_2 = \sigma_3 = 0.3$, value of T^* will be 1.4379 (see Fig. 9). However, if the intensity of stochastic perturbation is large, then stochastic permanence of all species and pest extinction of the

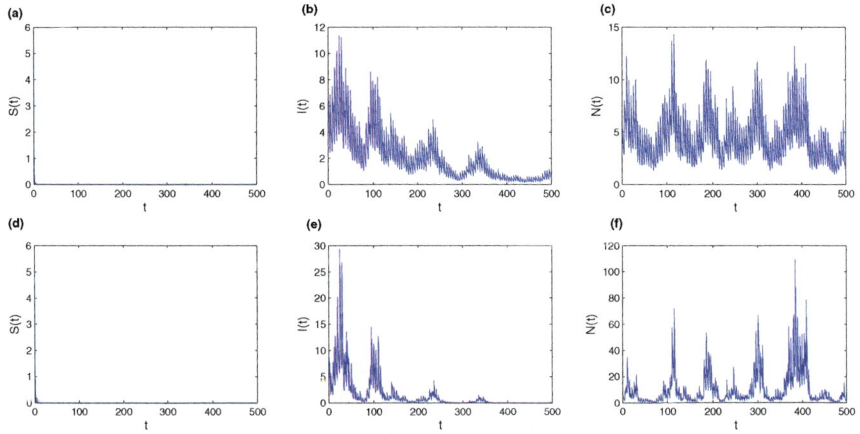

Fig. 7 Stability analysis of system (2.1), **a–c** with $\delta_n = 0.1$, $p_n = 1.8$, $q_n = 1.8$, $T = 5$, $\sigma_1 = \sigma_2 = \sigma_3 = 0.1$, Population dynamics in $t - S(t)$, $t - I(t)$, $t - N(t)$ planes, respectively and **d–f** with $\delta_n = 0.1$, $p_n = 2.5$, $q_n = 2.5$, $T = 5$, $\sigma_1 = \sigma_2 = \sigma_3 = 0.3$, Population dynamics in $t - S(t)$, $t - I(t)$, $t - N(t)$ planes, respectively

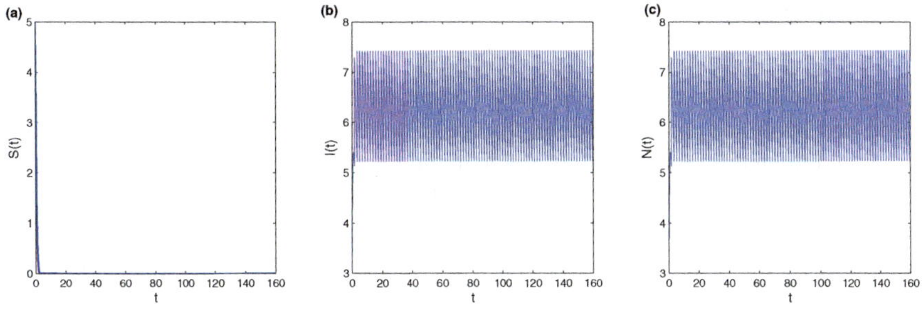

Fig. 8 Stability analysis of system (2.1) with $\delta_n = 0.1$, $p_n = 0.4208$, $q_n = 0.4208$, $T = 1.7557 < T^* = 1.8285$, $\sigma_1 = \sigma_2 = \sigma_3 = 0$, **a–c** Population dynamics in $t - S(t)$, $t - I(t)$, $t - N(t)$ planes, respectively

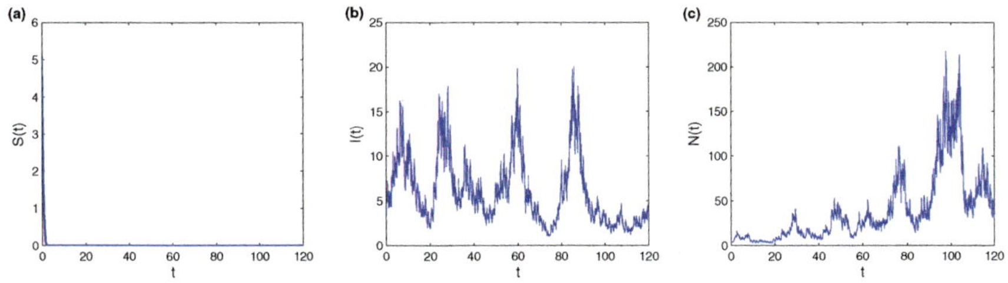

Fig. 9 Stability analysis of system (2.1) with $\delta_n = 0.1$, $p_n = 0.4208$, $q_n = 0.4208$, $T = 1.27 < T^* = 1.4379$, $\sigma_1 = \sigma_2 = \sigma_3 = 0.3$, **a–c** Population dynamics in $t - S(t)$, $t - I(t)$, $t - N(t)$, planes, respectively

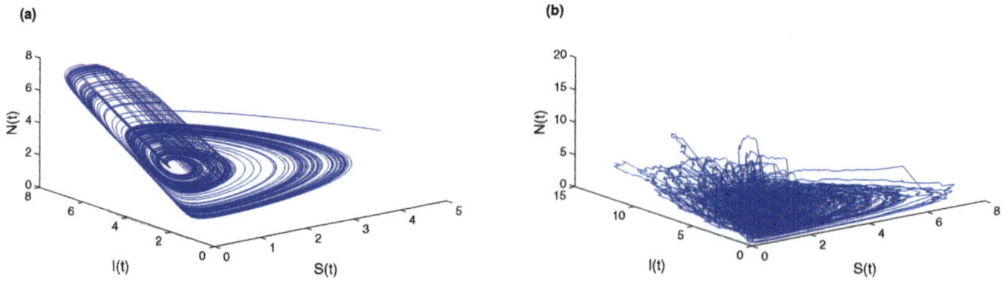

Fig. 10 3-D plot between: Susceptible pest, infected pest and natural enemy with $\delta_n = 0.1$, $p_n = 1$, $q_n = 1$, $T = 20$, and **a** with $\sigma_1 = \sigma_2 = \sigma_3 = 0$, **b** with $\sigma_1 = \sigma_2 = \sigma_3 = 0.1$

susceptible pest population could be changed. Further, we can also get that analysis of the system (2.1) becomes more complex when the impulsive period is greater than a certain critical value. A typical chaos oscillation occurs, and pulse phenomena are covered (see Fig. 10).

5 Conclusion

In this paper, a stochastic prey-predator system with hybrid impulses has been proposed and analyzed. In the modeling process, we considered that pests were completely eradicated by using both biological and chemical control strategies together impulsively. We released natural enemies, infected pests, and chemical pesticides to suppress the pest population. It is also noticed that real life is full of randomness. This random fluctuation occurs in a relatively short time interval due to some random environmental factors such as flood, rainfall, harvesting, fire, etc. Generally, the species' growth often suffers due to environmental fluctuations, and these fluctuations affect the long-term dynamics of the system. So, we can not neglect these random fluctuations, and it is essential to consider the prey-predator system with white noise. Keeping this in view, we have constructed a stochastic prey-predator model with disease in prey and a hybrid impulsive model for integrated pest management. We analyzed the dynamics of the system (2.1) and carried out some theoretical results. Theorem 1 shows that there exists a positive solution of the system (2.1). Similarly, Theorem 2 proves the permanence of the system (2.1). Moreover, By the result of Theorem 3, we obtained the condition for the pest extinction. Our analysis shows that these conditions mainly depend upon the parameters: impulsive releasing amount of chemical pesticides δ_n, impulsive releasing amount of infected pest p_n, impulsive releasing amount of natural enemies, impulsive time-period T and the magnitude of the intensities of external interferences $\sigma_1, \sigma_2, \sigma_3$. If the birth rate of susceptible pests is low, then susceptible pests can be controlled by using a small amount of chemical pesticides. However, the birth rate of the susceptible pest is high, and the intensity of stochastic perturbation is large, then susceptible pest population can be suppressed by releasing both infectious pests and natural enemies. Moreover, pulse releasing the amount of infected pests and natural enemies is also responsible for the susceptible pest extinction. The susceptible pest population can be suppressed by releasing a large amount of controlling agents. Also, if the controlling agents are released more frequently, the susceptible pest population can be eradicated completely. Our analysis also shows that pest control becomes more complex due to the high amplitude of impulsive period and higher intensities of external interference. Moreover, our analysis can help the farmer to understand the interactions of susceptible pests, infected pests, natural enemies, chemical pesticides, and environmental fluctuations to design the appropriate pest control strategies and make pest management decisions to control the pest.

References

1. Akman, O., Comar, T.D., Hrozencik, D.: On impulsive integrated pest management models with stochastic effects. Front. Neurosci. **9**, 119 (2015)
2. Comar, T., Akman, O., Hrozencik, D.: Model selection and permanence in a stochastic integrated pest management model (2017)
3. Feng, T., Meng, X., Zhang, T., Qiu, Z.: Analysis of the predator-prey interactions: a stochastic model incorporating disease invasion. Qual. Theory Dyn. Syst. **19**(2), 1–20 (2020)
4. Gao, W., Tang, S.: The effects of impulsive releasing methods of natural enemies on pest control and dynamical complexity. Nonlinear Anal. Hybrid Syst. **5**(3), 540–553 (2011)
5. Higham, D.J.: An algorithmic introduction to numerical simulation of stochastic differential equations. SIAM Rev. **43**(3), 525–546 (2001)
6. Huang, L., Chen, X., Tan, X., Chen, X., Liu, X.: A stochastic predator-prey model for integrated pest management. Adv. Differ. Equ. **2019**(1), 1–10 (2019)
7. Huang, Y., Shi, W., Wei, C., Zhang, S.: A stochastic predator-prey model with Holling II increasing function in the predator. J. Biol. Dyn. **15**(1), 1–18 (2021)
8. Jiao, J., Chen, L.: A pest management SI model with periodic biological and chemical control concern. Appl. Math. Comput. **183**(2), 1018–1026 (2006)
9. Jiao, J., Chen, L., Cai, S.: Impulsive control strategy of a pest management SI model with nonlinear incidence rate. Appl. Math. Model. **33**(1), 555–563 (2009)
10. Lakshmikantham, V., Simeonov, P.S., et al.: Theory of Impulsive Differential Equations, vol. 6. World scientific (1989)
11. Li, L., Zhao, W.: Deterministic and stochastic dynamics of a modified Leslie-Gower prey-predator system with simplified Holling-type IV scheme. Math. Biosci. Eng. **18**(3), 2813–2831 (2021)
12. Liang, J., Tang, S.: Optimal dosage and economic threshold of multiple pesticide applications for pest control. Math. Comput. Model. **51**(5–6), 487–503 (2010)
13. Shi, R., Jiang, X., Chen, L.: A predator-prey model with disease in the prey and two impulses for integrated pest management. Appl. Math. Model. **33**(5), 2248–2256 (2009)
14. Shi, R., Tang, S., Feng, W.: Two generalized predator-prey models for integrated pest management with stage structure and disease in the prey population. Abstr. Appl. Anal. **2013**. Hindawi (2013)
15. Tang, S., Cheke, R.A.: State-dependent impulsive models of integrated pest management (IPM) strategies and their dynamic consequences. J. Math. Biol. **50**(3), 257–292 (2005)
16. Tang, S., Cheke, R.A.: Models for integrated pest control and their biological implications. Math. Biosci. **215**(1), 115–125 (2008)
17. Tan, X., Qin, W., Tang, G., Xiang, C., Liu, X.: Models to assess the effects of nonsmooth control and stochastic perturbation on pest control: a pest-natural-enemy ecosystem. Complexity **2019** (2019)

18. Wang, L., Xie, Y., Fu, J.: The dynamics of natural mortality for pest control model with impulsive effect. J. Frankl. Inst. **350**(6), 1443–1461 (2013)
19. Xiao, Y., Van Den Bosch, F.: The dynamics of an eco-epidemic model with biological control. Ecol. Model. **168**(1–2), 203–214 (2003)
20. Zhang, H., Liu, X., Xu, W.: Threshold dynamics and pulse control of a stochastic ecosystem with switching parameters. J. Frankl. Inst. **358**(1), 516–532 (2021)
21. Zhao, W., Liu, Y., Zhang, T., Meng, X.: Geometric analysis of an integrated pest management model including two state impulses. Abstr. Appl. Anal. **2014**. Hindawi (2014)

Bifurcation Analysis of Longitudinal Dynamics of Generic Air-Breathing Hypersonic Vehicle for Different Operating Flight Conditions

Ritesh Singh, Om Prakash, Sudhir Joshi, and Yogananda Jeppu

Abstract The advancements in the Hypersonic Technology have taken us a step closer to the cost effective and reliable Hypersonic flight to Space. The paper outlines nonlinear dynamical model analysis of the Air-breathing Hypersonic Vehicle (AHV) using Bifurcation Method. It shows the Bifurcation Method implementation and its application to the nonlinear dynamics and stability investigation for 3DOF Longitudinal nonlinear dynamics of the Generic AHV. Bifurcation Analysis of AHV presents a quantifiable valuation with equilibrium states throughout the whole broad flight envelop and with dynamic stability analysis for Mach Number, $M = 0.9$. AUTO-07p platform is used here to demonstrate the AHV flight dynamics and control analysis using the Bifurcation Technique and Continuation approach. The Bifurcation Methodology is implemented for the AHV dynamic model using AUTO-07p for different operating flight conditions with elevator deflection, δ_e.

Keywords Bifurcation analysis · Longitudinal dynamics · Hypersonic vehicle

1 Introduction

Air-breathing Hypersonic Vehicle provides a way forward for the Hypersonic Technology in the coming decades to achieve tourism in space. Accomplishment of Hyper-X experimental vehicle during the last decade has made an increased interest in the AHV, which can lead the ultramodern dreams of rapid transportation across the world and to Space in the coming decades. Hypersonic Technology is now being developed around the world, with promising military and commercial applications, and

R. Singh (✉)
Manipal University Jaipur, Jaipur 303007, Rajasthan, India
e-mail: ritesh.singh23@gmail.com

R. Singh · O. Prakash · S. Joshi
University of Petroleum & Energy Studies, Dehradun 248007, Uttarakhand, India

Y. Jeppu
Honeywell Technology Solutions, Hyderabad 500032, Telangana, India

S. Banerjee and A. Saha (eds.), *Nonlinear Dynamics and Applications*,
Springer Proceedings in Complexity,
https://doi.org/10.1007/978-3-030-99792-2_97

for achieving Low Earth Orbit. The achievement with NASA Programs throughout the last seven decades has opened a new awareness for cost-effective Space entrance to many noncombatants and military missions, and provides renewed interest in Hypersonic Vehicles and its technology. The challenges with the aerodynamics of the AHV flight shows the dynamic interaction and nonlinear aerodynamics phenomena occurring at the different high speeds for the wide flight regime. Therefore, the flight dynamics of the AHV vehicle presents a challenge for the different operating flight conditions for the vehicle and shows that further dynamics analysis can be carried using the different nonlinear analysis methods. This gives a way forward to the Bifurcation Technique, which can be investigated and implemented due to the highly nonlinear AHV model, complicated aerodynamic characteristics over the whole flight regimes, large flight envelop, and significant coupling interaction.

Bifurcation approach provides a potential to significantly enhance the flight dynamics design process. The relevance of the bifurcation approach is that it may show global stability and improvements in aviation control design parameters. The nonlinear dynamical system with trim and stability analysis, may be investigated computationally using Bifurcation approaches. The bifurcation method can be used to find multiple trim points and different states which can co-exist utilizing a static categorization with control restrictions. It also forecasts system behavior utilizing several sorts of bifurcations that lead to the commencement of state dynamics, as well as information on phugoid stability for AHV dynamics analysis. Continuation algorithm implemented with the bifurcation technique delivers local stability evidence around all trim points, making the method valuable for analysis. Bifurcation method provides promising application for flight controls and its effective analysis of nonlinear phenomena occurring in the flight dynamics. When it comes to flight dynamics and control system design, using Bifurcation analysis, it proves a better nonlinear approach to be implemented, for the analysis of complex nonlinear dynamics and dynamical system with multiple trim states and control parameters.

The organization of the paper is categorized with the following sections. Section 2 presents the Bifurcation Method introduction, application and its brief literature overview. Section 3 discusses the implementation of the Bifurcation Method to the 3DOF Longitudinal AHV model using AUTO-07p platform. Section 4 discusses the Bifurcation Results of AHV for the Mach Number, $M = 0.9$ with elevator deflection, δ_e. And at the end, Sect. 5 deliberates the conclusion of the Method implemented.

2 Bifurcation Method

Bifurcation techniques are a valuable tool for numerically investigating the nonlinear effects of dynamical systems for trim and stability. With these analysis, nonlinear behavior may be better understood since it keeps track of all trim points at which equilibriums are created or destroyed, or at which the stability of equilibriums varies. This technique was first implemented for flight dynamics by [1–3] around 1982. Bifurcation analysis gives record of all critical bifurcation points for the equilibriums

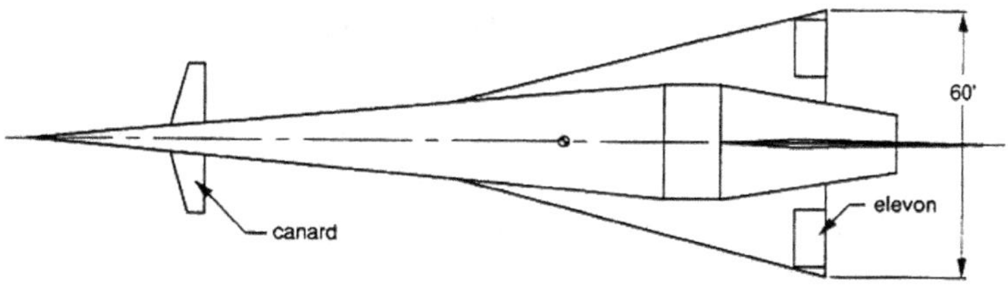

Fig. 1 AHV model

which shows the variation of the system stability change. Analysis and application of this technique is implemented for the aircraft stability, dynamics and control for high alpha effects in [4, 5]. The method is implemented for the 6DOF aircraft models with nonlinear dynamics for different flight conditions of level, straight and turn conditions for the flight dynamics analysis by [6–9]. Hence it can be shown that this method provides fruitful way to understand the nonlinear dynamics with all information regarding the equilibrium states and can be implemented to the flight dynamics area.

The vehicle considered here is Generic Hypersonic Vehicle known as Winged Cone Model from [10], developed by NASA is presented in the Fig. 1 [10]. The model developed is based on the rigid-body hypothesis and is used for various developments such for control systems, navigation and guidance, optimization and stability for Single-Stage-To-Orbit. The propulsion model used for the vehicle is the combination of the turbojet, ramjet and scramjet, and the rocket engine used for the entire flight envelop. The hypothetically proposed ramjet and scramjet engine, integrated to the vehicle makes the model to be called as Air-breathing Hypersonic Vehicle (AHV) [11]. The nonlinear ODE (Ordinary Differential Equation) of the 3DOF longitudinal AHV flight dynamic model, given by the Eqs. (1)–(5) using [11], are used for the bifurcation method analysis. The nonlinear behavior of the AHV with complicated aerodynamic model, large flight envelop and with strong dynamic coupling gives a reasonable way to implement and apply the bifurcation technique.

The bifurcation method uses first order ODE's represented by, $\dot{x}= f(x, u)$, to determine the steady states for the nonlinear systems. The method uses CBA (Continuation Based Algorithm) to determine the all-possible trim equilibrium points using the software platform AUTO-07p from [12]. The CBA computes, all potential trim solutions for the system while the other control parameters are held constant and by adjusting the free control parameter. At each of the trim points, the CBA computes the local stability information. The Bifurcation Diagram provides two-dimensional depiction with calculated trim results with the function in the variable control parameter that depicts the system's overall behavior. Bifurcation Points are the stability points where trim points calculated change with each calculation. These points resemble the movements of the eigenvalues from LHS to RHS in the complex plane, indicating and representing the dynamic behavior of the system stability as unstable in nature.

3 Bifurcation Method Study of 3DOF AHV Model

The 3DOF longitudinal nonlinear AHV model from [13] is considered for the Bifurcation methodology implementation with the nonlinear dynamic model given by the Eqs. (1)–(5). The geometric and aerodynamic data of the AHV model is used from [13], for the development of the nonlinear aerodynamic model. The aerodynamic model developed from [14] provides changes with Mach Number and with angle-of-attack for the entire AHV flight band. The propulsion model from [14] is considered here, which is combination of the engine of turbojet, ramjet and scramjet, and rocket for the complete AHV flight. The atmosphere model considering the altitude effects for the temperature and air density from [15] is considered for the AHV model. Combining the all-sub models, and the 3DOF AHV dynamic model given by the Eqs. (1)–(5) are combined and developed as 3DOF AHV longitudinal model for the bifurcation analysis. Bifurcation with SBA is implemented with the AHV model with states considered as $x = [V\ \alpha\ q\ \theta\ h]$ with the elevator deflection considered as input given by $u = [\delta_e]$ and PLA (pilot lever angle) is used as the parameter $p = [\text{PLA}]$. Here the velocity term V, of the AHV is used as M, called as Mach number, which is obtained by the ratio of the velocity to the speed of sound. The angle-of-attack of the AHV is directly affected by the parameter, and the parameters in the SBA analysis are fixed during the simulation run. For SBA implementation to this model, the results are depicted in the terms of Bifurcation Diagram (BD) which provides in interpreting and illustrating the AHV dynamic behavior. Bifurcation with EBA is applied with the longitudinal AHV model with the dynamics given by the Eqs. (1)–(5), for the level, climb/decent and straight flight paths.

$$m\dot{V} = \left(Tcos\alpha - F_y - mgsin\theta\right) \tag{1}$$

$$mV\dot{\alpha} = mVq - (Tsin\alpha + F_x - mgcos\theta) \tag{2}$$

$$I_{yy}\dot{q} = M_y \tag{3}$$

$$\dot{\theta} = q \tag{4}$$

$$\dot{h} = Vsin\theta \tag{5}$$

To initiate or run the developed AUTO-07p code, we need the initial, equilibrium or starting values for the bifurcation or simulation to run. For this the equilibrium values are obtained by solving the Eqs. (1)–(5), by equating the LHS of the Eqs. (1)–(5) to zero and solving for the state values. These equilibrium values are needed for the bifurcation to implement SBA and to start the simulation or to run the code. In order to simplify the calculation $\alpha = \theta = 0$, and the equilibrium solution is obtained for the different Mach number, M and the altitude $h = h_0$, and the state equilibrium solution is obtained and is given by the Eqs. (6)–(8).

$$T(h_0) = \overline{q}SC_D \tag{6}$$

$$V_0 = \overline{q}SC_L/(mg) \tag{7}$$

$$M_y = 0 \tag{8}$$

Here the term m, $\overline{q}$, S, T are the vehicle mass, dynamic pressure, reference area and vehicle thrust. And C_D and C_L, are the aerodynamic coefficient and M_y is the pitching moment. Using the solution, given by the Eqs. (6)–(8), the drag term is made equivalent to the thrust term, and this results in the initial thrust value which can be used for the initial run. And similarly, the V_0 start value for the initial run is used by the above relation $h = h_0$. Hence the bifurcation is carried with the states as $\alpha, qand\theta$, considering the other states as fixed and constant value for the whole bifurcation run.

The different bifurcation diagrams are obtained employing the bifurcation using CBA, and all the potential trims of the AHV are computed. Using AUTO-07p, the CBA method is used to compute different trims and elevator deflection, δ_e is varied in order to commence the continuation of the numerical simulation. Each of the bifurcation diagrams below shows the collection of equilibrium points that corresponds to the value of the specified parameter for each example. There is not at all variation with the state variables for any of the places on the curve where these solution branches are formed. When the Bifurcation Diagram is created for AUTO-07p platform, it provides altogether information of trim points in each variation with input variables. In this case, negative and positive eigen values represent the stability and instability of the dynamic system.

4 Bifurcation Results

Bifurcation analysis for the AHV at Mach number, $M = 0.9$ and altitude, $h = 10{,}000$ *ft*, is carried out and the Bifurcation Diagram obtained is shown in the Fig. 2, it illustrates how the equilibrium solution point shifts between –12° and 12° deflections when the elevator deflection, δ_e value changes. The AHV stability and bifurcation is examined by focusing on varying the value of the system parameter δ_e and making another parameter, thrust coefficient constant. Implementing the AUTO-07p code, we observe the numerous bifurcations occurrences for the AHV's states in consideration to the parameter variation. Here the angle-of-attack, α given in the Fig. 2a, when deflected for δ_e from 0° to 12° degrees, it indicates stable behaviour for the vehicle, and when deflected from 0° to –12° degrees down it shows stable action. The pitch angle, θ as shown in the Fig. 2b, for the elevator deflection, $\delta_{e,}$ 0° to –12° degrees the variation is stable between –4.5 to 1.5 radians and for the –1.5 to –4.25 radians. At each of the 24 possible equilibrium points, AUTO-07p generates result demonstrating stable dynamics for all the eigenvalues. For the pitch angle, θ, from the bifurcation

Fig. 2 Bifurcation diagram for $M = 0.9$

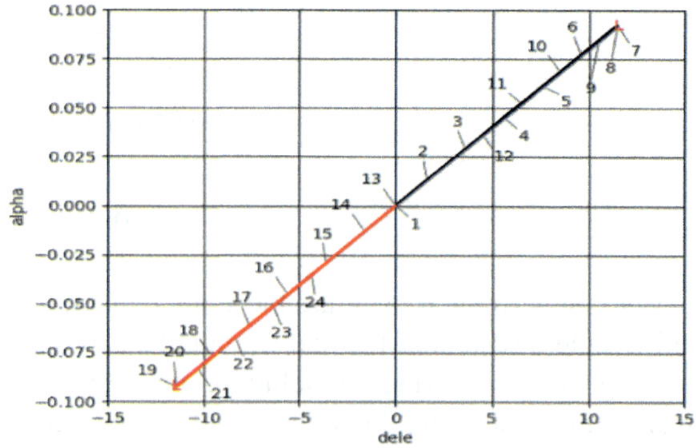

(a) Angle of attack with δ_e variation

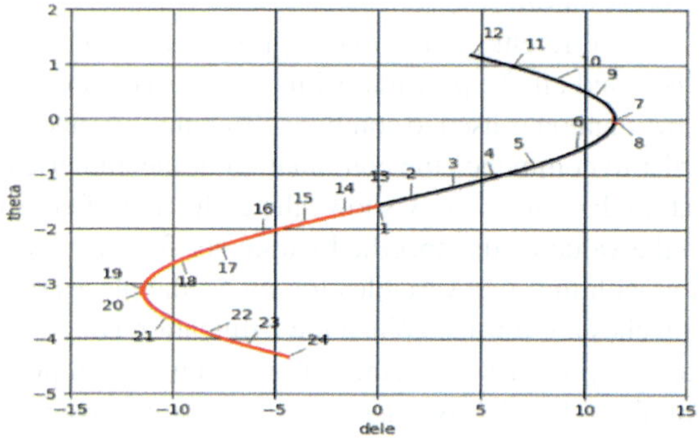

(b) Pitch with δ_e variation

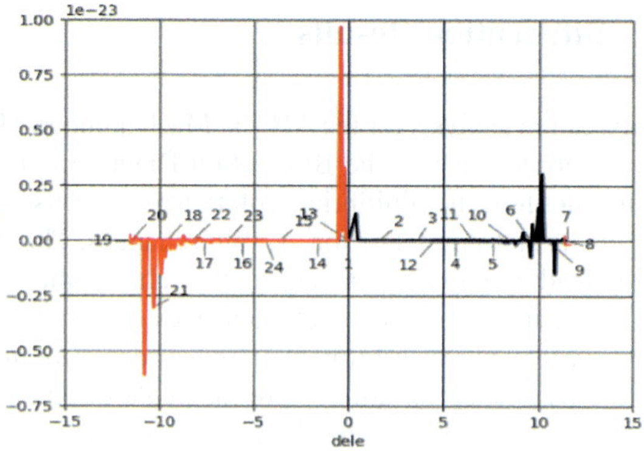

(c) Pitch rate with δ_e variation

diagram, the points 7 and 8 in the Fig. 2b show the nose going up, whereas points 19 and 20 show the nose going down. And for the elevator deflection 5°, the vehicle will have 60° nose-down and 60° nose-up directional changes at various periods of the vehicle's climb or fall. The pitch rate, q as shown in the Fig. 2c, for the elevator deflection, $\delta_{e,}$ 0° to −12° degrees the variation is stable and when deflected δ_e from 0° to 12° degrees, it also indicates stable behaviour for the vehicle, with ranging in a stable magnitude for the δ_e change.

Running the AUTO-07p code for the bifurcation of AHV using CBA, we obtain the bifurcation diagram as shown in the Fig. 2. We obtain the simulation results of forward run and backward run for the implemented bifurcation method and is shown in Table 1 and Table 2 respectively. These tabular results show the bifurcation diagram data points for the different runs with iterations, indicating the different values of α, θ and q. Here BR, PT, TY and LAB are the Bifurcation Result, Bifurcation Point, Bifurcation Type and Labelled Solution respectively. Here for different range (1 to 200) of PT's, the forward and backward run are carried out with different iterations and corresponding L2 Norm values for the different selected parameters are determined.

It shows that the result is obtained for 200 different points with different iterations and step size at each AUTO-07p run, with LP bifurcation type is obtained at 121 run; and this indicates the presence of bifurcation point as fold for the ordinary differential equation present in the AHV dynamics.

The eigen values determined from the forward and backward run are shown for the corresponding PT for 1, 121 and 200. This shows the different eigen values with respective to the different iteration and the stability of the AHV with the Bifurcation TY. It shows the presence of Hopf function at PT 1 and Iteration 1; and Fold, BP and Hopf function at PT 121, 200 and Iteration 5, 3; with the eigen values with each iteration. The stability information can be determined with the corresponding each

Table 1 Bifurcation result with forward run in AUTO-07p

BR	PT	TY	LAB	δe(rad)	L2-Norm	α(rad)	θ(rad)	q
1	1	EP	1	0.00000	1.57080	0.00E+00	– 1.570E+0	0.00E+00
1	20		2	1.61874	1.42901	1.30E-02	– 1.428E+0	2.90E-29
1	40		3	3.61065	1.25028	2.91E-02	– 1.25E+0	9.51E-31
1	60		4	5.60149	1.06042	4.51E-02	– 1.059E+0	– 2.70E-29
1	80		5	7.58995	0.84797	6.11E-02	– 8.46E-01	1.42E-27
1	100		6	9.57201	0.58508	7.71E-02	– 5.79E-01	– 1.191E-26
1	120		7	11.4416	0.09220	9.22E-02	– 3.05E-03	4.91E-27
1	121	LP	8	11.4416	0.09215	9.22E-02	1.38E-05	5.74E-28
1	140		9	10.4622	0.42539	8.43E-02	4.17E-01	7.73E-29
1	160		10	8.48822	0.73837	6.84E-02	7.35E-01	– 2.00E-27
1	180		11	6.50178	0.96809	5.24E-02	9.67E-01	– 4.068E-27
1	200	EP	12	4.51178	1.16626	3.63E-02	1.17E+00	– 2.19E-28

Table 2 Bifurcation result with backward run in AUTO-07p

BR	PT	TY	LAB	δe(rad)	L2-Norm	α(rad)	θ(rad)	q
1	1	EP	1	0.00E+00	1.57E+00	0.00E+00	– 1.57E+00	0.00E+00
1	20		2	– 1.62E+00	1.71E+00	– 1.30E-02	– 1.71E+00	9.76E-29
1	40		3	– 3.61E+00	1.89E+00	– 2.91E-02	– 1.89E+00	3.47E-29
1	60		4	– 5.60E+00	2.08E+00	– 4.51E-02	– 2.08E+00	7.11E-28
1	80		5	– 7.59E+00	2.30E+00	– 6.11E-02	– 2.29478	– 3.05E-27
1	100		6	– 9.57E+00	2.56E+00	– 0.077102	– 2.55934	– 9.77E-27
1	120		7	– 1.15E+01	3.12E+00	– 0.092300	– 3.11735	– 3.23E-26
1	121	LP	8	– 1.15E+01	3.14E+00	– 0.092327	– 3.14158	– 3.20E-29
1	140		9	– 1.03E+01	3.59E+00	– 0.083054	– 3.59342	– 3.03E-24
1	160		10	– 8.34E+00	3.90E+00	– 0.067136	– 3.89795	– 1.03E-25
1	180		11	– 6.35E+00	4.13E+00	– 0.051132	– 4.12518	– 1.55E-26
1	200	EP	12	– 4.36E+00	4.32E+00	– 3.51E-02	– 4.32E+00	4.09E-27

iteration with the eigen values. Considering the trim condition and iteration 1 the eigen values with the dynamic stability information are shown in Table 3 indicating the AHV stability information at the given Mach Number. It shows that there are three eigen values with one eigen value with real number and the two eigen values with complex in nature, as shown in the Table 3; and the corresponding damping ratio and the frequency at the given Mach Number is determined indicating the short period mode behaviour of the AHV.

For the eigen values considered the pole-zero plot is obtained and is shown in the Fig. 3, which outlines the stable behaviour of the AHV at the Mach Number, $M = 0.9$ as all the poles lying at the LHS plane of the stability axis. Thus, it can be said that using the Bifurcation Method the AHV's dynamically stability can be determined and at $M = 0.9$ it is dynamically stable. Considering the simulation carried out for the AHV at the Mach Number, $M = 0.9$ it shows, most stable behavior for it, using the Bifurcation Method; and their bifurcation diagrams are nonlinear in nature but are mirror images in the vertical plane and about $\delta_e = 0°$. This shows that for the different Mach number of the AHV's flight can be considered to determine the dynamical stability of the vehicle ranging from Mach number, $M = 0$ to 24.

Table 3 Dynamic stability using Bifurcation analysis for Mach Number, $M = 0.9$

Mach No. (M)	Eigen values	Damping ratio (ζ)	Frequency (ω_n)	Stability
	– 3.05831E-02			
	– 1.19248E+00			
0.9	3.12365E+00	0.356	3.342	Dynamically
	– 1.19248E+00			Stable
	– 3.12365E+00			

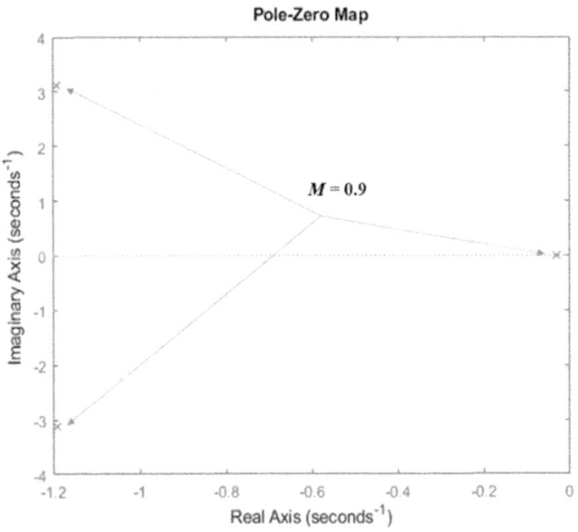

Fig. 3 Pole-zero plot for Mach Number, $M = 0.9$

5 Conclusion

Bifurcation analysis of Longitudinal dynamics for Generic AHV model considering CBA has been implemented for AHV dynamics at Mach Number, $M = 0.9$, for different choices of elevator deflection and with the aim to observe the control effects. Bifurcation technique is implemented with the 3DOF longitudinal AHV model using the AUTO-07p platform for the different elevator deflection, $\delta_e = -12°$ and $12°$. The Bifurcation Diagram obtained for the data points with different forward and backward runs with the different iterations, shows the parameter values of α, θ and q. The forward and backward run shows the presence of Hopf function at PT 1 with iteration 1, and Fold, BP and Hopf function at PT 121, 200 and iteration 5, 3; with the eigen values with each iteration. The stability information is determined with the corresponding iteration of the eigen values, showing the dynamic stability information. The eigen values determined, indicate the short period mode behaviour of the AHV indicating the stable behaviour at the Mach Number. The Method shows the AHV's dynamically stability determined at $M = 0.9$ is stable. This shows that for the different Mach Number of the AHV's flight, Bifurcation is promising method to determine the dynamical stability of the vehicle for the Mach number ranging from $M = 0$ to 24. Finally, the paper outlines Bifurcation Methodology application intended for the investigation of the dynamic stability of the Generic AHV.

References

1. Carroll, J.V., Mehra, R.K.: Bifurcation analysis of nonlinear aircraft dynamics. J. Guid. Control. Dyn. **5**(5), 529–536 (1982)
2. Zagainov, G.I., Goman, M.G.: Bifurcation analysis of critical aircraft flight regime. Int. Counc. Aeronaut. Sci. 217–223 (1984)
3. Jahnke, C.C., Culick, F.E.C.: Application of bifurcation theory to the high-angle-of-attack dynamics of the F-14. J. Aircr. **31**(1), 26–34 (1994)
4. Guicheteau, P.: Bifurcation theory a tool for nonlinear flight dynamics. Philos. Trans. R. Soc. Lond. Ser. A **356**(1745), 2181–2201 (1998)
5. Goman, M.G., Zagainov, G.I., Khramtsovsky, A.V.: Application of bifurcation methods to nonlinear flight dynamics problems. Prog. Aerosp. Sci. (Elsevier) **33**(9–10), 539–586 (1997)
6. Khatri, A.K., Singh, J., Sinha, N.K.: Aircraft design using constrained bifurcation and continuation method. J. Aircr. AIAA **51**(5), 1647–1652 (2014)
7. Thomas, S., Kwatny, H.G., Chang, B.C.: Bifurcation analysis of flight control systems. In: 16th Triennial World Congress, IFAC, Elsevier, Czech Republic (2005)
8. Khatri, A.K., Singh, J., Sinha, N.K.: Aircraft maneuver design using bifurcation analysis and sliding mode control techniques. J. Guid. Control Dyn. AIAA **35**(5), 1435–1449 (2012)
9. Prakash, O., Singh, R.: Flight dynamics analysis using high altitude & mach number for generic air-breathing hypersonic vehicle. In: AIAA 2021-3271, AIAA Propulsion and Energy 2021 Forum, Virtual Event USA (2021)
10. Shaughnessy, J.D., Pinckney, S.Z., Mcminn, J.D., Cruz, C.I., Kelley, M.L.: Hypersonic vehicle simulation model: Winged-cone configuration. NASA Langley Research Center, NASA-TM-102610, United States (1990)
11. Keshmiri, S., Colgren, R., Mirmirani, M.: Six DoF nonlinear equations of motion for a generic hypersonic vehicle. In: AIAA Atmospheric Flight Mechanics Conference and Exhibit, AIAA, Hilton Head, South Carolina (2007)
12. Doedel, E.J., Champneys, A.R., Fairgrieve, T.F., Kuznetsov, Y.A., Sandstede, B., Xang, X.: AUTO-07p: continuation and bifurcation software for ordinary differential equations. Ver. AUTO-07p 2007, Department of Computer Science, Concordia University, Montreal, Canada (2007)
13. Keshmiri, S., Colgren, R., Mirmirani, M.: Six-DOF modeling and simulation of a generic hypersonic vehicle for control and navigation purposes. In: AIAA Guidance, Navigation, and Control Conference and Exhibit, AIAA, Keystone, Colorado (2006)
14. Keshmiri, S., Colgren, R., Mirmirani, M.: Development of an aerodynamic database for a generic hypersonic air vehicle. In: AIAA Guidance, Navigation, and Control Conference and Exhibit, AIAA, San Francisco, California (2005)
15. Roskam, J., Lan, C.T.: Airplane Aerodynamics and Performance, Revised Darcorporation, Lawrence, Kansas (2000)

Multi-soliton Solutions of the Gardner Equation Using Darboux Transformation

Dipan Saha, Santanu Raut, and Prasanta Chatterjee

Abstract The Gardner equation is a particular version of the extended Korteweg-de Vries (KdV) equation which presents actually the same type of characters as the standard KdV model, but it extends the range of validity to a larger domain of the parameters of the wave motion for a dynamic system. The purpose of this article is to explore a new type of multi-soliton solutions for the Gardner equation. To confirm its integrability, first, we will construct the Lax pair of the Gardner equation using Ablowitz–Kaup–Newell–Segur (AKNS) approach and finally, derive the one-soliton and two-soliton wave solutions for the Gardner equation employing the Darboux transformation method (DTM). These solutions can be extended to the generalized multi-solitary for the Gardner equation by repetition of the transformation. Some numerical graphs of one-soliton and two-soliton solutions are drawn for a clear understanding of wave motion under Gardner equation.

Keywords Gardner equation · Wave motion · Multi-soliton solutions · Lax pair · Darboux transformation method

1 Introduction

For the last few decades, nonlinear evolution equations (NLEEs) have paid a lot of interest due to their wide applications in different branches of science and engineering fields. For instance, NLEEs are substantially employed to formulate distinct problems from fluid mechanics [1, 2], plasma physics [3–5], quantum field theory, quantum mechanics [6, 7], propagation of shallow water waves [8, 9], chemical kinetics etc.

D. Saha (✉)
Advanced Centre for Nonlinear and Complex Phenomena, Kolkata, India
e-mail: dipansaha12345@gmail.com

S. Raut
Department of Mathematics, Mathabhanga College, Coochbehar 736146, India

P. Chatterjee
Siksha Bhavana, Visva-Bharati, Santiniketan 731235, India

S. Banerjee and A. Saha (eds.), *Nonlinear Dynamics and Applications*,
Springer Proceedings in Complexity,
https://doi.org/10.1007/978-3-030-99792-2_98

The classical Korteweg-De Vries (KdV) equation is a particular type of generic evolutionary partial differential equation which was actually addressed by Boussinesq (A) and again rediscovered by the famous Dutch mathematician Diederik Johannes Korteweg (31 March 1848–10 May 1941) and his pupil Gustav de Vries (22 January 1866-16 December 1934). KdV equation is a solvable nonlinear model which may be exactly solved by using different analytical techniques such as, Inverse scattering transform method, Backlund transformation, Hirota's bilinear approach etc. In particular, the standard KdV equation is extensively used to model weakly nonlinear long waves. Many important nonlinear features, especially, in internal unsteady internal waves in oceanic water are explored through the KdV model. In remote sensing experiments and situ measurements, it was observed that long solitary type waves are commonly appeared in density layered shallow water. But, a number of experimental observations confirm that although the KdV model effectively defines the solitary waves for a vast range of parameters however there may arise some situations when the KdV framework miserably fails, for instance as, the critical values of a particular parameter which causes the dismiss of the coefficient of the nonlinear term in the said equation. To overcome this type of difficulty, it becomes necessary to extend the KdV equation by incorporating higher-order nonlinearity instead of quadratic nonlinearity. Often, these models become meaningful in the diverse fields of applications. Considering dual nonlinearity in the KdV model, the Gardner equation arises, which claims the validity to the larger parametric zone for internal wave motion in the different physical environment. The introduction of this equation is attributed to the famous mathematician Clifford Gardner in 1968 [10]. Some excellent works on Gardner mode for modeling of internal wave in the extensive parametric domain can be found in [11]. Gardner equation is actually combination form of KdV equation and modified KdV (mKdV) equation. The Gardner equation, or the combined KdV-mKdV equation, reads

$$u_t + auu_x + bu^2u_x + cu_{xxx} = 0, \tag{1}$$

where $u(x, t)$ presents the amplitude of the wave model and evaluates the time of the virtual displacement over the isopycnal surface, x signifies the scaled space variable along the direction of wave mode and t determines the scaled time. a and b are the coefficients of the quadratic and cubic nonlinear terms respectively whereas c presents a dispersive effect. In the present investigation, the constants a, b, c are taken as $a = k_1c$, $b = k_2c$ (k_1, k_2 are non-zero constants).

Many powerful methods, like Backlund transformation method [12], Inverse scattering transform method [13], Hirota method [14], pseudo spectral method [15], the tanh-sech method [16], Exp-function method [17] and the sine-cosine method [18] are used to investigate these types of equations. The Darboux transformation method (DTM) [19] is one of the most powerful and fruitful method for getting explicit solutions, which we will use in our present work.

So the outline of the present paper is as follows. In Sect. 2, we construct Lax pair of Eq. (1). Section 3 is approved for presenting the Darboux transformation of our proposed model. In Sects. 3.1 and 3.2, one-soliton and two-soliton solutions of the Gardner equation are explored respectively. Finally, the conclusions are briefly outlined in Sect. 5.

2 Lax Pair

In order to verify the integrability condition of a nonlinear partial differential equation, the Lax pair of a given equation is constructed. By Lax pair, we mean a set of two operators that, if they exist, implies that a nonlinear evolution equation is integrable. There is no general technique for finding the Lax pair of an integrable system. Gardner equation belongs to the class of integrable system. According to the AKNS approach [20], Lax pair of Eq. (1) is given by

$$\psi_x = U\psi = \begin{pmatrix} \lambda & u \\ -\frac{1}{6}(k_2 u + k_1) & -\lambda \end{pmatrix} \psi, \tag{2}$$

$$\psi_t = V\psi = \begin{pmatrix} A & B \\ C & -A \end{pmatrix} \psi, \tag{3}$$

where

$$\begin{aligned} A &= -4c\lambda^3 - \frac{1}{3}c(k_2 u^2 + k_1 u)\lambda - \frac{1}{6}k_1 c u_x, \\ B &= -(4cu\lambda^2 + 2cu_x\lambda + \frac{1}{3}k_2 c u^3 + \frac{1}{3}k_1 c u^2 + c u_{xx}), \\ C &= \frac{2}{3}c(k_1 + k_2 u)\lambda^2 - \frac{1}{3}k_2 c u_x \lambda + \frac{1}{18}k_2^2 c u^3 + \frac{1}{9}k_1 k_2 c u^2 \\ &\quad + \frac{1}{18}k_1^2 c u + \frac{1}{6}k_2 c u_{xx}. \end{aligned}$$

Here λ is a parameter independent of x and t. The Lax equation is

$$U_t - V_x + [U, V] = 0,$$

which is equivalent to Eq. (1).

3 Construction of Darboux Transformation (DT) and Soliton Solutions of Gardner Equation

In the late 1970s, V. B. Matveev showed that the spectral problem of second-order ordinary differential equations may be improved to some important soliton equations employing a technique addressed by G. Darboux about a century ago. This method is popularly known as DTM. A number of examples for describing the continuous and discrete spectrum management in quantum mechanics are demonstrated in [21, 22] where the elementary DT and binary DT (also known as twofold elementary DT) are expressed in detail. It is also found that the n-fold Darboux transformation [23] is a 2×2 matrix for the Kaup–Newell (KN) system. Employing n-fold DT, various types of wave solutions of the nonlinear Schrodinger (DNLS) equation, such as, periodic solution, rational traveling solution, breather solution, rogue wave, dark soliton, bright soliton, are derived explicitly from the different seed solutions. In particular, Darboux transformation provides new route to study the generalized Sawada–Kotera (SK) equation [24], the generalized TD equation, Kadomtsev–Petviashvili (KP) equation [25], the Gerdjikov–Ivanov (GI) equation etc. Here, we will employ the Darboux transformation to get multi-soliton solutions from an old solution of the Gardner equation. Starting with the trivial solution $u = 0$ of the Gardner equation (1), one can use the Darboux transformation to obtain the soliton solutions. For $u = 0$, the fundamental solution of the Lax pair can be obtained as

$$\psi(x, t, \lambda) = \begin{pmatrix} e^{-4c\lambda^3 t+\lambda x} & 0 \\ e^{\frac{2}{3}ck_1\lambda^2 t+\lambda x} & e^{4c\lambda^3 t-\lambda x} \end{pmatrix} \tag{4}$$

by integrating (3). Let λ_1, σ_1 be arbitrary real numbers and let

$$\gamma = \frac{e^{4c\lambda_1^3 t-\lambda_1 x} + \sigma_1 e^{\frac{2}{3}ck_1\lambda_1^2 t+\lambda_1 x}}{0 + \sigma_1 \cdot e^{-4c\lambda_1^3 t+\lambda_1 x}} \tag{5}$$

Take $\lambda_1 \neq 0$ and $\sigma_1 = \exp(-2\mu_1) > 0$, then (5) becomes

$$\gamma = \gamma_1 = e^{2\mu_1+8c\lambda_1^3 t-2\lambda_1 x} + e^{\frac{2}{3}ck_1\lambda_1^2 t+4c\lambda_1^3 t}. \tag{6}$$

Now consider the gauge transformation

$$\bar{\psi}(x, t, \lambda) = D(x, t, \lambda)\psi(x, t, \lambda), \tag{7}$$

where

$$D(x, t, \lambda) = \lambda I_2 - \frac{\lambda_1}{1+\gamma^2}\begin{pmatrix} 1-\gamma^2 & 2\gamma \\ 2\gamma & \gamma^2-1 \end{pmatrix}. \tag{8}$$

For such transformation, the spectral problems (2) and (3) are transformed into

$$\bar{\psi}_x = \bar{U}\bar{\psi}, \quad \bar{\psi}_t = \bar{V}\bar{\psi}, \tag{9}$$

where

$$\bar{U} = \begin{pmatrix} \lambda & u_1 \\ -\frac{1}{6}(k_2 u_1 + k_1) & -\lambda \end{pmatrix},$$

$$\bar{V} = \begin{pmatrix} \bar{A} & \bar{B} \\ \bar{C} & -\bar{A} \end{pmatrix},$$

with

$$u_1 = u + \frac{4\lambda_1 \gamma}{1+\gamma^2}, \tag{10}$$

$$\bar{A} = -4c\lambda^3 - \frac{1}{3}c(k_2 u_1^2 + k_1 u_1)\lambda - \frac{1}{6}k_1 c u_{1x},$$
$$\bar{B} = -(4cu_1\lambda^2 + 2cu_{1x}\lambda + \frac{1}{3}k_2 c u_1^3 + \frac{1}{3}k_1 c u_1^2 + cu_{1xx}),$$
$$\bar{C} = \frac{2}{3}c(k_1 + k_2 u_1)\lambda^2 - \frac{1}{3}k_2 c u_{1x}\lambda + \frac{1}{18}k_2^2 c u_1^3 + \frac{1}{9}k_1 k_2 c u_1^2 + \frac{1}{18}k_1^2 c u_1 + \frac{1}{6}k_2 c u_{1xx}.$$

3.1 One-Soliton Solution

A solitary wave is a localized "wave of translation" that arises from a balance between nonlinear and dispersive effects; and also preserves its shape upon collision. The ultimate aim of the present investigation is to finding solitary waves solution for the Gardner model and for this purpose we set, $\upsilon_1 = 2\lambda_1 x - 8c\lambda_1^3 t - 2\mu_1$ and $\upsilon_2 = -\frac{2}{3}ck_1\lambda_1^2 t - 4c\lambda_1^3 t$. Then from Eq. (6), $\gamma = \gamma_1 = e^{-\upsilon_1} + e^{-\upsilon_2}$.

By substituting this value of γ in (10), we obtain

$$u_1 = 0 + \frac{4\lambda_1(e^{-\upsilon_1} + e^{-\upsilon_2})}{1 + e^{-2\upsilon_1} + e^{-2\upsilon_2} + 2e^{-\upsilon_1-\upsilon_2}}$$

The equation

$$\boxed{u_1 = \frac{4\lambda_1(e^{-\upsilon_1} + e^{-\upsilon_2})}{1 + e^{-2\upsilon_1} + e^{-2\upsilon_2} + 2e^{-\upsilon_1-\upsilon_2}}} \tag{11}$$

is taken as the one-soliton solution of the Gardner equation.

3.2 Two-Soliton solution

Now if we take u_1 as a seed solution, a new Darboux matrix can be constructed from $\bar{\psi} = (\bar{\psi}_{ij})$ and a series of new solutions of the Gardner equation can be obtained.

Take constants $\lambda_2 \neq 0$ ($\lambda_2 \neq \lambda_1$) and $\sigma_2 = \exp(-2\mu_2)$. According to (5),

$$\bar{\gamma}_2 = \frac{\bar{\psi}_{22}(x,t,\lambda_2) + \sigma_2\bar{\psi}_{21}(x,t,\lambda_2)}{\bar{\psi}_{12}(x,t,\lambda_2) + \sigma_2\bar{\psi}_{11}(x,t,\lambda_2)}. \tag{12}$$

Substituting $\bar{\psi} = D\psi = \begin{pmatrix} D_{11} & D_{12} \\ D_{21} & D_{22} \end{pmatrix}\begin{pmatrix} \psi_{11} & \psi_{12} \\ \psi_{21} & \psi_{22} \end{pmatrix}$ into Eq. (12), we have

$$\begin{aligned}
\bar{\gamma}_2 &= \left.\frac{(D_{21}\psi_{12} + D_{22}\psi_{22}) + \sigma_2(D_{21}\psi_{11} + D_{22}\psi_{21})}{(D_{11}\psi_{12} + D_{12}\psi_{22}) + \sigma_2(D_{11}\psi_{11} + D_{12}\psi_{21})}\right|_{\lambda=\lambda_2} \\
&= \left.\frac{D_{21}(\psi_{12} + \sigma_2\psi_{11}) + D_{22}(\psi_{22} + \sigma_2\psi_{21})}{D_{11}(\psi_{12} + \sigma_2\psi_{11}) + D_{12}(\psi_{22} + \sigma_2\psi_{21})}\right|_{\lambda=\lambda_2} \\
&= \left.\frac{D_{21} + D_{22}\frac{(\psi_{22}+\sigma_2\psi_{21})}{(\psi_{12}+\sigma_2\psi_{11})}}{D_{11} + D_{12}\frac{(\psi_{22}+\sigma_2\psi_{21})}{(\psi_{12}+\sigma_2\psi_{11})}}\right|_{\lambda=\lambda_2} \\
&= \left.\frac{D_{21} + D_{22}\gamma_2}{D_{11} + D_{12}\gamma_2}\right|_{\lambda=\lambda_2},
\end{aligned} \tag{13}$$

where

$$\gamma_2 = \frac{\psi_{22}(x,t,\lambda_2) + \sigma_2\psi_{21}(x,t,\lambda_2)}{\psi_{12}(x,t,\lambda_2) + \sigma_2\psi_{11}(x,t,\lambda_2)}. \tag{14}$$

Using $\gamma = e^{-v_1} + e^{-v_2}$ in (8) and using (4), from (7), we obtain

$$\bar{\psi}(x,t,\lambda) = \begin{pmatrix} \bar{\psi}_{11} & \bar{\psi}_{12} \\ \bar{\psi}_{21} & \bar{\psi}_{22} \end{pmatrix} = \frac{1}{L}\begin{pmatrix} P & Q \\ R & S \end{pmatrix}, \tag{15}$$

where

$$\begin{aligned}
L &= 1 + e^{-2v_1} + e^{-2v_2} + 2e^{-v_1-v_2}, \\
P &= e^{-4c\lambda^3 t+\lambda x}(\lambda - \lambda_1) + e^{-4c\lambda^3 t+\lambda x}(\lambda + \lambda_1)(e^{-2v_1} + e^{-2v_2} + 2e^{-v_1-v_2}) \\
&\quad -2\lambda_1 e^{\frac{2}{3}ck_1\lambda^2 t+\lambda x}(e^{-v_1} + e^{-v_2}), \\
Q &= -2\lambda_1 e^{4c\lambda^3 t-\lambda x}(e^{-v_1} + e^{-v_2}), \\
R &= -2\lambda_1 e^{-4c\lambda^3 t+\lambda x}(e^{-v_1} + e^{-v_2}) + e^{\frac{2}{3}ck_1\lambda^2 t+\lambda x}(\lambda + \lambda_1) \\
&\quad + e^{\frac{2}{3}ck_1\lambda^2 t+\lambda x}(\lambda - \lambda_1)(e^{-2v_1} + e^{-2v_2} + 2e^{-v_1-v_2}), \\
S &= e^{4c\lambda^3 t-\lambda x}(\lambda + \lambda_1) + e^{4c\lambda^3 t-\lambda x}(\lambda - \lambda_1)(e^{-2v_1} + e^{-2v_2} + 2e^{-v_1-v_2})
\end{aligned}$$

and hence from Eq. (12), we have

$$\bar{\gamma}_2 = \frac{X}{Y}, \tag{16}$$

where

$$\begin{aligned}
X = {} & \frac{1}{L}[e^{4c\lambda_2^3 t-\lambda_2 x}(\lambda_2+\lambda_1) \\
& +e^{4c\lambda_2^3 t-\lambda_2 x}(\lambda_2-\lambda_1)(e^{-2v_1}+e^{-2v_2}+2e^{-v_1-v_2}) \\
& +e^{-2\mu_2}\{-2\lambda_1 e^{-4c\lambda_2^3 t+\lambda_2 x}(e^{-v_1}+e^{-v_2})+e^{\frac{2}{3}ck_1\lambda_2^2 t+\lambda_2 x}(\lambda_2+\lambda_1) \\
& +e^{\frac{2}{3}ck_1\lambda_2^2 t+\lambda_2 x}(\lambda_2-\lambda_1)(e^{-2v_1}+e^{-2v_2}+2e^{-v_1-v_2})\}], \\
Y = {} & \frac{1}{L}[-2\lambda_1 e^{4c\lambda_2^3 t-\lambda_2 x}(e^{-v_1}+e^{-v_2}) \\
& +e^{-2\mu_2}\{e^{-4c\lambda_2^3 t+\lambda_2 x}(\lambda_2-\lambda_1) \\
& +e^{-4c\lambda_2^3 t+\lambda_2 x}(\lambda_2+\lambda_1)(e^{-2v_1}+e^{-2v_2}+2e^{-v_1-v_2}) \\
& -2\lambda_1 e^{\frac{2}{3}ck_1\lambda_2^2 t+\lambda_2 x}(e^{-v_1}+e^{-v_2})\}].
\end{aligned}$$

$$\text{Let,}\quad v_3 = \mu_2 + 4c\lambda_2^3 t - \lambda_2 x, \quad v_4 = \mu_2 + 2c\lambda_2^3 t - \frac{1}{3}ck_1\lambda_2^2 t - \lambda_2 x. \tag{17}$$

Then by simple calculations from Eq. (16), we have

$$\bar{\gamma}_2 = \frac{\bar{X}}{\bar{Y}}, \tag{18}$$

where

$$\begin{aligned}
\bar{X} = {} & (\lambda_2+\lambda_1)+(\lambda_2-\lambda_1)(e^{-2v_1}+e^{-2v_2}+2e^{-v_1-v_2}) \\
& -2\lambda_1 e^{-2v_3}(e^{-v_1}+e^{-v_2}) \\
& +e^{-2v_4}(\lambda_2+\lambda_1)+e^{-2v_4}(\lambda_2-\lambda_1)(e^{-2v_1}+e^{-2v_2}+2e^{-v_1-v_2}), \\
\bar{Y} = {} & -2\lambda_1(e^{-v_1}+e^{-v_2})+e^{-2v_3}(\lambda_2-\lambda_1) \\
& +e^{-2v_3}(\lambda_2+\lambda_1)(e^{-2v_1}+e^{-2v_2}+2e^{-v_1-v_2}) \\
& -2\lambda_1 e^{-2v_4}(e^{-v_1}+e^{-v_2}).
\end{aligned}$$

According to Eq. (10),

$$u_2 = u_1 + \frac{4\lambda_2\bar{\gamma}_2}{1+\bar{\gamma}_2^2}$$

$$\boxed{u_2 = \frac{4\lambda_1(e^{-v_1}+e^{-v_2})}{1+e^{-2v_1}+e^{-2v_2}+2e^{-v_1-v_2}} + \frac{4\lambda_2\bar{\gamma}_2}{1+\bar{\gamma}_2^2}}, \tag{19}$$

where $\bar{\gamma}_2$ is given by Eq. (18). Equation (19) is a new solution of the Gardner equation, which is taken as the two-soliton solution of the Gardner equation. Proceeding in this way finally we obtain multi-soliton solutions of the Gardner equation, but the problem is that the calculation is too tedious.

4 Result and Discussion

We discuss the dynamics of exact solutions (11) and (19). Some of the furnished solutions in this paper are depicted graphically for their physical appearance which stands for different types of soliton, like, kink type soliton, bell-type soliton, two kink type soliton, etc. Figure 1a, b are showing distorted kink type soliton solutions. It is interesting to note that a small transition from kink profile to bell profile follows in Fig. 1b. Figure 1c, d indicates the propagation of bell soliton in space-time domain. On the other hand, Fig. 2a shows the interaction of a kink and a solitary wave. Figure 2b signifies the interaction of two kinks and one solitary wave via the Eq. (19). This type of nonlinear phenomenon appeases due to the presence of quadratic and cubic nonlinearity along with the Burgers term. After interaction two kink waves appear, which are moving with different amplitudes as shown in Fig. 2c, d.

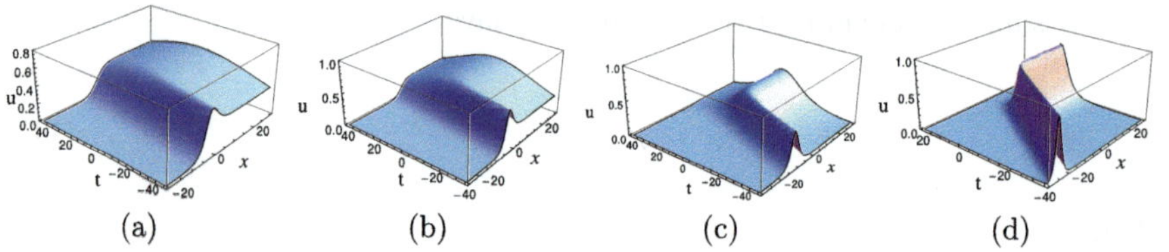

Fig. 1 3D profiles of the solution (11), **a** when $\lambda_1 = 0.25$; $c = 0.25$; $\mu_1 = 0.5$; $k_1 = 0.5$; **b** when $\lambda_1 = 0.3$; $c = 0.25$; $\mu_1 = 0.5$; $k_1 = 0.5$; **c** when $\lambda_1 = 0.25$; $c = 1$; $\mu_1 = 0.5$; $k_1 = 0.5$; **d** when $\lambda_1 = 0.4$; $c = 1$; $\mu_1 = 0.5$; $k_1 = 0.5$

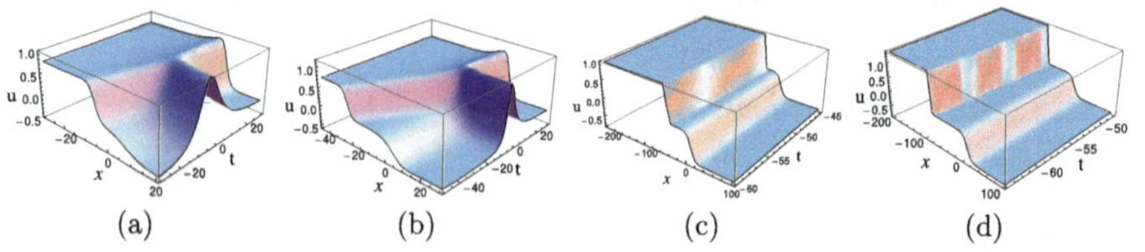

Fig. 2 3D Interaction of the two-soliton via solution (19) for **a** when $\lambda_1 = 0.5$; $c = 0.43$; $\mu_1 = 0.5$; $k_1 = 0.5$; $\mu_2 = 0.5$; $\lambda_2 = 0.4$; **b** when $\lambda_1 = 0.12$; $c = 0.8$; $\mu_1 = 0.5$; $k_1 = 0.25$; $\mu_2 = 0.5$; $\lambda_2 = 0.4$; **c** when $\lambda_1 = 0.1$; $c = 0.8$; $\mu_1 = 0.5$; $k_1 = 0.5$; $\mu_2 = 0.5$; $\lambda_2 = 0.5$; **d** when $\lambda_1 = 0.1$; $c = 0.7$; $\mu_1 = 0.5$; $k_1 = 0.5$; $\mu_2 = 0.5$; $\lambda_2 = 0.7$.

5 Conclusions

The present investigation provides a new class of effective solutions of the Gardner equation which may come as a bell-shaped soliton, kink type soliton, kink bell mixed soliton depending on the choosing of the values of the nonlinear and dispersive coefficients. To the best of our knowledge, for the first time, we derive the solution of the Gardner equation employing DTM. The one-soliton and two-soliton solutions are explicitly explored and repeating the same process, n folded multi-soliton solutions can be drawn from the previous seed. The obtained solutions acquired in the proposed scheme contain many free parameters and are claimed to be fresh and further general which might bear great importance in the research area. The numerical graphs emphasized that the method utilized in this observation is of significance in nature and the obtained solution could be utilized for modeling many natural phenomena such as wave motion in a plasma environment, water wave in the oceanic platform, etc.

References

1. Hosseinzadeh, E., Barari, A., Fouladi, F., Domairry, G.D.: Numerical analysis of forth-order boundary value problems in fluid mechanics and mathematics. Therm. Sci. **14**(4), 1101–1109 (2010)
2. Antontsev, S.N., Díaz, J.I., Shmarev, S.: Energy Methods for Free Boundary Problems: Applications to Nonlinear PDEs and Fluid Mechanics, vol. 48. Springer Science & Business Media (2012)
3. Saha, A., Pal, N., Chatterjee, P.: Dynamic behavior of ion acoustic waves in electron-positron-ion magnetoplasmas with superthermal electrons and positrons. Phys. Plasmas **21**(10), 102101 (2014)
4. Raut, S., Mondal, K.K., Chatterjee, P., Roy, A.: Propagation of dust-ion-acoustic solitary waves for damped modified Kadomtsev–Petviashvili–Burgers equation in dusty plasma with a q-nonextensive nonthermal electron velocity distribution. SeMA J. 1–23 (2021)
5. Raut, S., Mondal, K.K., Chatterjee, P., Roy, A.: Two-dimensional ion-acoustic solitary waves obliquely propagating in a relativistic rotating magnetised electron–positron–ion plasma in the presence of external periodic force. Pramana **95**(2), 1–13 (2021)
6. Moshinsky, M.: Canonical transformations and quantum mechanics. SIAM J. Appl. Math. **25**(2), 193–212 (1973)
7. Reinisch, G.: Nonlinear quantum mechanics. Phys. A: Stat. Mech. Appl. **206**(1–2), 229–252 (1994)
8. Thacker, W.C.: Some exact solutions to the nonlinear shallow-water wave equations. J. Fluid Mech. **107**, 499–508 (1981)
9. Kânoğlu, U., Synolakis, C.: Initial value problem solution of nonlinear shallow water-wave equations. Phys. Rev. Lett. **97**(14), 148501 (2006)
10. Griffiths, G., Schiesser, W.E.: Traveling Wave Analysis of Partial Differential Equations: Numerical and Analytical Methods with MATLAB and Maple. Academic (2010)
11. Holloway, P.E., Pelinovsky, E., Talipova, T.: A generalized Korteweg-de Vries model of internal tide transformation in the coastal zone. J. Geophys. Res.: Ocean. **104**(C8), 18333–18350 (1999)
12. Xiao, Z.-J., Tian, B., Zhen, H.-L., Chai, J., Xiao-Yu, W.: Multi-soliton solutions and Bäcklund transformation for a two-mode KDV equation in a fluid. Waves Random Complex Media **27**(1), 1–14 (2017)

13. Ji, J.-L., Zhu, Z.-N.: Soliton solutions of an integrable nonlocal modified Korteweg-de Vries equation through inverse scattering transform. J. Math. Anal. Appl. **453**(2), 973–984 (2017)
14. Hirota, R.: Exact solution of the Korteweg-de Vries equation for multiple collisions of solitons. Phys. Rev. Lett. **27**(18), 1192 (1971)
15. Fornberg, B.: The pseudospectral method: comparisons with finite differences for the elastic wave equation. Geophysics **52**(4), 483–501 (1987)
16. Malfliet, W., Hereman, W.: The tanh method: I. Exact solutions of nonlinear evolution and wave equations. Phys. Scr. **54**(6), 563 (1996)
17. He, J.-H., Xu-Hong, W.: Exp-function method for nonlinear wave equations. Chaos Solitons Fractals **30**(3), 700–708 (2006)
18. Wazwaz, A.-M.: A sine-cosine method for handling nonlinear wave equations. Math. Comput. Model. **40**(5–6), 499–508 (2004)
19. Wang, X., Wang, L.: Darboux transformation and nonautonomous solitons for a modified Kadomtsev-Petviashvili equation with variable coefficients. Comput. & Math. Appl. **75**(12), 4201–4213 (2018)
20. Liu, L.-J., Yu, X.: Solitons and breathers for nonisospectral mKDV equation with Darboux transformation (2017). arXiv:1710.05108
21. Matveev, V.B.: Darboux transformation and explicit solutions of the Kadomtcev-Petviaschvily equation, depending on functional parameters. Lett. Math. Phys. **3**(3), 213–216 (1979)
22. Doktorov, E.V., Leble, S.B.: A Dressing Method in Mathematical Physics, vol. 28. Springer Science & Business Media (2007)
23. Shuwei, X., He, J., Wang, L.: The Darboux transformation of the derivative nonlinear Schrödinger equation. J. Phys. A: Math. Theor. **44**(30), 305203 (2011)
24. He, G.L., Su, T.: Darboux transformation and explicit solutions for a generalized Sawada-Kotera equation. Int. Sch. Res. Not. **2013** (2013)
25. Nimmo, J.J.C.: Darboux transformations and the discrete KP equation. J. Phys. A: Math. Gen. **30**(24), 8693 (1997)

Optical Dark and Kink Solitons in Multiple Core Couplers with Four Types of Nonlinearity

Anand Kumar, Hitender Kumar, Fakir Chand, Manjeet Singh Gautam, and Ram Mehar Singh

Abstract In recent years, solitons in nonlinear couplers have acquired attention of the researchers. The switching of solitons is possible using any optical logic gate. Optical switching is of particular interest due to the possibility of extremely fast switching time within the femtosecond range. In this study the dark and kink type solitons to two different types of optical multiple core couplers with four types of nonlinearities viz Kerr law nonlinearity, power law nonlinearity, parabolic law nonlinearity and dual power-law nonlinearity are extracted using the Kudryashov integration algorithm. Coupling with nearest neighbors is one example, whereas in the other case the coupling with all nearest neighbors is considered. The parametric constraint conditions, also called integrability criteria are emerging with these novel solutions and reported.

1 Introduction

Nonlinear optical couplers are crucial appliances that route light from the main fibre in one or more section fibres. Optical couplers have also been employed as limiters and intensity-dependent switches. Optical couplers can be manufactured as planar semiconductor devices or as dual-core single-mode fibres with solitons conveying

A. Kumar (✉)
Department of Physics, Chaudhary Ranbir Singh University, Jind 126102, India
e-mail: anandkumar@crsu.ac.in

H. Kumar
Department of Physics, Government College for Women, Gharaunda 132114, India

F. Chand
Department of Physics, Kurukshetra University, Kurukshetra 136119, India

M. S. Gautam
Department of Physics, Government College, Alewa, Jind 126111, India

R. M. Singh
Department of Physics, Chaudhary Devi Lal University, Sirsa 125055, India

S. Banerjee and A. Saha (eds.), *Nonlinear Dynamics and Applications*,
Springer Proceedings in Complexity,
https://doi.org/10.1007/978-3-030-99792-2_99

in each core. Multiple-core optical fibers can be used with high-powered lasers and all-optical switching between the fiber cores [1].

The study of optical solitons in nonlinear optics has attracted a lot of attention, and it's played a big part in the development of all-optical systems. Studying the dynamical behaviour of propagation of soliton along optical fibers, couplers, dense wavelength division multiplexing (DWDM) systems, metamaterials, and metasurfaces is therefore crucial to improve soliton transmission performance across long-haul optical communication networks. Many efforts have been brought in the latter half of the 19th century to solve the difficulties of optical soliton transmission by appropriate control of the fiber group dispersion. Hasegawa et al. [2] presented adiabatic dispersion control to reduce dispersive wave radiation and collision-induced frequency shift in WDM systems by altering dispersion in percentage to the soliton power. Suzuki et al. [3] utilized non-adiabatic periodic dispersion compensation to minimize integrated dispersion and transmit a 10 Gbit/s soliton signal across the Pacific without using soliton management. Smith et al. [4] shows, even though the dispersion is almost zero, a nonlinear soliton-like pulse exist in a fiber with a periodic modulation of the dispersion. The nonlinearities such as Kerr law, power law, parabolic law, and dual-power law are four forms of nonlinear media used in this investigation. The governing model for multiple core fibers with STD (spatiotemporal dispersion) in addition to the standard GVD (group velocity dispersion) is the nonlinear Schrödinger's equation (NLSE). Hence, it is essential to investigate NLSE in couplers and in fibers with the STD term incorporated. The ansatz technique [5], the Jacobi elliptic function method [7], and other methods [6, 8–10] have all been used to study optical couplers previously.

In this study, we use the Kudryashov approach [11] to handle multiple-core couplers with four different types of nonlinearities, resulting in kink and dark soliton solutions that will serve the soliton community. Kink and dark optical soliton solutions will be found alongside their existing conditions that naturally occur from the solution profiles.

Outlined of the manuscript is given as. The brief idea of Kudryashov method is presented in Sect. 2. Section 3 deals with multiple core couplers with nearest-neighbor coupling for four forms of nonlinearity. In Sect. 4, we extend our study on multiple core couplers coupling with all neighbors for four forms of nonlinearity. Section 5 allotted to graphical results. Finally, conclusions are made in Sect. 6.

2 A Succinet Overview of the Kudryashov Method

To present the analysis more coherently, we highlight succinctly the prime aspects of the Kudryashov method.

We acknowledge a nonlinear partial differential equation (PDE), with a physical field q, which is a function of independent variables x, t as:

$$R(q_x, q_t, q_{xx}, q_{xt}, q_{tt}, \ldots) = 0, \tag{1}$$

where R is polynomial in $q(x, t)$ Here, we briefly highlighted the steps of the method:

Step 1: By using $\xi = k(x + vt)$, Eq. (1) modify to ODE (ordinary differential equation):

$$S(q, q_\xi, q_{\xi\xi}, q_{\xi\xi\xi}, \ldots) = 0 \tag{2}$$

Step 2: The solution of above equation can be specified in the more general form of physical field $q(\xi)$ as:

$$q(\xi) = \sum_{n=0}^{N} a_n [\Psi(\xi)]^n, \tag{3}$$

with $a_N \neq 0$ and the function $\Psi(\xi)$ satisfying the new equation

$$\Psi(\xi) = \frac{1}{1 + \exp(\xi + \xi_0)}, \tag{4}$$

which is the solution of a special kind of Riccati equation

$$\frac{d\Psi}{d\xi} = \Psi^2(\xi) - \Psi(\xi). \tag{5}$$

Step 3: The integer N is defined by homogeneous balance principal in Eq. (2).

Step 4: Replace Eq. (3) into Eq. (2), and determine all required derivatives q_ξ, $q_{\xi\xi}$, $q_{\xi\xi\xi}, \ldots$ as follows

$$q_\xi = \sum_{n=0}^{N} a_n n \Psi^n (\Psi - 1), \tag{6}$$

$$q_{\xi\xi} = \sum_{n=0}^{N} n \Psi^n (\Psi - 1)[(1 + n)\Psi - n] a_n, \tag{7}$$

and so on. Replacing Eqs. (3), (6) and (7) along with (5) into Eq. (2), we attain the polynomial form as

$$S[\Psi(\xi)] = 0. \tag{8}$$

Step 5: Using Eqs. (8) and (4) and symbolic computer packages such as Mathematica, we found the analytic exact solutions of Eq. (1).

In the succeeding sections, we implemented the Kudryashov method to obtain dark soliton solutions for multiple core couplers in which the coupling is held with the nearest neighbors and all neighbors respectively.

3 Multiple-Core Couplers (Coupling with Nearest Neighbors)

This types of problems can be designated by the N-coupled NLSE with the nearest neighbor linear coupling. The dictating NLSE for multiple-core couplers (coupling with nearest neighbors) is written as

$$iu_t^{(j)} + a_j u_{xx}^{(j)} + b_j u_{xt}^{(j)} + c_j F(|u^{(j)}|^2)u^{(j)} = Q[u^{(j-1)} - 2u^{(j)} + u^{(j+1)}], \quad (9)$$

where $1 \leq j \leq N$ and u^j denotes the optical field in the jth core. The Eq. (9) is distinguished as the coupled NLSE in which initial term represents the evolving soliton with time. The coefficients a_j are the GVD coefficients while b_j are the coefficients of STD and c_j denotes nonlinearity's coefficients. The sign of nonlinearity is given by functional F. Here F denotes algebraic function with real-valued. The constants Q in Eq. (9) denotes the coupling coefficients in optical fibres and signifies the strength of linear coupling. Also, coupled Eq. (9) possess three integrals of motion such as energy (E), the Hamiltonian (H) and linear momentum (M). To attend these coupled equations by the Kudryashov method, the following ansatz is appropriated.

$$u^j(x,t) = B_j(\xi)e^{i\phi(x,t)}, \quad (10)$$

Here, $B_j(j = 1, 2)$, nearest neighbour signify the amplitude component and v is the soliton's speed, while $\phi(x,t) = -Kx + wt + \theta$ is phase component, where K, w, θ are the soliton frequency, wave number and phase constant respectively. Plugging ansatz (10) into Eq. (9), then the real and imaginary parts are:

$$\begin{aligned} k^2(a_j - b_j v)B_j^{''} + (b_j wK - w - a_j K^2)B_j + c_j F(B_j^2)B_j \\ -Q[B_{j-1} - 2B_j + B_{j+1}] = 0, \end{aligned} \quad (11)$$

$$-(1 - b_j K)kvB_j^{'} + (b_j w - 2a_j K)kB_j^{'} = 0. \quad (12)$$

From the imaginary part (12), we determine soliton speed as

$$v = \frac{b_j w - 2a_j K}{1 - b_j K}, \quad (13)$$

whenever $1 \neq b_j K$. The balancing principle in (11) leads to $B_{j-1} = B_j = B_{j+1}$, as a result, the real part Eq. (11) transformed to

$$k^2(a_j - b_j v)B_j^{''} + (b_j wK - w - a_j K^2)B_j + c_j F(B_j^2)B_j = 0, \quad (14)$$

In the next subsections, we examine this equation for four different sorts of nonlinearity viz Kerr law, power law, parabolic law and dual power-law nonlinearities, respectively.

3.1 Kerr Law Nonlinearity

This is most basic kind of cubic nonlinearity which is originates from the fact that due to corresponding electric field, nonharmonic motion is shown by the bound electrons. Therefore, nonlinear responses exhibits by a light wave in an optical fiber. Hence, the induced polarization (P) is also not linear in the electric field (E), but includes higher-order factors in the amplitude of the electric field. $F(u) = u$ for Kerr law nonlinearity. For multiple-core couplers (coupling with nearest neighbors) with Kerr law form nonlinearity, the executive model Eq. (9) reduces to

$$iu_t^{(j)} + a_j u_{xx}^{(j)} + b_j u_{xt}^{(j)} + c_j(|u^{(j)}|^2)u^{(j)} = Q[u^{(j-1)} - 2u^{(j)} + u^{(j+1)}], \quad (15)$$

and Eq. (14) becomes

$$k^2(a_j - b_j v)B_j^{''} + (b_j wK - w - a_j K^2)B_j + c_j B_j^3 = 0. \quad (16)$$

With balancing $B_j^{''}$ and B_j^3 in Eq. (16), we have $N = 1$. Consequently, we reach $B_j(\xi) = a_0 + a_1\Psi(\xi)$, where $\Psi(\xi)$ satisfies the following general first and second order nonlinear differential equations: $B_j^{'}(\xi) = a_1\Psi(\xi)[\Psi(\xi) - 1]$ and $B_j^{''}(\xi) = a_1\Psi(\xi)[\Psi(\xi) - 1][2\Psi(\xi) - 1]$, where a_0, a_1 are constants to be determined later, such that $a_1 \neq 0$. Plugging the form of $B_j(\xi)$ and their derivatives into Eq. (16) and after collecting all the terms of Ψ^j $(j = 0, 1, 2, 3)$ gives a set of algebraic equations which on solving by aid of Maple, we have the results:

$$a_0 = \frac{k^2(a_j - b_j v)}{c_j}, a_1 = -2a_0, w = \frac{2k^4 b_j v a_j - k^4(a_j^2 - b_j^2 v^2) + a_j K^2 c_j}{c_j(b_j K - 1)}. \quad (17)$$

Substituting Eq. (17) in Eqs. (3) and (4), we get the optical solitary wave solution for multiple core coupler with Kerr law nonlinearity (coupling with nearest neighbours) as

$$B_j(\xi) = \frac{k^2(a_j - b_j v)}{c_j}\left[1 - \frac{2}{1 + \exp(\xi + \xi_0)}\right], \quad (18)$$

which can be equivalently written in more simplified form using the relation $\frac{1}{1+\exp(\xi)} = \frac{1}{2} - \frac{1}{2}\tanh(\xi/2)$ and using Eq. (10), the optical dark solitary wave solution is specified as

$$u^j(x,t) = \frac{k^2(a_j - b_j v)}{c_j} \tanh\left(\frac{k(x - \frac{b_j w - 2a_j K}{1 - b_j K} t)}{2} + \frac{\xi_0}{2}\right) e^{[i(-Kx + wt + \theta)]} \quad (19)$$

This optical dark solitary wave solution will exist provided that the constraint conditions $a_j \neq b_j v, b_j K \neq 1$ and the relation between wave number w and wave speed v from Eq. (17) can be understood as integrability condition with Kerr law nonlinearity.

3.2 Power Law Nonlinearity

Such type of nonlinearity is a generalized version of Kerr law and is commonly observed in nonlinear fiber optics and nonlinear plasmas. This nonlinearity can be seen in a variety of materials, such as semiconductor lasers. For this nonlinearity, $F(u) = u^n$ where n accounts for the power-law nonlinearity factor and have $n \neq 2$ and $0 < n < 2$ to avoid self-focussing effect. Henceforth, the Eq. (9), for multiple-core couplers (nearest-neighbor coupling) with power-law nonlinearity will now be changed to

$$iu_t^{(j)} + a_j u_{xx}^{(j)} + b_j u_{xt}^{(j)} + c_j(|u^{(j)}|^{2n})u^{(j)} = Q[u^{(j-1)} - 2u^{(j)} + u^{(j+1)}], \quad (20)$$

and Eq. (14) leads to

$$k^2(a_j - b_j v)B_j^{''} + (b_j wK - w - a_j K^2)B_j + c_j B_j^{2n+1} = 0. \quad (21)$$

On setting $B_j = U_j^{\frac{1}{n}}$ then Eq. (21) change into

$$k^2(a_j - b_j v)\left(nU_j U_j^{''} + (1-n)(U_j^{'})^2\right) + n^2 U_j^2(b_j wK - w - a_j K^2) + c_j n^2 U_j^4 = 0. \quad (22)$$

When the powers of $U_j U_j^{''}$ are compared to the powers of U_j^4, the homogeneous balancing in Eq. (22) leads to $N = 1$. We derive the following results using the Kudryashov method's solution approach:

$$a_0 = -a_1, a_1 = \frac{3}{2}\frac{n}{(n-1)}, w = \frac{4a_j K^2(n-1) + 3k^2(a_j - b_j v)}{4(n-1)(b_j K - 1)}. \quad (23)$$

Substituting Eq. (23) in Eqs. (3) and (4) and using the relation $B_j = U_j^{\frac{1}{n}}$, we get the optical dark solitary wave solution for multiple core coupler with power law nonlinearity as

$$u^j(x,t) = \left[\frac{3}{2}\frac{n}{(n-1)}\left\{1 + \tanh\left(\frac{k(x - \frac{b_j w - 2a_j K}{1 - b_j K}t)}{2} + \frac{\xi_0}{2}\right)\right\}\right]^{\frac{1}{n}} \times e^{[i(-Kx+wt+\theta)]}, \tag{24}$$

It is worth to see that the dark optical solitary wave solution (24) will satisfy the parametric constraint conditions $n \neq 0, 1$, $a_j \neq b_j v$, $b_j K \neq 1$, $b_j w \neq 0$ and the relation between wave number w and wave speed v from Eq. (23) can be understood as integrability condition with power law nonlinearity.

3.3 *Parabolic Law Nonlinearity*

This is known as cubic-quintic nonlinearity which is portrayed by $F(u) = \alpha u + \beta u^2$, where α and β are in overall constants that connects the two nonlinear forms. This sort of nonlinearity has significant attention after it occurs in materials such as para-toluene sulphonate that manifest fifth-order nonlinearity in response to extreme ultra-short optical pulses at 620 nm. With cubic-quintic nonlinearity, Eq. (9) for multiple-core couplers (coupling with nearest neighbors) is formulated as

$$iu_t^{(j)} + a_j u_{xx}^{(j)} + b_j u_{xt}^{(j)} + c_j(\alpha|u^{(j)}|^2 + \beta|u^{(j)}|^4)u^{(j)} = Q[u^{(j-1)} - 2u^{(j)} + u^{(j+1)}], \tag{25}$$

and Eq. (14) transformed to

$$k^2(a_j - b_j v)B_j'' + (b_j wK - w - a_j K^2)B_j + c_j(\alpha B_j^2 + \beta B_j^4)B_j = 0. \tag{26}$$

A transformation formula is used to obtain a closed form analytic solution as $B_j = U_j^{\frac{1}{2}}$ then Eq. (26) changes into

$$k^2(a_j - b_j v)\big(2U_j U_j'' - (U_j')^2\big) + 4U_j^2(b_j wK - w - a_j K^2) + 4\alpha c_j U_j^3 + 4\beta c_j U_j^4 = 0. \tag{27}$$

From Eq. (27), we have $N = 1$. Using the recipe of the Kudryashov method, we obtain the following results:

$$\begin{aligned} a_0 &= -a_1, a_1 = \frac{3k^2(b_j v - a_j) - 2\alpha c_j}{k^2(b_j v - a_j)}, \\ \beta &= \frac{(b_j v - a_j)(-k^2 a_j + 2c_j\alpha + k^2 b_j v)k^2}{4[3k^2(b_j v - a_j) - 2\alpha c_j]}, \\ w &= \frac{-3(b_j v - a_j)^2 k^4 + 4(a_j K^2 + 2c_j\alpha)(b_j v - a_j)k^2 - 4c_j^2\alpha^2}{4k^2(b_j K - 1)(b_j v - a_j)}. \end{aligned} \tag{28}$$

Replacing Eq. (28) in Eqs. (3) and (4) and setting the relation $B_j = U_j^{\frac{1}{2}}$, we acquire the dark optical solitary wave solution wave with parabolic law nonlinearity as

$$u^j(x,t) = \left[-\frac{3k^2(b_j v - a_j) - 2\alpha c_j}{2k^2(b_j v - a_j)} \left(1 + \tanh\left(\frac{k(x - \frac{b_j w - 2a_j K}{1 - b_j K} t)}{2} + \frac{\xi_0}{2} \right) \right) \right]^{\frac{1}{2}} \times e^{[i(-Kx + wt + \theta)]}. \tag{29}$$

It is pointed that the dark optical solitary wave solution (29) will exist under the parametric constraint conditions $a_j \neq b_j v$, $b_j K \neq 1$, $b_j w \neq 0$ and the relation between wave number w and wave speed v from Eq. (28) can be specified as integrability condition with parabolic law form nonlinearity.

3.4 Dual-Power Law Nonlinearity

Such type of nonlinearity is commonly used to explain spatial solitons in photovoltaic-photo refractive materials like as $LiNbO_3$. For dual-power law nonlinear media, the function $F(u) = \alpha u^n + \beta u^{2n}$, where α and β are defined as real-valued constants. Equation (9), account for multiple-core couplers (coupling with nearest neighbors) with dual-power law nonlinearity as

$$\begin{aligned} iu_t^{(j)} + a_j u_{xx}^{(j)} + b_j u_{xt}^{(j)} + c_j(\alpha |u^{(j)}|^{2n} + \beta |u^{(j)}|^{4n}) u^{(j)} \\ = Q[u^{(j-1)} - 2u^{(j)} + u^{(j+1)}], \end{aligned} \tag{30}$$

and Eq. (14) becomes

$$k^2(a_j - b_j v)B_j'' + (b_j wK - w - a_j K^2)B_j + c_j(\alpha B_j^{2n} + \beta B_j^{4n})B_j = 0. \tag{31}$$

It should be observed that in Eq. (31), $\beta = 0$ recovers power-law nonlinearity, and when $n = 1$ recovers Kerr law nonlinearity as well. However, if $\beta \neq 0$ and $n = 1$, one is back in the case of parabolic law nonlinearity, which was previously explored. On equating $B_j = U_j^{\frac{1}{2n}}$ then Eq. (31) change into

$$\begin{aligned} k^2(a_j - b_j v)\big(2nU_j U_j'' + (1 - 2n)(U_j')^2\big) + 4n^2 U_j^2 (b_j wK - w - a_j K^2) r \\ + 4n^2 \alpha c_j U_j^3 + 4n^2 \beta c_j U_j^4 = 0. \end{aligned} \tag{32}$$

Balancing the leading dispersive term $U_j U_j''$ with nonlinear term U_j^4 in Eq. (32), provides $N = 1$. Using the Kudryashov method's solution recipe, we obtain the value of a_0, a_1, w and β.

$$a_0 = -a_1, a_1 = \frac{n(3k^2(b_j v - a_j) - 2n\alpha c_j)}{k^2(2n-1)(b_j v - a_j)},$$
$$w = \frac{-3(b_j v - a_j)^2 k^4 + 8\left(K^2\left(n - \frac{1}{2}\right)a_j + n\alpha c_j\right)(b_j v - a_j)k^2 - 4n^2\alpha^2 c_j^2}{8k^2(b_j K - 1)\left(n - \frac{1}{2}\right)(b_j v - a_j)},$$
$$\beta = \frac{(a_j + 2b_j vn - b_j v - 2a_j n)(2n\alpha c_j - k^2 a_j + k^2 b_j v)k^2}{4n^2(-2n\alpha c_j + 3k^2 b_j v - 3k^2 a_j)c_j}. \tag{33}$$

Substituting these in Eqs. (3) and (4) and using the relation $B_j = U_j^{\frac{1}{2n}}$, we found the exact dark optical solitary wave solution as

$$u^j(x,t) = \left[-\frac{n(3k^2(b_j v - a_j) - 2n\alpha c_j)}{2k^2(2n-1)(b_j v - a_j)} \left(1 + \tanh\left(\frac{k(x - \frac{b_j w - 2a_j K}{1 - b_j K}t)}{2} + \frac{\xi_\theta}{2}\right)\right)\right]^{\frac{1}{2n}}$$
$$\times \exp[i(-Kx + wt + \theta)]. \tag{34}$$

It is pointed that dark optical solitary wave solution (34) will exist under the additional parametric constraint condition $n \neq \frac{1}{2}$ and the relation between wave number w and wave speed v from Eq. (33) can be specified as integrability condition with dual power law form nonlinearity. For $n = 1$ as a special case, we get the previously found solution (29) with parabolic form nonlinearity.

4 Multiple-Core Couplers (Coupling with All Neighbors)

For the pulses propagating through N coupled nonlinear fiber arrays, the governing equation for multiple-core couplers is provided in the dimensionless form as

$$iu_t^{(j)} + a_j u_{xx}^{(j)} + b_j u_{xt}^{(j)} + c_j F(|u^{(j)}|^2)u^{(j)} = \sum_{m=1}^{N} \lambda_{jm} u^m, \tag{35}$$

where $1 \leq j \leq N$. The Eq. (35) depicts the generic model for optical couplers wherever coupling with all neighbors is included with GVD and STD, which defines soliton passage via multiple-core optical fibers under balancing outcome of dispersion and nonlinearity. Here λ_{jm} in Eq. (35) renders the linear coupling coefficients in optical fibers. To approach this model by the Kudryashov method for the four types of nonlinear media, the first hypothesis is supposed to be

$$u^j(x,t) = B_j(\xi)e^{(i\phi(x,t))}, \tag{36}$$

where $\xi = k(x - vt)$. Replacing the hypothesis (36) into Eq. (35) and subsequently, splitting into real and imaginary parts results in

$$k^2(a_j - b_j v)B_j^{''} + (b_j wK - w - a_j K^2)B_j + c_j F(B_j^2)B_j - \sum_{m=1}^{N} \lambda_{jm} B_m = 0, \tag{37}$$

$$-(1 - b_j K)kvB_j^{'} + (b_j w - 2a_j K)kB_j^{'} = 0 \tag{38}$$

From the imaginary part (38), we deduce soliton speed as

$$v = \frac{b_j w - 2a_j K}{1 - b_j K}. \tag{39}$$

For all sorts of specified nonlinearity in question, the velocity of the soliton, given by (39), is constant. The balancing principle in Eq. (37) leads to $B_j = B_m$, as a result, the real part Eq. (37) reduces to

$$k^2(a_j - b_j v)B_j^{''} + (b_j wK - w - a_j K^2 - \sum_{m=1}^{N} \lambda_{jm})B_j + c_j F(B_j^2)B_j = 0 \tag{40}$$

From the Kudryashov method, this equation have the dark optical soliton with four types of nonlinearities.

5 Graphical Results and Discussion

In this section, we illustrate the graphic representation of several wave structures of the considered system. By utilizing Kudryashov method the soliton solution of optical couplers are retrieved and graphically represented in 3-D, and their contours with the selection of different parameters. The Kudryashov approach was used to examine solitary wave solutions, which are unique and different from what many other researchers have found using various methodologies.

The intensity distribution of dark solitary wave solution (19) is shown in Fig. 1 when parameter values $a_j = 1$, $b_j = 1$, $c_j = 1$, $k = 1$, $v = 0.2$, $K = 2$, $w = 3.44$, and $\xi_0 = 0$. In Fig. 2, the intensity distribution of kink solitary wave solution (24) when parameter values $n = 3$, $a_j = 1$, $b_j = 1$, $c_j = 1$, $k = 1$, $v = 0.2$, $K = 2$, $w = 3.44$, and $\xi_0 = 0$ is depicted. In nonlinear optics, these types of propagating structures are extremely significant. Soliton has been utilized to greatly improve the transmission capabilities of Telecom lines. Solitons are well-localized structures that can travel great distances without changing shape, and optical pulses propagate in the form of solitons.

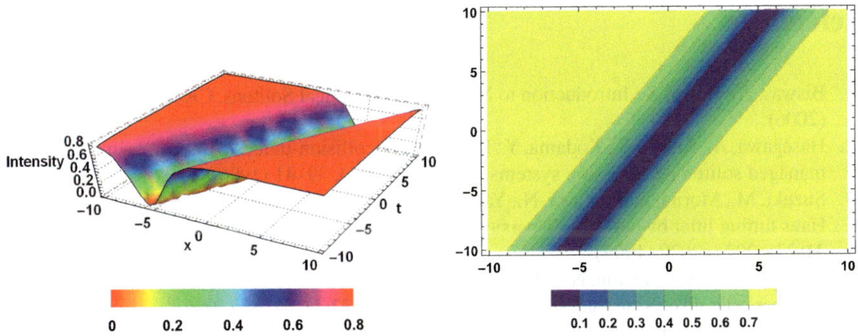

Fig. 1 The intensity distribution of solitary wave solution (19) with parameter values $a_j = 1$, $b_j = 1$, $c_j = 1$, $k = 1$, $v = 0.2$, $K = 2$, $w = 3.44$, and $\xi_0 = 0$

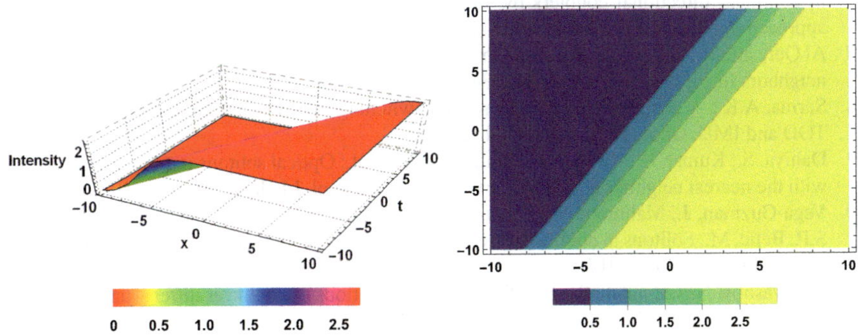

Fig. 2 The intensity distribution of kink solitary wave solution (24) with parameter values $n = 3$, $a_j = 1$, $b_j = 1$, $c_j = 1$, $k = 1$, $v = 0.2$, $K = 2$, $w = 3.44$, and $\xi_0 = 0$

6 Conclusions

Optical couplers and switches are indispensable components of optical communication systems. In the present work, we obtains dark optical solitons to nonlinear directional multiple cores couplers with four different kinds of nonlinearities using Kudryashov algorithm scheme. Here, two distinctive cases were analyzed. Coupling with nearest neighbors was addressed in the first instance, while coupling with all neighbors was addressed subsequently. To ensure the existence of these solitons, the necessary constraint conditions emerge naturally. The outcomes of this paper are certainly stimulating to focus study on a diverse possibility with couplers.

References

1. Biswas, A., Konar, S.: Introduction to Non-kerr Law Optical Solitons. CRC Press, Boca Raton (2006)
2. Hasegawa, A., Kumar, S., Kodama, Y.: Reduction of collision-induced time jitters in dispersion-managed soliton transmission systems. Opt. Lett. **21**(1), 39–41 (1996)
3. Suzuki, M., Morita, I., Edagawa, N., Yamamoto, S., Taga, H., Akiba, S.: Reduction of Gordon-Haus timing jitter by periodic dispersion compensation in soliton transmission. Electron. Lett. **31**(23), 2027–2029 (1995)
4. Smith, N., Knox, F., Doran, N., Blow, K., Bennion, I.: Enhanced power solitons in optical fibres with periodic dispersion management. Electron. Lett. **32**(1), 54–55 (1996)
5. Savescu, M., Bhrawy, A., Alshaery, A., Hilal, E., Khan, K.R., Mahmood, M., Biswas, A.: Optical solitons in nonlinear directional couplers with spatio-temporal dispersion. J. Mod. Opt. **61**(5), 441–458 (2014)
6. Mirzazadeh, M., Eslami, M., Zerrad, E., Mahmood, M.F., Biswas, A., Belic, M.: Optical solitons in nonlinear directional couplers by sine-cosine function method and Bernoulli's equation approach. Nonlinear Dyn. **81**(4), 1933–1949 (2015)
7. Al Qurashi, M.M., Ates, E., Inc, M.: Optical solitons in multiple-core couplers with the nearest neighbors linear coupling. Optik **142**, 343–353 (2017)
8. Sarma, A.K.: A comparative study of soliton switching in a two-and three-core coupler with TOD and IMD. Optik **120**(8), 390–394 (2009)
9. Dahiya, S., Kumar, H., Kumar, A., Gautam, M.S., et al.: Optical solitons in twin-core couplers with the nearest neighbor coupling. Part. Differ. Equ. Appl. Math. **4**, 100–136 (2021)
10. Vega-Guzman, J., Mahmood, M., Zhou, Q., Triki, H., Arnous, A.H., Biswas, A., Moshokoa, S.P., Belic, M.: Solitons in nonlinear directional couplers with optical metamaterials. Nonlinear Dyn. **87**(1), 427–458 (2017)
11. Kudryashov, N.A.: One method for finding exact solutions of nonlinear differential equations. Commun. Nonlinear Sci. Numer. Simul. **17**(6), 2248–2253 (2012)

Analysis of a Variable-Order Multi-scroll Chaotic System with Different Memory Lengths

N. Medellín-Neri, J. M. Munoz-Pacheco, O. Félix-Beltrán, and E. Zambrano-Serrano

Abstract This work presents a numerical analysis of the dynamical behavior of a multi-scroll chaotic system using variable-order calculus. In this scenario, we introduce the concept of variable-order from two approaches denominated herein as short-memory and full-memory, respectively. For the first one, the basic idea is to study the chaotic dynamics when the fractional-order changes abruptly like step-function with respect to time. The second approach is related to a smoother variation between the preceding order and the new fractional-order. To demonstrate the implications of using variable-order with distinct memory contributions, we show several numerical simulations of a multi-scroll chaotic system that contains a piecewise linear function. Numerical results are consistent with the underlying theory demonstrating the usefulness of the proposed study.

Keywords Variable-order · Fractional calculus · Chaos · Multi-scroll · PWL

1 Introduction

The chaotic behavior has remarkable characteristics such as extreme sensitivity to small variations of the initial conditions and parameters, limited trajectories in phase space, and at least, a positive Lyapunov exponent [1]. For instance, two initial trajectories that are extremely close between them, will diverge exponentially as time tends to infinite, and thereafter, have totally different evolutions. In this manner, the chaotic systems have been used in almost all fields of science and engineering, such as secure communications [2], cryptography [3], robotics [4], mechanics [5], etc.

N. Medellín-Neri (✉) · J. M. Munoz-Pacheco · O. Félix-Beltrán
Faculty of Electronics Sciences, Benemérita Universidad Autónoma de Puebla, 72570 Puebla, Mexico
e-mail: nadia.medellin@alumno.buap.mx

E. Zambrano-Serrano
Facultad de Ingeniería Mecánica y Eléctrica, Universidad Autónoma de Nuevo León, 66455 San Nicolás de los Garza, Nuevo León, Mexico

S. Banerjee and A. Saha (eds.), *Nonlinear Dynamics and Applications*,
Springer Proceedings in Complexity,
https://doi.org/10.1007/978-3-030-99792-2_100

For several decades, it was well known that chaotic behavior requires, at least, a three-dimensional system to emerge [6], where the dimension relates to the order of their derivatives. Therefore, a n-dimensional system is a system of mth order, $n = m, n \in \mathbb{Z}$. However, the exception to this rule is when the order of the derivative is no longer integer but fractional (from now on we will use q when $m \in \mathbb{R}$), like in $n \neq q$ and also $q \in \mathbb{R}$. Fractional calculus refers to the generalization of integrals and derivatives to arbitrary order. Although this topic was proposed more than 300 years ago, it has recently been noted that the fractional calculus has superior characteristics to the conventional calculus [7]. The main reason is that the fractional-order derivatives have memory properties, giving a more convenient way to describe living and nonliving phenomena [8–12].

The three main definitions for fractional derivatives are Riemann-Liouville [13], Caputo [14], and Grünwald-Letnikov [15–17] that are equivalent under some assumptions. In literature, we can find many excellent works using fractional calculus with constant values for the fractional-order, i.e., the fractional-order q is a positive real constant that remains unchanged throughout the simulation time. As a result, memory contributions follow a power-law evolution such as the Caputo derivative with singular kernel, where past events have lower implications than recent ones. In Caputo's definition, the memory kernel is expressed in the form of a convolution integral. As seen, the fractional-order differential equations accumulate the whole past history in a weighted form, this is called the "memory effect".

However, a less researched area and still exciting is fractional calculus with variable-order. It means that the value of the fractional-order can be updated as time evolves, i.e., $q(t)$. Then, the fractional-order can be defined as trigonometric, quadratic, polynomial, and constant piecewise linear functions with respect to time. Indeed, the fractional-order value can also depend on a pseudo-state of the underlying dynamical system. With the variable-order calculus, the memory contributions change with time, altering the strength of the effects of this memory. Many works have shown the importance of considering variable-order to increase the accuracy and degrees of freedom in many scientific areas. For instance, a variable-order susceptible-infected-recovered (SIR) model described the COVID19 evolution with better approximation to real data [18]. There, the memory contributions to epidemic spread were captured using a piecewise-linear fractional-order. Reference [19] offered a unified discussion of variable-order differential operators in anomalous diffusion modeling. Reference [20] reported block-based image encryption where each block has a different fractional-order using a short-memory variable-order. In mechanics, the effect of nonuniform viscoelastic frictional forces described applying variable-order to demonstrate that constant fractional-order cannot approximate the transition between the relevant dynamic regimes [21].

In this work, we present in the Eq. (5) a numerical analysis of the dynamical behavior of a multi-scroll chaotic system [22] using variable-order calculus.

A multi-scroll chaotic system can be defined as a nonlinear dynamical system that presents a chaotic attractor composed of many scrolls [23], contrary to the classical chaotic systems with only double-scroll attractors, like Lorenz, Chua, and Chen systems, to mention a few. The increased number of scrolls are originated from nonlinear functions in the form of piecewise-linear (PWL) functions such as saturated, hysteresis, sawtooth, Heaviside, and so on [24]. As a result, the number of equilibrium points also increases, and the dynamical system has the potential to generate a multi-scroll attractor. Recently, multi-scrolls chaotic systems continue being a hot topic of research. Wu et al. proposed a new multi-scroll system that produces three distinct hidden attractors with stable equilibria and without equilibrium points [25]. Escalante-Gonzalez and Campos-Canton also reported a method for switching between hidden and self-excited multi-scroll chaotic attractors in multi-stable systems [26]. Zhang et al. introduced a multi-scroll system based on a memristive approach along with a Hindmarsh-Rose neuron model [27]. Ahmad et al. studied how to transform a multi-scroll system to the fractional calculus domain, mainly using the Caputo fractal-fractional operator [28].

The concept of variable-order is shown from two approaches denominated herein as short-memory and full-memory, respectively. For the first one, the basic idea is to study the chaotic dynamics when the fractional-order changes abruptly like step-function with respect to time. The second approach is related to a smoother variation between the initial order and the new fractional-order. Therefore, Section 2 gives the mathematical foundations of variable-order calculus. Section 3 introduces the multi-scroll chaotic system, studies the stability of equilibrium points, and presents phase portraits. Section 4 presents the proposed numerical analysis with short-memory and full-memory approaches. Numerical simulations of a four-scroll chaotic attractor are in particular analyzed. Finally, Section 5 concludes the work.

2 Mathematical Preliminaries

Definition 1 Fractional calculus is a generalization of integration and differentiation to the non-integer fundamental operator ${}_aD_t^q$, where a and t are the limits of the operation and $q \in \mathbb{R}$. We have that the continuous integro-differential operator is defined as [29]:

$$
{}_aD_t^q = \begin{cases} \frac{d^q}{dt^q}, & q > 0, \\ 1, & q = 0, \\ \int_a^t (d\tau)^q, & q < 0. \end{cases} \tag{1}
$$

Additionally, The variable-order calculus is based on the fact that the order of the derivative is not constant, for the case of study, the variation is carried out with respect to time.

Definition 2 Let us consider $q(t) > 0$ as a function that is limited; the fractional derivative of Caputo in its fractional-order (FO) version is defined as [19]:

$$
{}_{t_0}^{C}D_t^{q(t)} f(t) := \begin{cases} \frac{1}{\Gamma(q(t)-m)} \int_{t_0}^{t} \frac{f^m(\tau)}{(t-\tau)^{q(t)+1-m}} \, \mathrm{d}\tau, & m-1 < q(t) \leq m, \\ \frac{d^m}{dt^m} f(t), & q = m. \end{cases} \tag{2}
$$

Definition 3 Let us consider the following general form of arbitrary-order differential equation described by

$$ {}_{t_0}^{C}D_t^q x(t) = Ax(t) + Bu(t), \tag{3} $$

where $x \in \mathbb{R}^n$, $u \in \mathbb{R}^m$, and $A \in \mathbb{R}^{n\times n}$, $B \in \mathbb{R}^{n\times m}$, $n, m \in \mathbb{N}$, and ${}_{t_0}^{C}D_t^q x(t) = [{}_{t_0}^{C}D_t^q x_1(t), \ldots, {}_{t_0}^{C}D_t^q x_n(t)]^T$, $q \in (0, 1]$ is the fractional-order, t and t_0 are the limits of operation. When the system (3) is autonomous, it can be rewritten as ${}_{t_0}^{C}D_t^q x(t) = Ax(t)$, with $x(0) = x_0$, $0 < q < 1$, and $x \in R^n$. Then, the stability analysis of the autonomous system can be expressed according to following conditions [29]:

- The system ${}_{t_0}^{C}D_t^q x(t) = Ax(t)$ is *asymptotically stable* if and only if $|\arg(\lambda)| > \frac{q\pi}{2}$ for all eigenvalues (λ) of matrix A. In this scenario, the solution $x(t)$ tends to 0 like t^{-q}.
- The system ${}_{t_0}^{C}D_t^q x(t) = Ax(t)$ is *stable* if and only if $|\arg(\lambda)| \geq \frac{q\pi}{2}$ for all eigenvalues (λ) of matrix A obeying that the critical eigenvalues must satisfy $|\arg(\lambda)| = \frac{q\pi}{2}$ and have geometric multiplicity of one.

Definition 4 The general numerical solution of the fractional differential equation ${}_aD_t^q w(t) = f(w(t), t)$ can be expressed as [29]:

$$ w(t_k) = f\left(w(t_{k-1}), t_{k-1}\right) h^q - \sum_{j=1}^{k} c_j^{(q)} w(t_{k-j}), \tag{4} $$

with $k = 1, 2, \ldots, n$, $n = \frac{T_f}{h}$, $n \in \mathbb{N}$, h the time step, and $c_j^{(q)}$ are binomial coefficients given by $c_0^{(q)} = 1$, $c_j^{(q)} = \left(1 - \frac{1+q}{j}\right) c_{j-1}^{(q)}$.

The numerical algorithm in Definition 4 is based on the fact that for a wide class of functions, the three definitions, Caputo, Riemann-Liouville and Grünwald-Letnikov, are equivalent under the conditions [13].

3 Fractional-Order Multi-scroll Chaotic Systems Based on PWL Functions

Based on the chaotic system proposed in [30], we introduce the multi-scroll chaotic system with variable-order given by

$$ \begin{aligned} {}_{t_0}^{C}D_t^{q(t)} x(t) &= y(t), \\ {}_{t_0}^{C}D_t^{q(t)} y(t) &= z(t), \\ {}_{t_0}^{C}D_t^{q(t)} z(t) &= -\alpha x(t) - \beta y(t) - \gamma z(t) + \phi f(x(t); k_s, h_s, p_s, q_s), \end{aligned} \tag{5} $$

Table 1 Equilibrium points (EP_i) of the system (5)

Function saturated series	Equilibrium points (EP_i)
$f(x; k_s, h_s, p_s, q_s) = k_s(x - ih_s) + 2ik_s$	$EP_1 = (0, 0, 0)$, $EP_{4,5} = (\pm 2k_s, 0, 0)$
$f(x; k_s, h_s, p_s, q_s) \neq k_s(x - ih_s) + 2ik_s$	$EP_{2,3} = (\pm k_s, 0, 0)$, $EP_{6,7} = (\pm 3k_s, 0, 0)$

where ${}_{t_0}^{C}D_t^{q(t)}$ is variable-order Caputo's derivative operator determined by definition (2), $q(t)$ variable-order; $x(t), y(t), z(t)$ are the state variables, $\alpha = 2$, $\beta = 1$, $\gamma = 0.6$, $\phi = 2$, and $f(x; k_s, h_s, p_s, q_s)$ is a PWL function, which consists of a set of linear relationships valid in different regions as follows:

$$f(x; k_s, h_s, p_s, q_s) = \begin{cases} (2q_s + 1)k_s, & x > k_s h_s + 1, \\ k_s(x - i_s h_s) + 2ik_s, & |x - ih_s, \leq 1|, \\ & -p_s \leq i \leq q_s, \\ (2i + 1)k_s, & ih_s + 1 < x < (i + 1)h_s - 1, \\ & -p_s \leq i \leq q_s - 1, \\ -(2q_s + 1)k_s, & x < -p_s h_s - 1, \end{cases} \quad (6)$$

where $k_s = 1$ and $h_s = 2k_s$ are the multiplicative factors for the slopes and saturated regions, $i \in \mathbb{Z}$, and x is the state variable. When $q_s = p_s = 1$, we obtain a PWL function with 7 segments (four saturated plateaus and three slopes), to generate a 4-scroll chaotic attractor in 1D orientation on phase space. The equilibrium points (EP_i) of system (5) are shown in Table 1. In particular, when the PWL function $f(x; k_s, h_s, p_s, q_s) \neq k(x - ih_s) + 2ik_s$ with $i = 1, 2, 3$ and $k_s = 1$, the roots of the equation (5), for the equilibrium points $EP_{2,3}$ and $EP_{6,7}$ are: $\lambda_1 = -1.1833$ and $\lambda_{2,3} = 0.2916 \pm 1.2669i$ respectively. In this manner, the minimum fractional-order q, where the multi-scroll system may present a chaotic behavior is $q > \frac{2}{\pi}\left(\arctan\left(\frac{|\pm 1.2669i|}{0.2916}\right)\right)$, i.e., $q > 0.8560$.

4 Numerical Analysis of the Variable-Order-Based Memory

For simulation purposes, we herein use the numerical algorithm (4). Thus, the solution of system (5) with variable-order is given in (7). The idea is that the elements of the summation operation will be adapted for short-memory and full-memory implications.

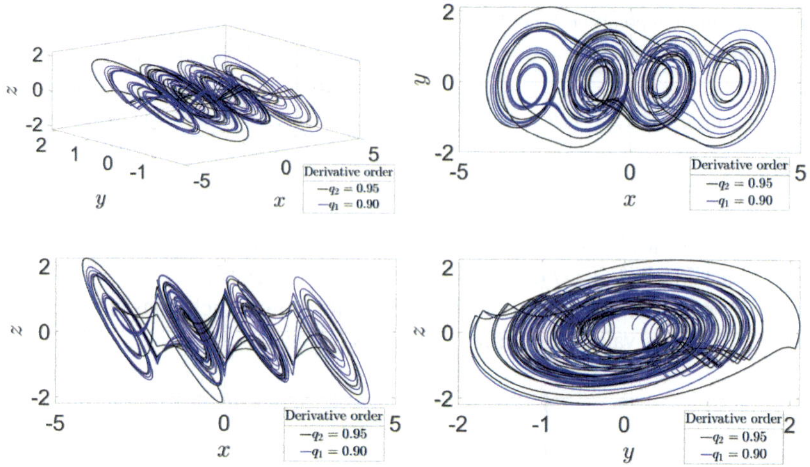

Fig. 1 Multi-scroll chaotic attractor of system (5) applying short-memory approach with variable order: $q_1 \in [t_0, t_1]$, and $q_2 \in [t_1, t_2]$

$$x(t_k) = [y(t_{k-1})]h^{q(t)} - \sum_{j=1}^{k} c_j^{q(t)} x(t_{k-j}),$$

$$y(t_k) = [z(t_{k-1})]h^{q(t)} - \sum_{j=1}^{k} c_j^{q(t)} y(t_{k-j}),$$

$$z(t_k) = [-\alpha x(t_{k-1}) - \beta y(t_{k-1}) - \gamma z(t_{k-1}) + \phi f(x(t_{k-1}))]h^{q(t)} - \sum_{j=1}^{k} c_j^{q(t)} z(t_{k-j}). \tag{7}$$

where x, y, z are the state variables, $\alpha = 2, \beta = 1, \gamma = 0.6, \phi = 2$, and $f(x; k_s, h_s, p_s, q_s)$ is the PWL function, h is the integration step, $q(t)$ is the variable-order, c_j are binomial coefficients, and k is the number of iterations.

4.1 Short-Memory

With the short-memory term, we mean that the simulation $t \in [t_0, t_n]$ is divided by n intervals where each interval associates to a specific fractional-order. Considering the initial time t_0 with initial condition x_0, we have $t \in [t_0, t_1]$ and the corresponding fractional differential equation ${}_{t_0}^{C}D_{t_1}^{q_0} x = f(x_0, t_0)$. For the next interval, only the memory from t_1 is considered without taking into account the initial condition t_0, i.e., the accumulated data are deleted. Thus, we obtain the updated interval $t \in [t_1, t_2]$ and fractional differential equation ${}_{t_1}^{C}D_{t_2}^{q_1} x = f(x_1, t_1)$; and so forth.

Figure 1 shows the resulting 4-scroll chaotic attractor using short-memory contributions with a constant PWL fractional-order q. In particular, the trajectory in *blue color* represents the chaotic oscillator with $q_1 = 0.90$, while in *black color* the other fractional-order $q_2 = 0.95$, with a simulation time of 300 s, and initial conditions $[0.1, 0.1, 0.1]$. The following pseudo-code gives the main instructions to implement the short-memory approach.

```
for i=1:NumberOfIterations
[t, y1]=GLMethod(Parameters, Order, AuxiliaryTime, ...
     InitialConditions);
cond=[y1(end,1), y1(end,2), y1(end,3)];
q_2=q+0.05;
Order=[q_2,q_2, q_2];
if (i==1)
yn=y1;
else
yn=[yn;y1];
end
end
```

4.2 *Full-Memory*

In the full-memory approach, we want to preserve all the values for the index of the summation in eq. (7). In this manner, the kth solution will depend on the whole previous values, including initial conditions $k(0)$. As a result, all the previously generated data since t_0 are preserved and used to compute the next values for the new fractional-order. Considering the initial time t_0 with initial condition x_0, we have $t \in [t_0, t_1]$ and the corresponding fractional differential equation ${}_{t_0}^{C}D_{t_1}^{q_0}x(t) = f(x_0, t_0)$. For the next interval, the full memory from t_0 is preserved, i.e., the data are accumulated as time evolves. Thus, we obtain the updated interval $t \in [t_1, t_2]$ and fractional differential equation ${}_{t_1}^{C}D_{t_2}^{q_1}x(t) = f(x_0, t_0)$. Next, we will obtain ${}_{t_2}^{C}D_{t_3}^{q_2}x(t) = f(x_0, t_0)$ with $t \in [t_2, t_3]$, and so forth. Based on this principle, we simulate again the multi-scroll chaotic system (5) with a constant PLW fractional-order $q_1 = 0.90$ for $t \in [0, 150\text{s}]$ and $q_2 = 0.95$ for $t \in [150\text{s}, 300\text{s}]$, and initial conditions $[0.1, 0.1, 0.1]$, as shown in Fig. 2. Similarly to the case of short-memory in the previous subsection, the following pseudo-code gives the main instructions to implement the full-memory approach:

```
function [t, Y]=GLMethod(parameter, order, simutime, Y0)
for i=2:n
    if i≤15000
        q=0.90;
    x(i)=y(i-1)*h^q - (memoria(x, c1, i));
    y(i)=z(i-1)*h^q - (memoria(y, c2, i));
    z(i)=(-alpha*x(i-1)-beta*y(i-1)-gamma*z(i-1)+...
        w*f_xpwl(x(i-1)))*h^q - (memoria(z, c3, i));
    else
        q=0.95;
    x(i)=y(i-1)*h^q - (memoria(x, c1, i));
    y(i)=z(i-1)*h^q - (memoria(y, c2, i));
    z(i)=(-alpha*x(i-1)-beta*y(i-1)-gamma*z(i-1)+...
        w*f_xpwl(x(i-1)))*h^q - (memoria(z, c3, i));
    end
end
```

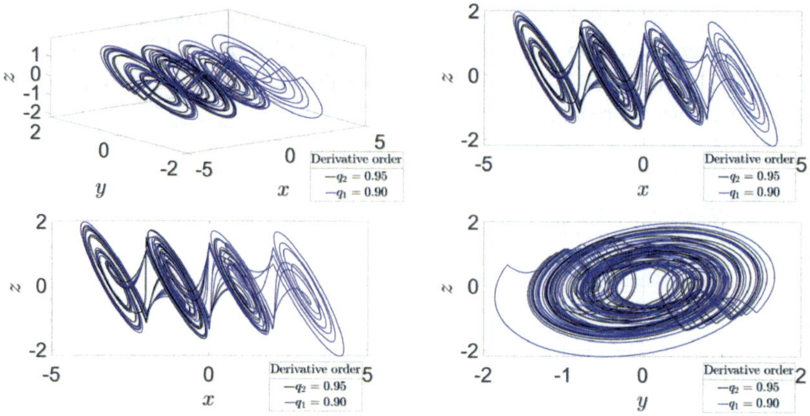

Fig. 2 Multi-scroll chaotic attractor of system (5) applying full-memory approach with variable-order: $q_1 \in [t_0, t_1]$, and $q_2 \in [t_1, t]$

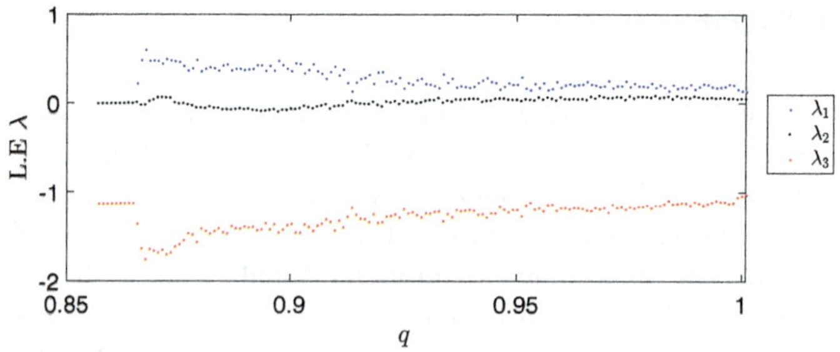

Fig. 3 The spectrum of the Lyapunov exponents for the system (5) with $q \in [0.8570, 0.9999]$

For the simulated chaotic attractors, we compute the Lyapunov exponent spectrum to demonstrate the chaos behavior. Figure 3 shows the Lyapunov exponents for $q \in [0.8570, 0.9999]$ based on the Wolf's algorithm [31]. The results confirm that in the interval $q \in [0.8656, 0.9999]$ there is a system of chaotic dynamics with a strange three-dimensional attractor.

4.3 Discussion

In Fig. 4, the time evolution $(x(t), y(t), z(t))$ of multi-scroll system (5) was computed with both short-memory and full-memory proposed approaches. While the evolution with short-memory is represented by the *blue color*, the full-memory implications are in *red color*. We observe that both solutions have a similar evolution due to the memory contributions of the fractional-order ($q = 0.9$) is the same for both aproaches during the first interval [0, 150s]. However, the evolution diverges for the

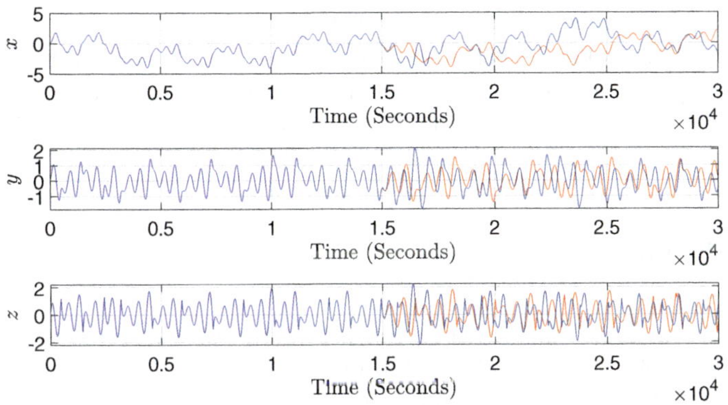

Fig. 4 Time evolution for (x, y, z) of variable-order multi-scroll chaotic system (5) with the short-memory (blue) and full-memory (red) proposed approaches

second time interval [150s, 300s]. The reason is that each approach (short-memory and full-memory) has different characteristics. For the short-memory-based chaotic system, the initial condition in the second interval is changed to the latter value of $(x(150), y(150), z(150))$. It means that previous data are deleted. In the full-memory-based chaotic system, the initial conditions remains fixed to its original value $(x(0), y(0), z(0))$, and all previous data of the state-variables vector are accumulated with the current and next solutions.

In Fig. 5, we plot the solution of (5) for constant orders $q = 0.9$ and $q = 0.95$, and variable-order $q(t) = [q(t_0), q(t_1)]$ with short-memory order and full-memory, respectively. As expected, the variable-order oscillator with short-memory (blue trajectory) has a strong correlation with constant order $q = 0.90$ when $t \in [0, 150]$,

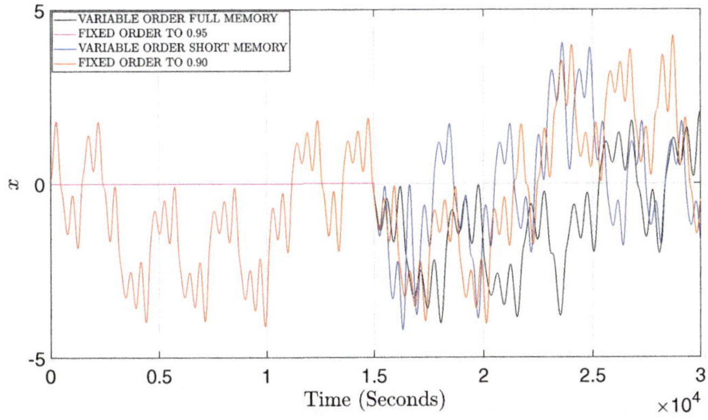

Fig. 5 Time evolution for state-variable x of multi-scroll chaotic system (5) with: constant order $q = 0.9$ (red), constant order $q = 0.95$ (pink), variable-order $q(t)$ with short-memory (blue), and variable-order $q(t)$ with full-memory (black)

and with $q = 0.95$ when $t \in [150, 300]$. On the other hand, the trajectory with full-memory (black color) has a behavior that diverges from the others once $t \geq 150$. This is the expected evolution because the full-memory approach considers all past values of $q = 0.9$ to compute the solutions of recent iterations with $q = 0.95$.

5 Conclusion

Based on the variable-order theory, in this paper the numerical analysis of a multi-scroll chaotic system using short- and full-memory implications has been presented. In particular, the proposed approach considered two scenarios. The first is related to splitting the simulation time into intervals with a specific fractional-order discarding past values, whereas the latter computes the solution considering both past values from the previous fractional-order and the current values for the new fractional-order. The numerical simulations confirmed that the short-memory approach is similar to compute the solution of the chaotic systems independently and just combine them. In contrast, the full-memory strategy evolved with a distinct dynamical behavior due to including not only current values but also the past values of the fractional-order.

Acknowledgements N. Medellín Neri thanks CONACYT-Mexico for the given support through the Master in Sciences scholarship, number: 1072803. The authors also thank VIEP-BUAP for the support through project 100519836-VIEP2021.

References

1. Wolf, A., Swift, J.B., Swinney, H.L., Vastano, J.A.: Determining Lyapunov exponents From a time series. Phys. 16D 285–317 (1985). North-Holland, Amsterdam, 18 October 1984
2. Wang, B., Zhong, S.M., Dong, X.C.: On the novel chaotic secure communication scheme design. Commun. Nonlinear Sci. Numer. Simul. **39**, 108–117 (2016)
3. Arroyo, D., Hernandez, F., Orúe, A.B.: Cryptanalysis of a classical chaos-based cryptosystem with some quantum cryptography features. Int. J. Bifurc. Chaos **27**(01), 1750004 (2017)
4. Zang, X., Iqbal, S., Zhu, Y., Liu, X., Zhao, J.: Applications of chaotic dynamics in robotics. Int. J. Adv. Rob. Syst. **13**(2), 60 (2016)
5. Sajid, M.: Recent developments on chaos in mechanical systems. Int. J. Theor. Appl. Res. Mech. Eng. **2**(3), 121–124 (2013)
6. Strogatz, S.: Nonlinear Dynamics and Chaos: With Applications to Physics, Biology, Chemistry, and Engineering (Studies in Nonlinearity) (2001)
7. Sun, H.G., Zhang, Y., Baleanu, D., Chen, W., Chen, Y.Q.: A new collection of real world applications of fractional calculus in science and engineering. Commun. Nonlinear Sci. Numer. Simul. **64**, 213–231 (2018)
8. Atangana, A., Vermeulen, P.D.: Analytical solutions of a space-time fractional derivative of groundwater flow equation. Abstr. Appl. Anal. **2014** (2014). Hindawi
9. Khan, A., Gómez-Aguilar, J.F., Abdeljawad, T., Khan, H.: Stability and numerical simulation of a fractional-order plant-nectar-pollinator model. Alex. Eng. J. **59**(1), 49–59 (2020)
10. Pandey, V.: Physical and geometrical interpretation of fractional derivatives in viscoelasticity and transport phenomena (2016)

11. Prathumwan, D., Sawangtong, W., Sawangtong, P.: An analysis on the fractional asset flow differential equations. Mathematics **5**(2), 33 (2017)
12. ElSafty, A.H., Tolba, M.F., Said, L.A., Madian, A.H., Radwan, A.G.: A study of the nonlinear dynamics of human behavior and its digital hardware implementation. J. Adv. Res. (2020)
13. Diethelm, K.: The Analysis of Fractional Differential Equations: An Application-Oriented Exposition Using Differential Operators of Caputo Type. Springer Science & Business Media (2010)
14. Garrappa, R., Kaslik, E., Popolizio, M.: Evaluation of fractional integrals and derivatives of elementary functions: overview and tutorial, vol. 7, Multidisciplinary Digital Publishing Institute (2019)
15. Podlubny, I.: Fractional differential equations. Mathematics in Science and Engineering, vol. 198. Academic, San Diego (1999)
16. Jacobs, B.A.: A new Grünwald-Letnikov derivative derived from a second-order scheme. Abstr. Appl. Anal. **2015** (2015). Hindawi
17. Wei, Y., Yin, W., Zhao, Y., Wang, Y.: A new insight into the Grünwald-Letnikov discrete fractional calculus. J. Comput. Nonlinear Dyn. **14**(4), 041008 (2019)
18. Jahanshahi, H., Munoz-Pacheco, J.M., Bekiros, S., Alotaibi, N.D.: A fractional-order SIRD model with time-dependent memory indexes for encompassing the multi-fractional characteristics of the COVID-19. Chaos Solitons & Fractals **143**, 110632 (2021)
19. Sun, H., Chen, W., Chen, Y.: Variable-order fractional differential operators in anomalous diffusion modeling. Phys. A **388**(21), 4586–4592 (2009)
20. Wu, G.C., Deng, Z.G., Baleanu, D., Zeng, D.Q.: New variable-order fractional chaotic systems for fast image encryption. Chaos: Interdiscip. J. Nonlinear Sci. **29**(8), 083103 (2019)
21. Coimbra, C.F.: Mechanics with variable-order differential operators. Ann. Phys. **12**(11–12), 692–703 (2003)
22. Deng, W., Lü, J.: Design of multidirectional multiscroll chaotic attractors based on fractional differential systems via switching control. Chaos: Interdiscip. J. Nonlinear Sci. **16**, 043120 (2006). https://doi.org/10.1063/1.2401061
23. Yalcin, M.E., Suykens, J.A.K., Vandewalle, J., Ozoguz, S.: Families of scroll grid attractors. Int. J. Bifurc. Chaos **12**(01), 23–41 (2002)
24. Lü, J., Han, F., Yu, X., Chen, G.: Generating 3-D multi-scroll chaotic attractors: a hysteresis series switching method. Automatica **40**(10), 1677–1687 (2004)
25. Wu, Y., Wang, C., Deng, Q.: A new 3D multi-scroll chaotic system generated with three types of hidden attractors. Eur. Phys. J. Spec. Top. **230**, 1863–1871 (2021)
26. Escalante-González, R.J., Campos-Canton, E.: Generation of self-excited and hidden multi-scroll attractors in multistable systems. Recent Trends in Chaotic, Nonlinear and Complex Dynamics, pp. 40–78 (2022)
27. Zhang, S., Zheng, J., Wang, X., Zeng, Z.: Multi-scroll hidden attractor in memristive HR neuron model under electromagnetic radiation and its applications. Chaos: Interdiscip. J. Nonlinear Sci. **31**(1), 011101 (2021)
28. Ahmad, S., Ullah, A., Akgül, A.: Investigating the complex behaviour of multi-scroll chaotic system with Caputo fractal-fractional operator. Chaos Solitons & Fractals **146**, 110900 (2021)
29. Petrás, I.: Fractional Order Non-Linear Systems. Modeling, Analysis and Simulation. Springer editorial (2011)
30. Deng, W., Lü, J.: Design of multidirectional multiscroll chaotic attractors based on fractional differential systems via switching control. Chaos: Interdiscip. J. Nonlinear Sci. **16**(4), 043120 (2006)
31. Wolf, A., Swift, J.B., Swinney, H.L., Vastano, J.A.: Phys. 16D 285–317 (1985). Department of Physics, University of Texas, Austin, Texas 78712, USA. North-Holland, Amsterdam, 18 October 1984

Effect of DEN-2 Virus on a Stage-Structured Dengue Model with Saturated Incidence and Constant Harvesting

Kunwer Singh Mathur and Bhagwan Kumar

Abstract The effect of the DEN-2 virus on dengue infection in children and adults plays an important role. This paper proposes and analyses a nonlinear stage-structured dengue model with a saturated incidence rate and constant harvesting with primary or secondary dengue infection. We analyze the local and global stability of disease-free and endemic equilibria of the system. The disease-free equilibrium is locally and globally asymptotic stable for $R_0 < 1$ and unstable for $R_0 > 1$. We also analyzed the stability of endemic equilibrium for $R_0 > 1$, but at $R_0 = 1$, the bifurcation exists, which is proven using the center manifold theory. Finally, numerical simulations are drawn to verify these theoretical results.

Keywords Dengue · Age-structure · Central manifold · Saturated incidence · Constant harvesting · Optimal control

1 Introduction

Dengue fever is the most common mosquito-borne acute arboviral (arthropod-borne viruses), caused by the bite of infected Aedes aegypti or Aedes Albopictus (also called Asian Tiger mosquito). Dengue virus infection is a leading cause of morbidity and mortality in the tropics and subtropics, mostly in urban and semi-urban areas of the world. The secondary infection of dengue in a person causes more severe complications. Sometimes, the secondary infection occurs in the form of Dengue haemorrhagic fever (DHF) or may have mild/moderate/high fever. It is also responsible for headaches, nausea, vomiting, pain in the muscles, bones, or common rashes on the skin, or most severe Dengue shock syndrome (DSS), which comprise rapid drops in blood pressure, a sudden weak pulse, suffering breathing problems, dilated pupils, cold, clammy skin, dry mouth, and restlessness. Once a patient goes into

K. S. Mathur (✉) · B. Kumar
Department of Mathematics and Statistics, Dr. Harisingh Gour Vishwavidyalaya, Sagar 470003, Madhya Pradesh, India
e-mail: ksmathur1709@gmail.com

S. Banerjee and A. Saha (eds.), *Nonlinear Dynamics and Applications*, Springer Proceedings in Complexity,
https://doi.org/10.1007/978-3-030-99792-2_101

DSS, it could be fatal within 12–24 h unless treatment is given immediately. There are four types of dengue virus (i.e., DEN-I, DEN-II, DEN-III, and DEN-IV), which are closely related to the serotypes of the virus that causes dengue infection. The four dengue serotypes, due to the infection, develop permanent immunity, probably lifelong to it, but this does not confer protective immunity against the other three serotypes. Thus, a person living in an endemic area can have as many as four dengue infections during his lifetime, one with each serotype. Moreover, the mosquitoes never recover from the infection and end their life during their infective period [7]. RT-PCR can detect the dengue viral genome in blood specimens up to day five simple precautions. There is no suitable vaccine and no immediate prospect of immunization prevention of the disease. Thus, one can say that dengue/severe dengue has no specific treatment, but the fatality rates can be lower below 1% by early detection of dengue infection and by providing proper medical care.

The transmission dynamics of infectious diseases through Mathematical modeling have been studied since a long time ago. However, the modeling of dengue disease still becomes a challenging question nowadays because of having more compartments in its mathematical model. Only a few researchers developed mathematical models for dengue disease transmission. Esteva and Vargas in [20], obtained the threshold value and the condition for the coexistence of two serotypes of dengue virus in a SIR model without including the age structure, which is not useful in the DHF outbreak of Thailand. The transmission of dengue fever is age-dependent (see, [9, 21]). Feng et al. analyzed the dengue transmission dynamics of the age–structured model [8]. Further, Suprianta is considered and investigated vaccination in a child age class models (see [1, 6]). Although, a severe manifestation of dengue infection develops in those who already have a primary infection [16]. This phenomenon should be considered in the modeling part to better present a realistic situation. Furthermore, the incidence rate will also play a key role in modeling infectious disease transmission. Many incidence rates are applicable in modeling. In the scenario of an epidemic, the bilinear incidence rate is also available, which is based on the law of mass action, which can't explain the disease dynamics [3]. Besides the bilinear incidence rate, the saturated incidence rate is more realistic in comparison to bilinear [5, 13, 18]. Therefore, we include the saturated incidence rate $\frac{\beta A(t)V(t)}{1+\alpha V}$ which tends to a level of saturation when V gets large, here α is the half-saturation constant.

Keeping in mind the above discussion, we will develop and analyze a dynamical system of nonlinear differential equations with two life stages in stage—structure form, and it is assumed that the adult has only primary dengue infection while the children under the age of 15 years have both primary and secondary infection. The presentation of the paper is as follows: In Sect. 2, a mathematical model is proposed, and in Sect. 3, the basic preliminary results including positivity and boundedness are proved. Further, the stability of equilibria is analyzed in Sect. 4 and an optimal control problem is discussed in Sect. 5. Finally, numerical simulations and conclusions are given respectively in Sects. 6 and 7.

2 Model Development

In this section, a mathematical model is proposed to study the effect of the DEN-2 virus in dengue disease transmission. The model is developed for two stages (e.g., children and adults) stage-structured population with a saturated incidence rate and constant harvesting. Let total density of host population be $N(t)$, in which, children population can be distinguised into two categories. Under the first category, children will not be infected by any dengue serotype DEN-I, DEN-III, DEN-IV, while in second category children may have an asymptomatic dengue infection. In this model, we will consider only second category's children population, which is divided into two compartment susceptible children $S(t)$ and infected children. Again, infected children is divided into primary infected $I_P(t)$ and secondary infected children $I_S(t)$. Here, the adult susceptible and infected populations are denoted by $A(t)$ and $I(t)$, respectively, all recovered popuplation including children and adult is represented by $R(t)$. Hence $N(t) = [S(t) + I_P(t) + I_S(t)] + [A(t) + I(t)] + R(t)$. let $U(t)$ be the susceptible vector population and $V(t)$ be the infected vector population such that $M(t) = U(t) + V(t)$. The function $\frac{\beta A(t)V(t)}{1+\alpha V}$ represents the saturated incidence rate. Further, it is assumed that:

(A1) The total susceptible children population is given by $S(t) = \nu S(t) + (1 - \nu)S(t)$; $(0 < \nu < 1)$, where $\nu S(t)$ represents the primary infection and $(1 - \nu)S(t)$ represents the secondary infection, when children population interact with infected (DEN-II) mosquitoes.

(A2) It is assumed that the probability of an adult being secondary infected is usually short. Therefore, we do not consider secondary infection for adults.

(A3) It is also assumed that only the DEN-II serotype is prevalent at time t.

Here, all parameters are assumed positive.

The complete transmission dynamics is given in Fig. 1, which leads to propose the following mathematical model:

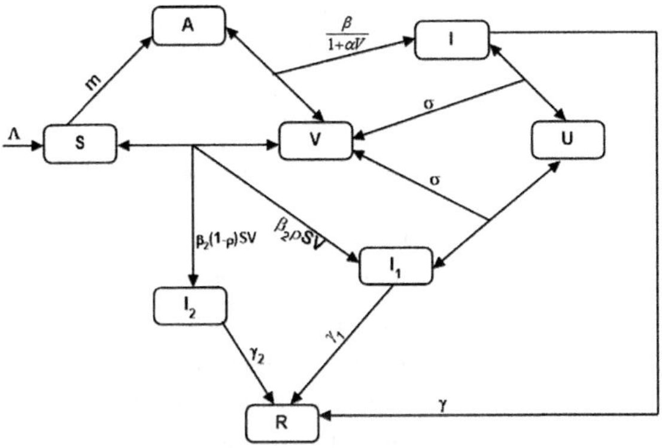

Fig. 1 The flow diagram for dengue disease transmission

$$
\begin{aligned}
\frac{dS}{dt} &= \Lambda - \beta_1 \nu SV - \beta_2(1-\nu)SV - mS - \mu_1 S,\\
\frac{dI_P}{dt} &= \beta_1 \nu SV - \gamma_1 I_P - \mu_1 I_P,\\
\frac{dI_S}{dt} &= \beta_2(1-\nu)SV - \gamma_2 I_S - \mu_1 I_S - \mu I_S,\\
\frac{dA}{dt} &= mS - \frac{\beta AV}{1+\alpha V} - \mu_1 A,\\
\frac{dI}{dt} &= \frac{\beta AV}{1+\alpha V} - \gamma I - \mu_1 I,\\
\frac{dR}{dt} &= \gamma_1 I_P + \gamma_2 I_S + \gamma I - \mu_1 R,\\
\frac{dU}{dt} &= \omega_1 - \sigma U(I_P + I) - \mu_2 U - pU,\\
\frac{dV}{dt} &= \sigma U(I_P + I) - \mu_2 V - qV.
\end{aligned}
\tag{2.1}
$$

The model possess the following non-negative initial conditions:

$$S(0) = S_0,\ I_P(0) = I_{10},\ I_S(0) = I_{20},\ I(0) = I_0,\ A(0) = A_0,\ R(0) = R_0,\ U(0) = U_0,\ V(0) = V_0.$$

Here, descriptions of other parameters are given in Table 1.

Table 1 Description of parameter used in model (2.1)

Parameter	Description
m	Maturation rate from child to adult
β_2	Transmission rate by which recovered children from prior asympomatics infection by heterologous serotype are further getting secondary infection by DEN-II serotype
γ_1	The rate of recovery for the primary infected children
γ_2	The rate of recovery for the secondary infected children
γ	The rate of recovery for adults
μ_1	The natural death rate for all human classes
μ_2	The natural death rate of mosquitoes
σ	Transmission rate of infection to mosquitoes by primary infected children and adults
μ	Disease-induced death rate for the secondary infected children
Λ	The constant recruitment rate
ω_1	The recruitment rate for vector
β_1	The rate of transmission of infection
β	The rate of transmission primary infection
p , q	Constant harvesting rate

3 Preliminary Results

3.1 Boundedness

Lemma 1 *The System* (2.1) *has a positively invariant and bounded solution in the closed set:*

$$\Gamma = \left\{(S, I_1, I_S, A, I, R, U, V) \in \mathbb{R}^8_+ : S + I_P + I_S + A + I + R \le \frac{\Lambda}{\mu_1}, U + V \le \frac{\omega_1}{\mu}\right\}.$$

Proof Consider the system of differential (2.1) in $\in \mathbb{R}^8_+$ as

$$\frac{dZ}{dt} = F(t, Z(t)), Z(0) = Z_0 \in \mathbb{R}^8_+ \tag{3.1}$$

Let $F_j(t, Z) \ge 0$ whenever $Z \in \mathbb{R}^8_+$ and $Z_j = 0$; $j = 1$ to 8. The system (3.1) has non-negative solution (see, [19]). Now, add the model equations of the host population and vector population separately, we can obtain that

$$\frac{dN}{dt} = \Lambda - \mu_1 N - \nu I_S, \frac{dM}{dt} \le \omega_1 - \mu M, \tag{3.2}$$

where $\mu = \min\{\mu_2 + p, \mu_2 + q\}$. Accordingly, $\limsup_{t\to\infty} N(t) = \frac{\Lambda}{\mu_1}$ and $\limsup_{t\to\infty} M(t) \le \frac{\omega_1}{\mu}$. Thus, Γ is positively invariant and bounded.

Since the state variables R and U are not playing any role in infection dynamics, hence we can exclude these variable. Thus, the following model is considered for further analysis:

$$\begin{aligned}
\frac{dS}{dt} &= \Lambda - \beta_1 \nu SV - \beta_2(1-\nu)SV - mS - \mu_1 S,\\
\frac{dI_P}{dt} &= \beta_1 \nu SV - \gamma_1 I_P - \mu_1 I_P,\\
\frac{dI_S}{dt} &= \beta_2(1-\nu)SV - \gamma_2 I_S - \mu_1 I_S - \mu I_S,\\
\frac{dA}{dt} &= mS - \frac{\beta AV}{1+\alpha V} - \mu_1 A,\\
\frac{dI}{dt} &= \frac{\beta AV}{1+\alpha V} - \gamma I - \mu_1 I,\\
\frac{dV}{dt} &= \sigma\left(\frac{\omega_1}{\mu_2} - V\right)(I_P + I) - \mu_2 V - qV,
\end{aligned} \tag{3.3}$$

3.2 Existence of Equilibria

The disease-free state $E_0 = (\tilde{S}, \tilde{I_P}, \tilde{I_S}, \tilde{A}, \tilde{I}, \tilde{V})$ always exists, where

$$\tilde{S} = \frac{\Lambda}{m+\mu_1}, \; \tilde{I_P} = 0, \; \tilde{I_S} = 0, \; \tilde{A} = \frac{m\Lambda}{(m+\mu_1)\mu_1}, \; \tilde{I} = 0, \; \tilde{V} = 0.$$

Let us denote $\beta_3 = (\beta_1 \nu + \beta_2(1-\nu))$. The endemic state $E^* = (S^*, {I_P}^*, {I_S}^*, A^*, I^*, V^*)$ is given as

$$S^* = \frac{\Lambda}{V^*\beta_3 + (m+\mu_1)}; \; I_P^* = \frac{\beta_1 \nu S^* V^*}{(\gamma_1+\mu_1)}; \; I_S^* = \frac{\beta_2(1-\nu)S^*V^*}{(\gamma_2+\mu_1+\nu)}; \; A^* = \frac{mS^*}{(\frac{\beta V^*}{1+\alpha V^*}+\mu_1)};$$

$$I^* = \frac{\beta A^* V^*}{(1+\alpha V^*)(\gamma+\mu_1)};$$

V^* is the root of following quadratic polynomial

$$\mathbf{a}V^{*2} + \mathbf{b}V^* + \mathbf{c} = 0, \tag{3.4}$$

$$\begin{aligned}
\mathbf{a} &= \sigma\nu\beta_1\Lambda(\beta+\alpha\mu_1)(\gamma+\mu_1) + \beta\beta_3(\mu_2+q)(\gamma_1+\mu_1)(\gamma+\mu_1) \\
&\quad + \mu_1\alpha\beta_3(\mu_2+q)(\gamma_1+\mu_1)(\gamma+\mu_1) \\
\mathbf{b} &= \sigma\beta_1\nu\Lambda\mu_1(\gamma+\mu_1) + m\beta\sigma\Lambda(\gamma_1+\mu_1) \\
&\quad + (\mu_2+q)(\gamma_1+\mu_1)(\gamma+\mu_1)(\mu_1\beta_3 + (m+\mu_1)(\beta+\mu_1\alpha)) \\
&\quad - \frac{\sigma\omega_1}{\mu}\beta_1\nu\Lambda(\gamma+\mu_1)(\beta+\mu_1\alpha) \\
\mathbf{c} &= \mu_1(\mu_2+q)(m+\mu_1)(\gamma_1+\mu_1)(\gamma+\mu_1)(1-R_0),
\end{aligned}$$

Let $V_\pm^*$ be the roots of (3.4) then

$$V_\pm^* = \frac{-\mathbf{b} \pm \sqrt{\mathbf{b}^2 - 4\mathbf{ac}}}{2\mathbf{a}}.$$

Since $\mathbf{a} > 0$. If $R_0 > 1$, $\mathbf{c} < 0$. Then it has one positive endemic state exits. If $R_0 < 1$, $\mathbf{c} > 0$. Then it has two real positive endemic state will exist provided $\mathbf{b} < 0$ and $\mathbf{b}^2 - 4\mathbf{ac} > 0$. And if $R_0 = 1$, $\mathbf{c} = 0$. Then one positive endemic state exist for $\mathbf{b} < 0$.

3.3 Basic Reproduction Number

We will determine the basic reproduction number by next generation approach [10], the jacobian matrix of the system (3.3)

$$F = \begin{pmatrix} 0 & 0 & 0 & \beta_1 \nu \tilde{S} \\ 0 & 0 & 0 & \beta_2(1-\nu)\tilde{S} \\ 0 & 0 & 0 & \dfrac{\beta \tilde{A}}{(1+\alpha \tilde{V})^2} \\ \dfrac{\sigma\omega_1}{\mu} & 0 & \dfrac{\sigma\omega_1}{\mu} & 0 \end{pmatrix}, V = \begin{pmatrix} \gamma_1+\mu_1 & 0 & 0 & 0 \\ 0 & \gamma_2+\mu_1+\nu & 0 & 0 \\ 0 & 0 & \gamma+\mu_1 & 0 \\ 0 & 0 & 0 & \mu_2+q \end{pmatrix}$$

clearly at E_0 the next generation matrix is

$$FV^{-1} = \begin{pmatrix} 0 & 0 & 0 & \dfrac{\beta_1 \nu \Lambda}{(m+\mu_1)(\mu_2+q)} \\ 0 & 0 & 0 & \dfrac{\beta_2(1-\nu)\Lambda}{(m+\mu_1)(\mu_2+q)} \\ 0 & 0 & 0 & \dfrac{\beta m \Lambda}{\mu_1(m+\mu_1)(\mu_2+q)} \\ \dfrac{\sigma\omega_1}{\mu(\gamma_1+\mu_1)} & 0 & \dfrac{\sigma\omega_1}{\mu(\gamma+\mu_1)} & 0 \end{pmatrix}.$$

We known that the largest eigenvalue of FV^{-1} is the basic reproduction number R_0. Which is computed as

$$R_0 = \frac{\sigma \Lambda \omega_1}{\mu(\mu_2+q)(m+\mu_1)}[\frac{\beta m}{\mu_1(\gamma+\mu_1)} + \frac{\nu\beta_1}{(\gamma_1+\mu_1)}].$$

4 Stability Analysis

Theorem 1 *The disease-free state is locally and globally asymptotically stable for* $R_0 < 1$ *and unstable for* $R_0 > 1$.

Proof The jacobian matrix[J] for the system (3.3) at E_0 is

$$J[E_0]=\begin{pmatrix} -(m+\mu_1) & 0 & 0 & 0 & 0 & -\dfrac{\beta_3\Lambda}{m+\mu_1} \\ 0 & -(\gamma_1+\mu_1) & 0 & 0 & 0 & \dfrac{\beta_1\nu\Lambda}{m+\mu_1} \\ 0 & 0 & -(\gamma_2+\mu_1+\nu) & 0 & 0 & \beta_2(1-\nu)\dfrac{\Lambda}{m+\mu_1} \\ m & 0 & 0 & -\mu_1 & 0 & -\dfrac{\beta m\Lambda}{\mu_1(m+\mu_1)} \\ 0 & 0 & 0 & 0 & -(\gamma+\mu_1) & \dfrac{\beta m\Lambda}{\mu_1(m+\mu_1)} \\ 0 & \dfrac{\omega_1\sigma}{\mu} & 0 & 0 & \dfrac{\omega_1\sigma}{\mu} & -(\mu_2+q) \end{pmatrix}$$

The eigenvalue of $[J(E_0)]$ are $-m-\mu_1$, $-\gamma_2-\mu_1-\nu$, $-\mu_1$ and other all the eigenvalue are given cubic polynomial $\lambda^3+B_1\lambda^2+B_2\lambda+B_3$. $Where$

$$B_1=2\mu_1+\gamma_1+\gamma+\mu_2+q,\ B_2=(\mu_2+q)(2\mu_1+\gamma_1+\gamma)$$
$$+(\gamma_1+\mu_1)(\gamma+\mu_1)-\frac{\sigma\omega_1}{\mu\mu_1}(\beta_1\nu\tilde{S}+\beta\mu_1\tilde{A})$$
$$B_3=(\mu_2+q)(\gamma+\mu_1)(\gamma_1+\mu_1)(1-R_0).$$

It is clear that $B_1>0\ and B_3>0\ for\ R_0<1.\ And$

$$B_1B_2-B_3=(2\mu_1+\gamma_1+\gamma+\mu_2+q)(\mu_2+q)(2\mu_1+\gamma_1+\gamma)$$
$$+(\gamma_1+\mu_1)(\gamma+\mu_1))+(\gamma+\mu_1)(\gamma_1+\mu_1)(\mu_2+q)(1-R_0)$$

is positive if $R_0<1$. By Routh-Hurwitz criterion the polynomial has eigenvalue with negative real part if $R_0<1$. This is showing the local stability at E_0. Now by Descartes Rule of signs it has one positive eigenvalue if $R_0>1$. Hence, at E_0 will be Unstable if $R_0>1$. Next we proof the system has global stability at E_0.
Let $L(I_P,I,V)=\frac{\sigma\Lambda I_P}{(\gamma_1+\mu_1)}+\frac{\sigma\Lambda I}{(\gamma+\mu_1)}+(\mu_2+q)V$ be the positive definite function

$$\dot{L}=\frac{\sigma\Lambda}{\gamma_1+\mu_1}(\beta_1\nu SV-\gamma_1I_P-\mu_1I_P)+\frac{\sigma\Lambda}{\gamma+\mu_1}(\frac{\beta AV}{1+\alpha V}-\gamma I-\mu_1 I)$$
$$+(\mu_2+q)(\sigma(\frac{2\omega_1}{\mu}-V)(I_P+I)-\mu_2V-qV)$$

$\dot{L}\le V\mu_2^2(R_0^2-1)\le 0$ for $R_0<1$. Therefore $L(I_P,I,V)$ is Lyapunov function for $R_0<1$. At $V=0$, $\dot{L}=0$. Because E_0 is the largest invariant set which is contains the subset in which V=0. by LaSalle's invariance principal [15], at E_0 Locally and globally stable for $R_0<1$.

Theorem 2 *The system* (3.3) *has a locally asymptotically stable endemic state* E^* *for* $R_0>1$ *and it has the forward bifurcation at* $R_0=1$.

Proof Let $\beta_1=\beta_1^c$ is bifurcation parameter corresponds to $R_0=1$,
$\beta_1^c=\frac{\gamma_1+\mu_1}{\nu}(\dfrac{\mu(\mu_2+q)(m+\mu_1)}{\sigma\Lambda\omega_1}-\dfrac{\beta m}{\mu_1(\gamma+\mu_1)})$

Let $\delta = (\delta_1, \delta_2, \delta_3, \delta_4, \delta_5, \delta_6)^T$ be a right eigenvector corresponding to the eigenvalue zero. It is given by

$$\begin{pmatrix} -(m+\mu_1) & 0 & 0 & 0 & 0 & -(\beta_1^c \nu + \beta_2(1-\nu))\tilde{S} \\ 0 & -(\gamma_1+\mu_1) & 0 & 0 & 0 & \beta_1^c \nu \tilde{S} \\ 0 & 0 & -(\gamma_2+\mu_1+\nu) & 0 & 0 & \beta_2(1-\nu)\tilde{S} \\ m & 0 & 0 & -\mu_1 & 0 & -\beta\tilde{A} \\ 0 & 0 & 0 & 0 & -(\gamma+\mu_1) & \beta\tilde{A} \\ 0 & \dfrac{\omega_1\sigma}{\mu} & 0 & 0 & \dfrac{\omega_1\sigma}{\mu} & -(\mu_2+q) \end{pmatrix} \begin{pmatrix} \delta_1 \\ \delta_2 \\ \delta_3 \\ \delta_4 \\ \delta_5 \\ \delta_6 \end{pmatrix} = 0$$

Solving above equation, the right eigenvector is

$$\left(\delta_1 = 0, \delta_2 = \frac{\beta_1^c \nu}{(\gamma_1+\mu_1))}, \delta_3 = 0, \delta_4 = 0, \delta_5 = \frac{\beta m}{\mu_1(\gamma_1+\mu_1)}, \delta_6 = \frac{(m+\mu_1)(\gamma+\mu_1)}{\Lambda(\gamma_1+\mu_1)}\right).$$

Further, the left eigenvector $\nu = (\nu_1, \nu_2, \nu_3, \nu_4, \nu_5, \nu_6)^T$ corresponding to the eigenvalue zero such that $\delta.\nu = 1$ is

$$\left(\nu_1 = 0, \nu_2 = \frac{\omega_1\sigma\Lambda}{(\mu_2+q)(m+\mu_1)(\gamma+\mu_1+\mu_2+q)}, \nu_3 = 0, \nu_4 = 0,\right.$$

$$\left.\nu_5 = \frac{\omega_1\sigma\Lambda\mu(\gamma_1+\mu_1)}{\mu(\mu_2+q)(m+\mu_1)(\gamma+\mu_1)(\gamma+\mu_1+\mu_2+q)}, \nu_6 = \frac{\mu\Lambda(\gamma_1+\mu_1)}{(\mu_2+q)(m+\mu_1)(\gamma+\mu_1+\mu_2+q)}\right).$$

Let $S = x_1$, $I_P = x_2$, $I_S = x_3$, $A = x_4$, $I = x_5$, $V = x_6$. The system (3.3) becomes,

$$\begin{aligned} \frac{dx_1}{dt} &= \Lambda - \beta_1\nu x_1 x_6 - \beta_2(1-\nu)x_1x_6 - mx_1 - \mu_1 x_1 := f_1 \\ \frac{dx_2}{dt} &= \beta_1\nu x_1x_6 - \gamma_1 x_2 - \mu_1 x_2 := f_2 \\ \frac{dx_3}{dt} &= \beta_2(1-\nu)x_1x_6 - \gamma_2 x_3 - \mu_1 x_3 - \mu x_3 := f_3 \\ \frac{dx_4}{dt} &= mx_1 - \frac{\beta x_4 x_6}{1+\alpha x_6} - \mu_1 x_4 := f_4 \\ \frac{dx_5}{dt} &= \frac{\beta x_4 x_6}{1+\alpha x_6} - \gamma x_5 - \mu_1 x_5 - p x_5 := f_5 \\ \frac{dx_6}{dt} &= \sigma(\frac{\omega_1}{\mu_2} - x_6)(x_2+x_5) - \mu_2 x_6 - q x_6 := f_6. \end{aligned} \tag{4.1}$$

The partial derivatives at (E_0) are

$$\frac{\partial^2 f_2}{\partial x_6 \partial \beta_1} = \frac{\nu\Lambda}{(m+\mu_1)}, \frac{\partial^2 f_5}{\partial x_6^2} = -\frac{2m\Lambda\alpha}{(m+\mu_1)\mu_1}, \frac{\partial^2 f_6}{\partial x_2 \partial x_6} = \frac{\partial^2 f_6}{\partial x_6 \partial x_2} = \frac{\partial^2 f_6}{\partial x_6 \partial x_5} = \frac{\partial^2 f_6}{\partial x_5 \partial x_6} = -\sigma$$

and remaining are zero. From Theorem 4.1 [4]. The coefficient **a** and **b** are computed as,

$$\mathbf{a} = 2v_6\delta_2\delta_6\frac{\partial^2 f_6}{\partial x_2 \partial x_6} + 2v_6\delta_5\delta_6\frac{\partial^2 f_6}{\partial x_5 \partial x_6} + v_5\delta_6\delta_6\frac{\partial^2 f_5}{\partial x_6 \partial x_6}, \mathbf{b} = v_2\delta_6\frac{\partial^2 f_2}{\partial x_6 \partial \beta_1}$$

Now substitue all the partial derivatives and left and right eigenvectors we get,

$$\mathbf{a} = -\frac{2\sigma(\gamma+\mu_1)}{\mu_1(\mu_2+q)(\gamma_1+\mu_1)(\gamma+\mu_1+\mu_2+q)}(\mu\mu_1\beta_1\nu + \mu\beta m + \omega_1 m\alpha),$$

$$\mathbf{b} = \frac{\omega_1\sigma\nu\Lambda(\gamma+\mu_1)}{(\mu_2+q)(m+\mu_1)(\gamma_1+\mu_1)(\gamma+\mu_1+\mu_2+q)}.$$

Clearly $\mathbf{a} < 0$ *and* $\mathbf{b} > 0$ is always. using Theorem 4.1 [4]. Forward bifurcation is possible and the E^* is found to be locally asymptotically stable for $R_0 > 1$.

5 Optimal Control Problem

In this section, we develop an optimal control problem for the model system (2.1), control the spread of an epidemic, it is imperative to propagate awareness amongst individuals, but a successful intervention strategy reduces the number of infective individuals with minimum cost [12] an effectual way of procuring the best strategy for information propagation is using optimal control theory [11, 14]

$$J[u(t)] = \int_0^T [z_1 I(t) + z_2 u^2(t)]dt, \tag{5.1}$$

$$\begin{aligned}
\frac{dS}{dt} &= \Lambda - \beta_1\nu SV - \beta_2(1-\nu)SV - mS - \mu_1 S,\\
\frac{dI_P}{dt} &= \beta_1\nu SV - \gamma_1 I_P - \mu_1 I_P,\\
\frac{dI_S}{dt} &= \beta_2(1-\nu)SV - \gamma_2 I_S - \mu_1 I_S - \mu I_S,\\
\frac{dA}{dt} &= mS - \frac{\beta AV}{1+\alpha V} - \mu_1 A,\\
\frac{dI}{dt} &= \frac{\beta AV}{1+\alpha V} - \frac{u(t)I}{1+\omega I} - \gamma I - \mu_1 I,\\
\frac{dR}{dt} &= \gamma_1 I_P + \gamma_2 I_S + \frac{u(t)I}{1+\omega I} + \gamma I - \mu_1 R,\\
\frac{dU}{dt} &= \omega_1 - \sigma U(I_P + I) - \mu_2 U - pU,\\
\frac{dV}{dt} &= \sigma U(I_P + I) - \mu_2 V - qV,
\end{aligned} \tag{5.2}$$

with initial condition

$$S(0) \geq 0,\ I_P(0) \geq 0,\ I_S(0) \geq 0,\ I(0) \geq 0,\ A(0) \geq 0,\ R(0) \geq 0, U(0) \geq 0,\ V(0) \geq 0$$

The coefficient z_1 and z_2 in the cost functional are balancing coefficients transforming the integral into currency expended. The first in the cost functional represent the cost due to infection caused by epidemic and last term represent cost associated with the implementation of awareness program. Quadratic expression of control indicates non-linear cost arising at high implementation level. find an optimal control $u^*(t)$ such that

$$J(u^*) = \min_{u \in \mathcal{U}} J(u) \tag{5.3}$$

where control set is defined as

$$Z = \{u(t) : 0 \leq u(t) \leq u_{\max};\ 0 \leq t \leq T,\ u(t) \text{ is Lebesgue measurable}\}.$$

Theorem 3 *There exist an optimal control* $u^* \in Z$ *such that* $J(u^*) = \min[J(u)]$ *corresponding to the control system* (5.1)–(5.2)

Proof The boundedness of solution of system (5.2) asserts the existence of solution to control system using results [17] Therefore, set of controls and corresponding state variables is nonempty. The control set is closed and convex by definition. The solution of system (5.1) are bounded above by a linear function in state and control. The integrand in the cost functional, $z_1 I(t) + z_2 u^2(t)$ is convex on control set Z. and also there exits $r_1, r_2 > 0$ *and* $m > 1$ such that, $z_1 I(t) + z_2 u^2(t) \geq r_1 + r_2 |u^2(t)|^m$ where r_1 depends upon the uper bound of $I(t)$ and $r_2 = z_2$ Hence, the existence of an optimal control is established.

5.1 Characterization of Optimal Control Function

The Pontryagin's Maximum Principle [2] can be used for the differential systems of adjoint variable and characterization of optimal control. we define the Hamiltoninan as,

$$\begin{aligned}
H(S, I_P, I_S, A, I, R, U, V, u, \lambda) &= L(S, I_P, I_S, A, I, R, U, V, u, \lambda) + \lambda_1 \dot{S} + \lambda_2 \dot{I_P} + \lambda_3 \dot{I_S} \\
+\lambda_4 \dot{A} + \lambda_5 \dot{I} + +\lambda_6 \dot{R} + \lambda_7 \dot{U} + \lambda_8 \dot{V} &= z_1 I(t) + z_2 u(t)^2 + \lambda_1 (\Lambda - \beta_1 \nu SV - \beta_2 (1-\nu) SV \\
-mS - \mu_1 S) + \lambda_2 (\beta_1 \nu SV - \gamma_1 I_P - \mu_1 I_P) &+ \lambda_3 (\beta_2 (1-\nu) SV - \gamma_2 I_S - \mu_1 I_S - \mu I_S) \\
+\lambda_4 (mS - \frac{\beta AV}{1+\alpha V} - \mu_1 A) &+ \lambda_5 (\frac{\beta AV}{1+\alpha V} - \frac{u(t) I}{1+\omega I} - \gamma I - \mu_1 I) \\
+\lambda_6 (\gamma_1 I_P + \gamma_2 I_S + \frac{u(t) I}{1+\omega I} &+ \gamma I - \mu_1 R) + \lambda_7 (\omega_1 - \sigma U (I_P + I) - \mu_2 U - pU) \\
&+ \lambda_8 (\sigma U (I_P + I) - \mu_2 V - qV)
\end{aligned} \tag{5.4}$$

where $\lambda = (\lambda_1, \lambda_2, \lambda_3, \lambda_4, \lambda_5, \lambda_6, \lambda_7, \lambda_8)$ is konow as adjoint variable. We obtain minimized cost functional subject to state variable for the given optimal control u^*

and state variable there exits adjoint variable λ_i satisfying the following canonical equations,

$$\begin{aligned}
\frac{d\lambda_1}{dt} &= (\beta_1 \nu V + m + \mu_1)\lambda_1 + (\beta_2(1-\nu)V - \beta_1 \nu V)\lambda_2 - \beta_2(1-\nu)V\lambda_3 - m\lambda_4,\\
\frac{d\lambda_2}{dt} &= (\gamma_1 + \mu_1)\lambda_2 - \gamma_1\lambda_6 - \sigma U\lambda_7 - \sigma U\lambda_8,\\
\frac{d\lambda_3}{dt} &= (\gamma_2 + \mu_1 + \mu)\lambda_3 - \gamma_2\lambda_6,\\
\frac{d\lambda_4}{dt} &= \left(\frac{\beta V}{1+\alpha V} + \mu_1\right)\lambda_4 - \frac{\beta V}{1+\alpha V}\lambda_5,\\
\frac{d\lambda_5}{dt} &= -z_1 + \left(\frac{u(t)}{(1+\omega I)^2} + \mu_1 + \gamma\right)\lambda_5 - \left(\frac{u(t)}{(1+\omega I)^2} + \gamma\right)\lambda_6 + \sigma U\lambda_7 - \sigma U\lambda_8,\\
\frac{d\lambda_6}{dt} &= \mu_1\lambda_6,\\
\frac{d\lambda_7}{dt} &= (\mu_2 + p)\lambda_7,\\
\frac{d\lambda_8}{dt} &= (\mu_2 + q)\lambda_8,
\end{aligned} \tag{5.5}$$

with transversality condition $\lambda_i(T) = 0$, i=1 to 8. The Hamiltonian is minimized with respect to u at the optimal value u^* Now from the optimality condition, we have $\frac{\partial H}{\partial u} = 0$ at $u = u^*$, so

$$H = z_2 u^2 + \lambda_5\left(\frac{-u(t)I}{1+\omega I}\right) + \lambda_6\left(\frac{u(t)I}{1+\omega I}\right) + \text{terms without u(t)}$$

differentiating H with respect to u gives:

$$\frac{\partial H}{\partial u} = 2z_2u^* - (\lambda_5 - \lambda_6)\frac{I}{1+\omega I^*} = 0. \text{ Thus we get, } u^* = \frac{I}{2z_2(1+\omega I^*)}(\lambda_5 - \lambda_6)$$

Now from the above findings along with the characteristics of control set U, we have

$$u^* = \begin{cases} 0 & \text{if } \frac{I^*}{2z_2(1+\omega I^*)} < 0,\\ \frac{I^*}{2z_2(1+\omega I^*)} & \text{if } 0 \le \frac{I^*}{2z_2(1+\omega I^*)} \le 1,\\ 1 & \text{if } \frac{I^*}{2z_2(1+\omega I^*)} > 1, \end{cases}$$

6 Numerical Simulation

We discuss the numerical simulation of the model as the parameters in the system (2.1) are varified in MATLAB using ODE45. We assume that arbitrary initial condition $Y_1 = [0.8, 0.5, 0.4, 0.07, 0.1, 0.5, 20, 5]$, $Y_2 = [0.6, 0.3, 0.6, 0.1, 0.6, 0.8, 22,$

12], $Y_3 = [0.4, 0.9, 0.2, 0.4, 0.8, 5, 30, 7]$, $Y_4 = [1.5, 0.1, 0.9, 2.9, 0.8, 2, 40, 20]$ and parameters value are $m = 0.00183$, $\beta_2 = 0.7$, $\gamma_1 = 0.3$, $\gamma_2 = 0.1428$, $\gamma = 0.5$, $\mu_1 = 0.0000457$, $\mu_2 = 0.0714$, $\sigma = 0.05$, $\mu = 0.001$, $\Lambda = 0.00001$, $\omega_1 = 50$, $\nu = 0.5$, $\beta_1 = 0.005$, $\beta = 0.003$, $\alpha = 0.09$, $p = 0.0001$, $q = 0.0001$. The basic reproducation number $R_0 = 0.76(< 1)$. So, E_0 exits and it is calculated as $E_0(S, I_P, I_S, A, I, R, U, V) = (0.0345, 0, 0, 0.1234, 0, 0, 512.280, 0)$. We can see from the projections of phase plot in $I_P - I_S - I$ hyperplanes drawn in Figs. 2 and 3, that the solution trajectories converge to the state E_0. This is showing that global stability at E_0. Now $\beta_1 = 0.1, \beta = 0.05$ the basic reproduction number $R_0 = 2.1432(> 1)$. So, E^* will exit. And it is calculated as $E^*(S, I_P, I_S, A, I, R, U,$

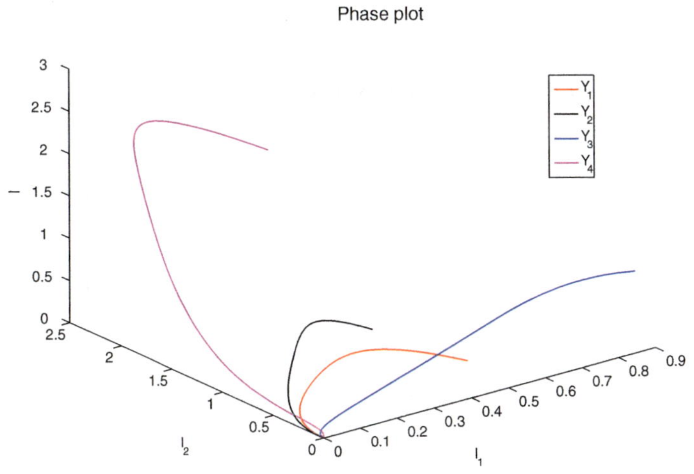

Fig. 2 A 3D $I_P - I_S - I$ this is showing that the convergence to E_0

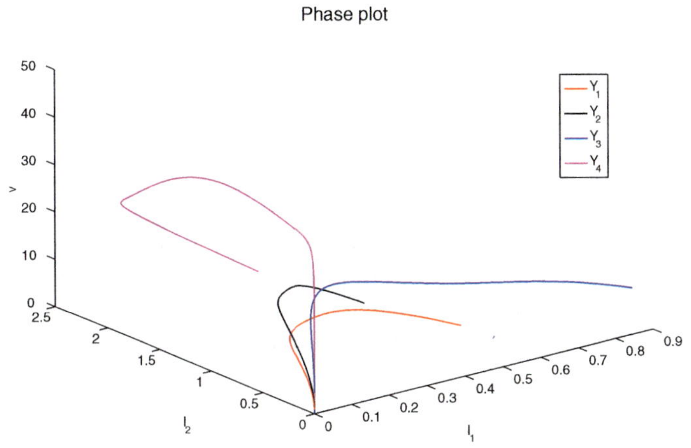

Fig. 3 A 3D $I_P - I_S - V$ this is showing that the convergence to E_0

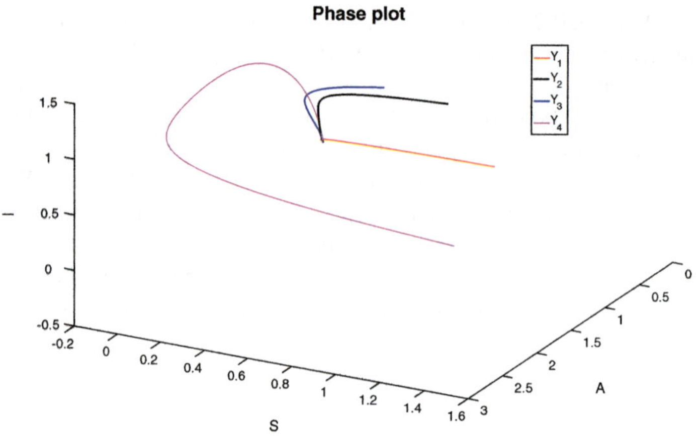

Fig. 4 A 3D $A - S - I$ this is showing that the convergence to E^*

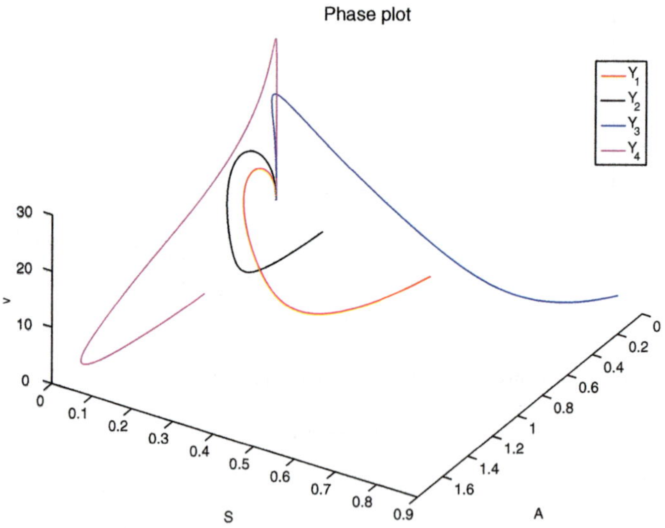

Fig. 5 A 3D $A - S - V$ this is showing that the convergence to E^*

$V) = (0.00463, 0.000001325, 0.00008542, 0.0022694, 0.0000012546, 0.24961, 512.267, 0.0032145)$. The projections of phase plot in I_P, I_S, V hyperplanes have been drawn in Figs. 4 and 5. The solution trajectories are found to converge to the state E^*, so the stability of the endemic equilibrium state.

7 Conclusion

We have constructed a stage-structured dengue model with a saturated incidence rate and constant harvesting to see the effect of the DEN-2 virus on the primary/secondary infection in children and adults. We have calculated the basic reproduction number R_0. It has been analyzed that the disease-free equilibrium point is locally and globally asymptotically stable for $R_0 < 1$ and unstable for $R_0 > 1$. Further, locally asymptotically stable endemic state E^* for $R_0 > 1$ and it has the forward bifurcation at $R_0 = 1$. An optimal control system has been constructed and by using the pontryagin maximum principle, we obtain the minimized cost function. From the numerical simulation, we verified some analytical results of the theorem. In disease dynamics, a basic reproduction number plays an essential role in controlling disease outbreaks. Here R_0 is the function of harvesting coefficient q only and not dependent on p, which suggests that the harvesting of susceptible mosquitos has no impact on the dynamics. However, the harvesting coefficient q plays a crucial role in controlling dengue infection. Hence, the Govt. policies can be regulated through the basic reproduction number obtained in this paper, and the scientists can make better policies to stop the outbreak. Moreover, the endemic equilibrium point became saturated and depended on the half-saturation constant α. Finally, it is concluded that the saturated incidence rate and vector harvesting rate have a more significant impact on the system's dynamics, which determines a more realistic analysis and is more practical from an epidemic point of view.

Acknowledgements The work is carried out under the project sponsored by Science and Engineering Research Board (SERB), New Delhi, India with file no. EMR/2017/001234.

References

1. Alera, M.T., Srikiatkhachorn, A., Velasco, J.M., Tac-An, I.A., Lago, C.B., Clapham, H.E., Fernandez, S., Levy, J.W., Thaisomboonsuk, B., Klungthong, C., et al.: Incidence of dengue virus infection in adults and children in a prospective longitudinal cohort in the philippines. PLoS Negl. Trop. Dis. **10**(2), e0004337 (2016)
2. Bellman, R.: The Mathematical Theory of Optimal Processes (1965)
3. Cai, L., Li, X., Ghosh, M., Guo, B.: Stability analysis of an HIV/AIDS epidemic model with treatment. J. Comput. Appl. Math. **229**(1), 313–323 (2009)
4. Castillo-Chavez, C., Song, B.: Dynamical models of tuberculosis and their applications. Math. Biosci. Eng. **1**(2), 361–404 (2004)
5. Chikaki, E., Ishikawa, H.: A dengue transmission model in thailand considering sequential infections with all four serotypes. J. Infection Dev. Ctries. **3**(09), 711–722 (2009)
6. Deen, J.L., Harris, E., Wills, B., Balmaseda, A., Hammond, S.N., Rocha, C., Dung, N.M., Hung, N.T., Hien, T.T., Farrar, J.J.: The who dengue classification and case definitions: time for a reassessment. Lancet **368**(9530), 170–173 (2006)
7. Enweronu-Laryea, C.C., Adjei, G.O., Mensah, B., Duah, N., Quashie, N.B.: Prevalence of congenital malaria in high-risk Ghanaian newborns: a cross-sectional study. Malaria J. **12**(1), 17 (2013)

8. Feng, W.J., Cai, L.M., Liu, K.: Dynamics of a dengue epidemic model with class-age structure. Int. J. Biomath. **10**(08), 1750109 (2017)
9. Guzmán, M.G., Kouri, G., Bravo, J., Valdes, L., Susana, V., Halstead, S.B.: Effect of age on outcome of secondary dengue 2 infections. Int. J. Infect. Diseas. **6**(2), 118–124 (2002)
10. Jones, J.H.: Notes on r0. Department of Anthropological Sciences, Califonia (2007)
11. Kandhway, K., Kuri, J.: How to run a campaign: optimal control of sis and sir information epidemics. Appl. Math. Comput. **231**, 79–92 (2014)
12. Kar, T., Jana, S.: A theoretical study on mathematical modelling of an infectious disease with application of optimal control. Biosystems **111**(1), 37–50 (2013)
13. Khan, M.F., Alrabaiah, H., Ullah, S., Khan, M.A., Farooq, M., bin Mamat, M., Asjad, M.I.: A new fractional model for vector-host disease with saturated treatment function via singular and non-singular operators. Alex. Eng. J. **60**(1), 629–645 (2021)
14. Kumar, B.: Role of optimal control in the smoking epidemic model with media awareness. Int. Res. J. Moderniz. Eng. Technol. Sci. **2**(10), 762–768 (2020)
15. LaSalle, J.P.: The Stability of Dynamical Systems, vol. 25. Siam (1976)
16. L'Azou, M., Moureau, A., Sarti, E., Nealon, J., Zambrano, B., Wartel, T.A., Villar, L., Capeding, M.R., Ochiai, R.L.: Symptomatic dengue in children in 10 Asian and Latin American countries. New Engl. J. Med. **374**(12), 1155–1166 (2016)
17. Lukes, D.L.: Differential Equations: classical to Controlled. Elsevier (1982)
18. Mathur, K.S., Narayan, P.: Dynamics of an [... formula...] epidemic model with vaccination and saturated incidence rate. Int. J. Appl. Comput. Math. **4**(5) (2018)
19. Perko, L.: Differential Equations and Dynamical Systems, vol. 7. Springer Science & Business Media (2013)
20. Pongsumpun, P., Tang, I.: Transmission of dengue hemorrhagic fever in an age structured population. Math. Comput. Modell. **37**(9–10), 949–961 (2003)
21. Sharp, T.W., Wallace, M.R., Hayes, C.G., Sanchez, J.L., DeFraites, R.F., Arthur, R.R., Thornton, S.A., Batchelor, R.A., Rozmajzl, P.J., Hanson, R.K., et al.: Dengue fever in us troops during operation restore hope, Somalia, 1992–1993. Am. J. Trop. Med. Hyg **53**(1), 89–94 (1995)

Modulational Instability Analysis in An Isotropic Ferromagnetic Nanowire with Higher Order Octopole-Dipole Interaction

T. Pavithra, L. Kavitha, Prabhu, and Awadesh Mani

Abstract We investigate the nonlinear spin excitation in a Heisenberg ferromagnetic nanowire with the higher order octupole-dipole interaction. In this study, the nonlinear dynamical equation of motion is obtained by a semi-classical limit employing Glauber's coherent state analysis along with Holstein-Primakoff bosonic representation for the spin operators. In the framework of linear stability analysis, we employ the Modulational Instability for the ferromagnetic nanowire with octupole-dipole interaction. It is found that the occurance of octupole-dipole exchange interaction systematically helps to localize the excitation which improve the growth of high amplitude localized robust solitons in the ferromagnetic nanowire lattice.

Keywords Nonlinear excitation · Octupole-dipole interaction · Modulational instability analysis · Heisenberg ferromagnetic nanowire

1 Introduction

It is more interesting to study the one dimensional classical Heisenberg ferromagnetic nanowire with several types of magnetic interactions which reveals the soliton spin excitation and integrability characteristics [1, 2]. In this view, the parallel symmetry of spins in the ferromagnetic nanostructures, especially in nanowire is controlled by bilinear spin-spin exchange interaction and higher order exchange interactions like magnetic octupole-dipole and biquadratic exchange interactions found to play a predominant role in recent years [3–6]. The intergability behaviour of ferromag-

T. Pavithra · L. Kavitha (✉)
Department of Physics, Central University of Tamil Nadu, Thiruvarur 610 005, Tamil Nadu, India
e-mail: lkavitha@cutn.ac.in

Prabhu
Department of Chemistry, Periyar University, Salem 636 011, Tamil Nadu, India

A. Mani
Condensed Matter Physics Division, Indira Gandhi Centre for Atomic Research, Kalpakkam 603102, Tamil Nadu, India

S. Banerjee and A. Saha (eds.), *Nonlinear Dynamics and Applications*,
Springer Proceedings in Complexity,
https://doi.org/10.1007/978-3-030-99792-2_102

netic systems in the view of biquadratic interaction is significantly established in both classically and quantum mechanically. Moriya draw out the concept of super exchange interaction to incorporate the spin-orbit coupling in a Heisenberg spin of Hamiltonian along with the biquadratic interaction, there look to be a third degree exchange interaction found to be $(S_i \cdot S_{i+1})(S_{i+1} \cdot \vec{k})^2$. Here '$k$' is fixed to be a constant vector [7–10]. Substantially, this interaction was elucidated as octupole-dipole interaction expressing the hyperfine construction of Γ_3 ions in cubic symmetry and additionally performing as fascinating nonlinear dynamical model with higher order interactions having much attention in field of nonlinear physics. Mostly, the research works are only related with magnetic systems have been concerned to the nearest-neighbor (NN) exchange interaction [11] and only very few research works handle with the interaction of next-nearest neighbor (NNN) one. However, for numerous materials with complex structure which is essential to take into account of the NNN spin-spin interactions in the ferromagnetic system. Daniel et al. [12] has investigated the octupole-dipole interaction through Painleve singularity structure analysis, where they explored the integrable spin chain model with nonlinear spin excitation in the Heisenberg ferromagnet. Further, Bing Tang et al. [13] studied effect of octupole-dipole interaction and localization of quantum breather and soliton in the anisotropic ferromagnetic chain through quantum approach whereas they consider the localised Hatree states as quantum breathers. And also, Kavitha et al. [15] informatively studied the effective nonlinear excitation of spin in anisotropic one dimensional Heisenberg ferromagnetic lattice by semi-classical approach in the presence of octupole-dipole interaction where using soliton flipping they analyzed the magnetic switching process under the influence of octupole-dipole interaction. Recently, Djoufack et al.[14] reported the localisation of bright intrinsic localised modes in an one dimensional isotropic ferromagnets with octupole-dipole magnetic interaction. In the connection with above research works, we planning to study the energy localisation in ferromagnetic nanowire of one dimensional with the higher order octupole-dipole magnetic exchange interactions of NN and NNN spins. This paper work is arranged in a following manner. First we introduce our dynamical model: a Heisenberg spin model of ferromagnetic nanowire with first and second neighbor exchange interaction, biquadratic and octupole-dipole interactions and we derived the corresponding equations of motion. The analytical study is attribute to the exploration of Modulational Instability (MI) analysis of a plane wave generating in a ferromagnetic system of discrete lattice of nanowire and also we illustrate the existence of symmetric solitons in the spin lattice of ferromagnetic nanowire.

2 Hamiltonian Model of the Dynamical System

We contemplate the motion of an isotropic one dimensional ferromagnetic nanowire of N spins which can be expressed by the Hamiltonian with nearest (NN) and next nearest neighbor(NNN) higher order magnetic spin interactions

$$H = -\sum_n \Big[2J_1\Big[(\vec{S}_n \cdot \vec{S}_{n+1}) + (\vec{S}_n \cdot \vec{S}_{n+2})\Big] + 2J_2\Big[(\vec{S}_n \cdot \vec{S}_{n+1})^2 + (\vec{S}_n \cdot \vec{S}_{n+2})^2\Big]$$
$$+2J_3\Big[(\vec{S}_n \cdot \vec{S}_{n+1})(\vec{S}_{n+1} \cdot \hat{k})^2 + (\vec{S}_n \cdot \vec{S}_{n+2})(\vec{S}_{n+2} \cdot \hat{k})^2\Big]\Big], \tag{1}$$

where $\vec{S}_n = (S_n^x, S_n^y, S_n^z)$ is the spin at the lattice site n and $\vec{S}_n \cdot \vec{S}_n = S(SH)$. There are three types of interactions are involved in (1):

(i) The first two expressions correspond to J_1 define the bilinear NN and NNN exchange interactions in the isotropic ferromagnetic nanowire.
(ii) The next two terms proportional to J_2 represents the biquadratic interactions between first and second neighbor spin.
(iii) The last term related with J_3 express the higher order octupole-dipole spin-spin exchange interactions.

In order to demonstrate the spin excitations in spin lattice, we incorporate the boson excitations using Holstein-Primakoff modification [16].

$$S_i^+ = \hat{a}_j^\dagger \sqrt{2S - \hat{a}_j^\dagger \hat{a}_j},$$
$$S_i^- = \sqrt{2S - \hat{a}_j^\dagger \hat{a}_j \hat{a}_j^\dagger},$$
$$S_i^z = \hat{a}_j^\dagger \hat{a}_j - S, \tag{2}$$

where the annihilation and creation operators are represented as $\hat{a}_i$ and $a_j^\dagger$ respectively which compensate the boson commutation correlation $[\hat{a}_i, \hat{a}_j{}^\dagger] = \delta_{ij}$. The number operator $\hat{n}_i = \hat{a}_i \hat{a}_i^\dagger$ designates the spin divergence from its esteem value (S). The ground state intimates the all magnetic spins are along in the direction of $+z$ or $-z$ in a ferromagnetic order . On that account, the magnetization begin to exist, which is elucidated as the nonvanishing average of $S^z = \sum_i S_i^z$. By developing $\sqrt{2S - \hat{n}_i} \simeq \sqrt{2S}[1 - \hat{n}_i/S]$ to a series around the truncated excitation $< \hat{n}_i >\ << 2S$, we can obtain the presiding equation for nonlinear excitations on ground states. Now, the (2) expanded as,

$$\hat{S}_i^+ = \sqrt{2}[\epsilon a_i - \frac{\epsilon^3}{4} a_i^\dagger a_i a_i - \frac{\epsilon^5}{32} a_i^\dagger a_i a_i^\dagger a_i a_i + O(\epsilon^7)],$$
$$\hat{S}_i^- = \sqrt{2}[\epsilon a_i^\dagger - \frac{\epsilon^3}{4} a_i^\dagger a_i^\dagger a_i - \frac{\epsilon^5}{32} a_i^\dagger a_i^\dagger a_i a_i^\dagger a_i + O(\epsilon^7)],$$
$$\hat{S}_i^z = [1 - \epsilon^2 a_i^\dagger a_i]. \tag{3}$$

Accordingly, behind the spin-wave approximation, we achieved the effective low-energy Hamiltonian (1) as

$$
\begin{aligned}
H = -\sum_n \Big[& \epsilon^2 (J_1+J_2+J_3)\big[a^\dagger_{n+1}a_n + a_{n+1}a^\dagger_n - a^\dagger_{n+1}a_{n+1} + a_n a^\dagger_{n+2} + a^\dagger_n a_{n+2} \\
& - a^\dagger_{n+2}a_{n+2} - a^\dagger_n a_n\big] - \frac{1}{4}\epsilon^4 \Big[(J_1+J_3)\big[a_n a^\dagger_{n+1} a_{n+1} + a^\dagger_n a_n a^\dagger_{n+1} a_n a^\dagger_{n+1} \\
& + a^\dagger_n a^\dagger_n a_n a_{n+2} - 4a^\dagger_n a^\dagger_{n+2} a_n a_{n+2} + a^\dagger_n a_n a_n a^\dagger_{n+2} + a^\dagger_n a^\dagger_{n+2} a_{n+2} a_{n+2} \\
& - 4a^\dagger_n a^\dagger_{n+1} a_{n+1} a_n + a^\dagger_{n+1} a^\dagger_n a_{n+1} a_{n+1} + a^\dagger_n a_{n+1} a^\dagger_n a_n + a^\dagger_{n+2} a_n a^\dagger_{n+2} a_{n+2}\big] \\
& + J_2\big[- 4a^\dagger_{n+1} a_n a_n a^\dagger_{n+1} - 4a^\dagger_n a_{n+1} a_{n+1} a^\dagger_n - 4a^\dagger_{n+1} a_n a_{n+1} a^\dagger_n - 4a^\dagger_n a_n a_{n+1} a^\dagger_{n+1} \\
& + 4a_n a^\dagger_{n+1} a_n a^\dagger_n + 4a_n a^\dagger_{n+1} a_{n+1} a^\dagger_{n+1} + a_n a^\dagger_{n+1} a^\dagger_{n+1} a_{n+1} + a^\dagger_n a_n a_n a^\dagger_{n+1} \\
& + a^\dagger_n a^\dagger_{n+1} a_{n+1} a_{n+1} + 4a^\dagger_n a_{n+1} a_n a^\dagger_n + a^\dagger_n a^\dagger_n a_n a_{n+1} + 4a^\dagger_n a_{n+1} a_{n+1} a^\dagger_{n+1} \\
& + 4a^\dagger_n a_n a_n a^\dagger_{n+1} + a^\dagger_{n+1} a_n a_{n+1} a^\dagger_{n+1} + a^\dagger_n a_n a_n a^\dagger_{n+1} + 4a^\dagger_{n+1} a_{n+1} a_n a^\dagger_{n+1} \\
& + 4a^\dagger_n a_n a_{n+1} a^\dagger_n + 4a^\dagger_{n+1} a_{n+1} a_{n+1} a^\dagger_n + a^\dagger_n a_{n+1} a_{n+1} a^\dagger_{n+1} + a^\dagger_n a^\dagger_n a_n a_{n+1} \\
& - 4a^\dagger_{n+2} a_n a_n a^\dagger_{n+2} - 4a^\dagger_n a_{n+2} a_{n+2} a^\dagger_n - 4a^\dagger_{n+2} a_n a_{n+2} a^\dagger_n - 4a^\dagger_n a_n a_{n+2} a^\dagger_{n+2} \\
& + 4a_n a^\dagger_{n+2} a_n a^\dagger_n + 4a_n a^\dagger_{n+2} a_{n+2} a^\dagger_{n+2} + a_n a^\dagger_{n+2} a^\dagger_{n+2} a_{n+2} + a^\dagger_n a_n a_n a^\dagger_{n+2} \\
& + a^\dagger_n a^\dagger_{n+2} a_{n+2} a_{n+2} + 4a^\dagger_n a_{n+2} a_{n+2} a^\dagger_{n+2} + a^\dagger_n a^\dagger_n a_n a_{n+2} + 4a^\dagger_n a_{n+2} a_n a^\dagger_n \\
& + a^\dagger_{n+2} a_n a_{n+2} a^\dagger_{n+2} + 4a^\dagger_{n+2} a_{n+2} a_n a^\dagger_{n+2} + a^\dagger_n a_n a_n a^\dagger_{n+2} + 14a^\dagger_n a_n a_n a^\dagger_{n+2} \\
& + 41a^\dagger_n a_n a_{n+2} a^\dagger_n + 4a^\dagger_{n+2} a_{n+2} a_{n+2} a^\dagger_n + a^\dagger_n a_{n+2} a_{n+2} a^\dagger_{n+2} + a^\dagger_n a^\dagger_n a_n a_{n+2} \\
& - 12a^\dagger_n a_n a_{n+1} a^\dagger_{n+1} - 4a^\dagger_{n+1} a_{n+1} a_n a^\dagger_n - a^\dagger_{n+1} a_{n+1} a_{n+1} a^\dagger_{n+1} - 4a^\dagger_n a_n a_n a^\dagger_n \\
& - 12a^\dagger_n a_n a_{n+2} a^\dagger_{n+2} - a^\dagger_{n+2} a_{n+2} a_{n+2} a^\dagger_{n+2} - 4a^\dagger_{n+2} a_{n+2} a_n a^\dagger_n\big]\Big]\Big].
\end{aligned}
\tag{4}
$$

Consequently, we introduce the spin-wave approximation with some nonlinear alterations. Let us consider the P representation also called the Glauber coherent-state representation [17] described by the multiplication of the multimode coherent states $|v>=\prod_i |v_i$, in which each element $|v_i>$ is an eigen state of the annihilation operator $\hat{a}_i$ i.e, $\hat{a}_i|v_i> v_i|v_i>$ where v_i the coherent amplitude. The field operator sandwiched by $|v>$ can be expressed only with their diagonal components because of the coherent states are normalized and over completed. The Glauber coherent-state representation of the nonlinear (4) can be written as

$$
i\hbar\frac{\partial a_i}{\partial t} = [a_i, H] = F(a_i^\dagger, a_{i+1}^\dagger, a_i, a_{i+1}). \tag{5}
$$

Further, the spin dynamics are demonstrated in the view of Glauber's coherent-state representations

$$
\begin{aligned}
-i\frac{dv_n}{dt} &= 2\epsilon^2\Big[(J'+J_2)\Big[v_{n+1}+v_{n-1}+v_{n+2}+v_{n-2}-4v_n\Big]-4J_3v_n\Big] \\
&-\frac{\epsilon^4}{4}\Big[J'\Big[2|v_n|^2(v_{n+1}+v_{n-1}+v_{n+2}+v_{n-2})+v_n^2(v_{n+1}^*+v_{n-1}^* \\
&+v_{n+2}^*+v_{n-2}^*)+|v_{n-2}|^2v_{n-2}+|v_{n+2}|^2v_{n+2}+|v_{n-1}|^2v_{n-1}+|v_{n+1}|^2v_{n+1} \\
&-4v_n(|v_{n+1}|^2+|v_{n-1}|^2+|v_{n+2}|^2+|v_{n-2}|^2)\Big]+J_2\Big[20|v_n|^2(v_{n+1}+v_{n-1} \\
&+v_{n+2}+v_{n-2})+10v_n^2(v_{n+1}^*+v_{n-1}^*+v_{n+2}^*+v_{n-2}^*)+10(|v_{n+1}|^2v_{n+1} \\
&+|v_{n-1}|^2v_{n-1}+|v_{n-2}|^2v_{n-2}+|v_{n+2}|^2v_{n+2})-8v_n^*(v_{n+1}^2+v_{n-1}^2+v_{n+2}^2 \\
&+v_{n-2}^2)-24v_n(|v_{n+1}|^2v_{n+1}+|v_{n-1}|^2v_{n-1}+|v_{n-2}|^2v_{n-2}+|v_{n+2}|^2v_{n+2}) \\
&-32|v_n|^2v_n\Big]+J_3\Big[8|v_{n+1}|^2v_{n+1}+|v_{n-2}|^2v_{n-2}+8|v_{n+2}|^2v_{n+2} \\
&-12v_n(|v_{n-1}|^2+|v_{n+1}|^2+|v_{n+2}|^2+|v_{n-2}|^2)+16(|v_n|^2v_{n-1}+|v_n|^2v_{n-2}) \\
&+8v_n^2(v_{n-1}^*+v_{n-2}^*)-48|v_n|^2v_n\Big]\Big], \qquad (6)
\end{aligned}
$$

where $J'=J_1+J_3$ and the above nonlinear differential equation is in the configuration of perturbed discrete nonlinear Schrödinger equation and finding the solution to this equation is extremely tough task because of its high discreteness and nonlinearity. In this curiosity we are attracted to examine the localization of energy in discrete NN and NNN spin chain of ferromagnetic nanowire, we carryout the MI analysis.

3 Modulational Instability Through Linear Stability Analysis

Our motivation of the current work is to study the MI of the extended nonlinear spin waves in a ferromagnetic system consisting higher order octupole-dipole interaction. Due to the interplay between the nonlinear onsite and intersite interactions as a mechanism for changing MI and rule the localized modes occur in a nonlinear lattice. let us assume the time periodic plane wave solution is $u_n(t)=u_0e^{i(kn+\omega_0t)}$, where ω_0 designates the frequency of the plane wave and k is a wave vector. The linear stability of the nonlinear plane wave explained by looking for solutions which is in the form of adding small perturbation $u_n(t)=u_0(1+B_n(t))e^{i(kn+\omega_0t)}$, where B_n is a perturbation of small magnitude of the carrier wave. Then we assume B_n in the configuration of $B_n(t)=B_1e^{i(Qn-\Omega t)}+B_2^*e^{-i(Qn-\Omega^*t)}$, here B_1 and B_2 represent the carrier wave amplitudes and taken as small as compared with the carrier wave parameters and the terms of complex conjugation is denoted by asterisk symbol. The frequency of modulation and wave vector are represented by Ω and Q respectively. Upon switching the modulated spin wave into the equation of motion we achieved a system of linearly coupled equations.

$$\begin{pmatrix} -\Omega + A^{+} & B \\ B & \Omega + A^{-} \end{pmatrix} \begin{pmatrix} B_1 \\ B_2 \end{pmatrix} = \begin{pmatrix} 0 \\ 0 \end{pmatrix} \tag{7}$$

If the determinant of the above matrix is vanishes, the equations yield only nontrivial solution, where B and $A^{\pm}$ are given as follows

$$\begin{aligned} B = & -\epsilon^4 u_0^2 \Big[J' \Big[\cos(k-Q) + \cos(2(k-Q)) + 2\cos(k) + 2\cos(2k) + \cos(k+Q) \\ & - 4\cos(2Q) - 3\cos(2(k+Q)) \Big] + 2J_2 \Big[6\cos(2k) + 10\cos(k) + 5\cos(k-Q) \\ & + 5\cos(2(k-Q)) + 5\cos(2(k+Q)) - 4\cos(4k) + 5\cos(k+Q) - 12\cos Q \\ & - 12\cos(2Q) - 8 \Big] + 32 J_3 \Big[\cos(2k) + \cos(k) - 1 \Big] \Big], \\ A^{\pm} = & -\omega_0 + 4\epsilon^2 \Big[(J' + J_2)(\cos(2(k \pm Q)) + \cos(k \pm Q) - 2) - 2J_3 \Big] \\ & - 2u0^2 \epsilon^4 \Big[J' \Big[2\cos(k \pm Q) + 2\cos(k) + 2\cos(2(k \pm Q)) - 2\cos(Q) - 4 \Big] \\ & + J_2 \Big[20\cos(k) + 20\cos(2k) + 20\cos(k \pm Q) + 20\cos(2(k \pm Q)) - 8\cos(2k \pm Q) \\ & - 8\cos(2(2k \pm Q)) - 12\cos(2Q) - 12\cos(Q) - 40 \Big] + J_3 \Big[8\cos(2k) + 8\cos(k) \\ & + 8\cos(k \pm Q) - 4\cos(2Q) + 8\cos(2(k \pm Q)) - 4\cos(Q) - 16 \Big] \Big]. \end{aligned}$$

where,

$$\begin{aligned} \omega_0 = & 4\epsilon^2 \Big[(J' + J_2)(\cos(2k) + \cos(k) - 2) - 2J_3 \Big] - 4u_0^2 \epsilon^4 \Big[J'[\cos(2k) + \cos(k) - 2] \\ & + 2J_2 \big[4\cos(2k) + 5\cos(k) - \cos(4k) - 8 \big] + 2J_3 \big[\cos(2k) + \cos(k) - 4 \big] \Big], \end{aligned}$$

Solving (7), we obtain

$$\begin{aligned} \Omega = & \frac{1}{2} \Big[4\epsilon^2 \Big[(J' + J_2) \big[\cos(k-Q) + \cos(k+Q) + \cos 2(k-Q) + \cos 2(k+Q) \\ & -2\cos k - 2\cos(2k) - 8 \big] - 6J_3 \Big] + 2\epsilon^4 u0^2 \Big[2J'(\cos(k+Q) + \cos 2(k+Q) \\ & + \cos(k-Q) + \cos 2(k-Q) - 2\cos Q + 4\cos(2k) + 6\cos k + 4\cos k + 4) \\ & - J_2(24\cos(2Q) + 16\cos(4k) - 104\cos(2k) - 20\cos 2(k+Q) - 80\cos k \\ & -20\cos(k-Q) - 20\cos(k+Q) + 8\cos(2k-Q) + 8\cos(2k+Q) \\ & -20\cos 2(k-Q) + 12\cos Q + 8\cos 2(2k-Q) - 68) + J_3(8\cos(2k) - 12\cos(2Q) \\ & -4\cos Q + 8\cos 2(k-Q) + 8\cos(k-Q) - 8\cos k - 16 - 8\cos(2Q) + 16\cos(2k) \\ & +32\cos(k+Q) + 32\cos 2(k+Q) + 32\cos k - 16\cos Q - 32) \Big] \\ & \pm \sqrt{M^2 - \epsilon^8 u_0^4 N^2} \Big], \end{aligned} \tag{8}$$

where

$$
\begin{aligned}
M &= 4\epsilon^2\Big[2J_3 - (J_2 + J')\big[-2 + \cos 2(k-Q) + \cos 2(k+Q) + \cos(k+Q) \\
&\quad + 2 - 2\cos k + \cos(k-Q) - 2\cos 2k\big]\Big] + 2u_0^2\epsilon^4\Big[J'\big[-4\cos Q - 10\cos k \\
&\quad - 40 + 2\cos 2(k-Q) + 2\cos(k-Q) + 2\cos(k+Q) + 2\cos(2Q+2k) - 16\cos 2k\big] \\
&\quad - J_2\big[-20\cos 2(k+Q) + 8\cos(2k-Q) + 24\cos(2Q) - 20\cos 2(k-Q) - 40\cos(2k) \\
&\quad + 8\cos 2(2k+Q) + 24\cos Q - 40\cos k + 8\cos(Q+2k) + 8\cos 2(2k-Q) \\
&\quad - 20\cos(k-Q) + 80 - 20\cos(Q+k)\big] + J_3\big[-8\cos(2k) - 8\cos(Q) \\
&\quad + 8\cos(2k-2Q) - 4\cos(2Q) + 8\cos(k-Q) - 16 + 8\cos(2k) + 8\cos(Q+k) \\
&\quad - 4\cos(2Q) + 48 + 8\cos(2Q+2k)\big]\Big], \\
N &= J'\big[-3\cos 2(Q+k) - 4\cos(2Q) + \cos 2(k-Q) + 2\cos k + \cos(k+Q) + \cos(k-Q) \\
&\quad + 2\cos(2k)\big] + 2J_2\big[6\cos 2k - 12\cos 2Q + 5\cos(k+Q) + 5\cos 2(k-Q) \\
&\quad - 4\cos(4k) + 5\cos 2(k+Q) - 8 + 5\cos(k-Q) - 12\cos Q + 10\cos k\big] \\
&\quad + 32J_3\big[\cos 2k\cos k - 1\big],
\end{aligned}
$$

$$
\begin{aligned}
g(\Omega) &\equiv Im(\Omega) \\
&\equiv \Bigg[\Big[J'\big[\cos(k+Q) - 4\cos(2Q) + \cos 2(k-Q) - 3\cos 2(Q+k) + 2\cos(2k) \\
&\quad + \cos(k-Q) + 2\cos k\big] + 2J_2\big[6\cos 2k - 4\cos(4k) - 12\cos 2Q \\
&\quad + 5\cos 2(k+Q) + 5\cos 2(k-Q) - 12\cos Q + 5\cos(k+Q) + 10\cos k \\
&\quad + 5\cos(k-Q) - 8\big] + 32J_3\big[\cos 2k - 1 + \cos k\big]\Big]^2 - \Big[5\epsilon^2\Big[\big[\cos 2(k+Q) + \cos(k+Q) \\
&\quad - 2 + \cos(k-Q) + \cos 2(k-Q) - 2\cos k + 2 + \cos 2(k-Q) - 2\cos 2k\big] \\
&\quad \times 2J_3 - (J' + J_2)\Big] + 2u_0^2\epsilon^4\Big[-J_2\big[24\cos(2Q) - 40\cos(2k) + 8\cos(2k-Q) \\
&\quad - 20\cos 2(k-Q) - 20\cos 2(k+Q) + 24\cos Q - 40\cos k + 8\cos 2(2k-Q) \\
&\quad - 20\cos(k-Q) + 8\cos(Q+2k) - 20\cos(Q+k) + 8\cos 2(2k+Q) + 80\big] \\
&\quad + J'\big[2\cos 2(k-Q) + 2\cos(k-Q) + 2\cos(k+Q) + 2\cos(2Q+2k) \\
&\quad - 10\cos k - 4\cos Q - 16\cos 2k - 40\big] + J_3\big[-8\cos(Q) - 4\cos(2Q) \\
&\quad - 4\cos(2Q) - 8\cos(2k) + 8\cos(2k) + 8\cos(k-Q) + 8\cos(2k-2Q) - 16 \\
&\quad + 8\cos(2Q+2k) + 8\cos(Q+k) + 48\big]\Big]\Big]^2\Bigg]^{1/2},
\end{aligned}
\tag{9}
$$

here Im represents the imaginary term and occurrence of localized constructions in the ferromagnetic spin lattice of nanowire which are only achievable if the steady amplitude of the solution becoming unstable. Whenever $\Omega < 0$, the steady state solution turn out to be unstable, whereas the perturbation exponentially expands with the appreciable intensity. The gain reveals more fascinating relation of Ω with the coupling exchange parameters J_1, J_2 and J_3 describing the exchange interaction and dispersive long range interactions for different values. Figure 1 depicts the stable/instable area in the plane of (k, Q) and be in tune with the influence of the

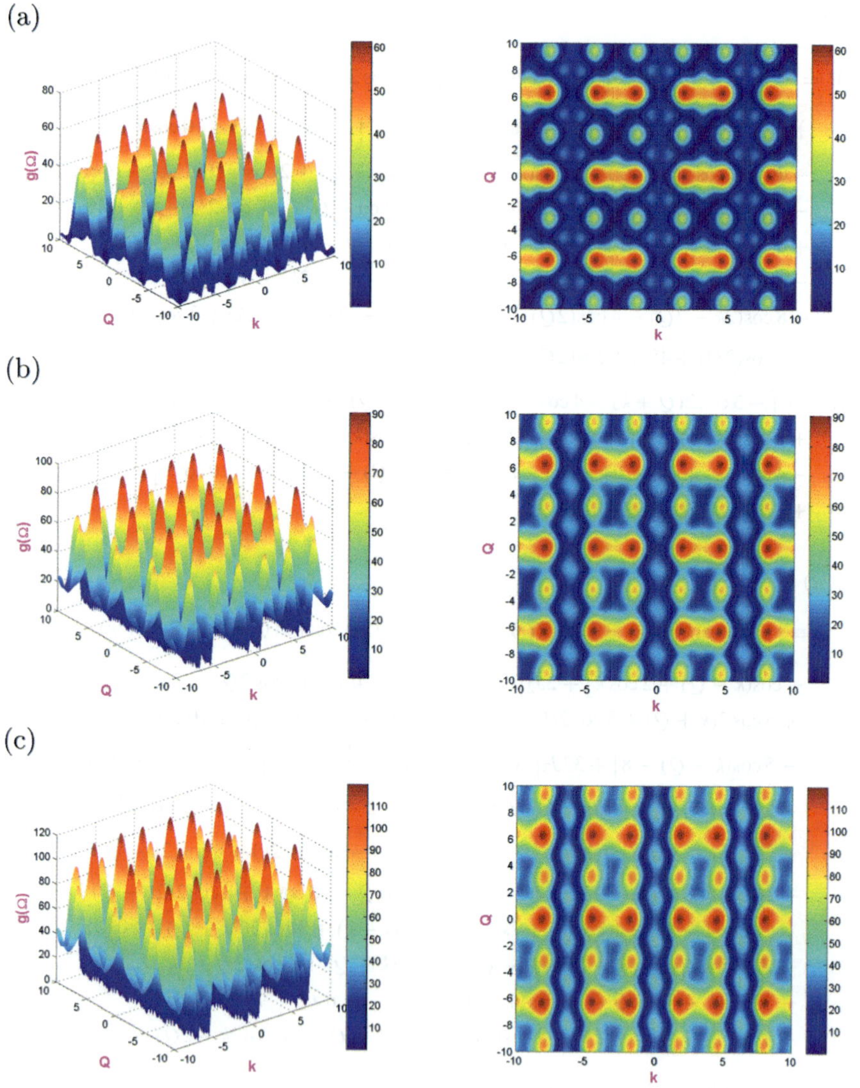

Fig. 1 MI gain profile for (a) $J_3 = 0.1$, (b) $J_3 = 0.5$, (c) $J_3 = 0.9$ and on all plots $J_1 = 0.1$, $J_2 = 0.2$

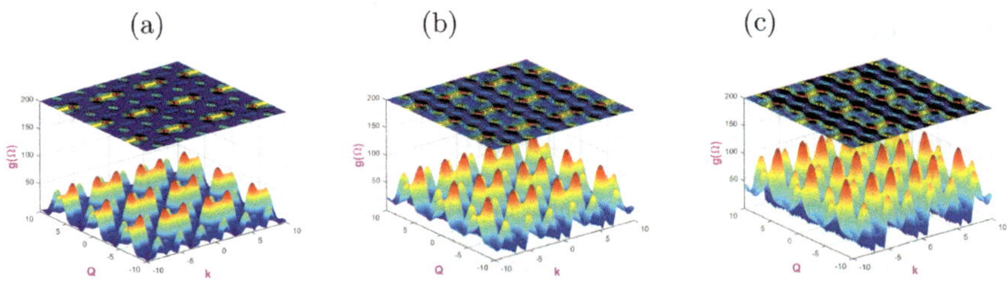

Fig. 2 MI gain cumulative profile for (a) $J_3 = 0.1$, (b) $J_3 = 0.5$, (c) $J_3 = 0.9$ and on all plots $J_1 = 0.1$, $J_2 = 0.2$

interaction of octupole-dipole coupling parameter denoted as J_3. In these figures, the nonlinear plane waves are reliable and stable in the dark bluish area by modulating the wavenumber Q and area with bright yellowish orange region where we expect the sudden exponential growth of the amplitude of wave which seems to be appreciable enhancement in the instability domain. Surprisingly, when we increase the octupole-dipole interaction parameter J_3 from 0.1 to 0.9 units (Fig. 1a–c), the stability region becomes faded out and the generation of high amplitude, robustic nonlinear solitonic waves are observed in the presence of biquadratic and bilinear interactions. Figure 2 depicts the cumulative representation of Fig. 2 as 2D and 3D plots.

4 Conclusion

In this investigation, we have effectively studied the modulational instability analysis of extended nonlinear solitary waves in isotropic ferromagnetic nanowire with NN-NNN higher order octupole-dipole magnetic spin interactions. The deliberated discrete nonlinear Schrödinger (DNLS) equation derived by using the spin wave approach of Glauber coherent approximation method. Then the highly discrete nonlinear complicated DNLS equation analysed through the modulational instability analysis and the result is depicted in the figures shows that control on the strength of octupole-dipole coupling exchange interaction of neighbouring spins, the place of stability/instability in the plane are predicted in which the instability region is increases with increasing of octupole-dipole interaction. The numerical simulations explore the existence of possible localized long-lived excitations of solitonic pulse in the ferromagnetic nanowire spin lattice that promote to nonlinear regime. This robustic soliton expected to play a potential application in manufacturing of magnetic recording and memory devices.

Acknowledgements L. K. delightedly acknowledge the financial aid from DST-SERB (MTR/2017/000 314/MS), India and CSIR (03 (1418)/17/EMR-II), India in the form of Major Research Projects (MRP). L. K. and T. P. gratefully acknowledge the financial assist from UGC-DAE CSR (CSR-KN/ CRS-102/2019-20) in the configuration of a MRP.

References

1. Huang, G., Shi, Z.P., Dai, X., Tao, R.: On soliton excitations in a one-dimensional Heisenberg ferromagnet. J. Phys.: Condens. Matter **42**(2), 8355 (1990)
2. Daumont, I., Dauxois, T., Peyrard, M.: Modulational instability: first step towards energy localization in nonlinear lattices. J. Nonlinearity **10**, 617 (1997)
3. Ivanov, B.A.: Nonlinear dynamics and two-dimensional solitons for spin-1 ferromagnets with biquadratic exchange. Phys. Rev. B **77**, 064402 (2008)
4. Alexey Kartsev: Biquadratic exchange interactions in two-dimensional magnets. NPJ. Comput. Mater. **6**, 150 (2020)
5. Christal Vasanthi, C., Latha, M.: Localized spin excitations in a disordered antiferromagnetic chain with biquadratic interactions. Euro. Phys. J. D **69**, 268 (2015)
6. Kavitha, L.: Modulational instability and nano-scale energy localization in ferromagnetic spin chain with higher order dispersive interactions. J. Magn. Magn. Mater. **404**(C), 91–118 (2016)
7. Anderson, P.W.: New approach to the theory of superexchange interactions. Phys. Rev. **115**, 2 (1959)
8. Anderson, P.W.: Theory of magnetic exchange interactions: exchange in insulators and semiconductors. Solid State Phys. **14**, 99–214 (1963)
9. Moriya, T.: Anisotropic superexchange interaction and weak ferromagnetism. Phys. Rev. **120**, 91 (1960)
10. Moriya, T.: New mechanism of anisotropic superexchange interaction. Phys. Rev. Lett. **4**, 228 (1960)
11. Huang, G., Zhang, S., Hu, B.: Nonlinear excitations in ferromagnetic chains with nearest- and next-nearest-neighbor exchange interactions. Phys. Rev. B **58**, 9194 (1998)
12. Daniel, M., Kavitha, L.: Soliton spin excitations in an anisotropic Heisenberg ferromagnet with octupole-dipole interaction. Phys. Rev. B **59**, 21 (1999)
13. Tang, Bing, Li, De-Jun., Tang, Yi.: Quantum solitons and breathers in an anisotropic ferromagnet with octupole-dipole interaction. Int. J. Theor. Phys. **53**, 359–369 (2014)
14. Djoufack, Z.I., Nguenang, J.P., Kenfack-Jiotsa, A.: Quantum breathers and intrinsic localized excitations in an isotropic ferromagnet with octupole-dipole interaction. Phys. B **598**, 412437 (2020)
15. Kavitha, L., Sathishkumar, P., Saravanan, M., Gopi, D.: Soliton switching in an anisotropic Heisenberg ferromagnetic spin chain with octupole-dipole interaction. Phys. Scr. **83**, 055701 (1963)
16. Holstein, T., Primakoff, H.: Field dependence of the intrinsic domain magnetization of a ferromagnet. Phys. Rev. **58**, 1098 (1940)
17. Roy, J.: Glauber: coherent and incoherent states of the radiation field. Phys. Rev. **131**, 2766 (1963)

Study of Nonlinear Dynamics of Vilnius Oscillator

Dmitrijs Pikulins, Sergejs Tjukovs, Iheanacho Chukwuma Victor, and Aleksandrs Ipatovs

Abstract Most chaotic dynamical systems exhibit non-robust chaos when small changes of system parameters lead to abrupt jumps from chaotic to periodic motions. These unstable regimes could not be successfully utilized in practice. The detailed systematic analysis of nonlinear dynamics becomes paramount for practical applications to identify regions of robust chaos and related transitions. This paper presents an in-depth analysis of the nonlinear dynamics of a Vilnius oscillator, providing a detailed numerical study of the operational regions of Vilnius oscillator's, constructing bifurcation map and brute-force bifurcation diagrams.

Keywords Chaotic Oscillator · Bifurcations · Robust Chaos

1 Introduction

For reliable and smooth communication, accurate data must be transmitted or received robustly and securely. This, in turn, requires the implementation of specific sophisticated data-coding algorithms. It has been revealed that the required complexity could be observed in very simple dynamical systems exhibiting chaotic oscillations. In recent years, an exponential rise in computer processing power allowed modelling, predicting, and exploiting irregular dynamics of nonlinear circuits.

What benefits could chaotic communications provide that sets it apart from existing conventional systems? The answer lies in the characteristics of chaos-based communication systems: a sensitivity to initial conditions, aperiodic and noise like time series, and wide frequency bandwidth. These properties allow chaotic transmissions to have reduced interception or even detection risk.

Chaotic oscillators are simple analogue or digital circuits producing chaotic signals. Such oscillators exhibit rich dynamics and offer a wide range of applications in engineering, such as pseudo-random number generators [1], radar and sonar

---

D. Pikulins (✉) · S. Tjukovs · I. Chukwuma Victor · A. Ipatovs
Institute of Radioelectronics, Riga Technical University, Azenes st.12, Riga, Latvia
e-mail: dmitrijs.pikulins@rtu.lv

S. Banerjee and A. Saha (eds.), *Nonlinear Dynamics and Applications*,
Springer Proceedings in Complexity,
https://doi.org/10.1007/978-3-030-99792-2_103

systems [2], switched-mode power supplies [3, 4], encryption for secure communication [5], and chaos-based communications [6–8]. Experiments on oscillatory systems using circuits simulations allow the in-depth study of phenomena and develop a wide range of practical applications.

The currently known dynamical systems frequently offer fragile chaos, meaning that the chaotic attractor has some periodic attractors in the nearby region in the parameter space [9]. This might be a concern for practical applications relying on continuous chaotic signals created by chaotic oscillators. The phase diagrams for several dynamic systems in [10] illustrate that small changes in the system parameters dramatically change its dynamics. If the chaotic attractor of a physical system meant to generate chaotic signals is fragile, the dynamics may not evolve as predicted. So, a given implementation may produce a system with parameters corresponding to a nonchaotic attractor. Even if the system starts in a chaotic state, slight changes in operational parameters caused by external stimuli or component faults may cause the transition to periodic oscillations. In practical applications, such issues may be avoided by using systems with robust chaos. Thus, it is necessary to study the parameters of oscillators in detail to determine how the system behaves in various conditions and know what regions of operation produce the most resilient chaos.

One of the chaotic systems has been presented in [11] and referenced as "Vilnius oscillator". Several papers were devoted to examining the nonlinear phenomena of this circuit [12–14]. But all of them were fragmentary and incomplete.

In this paper, an in-depth analysis of the Vilnius oscillator's nonlinear dynamics is performed to determine parameter ranges that reliably produce robust chaos.

This paper is organized as follows. The second section is devoted to the description of the schematic and analytical model of the Vilnius oscillator. The third section presents the nonlinear analysis of the dynamics of the system under study. The last section is devoted to the overall conclusions and suggestions on the applicability of this type of chaotic oscillator.

2 Vilnius Oscillator Model

The schematic of the Vilnius chaotic oscillator under study is shown in Fig. 1. It comprises affordable electronic components and can be implemented even on a breadboard, thus simplifying practical research. Values and part numbers of the components are summarized in Table 1.

The oscillator is built around a general-purpose operational amplifier. In the positive feedback loop, the RLC circuit is connected. A non-inverting amplifier configuration can be easily recognized in the negative feedback loop. Additional capacitor C2 and silicon diode D1 drastically change the oscillator's dynamics and transform it into a circuit with complex behaviour. The numerical study of the circuit is based on a system of differential equations, which is derived by applying general laws of circuit analysis and making several simplifications. Firstly, the equation of an ideal diode is used to describe current–voltage relations of D1:

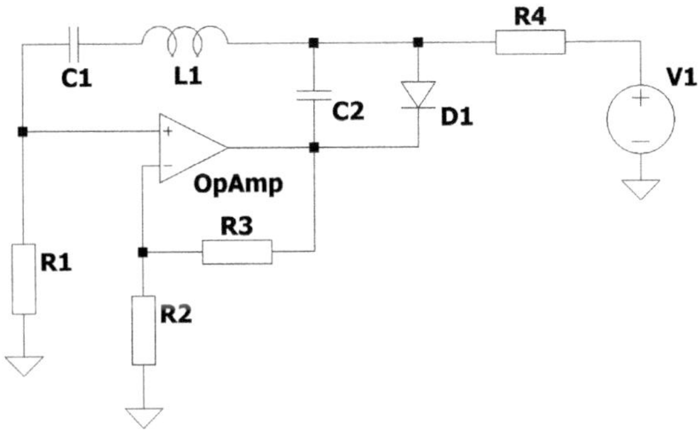

Fig. 1 The schematic diagram of the Vilnius chaos oscillator used in the study

Table 1 List of components used in the Vilnius oscillator

Component	Value or Part Number
R1	1 kΩ
R2	10 kΩ
R3	10 kΩ
R4	20 kΩ
C1	1 nF
C2	150 pF
L1	1 mH
D1	1N4148
OpAmp	TL082
V1	Laboratory power supply

$$I_D = I_S\left[exp\left(\frac{q \cdot V_D}{N \cdot k_B \cdot T}\right) - 1\right] \quad (1)$$

where I_S refers to saturation current, $q = 1.6 \times 10^{-19}$ C – the elementary charge, k_B – Boltzmann's constant, T – the temperature in Kelvin, N = 1 for an ideal diode, I_D and V_D are the current through a diode and voltage across it respectively. Equation 1 doesn't consider parasitic capacitances of the p–n junction and equivalent resistance Rs that models resistances of neutral regions and contacts associated with an actual device. Secondly, to express the current through the R4, it is assumed that R4 >> R1. As a result, according to Ohm's law:

$$I_{R4} = \frac{V1}{R4}. \quad (2)$$

Also, the gain of the non-inverting amplifier is expressed as:

$$k = 1 + \frac{R_2}{R_1}. \quad (3)$$

A system of ordinary differential equations describing the Vilnius oscillator is shown below:

$$C_1 \frac{dV_{C1}}{dt} = I_{L1} \tag{4}$$

$$L \frac{dI_{L1}}{dt} = (k-1) \cdot R_1 \cdot I_{L1} - V_{C1} - V_{C2} \tag{5}$$

$$C_2 \frac{dV_{C2}}{dt} = I_{R4} + I_{L1} - I_{D1} \tag{6}$$

Furthermore, in [11], dimensionless variables and parameters are suggested for conventional numerical analysis:

$$x = \frac{V_{C1} \cdot q}{k_B \cdot T} y = \frac{I_{L1} \cdot q \cdot \sqrt{\frac{L}{C_1}}}{k_B \cdot T} z = \frac{V_{C2} \cdot q}{k_B \cdot T} \tag{7}$$

$$a = \frac{(k-1) \cdot R_1}{\sqrt{\frac{L}{C_1}}} b = \frac{I_{R4} \cdot q \sqrt{\frac{L}{C_1}}}{k_B \cdot T} c = \frac{I_S \cdot q \cdot \sqrt{\frac{L}{C_1}}}{k_B \cdot T} \varepsilon = \frac{C_2}{C_1} \tag{8}$$

These substitutions allow to use of a simplified version of the system of differential equations:

$$\frac{dx}{dt} = y \tag{9}$$

$$\frac{dy}{dt} = ay - x - z \tag{10}$$

$$\varepsilon \frac{dz}{dt} = b + y - c(e^z - 1) \tag{11}$$

In forthcoming sections, a and ε will be used as bifurcation parameters. In practical experiments or circuit simulation software, it is possible to vary a by changing the value of R3 and ε by the corresponding adjustment of C2.

3 Analysis of Nonlinear Dynamics of Vilnius Oscillator

The study of nonlinear dynamics of the Vilnius oscillator requires the discrete-time model of the system. The model could be obtained numerically by sampling the variables x, y, z, defined by the system of nonlinear Eqs. (9)–(11). The oscillator under study is an example of an autonomous system, so it is impossible to obtain the stroboscopic map as no external clocking element is present. Thus, the Poincare mapping is derived by sampling the variables, as the trajectory hits the plane y = 0 from positive to the negative side. This mapping is used to construct bifurcation diagrams and provide the analysis of periodic or chaotic modes of operations.

First, the bifurcation map (also called the 2-parameter bifurcation diagram) is calculated for the predefined range of system parameters. This, essentially, is the graph depicting the periodic regions of regimes up to preset periodicity in the 2-parameter space. For the Vilnius oscillator, the bifurcation parameters of interest are a and ε, leaving all the other parameters fixed. The bifurcation map, depicted in Fig. 2, gives a variety of reference points to start the more detailed study of nonlinear dynamics of the oscillator, as we vary some parameters. The dashed areas depict periodic regimes, while the white area represents the periodicity >8 of chaotic modes of operation.

It could be seen that for small values of $a = 0.05–0.1$, the system exhibits stable period-1 operation as we change ε in the whole range of interest. As we increase the a for different values of ε, the dynamics of the system change from periodic to chaotic, and in some cases, returns to periodic motions.

To provide a more detailed analysis of the nonlinear dynamics of the system, the one-parameter brute-force bifurcation diagrams are obtained, essentially as cross-sections of the bifurcation map.

The first bifurcation diagram is constructed for ε=0.05 and shown in Fig. 3.

It could be noticed that for the defined parameter values, no classical period-doubling route to chaos could be observed. The stable period-1 solution suddenly becomes unstable through the period-doubling bifurcation and later recovers the stability. It could be shown that after the second period-doubling bifurcation, the

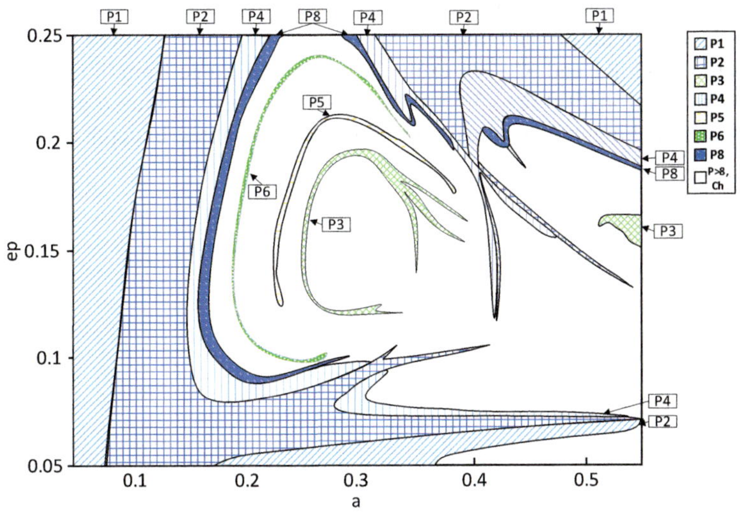

Fig. 2 The bifurcation map of Vilnius oscillator for $b = 40$; $a = 0.05–0.55$; $\varepsilon = 0.05–0.25$

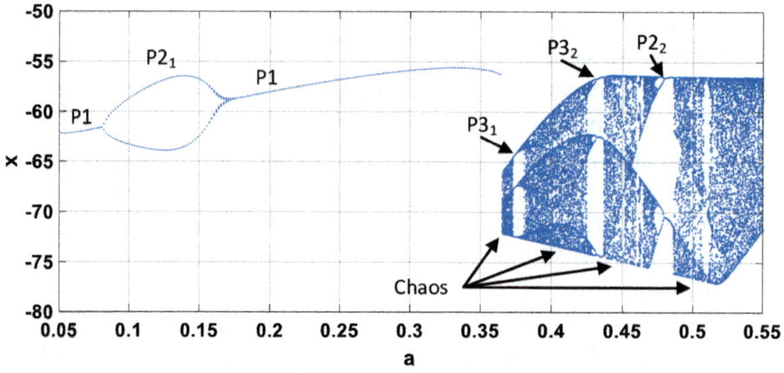

Fig. 3 Brute-force bifurcation diagram for $b = 40$; $\varepsilon = 0.05$; $a = 0.05$–0.55

period-1 solution is quantitatively different than that for smaller values of a. Figure 4 depicts the phase portraits in the x–y plane and the system's trajectory for $a = 0.05$ and $a = 0.34$.

While the phase portraits show only insignificant changes in the system's dynamics, as a changes, the study of the whole trajectory allows establishing the rapid growth of the amplitude of z (corresponding to the noticeable increase of voltage across the capacitor C2-see Fig. 1. and Eq. 8).

After $a = 0.37$, the system's dynamics become chaotic, exhibiting the transitions to periodic windows of period-3 and period-2. Thus, from the practical point of view-the most promising chaotic regions would be for $a = 0.38$–0.42 and $a = 0.51$–0.55.

To link obtained results to the physical implementation of Vilnius oscillator circuit one must keep in mind that according to Eq. 7 variable x is directly proportional to V_{C1}, y to I_{L1}, and z to V_{C2}. Also, Eqs. 3 and 8 state that in a real circuit parameter a can be changed in the defined range, either by sweep of value of R3 or simultaneous variation of both R1 and R3. For example, to cover a range from 0.05 to 0.55, R3 must be varied from 500 Ω to 5.5 kΩ, which can be easily achieved using variable resistor or switching matrix of resistors with different values. In depth experimental

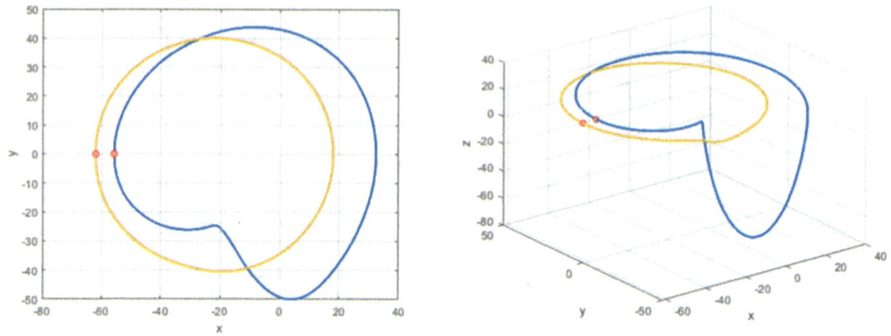

Fig. 4 Phase portrait and trajectory of period-1 regime: for $a = 0.05$ (yellow), $a = 0.34$ (blue)

verification is out of the scope of this paper. The main goal of current numerical study is to identify regions of interest for both possible implementations in secure communication systems and subsequent experimental verification of the obtained results.

The following bifurcation diagram has been obtained for $\varepsilon = 0.08$ and depicts the classical period-doubling route to chaos as a is varied from 0.05 to 0.55 (Fig. 5).

The system's dynamics drastically changes for large values of $a > 0.49$. It could be observed that there are two different chaotic attractors – one of them with a much larger amplitude than the other one (see Fig. 6.). In practical implementations, this could lead to the intermitted jumps from the main chaotic attractor that arose from the

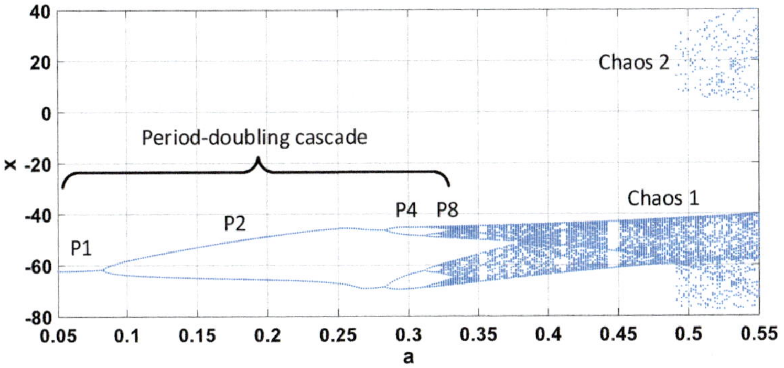

Fig. 5 Brute-force bifurcation diagram for $b = 40$; $\varepsilon = 0.08$; $a = 0.05$–0.55

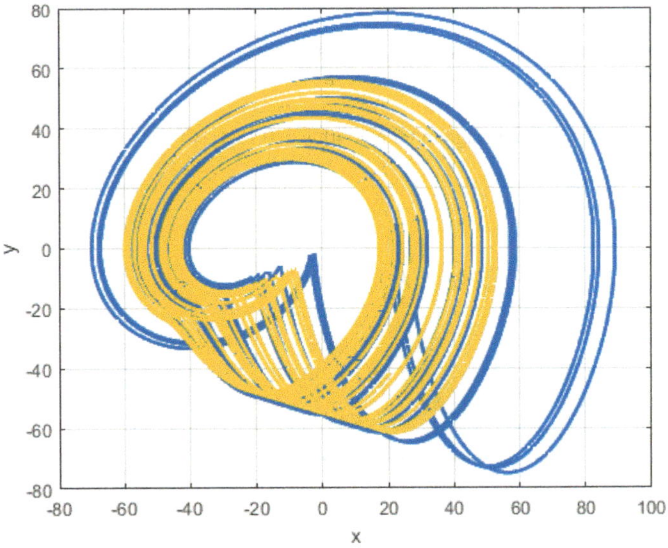

Fig. 6 Chaotic attractors for $b = 40$, $\varepsilon = 0.08$, $a = 0.46$ (yellow); $a = 0.55$(blue)

classical period-doubling cascade to the larger one, potentially leading to overvoltage and component failures. Thus, it would be desirable to avoid the operation of the Vilnius oscillator in this unsafe chaotic region.

The analysis of the bifurcation map in Fig. 2 allows the prediction of the relatively large chaotic region for a wide range of a values, as ε reaches 0.15. Figure 7 shows the brute-force bifurcation diagram, depicting chaotization of the system through the period-doubling cascade. However, unlike the chaotic region depicted, e.g. in Fig. 3, where a great variety of periodic windows is observed within the chaotic region, the diagram in Fig. 7 shows several intervals of robust chaotic oscillations (RCh_1-RCh_3). This means that setting system parameters with the defined ranges guarantee stable chaotic oscillations without the risk of shifting to periodic motion due to external noise or slight fluctuation in component nominal values.

Another two brute-force bifurcation diagrams were also obtained based on the bifurcation map in Fig. 2. In these cases, the value of parameter a was fixed ($a = 0.15$ in Fig. 8 and $a = 0.2$ in Fig. 9), and ε has been used as the bifurcation parameter of interest.

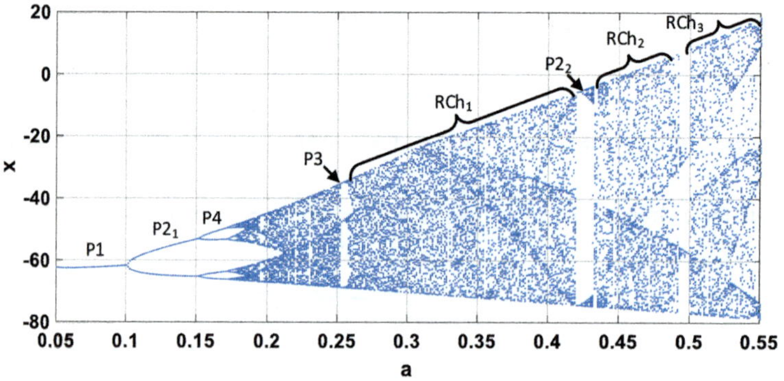

Fig. 7 Brute-force bifurcation diagram for $b = 40$; $\varepsilon = 0.15$; $a = 0.05 - 0.55$

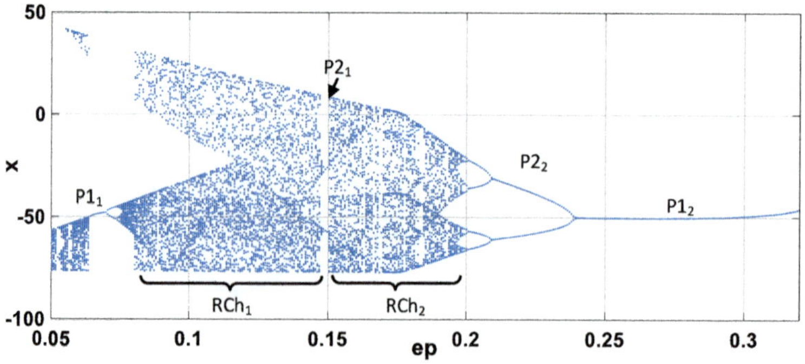

Fig. 8 Brute-force bifurcation diagram for $b = 40$; $a = 0.5$; $\varepsilon = 0.05 - 0.4$

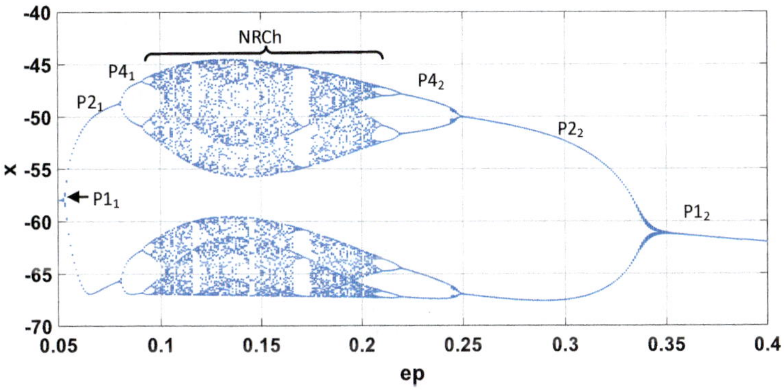

Fig. 9 Brute-force bifurcation diagram for $b = 40$; $a = 0.2$; $\varepsilon = 0.05$–0.4

The comparison of the diagrams allows us to draw important conclusions on the changes in the system's dynamics for two values of a. In Fig. 8, the Vilnius oscillator exhibits chaotic dynamics for small values of ε. Further, it operates in the narrow period-1 window, leading to the appearance of 2 regions of robust chaotic oscillations (RCh_1 and RCh_2) separated by a narrow period-2 window. For $\varepsilon > 0.18$, the transition to stable period-1 operation is observed through the inverse period-doubling cascade. Thus, the RCh_1 and RCh_2 could be used if the oscillator is supposed to provide reliable chaotic signals.

On the other hand, the similar transition from period-1 to chaos and back to period-1 is shown in Fig. 9, where $a = 0.2$. However, the formed chaotic region is full of stable periodic regimes (non-robust chaos- NRCh), making the oscillator not applicable as the source of chaotic oscillations.

4 Conclusions

As the scope of applications of chaotic systems expands, a growing number of different chaotic oscillators have been presented during several last decades. The main requirements include ease of implementation and rich nonlinear dynamics. The second point defines the necessity of the detailed numerical and experimental study of the oscillators, as frequently, even minor fluctuations of parameter values or external noises could lead to the transition from chaotic to periodic oscillations, compromising the robustness and security of the whole application.

This paper provided a detailed analysis of the nonlinear dynamics of the Vilnius oscillator in a wide parameter range. First, the usefulness of the bifurcation map has been proved, allowing the deliberate choice of parameter range, ensuring the required system's dynamics. Second, the mentioned map has been used as the keystone for constructing detailed bifurcation diagrams, revealing the exact transitions from periodic to chaotic motions and vice-versa. The analysis of the results shows that this oscillator could exhibit chaotic oscillations of different nature and properties: robust chaotic regions, non-robust chaotic motions, and even regions with coexisting chaotic attractors of various sizes. Thus, it has been proved that it is crucial to provide a comprehensive analysis of the system's dynamics to meet the requirements of the specific applications while utilizing the system within a feasible parameter range.

Acknowledgements This work has been supported by the European Regional Development Fund within the Activity 1.1.1.2 "Post-doctoral Research Aid" of the Specific Aid Objective 1.1.1 "To increase the research and innovative capacity of scientific institutions of Latvia and the ability to attract external financing, investing in human resources and infrastructure" of the Operational Programme "Growth and Employment" (No.1.1.1.2/VIAA/4/20/651).

References

1. Dantas, W.G., Rodrigues, L.R., Ujevic, S., Gusso, A.: Using nanoresonators with robust chaos as hardware random number generators. Chaos **30**, 043126 (2020)
2. Carroll, T.L., Rachford, F.J.: Target recognition using nonlinear dynamics. In: Leung, H. (ed.) Chaotic Signal Processing, pp. 23–48. SIAM, Philadelphia (2013)
3. Deane, J.H.B., Hamill, D.C.: Improvement of power supply EMC by chaos. Electron. Lett. **32**, 1045 (1996)
4. Pikulins, D.: Exploring types of instabilities in switching power converters: the complete bifurcation analysis. Elektronika ir Elektrotechnika **20**(5), 76–79 (2014)
5. Kocarev, L.: Chaos-based cryptography: a brief overview. IEEE Circuits Syst. Mag. **1**, 6–21 (2001)
6. Litvinenko, A., Aboltins, A.: Chaos based linear precoding for OFDM. RTUWO **2015**, 13–17 (2015)
7. Litvinenko, A., Bekeris, E.: Probability distribution of multiple-access interference in chaotic spreading codes based on DS-CDMA communication system. Elektronika Ir Elektrotechnika **123**(7), 87–90 (2012)
8. Babajans, R., Cirjulina, D., Grizans, J., Aboltins, A., Pikulins, D., Zeltins, M., Litvinenko, A.: Impact of the chaotic synchronization's stability on the performance of QCPSK communication system. Electronics **10**, 640 (2021)
9. Gusso, A., Ujevic, S., Viana, R.L.: Strong chaotification and robust chaos in the Duffing oscillator induced by two-frequency excitation. Nonlinear Dyn. **103**, 1955–1967 (2021)
10. Gallas, J.: The structure of infinite periodic and chaotic hub cascades in phase diagrams of simple autonomous flows. Int. J. Bifurc. Chaos **20**, 197–211 (2010)
11. Tamaševičius, A., Mykolaitis, G., Pyragas, V., Pyragas. K.: A simple chaotic oscillator for educational purposes. Eur. J Phys. **26**(1), 61 (2004)
12. Tamaševičius, A., Pyragienė, T., Pyragas, K.Ę.S.T.U.T.I.S., Bumelienė, S. and Meškauskas, M.: Numerical treatment of educational chaos oscillator. Int. J. Bifurc. Chaos **17**(10), 3657–3661 (2007)
13. Čirjuļina, D., Pikulins, D., Babajans, R., Anstrangs, D.D., Victor, I.C., Litvinenko, A: Experimental study of the impact of component nominal deviations on the stability of Vilnius Chaotic oscillator. In: 2020 IEEE Microwave Theory and Techniques in Wireless Communications (MTTW), vol. 1, pp. 231–236 (2020)
14. Babajans, R., Anstrangs, D.D., Cirjulina, D., Aboltins, A., Litvinenko, A.: Noise immunity of substitution method–based Chaos synchronization in Vilnius oscillator. In: 2020 IEEE Microwave Theory and Techniques in Wireless Communications (MTTW), vol. 1, pp. 237–242. IEEE, 2020

Classical Nonlinear Dynamics Associated with Prime Numbers: Non-relativistic and Relativistic Study

Charli Chinmayee Pal and Subodha Mishra

Abstract By mapping the system of prime numbers to a physical problem, it is possible to characterise the hidden nonlinear dynamics associated with it. In order to study the properties of primes, first the single particle Schrödinger equation is solved. The wave function used in this case is constructed from the prime counting function and their interaction potential is obtained. In the corresponding classical nonlinear system, the phase trajectories and the associated fixed points which happens to be half stable and half unstable are also studied. It is interesting to note that the Lambert W function appears in connection to solutions for the fixed points as a function of energy.

Keywords Quantum mechanics · Wave function · Nonlinear dynamics · Prime numbers

1 Introduction

One can represent a dynamical system using prime numbers [1, 2] e.g a system of gas molecules that are interacting with each other. Though the prime numbers are abstract points in the number universe, they can be represented by a one-particle system with an effective potential. Many interesting works have been published recently in this direction [3–8]. In the early 1970s Billingsley et al. [7] defined a random walk problem based on the fundamental theorem of arithmetic. Julia et al. [6] proposed the idea of a non-interacting gas where a single particle may have discrete energy equal to the logarithm of nth prime number. Berry and Keating developed [3] a theory where

Subodha Mishra deceased on 8th January 2022.

C. C. Pal (✉) · S. Mishra
Department of Physics, Siksha 'O' Anusandhan, Deemed to be University, Bhubaneswar, 751030, India
e-mail: charli.chinu@gmail.com
URL: http://www.springer.com/gp/computer-science/lncs

S. Banerjee and A. Saha (eds.), *Nonlinear Dynamics and Applications*, Springer Proceedings in Complexity,
https://doi.org/10.1007/978-3-030-99792-2_104

the zeros of the Riemann zeta function are related to the eigenvalues of the system. The dynamics of this classical system were reported to be chaotic which is represented by the Hamiltonian $H_{cl} = XP$ (X and P are position and momentum). Recently Bender et al. [4] constructed a Hamiltonian operator having eigenvalues as the nontrivial zeros of Riemann zeta function where the associated eigenfunctions obey certain boundary conditions. The classical limit of the operator is found to be exactly that predicted by Berry and Keating. All these seminal works beautifully connect and enrich quantum physics, dynamical chaos, and the prime number theory. In this work, we have taken a different approach to formulate a problem using the prime counting function [9] denoted by $\pi(x)$. The prime counting function $\pi(x)$ gives the number of primes below x as the number of particles. We construct the one dimension density as $\rho = \pi(x)/x$ and hence the corresponding wave function $\sqrt{\rho}(x)$. Any small smooth change in the parameter values (the bifurcation parameters) of a system causes a bifurcation in the system. It refers to a sudden qualitative change in the behavior of the dynamical system. The interaction potential between two bodies is obtained using the wave function. In the end, the relativistic and non-relativistic dynamics are studied by finding the flow trajectories, zeros, and bifurcation in that system.

2 Classical Dynamics: Non-relativistic

2.1 The Hamiltonian for Prime Number System

As discussed in the introduction, knowing the density function of primes, it is possible to construct the wave function and using the Schrodinger equation, one can map the prime number system to a dynamical system. The derivation of the interaction potential [10] is detailed in the apendix (10), which appears as $V(x) = \frac{\hbar^2}{m}\frac{1}{4x^2}\left(\frac{1}{ln(x)} + \frac{3}{2ln^2(x)}\right)$, The classical Hamiltonian [11] for a single particle system (taking $\hbar^2/m = 1$) using the derived potential appears as

$$H = \frac{p^2}{2m} + V(x) = \frac{p^2}{2} + \frac{1}{4x^2}\left(\frac{1}{ln(x)} + \frac{3}{2ln^2(x)}\right) = E \qquad (1)$$

The above classical Hamiltonian will be used to study the hidden classical nonlinear dynamics [12] of the system representing primes. The variation of potential $V(x)$ w.r.t x is plotted in Fig. 1 which shows $V(x) = 0$ at $x = e^{-3/2} = 0.223$ and is negative below this value of x up to $x = 0$. Also one can see that $V(x = 0) \rightarrow -\infty$ and $V(x = 1) \rightarrow \infty$. As x increases from 1, $V(x)$ decreases and goes to zero as $x \rightarrow \infty$.

Since the interaction potential (10) is determined, one can study the hidden classical dynamics characterizing prime numbers through the corresponding classical

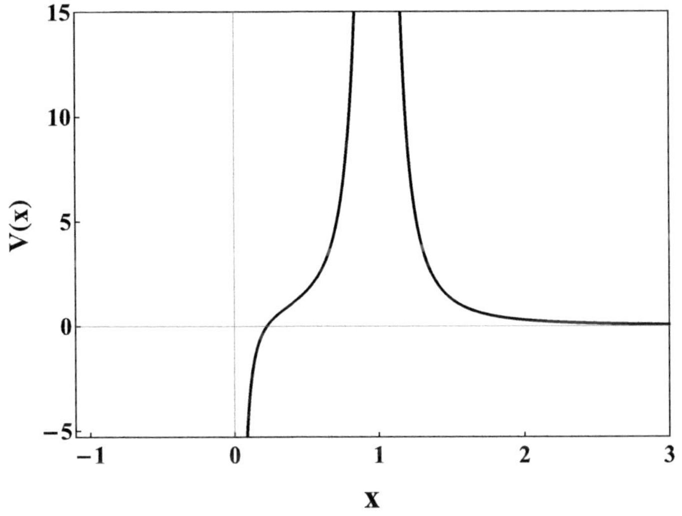

Fig. 1 Shows the variation of potential $V(x)$ with x

Hamiltonian system. This classical study will provide the connection of prime numbers with the Lambert W functions [13] as reported earlier [14].

2.2 *Phase Space, Trajectories and Zeroes*

Now we study the classical aspects of the problem in the phase space of x and p. Since the potential function in the above Hamiltonian is nonlinear in x, where the nonlinear dynamics [12] of this system is studied here. In nonlinear dynamics the trajectories and zeroes or fixed points are very important quantities which reveal the interesting dynamics peculiar to the system. We find $p(x)$ from the classical Hamiltonian given in (1) as

$$p(x) = \pm 2\sqrt{E - \frac{1}{4x^2}\left(\frac{1}{ln(x)} + \frac{3}{2ln^2(x)}\right)} \tag{2}$$

The plot between x and p for a given energy parameter E using Eq. (2) which is shown in Fig. 2. To be specific we choose $E = -10, 0, 1, 10, 100$ and analyze those trajectories on the x-p plane. We find there are only two zeros indicated by the circles (x is a zero or a fixed point if $\dot{x} = p = 0$ at that point). These two zeroes are half stable and half unstable fixed points. On the right branch of each of the x-p plot different E of Fig. 2, if a particle has positive velocity, then it will go away from the right side zero to ∞ and if it has negative velocity, it will come to that zero and stay there. On the other hand in the left branch, it is opposite in nature that is if the particle

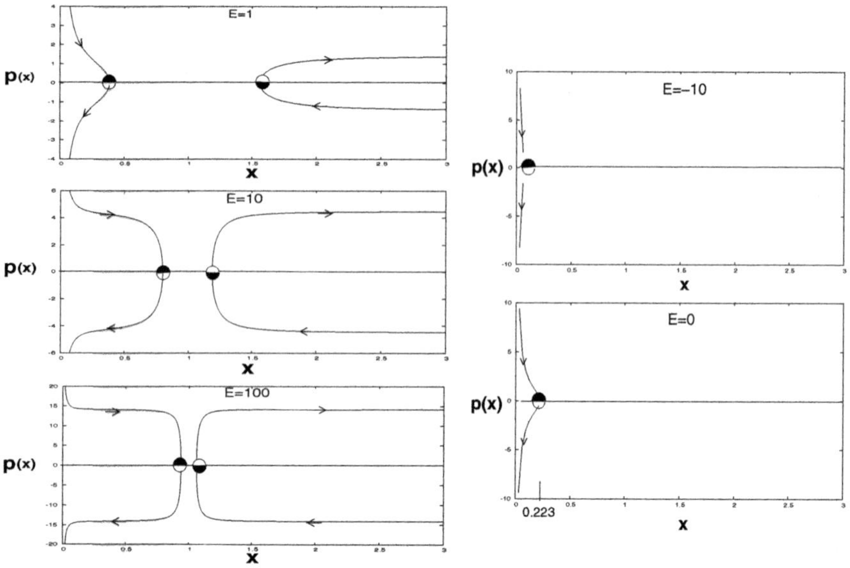

Fig. 2 Plot of momentum $p(x)$ as function of position x for five different values of energy E. The circles indicate the fixed points. A half full and half empty circle indicates half stable and half unstable nature of the fixed points. The two fixed points come closer as the energy increases. When energy is zero, one fixed point is at $x_2 = 0.223$ and other one is at infinity. For negative energy the left fixed point moves towards zero from 0.223

has positive velocity then it will move towards the fixed point and if it has negative velocity, then it will recede from it. As is done in standard nonlinear dynamics study, stability is shown in darkness and instability in emptiness of the small circles drawn at the fixed points. We also see that as the value of energy parameter E increases, the two zeroes come closer and when $E \to \infty$ they merge at $x = x_1 = x_2 = 1$. We also find, when $E = 0$, there is only the left branch cutting the x-axis at $x_2 = 0.223$, and the other one x_1 is moved to infinity. As E becomes negative, the left side zero is at a value greater than 0 and less than $x_2 = 0.223$.

2.3 Bifurcation with E as the Varying Parameter

As discussed in the Subsec-B, since the fixed points [12] are important in describing the dynamics, we make a detail study of these fixed points or zeroes. We see that in the Fig. 2, the two zeroes come closer as energy E increases. To find an analytic expression for the distance between the two zeroes or to study their bifurcation as a function of energy E, we first find out the positions of the two zeroes as functions of energy E. Putting $p = 0$ in (2) we obtain Eq. (3) which is plotted in Fig. 3

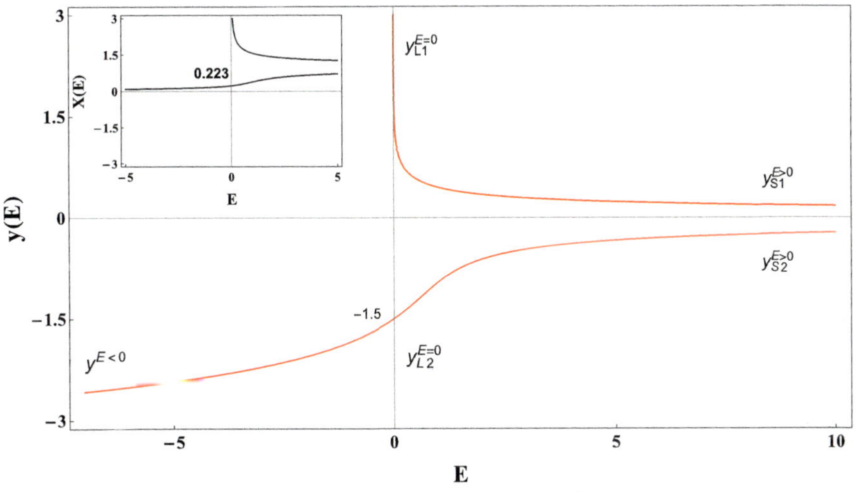

Fig. 3 The plot is the bifurcation diagram representing the positions of fixed points in x-space and y-space defined as $(y = ln(x))$ as functions of energy

$$\frac{1}{ln(x)} + \frac{1.5}{ln^2(x)} - 4Ex^2 = 0 \tag{3}$$

The solution of (3) when $x = e^y$ or $(ln(x) = y)$ and (large x or y) for any real E is given as $y = 0.5W_n\left(\frac{1}{2E}\right)$, ($W_n(x)$ being the famous Lambert W function [13] corresponding to $n = 0$ as the principal branch and other corresponds to the branch having $n = -1$).

The appearance of the Lambert W function in the prime number analysis is deep rooted at different levels of analogy. It has been shown [14] recently that the prime counting function $\pi(x)$ $(x \to \infty)$ is approximately equal to $exp(W_0(x))$ where $W_0(x)$ is the principal branch of the Lambert W function.

3 Classical Dynamics: Relativistic

In order to study the relativistic dynamics of the problem we construct the relativistic Hamiltonian using the Klein-Gordon equation [10]. The K-G equation is given as

$$H = \sqrt{p^2c^2 + m^2c^4} + V(x) = E \tag{4}$$

where the first term is the relativistic kinetic energy and the $V(x)$ is the potential used earlier in the non-relativistic analysis.

3.1 Phase Space, Trajectories and Zeroes

Now we study the classical aspects of the problem in the phase space of x and p when the above relativistic Hamiltonian is taken into consideration. From (4) we get

$$p(x) = \pm\sqrt{\left(\frac{E - V(x)}{c} + mc\right)\left(\frac{E - V(x)}{c} - mc\right)} \tag{5}$$

We will take $m = 1$ and $c = 1$ in our analysis without loss of generality. The flow shows that the dynamics is different than the corresponding non-relativistic case. The plot given in Fig. 4 shows that we have more number of trajectories and fixed points.

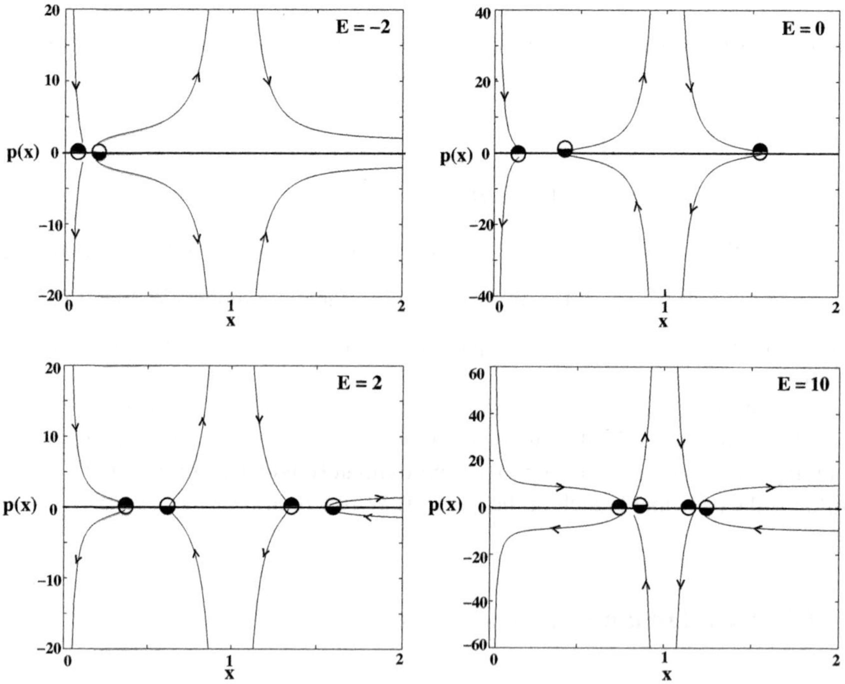

Fig. 4 The plot represents the trajectories and the positions of fixed points in x-space as functions of energy for the relativistic case. We see that when $E \leq -1$, there are two fixed points, $-1 < E < 1$, three fixed points and when $E > 1$ we have four fixed points but they pair up

3.2 Bifurcation with E as the Varying Parameter

We can also study the bifurcation in the system by taking E as the varying parameter. Hence putting $p = 0$ in (5) we get two equations which shows the bifurcation phenomenon in the system as,

$$\frac{1}{ln(x)} + \frac{1.5}{ln^2(x)} - 4x^2(E+1) = 0 \tag{6}$$

$$\frac{1}{ln(x)} + \frac{1.5}{ln^2(x)} - 4x^2(E-1) = 0 \tag{7}$$

These two equations are simultaneously valid. Now we have plotted the bifurcation diagram $x = x(E)$ in Fig. 5.

We see that when $E \leq -1$, there are two fixed points, for $-1 < E < 1$, three fixed points and when $E > 1$ we have four fixed points but they pair up (as the red and blue line).

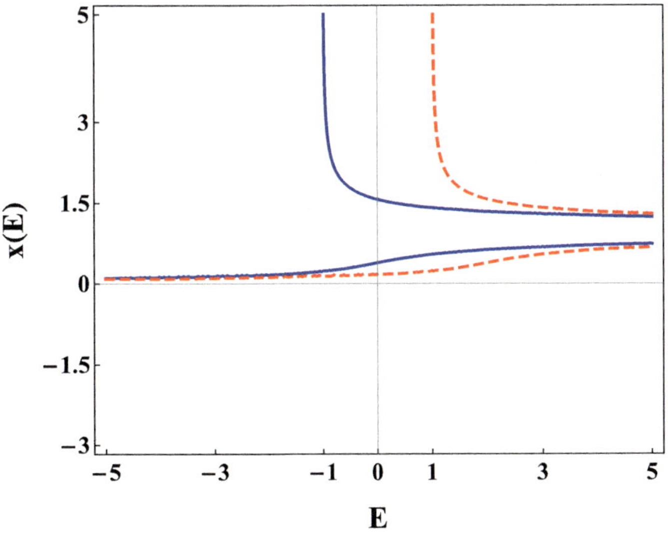

Fig. 5 The plot is the bifurcation diagram representing the positions of fixed points in x-space as functions of energy. We see that when $E \leq -1$, there are two fixed points, $-1 < E < 1$, three fixed points and when $E > 1$ we have four fixed points but they pair up (as the dotted red line for (6) and solid blue line for (7).)

4 Appendix: Derivation of the Potential

Here we construct the wave function $\psi(x)$ of Schrodinger equation (9) through the asymptotic form of prime counting function. The prime counting function denoted as $\pi(x)$ gives the number of primes below the real number x. if we take x as 4, then $\pi(x) = 2$ i.e. 2 and 3. $(\pi(x)/(x/ln(x)) \rightarrow 1 \;\; as \;\; x \rightarrow \infty)$ becomes $\frac{x}{ln(x)}$ in its asymptotic form [9]. The single particle local density for prime numbers in one dimension is found to be $\rho(x) = \frac{\pi(x)}{x} = \frac{1}{ln(x)}$. This gives a homogeneous distribution of the prime numbers below each value of x. In stead of considering $\pi(x)$ which is a stair case function [3], its asymptotic form $\frac{x}{ln(x)}$ is used. This consideration helps in obtaining a continuous and smooth distribution of density.

The wave function is obtained to be

$$\psi(x) = \sqrt{(}\rho(x)) = \frac{1}{\sqrt{ln(x)}} \tag{8}$$

We are concerned about a single particle system within a potential. So we write down the one particle Schrodinger equation [15] as

$$H\psi(x) = \left(\frac{-\hbar^2\nabla^2}{2m} + V(x)\right)\psi(x) = E\psi(x) \tag{9}$$

In this problem,since we know the asymptotic behaviour of prime counting function, we construct a probability density associated with primes as a function of x and from that we calculate the wave function. By using this single particle wave function, we find a formula for the effective potential through which we can study the classical aspect of the prime number system. We derive [10] the potential function by using the wave function $\psi(x)$ from the Schrodinger equation (9) (up to an additional constant and without loss of generality we have taken the constant $E = 0$ in the $V(x)$). Taking $\hbar^2/m = 1$, the interaction potential $V(x)$ appears as

$$V(x) = \frac{1}{4x^2}\left(\frac{1}{ln(x)} + \frac{3}{2ln^2(x)}\right) \tag{10}$$

5 Conclusion

In conclusion, a prime number system that is equivalent to an interacting quantum many-particle system is represented by a single particle Schrodinger equation. The interaction potential is obtained and the nonlinear classical dynamics associated with this novel system are studied. We show that finding large prime numbers, which is otherwise computationally challenging becomes easier. The fixed points associated with the classical trajectories are obtained to be half stable and half unstable. The Lambert W function appears in the solution of fixed points and is a function of energy. Thus by constructing the interaction potential for the prime number system, its properties have been investigated.

References

1. Wells ,D.: Prime Numbers: the Most Mysterious Figures in Math, p. 56. Wiley (2005)
2. Ribenboim, P.: The New Book of Prime Number Records, 3rd edn, pp. 252–253. Springer, New York, NY (1995)
3. Berry, M.V., Keating, J.P.: The Riemann zeros and eigenvalue asymptotics. SIAM Rev. **41**(2), 236 (1999)
4. Bender, C.M., Brody, D.C., Muller, M.P.: Hamiltonian for the zeros of the Riemann zeta function. Phys. Rev. Lett. **118**, 130201 (2017)
5. Julia, B.: Statistical theory of numbers. In: Luck, J.M., Moussa, P., Waldschmidt, M. (eds.) Number Theory and Physics, p. 276. Springer, Berlin (1990)
6. Julia, B.: Phys. A: Stat. Mech. Appl. 203(34), 425–436 (1994)
7. Billingsley, P.: Prime numbers and Brownian motion. Am. Math. Mon. **80**, 1099 (1973)
8. Okubo, S.: Lorentz-invariant Hamiltonian and Riemann hypothesis. J. Phys. A **31**, 1049 (1998)
9. Ingham, A.E.: The Distribution of Prime Numbers, pp. 1–3. Cambridge University Press, Cambridge (1932)
10. Griffiths, D.J.: Introduction to Quantum Mechanics, p. 19. Prentice Hall, New Jersey (1995)
11. Goldstein, H., Poole, C., Safko, J.: Classical Mechanics, 3rd edn, pp. 334–337. Addison Wesley, New York (2000)
12. Strogatz, S.H.: Nonlinear Dynamics and Chaos, p. 18. Perseus books, Massachusetts (1994)
13. Corless, R.M, et al.: On the LambertW function. Adv. Comp. Math. **5**, 329 (1996)
14. Visser, M.: Primes and the LambertW function. Mathematics **6**, 56 (2018)
15. Mishra, S., Pfeifer, P.: FAST TRACK COMMUNICATION: Schrdinger equation for the one-particle density matrix of thermal systems: an alternative formulation of Bose Einstein condensation. J. Phys. A.: Math. Theor. **4**(0), F243 (2007)

Other Fields of Nonlinear Dynamics

Dynamics of Chemical Excitation Waves Subjected to Subthreshold Electric Field in a Mathematical Model of the Belousov-Zhabotinsky Reaction

Anupama Sebastian, S. V. Amrutha, Shreyas Punacha, and T. K. Shajahan

Abstract We present a numerical study of the dynamics of spiral waves in a weak external electric field, using the Oregonator model of the Belousov-Zhabotinky (BZ) reaction. Both free and pinned spiral waves are studied in two types of electric fields: unidirectional (DC) and Circularly Polarised Electric Field (CPEF). Both free spirals and pinned spiral waves rotate faster in the DC field. The CPEF can help a free spiral to be spatially confined. A pinned spiral period can be controlled by varying the period of the CPEF. Both DC and CPEF can unpin the pinned spiral wave, but the minimum electric field required to unpin is much less with CPEF compared to DC. Thus, CPEF is more energy efficient to unpin a pinned spiral wave.

Keywords Excitable medium · Spiral wave · Belousov-Zhabotinsky reaction · Subthreshold stimulation · Unpinning · Critical threshold

1 Introduction

Excitable systems, in general, are non-equilibrium systems with a stable resting state. They can only be aroused to a transitory excited state after crossing a certain threshold. However, perturbations below the threshold go unnoticed since they cannot set off the system to an excited state. Following the excitation, the medium returns to their resting state after a certain time period called the refractory period, during which further perturbations cannot re-excite the system. Unlike the electromagnetic waves, which obey the superposition principle, excitation waves annihilate upon mutual

A. Sebastian · S. V. Amrutha · S. Punacha · T. K. Shajahan (✉)
Department of Physics, National Institute of Technology Karnataka, Surathkal, Mangalore 575025, Karnataka, India
e-mail: shajahan@nitk.edu.in

S. Banerjee and A. Saha (eds.), *Nonlinear Dynamics and Applications*,
Springer Proceedings in Complexity,
https://doi.org/10.1007/978-3-030-99792-2_105

collision. Many excitable systems exist in nature, including the heart muscles [1, 2], chicken retina [3, 4], slime-mold aggregates [5], Xenopus oocytes [6, 7], and Belousov-Zhabotinsky (BZ) reaction [8, 9].

One of the characteristic features of a two-dimensional excitable medium is the fascinating patterns such as rotating spiral and target waves. However, in the case of systems like the heart muscle, the high-frequency rotating spiral waves are known to disrupt the natural sinus rhythm. Furthermore, rotating spirals get stabilized once they get pinned to an obstacle. The pinned spiral waves play a key role in the progression of dynamical disorders, including cardiac arrhythmia [10, 11], and epileptic convulsions [12]. Understanding the dynamics of the pinned spiral waves is crucial to develop efficient techniques to control them.

The dynamics of spiral waves and their interaction with the external perturbations can be easily studied in the Belusov-Zhabotinsky reaction. It is an oscillating system that produces patterns by oxidizing malonic acid in the presence of a metal catalyst such as ferroin. External forcing, such as periodic illumination [13] and electric field application [14], are commonly used to study spiral dynamics. The electric field is known to influence the transport of ionic species in chemical media [15]. As a result, the spiral core drifts by forming a parallel and perpendicular component to the direction of the applied field [16]. Li et al. developed a theory of spiral wave drift caused by weak ac and polarised electric fields. Using response function theory, they derive the spiral drift velocity and direction [17]. Circularly Polarised Electric Field (CPEF) with rotational symmetry has been utilized to regulate spiral drift [18]. Frequency synchronization occurs if the CPEF and spiral are having a comparable frequency [19]. The CPEF has been used in cardiac models to investigate the control of both 2D [20] and 3D [21] pinned excitation waves. Punacha et al. proposed a theory for WEH-induced unpinning and showed that spirals can always be unpinned below a threshold time period of CPEF [22]. Furthermore, a unidirectional field is used for the electrically forced release of pinned spiral waves in the BZ medium when the field strength surpasses a critical threshold (i.e., for supra-threshold fields) [23]. Previous studies, however, have not looked into how a subthreshold field, one that is below a critical threshold, interacts with free and pinned spirals in a BZ medium.

In this article, we investigate the dynamics of both free and pinned spirals in the BZ medium exposed to the subthreshold electric field. We compare the behavior of spiral waves with and without an obstacle. We find that the subthreshold electric field generates a shift in the spiral period. At greater field strengths, this effect becomes more pronounced. Compared to DC, CPEF has a lower critical threshold for unpinning, making it a more energy-efficient method to control spiral waves.

2 Methodology

We model the BZ reaction system using the two-variable Oregonator model [24, 25]. The activator, 'u', initiates the reaction, and the inhibitor, 'v', with its slow dynamics, returns the system to the resting state. Their combined effect will steer the entire dynamics. In BZ reaction, u and v corresponds to the chemical concentration of $HBrO_2$ and catalyst respectively. In the presence of an electric field, E, the dynamics of u, and v are given by the following equations.

$$\frac{\partial u}{\partial t} = \frac{1}{\epsilon}(u(1-u) - \frac{fv(u-q)}{u+q}) + D_u \nabla^2 u + M_u \mathbf{E} \cdot \nabla u \tag{1}$$

$$\frac{\partial v}{\partial t} = (u - v) + D_v \nabla^2 v + M_v \mathbf{E} \cdot \nabla v \tag{2}$$

Electric field (E) is implemented in above equations by using an advection term $\mathbf{E} \cdot \nabla u$ and $\mathbf{E} \cdot \nabla v$ respectively. The model parameters are $q = 0.002$, $f = 1.4$, with diffusion coefficients $D_u = 1.0$, $D_v = 0.6$ and ionic mobilities $M_u = 1.0$ and $M_v = -2.0$. The value of ϵ is 0.01 and is used to explicitly determine the system's excitability.

The entire 300 $\times$ 300 computation domain is discretized in space into grids of uniform size $dx = dy = 0.1$ space units (s.u). The temporal evolution is studied using the explicit forward Euler technique with a timestep, $dt = 0.0001$ time units (t.u). Space and time are both measured in dimensionless units. A five-point Laplacian operator gives the coupling between the grids. No flux boundary conditions are imposed on the domain boundary. We use the phase-field method to implement them on the obstacle boundary [26].

An anticlockwise rotating spiral (ACW), either free or pinned to an obstacle of radius, 'r' s.u, is created at the center of the domain. The diffusion coefficient D_u = 0.0001 is set inside the obstacle. After five sustained rotations of the spiral wave with a period (T_s) in the medium, we apply (i) unidirectional DC and (ii) rotating CPEF of strength E.

The DC field along the x-axis is modeled as

$$\mathbf{E}_{DC} = E\hat{i} \tag{3}$$

and we implement an anticlockwise rotating CPEF in the form

$$\mathbf{E}_{CPEF} = E\left(\cos\left(\frac{2\pi t}{T}\right)\hat{i} + \sin(\frac{2\pi t}{T})\hat{j}\right) \tag{4}$$

where T denotes the rotational period of the CPEF. Later, we vary the pacing ratio, i.e., $p = \frac{T_s}{T}$ from 0.5 to 2 in the steps of 0.25 by altering the rotational period of the field. We apply an electric field until the critical threshold for pinned spirals, which is the lowest field strength for unpinning.

3 Results and Discussion

Our numerical study involves the interaction of the electric field of subthreshold amplitude with the rotating spiral. We consider both free and pinned spiral waves in this study.

We start with a unidirectional DC electric field in the medium oriented along the positive x-axis. A free spiral traces out a large core with increasing field strength and moves towards the positive electrode, as shown in Fig. 1a. We define the angle formed by the drift direction with the negative x-axis (represented as OX) as 'Θ', whereas the linear distance traveled by the spiral tip in one spiral rotation is called 'λ'. We calculate λ as the distance between two adjacent petals, which can also be referred to as the petal width. With the strength of the advective field, both Θ and λ increases (see Fig. 1b, c). With increasing field strength, E from 0.2 to 0.8, Θ rises quickly, reaching saturation at around 90 degrees at high field strength. Therefore, the spiral drifts precisely perpendicular to the direction of strong advective fields. The direction of spiral drift also depends on its chirality [16]. On the other hand, λ varies slowly with field strength. As extremely high fields lead to wave break, we limited our study up to $E = 0.8$.

In addition to the drift parameters, we measure the rotational period of the spiral in the presence of the DC field. The normalised spiral period, T_N, is defined as $\frac{T_f}{T_s}$, where T_s and T_f are the spiral periods before and after the forcing, respectively. We observe that the forcing induces an increase in T_N. When the field strength reaches

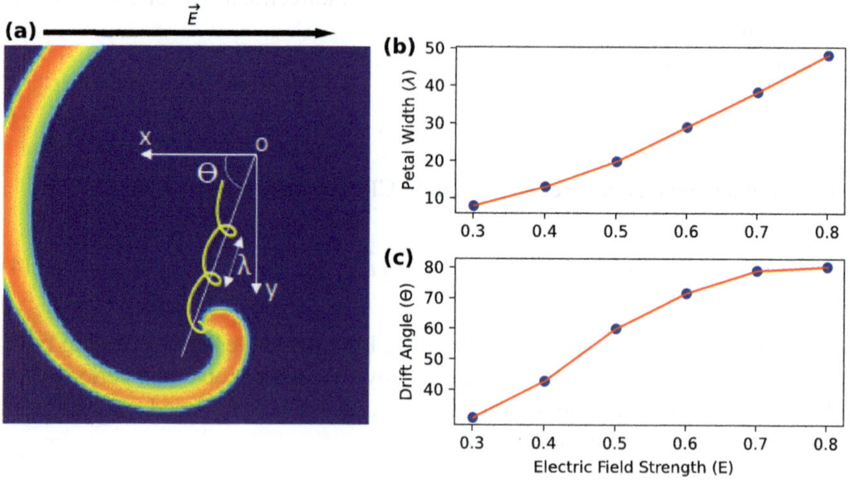

Fig. 1 **Spiral drift in the presence of DC electric field**. **a** An anticlockwise rotating free spiral drifts towards the positive direction of the field by forming an angle Θ and petal width, λ. The field, $\vec{E}$ is oriented in the positive x-direction, as indicated by the thick black arrow. **b**, **c** corresponds to the variation of λ and Θ with the field strength, respectively

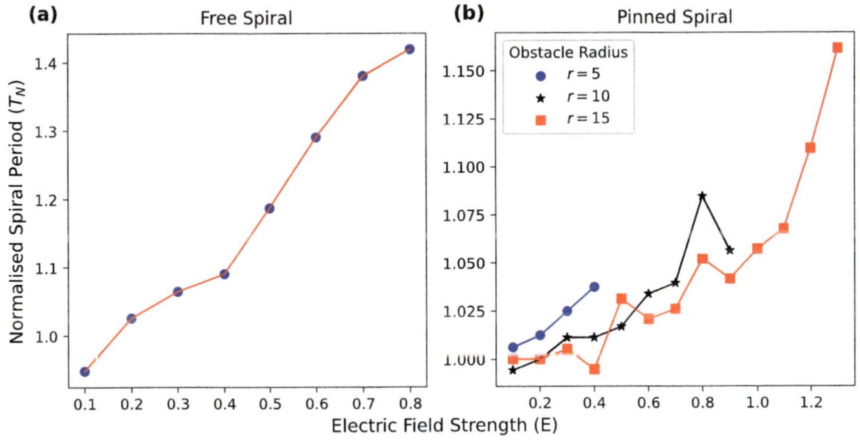

Fig. 2 Variation of spiral rotational period with the subthreshold field strength. a A free spiral's normalised time period (T_N), which increases with field strength. **b** T_N for a pinned spiral anchored to radius, r = 5, 10 and 15, respectively. The field strength (E) is varied till the wave was unpinned

E = 0.8, we see a 40% increase in the spiral period as in Fig. 2a. Hence, the field inhibits spiral rotation and slows it down.

To understand how stable pinned spirals interact with the applied field, we analyze spiral waves attached to obstacles with three different radii r = 5, 10, and 15. The spiral tip gets unpinned when the external forcing reaches a critical threshold [23]. We employ a field with a strength below the critical threshold to avoid unpinning. Like the free spiral, the pinned spiral attached to any obstacle of any size is slowed by the DC field. However, when compared to a free spiral, the increase in the rotational period is insignificant.

Unlike the DC field, CPEF is spatially uniform. In our simulations, we create an anticlockwise rotating CPEF with period T that has rotational symmetry with the spiral. The pacing ratio, $p = \frac{T}{T_s}$ indicates the extent of pacing. For each strength of the electric field, we vary p from 0.5 to 2 in the steps of 0.25. When subjected to CPEF, the spiral tip traces out meandering patterns if CPEF rotates too fast or too slow (see Fig. 3b). However, the motion of the free spiral tip could be spatially constrained for a range of p values closer to 1, as shown in Fig. 3c. In contrast to directed drift in the DC, the rotational symmetry of the CPEF causes the spiral tip to be confined to an area in the medium.

Surprisingly, the response of the pinned spiral to the electric field is strongly influenced by the pacing ratio, p. The spiral period increases with field amplitude for overdrive pacing ($p > 1$), meaning that the spiral slows down. The pacing ratio $p > 2$, on the other hand, does not affect the spiral dynamics (see Fig. 4). We infer that the spiral dynamics are unaffected by the exceedingly high pacing ratio. The spiral period is reduced, or the spiral advances faster when the field strength increases during underdrive pacing ($p < 1$). When the spiral period and the CPEF period are

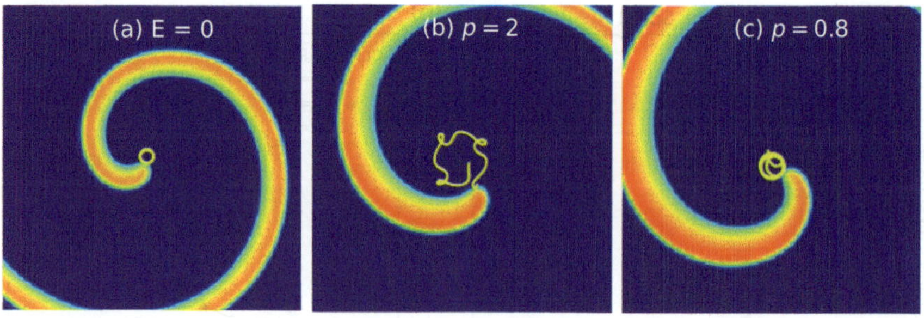

Fig. 3 The tip trajectory of the free spiral before and after applying CPEF. a A free spiral, in the absence of an electric field (i.e., E = 0), performs rigid rotation with a small core. From the same initial phase of the spiral (≈ 45°), CPEF with field strength, E =0.6, and pacing ratios of **b** $p = 2$ and **c** $p = 0.8$ is applied to the medium. Meandering occurs when $p = 2$, but $p = 0.8$ results in spatial confinement

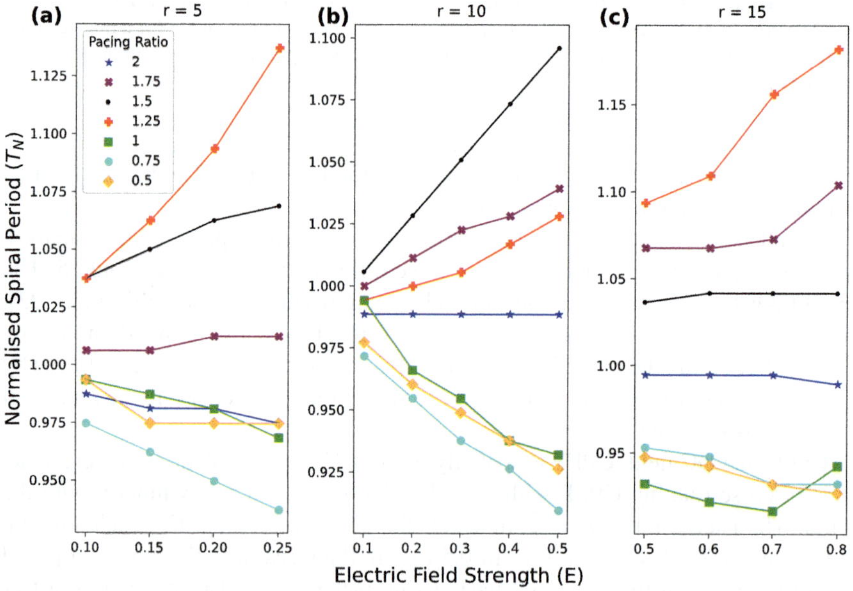

Fig. 4 Response of pinned spiral in the presence of CPEF with different strength and pacing ratio. a–c Shows the change in Normalised Spiral Period (T_N) as a function of field strength for pacing ratios p ranging from 0.5 to 2 for radii r = 5. **b** Is same as **a** for r = 10. **c** Is same as **a** for r = 15. T_N increases with the strength of the electric field for overdrive pacing and decreases for underdrive pacing

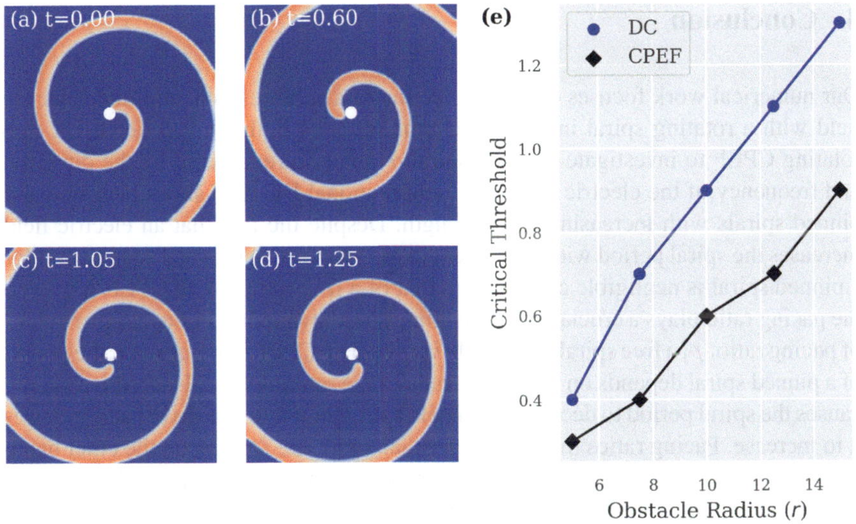

Fig. 5 **Unpinning at a critical threshold. a–d** Snapshots of DC electric field-driven unpinning from an obstacle of radius, $r = 7.5$ at the critical threshold of strength, $E = 0.7$. **e** For both DC (dots) and CPEF with overdrive pacing (diamond), the critical threshold for unpinning is linearly related to the obstacle size. The critical threshold for DC is higher than that of CPEF

both equal ($p \approx 1$), the electric field strength does not affect the T_N. Although the range of T_N varies, the trend mentioned above is followed for all obstacle sizes used in our simulation. Therefore, the spiral alters its rotational period as a response to the change in field frequency.

As the strength of the advective field increases, the pinned spiral's tip drifts away from the obstacle, a process known as unpinning. The critical threshold is the lowest field required to remove the spiral from the obstacle, which may also be thought of as the strength of the obstacle-spiral pinning interaction. In Fig. 5a–d, a DC electric field with a strength of $E = 0.7$ induces unpinning from an obstacle with a radius of r = 7.5 s.u. We investigate the critical thresholds for DC and CPEF for five obstacle sizes with radii ranging from 5 to 15. The pacing ratio is a crucial component in the case of CPEF. Underdrive pacing leads to unpinning only when the pacing ratio is very low. Moreover, unpinning at comparable frequencies ($p \approx 1$) is quite difficult. As a result, we estimate the critical threshold for overdrive pacing in our simulation.

As shown in Fig. 5e, the unpinning critical threshold is directly proportional to the obstacle radius in both DC and CPEF. Furthermore, unpinning may be influenced by the initial spiral phase at field initiation. Throughout this study, we have used the same initial phase ($\approx 45°$) of the spiral. In addition, the critical threshold for DC is higher than that for CPEF for a spiral fixed to any obstacle size. The low critical threshold is a benefit when comparing CPEF to DC. This result helps in the efficient selection of the best low-energy control method.

4 Conclusion

Our numerical work focuses on the interaction of a subthreshold amplitude electric field with a rotating spiral in a BZ medium. We used both unidirectional DC and rotating CPEF to investigate how the spiral dynamics are affected by the direction and frequency of the electric field. The unidirectional DC field slows both free and pinned spirals with increasing field strength. Despite the fact that an electric field increases the spiral period with field strength, the increase in the rotational period for a pinned spiral is negligible compared to that of a free spiral. With CPEF, however, the pacing ratio plays a crucial role in defining spiral dynamics. For a range of values of pacing ratio, p, a free spiral is spatially confined by CPEF. Furthermore, the period of a pinned spiral depends on the pacing ratio of the rotating field. Underdrive pacing causes the spiral period to decrease with field strength, while overdrive pacing causes it to increase. Pacing ratios that are greater than two (i.e., $p > 2$), on the other hand, have no effect. The critical electric field for unpinning by CPEF at overdrive pacing is much lower compared to that at the DC field.

We believe our work has implications for low energy control techniques used for spiral unpinning and control in cardiac dynamics. In particular, a lower critical threshold for CPEF compared to the DC field indicate that CPEF is more energy efficient for unpinning the pinned spiral waves.

5 Author Contributions

A.S and S.V.A conceived the idea and planned the simulations. S.P designed the numerical study as well as the computational framework. T.K.S supervised the project and provided constructive feedback. A.S performed numerical simulations, analyzed the data, and wrote the manuscript with contributions from all authors.

References

1. Davidenko, J.M., Pertsov, A.V., Salomonsz, R., Baxter, W., Jalife, J.: Stationary and drifting spiral waves of excitation in isolated cardiac muscle. Nature **355**(6358), 349–351 (1992)
2. Rostami, Z., Rajagopal, K., Khalaf, A.J.M., Jafari, S., Perc, M., Slavinec, M.: Wavefront-obstacle interactions and the initiation of reentry in excitable media. Phys. A: Stat. Mech. Appl. **509**, 1162–1173 (2018)
3. Gorelova, N.A., Bureš, J.: Spiral waves of spreading depression in the isolated chicken retina. J. Neurobiol. **14**(5), 353–363 (1983)
4. Fernandes de Lima, V.M., Hanke, W.: Relevance of excitable media theory and retinal spreading depression experiments in preclinical pharmacological research. Curr. Neuropharmacol. **12**(5), 413–433 (2014)
5. Siegert, F., Weijer, C.: Digital image processing of optical density wave propagation in Dictyostelium discoideum and analysis of the effects of caffeine and ammonia. J. Cell Sci. **93**(2), 325–335 (1989)

6. Lechleiter, J., Girard, S., Peralta, E., Clapham, D.: Spiral calcium wave propagation and annihilation in Xenopus Laevis oocytes. Science **252**(5002), 123–126 (1991)
7. Chatterjee, M., Sain, A.: Dynamic surface patterns on cells. J. Chem. Phys. (2022)
8. Zaikin, A.N., Zhabotinsky, A.M.: Concentration wave propagation in two-dimensional liquid-phase self-oscillating system. Nature **225**(5232), 535–537 (1970)
9. Tyson, J.J.: From the Belousov-Zhabotinsky reaction to biochemical clocks, traveling waves and cell cycle regulation. Biochem. J. **479**(2), 185–206 (2022)
10. Fenton, F.H., Cherry, E.M., Hastings, H.M., Evans, S.J.: Multiple mechanisms of spiral wave breakup in a model of cardiac electrical activity. Chaos: Interdiscip. J. Nonlinear Sci. **12**(3), 852–892 (2002)
11. Alonso, S., Bär, M., Echebarria, B.: Nonlinear physics of electrical wave propagation in the heart: a review. Rep. Progr. Phys. **79**(9), 096601 (2016)
12. Traub, R.D., Wong, R.K.: Cellular mechanism of neuronal synchronization in epilepsy. Science **216**(4547), 745–747 (1982)
13. Liu, G., Wu, N., Ying, H.: The drift of spirals under competitive illumination in an excitable medium. Commun. Nonlinear Sci. Numer. Simul. **18**(9), 2398–2401 (2013)
14. Muñuzuri, A.P., Gómez-Gesteira, M., Pérez-Muñuzuri, V., Krinsky, V.I., Pérez-Villar, V.: Mechanism of the electric-field-induced vortex drift in excitable media. Phys. Rev. E **48**(5), R3232 (1993)
15. Agladze, K.I., De Kepper, P.: Influence of electric field on rotating spiral waves in the Belousov-Zhabotinskii reaction. J. Phys. Chem. **96**(13), 5239–5242 (1992)
16. Steinbock, O., Schütze, J., Müller, S.C.: Electric-field-induced drift and deformation of spiral waves in an excitable medium. Phys. Rev. Lett. **68**(2), 248 (1992)
17. Li, T.C., Gao, X., Zheng, F.F., Pan, D.B., Zheng, B., Zhang, H.: A theory for spiral wave drift induced by ac and polarized electric fields in chemical excitable media. Sci. Rep. **7**(1), 1–9 (2017)
18. Chen, J.X., Zhang, H., Li, Y.Q.: Drift of spiral waves controlled by a polarized electric field. J. Chem. Phys. **124**(1), 014505 (2006)
19. Chen, J.X., Zhang, H., Li, Y.Q.: Synchronization of a spiral by a circularly polarized electric field in reaction-diffusion systems. J. Chem. Phys. **130**(12), 124510 (2009)
20. Feng, X., Gao, X., Pan, D.B., Li, B.W., Zhang, H.: Unpinning of rotating spiral waves in cardiac tissues by circularly polarized electric fields. Sci. Rep. **4**(1), 1–5 (2014)
21. Fu, Y.Q., Zhang, H., Cao, Z., Zheng, B., Hu, G.: Removal of a pinned spiral by generating target waves with a localized stimulus. Phys. Rev. E **72**(4), 046206 (2005)
22. Punacha, S., Shajahan, T.K.: Theory of unpinning of spiral waves using circularly polarized electric fields in mathematical models of excitable media. Phys. Rev. E **102**(3), 032411 (2020)
23. Sutthiopad, M., Luengviriya, J., Porjai, P., Tomapatanaget, B., Müller, S.C., Luengviriya, C.: Unpinning of spiral waves by electrical forcing in excitable chemical media. Phys. Rev. E **89**(5), 052902 (2014)
24. Keener, J.P., Tyson, J.J.: Spiral waves in the Belousov-Zhabotinskii reaction. Phys. D: Nonlinear Phenom. **21**(2–3), 307–324 (1986)
25. Field, R.J., Noyes, R.M.: Oscillations in chemical systems. IV. Limit cycle behavior in a model of a real chemical reaction. J. Chem. Phys. **60**(5), 1877–1884 (1974)
26. Fenton, F.H., Cherry, E.M., Karma, A., Rappel, W.J.: Modeling wave propagation in realistic heart geometries using the phase-field method. Chaos: Interdiscip. J. Nonlinear Sci. **15**(1), 013502 (2005)

Structural Transformation and Melting of the Vortex Lattice in the Rotating Bose Einstein Condensates

Rony Boral, Swarup Sarkar, and Pankaj K. Mishra

Abstract We numerically investigate the effect of the depth and lattice constant of square optical lattice on the vortex lattice structure of the rotating Bose-Einstein condensates. For a given angular velocity and lattice constant of the optical lattice, vortex lattice structure makes a transition from the hexagonal to the fully pinned square lattice upon increasing the pinning strength of the potential. A detailed quantitative analysis has been performed to understand the transition of the vortex lattice structure by changing the angular velocity, lattice constant and strength of the optical lattice potential. We find that the angular velocity at which the minimum potential strength is required to obtain fully pinned square vortex lattice increases upon decreasing the lattice constant of the potential with power law dependence. We also show the effect of random impurities along with pinning potential on the structure of the vortex lattice which triggers melting of the vortex lattice.

Keywords Optical lattice · Vortices · Abrikosov lattice · Random potential

1 Introduction

The experimental realization of Bose-Einstein condensates (BECs) in cold alkali-metallic gases has given an entirely new direction to the research in the cold atom physics [1]. One of the classical problem in ultracold system is to understand the genesis of the generation and dynamics of the vortices, a quantized circulation generated as a topological defect in the rotating BECs. Also the comprehensive understanding of different ordered lattice structure displayed by them upon trapping under the optical lattice has been the main thrust of the research. In last few decades the study related to the vortex lattice formation and its structural transformation in presence of different optical lattice, like square, hexagonal, rhomboid, etc. have caught a great

R. Boral (✉) · S. Sarkar · P. K. Mishra
Department of Physics, Indian Institute of Technology Guwahati,
Guwahati 781039, Assam, India
e-mail: rony176121108@iitg.ac.in

S. Banerjee and A. Saha (eds.), *Nonlinear Dynamics and Applications*,
Springer Proceedings in Complexity,
https://doi.org/10.1007/978-3-030-99792-2_106

attention of the scientific community due to the potential application of this in several other fields of the condensed matter Physics [2]. The experiments in BECs are highly tunable, so it helps to develop a better understanding of formation of pure vortex lattice and their melting in presence of the impurity. In this paper, we numerically investigate the effect of rotation and strength of the pinning potential on the lattice structure of the vortex lattice in rotating BECs.

After the first experimental realization of the vortices in BECs by the Madison Group [3] using laser stirring, the field has seen unprecedented growth at the theoretical and numerical level to unravel the underlying mechanism behind arrangements of the vortices on the lattice structure. Many theoretical investigations in BECs are based on the analysis of the vortex lattice structure and dynamics using the mean-field Gross-Pitaevskii equation. In recent years BECs trapped under the optical lattice have been used to understand many of interesting phenomena in many other fields of condensed matter physics [2]. As for as an example optical lattice are used to realize the Bloch Oscillations [4], Wannier-Stark Ladders [5], Josephson junction arrays [6], and quantum phase transition from a superfluid to a Mott-insulator [7]. There are several research groups mainly focused to investigate the effect of depth and periodicity of the optical lattice on the vortex lattice structure analytically and numerically [8, 9]. Tung et al. [10] experimentally observed the structural phase transition of the vortex lattice by rotating the condensates and investigated the effect of pinning strength of the pinning potential on transformation of the vortex lattice from the hexagonal to the square lattice structure. Pu et al. [11] showed that the structural phase transition of the vortex lattice is very highly responsive to the ratio of number of vortices with respect to the pinning sites when the condensate is trapped in harmonic and optical lattice. They also obtained that the vortex pinning increases when the vortex density matches the density of the lattice potential [12].

In recent years there are several studies that suggest a intricate nature of structural transition of the vortex lattice structures for the binary mixtures of rotating BECs trapped under optical lattice [13, 14]. William et al. [15] found that the dynamical regimes of vortex lattice depends on the depth and the periodicity of the optical lattice potential. In binary dipolar BECs such transitions between different vortex-lattice structures, like, Abrikosov lattice to square lattice, have been observed either by varying the angular velocity or by tuning the ratio of the inter- and intraspecies strengths even in absence of the optical lattice trapping [16]. In this direction Kumar et al. numerically investigated the transition of one vortex lattice to another for the dipolar rotating BECs trapped under square optical lattice and found the transition from the square to vortex sheet like structures upon tuning the ratio of the inter-species and intra-species interaction [17].

Apart from investigating the transformation between the different structure of the vortex in presence of the optical lattice, there are many studies that indicate the melting of the vortex lattice mainly induced by the random impurities present in the condensates. Mithun et al. numerically investigated the effect of the presence of Gaussian defects on the structure of vortex lattice for the rotating BECs trapped under the harmonic as well anharmonic potential and found that the beyond a certain threshold strength of the defect the vortex lattice gets melted [18]. Hu and Gu studied squared potential pinned vortex lattice in presence of random depth optical lattice and noticed the melting of the vortex lattice structure upon increasing the depth of the random potential [19].

So far we find that lots of work in the rotating BECs which have only focused on the different structure and their transition from one structure to another. However, a comprehensive quantitative picture the factors like interaction between the vortex lattice and optical lattice, interaction between the impurities and vortex lattice responsible for melting, etc. are lacking in the literature. In this paper, we first numerically investigate the effect of the the pinning potential strength and rotation on the structure of the vortex lattice. Further using the structure factor and lattice energy of the vortex lattice we compute the minimum rotation and minimum potential strength required to perfectly pin the lattice to the optical lattice which is also associated with the structural transformation of the lattice from the hexagonal lattice structure also know as Abrikosov lattice (AL) to the pinned lattice (PL) which follow the same square lattice structure as optical lattice have. We also identify the region in the potential and rotation parameter space where there is coexistence of both AL state and PL state. We extend our analysis for the vortex lattice structure in presence of the random lattice.

The paper is organized as follows. In Sect. 2, we introduce the basic formulation of the problem and review the optical lattice pinning potential. In Sect. 3 we present the numerical procedure and simulation details to solve GP equation. In Sect. 4 we present our numerical simulation results related to effect of change in the vortex lattice structure in presence of the square optical lattice. A detailed analysis related with different nature of the vortex lattice upon change in rotation rate and strength of the pinning potential is presented. It is followed by a phase diagram that show the different nature of the vortex lattice in the potential strength and rotation parameter space. Here we also discuss the effect of addition of impurities on the vortex lattice structure. In Sect. 5, we finally conclude our observations.

2 Governing Equations

We consider Bose-Einstein condensate gas confined in a harmonic potential and a periodic potential in a frame rotating with an angular velocity Ω along the z-axis. The macroscopic wave function of the condensate obeys the two-dimensional time-dependent Gross-Pitaevskii equation which is given by [12]:

$$(i-\gamma)\hbar\frac{\partial\psi}{\partial t} = [-\frac{\hbar^2}{2m}\nabla^2 + V_{ext} + g\,|\psi|^2 - \mu - \Omega L_z]\psi \tag{1}$$

where, ψ is the condensate wave function, $\nabla^2 = \partial_x^2 + \partial_y^2$, $L_z = -i\hbar(x\partial_y - y\partial_x)$ is the z-component of the angular momentum operator, $V_{ext}(x, y)$ the external optical potential which, we have chosen as a superposition of the harmonic and optical lattice potential, and Ω is the angular velocity of the condensate along the z-axis. Here, we have chosen both condensate and the potential rotates with the same angular velocity. The nonlinearity $g = \frac{4\pi\hbar^2 a_s}{m}$ is the strength of interaction with a_s being the s-wave scattering length and m is the mass of the atom. γ denotes the dissipation due to the presence of non-condensate at finite temperature and μ is the chemical potential of the condensate. The value of γ is set to be 0.03, as obtained via fitting the theoretical results with the experimental one [20]. For numerical simplification we have non-dimensionalized the above set of (1) using the scheme as $x = a_h\tilde{x}$, $t = \frac{\tilde{t}}{\omega_\perp}$, $\psi = \frac{\tilde{\psi}}{a_h^{3/2}}$, where $a_h = \sqrt{\frac{\hbar}{m\omega_\perp}}$, with $\omega_\perp$ is the transverse angular frequency of harmonic trap potential. Here the variables with tilde denote the non-dimensionalized variable. In what follows for simplicity we will remove the tilde from the non-dimensionalized variables. The above changes in the variables lead to the following non-dimensional GPE

$$(i - \gamma)\frac{\partial\psi}{\partial t} = [-\frac{1}{2}\Delta + V_{ext} + g \mid \psi \mid^2 - \mu - \Omega L_z]\psi \tag{2}$$

Where, $V_{ext} = \frac{1}{2}(x^2 + y^2) + V_{\text{lattice}}$, $\Delta = \frac{\partial^2}{\partial x^2} + \frac{\partial^2}{\partial y^2}$. The tilde is omitted for simplicity.

We have chosen the optical lattice as a superposition of the two perpendicular Gaussian beams which can be represented as

$$V_{\text{lattice}} = \sum_{n_1}\sum_{n_2} V_0 \exp(-\frac{|r - r_{n_1 n_2}|^2}{(\frac{\sigma}{2})^2})$$

Here $r_{n_1 n_2} = n_1 a_1 + n_2 a_2$ denotes the lattice vector with n_1 and n_2 number of lattice points in the x- and y- direction respectively, and V_0 is the strength of the laser beam. For the square optical lattice the two lattice unit vectors are given by $a_1 = a(1, 0)$ and $a_2 = a(0, 1)$. The width of the laser beam is considered to be $\sigma = 0.65$.

3 Simulation Details

We have used strange Alternate Direction Implicit-Time Splitting pseudo spectral (ADI-TSSP) method to solve the 2DGPE (2). In our numerical calculation, we consider spatial step as Δx=Δy=0.1860 and time step as Δt=0.001. All the simulation runs are performed for fixed non-linearity $g = 1000$. Due to the non-zero value of γ, the time development of (2) does not conserve the norm of the wave function. In order to preserve the norm of the condensate wave function, we calculate the correction to the chemical potential in each time step as

$$\Delta\mu = (\Delta t)^{-1} \ln \left[\int d^2 r \mid \psi(t) \mid^2 / \int d^2 r \mid \psi(t + \Delta t) \mid^2 \right]$$

and further run the simulation for the longer time until the stationary state is obtained. We used GPELab, a Matlab Toolbox, for the implementation of the above numerical procedure [21].

In simulation, we first determine the ground-state wave function of the condensate in the absence of rotation and then using this ground-state wave function as the initial state of the GP equation, the equilibrium state is obtained. We analyze the vortex lattice structure using the density profile of the equilibrium condensate. We identify the structural phase transition of the vortex lattice in the presence of optical lattice using: (i) density profile of the equilibrium condensate, (ii) density profile in reciprocal space, (iii) the lattice potential energy $< E_{\text{lattice}} >= \int dr \psi^*(\mathbf{r}) V_{\text{lattice}} \psi(\mathbf{r})$, and (iv) the structure factor of the vortex lattice $S(k) = \frac{1}{N_c} \sum_i n_i e^{ik \cdot r_i}$. Here n_i, r_i, and N_c are the winding number of individual vortices, the position of the ith vortex, and the total winding number, respectively. We calculate the above mentioned physical quantities for different optical lattice amplitude and observe the variation of them with the strength of the lattice potential.

4 Results and Discussion

4.1 Effect of Square Optical Lattice on the Vortex Lattice

In this section, we present our numerical results for the vortex lattice structure of a BECs in the presence of square optical lattice potential. In our numerical simulations, we first fix the lattice constant a of the potential and then vary the potential amplitude V_0 for different angular velocities. Figure 1 depicts the density profile of the condensate in real and Fourier space for lattice constant $a = 2.2$, at a constant angular velocity $\Omega = 0.8$ with increasing V_0.

As we analyze the ground state density of the condensate, we find that in the absence of optical lattice potential the single quantized vortices arranged in a Hexagonal pattern. These lattice structure is also known as Abrikosov lattice (AL) due to its similarity with the vortex lattice structure obtained in the type-II superconductor [22]. This is also observed by several other groups [23]. We notice slight distortion from the Hexagonal lattice for a sufficiently small V_0. At high pinning strength of the potential, all the vortices get pinned to the antinodes of the optical lattice potential and the vortex lattice gets transformed to the pinned lattice (PL) which is square for our case. We find that as the pinning strength of the potential is increased the resultant ground state of the vortex lattice gets transformed from AL state to the PL state.

In the bottom panel of Fig. 1, we show the reciprocal lattice structure of the vortex lattice for various pinning potential strength. In the absence of the optical

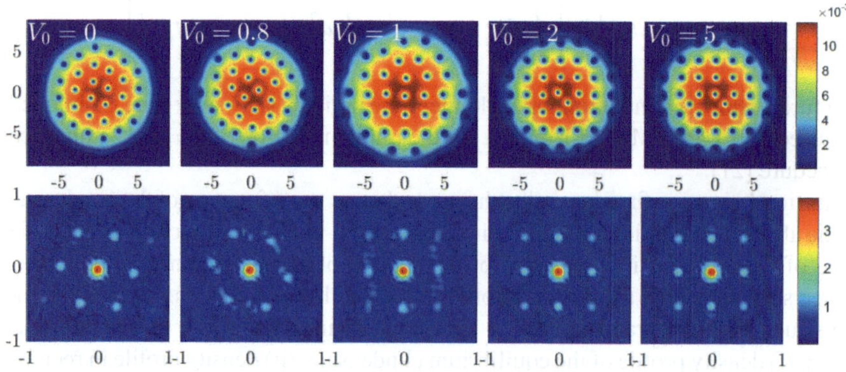

Fig. 1 Upper panel: Density profile in real space with the variation of pinning potential strength: $V_0 = 0.0, 0.8, 1.0, 2.0, 5.0$ (increasing strength from left to right). Bottom Panel: Fourier transform of the above density profile in k space. The density profile at $V_0 = 0.0$ shows the Abrikosov lattice (AL) lattice structure and with increment of the V_0 leads to pin the vortex lattice, for sufficient high pinning strength pinned lattice (PL) formed. Here, we used lattice constant $a = 2.2$ and $\Omega = 0.8$, $g = 1000$

lattice potential($V_0 = 0$), the reciprocal lattice is hexagonal, as quite evident from the density plot. Upon increase of the potential strength, we find that for low optical lattice strength (V_0) the vortex lattice gets distorted from the Hexagonal pattern at $V_0 = 0.8$. Upon further increase of V_0 leads to the formation of distorted square pattern which becomes perfect square at $V_0 = 2.0$. The square lattice structure of the vortex lattice becomes more pronounced at $V_0 = 5$.

In order to quantify the potential strength at which the structural transformation in the vortex lattice takes place, in Fig. 2, we show the variation of structure factor profile and lattice energy with respect to V_0 for different Ω. We notice that at small rotation ($\Omega = 0.55, 0.58, 0.62$), the structure factor profile and lattice energy show there exists an intermediate state where it is difficult to determine the structure of vortex lattice. This we term as the coexistence state (CS). We find that for $\Omega = 0.55$ the CS begins at $V_0 = 0.6$ and the vortex structure is completely pinned above $V_0 = 1.4$. For $\Omega = 0.58$ and $\Omega = 0.62$, the CS starts from $V_0 = 0.4$ and 0.8 respectively and the vortex lattice gets fully pinned above $V_0 = 1.4$ and 2.2, respectively.

The right panel of Fig. 2 shows the lattice energy and structure factor plots for relatively higher rotations ($\Omega = 0.725, 0.8, 0.85$) for the lattice constant $a = 2.2$. For $\Omega = 0.725$ we find that the lattice energy diminishes steadily upon increasing V_0, which suggests partial pinning of vortices. A sharp decrease in $E_{lattice}$ is noticeable at $V_0 = 0.6$ which indicates that the vortices are pinned above this pinning strength. Similarly, the structure factor $S(k)$ increases successively against V_0 and exhibits steady behavior above $V_0 = 0.6$ where all vortices get attached to the pinning sites. Similar trend has been observed for $\Omega = 0.8$ and 0.85 (See Fig. 2e, f).

In Fig. 3a we plot the phases of the vortex lattice in $\Omega - V_0$ plane. We find that at $\Omega = 0.725$ the minimum strength of potential ($V_{\min}$) required to get the perfectly

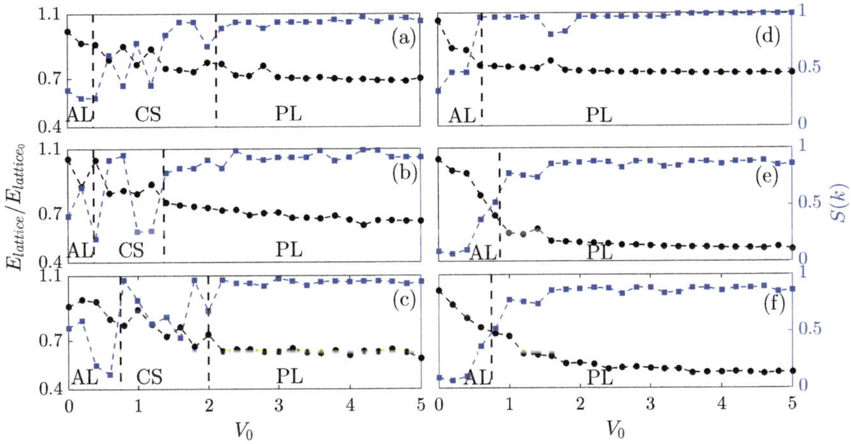

Fig. 2 Variation of structure factor $S(k)$ and lattice energy $E_{lattice}$ with pinning strength V_0 for different rotational frequencies: **a** $\Omega = 0.55$, **b** $\Omega = 0.58$, **c** $\Omega = 0.62$, **d** $\Omega = 0.725$, **e** $\Omega = 0.8$, and **f** $\Omega = 0.85$. The three regions are defined as Abrikosov lattice (AL), coexisting state (CS), and pinned lattice (PL). The faster rotation of the condensate can decrease the coexisting region and after a threshold value of Ω vortex lattice makes a sharp transition from Abrikosov lattice to pinned lattice. Here, other parameters are same as Fig. 1

pinned square lattice of the vortices is 0.6. The rotation at which the minimum strength of the potential is required to perfectly pin the lattice depends upon the lattice constant of the optical lattice. We vary the lattice constant between $a = 1.8 - 2.2$ and observe that the transformation of the vortex lattice structure against V_0 for different lattice constant a. The phase diagrams for other lattice constants (a) (does not present here) have the same nature as those for $a = 2.2$. In particular, there exists a rotational angular velocity at which the strength of the potential is minimum to get the pinned state. As we analyze the relation between the lattice constant a and the rotational velocity Ω at which minimum pinning strength is required to get the pinned lattice, we find that it follows

$$a = C\Omega^d \tag{3}$$

where constant $C = 1.76$ and d is the exponent of Ω as shown in Fig. 3b. We find that the magnitude of d obtained for the square lattice is ~ -0.73. This implies that the pinning strength V_0 takes the least value when a fulfills the above relation.

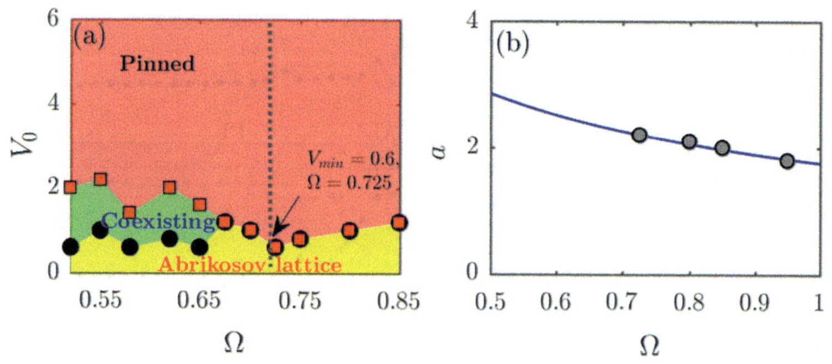

Fig. 3 **a** Phase diagram of different lattice structures depending upon the value of $S(k)$ obtained from Fig. 2 for different V_0 at lattice constant $a = 2.2$. Here (yellow, green, orange) region illustrates the AL, CS, and PL domains. Dashed line annotated for finding Ω corresponds to minimum pinning strength($V_{\min}$). **b** Optimum rotational frequencies Ω (obtained from phase diagram) for different lattice constant. Here, the solid line represents the fitted curve with (3)

4.2 Effect of Random Impurities on Pinned Square Lattice Structure

So far, we have presented the results in which the vortex lattice makes transformation from the Hexagonal to the square pinned lattice upon varying the strength of the potential. In this section, we present vortex lattice melting in BECs due to the random impurities or disorder in the system. Disorder is introduced in the system by the external potential associated with the random impurities. To introduce impurity, we consider a square optical lattice potential

$$V_{\text{impurity}} = \sum_{n_1}\sum_{n_2} V_{\text{imp}} \exp\left(-\frac{(x - n_1 a_{\text{imp}})^2 + (y - n_2 a_{\text{imp}})^2}{(\sigma/2)^2}\right). \tag{4}$$

Here n_1 and n_2 denotes the number of lattice points, respectively and V_{imp} is the strength of the impurity potential. $a_{\text{imp}} = 1$ is the lattice constant of the impurity potential, and $\sigma = 0.65$ is the width of the laser beam. For random impurities, the impurity potential or disorder is defined by an independent random variable uniformly distributed over $[-V_{\text{imp}}, V_{\text{imp}}]$ at each spatial position [18]. So the total lattice potential is given by

$$V_{\text{Total}} = V_{\text{lattice}} + V_{\text{impurity}} \tag{5}$$

Here V_{lattice} is the square optical lattice potential.

In our numerical simulation, we consider rotational angular velocity as $\Omega = 0.8$, lattice constant $a = 2.2$, and nonlinearity $g = 1000$. We first fix the strength of the lattice potential($V_0 = 5$), and then increase the strength of the impurity potential by

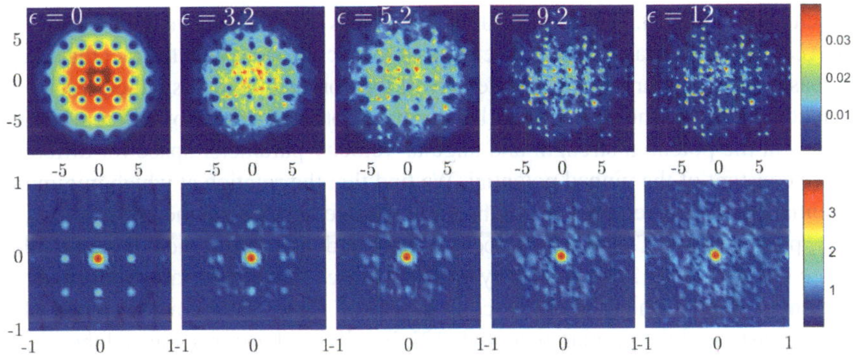

Fig. 4 (Upper panel) Density profile in real space with the variation of impurity potential strength ($\epsilon =$ 0.0, 3.2, 5.2, 9.2, 12). (lower panel) Fourier transform of the above density profile in k space. The density profile at $\epsilon = 0.0$ shows the square lattice (PL) lattice structure and with increment of the ϵ leads to disorders in the vortex lattice, for sufficient high ϵ pinned lattice(PL) shows melting behaviour. Here, we used lattice constant $a = 2.2$ $\Omega = 0.8$, $V_0 = 5$ and $g = 1000$

tuning the parameter ϵ. Here $\epsilon = \frac{V_{\mathrm{imp}}}{V_0}$ is the relative increase in the impurity potential strength with respect to the strength of square optical lattice potential.

In Fig. 4, we show the density profile in real and Fourier space for different ϵ. We observe that in the absence of impurity potential, vortices are arranged in a square pattern due to lattice potential, but the square geometry of the vortex lattice is slightly distorted for a sufficiently small ϵ. Upon further increase in ϵ we find that the lattice structure gets fully distorted and leads to the melting of the lattice structure. The same phenomenon has been observed from the condensate density in the Fourier space, where we can see that the periodic peaks of a square lattice distorted gradually as ϵ is increased. The detailed studies related to effect of different concentration of impurities (by varying the lattice constant of impurities) on the structure of the vortex lattice are underway and will be reported somewhere else.

5 Conclusion

Using the mean field model of the rotating BECs trapped in the square optical lattice we have obtained different form of vortex lattice structure like, Abrikosov, pinned square lattice, coexistence of both Abrikosov and square as the pinning strength of the potential and the rotational angular velocity of the condensate are varied. Using the structure factor and the lattice energy, we have been able to identify the minimum potential strength required to completely transform the vortex lattice from AL to PL state. We have chosen different lattice constant ($a = 2.2$, 2.0 and 1.8) of the optical lattice potential and different rotating speed ($\Omega = 0.55 - 0.95$) of the condensate. Our analysis indicate that the vortex lattice structure makes transition from the AL to PL state directly beyond a critical pinning potential strength for high angular velocity ($\Omega \gtrsim 0.725$). However, for lower angular velocity we have the presence of the AL, CL and PL state. This feature appears to be consistent with the fact that upon increasing the angular velocity the density of the vortices increases which results in perfectly pinning of the vortices. Further, we have obtained a phase of different vortex lattices in the potential strength and angular velocity parameter space for different lattice constant of the pinned potential. We find that the rotation at which minimum potential strength is required to pin the vortex lattice depends on the lattice constant of the pinned lattice. The angular velocity corresponding to that value is lowest for the lattice constant ($a = 2.2$) which physically signifies that structural transformation happens perfectly when the lattice constant of the vortex lattice matches with the lattice constant of the pinning potential [12]. Finally we have investigated the effect of the random impurities on the pinned lattice of the system. We have obtained that the perfect pinned vortex lattice gets melted upon increase in the impurities strength. Similar types of melting behaviour was shown to exist for the Abrikosov lattice. In this paper we have restricted our study for the mean-field model. Using the analysis performed in this paper in other direction it would be interesting to consider the effect of the quantum fluctuation on the vortex lattice structure transformation where the vortex lattice structure have the rich variety ranging from Abrikosov lattice, square, stripped, etc. upon tuning the quantum fluctuation [24].

References

1. Davis, K.B., Mewes, M.-O., Andrews, M.R., van Druten, N.J., Durfee, D.S., Kurn, D.M., Ketterle, W.: Bose-Einstein condensation in a gas of sodium atoms. Phys. Rev. Lett. **75**(22), 3969–3973 (1995)
2. Morsch, O., Oberthaler, M.: Dynamics of Bose-Einstein condensates in optical lattices. Rev. Mod. Phys. **78**(1), 179–215 (2006)
3. Madison, K.W., Chevy, F., Wohlleben, W., Dalibard, J.: Vortex formation in a stirred Bose-Einstein condensate. Phys. Rev. Lett. **84**(5), 806–809 (2000)
4. Ben, D., et al.: Bloch oscillations of atoms in an optical potential. Phys. Rev. Lett. **76**(24), 4508–4511 (1996)
5. Wilkinson, S.R., et al.: Observation of atomic Wannier-Stark ladders in an accelerating optical potential. Phys. Rev. Lett. **76**(24), 4512–4515 (1996)
6. Cataliotti, F.S., et al.: Josephson junction arrays with Bose-Einstein condensates. Science **293**(5531), 843–846 (2001)
7. Greiner, M., Mandel, O., Esslinger, T., Hansch, T.W., Bloch, I.: Quantum phase transition from a superfluid to a Mott insulator in a gas of ultracold atoms. Nature **415**, 39 (2002)
8. Reijnders, J.W., Duine, R.A.: Pinning of vortices in a Bose-Einstein condensate by an optical lattice. Phys. Rev. Lett. **93**(6), 060401 (2004)
9. Bhat, R., Holland, M.J., Carr, L.D.: Bose-Einstein condensates in rotating lattices. Phys. Rev. Lett. **96**(6), 060405 (2006)
10. Tung, S., Schweikhard, V., Cornell, E.A.: Observation of vortex pinning in Bose-Einstein condensates. Phys. Rev. Lett. **97**(24), 240402 (2006)
11. Pu, H., Baksmaty, L.O., Yi, S., Bigelow, N.P.: Structural phase transitions of vortex matter in an optical lattice. Phys. Rev. Lett. **94**(19), 190401 (2005)
12. Sato, T., Ishiyama, T., Nikuni, T.: Vortex lattice structures of a Bose-Einstein condensate in a rotating triangular lattice potential. Phys. Rev. A **76**(5), 053628 (2007)

13. Kasamatsu, K., Tsubota, M., Ueda, M.: Vortex phase diagram in rotating two-component Bose-Einstein condensates. Phys. Rev. Lett. **91**(15), 150406 (2003)
14. Mithun, T., Porsezian, K., Dey, B.: Pinning of hidden vortices in Bose-Einstein condensates. Phys. Rev. A **89**(5), 053625 (2014)
15. Williams, R.A., Al-Assam, S., Foot, C.J.: Observation of vortex nucleation in a rotating two-dimensional lattice of Bose-Einstein condensates. Phys. Rev. Lett. **104**(5), 050404 (2010)
16. Kumar, R.K., Tomio, L., Malomed, B.A., Gammal, A.: Vortex lattices in binary Bose-Einstein condensates with dipole-dipole interactions. Phys. Rev. A **96**(6), 063624 (2017)
17. Kumar, R.K., Tomio, L., Gammal, A.: Vortex patterns in rotating dipolar Bose-Einstein condensate mixtures with squared optical lattices. J. Phys. B: At. Mol. Opt. Phys. **52**, 0525302 (2019)
18. Mithun, T., Porsezian, K., Dey, B.: Disorder-induced vortex lattice melting in a Bose-Einstein condensate. Phys. Rev. A **93**(1), 013620 (2016)
19. Hu, P., Gu, Q.: Vortices in Bose-Einstein condensates with random depth optical lattice. J. Low Temp. Phys. **199**, 1314 (2020)
20. Choi, S., Morgan, S.A., Burnett, K.: Phenomenological damping in trapped atomic Bose-Einstein condensates. Phys. Rev. A **57**, 4057 (1998)
21. Antoine, X., Duboscq, R.: GPELab, a Matlab toolbox to solve Gross-Pitaevskii equations I: computation of stationary solutions. Comput. Phys. Commun. **185**(11), 2969–2991 (2014)
22. Abo-Shaeer, J.R., et al.: Observation of vortex lattices in bose-einstein condensates. Science **292**, 476 (2001); Haljan, P.C., et al.: Driving BEC vorticity with a rotating normal cloud. Phys. Rev. Lett. **87**, 210403 (2001)
23. Baksmaty, L.O., et al.: Tkachenko waves in rapidly rotating BEC. Phys. Rev. Lett. **92**, 160405 (2004); Mizushima, T., et al.: Collective oscillations of vortex lattices in rotating BEC. Phys. Rev. Lett. **92**, 060407 (2004)
24. Hsueh, C.H., Wang, C.W., Wu, W.C.: Vortex structures in a rotating Rydberg-dressed Bose-Einstein condensate with the Lee-Huang-Yang correction. Phys. Rev. A **102**, 063307 (2020)

Effect of Internal Damping on the Vibrations of a Jeffcott Rotor System

Raj C. P. Shibin, Amit Malgol, and Ashesh Saha

Abstract The vibration characteristics of a rotor system with a rigid disc mounted in the middle of a flexible shaft are investigated in order to determine the influence of the shaft's internal damping. A well-known bulk parameter model known as the Jeffcott rotor system (JRS) is considered for the analysis. The dynamics of the system are explored using time-displacement responses, phase-plane, and frequency response graphs. The effect of rotor eccentricity and internal damping on the vibration characteristics of JRS are analysed by numerically simulating the autonomous amplitude-phase equations under simultaneous resonance conditions derived using the method of multiple scales (MMS). Among the intriguing results reported from the amplitude-frequency responses are the multiple solutions, jump events, and multiple loops. The most important finding of this paper is the destabilizing effect of internal damping beyond a critical frequency resulting in the increase of amplitude of vibrations.

Keywords Internal damping · Cubic nonlinearity · Jeffcott rotor system · Jump phenomenon · Multiple solutions

1 Introduction

Rotors are used in a wide range of applications from toys to massive turbines. Unavoidable eccentricity or mass unbalance causes rotor system to whirl along its bearing axis. For the safe operation of the machinery it is necessary to study the effects of different parameters in these lateral vibrations. Jeffcott [1] proposed the most elementary discrete model for the study of rotor vibrations and found that

R. C. P. Shibin · A. Malgol (✉) · A. Saha
National Institute of Technology Calicut, Kattangal 673601, Kerala, India
e-mail: amitmalgol123@gmail.com

R. C. P. Shibin
e-mail: shibinraj009@gmail.com

A. Saha
e-mail: ashesh@nitc.ac.in

S. Banerjee and A. Saha (eds.), *Nonlinear Dynamics and Applications*,
Springer Proceedings in Complexity,
https://doi.org/10.1007/978-3-030-99792-2_107

system can be stable at frequencies higher than critical speed. Different parameters like damping, bearing clearance, gyroscopic and hydrodynamic effects are found to have considerable effect in shaft whirling in a rotor system [2].

Damping within the system itself, termed as internal damping was also found to have an effect on the nature of vibrations of a rotor system. It is found to be stabilizing the system while operating at speeds less than a critical speed and destabilizing at higher speeds [3]. Kimball was the first to discuss the physics of internal damping in his work [4]. Later on, different internal damping models are proposed by different authors such as Dimentberg [5], Vatta and Vigilani [6] and many others. However, an extensive analysis on the effect of internal damping on the vibrations of a rotor system with different nonlinear effects is lacking in literature. Inclusion of an internal damping model proposed by Yukio and Toshio [3] and analysing its effect on a nonlinear Jeffcott rotor model is the major contribution of this paper. For performing the nonlinear analysis, method of multiple scales (MMS) is used [7–9] to obtain the autonomous amplitude and phase equations.

The rest of the paper is organized as follows. In Sect. 2, we go over the mathematical model briefly. Section 3 describes the analytical procedure for getting closed form solutions. Section 4 discusses some of the findings, while Sect. 5 concludes.

2 Mathematical Modeling

As illustrated in Fig. 1, the rotor model consists of a rigid disc mounted on a flexible shaft supported by bearings. The disc's geometric centre and the shaft axis passes through point M, while the disc's centre of gravity is represented by point G. The eccentricity of the centre of gravity G away from the geometric centre M is e.

The governing equations of the rotor system for horizontal and vertical oscillations considering the nonlinear restoring forces (F_{rx} and F_{ry}), internal hysteretic damping forces (D_{ix} and D_{iy}) and external damping forces (F_{dx} and F_{dy}) are described as [1–3]

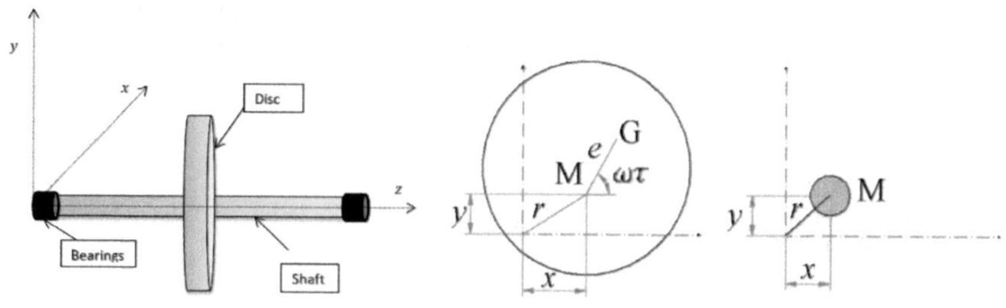

(a) (b) (c)

Fig. 1 **a** Schematic diagram of Jeffcott Rotor System. **b** Cross-sectional view of disc. **c** Cross-sectional view of the shaft

$$m\ddot{x} + F_{dx} + F_{rx} + D_{ix} = me\omega^2\cos(\omega\tau) \text{ and} \quad (1a)$$

$$m\ddot{y} + F_{dy} + F_{ry} + D_{iy} = me\omega^2\sin(\omega\tau) - mg \quad (1b)$$

respectively, where mg is the weight of the rotor and the overdots represent the derivatives with respect to the time τ.

The internal hysteretic damping terms along the horizontal (x-axis) and vertical directions (y-axis) are given by [3]

$$D_{ix} = \zeta^*(\dot{x} + \omega y), \text{ and} \quad (2a)$$

$$D_{iy} = \zeta^*(\dot{y} - \omega x) \quad (2b)$$

respectively, where ω is the angular velocity of the shaft and ζ^* is internal damping coefficent. The external damping forces along the horizontal (x-axis) and vertical directions (y-axis) as assumed to be similar to viscous damping terms represented by.

$$F_{dx} = c_x\dot{x}, \text{ and} \quad (3a)$$

$$F_{dy} = c_y\dot{y} \quad (3b)$$

respectively, where c_x and c_y are the damping coefficients.

The nonlinear restoring forces along the horizontal (x-axis) and vertical directions (y-axis) are expressed as [4]

$$F_{rx} = k_1x + k_2x\left(x^2 + y^2\right), \text{ and} \quad (4a)$$

$$F_{ry} = k_1y + k_2y\left(x^2 + y^2\right) \quad (4b)$$

respectively, where k_1 is linear stiffness coefficient and k_2 is the nonlinear stiffness coefficient.

Substituting Eqs. (2a, 2b) to (4a, 4b) in Eq. (1), we get.

$$m\ddot{x} + c_x\dot{x} + k_1x + k_2x\left(x^2 + y^2\right) + \zeta^*\dot{x} + \zeta^*\omega y = me\omega^2\cos(\omega\tau), \text{ and} \quad (5a)$$

$$m\ddot{y} + c_y\dot{y} + k_1y + k_2y\left(x^2 + y^2\right) + \zeta^*\dot{y} - \zeta^*\omega x = me\omega^2\sin(\omega\tau) - mg \quad (5b)$$

Substituting $\ddot{x} = \ddot{y} = \dot{x} = \dot{y} = \omega = 0$ in Eqs. (5a, 5b), we obtain the static equilibrium in x-direction as $x_{st} = 0$ and the static equilibrium in y-direction (y_{st})

is governed by the following nonlinear equation

$$k_1 y_{st} + k_2 y_{st}^3 = -mg. \tag{6}$$

Shifting the coordinates to the static equilibrium states as $x = x^*$, $y = y_{st} + y^*$, we obtain the governing equations in the shifted coordinates (x^*, y^*) as.

$$m\ddot{x}^* + (\zeta^* + c_x)\dot{x}^* + (k_1 + k_2 {y_{st}}^2)x^* + k_2 x^*({x^*}^2 + {y^*}^2) + 2k_2 x^* y^* y_{st} \\ + \zeta^* \omega(y_{st} + y^*) = me\omega^2 \cos(\omega\tau), \text{ and} \tag{7a}$$

$$m\ddot{y}^* + (c_y + \zeta^*)\dot{y}^* + (k_1 + 3k_2 {y_{st}}^2)y^* + k_2 y^*({x^*}^2 + {y^*}^2) \\ + 2k_2({x^*}^2 + 3{y^*}^2)y_{st} - \zeta^* \omega x^* = me\omega^2 \sin(\omega\tau). \tag{7b}$$

Introducing the dimensionless parameters.

$t = \omega_0 \tau$, $\omega_0 = \sqrt{\frac{k_1}{m}}$, $u = \frac{x^*}{y_{st}}$, $v = \frac{y^*}{y_{st}}$, $\zeta = \frac{\zeta^*}{\sqrt{k_1 m}}$, $f = \frac{e}{y_{st}}$, $\Omega = \frac{\omega}{\omega_0}$, ${\omega_1}^2 = 1 + \lambda$, ${\omega_2}^2 = 1 + 3\lambda$, $\mu_1 = \frac{c_x}{\sqrt{k_1 m}}$, $\mu_2 = \frac{c_y}{\sqrt{k_1 m}}$, $\lambda = \frac{k_2 {y_{st}}^2}{k_1}$,

the governing equations of the horizontally supported Jeffcott rotor system (JRS) for the horizontal and the vertical oscillations given as.

$$u'' + (\zeta + \mu_1)u' + u(1 + \lambda) + \lambda u(u^2 + v^2) + 2\lambda uv + \zeta\Omega(v + 1) = \\ f\Omega^2 \cos(\Omega \mathrm{t}), \textit{and} \tag{8a}$$

$$v'' + (\zeta + \mu_2)v' + v(1 + 3\lambda) + \lambda v\left(u^2 + v^2\right) + \lambda\left(u^2 + 3v^2\right) + 2\lambda uv - \zeta\Omega = \\ f\Omega^2 \sin(\Omega t) \tag{8b}$$

where 't' is dimensionless time, λ is the stiffness coefficient, ζ is an internal damping factor, Ω is rotor speed ratio, and f is the eccentricity ratio. For the horizontal and vertical directions, respectively, u and v are displacements of the JRS, μ_1 and μ_2 are the linear damping coefficients, and ω_1 and ω_2 are the natural frequencies.

As mentioned earlier, the nonlinear analysis by the method of multiple scales (MMS) is performed to obtain the autonomous amplitude and phase equations. This MMS is performed in next section.

3 Method of Multiple Scales (MMS) Analysis

In this section, amplitude and phase equations corresponding to the horizontal and vertical oscillations of the rotor are derived using the method of multiple scales (MMS). Simultaneous resonance condition is considered by defining two detuning parameters σ_1 and σ_2 as

$$\Omega = \omega_1 + \sigma_1, \sigma_2 = \omega_2 - \omega_1 \tag{9}$$

Accordingly, σ_1 represents the closeness of Ω to ω_1, and σ_2 represents the difference between ω_2 and ω_1.

We seek a series of the following form in the MMS:

$$\begin{aligned} u(T_0, T_1, T_2) &= \varepsilon u_1(T_0, T_1, T_2) + \varepsilon^2 u_2(T_0, T_1, T_2) \\ &\quad + \varepsilon^3 u_3(T_0, T_1, T_2) + O(\varepsilon^4), \textit{and} \end{aligned} \tag{10a}$$

$$v(T_0, T_1, T_2) = \varepsilon v_1(T_0, T_1, T_2) + \varepsilon^2 v_2(T_0, T_1, T_2) + \varepsilon^3 v_3(T_0, T_1, T_2) + O(\varepsilon^4). \tag{10b}$$

In Eqs. (10a, 10b), $T_0 = t$ is the fast time and $T_1 = \varepsilon t$, $T_2 = \varepsilon^2 t$, etc. are the slow times, where $0 < \varepsilon \ll 1$. The time derivatives then can be written in terms of fast and slow times as:

$$\frac{d}{dt} = \frac{\partial}{\partial T_0} + \varepsilon \frac{\partial}{\partial T_1} + \varepsilon^2 \frac{\partial}{\partial T_2}, \frac{d^2}{dt^2} = \frac{\partial^2}{\partial T_0^2} + 2\varepsilon \frac{\partial}{\partial T_0} \frac{\partial}{\partial T_1} + \varepsilon^2 \left(\frac{\partial^2}{\partial T_1^2} + 2 \frac{\partial}{\partial T_2} \frac{\partial}{\partial T_1} \right). \tag{11}$$

Other system parameters are scaled according to their orders as:

$$\mu_1 = \varepsilon^2 \mu_{h1}, \mu_2 = \varepsilon^2 \mu_{h2}, f = \varepsilon^2 f_h, \zeta = \varepsilon^2 D_h \tag{12}$$

Substituting Eqs. (10a, 10b) to (12) into Eqs. (8a, 8b) and collecting coefficients of ε yield

$$O(\varepsilon) : \frac{\partial^2 u_1}{\partial T_0^2} + \omega_1^2 u_1 = 0 \tag{13a}$$

$$\frac{\partial^2 v_1}{\partial T_0^2} + \omega_2^2 v_1 = 0 \tag{13b}$$

$$O(\varepsilon^2) : \frac{\partial^2 u_2}{\partial T_0^2} + \omega_1^2 u_2 = f_h \Omega^2 \cos(\Omega \mathrm{T}_0) - 2 \frac{\partial^2 (u_1)}{\partial T_0 \partial T_1} - 2\lambda u_1 v_1 - \Omega D_h \tag{14a}$$

$$\frac{\partial^2 v_2}{\partial T_0^2} + \omega_2^2 v_2 = f_h \Omega^2 \sin(\Omega T_0) - 2\frac{\partial^2 (v_1)}{\partial T_0 \partial T_1} - \lambda u_1^2 - 3\lambda v_1^2 \tag{14b}$$

$$O\left(\varepsilon^2\right) : \frac{\partial^2 u_3}{\partial T_0^2} + \omega_1^2 u_3 = -\frac{\partial^2 u_1}{\partial T_1^2} - 2\frac{\partial^2 (u_2)}{\partial T_0 \partial T_1} - 2\frac{\partial^2 (u_1)}{\partial T_0 \partial T_2}$$

$$+2\lambda(u_1 v_2 + u_2 v_1) + \lambda u_1 (u_1^2 + v_1^2) - \Omega D_h v_1 - D_h \frac{\partial u_1}{\partial T_0} - \mu_{h1} \frac{\partial u_1}{\partial T_0} \tag{15a}$$

$$\frac{\partial^2 v_3}{\partial T_0^2} + \omega_1^2 v_3 = -\frac{\partial^2 v_1}{\partial T_1^2} - 2\frac{\partial^2 (v_2)}{\partial T_0 \partial T_1} - 2\frac{\partial^2 (v_1)}{\partial T_0 \partial T_2} + 2\lambda(u_1 u_2 + 3 v_2 v_1)$$

$$- \lambda v b_1 (u_1^2 + v_1^2) + \Omega D_h u_1 - D_h \frac{\partial v_1}{\partial T_0} - \mu_{h2} \frac{\partial v_1}{\partial T_0} \tag{15b}$$

The solutions of Eqs. (13a) and (13b) can be written as:

$$u_1(T_0, T_1, T_2) = A(T_1, T_2) e^{i\omega_1 T_0} + \overline{A}(T_1, T_2) e^{-i\omega_1 T_0}, \tag{16a}$$

$$v_1(T_0, T_1, T_2) = B(T_1, T_2) e^{i\omega_2 T_0} + \overline{B}(T_1, T_2) e^{-i\omega_2 T_0}. \tag{16b}$$

Substituting these solutions and eliminating secular terms by equating coefficient of $e^{i\omega_1 t}$ to zero we get,

$$\frac{\partial A(T_1, T_2)}{\partial T_1} = -\frac{i e^{iT_1} \sigma_{h1}}{4\omega_2} f_h \Omega^2, \tag{17a}$$

$$\frac{\partial B(T_1, T_2)}{\partial T_1} = -\frac{e^{iT_1} (\sigma_{h1} - \sigma_{h2})}{4\omega_2} f_h \Omega^2. \tag{17b}$$

The remaining equations after the elimination of the secular terms are solved for u_2 and v_2. All the solutions foru_1, v_1, u_2 and v_2 are substituted in Eqs. (15a) and (15b). Again we will get secular terms which are eliminated to get expressions for $\frac{\partial A(T_1, T_2)}{\partial T_2}$ and $\frac{\partial B(T_1, T_2)}{\partial T_2}$. These expressions are quite large and not included in this work. The time derivatives of A and B are determined using the following relations:

$$\dot{A} = \frac{dA}{dt} = \varepsilon \frac{\partial A(T_1, T_2)}{\partial T_1} + \varepsilon^2 \frac{\partial A(T_1, T_2)}{\partial T_2}, \tag{18a}$$

$$\dot{B} = \frac{dB}{dt} = \varepsilon \frac{\partial B(T_1, T_2)}{\partial T_1} + \varepsilon^2 \frac{\partial B(T_1, T_2)}{\partial T_2}. \tag{18b}$$

The final expressions for $\dot{A}$ and $\dot{B}$ are quite large and not shown here. The polar forms of A and B are expressed as

$$A(T_1, T_2) = \frac{1}{2} \hat{a}_1 e^{i\delta_1} = \frac{1}{2\varepsilon} a_1 e^{i\delta_1}, \tag{19a}$$

$$B(T_1, T_2) = \frac{1}{2}\hat{a}_2 e^{i\delta_2} = \frac{1}{2\varepsilon} a_2 e^{i\delta_2}. \tag{19b}$$

where a_1 and a_2 are the steady state amplitudes in the horizontal and vertical directions of the JRS, respectively. The corresponding phases are δ_1 and δ_2, respectively. We can obtain the amplitude and phase equations from Eqs. (19a) and (19b) as

$$\dot{a}_1 = \frac{da_1}{dt} = 2\varepsilon \cdot \mathrm{Re}(\dot{A})e^{-i\delta_1}, \dot{\delta}_1 = \frac{d\delta_1}{dt} = \frac{2\varepsilon}{a_1} \cdot \mathrm{Im}(\dot{A})e^{-i\delta_1}, \tag{20a}$$

$$\dot{a}_2 = \frac{da_2}{dt} = 2\varepsilon \cdot \mathrm{Re}(\dot{B})e^{-i\delta_2}, \delta_2 = \frac{d\delta_2}{dt} = \frac{2\varepsilon}{a_2} \cdot \mathrm{Im}(\dot{B})e^{-i\delta_2}, \tag{20b}$$

The final expressions for the amplitude and phase equations are obtained as

$$\dot{a}_1 = -\frac{1}{4}\frac{(-3\omega_1 + \Omega) f \Omega^2}{\omega_1^2}\sin(\phi_1) + \frac{1}{2}\frac{(-\omega_1^2 + 2\lambda)\zeta a_2 \Omega}{\omega_1^3}\sin(\phi_1 - \phi_2) \tag{21a}$$
$$-\frac{\lambda a_2^2 a_1 (4\omega_1\lambda - 6\lambda\omega_2 + 2\omega_1\omega_2^2 - \omega_1^3)}{8\omega_1(2\omega_1 - \omega_2)\omega_2^2}\sin 2(\phi_1 - \phi_2) - \frac{(\mu_1 + \zeta)a_1}{2},$$

$$\dot{\phi}_1 = \frac{1}{4}\frac{(3\omega_1 - \Omega)\Omega^2 f}{\omega_1^2 a_1}\cos(\phi_1) + \frac{1}{2}\frac{(-\omega_1^2 + 2\lambda)\zeta a_2 \Omega}{\omega_1^3 a_1}\cos(\phi_1 - \phi_2) \tag{21b}$$
$$-\frac{1}{8}\frac{\lambda a_2^2(-6\lambda\omega_2 + 4\omega_1\lambda + 2\omega_1\omega_2^2 - \omega_1^3)}{\omega_1(2\omega_1 - \omega_2)\omega_2^2}\cos(2\phi_1 - 2\phi_2)$$
$$+\frac{1}{8(4\omega_1^2 - \omega_2^2)a_1\omega_2^2\omega_1^3}\begin{pmatrix} 16\omega_1^4 a_1^3 \lambda^2 - 12\omega_1^4 a_1^3 \lambda \omega_2^2 - 6\omega_1^2 a_1^3 \lambda^2 \omega_2^2 \\ +3\omega_1^2 a_1^3 \omega_2^4 \lambda + 48\omega_1^4 a_2^2 \lambda^2 a_1 + 2\omega_1^2 \lambda a_2^2 \omega_2^4 a_1 \\ -8\omega_1^4 \lambda a_2^2 \omega_2^2 a_1 + \omega_1^2 \lambda a_2^2 \omega_2^4 a_1 - 4\omega_1^4 a_2^2 \lambda^2 \omega_2^2 a_1 \end{pmatrix}$$

$$\dot{a}_2 = \frac{1}{4}\frac{(\Omega - 3\omega_2)\Omega^2 f}{\omega_2^2}\cos(\phi_2) - \frac{1}{2}\frac{(\omega_1^2 + 2\lambda)\zeta a_1 \Omega}{\omega_2\omega_1^2}\sin(\phi_1 - \phi_2) \tag{21c}$$
$$-\frac{\lambda a_1^2 a_2(-2\lambda\omega_2 + 8\omega_1\lambda - 4\omega_2\omega_1^2 + \omega_2^3)}{8(4\omega_1^2 - \omega_2^2)\omega_2^2}\sin(2\phi_1 - 2\phi_2) - \frac{1}{2}(\mu_2 + \zeta)a_2$$

$$\dot{\phi}_2 = \sigma_1 - \sigma_2 + \frac{\lambda a_1^2(-2\lambda\omega_2 + 8\omega_1\lambda - 4\omega_2\omega_1^2 - \omega_2^3)}{8(4\omega_1^2 - \omega_2^2)\omega_2^2}\cos(2\phi_1 - 2\phi_2)$$
$$-\frac{(\Omega - 3\omega_2)\Omega^2 f}{4\omega_2^2}\sin(\phi_2) + \frac{(\omega_1^2 + 2\lambda)\zeta a_1 \Omega}{2\omega_2 a_2 \omega_1^2}\cos(\phi_1 - \phi_2) \tag{21d}$$

$$+\frac{1}{8\left(4\omega_1^2-\omega_2^2\right)\omega_1^3}\begin{pmatrix} 120\omega_1^2a_2^2\lambda - 12\omega_1^2a_2^2\omega_2^2 - 30a_2^2\lambda\omega_2^2 \\ -4a_1^2\omega_2^2\lambda - 8\omega_1^2a_1^2\omega_2^2 + 2\omega_2^4a_1^2 \\ 3\omega_2^4a_2^2 + 48\omega_1^2\lambda a_1^2 \end{pmatrix}$$

It is to be noted here that $\dot{\phi}_1 = \sigma_1 - \dot{\delta}_1$, and $\dot{\phi}_2 = \sigma_1 - \sigma_2 - \dot{\delta}_2$.

4 Results and Discussion

All the results are discussed in this section. The following set of parameter values [8] are used for all the forthcoming analysis $\boldsymbol{\mu}_1 = 0.0154$ and $\boldsymbol{\mu}_2 = 0.0247$.

4.1 Comparison of Numerical and Analytical Results

In Fig. 2, the MMS results obtained by numerically simulating the amplitude and phase Eqs. (21a–21d) are compared with the numerical simulation results of the original JRS model given by the governing Eqs. (8a, 8b). The time-displacement responses and the phase-plane diagrams in Fig. 2 show that the MMS results correlate

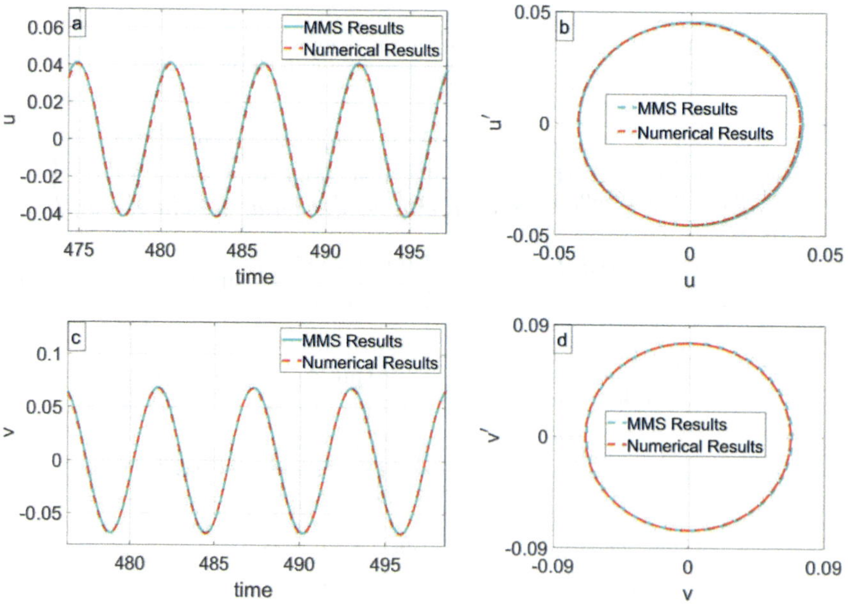

Fig. 2 Time-displacement responses and phase plane plots of the Jeffcott rotor system for $\boldsymbol{f} = 0.025$, $\boldsymbol{\zeta} = 0.005$ and $\boldsymbol{\lambda} = 0.05$. (**a**) and (**b**) for horizontal oscillations and (**c**) and (**d**) for vertical oscillations

very well with the results obtained from the original JRS model. In all the forthcoming analysis, we will use the amplitude and phase Eqs. (21a–21d) to obtain the frequency–response plots.

4.2 Non-Localised Oscillations

Non-Localized oscillations refer to the condition where $a_1 \neq 0$, $\phi_1 \neq 0$ $a_2 \neq 0$ and $\phi_2 \neq 0$, i.e. both horizontal and vertical oscillations are coupled with each other. In the case of non-localized oscillations, two amplitudes a_1 and a_2 are evaluated simultaneously.

4.2.1 Influence of Eccentricity Ratio *F*

The influence of the eccentricity ratio f on the horizontal and vertical oscillations of the JRS are shown in Figs. 3 and 4. In all of the figures, the limit point (LP) represents a change in the nature of the system solutions from stable to unstable, or vice versa. Blue colour indicates the stable solutions whereas red indicates unstable solutions. As shown in Fig. 3, a single jump phenomenon is observed for $f = 0.015$. The jump phenomena appears differently for horizontal and vertical oscillations. For horizontal oscillations, jump occurs close to the resonance peak, whereas for vertical oscillations, jump appears before the system reaches the resonance frequency. Furthermore, when compared to horizontal oscillations, this jump is substantially less for vertical oscillations.

Multivalued solutions with multiple loops are observed in the frequency response diagrams in Fig. 3 for the increased of the eccentricity ratio $f = 0.05$. The multi-valued solutions and jump phenomena disspear for $f = 0.015$ with the addition

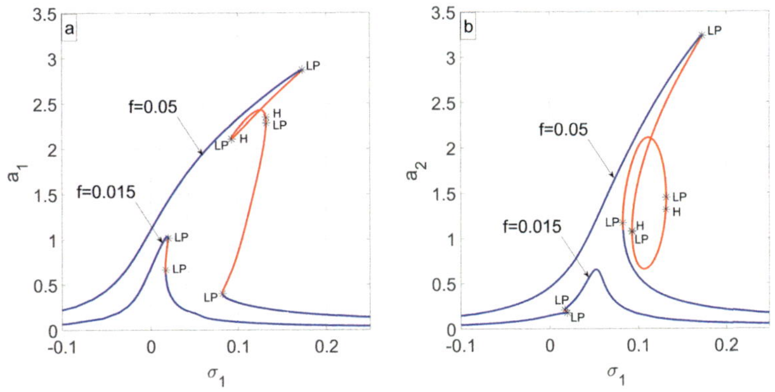

Fig. 3 Amplitude-frequency responses of the Jeffcott rotor system for different values of f, $\zeta = 0$ and $\lambda = 0.05$

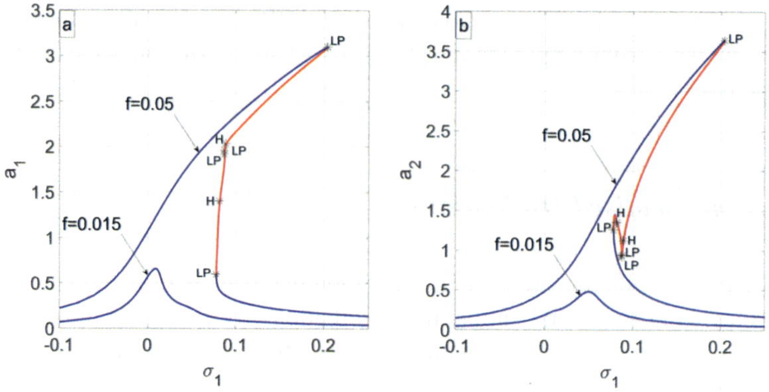

Fig. 4 Amplitude-frequency responses of the Jeffcott rotor system for different values of f, $\zeta = 0.01$ and $\lambda = 0.05$

of small amount of internal damping $\zeta = 0.01$ as shown in Fig. 4. Moreover, the multiple loops in the frequency response plots for $f = 0.05$ seem to disappear with the inclusion of internal damping $\zeta = 0.01$. However, the comparisons of Figs. 4 with Figs. 3 for $f = 0.05$ show that the internal damping causes the resonance peaks to shift towards higher frequencies with an increase in the amplitude at resonance peak.

4.2.2 Influence of Internal Damping Parameter

Frequency response diagrams for two different values of internal damping factor are shown in Fig. 5 to analyse the effect of internal damping on the dynamics of the JRS. Some complicated dynamics are observed from Fig. 5. For better clarity, the stable and unstable solutions for $\zeta = 0$ are shown by blue and red colors, and for $\zeta = 0.025$

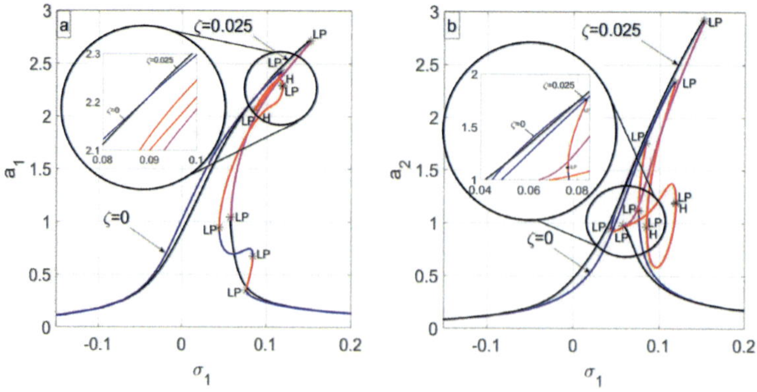

Fig. 5 Amplitude-frequency responses of the Jeffcott rotor system for different values of ζ, f=0.04 and λ=0.05

by black and magenta colors, respectively. Also, zoomed views of multiple crossings near $\sigma_1 = 0.1$ are shown at the insets of Fig. 5. It is observed from Fig. 5 that the multiple jump phenomenon and the multiple loops are eliminated with the increase of the internal damping factor from $\zeta = 0$ to $\zeta = 0.025$. However, the amplitude of oscillations at resonance peaks become larger with the inclusion of internal damping in the JRS model.

It is discussed in [3] that the internal damping can have destabilizing effect on the system after a threshold speed in a sense that the effective damping of the linearized system becomes negative beyond that speed. The negative damping of the linearized system might cause an increase in the amplitude of oscillations beyond that threshold speed. A similar characteristic is shown in Fig. 5a for horizontal oscillations. For lower frequencies, amplitude of horizontal oscillations is less for $\zeta = 0.025$ than that for $\zeta = 0$. The internal damping factor causes positive damping effect in this region by reducing the amplitudes of oscillations. However, with the increase in frequency beyond a certain critical value, the amplitude of horizontal oscillations become larger for $\zeta = 0.025$. The internal damping appear to have destabilizing effect beyond this critical value. The same conclusions cannot be drawn from the vertical oscillations shown in Fig. 5b as the amplitude of oscillations for $\zeta = 0.025$ appear to be larger in comparison to amplitude of oscillations for $\zeta = 0$ for the range of selected frequencies. It might happen that the critical frequency is already been crossed below the lower frequency bound shown in Fig. 5b.

5 Conclusions

The effects of internal damping (internal damping term ζ) on the vibration of a horizontally supported Jeffcott rotor system (JRS) are analyzed in this paper. Non-linear restoring force (with stiffness ratio λ) is also considered in the JRS model and the effect of eccentricity ratio f is also analysed. The method of multiple scales is used to derive autonomous amplitude-phase equations in the horizontal and vertical directions. The accuracy of the MMS are verified by comparing with the results produced from numerical simulation of the original JRS equations. The influence of different parameters are analysed from the frequency response curves obtained by numerically simulating the MMS equations in 'Matcont', a numerical bifurcation toolbox based on 'Matlab'.

Internal damping reduces the occurrence of multi-jump phenomena and multiple loops in the frequency response curves. However, the peak amplitude of vibration at resonance increases with the increase of ζ. The internal damping reduces the amplitude of horizontal vibrations upto a certain critical frequency. Above that critical frequency, the internal damping appears to be injecting energy into the system causing an increase in the amplitude of vibrations.

Acknowledgements This work has been supported by DST SERB - SRG under Project File no. SRG/2019/001445 and National Institute of Technology Calicut under Faculty Research Grant.

References

1. Jeffcott, H.H.: The lateral vibration of loaded shafts in the neighbourhood of a whirling speed—the effect of want of balance. Lond. Edinb. Dublin Philos. Mag. J. Sci. **37**(219), 304–314 (1919)
2. Cveticanin, L.: Free vibration of a Jeffcott rotor with pure cubic non-linear elastic property of the shaft. Mech. Mach. Theory **40**, 1330–1344 (2005)
3. Ishida, Y., Yamamoto, T.: Linear and Nonlinear Rotordynamics, 2nd edn. Wiley-VCH, Germany (2012)
4. Kimball, A.L.: Internal friction as a cause of shaft whirling. Philos. Mag. Ser. **649**(292), 724–727 (1925).
5. Dimentberg, M.: Vibration of a rotating shaft with randomly varying internal damping. J. Sound Vib. **285**, 759–765 (2005)
6. Vatta, F., Vigliani, A.: Internal damping in rotating shafts. Mech. Mach. Theory **43**(11), 1376–1384 (2008)
7. Shad, M.R., Michon, G., Berlioz, A.: Nonlinear dynamics of rotors due to large deformations and shear effects. Appl. Mech. Mater. **110**(116), 3593–3599 (2012)
8. Saeed, N.A., El-Gohary, H.A.: On the nonlinear oscillations of a horizontally supported Jeffcott rotor with a nonlinear restoring force. Nonlinear Dyn. **88**, 293–314 (2017)
9. Yabuno, H., Kashimure, T., Inoue, T., Ishida, Y.: Nonlinear normal modes and primary resonance of horizontally supported Jeffcott rotor. Non-linear Dyn. **66**, 377–387 (2011)

The Collective Behavior of Magnetically Coupled Neural Network Under the Influence of External Stimuli

T. Remi and P. A. Subha

Abstract We have analysed the collective behavior and synchronisation scenario under the combined effect of the magnetic coupling and external stimuli on a network of Hindmarsh Rose neurons. In the presence of constant input, the magnetic coupling induces synchrony and stabilises the equilibrium state. The periodically varying sinusoidal input enhances the synchrony in the network. The synchrony is inevitable for the signal transmission in the neurons. The external stimulus in the form of square wave has the capability to desynchronise the magnetically coupled network. The suppression of synchrony may find its relevance in clinical procedures to alleviate the symptoms of certain nervous system disorders.

Keywords Neural network · Linear chain · Memristor field effects · Different external stimuli

1 Introduction

A great effort has been carried out in studying and controlling the complex oscillatory networks in nonlinear science, with a number of practical applications [1–3]. The nonlinear models with high number of population is very common in Physics, Chemistry and Biology [4–6]. One of the many applications of nonlinear systems happens to be in neuro science as the brain comprises of several specialized areas with different functions, each being a complex network by itself and thus, is a perfect example of a complex dynamical system [7]. Different models have been proposed for biological neuronal networks which are capable of explaining all the intrinsic dynamics, response to external stimuli and collective behavior in different connection architectures [8].

T. Remi · P. A. Subha (✉)
Department of Physics, Farook College University of Calicut, Kozhikode 673632, Kerala, India
e-mail: pasubha@farookcollege.ac.in

T. Remi
e-mail: remi@farookcollege.ac.in

S. Banerjee and A. Saha (eds.), *Nonlinear Dynamics and Applications*,
Springer Proceedings in Complexity,
https://doi.org/10.1007/978-3-030-99792-2_108

The use of memristors to represent the electromagnetic field induced by the flow of ions in the neurons was proposed recently [9] and still continue to attract the researchers [10–13]. The dynamics and energy aspects of HR neurons with quadratic and cubic memrsitor effects have been analysed [14–16]. The impact of electromagnetic field on individual and collective dynamics of neural system is studied recently [17]. The utilisation of energy for the electrical activities in the presence of magnetic coupling and external stimuli have been studied on Hodgkin-Huxley model [18]. The hyperbolic tangent function was recognised to be easily implementable in electronic circuits and effective in numerical simulations [19, 20].

The suppression of oscillations and eventual asymptotic stabilisation of equilibrium, named as Amplitude Death (AD) has been achieved under the influence of memristor [14]. AD has been observed both theoretically and experimentally in neural networks [21, 22]. The dynamics of AD has been used to explain the temporal activity in the olfactory bulb [23]. One of the major causes of attaining AD is attributed to parameter mismatch [24–28].

The diverse firing patterns and synchronous oscillations are crucial features in neural dynamics of brains. The physiological conditions have a great influence on the electrical activity of nervous system. The collective synchrony plays an important role in the generation of both vital and pathological biological conditions. Both external and inherent stimuli constructively take part in the excitation of neural activity [29]. Zhijun Li, et al. investigated the coexistence of multiple firing patterns for different initial conditions and considering coupling strength as the sole control parameter [30]. Different methods have been employed to enhance the synchrony in the neural network owing to its vital role in signal transmission and signal processing [31, 32]. However, the unwanted synchrony among neurons which ought to have behaved independently, results in several pathological conditions [33–35]. Thus the enhancement and suppression of this synchrony is very necessary [36]. The phase synchrony in mean field coupled HR neurons with an external stimuli in form of spikes has been studied recently [37]. The external stimuli has the capability of controlling chaotic dynamics in neural network [38, 39]. However, the influence of square input on neural network is unexplored. In this work, we have analysed the capability of externally applied periodical inputs to enhance and suppress synchrony in a linear chain of HR neurons, coupled magnetically by a hyperbolic function.

The paper is organised as follows: The model is described in Sect. 2. The phase portraits and collective dynamics of the network with constant external stimulus is explained in Sect. 3.1. The control of phase space trajectories and synchrony by the combined effect of magnetic coupling and time varying external currents are presented in Sect. 3.2. Section 3.3 has been devoted to quantifying synchrony in the network with magnetic coupling and external stimuli. Section 4 concludes the study.

2 The Neural Network

The electromagnetic fields and externally applied stimuli are capable of altering collective electrical activities and signal propagation in a neural network. The current, induced by the change in magnetic flux across the membrane is represented by the memristor and that, due to the external stimuli are considered under three cases. The dynamical equations for a linear chain of HR neurons are described by [40, 41]

$$
\begin{aligned}
\dot{x_i} &= y_i + ax_i^2 - bx_i^3 - z_i + I - k_1\rho(\phi_i)x_i, \\
\dot{y_i} &= c - dx_i^2 - y_i, \\
\dot{z_i} &= r(s(x_i - x_e) - z_i), \\
\dot{\phi_i} &= k_2x_i - k_3\phi_i + D\left(\phi_i - \sum_{\substack{j=1 \\ j\neq i}}^{N} \frac{W}{|i-j|}\phi_j\right), \qquad i = 1, 2, \ldots, N.
\end{aligned} \tag{1}
$$

where, x_i, y_i and z_i represent the membrane potential, spiking variable and bursting variable of ith neuron, respectively. y_i is constituted by the flow of Na^+ and K^+ ions and the flow of Ca^+ ions constitute the z_i term. The fast oscillations of x_i and y_i correspond to spikes, whereas, the slow oscillations of the z_i variable cause burst [42]. Indices i, j and N, represent pre synaptic neuron, post synaptic neuron and total number of neurons in the network, respectively. The external current, I, is considered as a constant and two varying forms. The interaction between membrane potential, x_i and magnetic flux, ϕ_i is realised with the help of memristor, $k_1\rho(\phi_i)x_i$, where, k_1 is the magnetic coupling strength. We have considered the memory conductance term, $\rho(\phi_i)$ as a hyperbolic form, $tanh(\phi_i)$ [19]. This function can be easily approximated to linear or nonlinear forms. The term k_2x_i represents the change in magnetic flux induced by membrane potential of the cell and $k_3\phi_i$ denotes the leakage of the magnetic flux. 'D' describes the field interaction between neurons. 'W' represents the intensity of the field effect associated with distance between neurons. The parameters of the model are chosen as, $a = 3.0$, $b = 1.0$, $c = 1$, $d = 5$, $r = 0.006$, $s = 4.0$, $x_e = -1.61$ [43], $k_2 = 0.9$, $k_3 = 0.4$, $D = 0.0001$, $W = 1$ [14].

3 Collective Dynamics and Synchronisation Scenario

The changes in the firing pattern, phase portraits and synchrony level, under the combined effect of magnetic coupling and external stimulus have been analysed. We have chosen the external stimuli in different forms (i) a constant (I_1) and (ii) as time varying function. The time varying external inputs are considered in two different forms: (a) sinusoidal wave, given as $I_2 = \mathrm{A}(sin(\omega t) + cos(\omega t))$ with frequency, ω

and (b) square wave, represented as I_3 =A $\mathcal{H}\Big(sin\big(\frac{2\pi t}{\rho_s}\big)\Big)$, with an interval, ρ_s. $\mathcal{H}$ is the bi-valued Heaviside step function, whose value is zero, for negative arguments and one, for positive arguments.

3.1 Constant External Current

In this section, the dynamics, phase portraits and the synchrony pattern have been analysed under the memristor effect and constant external current. The dynamics and collective behavior of the network are analysed numerically by solving Eq. (1) for $x_i(t)$, $y_i(t)$, $z_i(t)$ and $\phi_i(t)$ using Runge–kutta method in step size of 0.1. Initial conditions for different neurons are chosen as, $x_0(t)$, $y_0(t)$ and $z_0(t)$ in the range $[-0.5, 0.5]$ and $\phi_0(t)$ in the range [0,1]. The study of variations in x_i reveals the dynamical change induced in each neuron under the influence of magnetic coupling and external stimuli. The average of x_i over the number of neurons in the network reveals the collective behavior of the network. The average of membrane potentials is given by

$$\bar{x}(t) = \frac{1}{N}\sum_{i=1}^{N} x_i(t) \tag{2}$$

The dynamical change induced in a single neuron and the collective synchrony is analysed in the memristive network for different values of k_1 in the presence of constant external current ($I_1 = 2.9$) and is as shown in Fig. 1.

For low values of k_1, the system exhibits square wave bursting with five spikes per burst as shown in Fig. 1a. The lack of synchrony is visible from difference of $\bar{x}$ from the individual membrane potential. We have validated the presence of plateau bursting dynamics with increase in the value of k_1, as shown in Fig. 1b. The dissimilarity in average membrane potential from the individual values represent the absence of synchrony. On further increasing the value of k_1, Fig. 1c shows AD state with no oscillations. From the studies, it is clear that magnetic coupling stabilises the system. The system is in complete synchrony as visible from the convergence of $\bar{x}$ with the individual values. The memritor drives the system from square wave bursting to plateau bursting and to AD state, where the system exhibits complete synchrony.

The synchrony pattern in the network is further justified by the phase portraits as shown in Fig. 2. The phase space has been plotted with average membrane potential $\bar{x}(t)$ on X-axis and membrane potential of a single neuron on Y-axis. In Fig. 2a, the phase space is dense at $k_1 = 0.1$, representing lack of synchrony. As the value of k_1 is increased, the phase space becomes more controlled and synchronised as shown in Fig. 2b. The phase space shrinks to a single point representing stabilisation of the states, on further increase in k_1, as in Fig. 2c. The enhancement in synchrony with the increase in strength of magnetic coupling is visible from the control and shrinking of phase space.

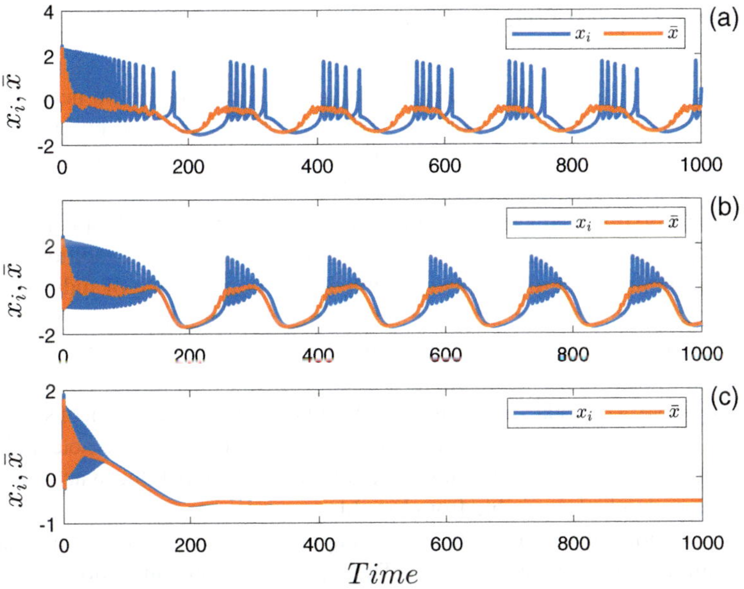

Fig. 1 Time series of discrete and average membrane potential of the magnetically coupled neural network with constant external current ($I = 2.9$). Here, $i = 25$ and the dynamics is consistent for any neuron in the network. **a** $k_1 = 0.1$, **b** $k_1 = 0.3$ and **c** $k_1 = 1.5$

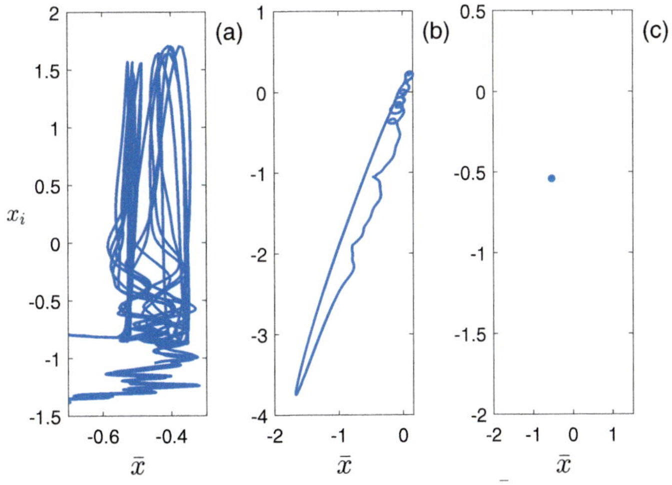

Fig. 2 Phase portraits of neural network with magnetic coupling and constant external current ($I = 2.9$). The average membrane potential is plotted in X-axis and membrane potential of single neuron on Y-axis. Here, $i = 25$. **a** $k_1 = 0.1$, **b** $k_1 = 0.3$ and **c** $k_1 = 1.5$

3.2 Time Varying External Currents

The variations in firing patterns and the collective dynamics of the network with magnetic coupling, under the influence of time varying external stimuli is analysed in this section. The time series of $\bar{x}(t)$ for the two different external stimulus is shown in Fig. 3. In the presence of memristor and sinusoidal wave, the system exhibits square wave bursting character, as shown in Fig. 3a, for low value of input frequency, ω. The initial transients lack complete synchrony, whereas, with evolution of time, the system attains synchrony. This is visible as the convergence of individual and average values of the membrane potential. With increase in frequency ω, the number of spikes per burst decreases to two as shown in Fig. 3b. But the synchrony pattern is maintained. Figure 3c shows the spiking behavior under the influence of memristor and square wave. The spikes are induced in the quiescent state. The lack of synchrony is visible from the figure, by the separation of individual membrane potential from the average value. With decrease in interval of the square input, the synchrony level is even reduced, which is visible from Fig. 3d.

The phase portraits of the system with magnetic coupling and the time varying external stimulus is as shown in Fig. 4. The top and bottom panel shows the phase space of the system with sinusoidal and square inputs, respectively. In both panels, the value of k_1 increases from left to right. The Fig. 4a present the influence of

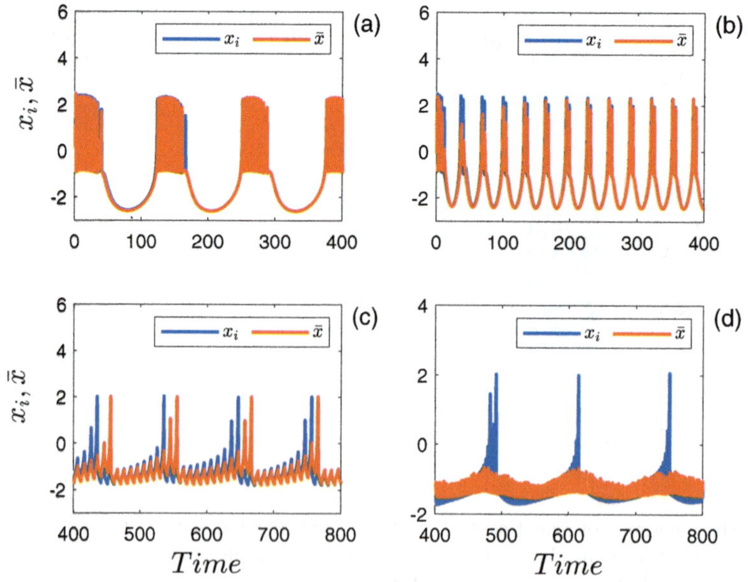

Fig. 3 Time series of discrete and average membrane potential of the magnetically coupled neural network with time varying external stimuli. Here $i = 25$, $k_1 = 0.1$. **a** sinusoidal input ($A = 3$, $\omega = 0.05$), **b** sinusoidal input ($A = 3$, $\omega = 0.2$), **c** square input ($A = 3$, $\rho_s = 10$), **d** square input ($A = 3$, $\rho_s = 3$)

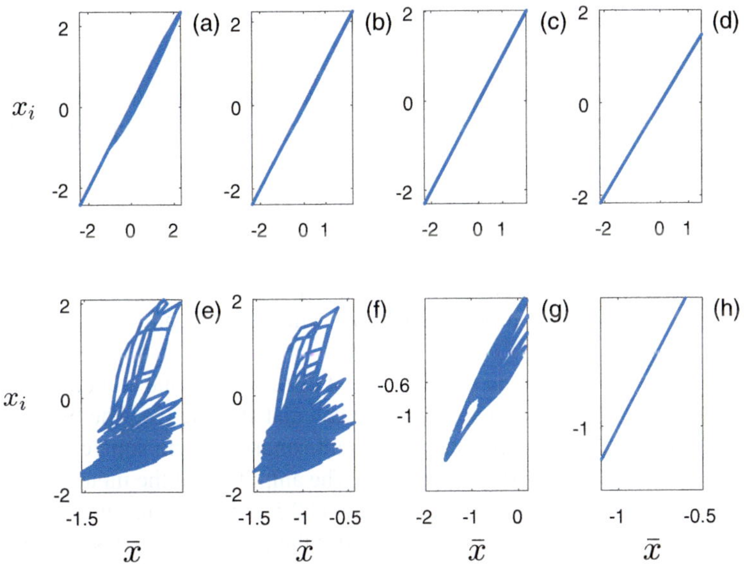

Fig. 4 Phase space with sinusoidal (top panel) and square input (bottom panel). Here, $i = 25$, $A = 3$, $\omega = 0.2$, $\rho_s = 3$. The value of k_1 used are 0.1 for **a** and **e**, 0.3 for **b** and **f**, 0.7 for **c** and **g**, 1.5 for **d** and **h**

input I_2 on a network with weak magnetic coupling. The phase space exhibits small dispersion. On increasing the strength of magnetic coupling, k_1, the phase space is reduced to a beeline as shown in Fig. 4b. On further increase of k_1, the synchrony in the network is preserved by the combined effect of magnetic coupling and sinusoidal input, (I_2), observed as the straight line in phase portrait of Fig. 4c. The synchrony attained is sustained at even higher value of magnetic coupling strength, as shown in Fig. 4d. The dispersed phase portrait for low magnetic coupling strength and under the influence of square input, I_3, represents the lack of synchrony in the network, as shown in Fig. 4e. With further increase in magnetic coupling, the phase portrait is still dispersed, in the presence of square input, as shown in Fig. 4f. A slight control in the phase portrait is visible for further increase in magnetic coupling strength, as shown in Fig. 4g. At high value of magnetic coupling strength, the phase portrait is reduced to a beeline in the presence of input, I_3, as shown in Fig. 4h. Thus, it is found that the magnetic coupling helps to control the phase space in the presence of time varying external inputs. However, the control is faster in the presence of sinusoidal input compared to the square input.

3.3 Statistical Factor of Synchronisation

The synchrony induced by the magnetic coupling, in the presence of external stimuli is quantified using statistical factor of synchronisation [41, 44], which has the form:

$$R = \frac{\langle \bar{x}(t)^2 \rangle - \langle \bar{x}(t) \rangle^2}{\frac{1}{N}\sum_{i=1}^{N}[\langle x(t)_i^2 \rangle - \langle x(t)_i \rangle^2]} \tag{3}$$

where, $\bar{x}(t)$ is given by Eq. (2) and '$\langle\ \rangle$' represent the average of the variable over time. The system attains complete synchrony when $R = 1$ and desynchrony when $R = 0$.

The variation in R with k_1 in the presence of different external stimuli is shown in Fig. 5a. The synchronisation factor has been calculated using (3). The blue, red and yellow lines represent the influence of constant, sinusoidal and square input on the magnetically coupled network, respectively. The amplitude of the inputs are fixed at $A = 3$. At low values of k_1, the value of R is about 0.5, representing partial synchrony. The synchrony obtained by the network, under the influence of constant current is found to be high compared to the square input. The high value of R obtained for even low value of magnetic coupling in the presence of sinusoidal input, represents the synchrony obtained by the network. For weakly coupled system, the synchrony obtained by the constant input lies in between square and sinusoidal input and for high values of k_1, the magnetic coupling overrides the input effects.

The variations in the R with the amplitude of the inputs is presented in Fig. 5b. The colour codes used for the inputs are similar to Fig. 5a. In the absence of any external input, the synchrony level is obtained to be about $R = 0.85$, as the value of k_1 is fixed as 0.3. For low intensities of the inputs, the synchrony of the network is reduced from the initial value in the presence of constant input. The synchrony is even low under the influence of square input. The presence of sinusoidal input increases the synchrony in the network. With increase in the intensities of the inputs, the differences

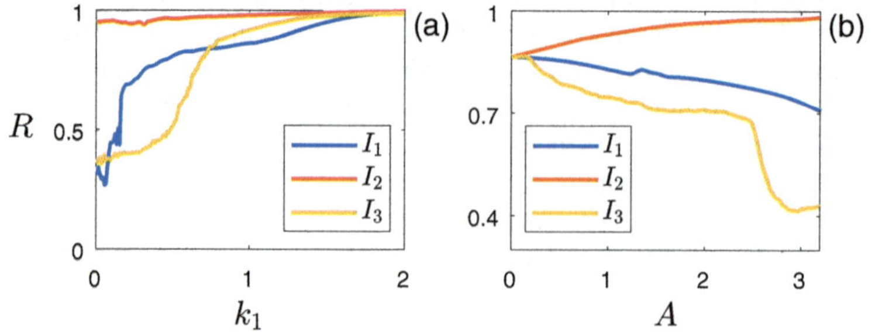

Fig. 5 Variation of statistical factor of synchronisation with **a** magnetic coupling strength and **b** amplitude of external input, under the influence of different external stimuli. The constant input, sinusoidal and square inputs are represented by blue line, red and yellow colours respectively. **a** $A = 3$ **b** $k_1 = 0.3$. Here, $\omega = 0.2$, $\rho_s = 3$

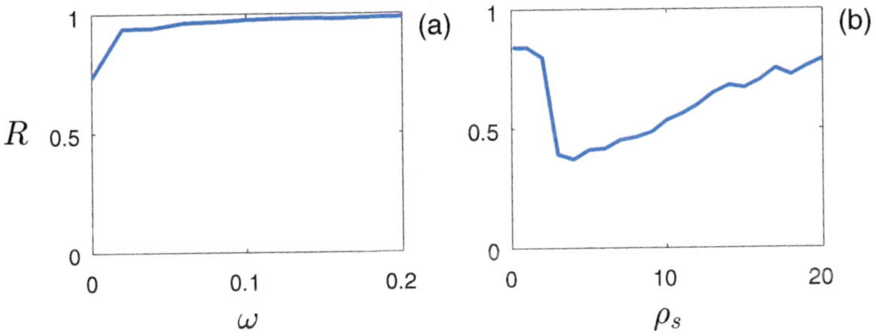

Fig. 6 Statistical factor of synchronisation for different parameters of the time varying input. Here, $k_1 = 0.3$, $A = 3$

in synchrony becomes more visible. The results obtained are in perfect agreement with the studies of Fig. 5a. At sufficiently high amplitudes of the external input, the sinusoidal input is capable of enhancing synchrony compared to the constant input. The desynchronising ability of the square input on the magnetically coupled network is also visible.

The study of synchrony pattern has been extended by varying frequencies of the inputs, I_2 and I_3. The variations in R with the frequency, ω of sinusoidal input is shown in Fig. 6a. For low value of frequency, the synchrony is quantified to be about $R = 0.7$. With further increase in ω, the synchrony is also increased. Thus, the input with high frequency has a greater synchronising ability. The desynchronising ability of the square input is analysed for the variations in interval of the square input as presented in Fig. 6b. For low values of the interval, ρ, (i.e.) high frequency, the synchrony is drastically reduced. But on further increasing the interval (reducing frequency), the synchrony is increasing, representing the inability of the low frequency input to desynchronise. Thus, it is inferred that high frequency sinusoidal input and square input has the capability to synchronise and desynchronise the magnetically coupled network, respectively.

4 Conclusions

We have analysed a linear chain of HR neurons with magnetic coupling and external stimuli of different forms. The magnetic coupling is capable of inducing plateau bursting and amplitude death in the HR neurons with constant external input. The phase space trajectories are controlled and shrinks to a single point with the increase in magnetic coupling strength, representing synchrony and amplitude death. The synchrony is quantified using statistical factor of synchrony. The HR neural network with magnetic coupling is also analysed in the presence of sinusoidal and square waves. The dynamics of the individual neurons are square wave bursting and spiking in the presence of sinusoidal and square inputs, respectively. In the presence of time varying external inputs, the phase space trajectories of the system are reduced to beeline, for high magnetic coupling strength. The statistical factor of synchronisation justifies the fact that, the sinusoidal input enhances synchrony and the square input suppresses the synchrony in the network, compared to the constant input. The inputs with high frequencies have greater ability to enhance and suppress the synchrony.

References

1. Strogatz, S.H.: From Kuramoto to Crawford: exploring the onset of synchronization in populations of coupled oscillators. Phys. D: Nonlinear Phenom. **143**(1–4), 1–20 (2000)
2. Batista, C.A.S., Lopes, S.R., Viana, R.L., Batista, A.M.: Delayed feedback control of bursting synchronization in a scale-free neuronal network. Neural Netw. **23**(1), 114–124 (2010)
3. Krylov, D., Dylov, D.V., Rosenblum, M.: Reinforcement learning for suppression of collective activity in oscillatory ensembles. Chaos: Interdiscip. J. Nonlinear Sci. **30**(3), 033126 (2020)
4. Kuramoto, Y.: Chemical oscillations, waves, and turbulence. Courier Corporation, 2003
5. Backwell, P., Jennions, M., Passmore, N., Christy, J.: Synchronized courtship in fiddler crabs. Nature **391**(6662), 31–32 (1998)
6. Mirollo, R.E., Strogatz, S.H.: Synchronization of pulse-coupled biological oscillators. SIAM J. Appl. Math. **50**(6), 1645–1662 (1990)
7. Thompson, R.F.: The Brain: a Neuroscience Primer. Macmillan (2000)
8. Izhikevich, E.M.: Dynamical Systems in Neuroscience. MIT press (2007)
9. Xu, Y., Jia, Y., Ma, J., Alsaedi, A., Ahmad, B.: Synchronization between neurons coupled by memristor. Chaos, Solitons Fractals **104**, 435–442 (2017)
10. Yang, J.Q., Wang, R., Wang, Z.P., Ma, Q.Y., Mao, J.Y., Ren, Y., Yang, X., Zhou, Y., Han, S.T.: Leaky integrate-and-fire neurons based on perovskite memristor for spiking neural networks. Nano Energy **74**, 104828 (2020)
11. Xu, L., Qi, G., Ma, J.: Modeling of memristor-based Hindmarsh-Rose neuron and its dynamical analyses using energy method. Appl. Math. Model. **101**, 503–516 (2022)
12. Rajagopal, K., Karthikeyan, A., Jafari, S., Parastesh, F., Volos, C., Hussain, I.: Wave propagation and spiral wave formation in a Hindmarsh-Rose neuron model with fractional-order threshold memristor synaps. Int. J. Mod. Phys. B **34**(17), 2050157 (2020)
13. Han, B., Yunzhen, Z., Liu, W., Bocheng, B.: Memristor synapse-coupled memristive neuron network: synchronization transition and occurrence of chimera. Nonlinear Dyn. **100**(1), 937–950 (2020)
14. Usha, K., Subha, P.A.: Hindmarsh-Rose neuron model with memristors. Biosystems **178**, 1–9 (2019)
15. Usha, K., Subha, P.A.: Energy feedback and synchronous dynamics of Hindmarsh-Rose neuron model with memristor. Chin. Phys. B **28**(2), 020502 (2019)
16. Usha, K., Subha, P.A.: Collective dynamics and energy aspects of star-coupled Hindmarsh-Rose neuron model with electrical, chemical and field couplings. Nonlinear Dyn. **96**(3), 2115–2124 (2019)
17. Zandi-Mehran, N., Jafari, S., Hashemi Golpayegani, S.M.R., Nazarimehr, F., Perc, M.: Different synaptic connections evoke different firing patterns in neurons subject to an electromagnetic field. Nonlinear Dyn. **100**(2), 1809–1824 (2020)
18. Wu, F.Q., Ma, J., Zhang, G.: Energy estimation and coupling synchronization between biophysical neurons. Science China Technol. Sci. **63**(4), 625–636 (2020)
19. Tan, Y., Wang, C.: A simple locally active memristor and its application in hr neurons. Chaos: Interdiscip. J. Nonlinear Sci. **30**(5), 053118 (2020). https://doi.org/10.1063/1.5143071
20. Chen, M., Chen, C.J., Bao, B.C., Xu, Q.: Multi-stable patterns coexisting in memristor synapse-coupled Hopfield neural network. In: Mem-elements for Neuromorphic Circuits with Artificial Intelligence Applications, pp. 439-459. Academic Press (2021)

21. Saxena, G., Prasad, A., Ramaswamy, R.: Amplitude death: the emergence of stationarity in coupled nonlinear systems. Phys. Rep. **521**(5), 205–228 (2012)
22. Herrero, R., Figueras, M., Rius, J., Pi, F., Orriols, G.: Experimental observation of the amplitude death effect in two coupled nonlinear oscillators. Phys. Rev. Lett. **84**(23), 5312 (2000)
23. Monteiro, L.H.A., Filho, A.P., Chaui-Berlinck, J.G., Piqueira, J.R.C.: Oscillation death in a two neuron network with delay in a self connection. J. Biol. Syst. **15**, 49–61 (2007)
24. Koseska, A., Volkov, E. and Kurths, J.: Parameter mismatches and oscillation death in coupled oscillators. Chaos: Interdiscip. J. Nonlinear Sci. **20**(2), 023132 (2010)
25. Prasad, A.: Universal occurrence of mixed-synchronization in counter-rotating nonlinear coupled oscillators. Chaos Solitons Fractals **43**(1–12), 42–46 (2010)
26. Sharma, A., Shrimali, M.D.: Amplitude death with mean-field diffusion. Phys. Rev. E **85**(5), 057204 (2012)
27. Gjurchinovski, A., Zakharova, A., Schöll, E.: Amplitude death in oscillator networks with variable-delay coupling. Phys. Rev. E **89**(3), 032915 (2014)
28. Teki, H., Konishi, K., Hara, N.: Amplitude death in a pair of one-dimensional complex Ginzburg-Landau systems coupled by diffusive connections. Phys. Rev. E **95**(6), 062220 (2017)
29. Eteme, A.S., Tabi, C.B., Mohamadou, A.: Firing and synchronization modes in neural network under magnetic stimulation. Commun. Nonlinear Sci. Numer. Simul. **72**, 432–440 (2019)
30. Li, Z., Zhou, H., Wang, M., Ma, M.: Coexisting firing patterns and phase synchronization in locally active memristor coupled neurons with HR and FN models. Nonlinear Dyn. **104**(2), 1455–1473 (2021)
31. Dhamala, M., Jirsa, V.K., Ding, M.: Enhancement of neural synchrony by time delay. Phys. Rev. Lett. **92**(7), 074104 (2004)
32. Sakurai, Y., Song, K., Tachibana, S., Takahashi, S.: Volitional enhancement of firing synchrony and oscillation by neuronal operant conditioning: interaction with neurorehabilitation and brain-machine interface. Front. Syst. Neurosci. **8**, 11 (2014)
33. Uhlhaas, P.J., Singer, W.: Neural synchrony in brain disorders: relevance for cognitive dysfunctions and pathophysiology. Neuron **52**(1), 155–168 (2006)
34. Buzsaki, G.: Rhythms of the Brain. Oxford University Press (2006)
35. Schnitzler, A., Gross, J.: Normal and pathological oscillatory communication in the brain. Nat. Rev. Neurosci. **6**(4), 285–296 (2005)
36. Rosenblum, M.: Controlling collective synchrony in oscillatory ensembles by precisely timed pulses. Chaos: Interdiscip. J. Nonlinear Sci. **30**(9), 093131 (2020)
37. Remi, T., Subha, P.A., Usha, K.: Controlling phase synchrony in the mean field coupled Hindmarsh-Rose neurons. Int. J. Mod. Phys. C 2250058 (2021)
38. Lin, H., Wang, C., Yao, W., Tan, Y.: Chaotic dynamics in a neural network with different types of external stimuli. Commun. Nonlinear Sci. Numer. Simul. **90**, 105390 (2020)
39. Eteme, A.S., Tabi, C.B., Beyala Ateba, J.F., Ekobena Fouda, H.P., Mohamadou, A. and Crepin Kofane, T.: Chaos break and synchrony enrichment within Hindmarsh-Rose-type memristive neural models. Nonlinear Dyn. **105**(1), 785–795 (2021)
40. Xu, Y., Ying, H., Jia, Y., Ma, J., Hayat, T.: Autaptic regulation of electrical activities in neuron under electromagnetic induction. Sci. Rep. **7**, 43452 (2017). https://doi.org/10.1038/srep43452
41. Xu, Y., Jia, Y., Ma, J., Hayat, T., Alsaedi, A.: Collective responses in electrical activities of neurons under field coupling. Sci. Rep. **8**, 1349 (2018). https://doi.org/10.1038/s41598-018-19858-1
42. Buric, N., Todorovic, K., Vasovic, N.: Synchronization of bursting neurons with delayed chemical synapses. Phys. Rev. E **78**(3), 036211 (2008). https://doi.org/10.1103/physreve.78.036211
43. Shi, X., Wang, Z.: Adaptive synchronization of time delay Hindmarsh-Rose neuron system via self-feedback. Nonlinear Dyn. **69**(4), 2147–2153 (2012)
44. Qin, H., Ying, W., Wang, C., Ma, J.: Emitting waves from defects in network with autapses. Commun. Nonlinear Sci. Numer. Simul. **23**(1–3), 164–174 (2015). https://doi.org/10.1016/j.cnsns.2014.11.008

Excitation Spectrum of Repulsive Spin-Orbit Coupled Bose-Einstein Condensates in Quasi-one Dimension: Effect of Interactions and Coupling Parameters

Sanu Kumar Gangwar, R. Ravisankar, and Pankaj K. Mishra

Abstract We investigate the stability of the repulsive spin-orbit (SO) coupled Bose-Einstein condensates with linear Rabi mixing by employing the Bogoliubov-de-Gennes theory. We analytically compute the eigenenergy spectrum for both non-interacting and interacting cases. The magnitude of the imaginary part of the eigenenergy has been used to characterize the dynamical instability of the condensate. We find that increase in the SO coupling (k_L) leads to the transformation from a stable state to a single instability band then a multiple instability band. However, the effect of increase in Rabi coupling (Ω) is the opposite, which makes the system more stable. Further, we perform a systematic analysis to understand the effect of the variations of the interaction parameters on the instability of the spectrum. Finally, a stability phase diagram in the interspecies and intraspecies parameter plane and $\Omega - k_L$ plane has been obtained.

Keywords Bose-Einstein condensates · Bogliubov-de-Gennes theory · Spin-Orbit coupling · Gross-Pitaevskii equation · Instability

1 Introduction

After the successful achievement of experimental realization of Bose-Einstein condensates (BECs) in the dilute atomic gases, with a lot of free parameters to control the system, it has totally given a new direction to the research in ultracold Physics.

WWW home page: https://www.iitg.ac.in/pankaj.mishra/.

S. K. Gangwar · R. Ravisankar · P. K. Mishra (✉)
Department of Physics, Indian Institute of Technology, Guwahati 781039, Assam, India
e-mail: pankaj.mishra@iitg.ac.in

S. Banerjee and A. Saha (eds.), *Nonlinear Dynamics and Applications*,
Springer Proceedings in Complexity,
https://doi.org/10.1007/978-3-030-99792-2_109

In BEC, the nonlinear interaction evolves from the atomic interactions, which are revealed as nonlinear terms in the Gross-Pitaevskii equation (GPE) [1]. The nonlinear interactions may be attractive or repulsive interatomic interactions.

In recent years the coupled BECs [2] have attracted the attention of the scientific community owing to its unique features of having two internal atomic states, which can be tuned to generate the multiple component BECs. Moreover, at ultralow temperatures, quantum gases allow the realizations of multi-component BECs [2–4], whose behavior is very different with respect to that of a single component BEC. The possibility of tuning a number of system parameters makes such systems ideal for studying the structure of the various phases and the nature of the phase transitions. There are lots of studies in coupled BECs system that focus on the stability region [3, 5], chaotic and unstable cycle behavior [4], etc. However, after a breakthrough experiment by Spielman and his group at NIST, who have been able to engineer a synthetic spin-orbit (SO) coupling in BECs [6], the coupled BEC field has generated renewed interests among the community. In this experiment, two Raman laser beams were used to couple hyperfine states of BECs, where the momentum transfer between atoms and lasers leads to the synthetic SO coupling [7].

In 1941, to explain the superfluid behaviour in ^{4}He Landau developed the fundamental theory of elementary excitation. On similar line further Bogoliubov in 1947 [8] performed the analytical derivation of the excitation spectrum for the Bose gases. The collective excitation spectrum of BECs gives the basic information about the dynamics of the ultracold gases [9]. Roton-phonon-Maxon modes were found in the quasi-1D SO coupled BECs experimentally [10], and analytically [11], also the dynamical and energetic instability were investigated numerically [12]. Although there are recent works that highlight the stability/instability of the excitation spectrum of SO coupled BECs in two dimension such analysis has been lacking for 1D system where the systems exhibit interesting dynamical behaviour [13]. In this paper, we focus on the instability regions for different SO and Rabi coupling parameters while considering the repulsive inter-and intraspecies interactions between the components of condensates.

We arrange the paper in the following sequence. In Sect. 2, we develop the mathematical model and corresponding coupled Gross-Pitaevskii equations for spin-$1/2$ BECs. In Sect. 3 we discussed the single-particle spectrum. Section 4 we divide in three different subsections. We develop the analytical model to evaluate the excitation spectrum in the first subsection. In the next subsection, we focus on the effect of SO and Rabi couplings on the excitation spectrum of spin-$1/2$ BECs, and further, the effect of intra- and interspecies interactions. In Sect. 5 we present the stability phase diagrams for both cases as discussed in Sect. 4. Finally, we provide the summary of our work in Sect. 6.

2 The Model: Coupled Gross-Pitaevskii Equations

This section illustrates the mean-field model of spin-orbit (SO) coupled quasi-one-dimensional pseudospin-$1/2$ BECs. Theoretically, we can study the properties of these kinds of systems with the help of the coupled Gross-Pitaevskii equations given below (as in dimensionless form) [13]:

$$\mathrm{i}\frac{\partial\psi_\uparrow}{\partial t} = \left[-\frac{1}{2}\frac{\partial^2}{\partial x^2} - \mathrm{i}k_L\frac{\partial}{\partial x} + V(x) + \alpha|\psi_\uparrow|^2 + \beta|\psi_\downarrow|^2\right]\psi_\uparrow + \Omega\psi_\downarrow, \quad (1a)$$

$$\mathrm{i}\frac{\partial\psi_\downarrow}{\partial t} = \left[-\frac{1}{2}\frac{\partial^2}{\partial x^2} + \mathrm{i}k_L\frac{\partial}{\partial x} + V(x) + \beta|\psi_\uparrow|^2 + \alpha|\psi_\downarrow|^2\right]\psi_\downarrow + \Omega\psi_\uparrow, \quad (1b)$$

Here, $\psi_\uparrow$ and $\psi_\downarrow$ are the wavefunction corresponding to the spin-up and spin-down component of the condensates, $V(x)$ is the trapping potential, α and β are the intra- and interspecies components of the condensates respectively, k_L is SO coupling and Ω is the Rabi coupling parameters. In the above equations (1), length is measured in units of harmonic oscillator length $a_0 = \sqrt{\hbar/(m\omega_\perp)}$, time in the units of $\omega_\perp^{-1}$, and energy in the units of $\hbar\omega_\perp$, where $\omega_\perp$ is the transverse direction frequency of the harmonic confinement. The SO coupling and the Rabi coupling parameters have been rescaled as $k_L = k'_L/a_0\omega_\perp$ and $\Omega = \Omega'/\omega_\perp$, respectively, while the wave function is rescaled as $\psi_{\uparrow,\downarrow} = \psi_{\uparrow,\downarrow}a_0^{3/2}/\sqrt{N}$. We consider the Rabi coupling as $\Omega = |\Omega|e^{\mathrm{i}\theta}$ that minimizes the energy when $\Omega = -|\Omega|$ for $\theta = \pi$ [13]. The wave functions are subjected to the following normalization condition $\int_{-\infty}^{\infty}\left(|\psi_\uparrow|^2 + |\psi_\downarrow|^2\right)dx = 1$.

3 Single-particle Spectrum

In this section, we study the single-particle spectrum of the spin-orbit coupled binary BECs, which arises by solving the coupled trapless GP equations in the absence of inter- and intraspecies interactions.

Let us consider the plane wave solution as $\psi_{\uparrow,\downarrow} = \phi_{\uparrow,\downarrow}e^{\mathrm{i}(k_x x - \omega t)}$, and also $\alpha = \beta = V = 0$. Therefore we can write the eigenvalue problem as,

$$\omega\left(\phi_\uparrow\ \phi_\downarrow\right)^T = \mathcal{L}_{sp}\left(\phi_\uparrow\ \phi_\downarrow\right)^T \quad (2)$$

where,

$$\mathcal{L}_{sp} = \begin{pmatrix} \frac{1}{2}k_x^2 + k_Lk_x & \Omega \\ \Omega & \frac{1}{2}k_x^2 - k_Lk_x \end{pmatrix} \quad (3)$$

Which gives the single-particle spectrum as,

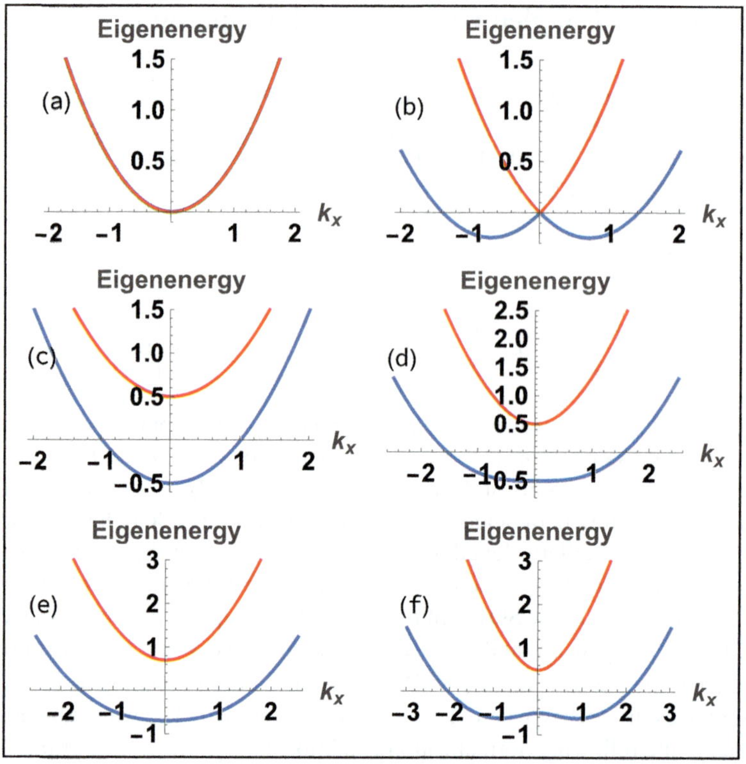

Fig. 1 Single-particle spectrum for different combination of spin-orbit and Rabi coupling parameters **a** $k_L = 0.0, \Omega = 0.0$ **b** $k_L = 0.7, \Omega = 0.0$, **c** $k_L = 0.0, \Omega = 0.5$ **d** $k_L = 0.7, \Omega = 0.5$, **e** $k_L = 0.7, \Omega = 0.7$ **f** $k_L = 1.0, \Omega = 0.5$

$$\omega_{\pm} = \frac{1}{2}\left(k_x^2 \pm 2\sqrt{k_x^2 k_L^2 + \Omega^2}\right) \tag{4}$$

The single-particle spectrum has two solutions positive branch (ω_+) and negative branch (ω_-). The negative branch shows the transition from single minima to double minima on the variation of SO coupling for finite Rabi coupling strength, while the positive branch has single minima throughout. So mainly, in this part, we will focus on the negative branch.

For zero spin-orbit ($k_L = 0$) and Rabi ($\Omega = 0$) couplings, the non-degenerate parabolic spectrum occurs (see Fig. 1a). For further analysis, we introduce finite value to $k_L = 0.7$ in the absence of Rabi ($\Omega = 0$), the negative branch of the spectrum makes a transition from parabolic to double minima, which locates at the position $k_x = \pm k_L$ (see Fig. 1b). Now we fix finite value to Rabi ($\Omega = 0.5$) in the absence of the SO coupling parameter ($k_L = 0$), double minima disappears, and the depth of the minima increases. We also investigate that the energy gap between negative and positive branches is 2Ω (see Fig. 1c). Since we are looking forward to the effect of variation of SO (Rabi) with the other coupling term Rabi (SO) at some finite value, on the negative branch of the spectrum. On increasing Rabi term (see Fig. 1d and e) from $\Omega = 0.5$ to $\Omega = 0.7$ at fix value of SO $k_L = 0.7$, we analyze that the depth of minima increases, this corresponds to lowering in the energy of the system. It indicates that the system is more stable upon increase in Ω. Again on increasing SO term from $k_L = 0.7$ to $k_L = 1.0$ at the fixed value of Rabi term $\Omega = 0.5$, we observe that the phase transition from single minimum to double minima in the negative branch of the spectrum (see Fig. 1d and f). This observation confirms the phase transition, and it will occur only when $\Omega < k_L^2$.

4 Collective Excitation Spectrum

The excitation spectrum of the condensates provides an important clue about their dynamical stability. In this section, first, we discuss the analytical results of the eigenenergy of the excitation spectrum using Bogoliubov-de Gennes (BdG) theory. It is followed by the effect of the coupling parameters on the stability of the energy spectrum.

4.1 Analytical Study of Excitation Spectrum

In this section, we present the analytical study of the excitation spectrum of coupled Binary BECs in the presence of SO and Rabi couplings by applying the BdG theory given by Bogoliubov in 1947. Initially, we perturb our ground state wave function $\phi_{\uparrow,\downarrow}$ by adding a small perturbation term $\delta\phi_{\uparrow\downarrow}$. After inserting this modified wave function in the coupled GP Equation, we get four different branches of the excitation spectrum. Out of four, two are positive, and two are negative. We analyze the behavior of positive and negative energy branches in the presence of SO and Rabi coupling. Elaborately, we vary the SO (Rabi) term at the fixed Rabi (SO) term. The excitation wave function to get the BdG transformation matrix is given by,

$$\psi_j = e^{-\mathrm{i}\mu t}[\phi_j + \delta\phi_j] \tag{5}$$

$$\delta\phi_j = u_j e^{\mathrm{i}(k_x x - \omega t)} + v_j^* e^{-\mathrm{i}(k_x x - \omega^* t)} \tag{6}$$

Where, $\phi_j = \sqrt{n_j} e^{i\varphi_j}$, $(j = \uparrow, \downarrow)$ is the ground state wavefuction, $u_j' s$ and $v_j' s$ are the BdG amplitudes. Also n_j and φ_j is the density and phase respectively. We could calculate BdG amplitudes by substituting equation Eq. (5) in Eq. (1). Therefore,

$$\omega \left(u_\uparrow \; v_\uparrow \; u_\downarrow \; v_\downarrow\right)^T = \mathcal{L} \left(u_\uparrow \; v_\uparrow \; u_\downarrow \; v_\downarrow\right)^T \tag{7}$$

where the superscript T denotes the transpose of the matrix and $\mathcal{L}$ is given by,

$$\mathcal{L} = \begin{pmatrix} H_1 - \mu & \alpha n_\uparrow & \beta\sqrt{n_\uparrow n_\downarrow} - \Omega & \beta\sqrt{n_\uparrow n_\downarrow} \\ -\alpha n_\uparrow & -H_2 + \mu & -\beta\sqrt{n_\uparrow n_\downarrow} & -\beta\sqrt{n_\uparrow n_\downarrow} + \Omega \\ \beta\sqrt{n_\uparrow n_\downarrow} - \Omega & \beta\sqrt{n_\uparrow n_\downarrow} & H_3 - \mu & \alpha n_\downarrow \\ -\beta\sqrt{n_\uparrow n_\downarrow} & -\beta\sqrt{n_\uparrow n_\downarrow} + \Omega & -\alpha n_\downarrow & -H_4 + \mu \end{pmatrix}$$

Where, the matrix element H_1, H_2, H_3, H_4 and μ (chemical potential) is given in the Appendix. A. The normalization condition yields $\int(|u_j|^2 - |v_j^*|^2)dx = 1$. The Simplified form of the BdG equation can be obtain by substituting $det(\mathcal{L})$ equal to zero with $n_\uparrow = n_\downarrow = 1/2$. Therefore we get,

$$\omega^4 + a\omega^2 + b = 0 \tag{8}$$

where the coefficients of the dispersion relation a, b are given in Appendix. B.

We calculate the eigenvalue of the matrix $\mathcal{L}$ (see Eq. (7)). As discussed in the last section, the single-particle spectrum ($\alpha = \beta = 0$) generally has a positive branch (ω_+) and a negative branch (ω_-). The positive branch exhibits single minima throughout, while the negative branch shows the phase transition from single minimum to double minima upon the variation of spin-orbit coupling at some finite Rabi coupling term. Here, we will focus on the negative branch. Note that the imaginary or complex eigenenergies indicate the dynamical instability, while the negative eigenenergy of the excitation spectrum implies that the system is energetically unstable [14]. As we are interested in investigating the effect of the different interaction and coupling parameters on the dynamical instability, we will be mainly interested in looking at the nature of the negative branch of the eigenenergy. We define the instability factor $G = |\Im(\omega_-)|$.

In the following, we present the effect of the couplings on the stability of the excitation spectrum.

4.2 *Effect of SO and Rabi Coupling Parameters on the Stability of Spectrum*

Recently, Ravisankar et al. [14] have analyzed the effect of the SO and Rabi couplings on the stability of the excitation spectrum for quasi two-dimension binary BECs. It has been observed while the increase in SO coupling leads the instability, increment in Rabi couplings brings the stability. In a similar line, we want to analyze the effect of variation in the couplings for a quasi-one dimension where the SO coupling acts in the same way. Figure 2 exhibits the variation of the instability (G) along the wavenumber for different k_L with fixed $\alpha = \beta = \Omega = 1$. We find that the condensate is stable for $k_L = 0$ for all ranges of the wavenumber. For small k_L ($= 1.5, 1.75$) the instability appears for a band of the wavenumber. Such kind of phase transition will

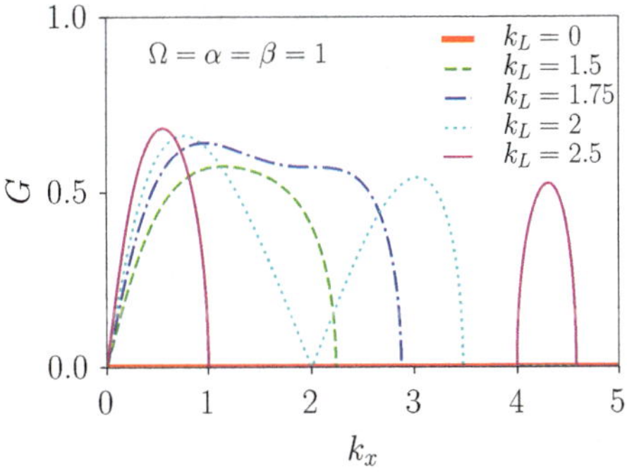

Fig. 2 Variation of the imaginary part of the eigenvalue of the excitation spectrum with k_x for different k_L with fixed $\Omega = \alpha = \beta = 1$. The instability region expands along k_x with increase in k_L

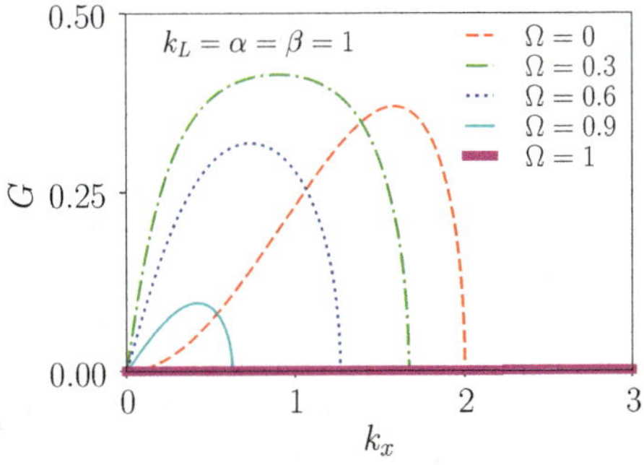

Fig. 3 Variation of the imaginary part of the eigenvalue of the BdG spectrum with k_x for different Ω with fixed $k_L = \alpha = \beta = 1$. The instability region reduce along k_x with increase in Ω

occur only for $\Omega > k_L^2$ in the weak repulsive BECs. In particular, when $k_L \gtrsim 1.75$ we noticed the single instability band transforms to multi-band. Upon further increase in k_L $(= 2.0, 2.5)$ leads to the generation of two or more wavenumber bands in which the spectrum is unstable. Note that for simplicity here we have shown the instability factor (G) for the positive wavenumber. The behavior is symmetric about $k_x = 0$.

In Fig. 3 we show the variation of the instability factor (G) along the wavenumber for different Rabi coupling (Ω) with fixed interaction parameters $\alpha = \beta = 1$ and $k_L = 1$. For zero Rabi coupling ($\Omega = 0$), the condensates lie in the stripe wave regime; thus, it has an instability band [13]. Further increase in the Rabi coupling to $\Omega = 0.3$, we observed that the instability bandwidth gets reduced with enhanced amplitude. However, increment in $\Omega (= 0.6, 0.9)$ resulted reduction in the bandwidth as well as the amplitude of the instability. Finally, we obtained the stable regime for $\Omega \gtrsim 1$. From the single-particle spectrum, when $\Omega > k_L^2$, we have a plane-wave phase which is a more stable one. Here we also confirm this through our stability analysis.

In the next section, we discuss the effect of variations of interaction parameters on the instability of the spectrum.

4.3 *Effect of Intra- And Interspecies Interactions on the Stability of Spectrum*

In the last subsection, we discussed the effect of SO and Rabi coupling parameters on the stability of the excitation spectrum; here, we present our analysis attributed to the effect of the intra- and interspecies nonlinear interaction strengths on the stability of the excitation spectrum. First we analyze the effect of intraspecies interactions (α) on the BdG spectrum for fixed SO coupling $k_L = 2$, and $\Omega = \beta = 1$ (see Fig. 4). For $k_L = 1$, and $\Omega = 1$, we know that the system should be in the stable regime because it is plane wave phase as discussed in Fig. 3. Here as we are interested in analyzing the instability of the system, we choose the stripe wave regime, which is the more unstable regime. For $\alpha = 1$, we find two instability bands, which are maintained up to $\alpha \lesssim 2$. When $\alpha > 2$, we noticed that the two instability bands converted into a single band accompanied by a reduction in the instability amplitude. It is quite interesting that we obtain the stable regime for more repulsive intraspecies interaction strength $\alpha = 8$ (see magenta thick solid line in Fig. 4). Overall we conclude that as we select the unstable regime, increasing α makes the system enter into the stable regime.

Next, we analyze the role of interspecies interaction strength (β) on the stability of the system with fixed parameters as $k_L = \Omega = \alpha = 1$. Here we consider the plane wave regime, which is the more stable regime, then we show the effect of increasing β on the stability of the system. For finite repulsive intraspecies interaction strength, we have a more stable regime (see red thick solid line in Fig. 5). When $\beta > 1$, we observe the appearance of the instability band. Further, we notice that the single instability band gets converted to multi-band when $\beta > 5$.

So far, we have discussed the effect of interactions and coupling parameters on the stability of the quasi-1D SO coupled BECs. In the next section, we will present the stability phase diagrams that we have obtained from our analysis.

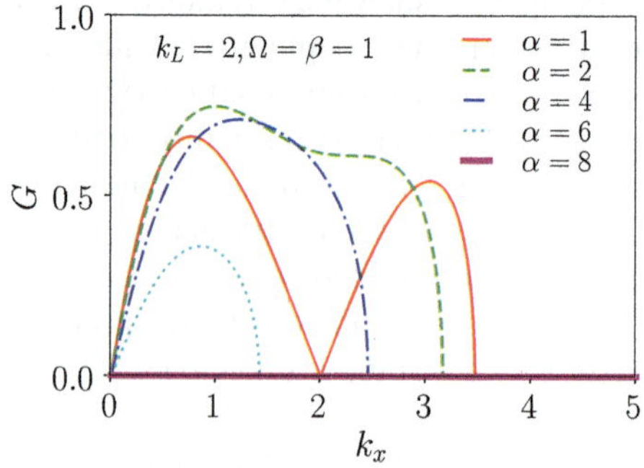

Fig. 4 Variation of the imaginary part of the eigenvalue of the excitation spectrum with k_x for different α with fixed $k_L = 2$, $\Omega = \beta = 1$. The instability region reduce along k_x with increase in α

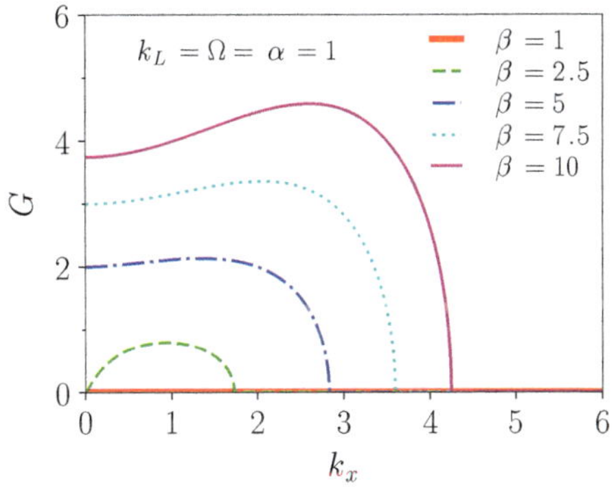

Fig. 5 Variation of the imaginary part of the eigenvalue of the BdG spectrum with k_x for different β with fixed $k_L = \Omega = \alpha = 1$. The instability region expands along k_x with increase in β

5 Stability Phase Diagrams

For better visualization of the stability of the system, we plot two different stability phase diagrams in Fig. 6. The left panel is in the $\alpha - \beta$ plane for $\Omega = k_L = 1$ and right panel is in the $\Omega - k_L$ plane for $\alpha = \beta = 1$. Both phase diagram is for the G value at $k_x = 1$. First, let us consider the $\alpha - \beta$ stability phase diagram. When $\beta < 0.9$, we found a stable regime for all ranges of the α. Afterward, we observed that the increase in α with fixed β leads to the transformation of the unstable phase to the stable phase. However, increment of β for fixed α results in the transformation from stable to unstable. The $\alpha - \beta$ phase diagram clearly indicates the emergence of the more stable regime for repulsive intraspecies interaction strengths (α). On the other hand, the $k_L - \Omega$ phase diagram shows the instability regimes (see right side Fig. 6). For zero Rabi coupling ($\Omega = 0$) as k_L is increased initially, the stable

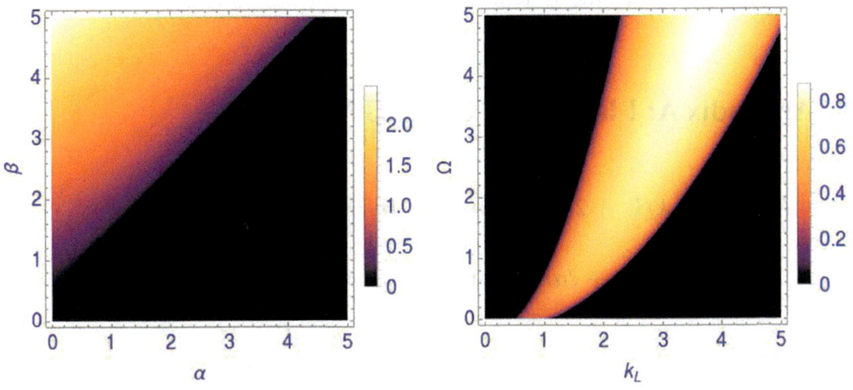

Fig. 6 Stability and instability phases in the $\alpha - \beta$ (left) and $k_L - \Omega$ (right) parameter space. The fixed parameters in the left are $\Omega = k_L = 1$ and in the left are $\alpha = \beta = 1$. The pseudo colour represents the G value

phase is observed, which becomes unstable for $0.5 < k_L < 1.0$. For $k_L > 1.0$, the system returns back to the stable regime. In the same line, all range of Rabi coupling behaves, which suggests that upon increasing SO coupling with fixed Rabi coupling we will have some intermediate instability regime for a short range. Recently such types of intermediate ground state phase is reported in the Ref. [15].

6 Conclusion

Using the Bogoliubov-de-Gennes analysis, we have analyzed the effect of the interaction and coupling parameters on the stability of the collective excitation spectrum of the SO and Rabi coupled binary BECs in quasi-one dimension. We have computed the single-particle as well the collective excitation spectrum. The absolute value of the imaginary part of the negative eigenenergy of the excitation spectrum has been used to characterize the stability of the spectrum. Our analysis shows that the increase in the SO coupling (k_L) leads to the conversion of the single and multi-instability bands to the stable state. However, increase in Rabi coupling resulted the instability having a single-band nature. The increase in α also changes the multiple instability band into single band which finally converts into stable one for strong repulsion. However, increase in β keeping other parameters fixed yields multiple unstable bands from a stable state. Only the amplitude to instability increases. Finally, we have obtained a stability phase diagram in $\alpha - \beta$ and $\Omega - k_L$ parameter space. We found that increase in β for a fixed value of α leads the system to enter from a stable to unstable region while the effect of the increase in α for fixed β is opposite. Similarly, we find that increase in k_L for fixed Ω brings the system from stable to unstable phase while behavior is opposite as Ω is increased for fixed k_L. The essential observation is that upon the increase in Ω and α alone makes the system more stable, however, increase in k_L and β transformed the system from stable to unstable one.

7 Appendix A: Elements of the BdG Matrix

$$H_1 = \frac{k_x^2}{2} + k_L k_x + 2\alpha n_\uparrow + \beta n_\downarrow; \qquad H_2 = \frac{k_x^2}{2} - k_L k_x + 2\alpha n_\uparrow + \beta n_\downarrow$$

$$H_3 = \frac{k_x^2}{2} - k_L k_x + 2\alpha n_\downarrow + \beta n_\uparrow; \qquad H_4 = \frac{k_x^2}{2} + k_L k_x + 2\alpha n_\downarrow + \beta n_\uparrow$$

and

$$\mu = \frac{1}{2}\left[\alpha n + \beta n - \Omega \frac{n}{\sqrt{n_\uparrow n_\downarrow}}\right]$$

8 Appendix B: Coefficients of the BdG Excitation Spectrum

$$a = -\frac{k_x^4}{2} - 2\Omega\Big(\alpha - \beta + 2\Omega\Big) - k_x^2\Big(\alpha + 2(k_L^2 + \Omega)\Big)$$
$$b = \frac{1}{16}k_x^2\Big(k_x^2 + 2\alpha - 2\beta - 4k_L^2 + 4\Omega\Big)\Big(k_x^4 + 8(\alpha + \beta)\Omega + 2k_x^2(\alpha + \beta - 2k_L^2 + 2\Omega)\Big)$$

Acknowledgements Supported by DST-SERB (Department of Science and Technology - Science and Engineering Research Board) for the financial support through Project No. ECR/2017/002639.

References

1. Pethick, C.J., Smith, H.: Bose-Einstein Condensation in Dilute Gases, 2nd edn. Cambridge University Press, Cambridge (2008)
2. Ho, T.-L.: Spinor Bose condensates in optical traps. Phys. Rev. Lett. **81**, 742 (1998)
3. Li, Y., Hai, W.: Three-body recombination in two coupled Bose-Einstein condensates. J. Phys. A: Math. Gen. **38**, 4105–4114 (2005)
4. Wang, Z., Zhang, X.: Ke Shen, Unstable cycle of two coupled bose-einstein condensates with three-body interaction. J. Low Temp. Phys. **152**, 136–146 (2008)
5. Sudharsan, J.B., Radha, R., Muruganandam, P.: Collisionally inhomogeneous Bose-Einstein condensates with binary and three-body interactions in a bichromatic optical lattice. J. Phys. B: At. Mol. Opt. Phys. **46**, 155302 (2013)
6. Lin, Y.J., García, K.J., Spielman, I.B.: Spin-orbit coupled Bose-Einstein condensates. Nature (London) **471**, 83 (2011)
7. Campbell, D.L. , Juzeliūnas, G., Spielman, I.B.: Realistic spin-orbit and Dresselhaus spin-orbit coupling for neutral atoms. Phys. Rev. A **84**, 025602 (2011)
8. Bogolyubov, N.N.: On the theory of superfluidity. J. Phys. (USSR) **11**, 23 (1947)
9. Zilsel, P.R.: Liquid Helium II: the hydrodynamics of the two-fluid model. Phys. Rev. **79**, 309 (1950)
10. Khamehchi, M.A., Zhang, Y., Hamner, C., Busch, T., Engels, P.: Measurement of collective excitations in a spin-orbit-coupled Bose-Einstein condensate. Phys. Rev. A **90**, 063624 (2014). https://doi.org/10.1103/PhysRevA.90.063624
11. Li, Y., Martone, G.I., Pitaevskii, L.P., Stringari, S.: Superstripes and the excitation spectrum of a spin-orbit-coupled Bose-Einstein condensate. Phys. Rev. Lett. **110**, 235302 (2013). https://doi.org/10.1103/PhysRevLett.110.235302
12. Ozawa, T., Pitaevskii, L.P., Stringari, S.: Supercurrent and dynamical instability of spin-orbit-coupled ultracold Bose gases. Phys. Rev. A **87**, 063610 (2013)
13. Ravisankar, R., Sriraman, T., Salasnich, L., Muruganandam, P.: Quenching dynamics of the bright solitons and other localized states in spin-orbit coupled Bose-Einstein condensates. J. Phys. B: At. Mol. Opt. Phys. **53**, 195301 (2020). https://doi.org/10.1088/1361-6455/aba661
14. Ravisankar, R., Fabrelli, H., Gammal, A., Muruganandam, P., Mishra, P. K.: Effect of Rashba spin-orbit and Rabi couplings on the excitation spectrum of binary Bose-Einstein condensates. Phys. Rev. A **104**, 053315 (2021). https://doi.org/10.1103/PhysRevA.104.053315
15. Ravisankar, R., Sriraman, T., Kumar, R.K., Muruganandam, P., Mishra, P.K.: Influence of Rashba spin-orbit and Rabi couplings on the miscibility and ground state phases of binary Bose-Einstein condensates. J. Phys. B: At. Mol. Opt. Phys. **54**, 225301 (2021). https://doi.org/10.1088/1361-6455/ac41b2

Empirical Models for Premiums and Clustering of Insurance Companies: A Data-Driven Analysis of the Insurance Sector in India

Rakshit Tiwari and Siddhartha P. Chakrabarty

Abstract The article deals with the modeling of insurance premiums and their clustering, from the paradigm of the Indian insurance sector, during a 15 year period, using data-driven regression and clustering techniques. Among three approaches considered for the predictive modeling of insurance premiums, the most effective method was determined to be the random forest approach. Interesting insights for the pre and post 2008 (financial crisis) period, revealed distinct clustering characteristics between the private and public sector insurance companies operating in India, especially in terms of consumer behavior.

Keywords Insurance premium · Machine learning · Regression · Clustering

1 Introduction

Since the solvency of an insurance company is driven by its ability to generate premiums, which must exceed the expected claims payout and other liabilities, therefore it is essential to develop predictive models of both the premium receipts, as well as the claims losses. Our analysis in this article is motivated by the somewhat minimal literature available in case of the former, as compared to the latter. The role of actuaries, while developing pricing strategies, is to assess a fair price for the insurance products they wish to sell. This however, has to be done in a manner so as to fulfill the outstanding liabilities of the insurance company, while safeguarding its solvency and reserve capital. Consequently, the actuaries must predict, with maximum possible accuracy, the total amount required to meet the claims payout. These reserves form the principal item on the insurance company's balance sheet's liability side and thus have a significant economic impact.

R. Tiwari · S. P. Chakrabarty (✉)
Indian Institute of Technology Guwahati, Guwahati 781039, Assam, India
e-mail: pratim@iitg.ac.in

R. Tiwari
e-mail: rakshit10@alumni.iitg.ac.in

S. Banerjee and A. Saha (eds.), *Nonlinear Dynamics and Applications*,
Springer Proceedings in Complexity,
https://doi.org/10.1007/978-3-030-99792-2_110

In this article, we perform statistical and data analysis, on the premium data collected from the website of the Insurance Regulatory Development Authority of India (IRDAI) [1], with the objective of developing predictive models for the monthly premium collection of the insurance companies included in our dataset. The dataset available on the website of IRDAI were for 13 different companies, namely, *Royal Sundaram, Tata-AIG, Reliance General, IFFCO-Tokio, ICICI-lombard, Bajaj Allianz, HDFC CHUBB, Cholamandalam, New India, National, United India, Oriental and ECGC.*

The data that we use in this article, consists of the monthly premium amounts collected during the period of April 2003 to Dec 2017, for these 13 insurance companies mentioned above. We note here, that more detailed data was not available. Also, the issue of some anomalies and errors in some of the cells of the considered dataset, was addressed, by removing the erroneous data, so as to achieve the best possible results. Further, during the process of building a machine learning model, we followed the standard practice of training and then back-testing the model on the untrained part of the dataset. Accordingly, the training set comprised of 80% of the total data points, while the remaining 20% of the data points was used for back-testing the results, as well as finding the mean squared and root mean squared error, for comparative analysis across several types of models. The key aspect in this analysis is not just about building mathematical models, for finding results using the same, but to explain those results with the help of real life comparisons which have been discussed towards the end of the article.

Most of the modeling approaches in actuarial mathematics focus on the determination of the distribution that best fits the data. In a recent work [2], a dependent modeling framework is adopted for predictive distribution with accuracy in case of frequencies, as well as claims score, pertaining to insurance claims. In another distribution driven article [3], the authors make use of a tweedie compound Poisson model to achieve a robust prediction performance for premiums. The classification of applicants, for risk prediction in life insurance sector, is carried out by way of the supervised learning approach in [4].

2 Predicting the Monthly Premium Amount

We will approach the problem of predicting the monthly premium amount received by each insurance company, making use of three commonly used machine learning models. Further, we implement the models in terms of its increasing level of complexity, in order to understand whether this results in the concurrent improvement in terms of the backtesting results. The eventual goal is to ascertain whether the models and their predictive results obtained here, would be helpful in obtaining a viable prediction mechanism, for forecasting the premium amounts in the future. Accordingly, we will make use of three models, as outlined in the following subsections.

2.1 Linear Regression

The basis for *linear regression* model is the assumption that there exists a linear relationship between a dependent (output) variable y and several independent (input) variables $x_1, x_2, \ldots, x_n$ [5, 6]. A univariate or simple linear regression model is one where there is only one input variable, while linear regression models with multiple input variables is termed as multiple linear regression model. The most common approach of obtaining a linear regression equation, from observed data is the method of *least squares regression*. The coefficients in case of both multiple, as well single regression model, can be obtained by minimizing the sum of the least squared errors, which can be defined as $\epsilon_i = y_i - \widehat{y_i}$, where the variable y has n observed (predicted) values, say y_i $(\widehat{y_i})$, $i = 1, 2, \ldots, n$. For the sake of brevity, we state the results for multiple linear regression which can subsequently be used in case of polynomial regression model. Accordingly, we let,

$$\vec{y} = \begin{pmatrix} y_1 \\ y_2 \\ \vdots \\ y_n \end{pmatrix}, \; X = \begin{pmatrix} 1 & x_{11} & x_{12} & \cdots & x_{1k} \\ 1 & x_{21} & x_{22} & \cdots & x_{2k} \\ \vdots & \vdots & \vdots & \vdots & \\ 1 & x_{n-11} & x_{n-12} & \cdots & x_{n-1k} \end{pmatrix}, \; \vec{\beta} = \begin{pmatrix} \beta_1 \\ \beta_2 \\ \vdots \\ \beta_k \end{pmatrix} \text{ and } \vec{\epsilon} = \begin{pmatrix} \epsilon_1 \\ \epsilon_2 \\ \vdots \\ \epsilon_n \end{pmatrix}.$$

The multiple linear regression thus can be represented in the form,

$$\vec{y} = X\vec{\beta} + \vec{\epsilon}.$$

The least-squares regression approach is used to estimate the parameter $\vec{\beta}$ by minimizing,

$$\sum_{i=1}^{n} \epsilon_i^2 = \left(\vec{y} - X\vec{\beta}\right)^{\top} \left(\vec{y} - X\vec{\beta}\right),$$

which then results in,

$$\widehat{\vec{\beta}} = \left(X^{\top}X\right)^{-1} X^{\top}\vec{y}.$$

Note that the methods of regularization, is used to modify the learning algorithm (so as to achieve reduction in complexity of the regression models) by placing pressure on the absolute size of the coefficients, thereby reducing some of the coefficients to zero.

2.2 Polynomial Regression

We next, seek to address a key shortcoming of the linear model, namely the underfitting, by increasing the complexity of the model, and the natural choice for the same is to add higher powers, resulting in higher order equations [6]. Accordingly, the linear model obtained above is now represented as,

$$y_i = \sum_{j=1}^{m} x_i^j + \epsilon_i, \ i = 1, 2, \ldots, n.$$

Even though terms like x_i^j are present in the model, the coefficients are still linear. However, the curve we fit is going to be polynomial in nature. Therefore, *polynomial regression* is a type of linear regression in which the relationship between the input variable and output variable, is modeled as a polynomial of m-th degree. Polynomial regression matches a nonlinear relationship between the value of the input variable x and the corresponding conditional mean of the output variable y, denoted as $E\ (y|x)$. An interesting point that was observed while working with polynomial regression model, is the accuracy of the results (or equivalently, the reduction in absolute error) while back-testing results, as we select higher degree polynomials. A naive approach of selecting degree 2 polynomials till degree 10 polynomials was tested, and the same was confirmed.

2.3 *Random Forest Regression*

Random forest, as the nomenclature suggests, comprises of a large collection of decision trees which act collectively [7]. Each individual tree of the random forest outputs a prediction of class, with the one gaining the most votes, taken as the prediction of the model. The rationale for this impact is that the trees protect one other from their individual incorrectness (provided they do not all fail in a concurrent and consistent manner). While some trees might be incorrect, many other trees would be correct, and consequently, the trees will move in the correct direction, collectively. In our case below, we started with 10 decision trees, since this is the default value in sklearn.ensemble library [8] and slowly increased the value so that the model can learn and adapt from the training set and make better decisions.

3 Procedure

In this Section, we present the approach for predicting the monthly premium as elaborated in Sects. 2.1, 2.2 and 2.3.

3.1 Predicting Monthly Premiums

We will discuss the methods that were used primarily to obtain the results reported in this article. The first step was the filtering of data containing approximately 2000 data points, in order to remove the anomalies. Our main focus then was to judge the fitness of models on the test data and to suggest ways that can be used by the actuaries for projecting the growth of premiums received by the insurance firms in the future. Three modeling methods have been used in this article as described above in Sect. 2, with one of them distinctly emerging as the best choice, because of a substantial difference obtained in the mean absolute error while back-testing. Some of the standard practices of machine learning that have been used in this article are:

1. The data considered was divided in the ratio 4 : 1, between the training and the testing sets which is the standard approach adopted for machine learning programs. Also, due to the lack of availability of extensive data, as well as for the generation of a fully functional model, it is important to prioritize our limited resources, for training the model first.
2. As already noted, the data provided, comprised of monthly premiums collected across a span of 15 years. For regression purposes, we have taken the help of date ordinals, which are the standard hashed values of dates used in machine learning.
3. One of the other challenges while modeling, pertained to the choice of the degree of polynomial being used in the polynomial regression model. All polynomials from degree 2 to degree 10, were chosen one by one, and the degree which gave the least absolute error during back-testing was finally chosen.
4. Number of estimators/decision trees chosen in the random forest model was equal to 1000.

3.2 Predicting the Optimum Number of Clusters for Clustering

1. An appropriate data-frame of monthly premiums was formed to carry out the basic *K-Means clustering* [9].
2. The purpose of going forward with a basic clustering algorithm and not any advanced clustering algorithms, lies in our regression results. Since the data did not exhibit any absurd patterns, it was an optimum choice for carrying out the clustering experiment with basic algorithms.
3. In order to determine the optimum number of clusters, we used the standard *Elbow Method* [9]. The results obtained during this experiment can be seen in Sect. 4. The value of optimum clusters obtained is 2.
4. Using the same result, going forward and applying K-Means algorithm, we obtained 2 group of companies. The results obtained were on expected lines, as it segregated the 13 companies into 2 groups primarily based on their positive or negative growth of monthly premiums.

5. In order to examine the results of this experiment in a more elaborate manner, we worked around with cluster size of 3 and 4 also. This helped us in segregating organizations even further on their growth/product quality.

4 Results

This Section is entirely dedicated to the presentation of the results that were obtained from the data considered, and the models we have enumerated in Sect. 3. The results of the regression modeling using the random forest algorithm are presented in Fig. 1a–m, where the consistently superior performance of the random forest algorithm can be easily observed.

Using K-Means Clustering and Elbow method for finding optimum number of clusters, we were assured of that there are two optimum number of clusters present in the dataset considered, into which, we can segregate the enumerated insurance companies. Using the Elbow method, we obtained the first cluster comprising of eight companies, namely, Royal Sundaram, Tata-AIG, Reliance General, IFFCO-Tokio, ICICI-Lombard, Bajaj Allianz, HDFC CHUBB and Cholamandalam, all of which are private sector companies. The second cluster comprises of the remaining five companies, that is, New India, National, United India, Oriental and ECGC, all of which are public sector companies. The sum of the squared distances against the number of clusters are presented in Fig. 2.

When we experimented with the clustering of the companies, using different cluster sizes, we observed that the existing partition of size two, in the preceding paragraph, gets further sub-divided, on the basis of better/good performance or better/worse performance. With the number of clusters being three, the first cluster includes, Royal Sundaram, Tata-AIG, Reliance General, IFFCO-Tokio, ICICI-Lombard, Bajaj Allianz, HDFC CHUBB and Cholamandalam. The eight companies, included in the first list for three clusters, were all included in the first list for two clusters. These eight are the group of companies, who have shown an impeccable growth in period of 15 years that have been considered. The second cluster, now, comprises of four companies, namely, New India, National, United India and Oriental. These can be described as those players who suffered a major setback in 2008 but are showing signs of revival, since then. Finally, the third cluster in this case, includes just one company, that is, ECGC. It is observed from the data, that ECGC is the only company that has failed to achieve visible revival, after going through setbacks as a result of the 2008 financial crisis.

In case of cluster of size four, the first cluster includes ICICI-Lombard and Bajaj Allianz. These two companies are the ones who have an outstanding growth chart when compared to the other six companies that are in the positive growth cluster. The remaining six companies namely Royal Sundaram, Tata-AIG, Reliance General, IFFCO-Tokio, HDFC CHUBB, Cholamandalam are now a part of the second cluster out of the four clusters. The third cluster, comprises of the same cluster of public comprises, that formed the second cluster in case of three clusters, namely, New

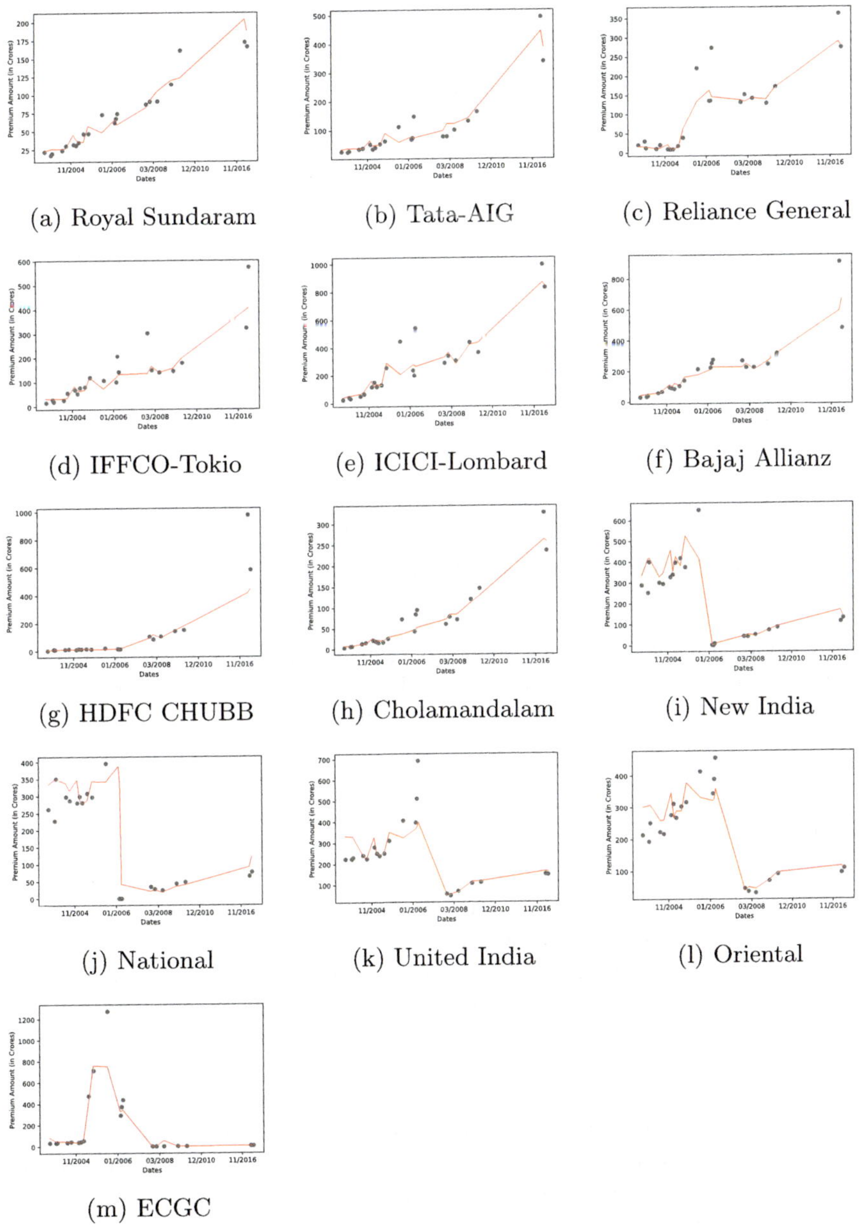

(a) Royal Sundaram (b) Tata-AIG (c) Reliance General

(d) IFFCO-Tokio (e) ICICI-Lombard (f) Bajaj Allianz

(g) HDFC CHUBB (h) Cholamandalam (i) New India

(j) National (k) United India (l) Oriental

(m) ECGC

Fig. 1 The fit to data of all the 13 insurance companies using the Random Forest Algorithm.

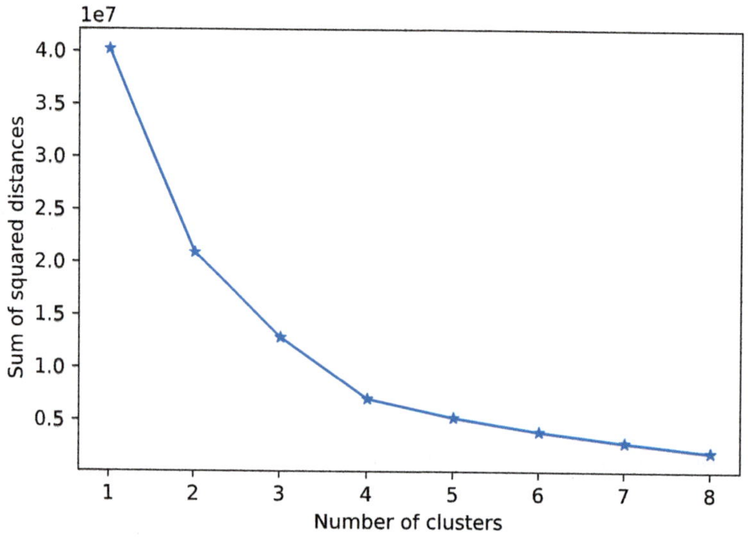

Fig. 2 The sum of squared distances against the number of clusters using the Elbow Method

India, National, United India, Oriental. Consequently, ECGC is the only company in the last of the four clusters. We observe that, the third and fourth cluster in this case, is identical to the last two clusters for the previous case of three clusters.

5 Discussion

It is clear that in the 15 years, for which the data was considered, the first group of companies (**for the two-cluster case**) have shown a tremendous growth, while the opposite has happened in case of the second group of companies. There are several reasons to which this happenings can be attributed to, which will be elaborated upon, in the following discussion.

One of the things that we can look at, with the help of the models, is the accuracy of our back-testing results which exhibits improvement with the increasing complexity of the model. Table 1 depicts the mean absolute error obtained using the Linear Regression (LR), Polynomial Regression (PR) and the Random Forest (RF), model in Crores.[1] The error data obtained from the first eight companies confirms that random forest is indeed the best model being used to predict premium amount collections in our case. Meanwhile, when we first look at the error data of last five columns in Table 1, it suggests a completely different narrative, which is actually not the case. A close look at Fig. 3 above for these companies itself, one can see a huge spike (downfall) right around the 2007–09 period, which results in the outliers. In the context of

[1] 1 Crore=10 Million.

Table 1 Mean Square Error of the Three Regression Algorithms (in Crores) for all the thirteen insurance companies

Company name	Linear regression	Polynomial regression	Random forest
Royal Sundaram	14.15	14.13	10.33
Tata-AIG	38.09	14.895	17.89
Reliance General	40.09	26.25	23.07
IFFCO-Tokio	43.61	33.15	30.79
ICICI-lombard	77.86	41.51	40.54
Bajaj Allianz	52.93	47.88	30.14
HDFC CHUBB	92.70	32.47	37.43
Cholamandalam	25.52	12.05	11.57
New India	120.23	294.25	321.45
National	86.98	258.92	253.53
United India	105.25	197.25	197.71
Oriental	88.52	195.62	200.93
ECGC	243.73	261.99	313.92

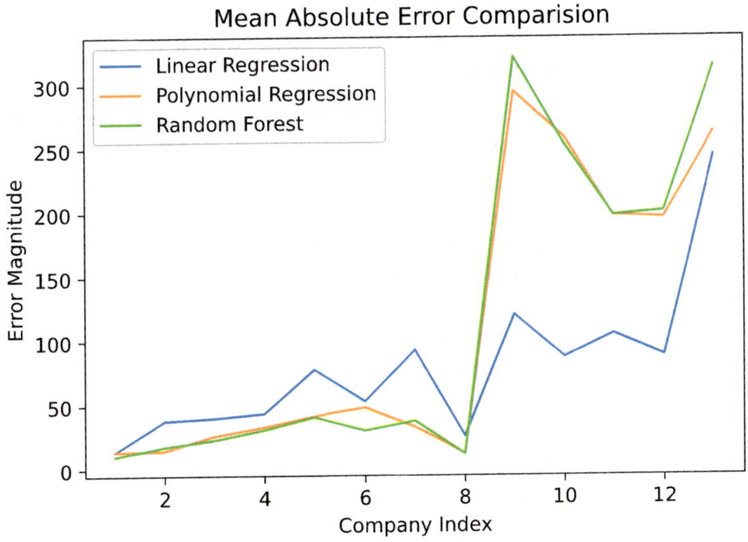

Fig. 3 Comparison of the Mean Absolute Error using the three methods

the discussion, outliers comprise of values which exhibit significant deviation from the other data points. This deviation could be attributed to measurement inaccuracies and flaws in the experimental design, among other considerations. An outlier can be viewed as a deviation from the general trend. It is because of this reason, the data analysis obtained provides an inaccurate picture and we should rather look the graphs for judging the fitness of data, in the case of Random Forest. Accordingly, the graphical comparison is shown in Fig. 3, which highlights the heavy downfall experienced by the second group of companies (for the indices $9-13$ in the figure), where because of outliers (as discussed above), one can observe a large deviation from truth. Also, the errors presented in Table 1 looks large, when someone discusses all these figures in Crores. The point that should be noted along with this, is that the typical figures for the monthly premium being collected is in the range of $300-800$ Crores (as provided in our dataset), which, consequently, gives us a relative error close to $1-3\%$.

The purpose of this article is not only restricted to the determination of mathematical validity of the model to ascertain the best one, but also to effectively we can correlate the right real world reasons with what can be ascribed to the dataset. Talking first about the companies which belonged to the second cluster, one can note that they are being differentiated from other companies primarily because of the lack of growth in the last 15 years. Once we started taking a deep dive into it, which in our observation can be attributed to:

1. **Marketing:** For organizations that extensively require new customers and making sure that old customers stay onboard, one can clearly see the companies in first cluster spending extensively on marketing especially in metropolitan states which is a huge reason for their growth post the 2008 crisis. These organizations have gone out of their way, post the financial crisis to instill a deep trust between the people for using their insurance advisory.
2. **Technology:** The world is run on applications today. It is important for all the banks and financial businesses to have an online presence. Those organizations who are not having an application/website where customers can come and get the job done in one click, have faced a huge downfall must similar to what is being faced by the second cluster of companies. It may be noted that, ICICI Lombard and Bajaj Allianz are leaders, when it came to adoption of technology into their operation, especially the customer interface. In addition, both these organizations have also invested substantially on social marketing to propel their businesses online.

References

1. IRDAI monthly insurance data. https://www.irdai.gov.in/ADMINCMS/cms/frmGeneral_List.aspx?DF=MBFN&mid=3.2.8
2. Lee, G.Y., Shi, P.: A dependent frequency-severity approach to modeling longitudinal insurance claims. Insur.: Math. Econ. **87**, 115–129 (2019)

3. Yang, Y., Qian, W., Zou, H.: Insurance premium prediction via gradient tree-boosted Tweedie compound Poisson models. J. Bus. Econ. Stat. **36**(3), 456–470 (2018)
4. Boodhun, N., Jayabalan, M.: Risk prediction in life insurance industry using supervised learning algorithms. Complex Intell. Syst. **4**(2), 145–154 (2018)
5. Schneider, A., Hommel, G., Blettner, M.: Linear regression analysis: part 14 of a series on evaluation of scientific publications. Deutschesrzteblatt Int. **107**(44), 776 (2010)
6. Ostertagova, E.: Modelling using polynomial regression. Procedia Engineering **48**, 500–506 (2012)
7. Liaw, A., Wiener, M.: Classification and regression by randomForest. R news **2**(3), 18–22 (2002)
8. Pedregosa, F., Varoquaux, G., Gramfort, A., Michel, V., Thirion, B., Grisel, O., Duchesnay, E.: Scikit learn: machine learning in python. J. Mach. Learn. Res. **12**, 2825–2830 (2011)
9. Yuan, C., Yang, H.: Research on K-value selection method of K-means clustering algorithm. Journal **2**(2), 226–235 (2019)

Variations in the Scroll Ring Characteristics with the Excitability and the Size of the Pinning Obstacle in the BZ Reaction

Puthiyapurayil Sibeesh, S V Amrutha, and T K Shajahan

Abstract We report the experimental results of the effects of excitability on the wave characteristics of free rotating and pinned scroll rings in the Belousov-Zhabotinsky (BZ) reaction. The experiments show that the stability of the scroll ring depends on the excitability of the medium. At low excitability, the scroll ring becomes less stable and eventually breaks up. As we increase the excitability of the medium, the time period (T) and wavelength (λ) of the excitation wave decrease while wave velocity (v) increases. Properties of both free and pinned scroll rings change in the same way. However, at a given excitability, both the λ and v of a pinned scroll ring increase with the size of the obstacle. For the range of parameters chosen in our experiments, the excitability changes brought by varying reactant concentrations have a higher impact on the scroll ring properties than those induced by the size of the pinning obstacle.

Keywords BZ reaction · Scroll wave · Excitability · Pinning

1 Introduction

Excitable media support nonlinear waves that propagate as target, spiral, or scroll waves. Such waves are observed in many diverse systems including in the aggregation of Dictyostelium discoideum amoeba [1], the chicken retina [2], the brain [3] and the cardiac tissues [4], the Belousov-Zhabotinsky (BZ) chemical reaction [5], and the oxidation of CO on Pt surfaces [6]. In physiological tissue such self sustained spiral and scroll waves can lead to life-threatening dynamical diseases such as cardiac arrhythmias[7] or epilepsy [8].

A scroll wave is a three-dimensional manifestation of a rotating spiral wave found in the two-dimensional excitable media. While the spiral wave rotates around a single point at the tip of the spiral, the scroll wave rotates around a one-dimensional filament.

P. Sibeesh · S. V. Amrutha · T. K. Shajahan (✉)
Department of Physics, National Institute of Technology Karnataka Surathkal, 575025 Mangalore, India
e-mail: shajahan@nitk.edu.in

S. Banerjee and A. Saha (eds.), *Nonlinear Dynamics and Applications*, Springer Proceedings in Complexity,
https://doi.org/10.1007/978-3-030-99792-2_111

These filaments take the shape of a straight line or a circular ring. A scroll wave with a straight filament is a stack of two-dimensional spiral waves in parallel [9, 10]. A scroll wave with a circular closed loop as the filament is called a scroll ring. The curvature of the ring induces motion of the filament such that its radius becomes smaller and leads to the self-annihilation of the scroll ring. The collapse or shrinkage of a free scroll ring can be delayed by modifying the medium parameters. In an excitable medium, the life span of a scroll ring is short. Pinning the scroll filament to medium heterogeneities can elongate the lifetime of scroll waves [11].

The BZ reaction is one of the simplest laboratory models used to the study excitable media in which the excitation waves can be observed with naked eyes. Despite the fact that the BZ reaction is far less complex than heart tissue or any other excitable medium, they all possess similar dynamical behavior. Because of this similarity in the dynamics of excitation waves, we employ the BZ reaction as a model medium to investigate the dynamics of scroll rings. Previous studies reported that the properties of a spiral wave in a two-dimensional BZ medium modify according to the medium excitability [12], the size, and the shape of the pinning obstacle [13, 14]. The excitability of the BZ medium can be controlled by varying the initial concentrations of reactants [12]. Simulations with the Oregonator model have shown that both wave period and wavelength of a scroll wave decrease with increasing the excitability [15]. To the best of our knowledge, the dynamics of the scroll ring with the excitability and the size of the pinning obstacle is not reported so far. This article reports the dynamics of both free and pinned scroll rings in the BZ reaction by varying the excitability and the size of the pinning obstacle. In our experiments, we observed a breakup of a scroll ring filament at very low excitability. Our experimental observations show that the frequency and the wave velocity of a scroll ring increase with the excitability, whereas the wavelength decreases. As the size of the pinning obstacle increases, the time period (T), wavelength (λ), and wave velocity (v) of a scroll ring increase for a given excitability.

2 Experimental Methods

We conducted experiments in chemically identical double layers of ferroin - catalyzed BZ reaction as given in [9]. Each layer of thickness 4 mm is embedded in 1.4 % w/v of agar. We performed experiments with initial concentrations of reactants as following: $[H_2SO_4] = 0.25$–0.75 M, $[NaBrO_3] = 1$ M, $[MA] = 1$ M, $[SDS] = 0.0245$ M, $[Ferroin] = 0.025$ M. All the experiments were carried out at constant room temperature.

Spherical glass beads of diameters varying from 2 mm to 5 mm were used to pin the scroll waves. During the gelation of the first layer, two glass beads of the same size were symmetrically inserted halfway into the surface. A half-spherical wave was initiated at the center of the line joining the two glass beads by inserting the tip of a properly cleaned silver wire for a few seconds. The second layer is added above

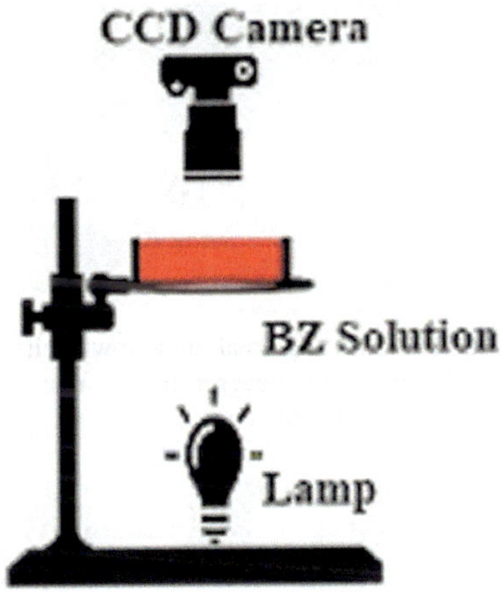

Fig. 1 Schematic diagram of the experiment setup: The BZ reaction medium is represented in red colour. A charge-coupled device camera kept above the medium captures the images of the reaction medium. Light source placed below enhances the imaging

the first layer when the wavefront touches the glass beads. The half-spherical wave at the lower layer curl upwards and forms a scroll ring pinned to both the beads.

Ferroin indicator undergoes striking color difference during each excitation cycle, which allows for the optical detection of the excitation waves. The experimental medium was illuminated using a diffused white light placed below to monitor the chemical waves. The images were captured by a charge-coupled device (CCD) camera (mvBlueCougarx 120bc) positioned above the medium. A blue filter (MidOpt BP470-27) was mounted on the camera to increase the contrast between the excitation waves and the unexcited medium. The images were recorded onto a computer at 2 frames/seconds with the help of LabVIEW and the data analyzed using software developed in Python. Figure 1 depicts the schematic representation of the experiment setup.

3 Results and Discussion

We explored the behavior of free rotating and pinned scroll rings by increasing the excitability of the BZ reaction medium. The excitability of the BZ reaction medium can be controlled by varying the concentration of the reactants ($[H_2SO_4]$ or $[NaBrO_3]$).

To vary the excitability of the medium in our experiments, we adjusted the concentration of H_2SO_4 from 0.25 M to 0.75 M while maintaining the concentrations of the other reactants constant. According to the relationship suggested by Jahnke et al. [16] the excitability of the medium increases with an increase in the concentration of H^+. In a low excitable medium with $[H_2SO_4] = 0.25$ M, the filament of a scroll ring breaks up within a short duration of time. Figure 2 represents the time evolution of a free scroll ring induced in a low excitable medium. The scroll ring breaks by the time t=1025s as shown in Fig. 2c.

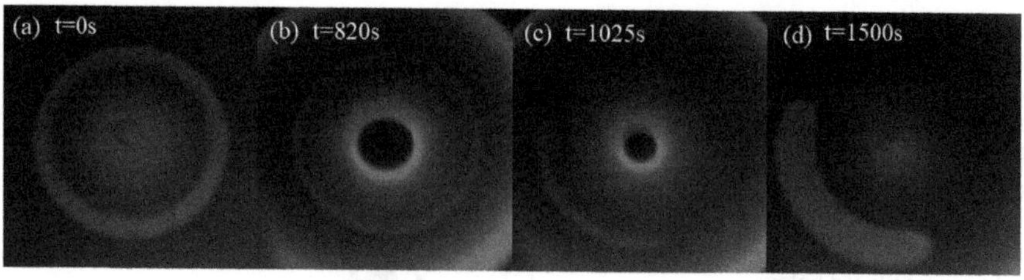

Fig. 2 Scroll ring break up at low excitability ([H_2SO_4] = 0.25 M): Snapshots of a free rotating scroll ring. A complete scroll ring (**a**) at t = 0s (**b**) at t = 820s (**c**) the scroll ring breaks at t = 1025 s (**d**) the broken scroll ring at = 1500s

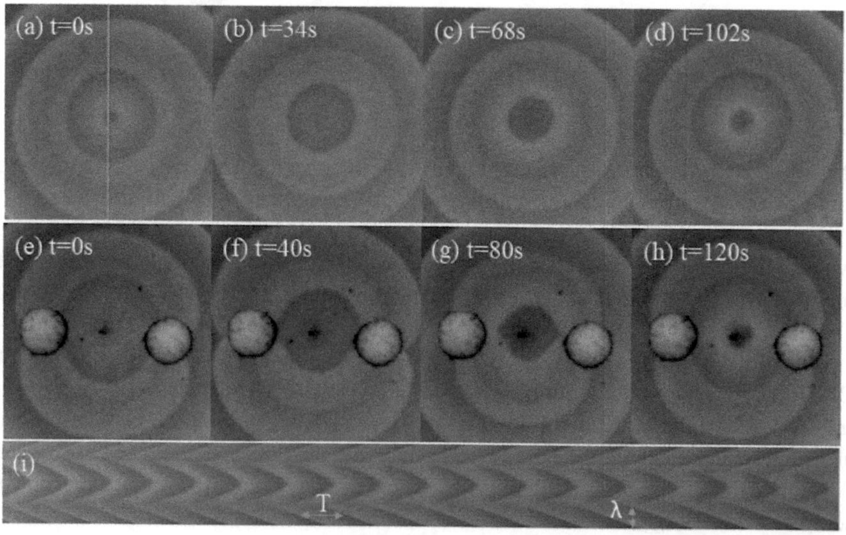

Fig. 3 Time evolution of Scroll rings and the space time plot of the free scroll ring in BZ reaction medium ([H_2SO_4] = 0.7 5 M). (**a**)–(**d**) Free rotating, scroll ring (**e**)–(**h**) Scroll ring pinned to two spherical glass beads of diameter 3.5 mm (i) Space-time plot generated for the free rotating scroll ring along a vertical line at the center of each image in the experiment. It spans a time interval of 41 min

Figure 3a–d show the snapshots of the time evolution of a free rotating scroll ring and Fig. 3e–f show snapshots of the time evolution of a scroll ring pinned to two glass beads of diameter 3.5 mm at [H_2SO_4] = 0.75 M. Figure 3i is a space-time plot that is obtained for the free rotating scroll ring along the central vertical line (orange line in a) of the captured images for a duration of 41 min. T and wavelength λ of the scroll rings are calculated from the corresponding space-time plot as described in [12].

As shown in Fig. 4a (green diamond line), the frequency of the scroll rings increases as the excitability increases, and that the wavelength of the scroll rings decreases as shown in Fig. 5a (green diamond curve). Similar changes are also found

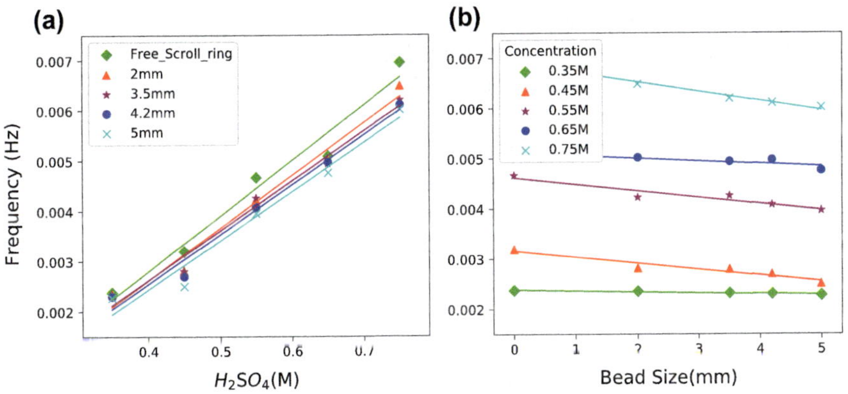

Fig. 4 Variation of scroll ring frequency with concentration of H_2SO_4 and obstacle size: (**a**) Frequency of the scroll ring increases with excitability. Each line corresponds to scroll ring pinned to different obstacles: free rotating scroll ring (green diamond), 2 mm glass bead (red triangle), 3.5 mm glass bead (purple star), 4.2 mm glass bead (blue dot) and 5 mm glass bead (cyan cross). (**b**) Frequency of the scroll ring decreases with increase in obstacle size. Each line corresponds to different excitability: 0.35 M (green diamond), 0.45 M (red triangle), 0.55 M (purple star), 0.65 M (blue dot) and 0.75 M (cyan cross)

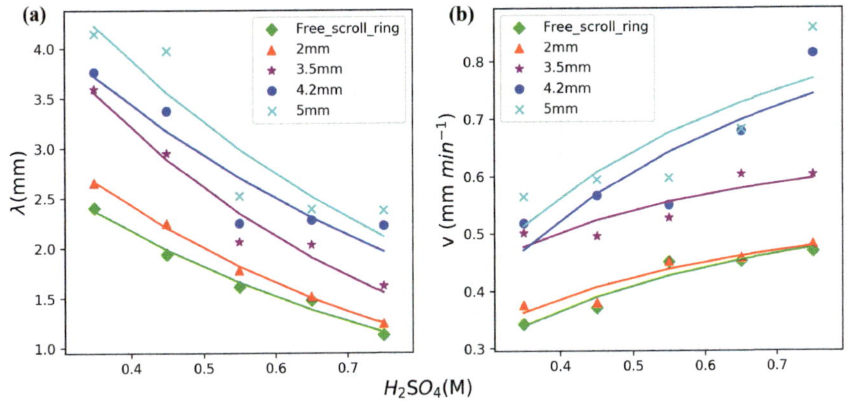

Fig. 5 Properties of scroll ring with concentration of H_2SO_4 and obstacle size: (**a**) Wavelength (λ) of the scroll ring decreases and (**b**) wave velocity (v) increases. Both quantities increases with increase in obstacle size for a particular excitability. Each curve corresponds to scroll ring pinned to different obstacles: free rotating scroll ring (green diamond), 2 mm glass bead (red triangle), 3.5 mm glass bead (purple star), 4.2 mm glass bead (blue dot) and 5 mm glass bead (cyan cross)

in two-dimensional spiral waves [12, 17]. The wave velocity is estimated as λ/T [13]. As excitability increases, the wave velocity of the free scroll ring increases as shown in Fig. 5b (green diamond curve). A numerical study on a scroll wave with a straight filament pinned to a cylindrical obstacle with fixed diameter and length shows a similar trend in wave characteristics [15].

The size of the pinning obstacle influences the scroll wave dynamics. According to the experimental studies in the BZ reaction, the T, λ and v of a spiral wave increases with the obstacle size [13, 14]. We used glass beads of different diameters to pin the scroll rings.

As illustrated in Fig. 4 a and b, the frequency of the pinned scroll ring increases with excitability for a certain bead size but decreases as the size of the obstacle increases. For a given excitability, the λ and v of a pinned scroll wave increase in proportion to the size of the obstacle. On the other hand, for a scroll ring attached to a specific obstacle size, λ decreases as excitability rises, while v increases (Fig. 5a and b). The change in frequency associated with the change in excitability for a given bead size is more significant than the same observed with the change in bead size for a given excitability.

4 Conclusions

We investigated the variations in the wave properties of the three-dimensional scroll rings by varying the concentration of sulphuric acid. We looked at the dynamics of free scroll rings and scroll rings pinned to various spherical glass beads. A scroll ring is not stable in a low excitable medium as it breaks up within a short period of time after wave initiation. When excitability increases, frequency and wave velocity increases, while wavelength decreases. Previous reports indicate that the wave properties of the spirals in the two-dimensional BZ system are modified in the same manner with changes in the concentrations of sulphuric acid and sodium bromate [12]. We also studied the scroll ring properties by varying the size of the spherical glass beads that serve as the pinning obstacle. With an increase in the size of the glass beads, the scroll ring takes a longer time to complete one rotation. This delay leads to a decrease in the frequency of the scroll ring. As a result, λ and v increase. Our experiments reveal that the behavior of scroll rings with variations in excitability and size of pinning obstacle is similar to that of spiral waves as previously observed. However, compared to spiral waves, scroll rings are less stable in a low excitable medium. A clear understanding and quantification of variations occurring in the wave dynamics with medium inhomogeneities will have a wide range of applications in different excitable media. Because of the similarity in the mathematical equations, the excitation waves in a wide variety of excitable media behave in similar ways. We believe our findings about scroll waves in chemical excitation waves are also applicable to other excitation waves, including the excitation waves in the cardiac tissue.

Acknowledgements This research was partially supported by SERB (DST) early career research grant.

References

1. Tan, T.H., Liu, J., Miller, P.W., Tekant, M., Dunkel, J., Fakhri, N.: Topological turbulence in the membrane of a living cell. Nat. Phys. **16**(6), 657–662 (2020)
2. Yu, Y., Santos, L.M., Mattiace, L.A., Costa, M.L., Ferreira, L.C., Benabou, K., Rozental, R.: Reentrant spiral waves of spreading depression cause macular degeneration in hypoglycemic chicken retina. Proc. Natl. Acad. Sci. **109**(7), 2585–2589 (2012). https://doi.org/10.1073/pnas.1121111109
3. Rostami, Z., Jafari, S.: Defects formation and spiral waves in a network of neurons in presence of electromagnetic induction. Cogn. Neurodynamics **12**(2), 235–254 (2018)
4. Kappadan, V., Telele, S., Uzelac, I., Fenton, F., Parlitz, U., Luther, S., Christoph, J.: High-resolution optical measurement of cardiac restitution, contraction, and fibrillation dynamics in beating vs. blebbistatin-uncoupled isolated rabbit hearts. Front. Physiol. **11**, 464 (2020)
5. Bhattacharya, S., Iglesias, P.A.: Controlling excitable wave behaviors through the tuning of three parameters. Biol. Cybern. **113**(1), 61–70 (2019)
6. Kundu, S., Muruganandam, P., Ghosh, D., Lakshmanan, M.: Amplitude-mediated spiral chimera pattern in a nonlinear reaction-diffusion system. Phys. Rev. E **103**(6), 062209 (2021)
7. Punacha, S., Shajahan, T.K.: Theory of unpinning of spiral waves using circularly polarized electric fields in mathematical models of excitable media. Phys. Rev. E **102**(3), 032411 (2020)
8. Kalitzin, S., Petkov, G., Suffczynski, P., Grigorovsky, V., Bardakjian, B.L., da Silva, F.L., Carlen, P.L.: Epilepsy as a manifestation of a multistate network of oscillatory systems. Neurobiol. Dis. **130**, 104488 (2019)
9. Jiménez, Z.A.: Dynamical behavior of scroll rings in the presence of heterogeneities in the Belousov-Zhabotinsky excitable medium (Doctoral dissertation, The Florida State University) (2012)
10. Luengviriya, C., Storb, U., Lindner, G., Müller, S.C., Bär, M., Hauser, M.J.: Scroll wave instabilities in an excitable chemical medium. Phys. Rev. Lett. **100**(14), 148302 (2008). https://link.aps.org/doi/10.1103/PhysRevLett.100.148302
11. Das, N.P., Mahanta, D., Dutta, S.: Unpinning of scroll waves under the influence of a thermal gradient. Phys. Rev. E **90**(2), 022916 (2014). https://link.aps.org/doi/10.1103/PhysRevE.90.022916
12. Mahanta, D., Das, N.P., Dutta, S.: Spirals in a reaction-diffusion system: dependence of wave dynamics on excitability. Phys. Rev. E **97**(2), 022206 (2018). https://link.aps.org/doi/10.1103/PhysRevE.97.022206
13. Sutthiopad, M., Luengviriya, J., Porjai, P., Phantu, M., Kanchanawarin, J., Müller, S.C., Luengviriya, C.: Propagation of spiral waves pinned to circular and rectangular obstacles. Phys. Rev. E **91**(5), 052912 (2015). https://link.aps.org/doi/10.1103/PhysRevE.91.052912
14. Phantu, M., Sutthiopad, M., Luengviriya, J., Müller, S.C., Luengviriya, C.: Robustness of free and pinned spiral waves against breakup by electrical forcing in excitable chemical media. Phys. Rev. E **95**(4), 042214 (2017). https://link.aps.org/doi/10.1103/PhysRevE.95.042214
15. Wattanasiripong, N., Kumchaiseemak, N., Porjai, P.: Behavior of pinned scroll waves with different excitability in a simulated excitable media. Prog. Appl. Sci. Technol. **8**(2), 80–85 (2018). https://ph02.tci-thaijo.org/index.php/past/article/view/243029
16. Jahnke, W., Skaggs, W.E., Winfree, A.T.: Chemical vortex dynamics in the Belousov-Zhabotinskii reaction and in the two-variable Oregonator model. J. Phys. Chem. **93**(2), 740–749 (1989). https://pubs.acs.org/doi/10.1021/j100339a047
17. Luengviriya, J., Sutthiopad, M., Phantu, M., Porjai, P., Kanchanawarin, J., Müller, S.C., Luengviriya, C.: Influence of excitability on unpinning and termination of spiral waves. Phys. Rev. E **90**(5), 052919 (2014). https://link.aps.org/doi/10.1103/PhysRevE.90.052919

Periodic Amplifications of Attosecond Three Soliton in an Inhomogeneous Nonlinear Optical Fiber

M. S. Mani Rajan, Saravana Veni, and K. Subramanian

Abstract We show the periodic amplification of attosecond three solitons in an inhomogeneous optical fiber which is can be governed by nonlinear Schrödinger equation with higher order linear and nonlinear effects. Through AKNS method, we construct the Lax pair for higher order inhomogeneous nonlinear Schrödinger equation. Based on Lax pair, three soliton solutions are attained by means of Darboux transformation (DT) method. By properly tailoring the dispersion and nonlinear profiles, periodic amplifications of three solitons are demonstrated through some graphical illustrations. Especially, three soliton interactions are portrayed via 2D and 3D plots. Results attained through this work will be potentially useful in the field of soliton amplifications by soliton control and management. Also, we clearly observed that the impact of variable coefficients on attosecond soliton dynamics.

Keywords Optical solitons · Attosecond soliton · Lax pair · Darboux transformation · Soliton management · Amplification

1 Introduction

The controlling of optical solitons and their management in an inhomogeneous optical fiber system have been received huge attention in both theoretical and experimental research due to their enormous potential applications in long-haul communication and ultrafast signal routing systems [1–3]. The study on optical soliton shaping

M. S. M. Rajan (✉)
Anna University, University College of Engineering, Ramanathapuram 623513, India
e-mail: senthilmanirajanofc@gmail.com

S. Veni
Department of Physics, Amirta College of Engineering and Technology, Erachakulam Campus, Nagercoil 629901, India

K. Subramanian
Department of Physics, SRM Institute of Science and Technology, Ramapuram Campus, Chennai, India
e-mail: msmanirajan@aucermd.edu.in

S. Banerjee and A. Saha (eds.), *Nonlinear Dynamics and Applications*,
Springer Proceedings in Complexity,
https://doi.org/10.1007/978-3-030-99792-2_112

and management is very important because of rapid growth of modern communication technology. In the picosecond regime, optical soliton pulse propagation in an optical fiber which governs the nonlinear Schrödinger (NLS) equation with constant coefficients [4, 5]. NLS equations are most significant models in modern nonlinear Science. However, real fiber cannot be a homogeneous in nature and because of various factors such as variation in diameter of core and lattice points, fiber medium becomes as inhomogeneous [6]. In such case, soliton propagation can be described by generalized variable-coefficient NLS equation [7]. Furthermore, there have been paid great attention on the investigation of generalized variable-coefficient NLS equations which contains dispersion, nonlinearity and some inhomogeneous terms with varying nature along the propagation axis [8]. The rapid growth of computational methods leads to the investigation for the bountiful models of nonlinear Schrödinger (NLS) with inhomogeneous terms by several researchers in various aspects like nonlinear fiber optics and nonlinear science [9, 10]. In a nonlinear fiber medium, soliton control technique can be theoretically represented by nonlinear Schrödinger model with dispersion, nonlinearity and gain or loss terms. Recently, this method of soliton control is new and important developments in the application of optical solitons for optical transmission systems which have been discussed in detail by Serkin et al. [11]. On the other hand, attosecond soliton pulses has a better performance on the transmission characteristics where higher-order linear and nonlinear effects should be taken into consideration [12].

2 Inhomogeneous NLS Equation with Higher Order Linear and Nonlinear Effects

To our knowledge, in real fiber systems, inhomogeneous nonlinear Schrödinger models with higher order linear and nonlinear effects are considered to describe the attosecond optical pulse transmission in an inhomogeneous fiber system. For example, in modern optical fiber communication systems, inhomogeneous profiles are varying with respect to propagation distance. Hence, we address the following generalized higher order nonlinear Schrödinger equation with variable coefficients as given below

$$i\frac{\partial E(z,t)}{\partial z} + \frac{1}{2}D_1 + \chi(z)D_2 + \beta(z)D_3 - i\gamma(z)D_4 + \delta(z)D_5 + iG(z) = 0 \quad (1)$$

where

$$D_1 = D(z)E_{tt} + 2R(z)|E|^2 E$$

$$D_2 = E_{ttt} + 6|E|^2 E_t$$

$$D_3 = E_{tttt} + 8|E|^2E_{tt} + |E|^4E + 4|E_t|^2E + 6{E_t}^2E^* + 2E^2E^*_{tt}$$

$$D_4 = E_{ttttt} + 10|E|^2E_{ttt} + 10(|E|^2E)_t + 20E^*E_tE_{tt} + 6{E_t}^2E^* + 2E^2E^*_{tt}$$

$$\begin{aligned}D_5 = E_{tttttt} &+ \left[60E^*|E_t|^2 + 50\left(E^*\right)^2E_{tt} + 2E^*_{tttt}\right]E^2 \\ &+ E\left[12E^*E_{tttt} + 8E_tE^*_{ttt} + 22|E_{tt}|^2\right] + E\left[18E_{ttt}E^*_t + 70\left(E^*\right)^2(E_t)^2\right] \\ &+ 20(E_t)^2E^*_{tt} + 10E^3\left[\left(E^*_t\right)^2 + 2E^*E^*_{tt}\right] + 20|E|^6E\end{aligned}$$

$$G(z) = \frac{1}{2}\frac{W[R(z), D(z)]}{R(z), D(z)}$$

$$W[R(z), D(z)] = R(z)\frac{dD(z)}{dz} - D(z)\frac{dR(z)}{dz}$$

where E (z, t) denotes the complex envelope of incident light filed, the subscripts z and t denote respectively the partial derivatives with respect to the normalized propagation distance and retarded time. $D(z)$ represents the GVD coefficient, $R(z)$ is arise due to Kerr nonlinearity especially self-phase modulation which is particularly cubic nonlinearity coefficient and $G(z)$ represents the loss (attenuation) or gain (amplification) profile. D_2, D_3, D_4 and D_5 are inhomogeneous coefficients of higher order dispersion and nonlinear terms. It should be emphasized that Eq. (1) not only describing the attosecond soliton propagation but also soliton control and management in an inhomogeneous fiber system. In different domain, various researchers solved many kinds of NLS equations using some mathematical techniques. In the present work, we aimed to solve inhomogeneous HNLS equations in attosecond regime.

3 Lax Pair

With the aid of AKNS formalism [13], matrices M and N are constructed for the Eq. (1) which can be derived from the zero-curvature equation. In order to apply the Darboux transformation, the following 2×2 eigenvalue problem is considered. In the obtaining of soliton solutions via Darboux transformation, Lax pair plays a vigorous role.

$$\psi_t = M\psi \tag{2}$$

$$\psi_z = N\psi$$

M and N matrices are can be expressed as given below for the linear eigen value problem (2)

$$M = i\begin{pmatrix} \lambda & E^*(z,t) \\ E(z,t) & -\lambda \end{pmatrix} \tag{3}$$

$$N = \sum_{j=0}^{6} \lambda^j V_j$$

$$M_j = i\begin{pmatrix} A_j & B_j^* \\ B_j & -A_j \end{pmatrix} \tag{4}$$

$$A_6 = 32\,\delta(z)$$

$$B_6 = 0$$

$$A_5 = 16\,\gamma(z)$$

$$B_5 = 32\delta(z)E$$

$$A_4 = -8\gamma(z) - 16\delta(z)|E|^2$$

$$B_4 = 16\gamma(z)E + 16i\delta(z)E_t$$

$$A_3 = -4\chi(z) - 8\gamma(z)|E|^2 - 8i\delta(z)|E|_t^2$$

$$B_3 = -8\beta(z)E + 8i\gamma(z)E_t - 8E_{tt} - 16\delta(z)|E|^2E$$

$$A_2 = 1 + 4\beta(z)|E|^2 + 4i\gamma(z)(E_t^*E - E_tE^*) + 12\delta(z)|E|^4 - 8\delta(z)|E_t|^2 + 4\delta(z)(E_t^*E - EE^*)_t$$

$$B_2 = -4\chi(z)E - 8\gamma(z)|E|^2E - 24i\delta(z)|E|^2E_t - 4i\beta(z)E_t - 4i\gamma(z)E_{tt} - 4i\delta(z)E_{ttt}$$

$$A_1 = 2\chi(z)|E|^2 + 6\gamma(z)|E|^4 - 2i\beta(z)(E_t^*E - E_tE^*) + 12i\delta(z)|E|^2(E_tE^* - E_t^*E) - 2\gamma(z)|E_t|^2 + 2\gamma(z)(E_{tt}^*E + E_{tt}E^*) + 2i\delta(z)(E_tE_{tt}^* - E_t^*E_{tt} + E^*E_{ttt} - E_{ttt}^*E)$$

$$B_1 = E + 4\beta(z)|E|^2E - 2i\chi(z)E_t - 12i\gamma(z)|E|^2E_t + 12\delta(z)E^*E_t^2 + 16\delta(z)|E|^2E_{tt}$$

$$+\,4\delta(z)E^2E_{tt}^{*}-2i\gamma(z)E_{ttt}+2\beta(z)E_{tt}+2\delta(z)E_{tttt}+12\delta(z)|E|^4E+8\delta(z)|EE_t|^2$$

The Eq. (1) can be obtained directly from zero-curvature equation i.e., $M_z - N_t + [M, N] = 0$. In Eq. (3), λ being the spectral parameter in Lax pair.

4 Three Soliton Solutions

Using Darboux matrix $D = \lambda\,\mathrm{I} - \mathrm{S}$ in the process of Darboux transformation method [14], explicit multi-soliton solutions can be obtained. To demonstrate the propagation of three solitons in a real fiber system, three soliton solutions with arbitrary control parameters are attained in this section. Consequently, three soliton solutions are computed via Darboux transformation technique as described below

$$E^{(3)} = 2i\,\frac{N_3}{D_3} \tag{5}$$

$$\begin{aligned}
N_3 = {} & e^{2i(\theta_1+\xi_1 t)+\sigma_1} + e^{2i(\theta_2+\xi_2 t)+\sigma_2} + e^{2i(\theta_3+\xi_3 t)+\sigma_3} \\
& + e^{2i\left(\theta_1+\theta_1^*+\theta_2+\xi_1 t+\xi_1^* t+\xi_2 t\right)+\lambda_{123}} + e^{2i\left(\theta_1+\theta_1^*+\theta_3+\xi_1 t+\xi_1^* t+\xi_3 t\right)+\lambda_{132}} \\
& + e^{2i\left(\theta_1+\theta_2+\theta_2^*+\xi_1 t+\xi_2 t+\xi_1^* t\right)+\lambda_{213}} + e^{2i\left(\theta_1+\theta_3+\theta_3^*+\xi_1 t+\xi_3 t+\xi_3^* t\right)+\lambda_{312}} \\
& + e^{2i\left(\theta_2+\theta_2^*+\theta_3+\xi_2 t+\xi_2^* t+\xi_3 t\right)+\lambda_{231}} + e^{2i\left(\theta_2+\theta_3+\theta_3^*+\xi_2+\xi_3 t+\xi_3^* t\right)+\lambda_{321}} \\
& + e^{2i\left(\theta_1+\theta_2+\theta_3^*+\xi_1 t+\xi_2 t+\xi_3^* t\right)+\upsilon_{123}} + e^{2i\left(\theta_1+\theta_3+\theta_2^*+\xi_1 t+\xi_3 t+\xi_2^* t\right)+\upsilon_{312}} \\
& + e^{2i\left(\theta_2+\theta_3+\theta_1^*+\xi_2 t+\xi_3 t+\xi_1^* t\right)+\upsilon_{231}} \\
& + e^{2i\left(\theta_1+\theta_2+\theta_2^*+\theta_3+\theta_3^*+\xi_1 t+\xi_2 t+\xi_2^* t+\xi_3 t+\xi_3^* t\right)+\varphi_{123}} \\
& + e^{2i\left(\theta_1+\theta_1^*+\theta_2+\theta_3+\theta_3^*+\xi_1 t+\xi_1^* t+\xi_2 t+\xi_3 t+\xi_3^* t\right)+\varphi_{231}} \\
& + e^{2i\left(\theta_1+\theta_1^*+\theta_2+\theta_2^*+\theta_3+\xi_1 t+\xi_1^* t+\xi_2 t+\xi_2^* t+\xi_3 t\right)+\varphi_{321}}
\end{aligned}$$

$$\begin{aligned}
D_2 = {} & 1 + e^{2i\left(\theta_1+\theta_1^*+\xi_1 t+\xi_1^* t\right)+\varkappa_{123}} + e^{2i\left(\theta_2+\theta_2^*+\xi_2 t+\xi_2^* t\right)+\varkappa_{231}} + e^{2i\left(\theta_3+\theta_3^*+\xi_3 t+\xi_3^* t\right)+\varkappa_{321}} \\
& + e^{2i\left(\theta_1^*+\theta_2+\xi_1^* t+\xi_2 t\right)+\kappa_{123}} + e^{2i(\theta_1+\theta_2^*+\xi_1 t+\xi_2^* t)+\kappa_{213}} + e^{2i\left(\theta_1^*+\theta_3+\xi_1^* t+\xi_3 t\right)+\kappa_{132}} \\
& + e^{2i(\theta_1+\theta_3^*+\xi_1 t+\xi_3^* t)+\kappa_{312}} + e^{2i(\theta_3+\theta_2^*+\xi_3 t+\xi_2^* t)+\kappa_{231}} + e^{2i(\theta_2+\theta_3^*+\xi_2 t+\xi_3^* t)+\kappa_{321}} \\
& + e^{2i\left(\theta_1+\theta_1^*+\theta_2+\theta_2^*+\xi_1 t+\xi_1^* t+\xi_2 t+\xi_2^* t\right)+\omega_{123}} + e^{2i\left(\theta_1+\theta_1^*+\theta_3+\theta_3^*+\xi_1 t+\xi_1^* t+\xi_3 t+\xi_3^* t\right)+\omega_{312}} \\
& + e^{2i\left(\theta_2+\theta_2^*+\theta_3+\theta_3^*+\xi_1 t+\xi_1^* t+\xi_3 t+\xi_3^* t\right)+\omega_{321}} + e^{2i\left(\theta_1+\theta_1^*+\theta_2+\theta_3^*+\xi_1 t+\xi_1^* t+\xi_2 t+\xi_3 t+\xi_3^* t\right)+\varrho_{321}} \\
& + e^{2i\left(\theta_1+\theta_1^*+\theta_3+\theta_2^*+\xi_1 t+\xi_1^* t+\xi_2 t+\xi_3 t+\xi_2^* t\right)+\varrho_{231}} + e^{2i\left(\theta_1+\theta_2+\theta_2^*+\theta_3^*+\xi_1 t+\xi_2 t+\xi_2^* t+\xi_3^* t\right)+\varrho_{312}} \\
& + e^{2i\left(\theta_1+\theta_3+\theta_2^*+\theta_3^*+\xi_1 t+\xi_3 t+\xi_2^* t+\xi_3^* t\right)+\varrho_{213}} + e^{2i\left(\theta_2+\theta_3+\theta_1^*+\theta_2^*+\xi_2 t+\xi_3 t+\xi_1^* t+\xi_2^* t\right)+\varrho_{132}} \\
& + e^{2i\left(\theta_2+\theta_3+\theta_1^*+\theta_3^*+\xi_2 t+\xi_3 t+\xi_1^* t+\xi_3^* t\right)+\varrho_{123}}
\end{aligned}$$

$$+ e^{2i(\theta_1+\theta_1^*+\theta_2+\theta_2^*+\theta_3+\theta_3^*+\xi_1 t+\xi_1^* t+\xi_2 t+\xi_2^* t+\xi_3 t+\xi_3^* t)+\Gamma}$$

$$\theta_1 = \xi_1^2 \int (D(z) - 4\xi_1 R(z) - 8\xi_1^2 \beta(z) + 16\xi_1^3 \gamma(z) + 32\xi_1^4 \delta(z))dz$$

$$\theta_2 = \xi_2^2 \int (D(z) - 4\xi_2 R(z) - 8\xi_2^2 \beta(z) + 16\xi_2^3 \gamma(z) + 32\xi_2^4 \delta(z))dz$$

$$\theta_3 = \xi_3^2 \int (D(z) - 4\xi_3 R(z) - 8\xi_3^2 \beta(z) + 16\xi_3^3 \gamma(z) + 32\xi_3^4 \delta(z))dz$$

$$\xi_1 = \mu_1 + i\eta_1$$

$$\xi_2 = \mu_2 + i\eta_2$$

$$\xi_3 = \mu_3 + i\eta_3$$

$$e^{\sigma_i} = -\frac{C_{1i}\prod_j(\xi_j^* - \xi_i)}{C_{2i}\prod_{i\neq j}(\xi_j - \xi_i)}$$

$$e^{\lambda_{ijk}} = -\frac{C_{1i}^2 C_{1j}(\xi_j^* - \xi_j)(\xi_j^* - \xi_i)(\xi_k^* - \xi_i)(\xi_k^* - \xi_j)(\xi_i^* - \xi_k)}{C_{2i}^2 C_{2j}(\xi_i^* - \xi_j^*)(\xi_i^* - \xi_k^*)(\xi_i - \xi_k)(\xi_k - \xi_j)}$$

$$e^{\upsilon_{ijk}} = \frac{C_{1i}C_{1j}C_{1k}(\xi_i^* - \xi_i)(\xi_j^* - \xi_i)(\xi_k^* - \xi_k)(\xi_j^* - \xi_i)(\xi_i^* - \xi_j)}{C_{2i}C_{2j}C_{2k}(\xi_i^* - \xi_k^*)(\xi_k^* - \xi_j^*)(\xi_i - \xi_k)(\xi_k - \xi_j)}$$

$$e^{\varphi_{ijk}} = -\frac{C_{1i}C_{1j}^2 C_{1k}^2(\xi_i^* - \xi_i)(\xi_i^* - \xi_j)(\xi_j^* - \xi_k^*)(\xi_i^* - \xi_k)}{C_{2i}C_{2j}^2 C_{2k}^2(\xi_i^* - \xi_j^*)(\xi_i^* - \xi_k^*)(\xi_k^* - \xi_j^*)}$$

$$e^{\varkappa_{ijk}} = -\frac{C_{1i}^2(\xi_j^* - \xi_i)(\xi_i^* - \xi_j)(\xi_k^* - \xi_i)(\xi_i^* - \xi_k)}{C_{2i}^2(\xi_i^* - \xi_j^*)(\xi_i^* - \xi_k^*)(\xi_j - \xi_i)(\xi_i - \xi_k)}$$

$$e^{\kappa_{ijk}} = -\frac{C_{1i}C_{1j}(\xi_i^* - \xi_i)(\xi_j^* - \xi_j)(\xi_k^* - \xi_j)(\xi_i^* - \xi_k)}{C_{2i}C_{2j}(\xi_i^* - \xi_j^*)(\xi_i^* - \xi_k^*)(\xi_j - \xi_i)(\xi_k - \xi_j)}$$

$$e^{\omega_{ijk}} = \frac{C_{1i}^2 C_{1j}^2(\xi_k^* - \xi_i)(\xi_k^* - \xi_j)(\xi_i^* - \xi_k)(\xi_j^* - \xi_k)}{C_{2i}^2 C_{2j}^2(\xi_i^* - \xi_k^*)(\xi_k^* - \xi_j^*)(\xi_k - \xi_j)(\xi_i - \xi_k)}$$

$$e^{\varrho_{ijk}} = \frac{C_{1i}C_{1j}C_{1k}^2(\xi_i^* - \xi_i)(\xi_j^* - \xi_j)(\xi_k^* - \xi_i)(\xi_j^* - \xi_k)}{C_{2i}C_{2j}C_{2k}^2(\xi_i^* - \xi_j^*)(\xi_k^* - \xi_j^*)(\xi_i - \xi_k)(\xi_i - \xi_j)}$$

$$e^{\Gamma} = \frac{C_{11}^2 C_{12}^2 C_{13}^2}{C_{21}^2 C_{22}^2 C_{23}^2}, \text{j} = 1, 2, 3.$$

By adopting particular profile for inhomogeneous functions that are present in the obtained attosecond three soliton solutions, various transmission characteristics can be investigated. Especially, with the expression of three soliton solutions (5), periodic amplification properties of attosecond three solitons to be observed by adopting periodic functions for inhomogeneous profiles in the next section.

5 Periodic Amplifications of Three Attosecond Solitons

The present study focused on three attosecond optical bright soliton transmission with the periodic amplification characteristics under the dispersion and nonlinearity management scheme (DNMS). Furthermore, by studying the three soliton interactions in an inhomogeneous fiber system, we can enrich the capacity of optical fiber communication system. In order to exemplify the behavior of three solitons which are transmitted under periodic amplification system, we chose the variable coefficients especially GVD parameter $D(z)$ and Kerr nonlinearity parameter $R(z)$ are adopted as given in the references [15–18]

$$D(z) = \frac{1}{d_0}\exp(kz)\,R(z) \tag{6}$$

$$R(z) = R_0 + R_1 \sin(gz)$$

Here, R_0, R_1 and g are the control parameters describing Kerr nonlinearity and d_0 represent the initial peak power of the system, respectively. Here, for the sake of simplicity, control parameters are considered as $R_0 = 0$, $d_0 = 1$ and $g = 1$. Specifically, when $k = 0$ in the expression (6) which corresponds to the loss less optical fiber medium (without any loss or gain).

As depicted in Fig. 1a, the evolution of three solitons is displayed which is obtained through the adopting of dispersion and nonlinearity coefficients as periodic functions. At the beginning of propagation along the z-axis, the period of oscillation is small while it is increasing significantly during the transmission of optical solitons in attosecond regime. Three solitons are got highly amplifications when propagation distance is very long. From the density plot Fig. 1b, one can conclude that three attosecond solitons are getting not only amplification but also the compressed and broadened width periodically. The snake like propagation trajectory of optical solitons is observed and this type of solitons are namely as "snake solitons". In real optical communication network, amplified solitons may be attained through the technique of appropriately tailoring the dispersion and nonlinearity profiles which is called dispersion and nonlinearity management (DNLM) scheme. Recently, in literature, various periodic soliton amplification schemes through several methods have been suggested for numerous applications [19–23].

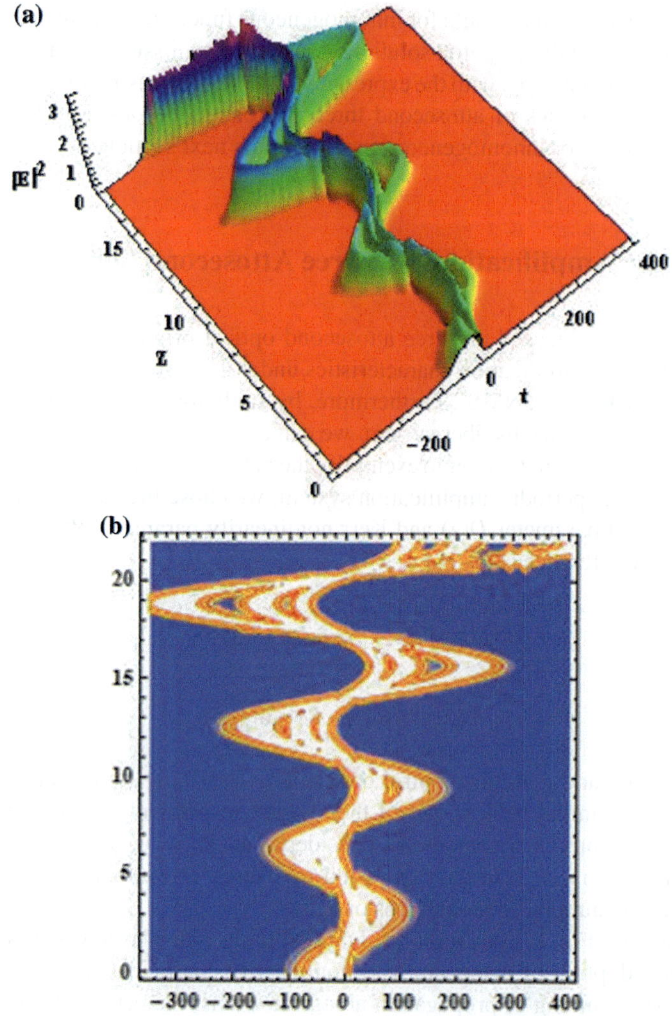

Fig. 1 **a** Intensity profile of attosecond three solitons. The parameters are $\mu_1 = 0.07$, $\mu_2 = 0.01$, $\mu_3 = 0.08$, $\eta_1 = 0.08$, $\eta_2 = 0.09$, $\eta_3 = 0.07$. **b** Corresponding contour plot for **a**

6 Conclusions

We proposed a theoretical model i.e., variable coefficient inhomogeneous NLS equation with the presence of higher order linear and nonlinear effects which governs the propagation of attosecond soliton in an inhomogeneous nonlinear optical fiber medium. Bright three soliton solutions are derived by employing the Darboux transformation method based on the constructed Lax pair. Finally, three attosecond soliton transmissions in an inhomogeneous fiber have been demonstrated graphically with the aid of *Mathematica* software tool. By careful choices of trigonometric functions for GVD and nonlinearity parameters, periodic soliton amplification have been attained. Moreover, we observed that one can control the soliton transmission and its characteristics through adopting the dispersion—nonlinearity management scheme.

The study on simultaneous propagation of three attosecond solitons in an inhomogeneous fiber optic communication system will be used for the development of optical soliton-based switching devices and soliton shaping or management by tailoring the inhomogeneous profiles properly. The elastic interaction among three solitons offers a new way for controllable optical soliton amplification in multi soliton transmission system where ultrahigh capacity can be easily achieved. Especially, this work is useful in the study of propagation properties of attosecond soliton in an inhomogeneous fiber with the modulation of dispersion and nonlinearity. We realized that three soliton interactions effectively controlled by properly tailoring the inhomogeneous profiles in attosecond regime.

References

1. Hasegawa, A., Kodama, Y.: Solitons in Optical Communications. Clarendon, Oxford, UK (1995)
2. Toda, H., Furukawa, Y., Kinoshita, T., Kodama, Y., Hasegawa, A.: Optical soliton transmission experiment in a comb-like dispersion profiled fiber loop. IEEE Photon. Technol. Lett. **9**, 1415–1417 (1997)
3. Liu, W.J., Zhang, Y., Wazwaz, A.M., Zhou, Q.: Analytic study on triple-S, triple-triangle structure interactions for solitons in inhomogeneous multi-mode fiber. Appl. Math. Comput. **361**, 325–331 (2019)
4. Agrawal, G.P.: Nonlinear Fiber Optics. Academic Press, San Diego (1995)
5. Tian, B., Gao, Y.T., Zhu, H.W.: Variable-coefficient higher order nonlinear Schrödinger model in optical fibers: variable-coefficient bilinear form, Bäcklund transformation, brightons and symbolic computation. Phys. Lett. A **366**, 223–229 (2007)
6. Serkin, V.N., Hasegawa, A.: Exactly integrable nonlinear Schrodinger equation models with varying dispersion, nonlinearity and gain: application for soliton dispersion management. IEEE J. Sel. Top. Quant. Electron. **8**, 418–431 (2002)
7. Serkin, V.N., Hasegawa, A., Belyaeva, T.L.: Nonautonomous solitons in external potentials. Phys. Rev. Lett. **98**, 074102 (2007)
8. Kruglov, V.I., Peacock, A.C., Harvey, J.D.: Exact self-similar solutions of the generalized nonlinear Schrödinger equation with distributed coefficients. Phys. Rev. Lett. **90**, 113902–113905 (2003)
9. Vijayalekshmi, S., Mani Rajan, M.S., Mahalingam, A., Uthayakumar, A.: Hidden possibilities in soliton switching through tunneling in erbium doped birefringence fiber with higher order effects. J. Mod. Opt. **62**, 278–287 (2015)
10. Serkin, V.N., Hasegawa, A., Belyaeva, T.L.: Solitary waves in nonautonomous nonlinear and dispersive systems. J. Mod. Opt. **57**, 1456 (2010)
11. Serkin, V.N., Belyaeva, T.L.: Optimal control of optical soliton parameters: part 1. The Lax representation in the problem of soliton management. Quantum Electron. **31**, 1007 (2001)
12. Prathap, N., Arunprakash, S., Mani Rajan, M.S., Subramanian, K.: Multiple dromion excitations in sixth order NLS equation with variable coefficients. Optik **158**, 1179–1185 (2018)
13. Ablowitz, M.J., Kaup, D.J., Newell, A.C.: Nonlinear-evolution equations of physical significance. Phys. Rev. Lett. **31**, 125 (1973)

14. Gu, H., Hu, H.S., Zhou, Z.X.: Darboux transformation in Soliton Theory and Its Geometric Applications. Shanghai Science and Technology Publishers, Shanghai (2005)
15. Mahalingam, A., Porsezian, K., Mani Rajan, M.S., Uthayakumar, A.: Propagation of dispersion nonlinearity managed solitons in an inhomogeneous erbium doped fiber system. J. Phys. A Math. Theor. **42**, 165101 (2009)
16. Mani Rajan, M.S.: Dynamics of optical soliton in a tapered erbium-doped fiber under periodic distributed amplification system. Nonlinear Dyn. **85**, 599–606 (2016)
17. Karthikeyaraj, G., Mani Rajan, M.S., Tantawy, M., Subramanian, K.: Periodic oscillations and nonlinear tunneling of soliton for Hirota-MB equation in inhomogeneous fiber. Optik **181**, 440–448 (2019)
18. Yao,Y., Ma, G., Zhang, X., W.J. Liu, W.J: M-typed dark soliton generation in optical fibers. Optik **193**, 162997(2019)
19. Yu, W.T., Zhou, Q., Mirzazadeh, M., Liu, W.J.: Phase shift, amplification, oscillation and attenuation of solitons in nonlinear optics. J. Adv. Res. **15**, 69–76 (2019)
20. Mani Rajan, M.S., Mahalingam, A., Uthayakumar, A.: Observation of two soliton propagation in an erbium doped inhomogeneous lossy fiber with phase modulation. Commun. Nonlinear Sci. Numer. Simulat. **18**, 1410–1432 (2013)
21. Maddouri, K., Azzouzi, F., Triki, H., Bouguerra, A., Korba, S.A.: Dark-managed solitons in inhomogeneous cubic–quintic–septimal nonlinear media. Nonlinear Dyn. **103**, 2793–2803 (2021)
22. Liu, X., Triki, H., Zhou, Q., Liu, W.J., Anjan B.: Analytic study on interactions between periodic solitons with controllable parameters. Nonlinear Dyn. 94, 703–709 (2018)
23. Wang, L., Luan, Z., Zhou, Q., Biswas, A., Alzahrani, A.K., Liu, W.J.: Effects of dispersion terms on optical soliton propagation in a lossy fiber system. Nonlinear Dyn. **104**, 629–637 (2021)

Analysis of Flexoelectricity with Deformed Junction in Two Distinct Piezoelectric Materials Using Wave Transmission Study

Abhinav Singhal, Rakhi Tiwari, Juhi Baroi, and Chandraketu Singh

Abstract Analysis of flexoelectricity in distinct piezoelectric (PE) materials bars (PZT-7A, PZT-6B) with deformed interface in stick over Silicon oxide layer is studied analytically with the help of Love-type wave vibrations. Using the numerical data for PE material, then research achieves the noteworthy fallouts of flexoelectric effect (FE) and PE. The effect of flexoelectricity is compared first between biomaterials of piezoelectric ceramics. Dispersion expressions are procured logically for together electrically unlocked/locked conditions under the influence of deformed interface in the complex form which is transcendental. Fallouts of the research identify that contexture consisting of FE has a noteworthy impact on the acquired dispersion expressions. Existence of FE displays that the unreal section of the phase velocity rises monotonically. Competitive consequences are displayed diagrammatically and ratified with published outcomes. The outcomes of the present research done on both the real and imaginary section of the wave velocity. The comparative study between the two piezo-ceramics bars helps us to understand the properties of one piezo-material over the another and as an outcomes the significance of the present study helps in structural health monitoring, bioengineering for optimizing the detection sensitivity in the smart sensors.

Keywords Flexoelectricity · PZT-7A · Vibrations · Deformed Interface · PZT-6B · Deformed Interface

Nomenclature

σ Stress tensor

A. Singhal (✉) · J. Baroi · C. Singh
School of Sciences, Christ (Deemed to Be University) Delhi NCR, Ghaziabad 201003, India
e-mail: ism.abhinav@gmail.com

R. Tiwari
Department of Mathematics, Babasaheb Bhimrao Ambedkar Bihar University, Nitishwar college, Muzaffarpur 842002, India

S. Banerjee and A. Saha (eds.), *Nonlinear Dynamics and Applications*,
Springer Proceedings in Complexity,
https://doi.org/10.1007/978-3-030-99792-2_113

τ	Higher order stress (moment stress) tensor
ε	Strain
E^o	Electric field intensity
V^o	Electric field gradient
w	Strain gradient
D	Electrical displacement vector
Q^o	Electric quadrupole tensor
c	Elastic tensors
a	Permittivity tensors
e	Piezoelectric tensors
f	Direct piezoelectric tensors
d	Converse piezoelectric tensors
u	Particle displacements
ϕ	Electric potential
u_i	Mechanical displacement
$k = 2\pi/\lambda$	Wave number
λ	Wavelength
$i = \sqrt{-1}$	Imaginary unit
c_{44}^e	Shear modulus of the lower plate
ρ^e	Density of the lower plate
σ^o	Initial stress
u^e	Mechanical displacement of the lower plate
ϕ^e	Electric potential of the lower plate

1 Introduction

The study of surface acoustic wave (SAW) propagation in piezoelectric materials based smart devices has attracted significant attention due to its distinctive applications. The SAW devices can be used as biosensors or for liquid sensing, for which the sensing layer needs to be paired with a liquid medium. To generate the SAW, an oscillating electric signal is applied to the interdigital transducer (IDT) which is designed on the surface of piezoelectric layer. Subsequently, piezoelectric crystal converts the electric signal into mechanical vibration, which another IDT receives for further processing. To fabricate metal electrodes in IDT's, complex procedure involving lithographic patterning and metal deposition along with other special equipment. Chu et al. [1] studied the vibrations in smart materials following the structures of MEMS. Singhal et al. [2] and Nathankumar kumar et al. [3] elobrated the research of seismic wave vibrations in smart materials structure where the flexoelectricty and piezoelectricity charachterstics and Sahu et al. [4] extend the study of dispresion relation of wave transmission in intelligent structures.

Lately for designing composite structure, functional grading of the piezoelectric material is preferred over conventional material of same thickness due to the brittle nature and mechanical stiffness of the material. Functional grading assists to avoid the local stress concentration and increases the bonding strength due to smooth variation of material properties. Now, currently, Othmani et al. [5] explored the results of frequency equations following the legendres polynomial approach in the distinct materials, Li et al. [6] extended the new results on the same materials and Barati [7] studied the wave transference in the nonporous materials. Arani et al. [8] mentioned the nonlinear analysis of vibrations of the microbeams rubbery soldered with a PE beam using strain gradient theory. Moreover, Singhal et al. [9] covered and extended the topic of PE material variables on shear horizontal (SH) waves continuance in multiferroic structure.

The effect of various parameter such as material gradient parameter in electrically open and short case has been studied for both FE and PE. Although several modes existing in the given range have been shown through graphs for both the dispersion and attenuation curve. The dispersion equation for considered structure for electrically open and short circuit condition have also been obtained.

2 Mathematical Brief Implications

Governing material equations for piezoelectricity and flexoelectricity are:

$$\sigma_{ij} = c_{ijkl}\varepsilon_{kl} - d_{ijkl}V^o_{kl} - e_{ijk}E^o_k \tag{1}$$

$$\tau_{ijm} = -f_{ijkm}E^o_k \tag{2}$$

$$D_i = a_{ij}E^o_j + e_{jki}\varepsilon_{jk} + f_{jkil}w_{jkl} \tag{3}$$

$$Q^o_{ij} = d_{klij}\varepsilon_{kl} \tag{4}$$

Here $\tau_{ijm} = \tau_{jim}, \sigma_{ij} = \sigma_{ji}$, and$Q^o_{ij} = Q^o_{ji}$. The equation $d = -f$ is used to derive the results. Hence,

$$\varepsilon_{ij} = \frac{1}{2}\left(u_{i,j} + u_{j,i}\right) \tag{5}$$

$$E^o_i = -\phi_{,i} \tag{6}$$

The Eqs. (2)–(6) are the main fundamental equations for the piezoelectric materials. The physical significance of the above equations in mechanical systems displays the relationship between the elastic tensors, piezoelectric tensors, and inverse piezoelectric tensors. The equilibrium is set up among all the above equations by electric field intensity and electric potential. Now:

$$w_{jkl} = \varepsilon_{jk,l} \tag{7}$$

$$V_{ij}^{o} = E_{i.j}^{o} \tag{8}$$

Also, $V_{ij}^{o} = V_{ji}^{o}$, $w_{jkl} = w_{kjl}$, $\varepsilon_{ij} = \varepsilon_{ji}$.
The assumption equations for the Love wave propagation are:

$$u_1^p = u_2^p = 0, u_3^p = u_3^p(x, y, t), \phi^p = \phi^p(x, y, t), \tag{9}$$

Following some considered circumstances the equation yields

$$\frac{\partial}{\partial x}\left(\sigma_{31} - \frac{\partial \tau_{311}}{\partial x} - \frac{\partial \tau_{312}}{\partial y}\right) + \frac{\partial}{\partial y}\left(\sigma_{32} - \frac{\partial \tau_{321}}{\partial x} - \frac{\partial \tau_{322}}{\partial y}\right) = \rho^p \frac{\partial^2 u_3^p}{\partial t^2} \tag{10}$$

$$\frac{\partial}{\partial x}\left(D_1^o - \frac{\partial Q_{11}^o}{\partial x} - \frac{\partial Q_{12}^o}{\partial y}\right) + \frac{\partial}{\partial y}\left(D_2^o - \frac{\partial Q_{21}^o}{\partial x} - \frac{\partial Q_{22}^o}{\partial y}\right) = 0 \tag{11}$$

The most important strain–stress relations are for the considered materials:

$$\varepsilon_{23} = \frac{1}{2}\left(\frac{\partial u_3^p}{\partial y}\right), \varepsilon_{31} = \frac{1}{2}\left(\frac{\partial u_3^p}{\partial y}\right) \tag{12}$$

$$w_{231} = \frac{1}{2}\left(\frac{\partial^2 u_3^p}{\partial x \partial y}\right), w_{232} = \frac{1}{2}\left(\frac{\partial^2 u_3^p}{\partial y^2}\right), w_{311} = \frac{1}{2}\left(\frac{\partial^2 u_3^p}{\partial x^2}\right), w_{312} = \frac{1}{2}\left(\frac{\partial^2 u_3^p}{\partial x \partial y}\right) \tag{13}$$

$$E_1^o = -\frac{\partial \phi^p}{\partial x}, E_2^o = -\frac{\partial \phi^p}{\partial y} \tag{14}$$

$$V_{11}^o = -\frac{\partial^2 \phi^p}{\partial x^2}, V_{12}^o = -\frac{\partial^2 \phi^p}{\partial x \partial y}, V_{21}^o = -\frac{\partial^2 \phi^p}{\partial x \partial y}, V_{22}^o = -\frac{\partial^2 \phi^p}{\partial y^2} \tag{15}$$

Now using the Eqs. (12)–(15) in Eqs. (1)–(4), It is obtained:

$$\sigma_{31} - \frac{\partial \tau_{311}}{\partial x} - \frac{\partial \tau_{312}}{\partial y} = c_{44}\frac{\partial u_3^p}{\partial x} + e_{15}\frac{\partial \phi^p}{\partial x} \tag{16}$$

$$\sigma_{32} - \frac{\partial \tau_{321}}{\partial x} - \frac{\partial \tau_{322}}{\partial y} = c_{44}\frac{\partial u_3^p}{\partial y} + e_{15}\frac{\partial \phi^p}{\partial y} - h_{41}\frac{\partial^2 \phi^p}{\partial x^2} + h_{41}\frac{\partial^2 \phi^p}{\partial y^2} \tag{17}$$

$$D_1^o - \frac{\partial Q_{11}^o}{\partial x} - \frac{\partial Q_{12}^o}{\partial y} = -a_{11}\frac{\partial \phi^p}{\partial x} + e_{15}\frac{\partial u_3^p}{\partial x} + (h_{41} + h_{52})\frac{\partial^2 u_3^p}{\partial x \partial y} \tag{18}$$

$$D_2^o - \frac{\partial Q_{21}^o}{\partial x} - \frac{\partial Q_{22}^o}{\partial y} = -a_{11}\frac{\partial \phi^p}{\partial y} + e_{15}\frac{\partial u_3^p}{\partial y} - h_{52}\frac{\partial^2 u_3^p}{\partial x^2} - h_{41}\frac{\partial^2 u_3^p}{\partial y^2} \tag{19}$$

Hence, the fundamental expressions of PE material bars are:

$$c_{44}\frac{\partial^2 u_3^p}{\partial x^2} + c_{44}\frac{\partial^2 u_3^p}{\partial y^2} + e_{15}\frac{\partial^2 \phi^p}{\partial x^2} + e_{15}\frac{\partial^2 \phi^p}{\partial y^2} - h_{41}\frac{\partial^3 \phi^p}{\partial x^2 \partial y} + h_{41}\frac{\partial^3 \phi^p}{\partial y^3} = \rho^p \frac{\partial^2 u_3^p}{\partial t^2} \quad (20)$$

$$-a_{11}\frac{\partial^2 \phi^p}{\partial x^2} - a_{11}\frac{\partial^2 \phi^p}{\partial y^2} + e_{15}\frac{\partial^2 u_3^p}{\partial x^2} + e_{15}\frac{\partial^2 u_3^p}{\partial y^2} + h_{41}\frac{\partial^3 u_3^p}{\partial x^2 \partial y} - h_{41}\frac{\partial^3 u_3^p}{\partial y^3} = 0 \quad (21)$$

The following equations are arrived in the absence of flexoelectric affect:

$$u_3^p(x, y, t) = U(x)e^{ik(y-ct)}, \phi^p(x, y, t) = \Phi(x)e^{ik(y-ct)} \quad (22)$$

So, by solving Eq. (22) with mentioned below boundary conditions, dispersion relation is obtained.

3 Boundary Conditions (States) and Dispersion Expressions

(1) Mechanically and electrically constraint for electrically unlocked alliance at $x = -h_1$

$$\left[\left(\sigma_{zx} - \tau_{zxx,x} - \tau_{zxy,y}\right) - \tau_{zyx,y}\right]_{upperplate} = 0$$
$$\left[\left(D_x - Q_{xx,x} - Q_{xy,y}\right) - Q_{yx,y}\right]_{upperplate} = 0$$

(2) Mechanically and electrically constraint for electrically locked alliance at $x = -h_1$

$$(a)\left[\left(\sigma_{zx} - \tau_{zxx,x} - \tau_{zxy,y}\right) - \tau_{zyx,y}\right]_{upperplate} = 0$$
$$(b)\left[\phi_1^P(x, y)\right]_{upperplate} = 0$$

(3) At the interface, the continuous conditions and impedance boundary condition is given at $x = 0$ as follows

$$(a)\left[\left(\sigma_{zx} + \omega Z_1 u^P\right)\right]_{upperplate} = \left[\left(\tau_{zx} + \omega Z_1 u^e\right)\right]_{lowerplate}$$

$$(b)[D_x]_{upperplate} = [D_x]_{lowerplate}$$

$$(c)\left[u_3^P\right]_{upperplate} = \left[u^e\right]_{lowerplate} \quad (d)\left[\phi^P\right]_{upperplate} = \left[\phi^e\right]_{upperplate}$$

(4) Mechanically and electrically state for electrically unlocked condition at $x = -h_2$

$$(a)\left[\sigma_{zx}\right]_{lowerplate} = 0(b)[D_x]_{lowerplate} = 0$$

(5) Mechanically and electrically state for electrically locked case at $x = -h_2$

$$(a)\left[\sigma_{zx}\right]_{lowerplate} = 0(b)\left[\phi_1^e(x, y)\right]_{lowerplate} = 0$$

3.1 Dispersion Expressions

The equation called dispersion relation is obtained

$$\boldsymbol{Det}[\Lambda_i]\mathrm{i} = 1, 2, \ldots, 8, \tag{23}$$

Therefore, to solve the Eq. (23), obtained equation can be equal to zero to determine the unknowns and hence the closed form of dispersion equation can be achieved. The values of Eq. (23) are given in Appendix 2.

The equation called dispersion relation is obtained

$$\boldsymbol{Det}[\Lambda_i]\mathrm{i} = 1, 2, \ldots, 8, \tag{24}$$

Therefore, to solve the Eq. (24), obtained equation can be equal to zero to determine the unknowns and hence the closed form of dispersion equation can be achieved. The values of Eq. (24) are given in Appendix 4.

4 Numerical Discussion

Present section gives the insight of Love propagation in the piezo-composite structure. The dispersion equation is obtained in closed form. Typically, dispersion relation defines the relationship between phase velocity and wave number, but it also displays the impact of material gradient parameters on the phase velocity and attenuation (Table 1).

The consecutive Fig. 1a, b are plotted to study the results of material gradients coefficients on the Love wave phase velocity. The dimensionless variation of Love-type phase velocity increases with the increment of material gradient coefficients in both electrically open and short cases in the Fig. 1a, b. Increment in the material gradients coefficients decreases monotonically with the angular frequency ($\omega = k_1 c$). The Fig. 1a, b shown the phase velocity increment in the electrically closed case, but there

Table 1 Materials values are given

	PZT-6B	PZT-7A
$C_{44}(10^9 \text{Nm}^{-2})$	2.71	25.4
$\rho(10^3 \text{kgm}^{-3})$	5.8	76,000
$\varepsilon_{11} = \varepsilon_{22}(10^{-9} \text{Fm}^{-1})$	3.6	−2.1
$\mu_{11}(10^{-6} \text{Ns}^2\text{C}^{-2})$	5	5
$e_{15}(\text{Cm}^{-2})$	4.6	9.2

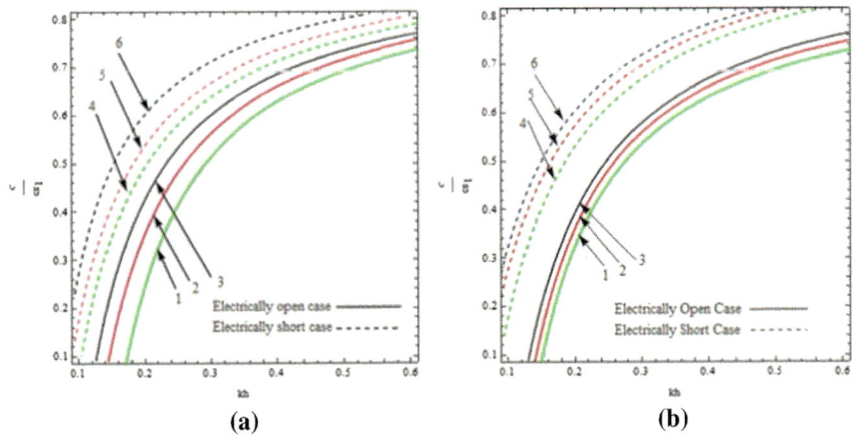

Fig. 1 **a** and **b** Variation of dimensionless phase velocity against dimensionless wave number for different values of material gradient of PE material gradient β under the influence of FE for electrically open and short cases in PZT-6B and PZT-7A materials respectively

is a less phase velocity in the electrically open case under the increment of the piezoelectric material gradient. This leads the main outcomes of the study which will be very beneficial for the material engineering.

For varied material gradients bars, the graph lines in the dispersive curves are shown in Fig. 2a, b. From the both the figs. It is concluded that the dispersive curves run towards the right and the gap between the dispersive curves of mode increases as the mode order increasing when both material gradients increase this will leads to the difference between in PZT-6B and PZT-7A. Means, phase velocity curves rise with the increase of the material plates gradients for selected wavelength and wavenumber. The graphs shown in present study quantitatively characterize the significance of individual parameters of the liquid layer as well as other parameters of piezoelectric composite on the propagation behaviour of the Love wave. It also allows the integration of SAW devices for Lab-on-chip systems. The present study may apply in the theoretical study of acoustic wave-based smart materials and non-invasive analysis of SAW devices. This model could be used in tailoring and analysing acoustic wave sensors based on Love wave and piezoelectric materials. For designing the device, the parameters like thickness of porous piezoelectric plate, the operating frequency

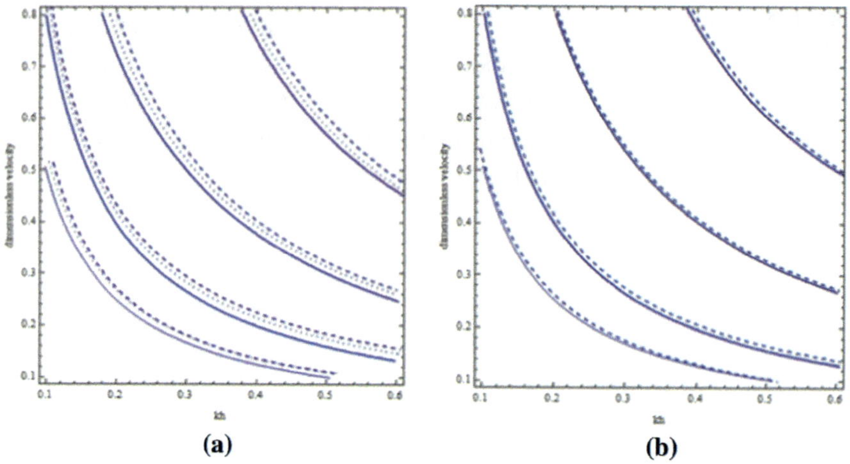

Fig. 2 **a** and **b** Dispersive curves of Love-type wave number for different values PE material gradient β under the influence of FE for electrically open and short cases in PZT-6B and PZT-7A materials respectively

which depends on wave number, the interface of two medium of composite structure are changed to optimize the structure.

5 Conclusions

The present article comprises the investigation of Love wave propagating in a piezoelectric biomaterial. Present study primarily focuses on the flexoelectricity and piezoelectricity, although the dispersion equation for the the materials is also established for both electrically open and short circuit conditions. The conclusions based on present study and graphical interpretation is encapsulated as follows:

- Flexoelectricity depends on the piezoelectric material gradient.
- Between piezoelectricity and flexoelectricity, from the results the flexoelectricity depends more on the materials width. This will help in enchaining the efficiency of piezo sensors.
- Higher is the material gradient value in the electric short case, higher is the phase velocity.
- The finer the functional grading, the higher is the phase velocity.
- Phase velocity and attenuation portrays completely opposite trend pertaining to a parameter when plotted against wave number.
- As an important remarks, waves phase velocity jumps high in the absence of electricity.
- The obtained results may be useful to enhance the efficiencies of IDT and Actuators.

This model could be used in tailoring and analysing acoustic wave sensors based on Love wave and piezoelectric materials. For designing the device, the parameters like thickness of porous piezoelectric plate, the operating frequency which depends on wave number, the interface of two medium of composite structure are changed to optimize the structure.

Acknowledgements The authors convey their sincere thanks "Department of Computational Sciences, School of Sciences, Christ (Deemed to be University) Delhi NCR Campus, Ghaziabad-201003, Uttar Pradesh, India." for providing all necessary research facility.

Compliance with Ethical Standards

Conflict of Interest: The authors declare that they have no conflict of interest.

Appendix 1 Secular Equation for Electrically Open Case

$$\Lambda_{11} = \left(Sc_{44} + e_{15} - \frac{ikh_{41}}{2}\right)s_1 e^{-s_1 h_1},$$

$$\Lambda_{12} = -\left(Sc_{44} + e_{15} - \frac{ikh_{41}}{2}\right)s_1 e^{s_1 h_1},$$

$$\Lambda_{13} = \left(Tc_{44} + e_{15} - \frac{ikh_{41}}{2}\right)s_2 e^{-s_2 h_1}$$

$$\Lambda_{14} = -\left(Tc_{44} + e_{15} - \frac{ikh_{41}}{2}\right)s_2 e^{s_2 h_1},$$

$$\Lambda_{21} = \left[-a_{11} + Se_{15} + ikS\left(\frac{h_{52}}{2} + h_{41}\right)\right]s_1 e^{-s_1 h_1}$$

$$\Lambda_{22} = -\left[-a_{11} + Se_{15} + ikS\left(\frac{h_{52}}{2} + h_{41}\right)\right]s_1 e^{s_1 h_1},$$

$$\Lambda_{23} = \left[-a_{11} + Te_{15} + ikT\left(\frac{h_{52}}{2} + h_{41}\right)\right]s_2 e^{-s_2 h_1},$$

$$\Lambda_{24} = -\left[-a_{11} + Te_{15} + ikT\left(\frac{h_{52}}{2} + h_{41}\right)\right]s_2 e^{s_2 h_1},$$

$$\Lambda_{31} = \left(Sc_{44} + e_{15} - \frac{ikh_{41}}{2}\right)s_1,$$

$$\Lambda_{32} = -\left(Sc_{44} + e_{15} - \frac{ikh_{41}}{2}\right)s_1,\ \Lambda_{33} = \left(Tc_{44} + e_{15} - \frac{ikh_{41}}{2}\right)s_2,$$

$$\Lambda_{34} = -\left(Tc_{44} + e_{15} - \frac{ikh_{41}}{2}\right)s_2,\ \Lambda_{35} = -k\alpha e_{15},\ A_{36} = k\alpha e_{15},$$

$$\Lambda_{41} = \left[-a_{11} + Se_{15} + ikS\left(\frac{h_{52}}{2} + h_{41}\right)\right]s_1,$$

$$\Lambda_{42} = -\left[-a_{11} + Se_{15} + ikS\left(\frac{h_{52}}{2} + h_{41}\right)\right]$$

$$s_1\Lambda_{43} = \left[-a_{11} + Te_{15} + ikT\left(\frac{h_{52}}{2} + h_{41}\right)\right]s_2,$$

$$\Lambda_{44} = -\left[-a_{11} + Te_{15} + ikT\left(\frac{h_{52}}{2} + h_{41}\right)\right]s_2,\ \Lambda_{47} = \varepsilon_{11}^e k,\ A_{48} = -\varepsilon_{11}^e k$$

$$\Lambda_{51} = K_T S,\ \Lambda_{52} = SK_T,\ \Lambda_{53} = TK_T,\ \Lambda_{54} = TK_T,$$

$$\Lambda_{55} = -\left(K_T + k\alpha c_{44}^e\right),\ \Lambda_{56} = \left(-K_T + k\alpha c_{44}^e\right)$$

$$\Lambda_{61} = K_L,\ \Lambda_{62} = K_L,\ \Lambda_{63} = K_L,\ \Lambda_{64} = K_L,$$

$$\Lambda_{65} = -K_L,\ \Lambda_{66} = -K_L,\ \Lambda_{67} = k\varepsilon_{11}^e,\ \Lambda_{77} = -k\varepsilon_{11}^e,$$

$$\Lambda_{75} = k\alpha e^{k\alpha h_2},\ \Lambda_{76} = -k\alpha e^{-k\alpha h_2},\ \Lambda_{77} = ke^{kh_2},\ \Lambda_{78} = -ke^{-kh_2}.$$

Appendix 2 Secular Expressions for Electrically Unlocked Case

$$\Lambda_{11} = \left(Sc_{44} + e_{15} - \frac{ikh_{41}}{2}\right)s_1e^{-s_1h_1},\ \Lambda_{12} = -\left(Sc_{44} + e_{15} - \frac{ikh_{41}}{2}\right)s_1e^{s_1h_1},$$

$$\Lambda_{13} = \left(Tc_{44} + e_{15} - \frac{ikh_{41}}{2}\right)s_2e^{-s_2h_1}$$

$$\Lambda_{14} = -\left(Tc_{44} + e_{15} - \frac{ikh_{41}}{2}\right)s_2e^{s_2h_1},$$

$$\Lambda_{21} = \left[-a_{11} + Se_{15} + ikS\left(\frac{h_{52}}{2} + h_{41}\right)\right]s_1e^{-s_1h_1}$$

$$\Lambda_{22} = -\left[-a_{11} + Se_{15} + ikS\left(\frac{h_{52}}{2} + h_{41}\right)\right]s_1e^{s_1h_1},$$

$$\Lambda_{23} = \left[-a_{11} + Te_{15} + ikT\left(\frac{h_{52}}{2} + h_{41}\right)\right]s_2e^{-s_2h_1},$$

$$\Lambda_{24} = -\left[-a_{11} + Te_{15} + ikT\left(\frac{h_{52}}{2} + h_{41}\right)\right]s_2e^{s_2h_1},$$

$$\Lambda_{31} = \left(Sc_{44} + e_{15} - \frac{ikh_{41}}{2}\right)s_1,$$

$$\Lambda_{32} = -\left(Sc_{44} + e_{15} - \frac{ikh_{41}}{2}\right)s_1,$$

$$\Lambda_{33} = \left(Tc_{44} + e_{15} - \frac{ikh_{41}}{2}\right)s_2$$

$$\Lambda_{31} = \left(Sc_{44} + e_{15} - \frac{ikh_{41}}{2}\right)s_1,\ \Lambda_{32} = -\left(Sc_{44} + e_{15} - \frac{ikh_{41}}{2}\right)s_1,$$

$$\Lambda_{33} = \left(Tc_{44} + e_{15} - \frac{ikh_{41}}{2}\right)s_2$$

$$\Lambda_{34} = -\left(Tc_{44} + e_{15} - \frac{ikh_{41}}{2}\right)s_2,\ \Lambda_{35} = -k\alpha e_{15},$$

$$\Lambda_{36} = k\alpha e_{15}\Lambda_{43} = \left[-a_{11} + Te_{15} + ikT\left(\frac{h_{52}}{2} + h_{41}\right)\right]s_2,$$

$$\Lambda_{44} = -\left[-a_{11} + Te_{15} + ikT\left(\frac{h_{52}}{2} + h_{41}\right)\right]s_2,$$

$$\Lambda_{47} = \varepsilon_{11}^{e}k,\ \Lambda_{48} = -\varepsilon_{11}^{e}k\Lambda_{51} = K_T S,\ \Lambda_{52} = SK_T,$$

$$\Lambda_{53} = TK_T,\ \Lambda_{54} = TK_T,\ \Lambda_{55} = -\left(K_T + k\alpha c_{44}^{e}\right),$$

$$\Lambda_{56} = \left(-K_T + k\alpha c_{44}^{e}\right)\Lambda_{61} = K_L,$$

$$\Lambda_{62} = K_L,\ \Lambda_{63} = K_L,\ \Lambda_{64} = K_L,$$

$$\Lambda_{65} = -K_L,\ \Lambda_{66} = -K_L,\ \Lambda_{67} = k\varepsilon_{11}^{e},$$

$$\Lambda_{77} = -k\varepsilon_{11}^{e},\ \Lambda_{75} = k\alpha e^{k\alpha h_2},$$

$$\Lambda_{76} = -k\alpha e^{-k\alpha h_2},$$

$$\Lambda_{77} = ke^{kh_2},\ \Lambda_{78} = -ke^{-kh_2}.$$

References

1. Chu, L., Dui, G., Ju, C.: Flexoelectric effect on the bending and vibration responses of functionally graded piezoelectric nanobeams based on general modified strain gradient theory. Compos. Struct. **186**, 39–49 (2018)
2. Singhal, A., Tiwari, R., Baroi, J., Kumhar, R.: Perusal of flexoelectric effect with deformed interface in distinct (PZT-7A, PZT-5A, PZT-6B, PZT-4, PZT-2) piezoelectric materials. Waves Random and Complex Media (2022). 10.1080/17455030.2022.2026522
3. Nanthakumar, S.S., Zhuang, X., Park, H.S., Rabczuk, T.: Topology optimization of flexoelectric structures. J. Mech. Phys. Solids **105**, 217–234 (2017)
4. Sahu, S.A., Singhal, A., Chaudhary, S.: Surface wave propagation in functionally graded piezoelectric material: an analytical solution. J. Intell. Mater. Syst. Struct. **29**(3), 423–437 (2018)
5. Othmani, C., Takali, F., Njeh, A., Ghozlen, M.H.B.: Numerical simulation of Lamb waves propagation in a functionally graded piezoelectric plate composed of GaAs-AlAs materials using Legendre polynomial approach. Optik **142**, 401–411 (2017)
6. Li, X.Y., Wang, Z.K., Huang, S.H.: Love waves in functionally graded piezoelectric materials. Inter. J. Sol. Struct. **41**(26), 7309–7328 (2004)
7. Barati, M.R.: On wave propagation in nanoporous materials. Int. J. Eng. Sci. **116**, 1–11 (2017)
8. Arani, A.G., Abdollahian, M., Kolahchi, R.: Nonlinear vibration of a nanobeam elastically bonded with a piezoelectric nanobeam via strain gradient theory. Int. J. Mech. Sci. **100**, 32–40 (2018)
9. Singhal, A., Baroi, J., Sultana, M., Baby, Riya.: Analysis of SH-waves propagating in multiferroic structure with interfacial imperfections. Mech. Adv. Compos. Struct. **9**(1), 1–10 (2022)

A Review on the Reliability Analysis of Point Machines in Railways

Deb Sekhar Roy, Debajyoti Sengupta, Debraj Paul, Debjit Pal, Aftab Khan, Ankush Das, Surojit Nath, Kaushik Sinha, and Bidhan Malakar

Abstract Railways, being one of the most eminent modes of transportation used worldwide. It requires proper maintenance ensuring maximum reliability to conduct a safe journey for the passengers. Point Machines or Railway Turnouts are the vital safety assets that play a critical role in maintaining the flexibility of the rail networks. This paper provides an insightful review on the fault analysis, health monitoring and reliability analysis of point machines. A detailed review of the different technologies and algorithms proposed by the researchers, worldwide has been presented along with their current modifications for the adaptation in modern technologies. This review enables the researchers a basis for an enhanced quality and reliability in the working of point machines.

D. Sekhar Roy (✉) · D. Sengupta · D. Paul · D. Pal · A. Khan · A. Das · S. Nath · K. Sinha · B. Malakar
Department of Electrical Engineering, JIS College of Engineering, Kalyani, West Bengal, India
e-mail: dsroy.96.00@gmail.com

D. Sengupta
e-mail: debajyotisengupta98@gmail.com

D. Paul
e-mail: debrajpaul15@gmail.com

D. Pal
e-mail: debjitpal741507@gmail.com

A. Khan
e-mail: ashifkhankatwa@gmail.com

A. Das
e-mail: ankush2001das@gmail.com

S. Nath
e-mail: surojitnath17@gmail.com

K. Sinha
e-mail: ksin08546@gmail.com

B. Malakar
e-mail: bidhan.malakar@jiscollege.ac.in

S. Banerjee and A. Saha (eds.), *Nonlinear Dynamics and Applications*, Springer Proceedings in Complexity,
https://doi.org/10.1007/978-3-030-99792-2_114

Keywords Point machine · Reliability · Rail networks · Monitoring · Algorithms · Fault analysis

1 Introduction

Over the last few decades railway transport has evolved into a busiest network worldwide. Vast growing railway networks and fast-moving railways has become a mass transportation system, transporting billions of passengers and cargoes daily. Enormous safety, reliability and stability is required within railway networks and transportation keeping in view of its increasing demand. This vast grown railway network requires a well driven railway traffic management system for its smooth working. Point machines plays an effective role in railway trafficking that are operated in railway turnouts and are critical elements in railway tracks. So, encountering faults of point machines and its diagnosis is an important issue to guarantee safer train journeys and has a remarkable sense in railway monitoring [1–6].

For the last few decades several researches have been done in the field of fault detection and health monitoring processes on point machines in aspects of the different railway system. The researchers from different countries of the world are united to develop effective techniques and methodologies to increase the efficiency of the reliability of a point machine. For the sake of simplicity in understanding this paper has given a brief theory on point machines, its types and working principles with technical detailing in Sect. 2. Section 3 of this paper focuses on different analysis methods of point machines which are classified into three categories: Reliability analysis, Condition monitoring and Fault analysis of point machines.

In the reliability analysis of point machine wide range of methodologies and approaches that has been proposed by the researchers worldwide are discussed. This includes methodologies like Remote Condition Monitoring (RCM) system which uses Kalman Filter method [1], Functional Redundancy Approach (FRA) for reliability estimation and Monte Carlo simulation for model designing [3], Qualitative Trend Analysis (QTA) for incipient fault detection [5], etc. The monitoring and management of point machine gives brief survey of the sensors, actuators and modelling methods used to monitor the health of point machines including condition monitoring using Discrete Wavelet Transform (DWT) [4], Principal Component Analysis (PCA) technique etc. The fault detection of point machine provides different technologies and advancements done in the field of failure detection which includes Failure Mode and Effects Analysis (FMEA) [2], Autoregressive Integrated Moving Average (ARIMA) and Autoregressive Kalman (AR-Kalman) to diagnosis and detect faults [6], etc. Therefore, this paper presents a detailed and comprehensive survey of some works done within a decade on point machine utilizing past and present technologies.

2 A Brief Theory About Point Machines Used in Railways

The point machine, a piece of electro-mechanical equipment with potential failure mode, is commonly used in signaling systems combining crossings, cross-overs, stock rails, rods, cranks, levers, and locking arrangements. For smooth changes in railway tracks and avoiding hazardous accidents, the role of the point machine has been considered as one of the main components in the context of the world railway system.

In detail, the function of the point machine is to unlock and operate the point switches in the exact position for detection of the correct setting of the point switch. The system mechanism includes various subsystems: motor unit including a contactor control arrangement and a terminal area, gear unit including spur-gears, worm reduction unit [7].

The inner parts of the point machine used in Railways is shown in the following Fig. 1.

2.1 *Types of Point Machines*

Point machines are broadly classified on the basis of different parameters, are as follows [7, 8]:

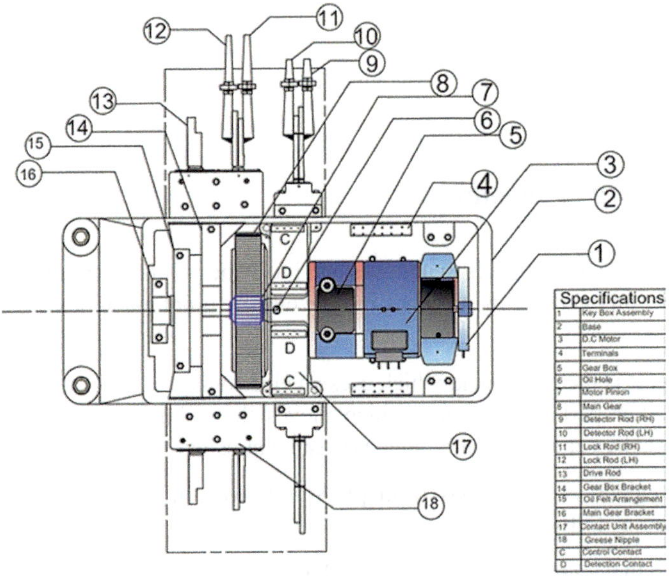

Fig. 1 Top view of inner portion of the point machine used in Railways

i. Type of point machine on the basis of field arrangement:

   a. Single field type point machine.
   b. Split field type point machine.

ii. Types of point machine on the basis of voltage rating:

   a. High voltage type (Operating voltage of 110 V DC).
   b. Low voltage type (Operating voltage of 24 V/ 36 V).
   c. High Voltage AC-380 AC.

iii. Type of point machine on the basis of speed:

   a. High speed machine – Operating speed of 3 s.
   b. Low speed machine – Operating speed of 5 s.

iv. Type of point machine on the basis of locking:

   a. Clamp type point machine.
   b. Rotary type point machine (Siemens).

v. Type of point machine on the basis of machine type:

   a. Combined type point machine.
   b. Separate type point machine.

vi. Type of point machine on the basis of machine type:

   a. Electro-mechanical type.
   b. Electro-pneumatic type.
   c. Electro-hydraulic type.

The construction of the Point Machines is shown in Fig. 2.

2.2 *Working Principles*

A point machine is the device that moves and locks the points of turnout remotely. It works under an interlocking mechanism. The electric point machine consists of two parts:

i. One part is the point lock, which uses an electromagnet to release the lock.
ii. The other part is the motor which moves the point blades. Both parts are electrically separated but mounted on same housing.

The sequence of operations from [7, 8] followed in the process of Railway Turnouts are as follows:

i. When the control operator wants to change a certain switch from "straight" to "thrown" or vice versa, a signal is sent to the signaling post.
ii. The electromagnet of the machine is energized and it releases the lock of it.

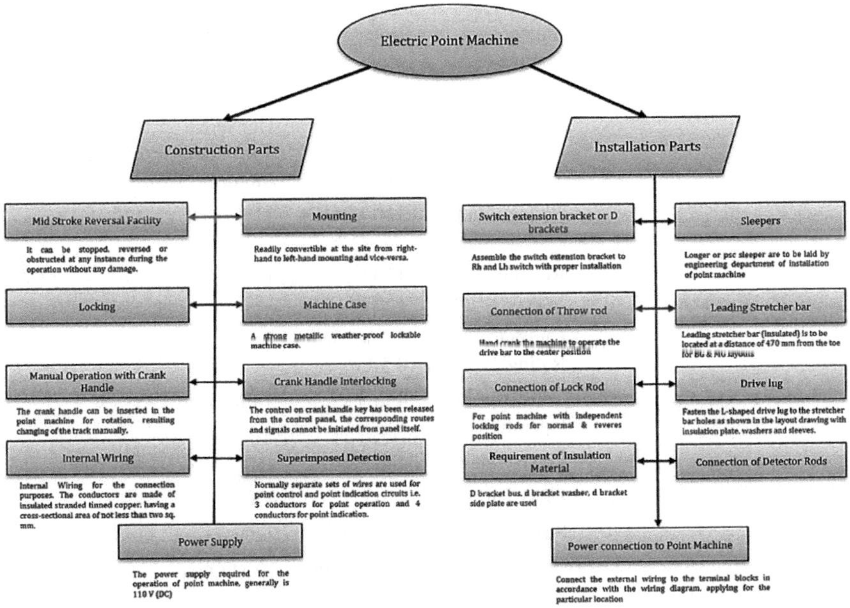

Fig. 2 Components of the point machine in Railways

iii. The motor is powered in the desired direction and lets the points move.
iv. At the end position, the motor gets turned off.
v. The electromagnet gets de-energized and points in the machine gets locked.
vi. The control operator gets confirmation of success about the change in track positions.

Henceforth, this Section briefly describes the point machines, its types and working principles used in Railways. The following table shows the technical data of motors used in different railway point machines in Table 1.

3 A Brief Theory About Point Machines Used in Railways

Different analyses carried out till now for the point machines are divided into the following 3 categories. Apart from these 3 categories, there may be some more categories but to keep this study simple only these three categories are considered.

Table 1 Table consisting of technical data of motors used in different railway point machines

Salient features	Rotary type point motor	Clamp type point motor	Siemens electric point machine
Type of Motor	DC Series Split Field Motor	DC Series Split Field Motor	DC Series Split Field Motor
Voltage Rating	110 V (DC) + - 25%	110 V (DC)	110 V (DC)
Revolution Speed (RPM)	1700 + - 15%	1700 + - 15%	1700 + - 15%
Current Rating	5.3A (Max-8.5A)	5.5A	5.3A
Time of Operation	4 to 5 s	5 s Max	3–5 s
Power Rating	440 Watts	440 Watts	440 Watts
EPM Thrust	450 kg	450 kg	-
Stroke	143 mm	220 mm	143 mm
Insulation	10 M Ohm	10 M Ohm	10 M Ohm
Gear oil	SAE 30	-	-

3.1 Reliability Analysis of Point Machines

Before developing any hardware or methods, the priority retains in the current developments and methodology used in the twenty-first century models. In 2003, researchers from Universidad de Castilla-La Mancha, Spain in collaboration with the University of Sheffield, UK discussed a new detection technique of the gradual failure in Railway Turnouts or Point Machines, which can be managed by a RCM system, by the help of a methodology known as the Kalman Filter. By the usage of the methodology, the faults from Reverse to Normal (RN) detected 100% of faults whereas it showed 97.1% of faults in Normal to Reverse (NR) of a Point Machine [1]. In 2012, researchers from Loughborough University, UK proposed a FRA to estimate the increase in the reliability in the Railway Point Machines. The paper mainly contributes to the usage of 2P-Weibull failure distributions for the collection of field data over a long period of time and using it to model engineering fault tolerance into the existing systems [3]. In 2013, researchers from the University of Nottingham, UK, introduces research that describes the obtaining the field data of lifetime distributions of components of Switches and Crossings (S&C) and as well as on the basis of historical failure data. The paper selected the Weibull distribution as the suitable probabilistic model which was used for the collection of both time-based and utilization-based failure data resulting in efficient and cost-effective asset management [9].

In 2014, researchers from India proposed a parametric model for estimating the probability of the reliability of the Point Machines in Railways which is capable of predicting the percentage of its failure rate. This paper frames the different types of reliability models and the procedures in the context of parametric analysis which can be applied in Indian Railway (IR) signaling systems [4]. In 2015, researchers from the University of Birmingham, UK introduces a novel algorithm that utilizes QTA for the detection and diagnosis of various incipient faults in the Point Machines, proving to be much reliable than the currently used commercial methods. The research aims to introduce various methods which are capable of reducing 30% of track Life Cycle Costs (LCCs) in the area of Condition Based Maintenance (CBM) where infrastructure operators aim to limit their maintenance costs as low as possible [5]. In 2021, researchers from the Institute of Engineering & Management, India in collaboration with the University of Greenwich, UK researched the rate of failures of different components of signaling systems, including Point Machines using analytical methods. The signaling system components (Point Machine, Track Circuit, and Signal Unit) was analyzed and developed with the help of mathematical modeling that required to increase the reliability of the signaling systems and to achieve the optimal interval inspection for the system [10].

3.2 *Case Studies on the Monitoring and Management of Point Machines*

In 2002, researchers from Oxford University discussed about the various sensors that are used for measuring motor driving force, driving current and voltage, temperatures and state changes etc. The data are directly collected from the sensors and processed through the operating system which remotely monitors, control and calibrates the extracted data [11]. In 2007, a paper describes about the remote condition monitoring, reliability, safety of the point machine using the Kalman filter approach. This special type of filter was basically a recursive data processing algorithm that is executed in the form of an equation. The main purpose of the Kalman filter is to filter the linear discrete data using the current sensor data in a point condition monitoring system [12]. In 2010, the authors of the paper mainly focus on the railway infrastructure and more specifically on railways point machines. By using smart techniques of the monitoring system, the problems can be predicted and be able to recover quickly. By using Artificial Intelligence (AI) and signal processing, the faults can be used to detect the faults of the point machines [13]. In 2011, [4] shows a new approach for the fault detection and diagnosis of the AC point machines used in Railways, from the University of Birmingham, UK. As electric active power can be shown as a parameter for the condition monitoring system of the AC point machines i.e., by means of the DWT. By using DWT, the original waveform is converted to multiple levels of resolution, sustaining local time information in each level of resolution. In 2012, a paper focuses on health management and condition monitoring of the point machine in the Railway [3]. This paper shows different strategies and technical architectures for the health management of the electro-mechanical point machine.

The Principal Component Analysis (PCA) was used to access the health of the point machine. In 2013, unique approach has been presented for fault detection and diagnosis for electric DC point machines operated on the railways in the UK. Here, the electric current is used as a main parameter for analyzing and acquiring for the DC point machine. The proposed method is dependable upon wavelet transform and Support Vector Machine (SVM). The DWT is used for feature extraction and SVM is used for the classifiers.

In 2015, a group of researchers from China had presented the improvement of the reliability, availability and maintainability, which necessitates the development of a predictive monitoring system for the fleet of point machines in the Railway industry. In the proposed monitoring system, the appropriate signals had been included and the advanced pattern was used to recognize for finding the primary indication of the point machine degradation, and a user interface for displaying and reporting the health results of the point machine [14]. In 2017, a paper describes electric current shape analysis used for condition detection of the point machine in Railway [15]. After analyzing the replacement data and labelling the shape of each replacement data, any further procedures can be undertaken. On the basis of replacement results with field replacement data, we can find the accuracy. In 2018, a paper represents a unique degradation detection method that can be used for mining and identifying the degradation state of the point machine used in Railways, researched in China. Power data is to be processed for extracting the set of features which can be used to describe the point machine properties effectively. Basically, a SVM was used to build the state classifier which can be used to identify the degradation state of the point machine in a combination with a featured map known as Self-Organizing Map (SOM). In this paper, time– frequency-based featured data were mined and a few of them were used to detect and diagnose faults based on a SVM classifier, which is used to build a preferable autonomous fault detection method, where prior knowledge of the data in CM is not vital [16]. Again, within that same year, the authors of [17] had shown their contribution by proposing a methodology known as Machine Health Assessment (MHA), which comprises both offline and online segments, for the analysis of sliding-chair degradation of Point Machines in Railways. In the offline phase, the extraction and data labeling methods were developed for diagnosing the feature selection and time-series segmentation-based fault severity whereas, in the online phase, a sliding-chair fault severity classification was studied by using supervised machine learning tools. Nevertheless, the proposed approach is only applicable if the monitoring data is readily available and if not, appropriate data imputation algorithms are to be used for the extraction of the complete data before using them for further fault detection and severity extraction. In 2019, a research paper was published that discussed the proposal of an initial fault diagnosis method of Railway Point Machines which emphasizes the mining of non-fault data and recognizing them from the assigned degradation states under diverse fault modes. Based upon extraction of data, a self-organizing feature map (SOM)-based degradation state mining method and a Particle Swarm Optimization (PSO)-SVM based classification model was proposed for various degradation conditions and to accurately identify different degradation states respectively. From the application of the esteemed method and model on the 'SIEMENS S700K' Point Machine, it is found that the method can be used to mine different degradation level states under various fault modes and the model can easily identify the types of states with an accuracy of 97.73% [18]. In 2020, a researcher from Embedded Systems Group, CMC Americas, Inc from the United States of America discussed a new method for maintenance and diagnosis of Railway Point Machines. The method results in the reduction of costs in the examination of manual checks. Also, the method utilizes various sensors to monitor the applicable parameters of Point Machines which can be used to understand and identify the cause of a fault by studying the current situation and comparing its problems from the history of faults that occurred in it [19].

3.3 Case Studies on the Fault Detection of Point Machines

As continuous growth in railways demands target of performance with proper maintenance and failure management. In 2007, to fulfil the demand a group of researchers had published a paper focusing on maintenance and diagnosis of failure in railway track side components especially on M63 type Point Machine in the UK. They had analysed fault by signal analysis by Kalman filter model and moving average filter and constructed a comparative study about it. This study would come to a decision that moving to the average filter is a more preferable method to the Kalman filter model [12]. In 2009, researchers from Kolkata, India had developed research work on failure mode and effect analysis (FMEA) of the railway signaling systems in working mode to avoid hazardous accidents that occurred in railway tracks. This analysis would cause gradual improvements in design, classifications on a ten-grade scale had been developed in the research [2]. In 2016, a methodology had been proposed to detect the fault in point operating equipment (POE) machine in advance by measuring the current consumption of point motor. One-class Support Vector Machine (SVM) method was used in the technique to get more accuracy than other threshold-based techniques [5]. In 2016, researchers had proposed a data mining solution using audio data to diagnose and detect faults to avoid accidents. This process had gone through four different modules- two online modules and another is offline. The online module includes feature extraction and fault detection; attribute subset selection and vector machines (SVM) training are included in the offline module. To detect the fault and classify it by audio analysis, two SVM had been used and this results in cost-effectiveness and automatic detection [14].

In 2018, a group of researchers from Iran had proposed a new methodology using Stacked Autoencoders (SAE) to analyse the fault. Initially, integration of feature extraction was performed, followed by failure detection. To diagnose the fault, a trained SAE had gone through the phase of final tuning [20]. In the same year, a new method using ARIMA and AR-Kalman had been presented to diagnose and detect the faults in the point machine (S700K). In this method, signal processing including wavelet transform and statistical analysis had been presented to acquire the related data of faults in the point machine [6]. In 2019, a group of researchers from China had presented a methodology using current signals and feature extraction approach based on locally connected autoencoder. The proposed methodology automatically acquires the informative features from raw electrical signals. For reducing the impact of the non-informative information in raw electrical signals a weighting strategy had been proposed to develop the usefulness and robustness in different features [21]. In the same year, using data from Centralized Traffic Control (CTC) systems, along with meteorological data, a methodology had been proposed to diagnose the risk factor in Point machine. In this methodology, extracting information from railway logs, presented a compact numerical representation along with algorithms of machine learning [22]. In the year 2020, a group of researchers from the UK had proposed a methodology utilizing unlabeled signal sensor data. It was claimed that it can deal with real-time faults using data pre-processing; also suitable for smart city infrastructure [19].

Henceforth, this Section describes about the existing technologies used in Point Machines in Railways in the recent decade, on the basis of Reliability, monitoring and the fault detection methods.

4 Conclusions

As Point Machines or Railway Turnouts are vital safety assets that play a critical role in maintaining the flexibility of rail networks, hence requiring proper maintenance procedures for reliable productivity in the field of Railways. There are different technologies developed and used currently as discussed in this paper to overcome the problem of reliability of Point Machines in Railways. In this paper the current, as well as past research, works on point machine is carefully reviewed. More specifically, the paper presents and reviews the previously published works based on the technologies implemented, application of various technologies, algorithms and methodologies, surveyed for the detection of faults, allow feasible condition monitoring and is used for increasing the reliability of the point machine.

References

1. Márquez, F.P.G., Schmid, F., Collado, J.C.: A reliability centered approach to remote condition monitoring. A railway points case study. Reliab. Eng. & Syst. Saf. (ELSEVIER, UK) **80**(1), 33–40 (2003)
2. Panja, S.C., Ray, P.K.: Failure mode and effect analysis of Indian railway signalling system. Int. J. Perform. Eng. (India) **5**(2), 131 (2009)
3. Ardakani, H.D., Lucas, C., Siegel, D., Chang, S., Dersin, P., Bonnet, B., Lee, J.: PHM for railway system—a case study on the health assessment of the point machines. In: 2012 IEEE Conference on Prognostics and Health Management pp. 1–5. IEEE, USA (2012)
4. Atamuradov, V., Camci, F., Baskan, S., Şevkli, M.: Failure diagnostics for railway point machines using expert systems. IEEE, Spain (2011)

5. Vileiniskis, M., Remenyte-Prescott, R., Rama, D.: A fault detection method for railway point systems. Proc. Inst. Mech. Eng. Part F: J. Rail Rapid Transit (UK) **230**(3), 852–865 (2016)
6. Atamuradov, V., Medjaher, K., Camci, F., Dersin, P., Zerhouni, N.: Railway point machine prognostics based on feature fusion and health state assessment. IEEE Trans. Instrum. Meas. **68**(8), 2691–2704 (2018)
7. Handbook on Installation and Maintenance of Electric Point Machine, Indian Railways Centre for Advanced Maintenance Technology, India (2010)
8. Signal Directorate Research Designs & Standards Organisation, Indian Railways Standard Specification For Motors For Electric Point Machine, Specification No.: Irs: S 37, India (2020)
9. Rama, D., Andrews, J.D.: A reliability analysis of railway switches (NG7 2RD. UK). Proc. Inst. Mech. Eng. Part F: J. Rail Rapid Transit **227**(4), 344–363 (2013)
10. Kumar, N., Tee, K.F.: Reliability and inspection modelling of railway signalling systems. Modelling (MDPI, India) **2**(3), 344–351 (2021)
11. Zhou, F.B., Duta, M.D., Henry, M.P., Baker, S., Burton, C.: Remote condition monitoring for railway point machine. In ASME/IEEE Joint Railroad Conference, pp. 103–108. IEEE, USA (2002)
12. Marquez, F.P.G., Weston, P., Roberts, C.: Failure analysis and diagnostics for railway trackside equipment. Eng. Fail. Anal. (ELSEVIER, UK) **14**(8), 1411–1426 (2007)
13. García Márquez, F.P., Roberts, C., Tobias, A.M.: Railway point mechanisms: condition monitoring and fault detection. Proc. Inst. Mech. Eng., Part F: J. Rail Rapid Transit (Spain) **224**(1), 35–44 (2010)
14. Lee, J., Choi, H., Park, D., Chung, Y., Kim, H.Y., Yoon, S.: Fault detection and diagnosis of railway point machines by sound analysis. Sensors (MDPI, Korea) **16**(4), 549 (2016)
15. Sa, J., Choi, Y., Chung, Y., Kim, H.Y., Park, D., Yoon, S.: Replacement condition detection of railway point machines using an electric current sensor. Sensors (MDPI, Korea) **17**(2), 263 (2017)
16. Guo, Z., Ye, H., Dong, W., Yan, X., Ji, Y.: A fault detection method for railway point machine operations based on stacked autoencoders. In: 2018 24th International Conference on Automation and Computing (ICAC), pp. 1–6. IEEE, UK (2018)
17. Bian, C., Yang, S., Huang, T., Xu, Q., Liu, J., Zio, E.: Degradation detection method for railway point machines, China. arXiv:1809.02349 (2018)
18. Atamuradov, V., Medjaher, K., Camci, F., Zerhouni, N., Dersin, P., Lamoureux, B.: Feature selection and fault-severity classification–based machine health assessment methodology for point machine sliding-chair degradation. Qual. Reliab. Eng. Int. (UK) **35**(4), 1081–1099 (2019)
19. Mistry, P., Lane, P., Allen, P.: Railway point-operating machine fault detection using unlabeled signaling sensor data. Sensors (MDPI, UK) **20**(9), 2692 (2020)
20. Abbasnejad, S., Mirabadi, A.: Predicting the failure of railway point machines by using Autoregressive Integrated Moving Average and Autoregressive-Kalman methods. Proc. Inst. Mech. Eng. Part F: J. Rail Rapid Transit (Iran) **232**(6), 1790–1799 (2018)
21. Bian, C., Yang, S., Huang, T., Xu, Q., Liu, J., Zio, E.: Degradation state mining and identification for railway point machines. Reliab. Eng. & Syst. Saf. (China) **188**, 432–443 (2019)
22. Doboszewski, I., Fossier, S., Marsala, C.: Data driven detection of railway point machines failures. HAL Id: hal-02407540, China (2019)
23. Panja, S.C., Ray, P.K.: Reliability analysis of a 'point-and-point machine' of the Indian railway signaling system. Qual. Reliab. Eng. Int. (Wiley InterScience, India) **23**(7), 833–848 (2007)
24. Atamuradov, V., Camci, F., Baskan, S., Sevkli, M.: Failure diagnostics for railway point machines using expert systems. In: 2009 IEEE International Symposium on Diagnostics for Electric Machines, Power Electronics and Drives, pp. 1–5. IEEE, France (2009)
25. Asada, T., Roberts, C., Koseki, T.: An algorithm for improved performance of railway condition monitoring equipment: Alternating-current point machine case study. Transp. Res. Part C: Emerg. Technol. (ELSEVIER, Japan) **30**, 81–92 (2013)
26. Asada, T., Roberts, C.: Improving the dependability of DC point machines with a novel condition monitoring system. Proc. Inst. Mech. Eng. Part F: J. Rail Rapid Transit (UK. SAGE) **227**(4), 322–332 (2013)

27. Li, Z., Yin, Z., Tang, T., Gao, C.: Fault diagnosis of railway point machines using the locally connected autoencoder. Appl. Sci. (China) **9**(23) 5139 (2019)
28. Bemment, S.D., Goodall, R.M., Dixon, R., Ward, C.P.: Improving the reliability and availability of railway track switching by analysing historical failure data and introducing functionally redundant subsystems. Proc. Inst. Mech. Eng. Part F: J. Rail Rapid Transit (SAGE, UK) **232**(5), 1407–1424 (2018)
29. Kumar, N., Tee, K. F.: Reliability and inspection modelling of railway signalling systems. Modelling (MDPI, UK) **2**(3), 344–354 (2021)

Application of a Measure of Noncompactness in *cs*-Solvability and *bs*-Solvability of an Infinite System of Differential Equations

Niraj Sapkota, Rituparna Das, and Santonu Savapondit

Abstract The theory of measure of noncompactness has been a very helpful tool in non-linear functional analysis over the years. In this paper we have examined the solvability of an infinite system of third order differential equations in the sequence space of convergent series and sequence space of bounded series using the Hausdorff measure of noncompactness. We have also used the concept of Meir-Keeler condensing operator to utilize the fixed point theory in our approach and have analysed the approach for each sequence space with suitable examples.

Keywords Infinite system of third-order differential equations · Measures of noncompactness · Convergent and bounded sequences

1 Introduction

The concepts of measure of noncompactness (MNC), in the non-linear functional analysis is used in various applications of operator theory, fixed point theory and extensively to investigate the theories of differential equations, functional integral equations and in characterizing compact operators between sequence spaces. Kuratowski [1] was first to introduce the concept of measure of noncompactness in 1930. Later Darbo [2] used this in generalising the classical schauder fixed point principle and Banach's contraction mapping principle(special variant) for condensing operators. Goldenštein and Markus [3] gave Hausdorff MNC and Istrăţescu [4]

N. Sapkota (✉) · S. Savapondit
Department of Mathematics, Sikkim Manipal Institute of Technology,
Sikkim Manipal University, Sikkim 737136, India
e-mail: niraj.sapkota13@gmail.com

S. Savapondit
e-mail: sspondit@gmail.com

R. Das
Department of Mathematics, Pandu College, Guwahati, India
e-mail: ri2p.das@gmail.com

S. Banerjee and A. Saha (eds.), *Nonlinear Dynamics and Applications*,
Springer Proceedings in Complexity,
https://doi.org/10.1007/978-3-030-99792-2_115

provided the Istrăţescu MNC. It was only in 1980 that Banás and Goebel [5] gave the axiomatic definition of the measure of noncompactness. Over the years measure of noncompactness has proven itself to be a great tools in various branches of mathematics.

In the literature, Infinite system of differential equations has been introduced by Persidskii in 1959 [6], in 1961 [7] and in 1976 [8]. Here he has coined the term "Countable systems of Differential equations". Following this, series of other works came into existence for the solvability and existence theorems for different kinds of infinite system of calculus equations in line with our work (see, [9–16, 20–22]).

In 2011 Yueli Chen, Jingli Ren and Stefan Siegmund [17] showed the "existence of positive periodic solutions of third-order differential equations". For this they have established a particular array of Green's functions. The equation used here was a single equation. Borrowing the same later R. Saadati, E. Pourhadi and M. Mursaleen in 2019 [16] showed solvability for infinite set of equations of type used by Chen et al. [17] occurring simultaneously. Solvability of that infinite system was established in c_0 space. Our aim is to establish the solvability conditions in the sequence space of convergent series (cs) and bounded series (bs) for the infinite system of third-order differential equations (ISTODE) define in section (2). For various concepts and types of MNCs one can refer to [11].

2 Infinite System of Third-Order Differential Equations

We introduce the following ISTODE for further work

$$\frac{\mathrm{d}^3 k_n(\theta)}{\mathrm{d}\theta^3} + p\frac{\mathrm{d}^2 k_n(\theta)}{\mathrm{d}\theta^2} + q\frac{\mathrm{d}k_n(\theta)}{\mathrm{d}\theta} + rk_n(\theta) = h_n(\theta, k_1(\theta), k_2(\theta), \ldots), \quad (n \in \mathbb{N}) \tag{1}$$

Here we take $h_n \in C(\mathbb{R} \times \mathbb{R}^\infty, \mathbb{R})$ as family of ω-periodic functions with respect to θ; $p, q, r \in \mathbb{R}$ are constants and $r \neq 0$. In this paper we will investigate the solvability conditions for ω-periodic solutions of ISTODE (1) in cs and bs space (for definition see Sect. 3 and Sect. 4 below).

Homogeneous equations for the system (1) is $\frac{\mathrm{d}^3 k_n(\theta)}{\mathrm{d}\theta^3} + p\frac{\mathrm{d}^2 k_n(\theta)}{\mathrm{d}\theta^2} + q\frac{\mathrm{d}k_n(\theta)}{\mathrm{d}\theta} + rk_n(\theta) = 0$, $(n \in \mathbb{N})$ and the corresponding characteristic equation is

$$\alpha^3 + p\alpha^2 + q\alpha + r = 0. \tag{2}$$

The eigen value for the Polynomial Eq. 2 can have following four possibilities. (given $r \neq 0$): (i) $\alpha_1 \neq \alpha_2 \neq \alpha_3$, (ii) $\alpha_1 = \alpha_2 \neq \alpha_3$, (iii) $\alpha_1 = \alpha_2 = \alpha_3 = \alpha$, (iv) $\alpha_1 = u + iv$, $\alpha_2 = u - iv$, $\alpha_3 = \alpha$, for $u, v, \alpha \in \mathbb{R}$. We shall seek the solvability for each case. Further, with the concept of Green's functions one can claim that $k \in C^3(\mathbb{R}, \mathbb{R}^\infty)$ serves as a solution of infinite system of differential equations (1) if and only if $k \in C(\mathbb{R}, \mathbb{R}^\infty)$ acts as a solution of following integral equations

$$k_n(\theta) = \int_{\theta}^{\theta+\omega} \mathcal{K}(\theta, \eta) h_n(\eta, k(\eta)) d\eta \quad (n \in \mathbb{N}). \tag{3}$$

Equation (3) is an infinite system of integral equations. Here $\mathcal{K}(\theta, \eta)$ is a Green's function which takes up different form in different cases of eigen values. For the varying behaviour of $\mathcal{K}(\theta, \eta)$ for each case one may refer to [17].

3 Solvability of the Infinite System of Differential Equations in *cs* Space

$cs := \left\{ k = \{(k_i)\}_{i=0}^{\infty} : \lim_{n\to\infty} \sum_{i=0}^{n} k_i < \infty \right\}$. By definition cs is a space of sequences with BK and AK and also is equipped with monotone norm $||k||_{cs} = \sup_n \left|\sum_{i=0}^{n} k_i\right|$ (see Sect. 7.3. [19]). Thus with the application of formula given by Banaś and Mursaleen [11] we see that the Hausdorff measure of noncompactness for cs is

$$\chi(Q) = \lim_{n\to\infty} \left\{ \sup_{k\in Q} \left(\sup_{m} \left| \sum_{i=n}^{m} k_i \right| \right) \right\}, \quad m \geq n$$

for any $Q \in \mathfrak{M}_{cs}$. Further, let us state two hypothesis under which solvability of system (1) can be established:

(H1) The operator $h : \mathbb{R} \times cs \to cs$ defined as $(\theta, k) \rightarrowtail (hk)(\theta) = (h_1(\theta, k), h_2(\theta, k), ...)$ constitutes a family $\{(hk)(\theta)\}_{\theta\in\mathbb{R}}$ which is equi-continuous at each point in cs space, given that, each $h_n : \mathbb{R} \times \mathbb{R}^{\infty} \to \mathbb{R}$ is ω-periodic with respect to θ.

(H2) The inequality: $|h_n(\theta, k_1, k_2, \cdots)| \leq \phi_n(\theta) + \psi_n(\theta).k_n(\theta)$ holds pertaining to following conditions on $\phi_n(\theta)$ and $\psi_n(\theta)$; $\phi_n(\theta) \rightarrowtail \mathbb{R}$ and $\psi_n(\theta) \rightarrowtail \mathbb{R}$ are continuous, and $\phi(\theta)$ defined by $\phi(\theta) = \sum_{k\geq 1} \phi_k(\theta)$ for each θ, is uniformly convergent to a function that vanishes identically in $\mathbb{R}$. And the family $\{(\psi_n(\theta))\}_{n\in\mathbb{N}}$ in $\mathbb{R}$ is equibounded.

Consequent upon $\Phi := \sup\{\phi(\theta) : \theta \in \mathbb{R}\}$ and $\Psi := \sup\{\psi_n(\theta) : n \in \mathbb{N}, \theta \in \mathbb{R}\}$ exists.

In the following subsections we shall utilise hypothesis (H1)–(H2) as a conditions for solvability in each cases.

3.1 Solvability in cs for Case (i)

For investigating solvability of the solution of the system 1 for the condition $\alpha_1 \neq \alpha_2 \neq \alpha_3$ the Green's function appearing in 3 is given by [17], for $\eta \in [\theta, \theta + \omega]$

$$\mathcal{K}_1(\theta, \eta) = \frac{\exp(\alpha_1(\theta + \omega - \eta))}{(\alpha_1 - \alpha_2)(\alpha_1 - \alpha_3)(1 - \exp(\alpha_1 \omega))}$$

$$+ \frac{\exp(\alpha_2(\theta + \omega - \eta))}{(\alpha_2 - \alpha_1)(\alpha_2 - \alpha_3)(1 - \exp(\alpha_2 \omega))} + \frac{\exp(\alpha_3(\theta + \omega - \eta))}{(\alpha_3 - \alpha_1)(\alpha_3 - \alpha_2)(1 - \exp(\alpha_3 \omega))} \quad (4)$$

Further, considering $0 \leq \theta + \omega - \eta \leq \omega$ and relations between α_i, for $i = 1, 2, 3$ we get, $\sup_{\eta \in [\theta, \theta+\omega]}(\exp(\alpha_i)(\theta + \omega - \eta)) = \max\{1, \exp(\omega|\alpha_i|)\} = \exp(\omega|\alpha_i|)$, and hence we can deduce

$$\sup_{\eta \in [\theta, \theta+\omega]} |\mathcal{K}_1(\theta, \eta)| = (\text{let})\mathcal{Q}_1 = \frac{\exp(\omega|\alpha_1|)}{|(\alpha_1 - \alpha_2)(\alpha_1 - \alpha_3)(1 - \exp(\alpha_1 \omega))|}$$

$$+ \frac{\exp(\omega|\alpha_2|)}{|(\alpha_2 - \alpha_1)(\alpha_2 - \alpha_3)(1 - \exp(\alpha_2 \omega))|} + \frac{\exp(\omega|\alpha_3|)}{|(\alpha_3 - \alpha_1)(\alpha_3 - \alpha_2)(1 - \exp(\alpha_3 \omega))|} \quad (5)$$

Theorem 1 *The ISTODE of the form given in Eq. 1 satisfying (H1)-(H2) falling in case $\alpha_1 \neq \alpha_2 \neq \alpha_3$ has at least one ω-periodic solution $k(\theta) = (k_i(\theta)) \in cs$ whenever $0 < \omega \mathcal{Q}_1 \Psi < 1$, for all $\theta \in \mathbb{R}$.*

Proof From the relation (3), (4), (5) and the conditions (H2), for any $\theta \in \mathbb{R}$ we get

$$||k(\theta)||_{cs} = \sup_n \left| \sum_{i=0}^{n} \int_{\theta}^{\theta+\omega} \mathcal{K}_1(\theta, \eta) h_i(\eta, k(\eta)) d\eta \right|$$

$$\leq \sup_n \int_{\theta}^{\theta+\omega} |\mathcal{K}_1(\theta, \eta)| \sum_{i=1}^{n} (\phi_i(\eta) + \psi_i(\eta).k_i(\eta)) d\eta \leq \omega \mathcal{Q}_1(\Phi + \Psi ||k||_{cs}).$$

Hence, we get $||k||_{cs} \leq \frac{\omega \mathcal{Q}_1 \Phi}{1 - \omega \mathcal{Q}_1 \Psi} = l_0$.

Let $k^0(\theta) = (k_i^0(\theta))$ where $k_i^0(\theta) = 0$. Then a closed ball $U_0 = U(k_0, l_0)$ contains k. Now, let us consider the operator $T = (T_i)$ on $C(\mathbb{R}, U_0)$ defined as follows: For $\theta \in \mathbb{R}$,

$$(Tk)(\theta) = (T_i k)(\theta) = \left\{ \int_{\theta}^{\theta+\omega} \mathcal{K}_1(\theta, \eta) h_i(\eta, k(\eta)) d\eta \right\}, \quad (6)$$

where $k(\theta) = (k_i(\theta)) \in U_0$ and $k_i(\theta) \in C(\mathbb{R}, \mathbb{R})$, $\theta \in \mathbb{R}$. Since, $(h_i(\theta, k(\theta))) \in cs$ for each $\theta \in \mathbb{R}$, so we have

$$\sup_{n\to\infty}\left(\sum_{i=0}^{n}(T_ik)(\theta)\right) \leq \omega\mathcal{Q}_1(\Phi + \Psi||k||_{cs}) < \infty.$$

Therefore, $(Tk)(\theta) = \{(T_ik)(\theta)\} \in cs$ for all $\theta \in \mathbb{R}$. Also,

$$\begin{aligned}(T_ik)(\theta+\omega) &= \int_{\theta+\omega}^{\theta+2\omega} \mathcal{K}_1(\theta+\omega,\eta)h_i(\eta,k(\eta))d\eta \\ &= \int_{\theta}^{\theta+\omega} \mathcal{K}_1(\theta+\omega,\zeta+\omega)h_i(\zeta+\omega,k(\zeta+\omega))d\zeta \\ &= \int_{\theta}^{\theta+\omega} \mathcal{K}_1(\theta,\zeta)h_i(\zeta,k(\zeta))d\zeta = (T_ik)(\theta).\end{aligned}$$

i.e., each $(T_ik)(\theta)$ is ω-periodic whenever $k(\theta)$ is ω-periodic. Since $||(Tk)(\theta) - k^0(\theta)||_{cs} = ||(Tk)(\theta)||_{cs} \leq l$, thus T self maps on U_0. Also, by (H1), T and hence Tk is continuous. Now, to establish T as a Meir-Keeler condensing operator we do the following. Firstly, for any given $\varepsilon > 0$, we need to find $\delta > 0$ such that

$$\varepsilon \leq \chi(U_0) < \varepsilon + \delta \implies \chi(TU_0) < \varepsilon.$$

Using Eq. (4), (3), (3), we get

$$\begin{aligned}\chi(TU_0) &= \lim_{n\to\infty}\left\{\sup_{k(\theta)\in U_0}\left|\sum_{k\geq n}\int_{\theta}^{\theta+\omega}\mathcal{K}_1(\theta,\eta)h_k(\eta,k(\eta))d\eta\right|\right\} \\ &\leq \mathcal{Q}_1\lim_{n\to\infty}\left\{\sup_{k(\theta)\in U_0}\sum_{k\geq n}\int_{\theta}^{\theta+\omega}(|\phi_k(\eta)| + |\psi_k(\eta)||k_k(\eta)|)\right\} \\ &\leq \omega\mathcal{Q}_1\Psi\lim_{n\to\infty}\left\{\sup_{k(\theta)\in U_0}\sum_{k\geq n}|k_k|\right\} = \omega\mathcal{Q}_1\Psi\chi(U_0)\end{aligned}$$

Thus, we get $\chi(TU_0) < \omega\mathcal{Q}_1\Psi\chi(U_0) < \varepsilon \implies \chi(U_0) < \frac{\varepsilon}{\omega\mathcal{Q}_1\Psi}$.

Taking $\delta = \frac{(1-\omega\mathcal{Q}_1\Psi)}{\omega\mathcal{Q}_1\Psi}\varepsilon$, we get $\varepsilon \leq \chi(U_0) < \varepsilon + \delta$. Thus T is a Meir-Keeler condensing operator defined on a set $U_0 \subset cs$. T also satisfies all the hypothesis of Fixed point Theorem given by Aghajani et al. [18]. This shows that T has a fixed point in U_0, which is a solution of the system (1). □

$\mathcal{K}_1(\theta,\eta)$ may have different bounds depending upon the constraints on the value of α_i, for $i = 1, 2, 3$. as described in [17]. Then,

(E1) If $f_1 < g_1$, and either $\alpha_1 > \alpha_2 > \alpha_3 > 0$ or $\alpha_3 < \alpha_2 < 0 < \alpha_1$ holds, then $F_3 \leq \mathcal{K}_1(\theta, \eta) \leq G_3 < 0$.
(E2) If $f_2 > g_2$, and either $\alpha_3 < \alpha_2 < \alpha_1 < 0$ or $\alpha_3 < 0 < \alpha_2 < \alpha_1$ holds, then $0 < F_3 \leq \mathcal{K}_1(\theta, \eta) \leq G_3$.

We can easily refine our Theorem 1 using the above conditions (E1)–(E2) and hypothesis (H1)–(H2) by referring to the results by Chen et al. [17] as follows:

Theorem 2 *The ISTODE of the form given in equation (1) satisfying assumptions (H1)–(H2) along with hypothesis (E1) (resp. (E2)) has atleast one ω-periodic solution $k(\theta) = (k_i(\theta)) \in cs$ whenever $\omega\Psi|F_3| < 1$ (resp. $\omega\Psi G_3 < 1$), for all $\theta \in \mathbb{R}$.*

3.2 Solvability in cs for Case (ii)

For the second case where $\alpha_1 = \alpha_2 \neq \alpha_3$, Green's function $\mathcal{K}(\theta, \eta) = \mathcal{K}_2(\theta, \eta)$ (let) appearing in equation (3) for investigating solvability of the solution of the system (1) is given by [17]

$$\mathcal{K}_2(\theta, \eta) = \frac{\exp(\alpha_1(\theta + \omega - \eta))[(1 - \exp(\alpha_1\omega))((\eta - \theta)(\alpha_3 - \alpha_1) - 1) - (\alpha_3 - \alpha_1)\omega]}{(\alpha_1 - \alpha_3)^2(1 - \exp(\alpha_1\omega))^2} + \frac{\exp(\alpha_3(\theta + \omega - \eta))}{(\alpha_1 - \alpha_3)^2(1 - \exp(\alpha_3\omega))}, \quad \eta \in [\theta, \theta + \omega] \tag{7}$$

Then following conditions holds true

(E3) If $\alpha_3 < 0 < \alpha_1 = \alpha_2$, then $0 < F_4 \leq \mathcal{K}_2(\theta, \eta) \leq G_4$.
(E4) If $\alpha_1 = \alpha_2 < 0 < \alpha_3$, then $F_4 \leq \mathcal{K}_2(\theta, \eta) \leq G_4 < 0$.
(E5) If $0 < \alpha_1 = \alpha_2 < \alpha_3$, $\exp(\alpha_1\omega) < 1 + (\alpha_3 - \alpha_1)\omega$ then $F_5 \leq \mathcal{K}_2(\theta, \eta) \leq G_5 < 0$.
(E6) If $\alpha_1 = \alpha_2 < \alpha_3 < 0$, and $f_3 > 1$ then $0 < F_4 \leq \mathcal{K}_2(\theta, \eta) \leq G_4$.
(E7) If $0 < \alpha_3 < \alpha_2 = \alpha_1$, and $g_4 < 1$ then $F_6 \leq \mathcal{K}_2(\theta, \eta) \leq G_6 < 0$.

For, each notations refer to [17].
We refine our Theorem 1 for case (ii) using the above conditions (E3)–(E5) and hypothesis (H1)–(H2) by referring to the results by Chen et al. [17] as follows:

Theorem 3 *The ISTODE of the form given in Eq. (1) satisfying assumptions (H1)-(H2) along with hypothesis (E3) (resp. (E4), (E5), (E6), (E7)) has atleast one ω-periodic solution $k(\theta) = (k_i(\theta)) \in cs$ whenever $\omega\Psi G_4 < 1$ (resp. $\omega\Psi|F_4|) < 1$, $\omega\Psi G_4 < 1$, $\omega\Psi|F_6| < 1$), for all $\theta \in \mathbb{R}$.*

3.3 Solvability in cs for Case (iii)

For the third case where $\alpha_1 = \alpha_2 = \alpha_3 = \alpha$ the Green's function $\mathcal{K}(\theta, \eta) = \mathcal{K}_3(\theta, \eta)$(let) is represented by.

$$\mathcal{K}_3(\theta, \eta) = \frac{[(\eta - \theta)\exp(\alpha\omega) + \omega - \eta + \theta]^2 + \omega^2 \exp(\alpha\omega)}{2(1 - \exp(\alpha\omega))^3} \exp(\alpha(\theta + \omega - \eta)),$$
$$\eta \in [\theta, \theta + \omega] \quad (8)$$

Then

(E8) If $\alpha > 0$, then $F_7 \leq \mathcal{K}_3(\theta, \eta) \leq G_7 < 0$.
(E9) If $\alpha < 0$, then $0 < F_7 \leq \mathcal{K}_3(\theta, \eta) \leq G_7$.

Theorem 4 *The ISTODE of the form given in Eq. (1) satisfying assumptions (H1)–(H2) along with hypothesis (E8)(resp. (E9)) has atleast one ω-periodic solution $k(\theta) = (k_i(\theta)) \in cs$ whenever $\omega\Psi|F_7| < 1$(resp. $\omega\Psi G_7 < 1$), for all $\theta \in \mathbb{R}$.*

3.4 Solvability in cs for Case (iv)

For the fourth case where $\alpha_1 = u + iv$, $\alpha_2 = u - iv$, $\alpha_3 = \alpha$ the Green's function $\mathcal{K}(\theta, \eta) = \mathcal{K}_4(\theta, \eta)$ (let) is represented by.

$$\mathcal{K}_4(\theta, \eta) = \frac{\exp(u(\theta + \omega - \eta))[(u - \alpha)B_2(\theta) - vA_2(\theta)]}{v[(u - \alpha)^2 + v^2](1 + \exp(2u\omega) - 2\cos(v\omega)\exp(u\omega))} + \frac{\exp(\alpha(\theta + \omega - \eta))}{(1 - \exp(\alpha\omega))[(u - \alpha)^2 + v^2]} \quad \text{for } \eta \in [\theta, \theta + \omega] \quad (9)$$

Then following conditions holds true.

(E10) If $\alpha < 0 < u, v$ and $\frac{1+\exp(2u\omega)-2\cos(v\omega)\exp(u\omega)}{\exp(2u\omega)} > \frac{[(u-\alpha)^2+v^2](1-\exp(\alpha\omega))^2}{v^2\exp(2\alpha\omega)}$, then $0 < F_8 \leq \mathcal{K}_4(\theta, \eta) \leq G_8$.
(E11) If $u, \alpha < 0 < v$ and $1 + \exp(2u\omega) - 2\cos(v\omega)\exp(u\omega) > \frac{[(u-\alpha)^2+v^2](1-\exp(\alpha\omega))^2}{v^2\exp(2\alpha\omega)}$, then $0 < F_9 \leq \mathcal{K}_4(\theta, \eta) \leq G_9$.
(E12) If $u, v, \alpha > 0$ and $\frac{1+\exp(2u\omega)-2\cos(v\omega)\exp(u\omega)}{\exp(2u\omega)} > \frac{[(u-\alpha)^2+v^2](1-\exp(\alpha\omega))^2}{v^2}$, then $F_8 \leq \mathcal{K}_4(\theta, \eta) \leq G_8 < 0$.
(E13) If $u < 0 < v, \alpha$ and $1 + \exp(2u\omega) - 2\cos(v\omega)\exp(u\omega) > \frac{[(u-\alpha)^2+v^2](1-\exp(\alpha\omega))^2}{v^2}$, then $F_9 \leq \mathcal{K}_4(\theta, \eta) \leq G_9 < 0$.

We refine our Theorem 1 for case (iv) using the above conditions (E10)-(E13) and hypothesis (H1)-(H2) by referring to the results by Chen et al. [17] as follows:

Theorem 5 *The ISTODE of the form given in Eq. (1) satisfying assumptions (H1)–(H2) along with hypothesis (E10) (resp. (E11), (E12), (E13)) has atleast one ω-periodic solution $k(\theta) = (k_i(\theta)) \in cs$ whenever $\omega\Psi G_8 < 1$ (resp. $\omega\Psi G_9 < 1, \omega\Psi|F_8| < 1, \omega H|F_9| < 1$), for all $\theta \in \mathbb{R}$.*

4 Solvability of the Infinite System of Diiferential Equations in *bs* Space

$bs := \left\{k' = \{(k_i')\}_{i=0}^{\infty} : \sup_n \left|\sum_{i=0}^{n} k_i'\right| < \infty\right\}$. By definition bs is a sequence space of bounded series also equipped with the norm $||k'||_{bs} = \sup_n |\sum_{i=1}^{n} k_i'|$ which makes it a Banach space. Also, comparing the definition between cs and bs space it becomes clear that cs is a closed subspace of bs. For the sake of clarity we shall add superscript $*'$ for anything related to bs space that are equivalent to functions and elements related to cs space.
We again state the following two conditions so as to show the solvability of the system (1) in bs space:

(H3) The operator $h = h' : \mathbb{R} \times bs \to bs$ defined as $(\theta, k) \rightarrowtail (h'k)(\theta) = (h_1'(\theta, k), h_2'(\theta, k), \cdots)$ constitutes a family $\{(h'k)(\theta)\}_{\theta\in\mathbb{R}}$ which is equi-continuous at each point in bs space, given that, each $h_i' : \mathbb{R} \times \mathbb{R}^{\infty} \to \mathbb{R}$ is ω-periodic with respect to θ.

(H4) The inequality: $\left|h_n'(\theta, k_1, k_2, \ldots)\right| \le \phi_n'(\theta) + \psi_n'(\theta).k_n'(\theta)$ holds pertaining to following conditions on $\phi_n'(\theta)$ and $\psi_n'(\theta)$; $\phi_n'(\theta) \rightarrowtail \mathbb{R}$ and $\psi_n'(\theta) \rightarrowtail \mathbb{R}$ are continuous, and $\phi'(\theta)$ defined by $\phi'(\theta) = \sum_{k\ge1} \phi_k'(\theta)$ for each θ, is uniformly convergent to a function that vanishes identically in $\mathbb{R}$. And the family $\left\{(\psi_n'(\theta))\right\}_{n\in\mathbb{N}}$ in $\mathbb{R}$ is equibounded.

Consequent upon $\Phi' := \sup\{\phi'(\theta) : \theta \in \mathbb{R}\}$ and $\Psi' := \sup\{\psi_n'(\theta) : n \in \mathbb{N}, \theta \in \mathbb{R}\}$ exists. Do take note that the conditions for solvability in bs space is analogous to cs space. This is due to the fact that cs and bs enjoys the same norm. This fact greatly helps in shortening of proof for all the solvability cases below.

4.1 Solvability in bs for Different Cases

We shall reuse all the notations that we have used to shown solvability in cs. Hence, we will go straight to the theorem.

Theorem 6 *The ISTODE of the form given in Eq. 1 satisfying (H1)-(H2) falling in case $\alpha_1 \neq \alpha_2 \neq \alpha_3$ has at least one ω-periodic solution $k'(\theta) = (k_i'(\theta)) \in bs$ whenever $0 < \omega Q_1 \Psi < 1$, for all $\theta \in \mathbb{R}$.*

As mentioned earlier norm for both cs and bs spaces are same and hence it becomes obvious that the proof will look similar to proof in theorem (1). Only thing noteworthy here is that the set of all solutions in cs space is also a subset of the set of all solutions in bs space. $\mathcal{K}_1(\theta, \eta)$ may have different bounds depending upon the constraints on the value of α_i for $i = 1, 2, 3$ as described in Chen et al. [17]. By using those conditions and the hypothesis (H3)-(H4) we will refine our theorem for each case as follows:

Theorem 7 *The ISTODE of the form given in equation (1) satisfying assumptions (H1)-(H2) along with hypothesis (E1)(resp. (E2)) has atleast one ω-periodic solution $k'(\theta) = (k_i'(\theta)) \in bs$ whenever $\omega\Psi'|F_3| < 1$ (resp. $\omega\Psi'G_3 < 1$), for all $\theta \in \mathbb{R}$.*

Theorem 8 *The ISTODE of the form given in Eq. 1 satisfying assumptions (H1)-(H2) along with hypothesis (E3) (resp. (E4), (E5),(E6),(E7)) has atleast one ω-periodic solution $k'(\theta) = (k_i'(\theta)) \in bs$ whenever $\omega\Psi'G_4 < 1$ (resp. $\omega\Psi'|F_4|) < 1$, $\omega\Psi'G_4 < 1$, $\omega\Psi'|F_6| < 1$), for all $\theta \in \mathbb{R}$.*

Theorem 9 *The ISTODE of the form given in Eq. (1) satisfying assumptions (H1)-(H2) along with hypothesis (E8)(resp. (E9)) has atleast one ω-periodic solution $k'(\theta) = (k_i'(\theta)) \in bs$ whenever $\omega\Psi'|F_7| < 1$(resp. $\omega\Psi'G_7 < 1$), for all $\theta \in \mathbb{R}$.*

Theorem 10 *The ISTODE of the form given in Eq. (1) satisfying assumptions (H1)-(H2) along with hypothesis (E10)(resp. (E11), (E12), (E13)) has atleast one ω-periodic solution $k'(\theta) = (k_i'(\theta)) \in bs$ whenever $\omega\Psi'G_8 < 1$ (resp. $\omega\Psi'G_9 < 1, \omega\Psi'|F_8| < 1, \omega\Psi'|F_9| < 1$), for all $\theta \in \mathbb{R}$.*

5 Examples

Let us take examples and contrast the result that we got from two different spaces cs and bs space.

Example 1 Consider the differential equation for ($n \in \mathbb{N}$) and $\theta \in \mathbb{R}$:

$$\frac{d^3k_n(\theta)}{d\theta^3} - 2.1\frac{d^2k_n(\theta)}{d\theta^2} + 5.2\frac{dk_n(\theta)}{d\theta} - 0.5k_n(\theta) = \frac{\sin^n(\theta)}{(n+1)^3} + \sum_{j=n}^{\infty} \frac{\cos^n(\theta)k_j(\theta)}{(j^2+n^2)(nj+\pi^2)}, \tag{10}$$

Solution 1 With the application of theory of differential equations we can find the roots of the homogeneous equation associated with equation (10) are $\alpha_1 = 1 + 2i$, $\alpha_2 = 1 - 2i$, $\alpha_3 = 0.1$. Now we will check $h_n(\theta, k)$ of Eq. 10, Here, $h_n(\theta, k) = \frac{\sin^n(\theta)}{(n+1)^3} + \sum_{j=n}^{\infty} \frac{\cos^n(\theta)k_j(\theta)}{(j^2+n^2)(nj+\pi^2)}$. Considering $k = (k_n) \in cs$, we get

$$\sup_n \left| \sum_{j=1}^{n} h_j(\theta, k(\theta)) \right| \leq \frac{\pi^2}{6} + \frac{\pi^4}{180} ||k||_{cs}.$$

Further, check that (H1) holds by letting $\varepsilon > 0$ arbitrarily. Take $k'(\theta) = (k'_n(\theta)) \in cs$ that satisfies $||k(\theta) - k'(\theta)||_{cs} \leq \delta(\varepsilon) = 2(1+\pi^2)\varepsilon$, then

$$|h(\theta, k(\theta)) - h(\theta, k'(\theta))| \leq \frac{1}{2}(1+\pi^2)^{-1}\delta$$

This ensures the continuity in accord with (H1).

Conditions of (H2): $|h_n(\theta, k_1, k_2, \ldots)| \leq \phi_n(\theta) + \psi_n(\theta).k_n(\theta)$ can be established by taking $\phi_n(\theta) = (n+1)^{-3}$ and $\psi_n(\theta)$, the series sum from j from 1 to ∞ of the function $\frac{1}{(j^2+n^2)(nj+\pi^2)}$. Ofcourse, the series for $\psi_n(\theta)$ is convergent for each n.

$$|h_n(\theta, k(\theta))| \leq \frac{1}{(n+1)^3} + \sum_{j=n}^{\infty} \frac{|k_j(\theta)|}{(j^2+n^2)(nj+\pi^2)} \leq \phi_n(\theta) + \psi_n(\theta)|k_n(\theta)|$$

Further, series $\sum_{n\geq 1} \phi_n(\theta)$ converges to $-1+\zeta(3)$ and each $\{\psi_n(\theta)\}_n$ is independent of θ and hence the equiboundedness. Also $\Psi = \sup\{\psi_n(\theta) : n \in \mathbb{N}, \theta \in \mathbb{R}\} \approx 0.0606$ With these results both conditions (H1) and (H2) is satisfied.

Now as mentioned earlier, roots of characteristic equations associated with homogeneous equation pertaining to differential equation (10) are $\alpha_1 = 1+2i$, $\alpha_2 = 1-2i$, $\alpha_3 = 0.1$. This implies that $u, v, \alpha > 0$. Consequently, given example falls under case (iv)(given in Sect. 3.4). i.e., Green's function related to this example is given by Eq. (9). Further, since $u, v, \alpha > 0$ and $\omega = 2\pi$ satisfies all the conditions of (E12), we get $F_8 \leq \mathcal{K}_4(\theta, \eta) \leq G_8 < 0$ (for all notations ref. Section 3.4). Also, it is clear that $h_n(\theta, k)$ is $(\omega =)2\pi$-periodic with respect to θ.

Upon certain approximations we also get $\omega\Psi|F_8| \approx 0.48897 < 1$. This would mean that all the conditions of Theorem (5) is being satisfied, and hence the infinite system of differential equations (10) has at least one ω-periodic solution $k(\theta) = (x_j(\theta)) \in cs$.

Now, we are going to consider a example for bs space.

Example 2 For bs space we consider an ISTODE as follows: for ($n \in \mathbb{N}$) and $\theta \in \mathbb{R}$

$$\frac{d^3k_n(\theta)}{d\theta^3} + 4.5\frac{d^2k_n(\theta)}{d\theta^2} - 3\frac{dk_n(\theta)}{d\theta} - 2.5k_n = \frac{e^{(\sin(2\theta+1))}}{(n+1)^2} + \sin(\cos(2\theta))\sum_{j=n}^{\infty}\frac{k_j(\theta)}{j^3}. \tag{11}$$

Solution 2 In this example we have $h_n(\theta, k) = \frac{e^{(\sin(2\theta+1))}}{(n+1)^2} + \sin(\cos(2\theta))\sum_{j=n}^{\infty}\frac{k_j(\theta)}{j^3}$ Ofcourse, $k = (k_n) \in bs$ here. Now, Checking for $h_n(\theta, k)$ we get,

$$\sup_n \left|\sum_{i=1}^{n} h_i(\theta, k(\theta))\right| \leq \frac{\pi^2}{6}(r_1 + r_2||k||_{bs})$$

Here, $r_1 = \sup_\theta \left|e^{(\sin(2\theta+1))}\right| \approx 2.718$ and $r_2 = \sup_\theta |\sin(\cos(2\theta))| \approx 0.841$ are constants. One more thing to notice here is that $h_n(\theta, k(\theta))$ is π-periodic with respect to θ. Now, take $||k(\theta) - k'(\theta)||_{bs} \leq \delta(\varepsilon) = [r_2\zeta(3)]^{-1}\varepsilon$, where, both $k(\theta)$ and $k'(\theta)$ belongs to bs space, then $|h(\theta, k(\theta)) - h(\theta, k(\theta))| \leq r_2\zeta(3)\delta(\varepsilon) < \varepsilon$. Again, for $|h_n(\theta, k(\theta))|$, we get

$$|h_n(\theta, k(\theta))| \leq r_1(n+1)^{-2} + r_2 \sum_{j=n}^{\infty} \frac{1}{j^3} |k_j(\theta)| \leq \phi_n(\theta) + \psi_n(\theta)|k_n|$$

By taking $\phi_n(\theta) = r_1(n+1)^{-2}$ and family $(\psi_n(\theta))$ equibounded by $r_2\zeta(3)$ we get a series $\sum_{n\geq 1} \phi_n(\theta)$ convergent to a limit $\frac{\pi^2 r_2}{6}$ and each member of $(\psi_n(\theta))$ continuous and in fact a constant.

Summarizing all the results achieved so far we have $h_n(\theta, k(\theta)) \in bs$ a π-periodic functions with respect to first coordinate and also satisfies conditions (H3) and (H4). Now, we look upon the homogeneous equations associated with infinite system of differential equations (11). We get roots as $\alpha_1 = 1, \alpha_2 = -0.5$ and $\alpha_3 = -5$. I.e., the examples falls under case (i)(ref. Sect. 4.1). This makes clear that the Green's function of concern is given by equation (4). Also $37.206 = f_1 < g_1 = 42.079$, i.e., $\alpha_1, \alpha_2, \alpha_3$ satisfies (E1). Hence, $F_3 \leq \mathcal{K}_1(\theta, \eta) \leq G_3 < 0$. Upon certain approximation we get $\omega\Psi|F_3| = 0.9627 < 1$. All results combined together assures that all the conditions for theorem (7) are satisfied, which in turn ensures that there exist atleast one π-periodic solution $k(\theta) = (k_j(\theta)) \in bs$ for infinite system (11).

Remark Infinite systems of differential equations, system of Eq. (10) and system of Eq. (11) is interchangeable as an example for cs and bs space, if associated $k(\theta) \in$(cs or bs) in $h_n(\theta, k)$ is interchanged accordingly.

Acknowledgements Supported by TMA Pai University Research Fund.

References

1. Kuratowski, K.: Sur les espaces complets. Fundamenta mathematicae **1**(15), 301–309 (1930)
2. Darbo, G.: Punti uniti in trasformazioni a codominio non compatto. Rendiconti del Seminario matematico della Università di Padova **24**, 84–92 (1955)
3. Goldenstein, L., Markus, A.: On a measure of noncompactness of bounded sets and linear operators. Studies in Algebra and Mathematical Analysis, Kishinev, pp. 45–54 (1965)
4. Istrăţescu, V.I.: On a measure of noncompactness. Bulletin mathématique de la Société des Sciences Mathématiques de la République Socialiste de Roumanie **16**(2), 195–197 (1972)
5. Banaś, J., Goebel, K.: Measure of noncompactness in Banach space. Lecture Notes in Pure and Applied Mathematics, vol. 60 (1980)
6. Persidskii, K.: Countable systems of differential equations and stability of their solutions. Izv. Akad. Nauk Kazach. SSR **7**, 52–71 (1959)

7. Persidskii, K.: Countable systems of differential equations and stability of their solutions iii: Fundamental theorems on stability of solutions of countable many differential equations. Izv. Akad. Nauk Kazach. SSR **9**, 11–34 (1961)
8. Persidskii, K.: Infinite systems of differential equations, izdat (1976)
9. Mursaleen, M., Mohiuddine, S.: Applications of measures of noncompactness to the infinite system of differential equations in ℓ_p spaces. Nonlinear Anal.: Theory Methods Appl. **75**(4), 2111–2115 (2012)
10. Mursaleen, M., Alotaibi, A.: Infinite system of differential equations in some spaces. In: Abstract and Applied Analysis, vol. 2012. Hindawi (2012)
11. Banaś, J., Mursaleen, M.: Sequence Spaces and Measures of Noncompactness with Applications to Differential and Integral Equations. Springer (2014)
12. Arab, R., Allahyari, R., Haghighi, A.S.: Existence of solutions of infinite systems of integral equations in two variables via measure of noncompactness. Appl. Math. Comput. **246**, 283–291 (2014)
13. Aghajani, A., Pourhadi, E.: Application of measure of noncompactness to ℓ_1-solvability of infinite systems of second order differential equations. Bull. Belg. Math. Soc.-Simon Stevin **22**(1), 105–118 (2015)
14. Mursaleen, M., Rizvi, S.: Solvability of infinite systems of second order differential equations in c_0 and ℓ_1 by Meir-Keeler condensing operators. Proc. Am. Math. Soc. **144**(10), 4279–4289 (2016)
15. Srivastava, H., Das, A., Hazarika, B., Mohiuddine, S.: Existence of solutions of infinite systems of differential equations of general order with boundary conditions in the spaces c_0 and ℓ_1 via the measure of noncompactness. Math. Methods Appl. Sci. **41**(10), 3558–3569 (2018)
16. Saadati, R., Pourhadi, E., Mursaleen, M.: Solvability of infinite systems of third-order differential equations in c_0 by Meir-Keeler condensing operators. J. Fixed Point Theory Appl. **21**(2), 1–16 (2019)
17. Chen, Y., Ren, J., Siegmund, S.: Green's function for third-order differential equations. Rocky Mt. J. Math. 1417–1448 (2011)
18. Aghajani, A., Mursaleen, M., Haghighi, A.S.: Fixed point theorems for Meir-Keeler condensing operators via measure of noncompactness. Acta Math. Sci. **35**(3), 552–566 (2015)
19. Wilansky, A.: Summability Through Functional Analysis. Elsevier (2000)
20. Hazarika, B., Rabbani, M., Agarwal, R.P., Das, A., Arab, R.: Existence of solution for infinite system of nonlinear singular integral equations and semi-analytic method to solve it. Iranian J. Sci. Technol. Trans. A: Sci. **45**, 235–245 (2021)
21. Banaś, J., Woś, W.: Solvability of an infinite system of integral equations on the real half-axis. Adv. Nonlinear Anal. **10**(1), 202–216 (2021)
22. Nashine, H.K., Das, A.: Extension of Darbo's fixed point theorem via shifting distance functions and its application. Nonlinear Anal.: Model. Control **27**, 1–14 (2022)

Instabilities of Excitation Spectrum for Attractive Spin-Orbit Coupled Bose-Einstein Condensates in Quasi-one Dimension

Sonali Gangwar, R. Ravisankar, and Pankaj K. Mishra

Abstract In the paper, we present our analytical results to investigate the effect of spin-orbit (SO) and Rabi couplings on the excitation spectrum of attractive quasi-one dimensional binary Bose-Einstein condensates. We use Bogoliubov-de Gennes theory to analytically derive the spectrum for the non-interacting and interacting cases. The eigenvalues of the spectrum are used to identify the stability of the spectrum. First, we analyze the effect of attractive nonlinear interactions on the instability by fixing other coupling parameters. We obtain the appearance of multiple instability bands upon increase of intraspecies interaction. Similar observation is made as SO coupling strengths are increased for fixed Rabi coupling strength as $\Omega < \Omega_c$. For $\Omega > \Omega_c$ we have a phase transition from unstable state to stable state. While increase in Rabi coupling with fixed k_L, the multi-band instability gets transformed to single-band instability. However, the effect of variation of interspecies interaction does not yield multiple bands. Finally, we obtain a stability phase diagram of the excitation spectrum in the coupling parameters space.

Keywords Bose Einstein condensates · Bogoliubov-de Gennes theory · Spin-Orbit coupling · Instability

1 Introduction

Since its first realization in the laboratory experiment in 2011 [1], the spin-orbit (SO) coupled Bose- Einstein condensates (BECs) have triggered unprecedented growth in the research in ultracold matter [2–4]. Owing to its highly controllable nature of the experiment it has also opened new avenues for exploration for the other fields of Physics. One of the interesting question in the field is to ascertain the different kind of excitations of the condensates which characterize the overall nature and dynamical

S. Gangwar · R. Ravisankar · P. K. Mishra (✉)
Department of Physics, Indian Institute of Technology, Guwahati 781039, Assam, India
e-mail: pankaj.mishra@iitg.ac.in
URL: https://www.iitg.ac.in/pankaj.mishra/

S. Banerjee and A. Saha (eds.), *Nonlinear Dynamics and Applications*,
Springer Proceedings in Complexity,
https://doi.org/10.1007/978-3-030-99792-2_116

behaviour of the ground state of the system. The stability of the coupled BECs could be well understood by analyzing the elementary excitation spectrum. For example, the Bogoliubov-de Gennes (BdG) elementary excitation spectrum, much related to macroscopic quantum phenomena, namely, superfluidity and superconductivity, provides the fundamental information about the condensate dynamics. Excitation spectrum for the equal combination of Rashba-Dresselhaus 1D SO coupling [5, 6] in BECs gives rise roton-maxon structures theoretically [7, 8] and experimentally (see [9–11]). Effective study on SO coupling has played an important role in many exotic phenomena such as superfluidity [12], flat band structure in optical lattice potential [13] and ground state phase diagram. There are some works that indicate the presence of zero momentum phase [14] in quasi-1D SO coupled BECs. So far there are few studies of collective excitation study on SO coupled BEC in 2D [15]. However, such detailed studies are lacking for quasi-1D system. In this paper, we present the excitation spectrum of SO coupled BECs in quasi-1D.

SO coupled BECs can be described by coupled Gross-Pitaevskii equations. Using this we can find the single particle dispersion relation. It has two different distinct structures those are single minimum and double minima of parabola. We find that single minimum represents plane wave phase, while, the double minima corresponds to the stripe wave phase. These minima can be achieved with the help of free parameters namely SO and Rabi coupling strengths. Excitation spectrum is different from the single particle spectrum. We vary the contact inter- and intraspecies interactions and analyze their effect on the excitation spectrum. In this paper, we focus on solving the excitation spectrum and analyze its phase transition from stable to unstable state upon variation of several parameters.

The paper is organized as follows. In Sect. 2 we begin by describing the theoretical model, coupled Gross-Piteavskii equations. In Sect. 3 we present the analytical results of the Bogoliubov-de Gennes matrix and explain its excitation spectrum. The effect of interactions and coupling parameters on the stability of the excitation spectrum is demonstrated in Sect. 4, which is followed by the discussion on the stability phase diagram in Sect. 5. Finally, we conclude our paper in Sect. 6.

2 The Model

In this section we illustrate the mean-field model of pseudo spin-$1/2$ BECs in a quasi-one dimensional setting with strong transverse trap confinement. In experiments, the spin-orbit coupled BECs are created, for instance, by choosing two internal spin states of ^{87}Rb atoms within the $^5S_{1/2}$, $F = 1$ ground electronic manifold, which are designated as pseudo spin-up, $|\uparrow\rangle = |F = 1, m_F = 0\rangle$ and spin-down, $|\downarrow\rangle = |F = 1, m_F = -1\rangle$. Then a pair of counter propagating Raman lasers with strength Ω (Rabi coupling) is used to couple the two states. The properties of such SO coupled BECs can be described by a set of coupled Gross-Pitaevskii equations (in dimensionless form) as [16]:

$$i\frac{\partial \psi_\uparrow}{\partial t} = \left[-\frac{1}{2}\frac{\partial^2}{\partial x^2} - ik_L\frac{\partial}{\partial x} + V(x) + \alpha|\psi_\uparrow|^2 + \beta|\psi_\downarrow|^2 \right]\psi_\uparrow + \Omega\psi_\downarrow, \quad \text{(1a)}$$

$$i\frac{\partial \psi_\downarrow}{\partial t} = \left[-\frac{1}{2}\frac{\partial^2}{\partial x^2} + ik_L\frac{\partial}{\partial x} + V(x) + \beta|\psi_\uparrow|^2 + \alpha|\psi_\downarrow|^2 \right]\psi_\downarrow + \Omega\psi_\uparrow, \quad \text{(1b)}$$

Here, $\psi_\uparrow$ and $\psi_\downarrow$ are the wavefunction corresponding to the spin-up and spin-down component of the condensates, $V(x)$ is the one dimensional harmonic trapping potential, α and β are the intra- and respectively interspecies nonlinear interaction strengths respectively, k_L is spin-orbit and Ω is the Rabi coupling parameters. In the above equations (1), length is measured in units of harmonic oscillator length $a_0 = \sqrt{\hbar/(m\omega_\perp)}$, time in the units of $\omega_\perp^{-1}$, and energy in the units of $h\omega_\perp$, where $\omega_\perp$ is the transverse direction frequency of the harmonic confinement. The SO and Rabi coupling parameters have been rescaled as $k_L = k'_L/a_0\omega_\perp$ and $\Omega = \Omega'/\omega_\perp$, respectively, while the wave function is rescaled as $\psi_{\uparrow,\downarrow} = \psi_{\uparrow,\downarrow}a_0^{3/2}/\sqrt{N}$. We consider the Rabi coupling as $\Omega = |\Omega|e^{i\theta}$ that minimizes the energy when $\Omega = -|\Omega|$ for $\theta = \pi$ [16]. The wave functions are subjected to the following normalization condition,

$$\int_{-\infty}^{\infty} \left(|\psi_\uparrow|^2 + |\psi_\downarrow|^2\right) dx = 1, \quad (2)$$

3 Excitation Spectrum

The excitation spectrum provides the information about the dynamics of the condensates [15]. In this section, we present our analytical analysis of the stability of the SO coupled BECs with help of excitation spectrum. Here to understand the stability of different phases, we wish to study the excitation spectrum of plane wave solutions. This will be done by applying Bogoliubov theory. For this purpose, we assumed that homogeneous BECs with the total density of the system $n = n_\uparrow + n_\downarrow$ is conserved and the chemical potential μ is same for the both components [15]. Then the stationary state evolves as

$$\Psi_j = e^{-i\mu t}\left[\psi_j + u_j e^{i(k_x x - \omega t)} + v_j^* e^{-i(k_x x - \omega^* t)}\right], \quad (3)$$

where $\psi_j = \sqrt{n_j}e^{i\phi_j}$, $j = (\uparrow, \downarrow)$ is the ground state wave functions and u_j, v_j are the amplitudes of the two plane waves, n_j, ϕ_j are density and phase respectively of ground state wave function. The Bogoliubov coefficients u's and v's are obtained by substituting Eq. (3) in Eq. (1) and written as

$$\mathbf{L}\begin{pmatrix} u_\uparrow \\ v_\uparrow \\ u_\downarrow \\ v_\downarrow \end{pmatrix} = \hbar\omega \begin{pmatrix} u_\uparrow \\ v_\uparrow \\ u_\downarrow \\ v_\downarrow \end{pmatrix}, \tag{4}$$

where,

$$\mathbf{L} = \begin{pmatrix} f_1(n_\uparrow, n_\downarrow) & \alpha n_\uparrow & \beta\sqrt{n_\uparrow n_\downarrow} - \Omega & \beta\sqrt{n_\uparrow n_\downarrow} \\ -\alpha n_\uparrow & -f_2(n_\uparrow, n_\downarrow) & -\beta\sqrt{n_\uparrow n_\downarrow} & -\beta\sqrt{n_\uparrow n_\downarrow} + \Omega \\ \beta\sqrt{n_\uparrow n_\downarrow} - \Omega & \beta\sqrt{n_\uparrow n_\downarrow} & f_2(n_\downarrow, n_\uparrow) & \alpha n_\downarrow \\ -\beta\sqrt{n_\uparrow n_\downarrow} & -\beta\sqrt{n_\uparrow n_\downarrow} + \Omega & -\alpha n_\downarrow & -f_1(n_\downarrow, n_\uparrow) \end{pmatrix}, \tag{5}$$

where,

$$f_1(n_\uparrow, n_\downarrow) = \frac{k_x^2}{2} + k_L k_x + 2\alpha n_\uparrow + \beta n_\downarrow - \frac{1}{2}\left[\alpha n + \beta n - \frac{n\Omega}{\sqrt{n_\uparrow n_\downarrow}}\right],$$
$$f_2(n_\uparrow, n_\downarrow) = \frac{k_x^2}{2} - k_L k_x + 2\alpha n_\uparrow + \beta n_\downarrow - \frac{1}{2}\left[\alpha n + \beta n - \frac{n\Omega}{\sqrt{n_\uparrow n_\downarrow}}\right], \tag{6}$$

We consider $n_\uparrow = n_\downarrow = 1/2$. So using $n = n_\uparrow + n_\downarrow = 1$, the above quantities can be written as

$$f_1 = \frac{k_x^2}{2} + k_L k_x + \frac{\alpha}{2} + \Omega, \qquad f_2 = \frac{k_x^2}{2} - k_L k_x + \frac{\alpha}{2} + \Omega,$$

The simplified form of the Bogoliubov-de Gennes equation can be obtained by substituting det $\mathbf{L}$ equal to zero and written as

$$\omega^4 + \Lambda\omega^2 + \Delta = 0. \tag{7}$$

where,

$$\Lambda = \frac{1}{2}\Big[-k_x^4 - 4\Omega(\alpha - \beta + 2\Omega) - 2k_x^2\left(\alpha + 2\left(k_L^2 + \Omega\right)\right)\Big] \tag{8}$$

$$\Delta = \frac{1}{16}k_x^2\left(k_x^2 + 2\alpha - 2\beta - 4k_L^2 + 4\Omega\right)\left(k_x^4 + 8(\alpha+\beta)\Omega + 2k_x^2\left(\alpha + \beta - 2k_L^2 + 2\Omega\right)\right) \tag{9}$$

4 Stability Analysis of the Excitation Spectrum

The single-particle spectrum ($\alpha = \beta = 0$) generally have a positive branch (ω_+) and a negative branch (ω_-). Positive branch exhibits single minima throughout, while, the negative branch shows the transition from single minima to double minima upon the variation of spin-orbit coupling at some finite Rabi coupling term. We calculate the eigenvalue of the matrix $\mathbf{L}$ (Eq. (5)). we have two positive and two negative eigen energy spectrum.

Here, we will focus on the negative branch of the elementary excitation spectrum, which plays a vital role in the transition between the different phases of the condensates. Note that the imaginary or complex eigenenergies indicate the dynamical instability, while, negative eigenenergy of the excitation spectrum implies that the system is energetically unstable [15]. As we are interested in investigating the effect of the different interaction and coupling parameters on the dynamical instability, we will be mainly interested in looking at the nature of the negative branch of the eigenenergy. We define $G = |\Im(\omega_-)|$, where $\Im$ represents the imginary part of the variable.

4.1 *Effect of Interactions on the Stability of Excitation Spectrum*

First, we begin our discussion by analyzing the effect of the variation of the attractive nonlinear interaction on the excitation spectrum stability. In Fig. 1 we show the variation of the imaginary part of the negative branch of the eigenspectrum ($G = |\Im(\omega_-)|$) along the wavenumber by fixing the other parameters as $k_L = \Omega = 1$ and $\beta = -1$ for different attractive intraspecies (α) interactions. For lower α ($\alpha < -4$) the spectrum is stable (as $G = 0$) at $k_x = 0$, however, for $\alpha = -4$ the spectrum shows instability at $k_x = 0$. The instability of the spectrum even for $k_x = 0$ can

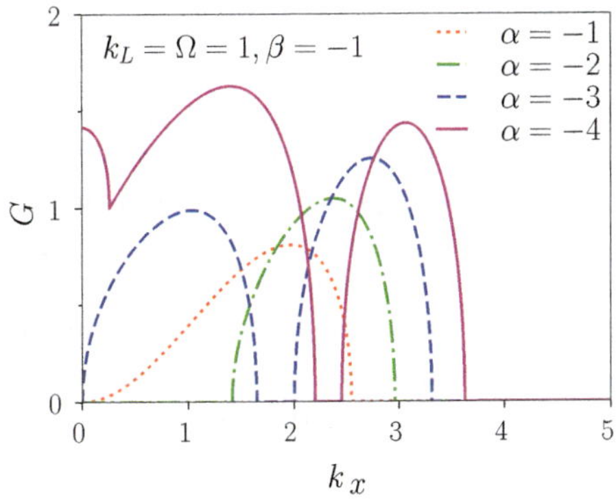

Fig. 1 Variation of the imaginary part of the eigenvalue ($G = |\Im(\omega_-)|$) with k_x for different range of α with fixed $k_L = \Omega = 1$ and $\beta = -1$. Multiple unstable bands starts appearing upon increase of attractive interaction

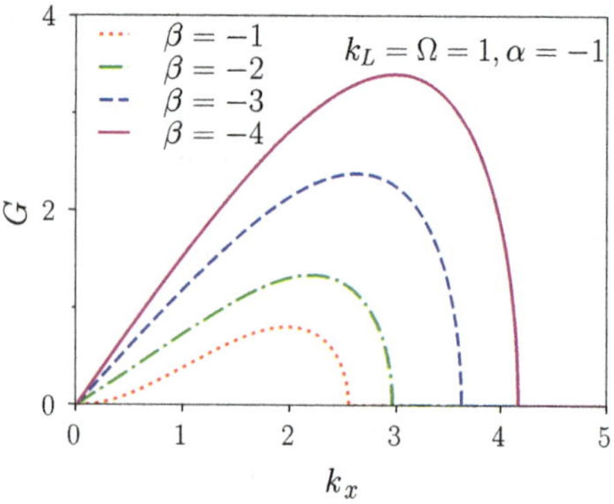

Fig. 2 Variation of the imaginary part of the eigenfrequency (G) with k_x for different range of β with fixed $k_L = \Omega = 1$ and $\alpha = -1$. The instability bands become more wider upon increase in β

be attributed to the collapsing nature of condensates at higher attractive value of α. At lower $\alpha(= -1, -2)$ the system have single instability band while for higher $\alpha(= -3, -4)$ multiple instability band seems to appear in along the wavenumber. Note that the instability band is symmetric about $k_x = 0$. However, as we analyze the effect of variation in the interspecies interaction (β) on the stability of the excitation spectrum we find that for all the values of β the system have one instability band. The wavenumber range of the instability band width and amplitude becomes wider upon increase in β for $k_L = \Omega = 1$ and $\alpha = -1$ (See Fig. 2). So we find that effect of change in α and change in β on the stability of the excitation is different.

After discussing the effect of the interactions of the stability of the excitation spectrum now in the following we will present the effect of SO and Rabi couplings on the instability band of the excitation spectrum.

4.2 *Effect of Coupling Parameters on the Stability of Excitation Spectrum*

In Fig. 3 we show the plots of the instability strength G in the wavenumber space for different $k_L(= 0, 1, 2, 4)$ by keeping the other coupling and interaction parameters fixed ($\Omega = 0.25, \alpha = \beta = -1$). We find the presence of single instability at $k_L = 0$. At finite $k_L(= 1, 2, 4)$ the spectrum shows multiple instability bands. However, as we analyze the effect of the Rabi coupling on the instability bands, we find that the multi band instability gets modified into single instability band that exhibits expansion along the wave number upon increase of Ω. They do not split into multiple instability bands (See Fig. 4). Overall we find that the effect of Rabi coupling on the instability is opposite to those due to the SO coupling.

After discussing both interactions and coupling parameters effects on the stability of the excitation spectrum, next we move to obtain a full stability phase diagrams in interactions and coupling parameters plane.

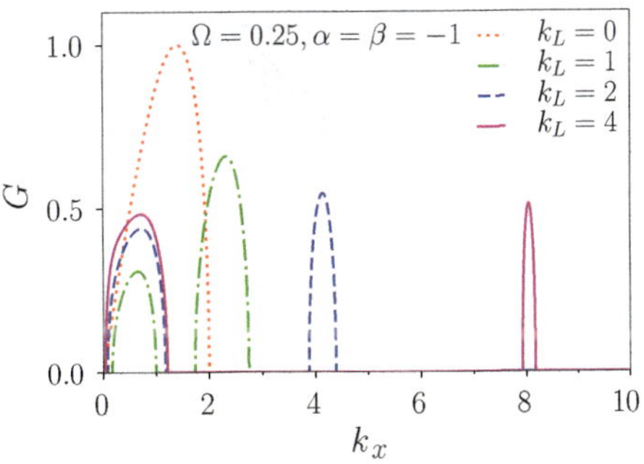

Fig. 3 Variation of the imaginary part of the eigenfrequency (G) with k_x for different range of k_L with fixed $\alpha = \beta = -1$ and $\Omega = 0.25$

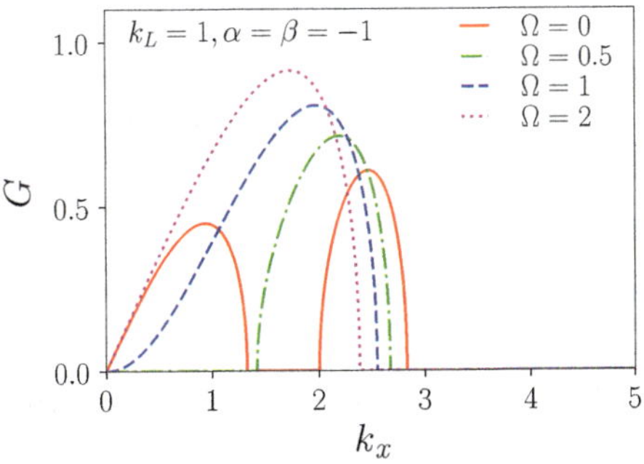

Fig. 4 Variation of the imaginary part of the eigenfrequency (G) with k_x for different range of Ω with fixed $\alpha = \beta = -1$ and $k_L = 1$

5 Stability Phase Diagrams

In order to get a complete picture of the stability of excitation spectrum for the different ranges of the coupling parameters in Fig. 5 we show the imaginary part of the eigenfrequency (G) in the $\alpha - \beta$ plane (in left panel) and in the $\Omega - k_L$ plane (right panel). For the left panel we fixed the parameters $\Omega = k_L = 1$ and for right panel $\alpha = \beta = -1$. We have chosen $k_x = 1$. We have the notion that the system in the strong attractive regime of condensates gets collapsed, however, here we find the presence of a small narrow stable regime in the $\alpha - \beta$ plane, which may be quite interesting for the experimental research. As we analyze the phase diagram in the $\Omega - k_L$ we find that at lower $k_L (\sim 0)$ the excitation spectrum exhibits instability for all the ranges of Ω. However, for $k_L \gtrsim 1$ the spectrum is unstable for small $\Omega (\sim 0)$. We find that the condensates become unstable upon increasing Ω for a fixed k_L. However, the effect of increase in k_L for a fixed Ω ($\gtrsim 0.3$) shows the transition from unstable to the stable phase. Overall, we find that the effect of k_L and Ω on the excitation spectrum is quite opposite to those observed in quasi-two dimension [15]. This is due to the SO coupling present in the same components of the quasi-1D binary BECs.

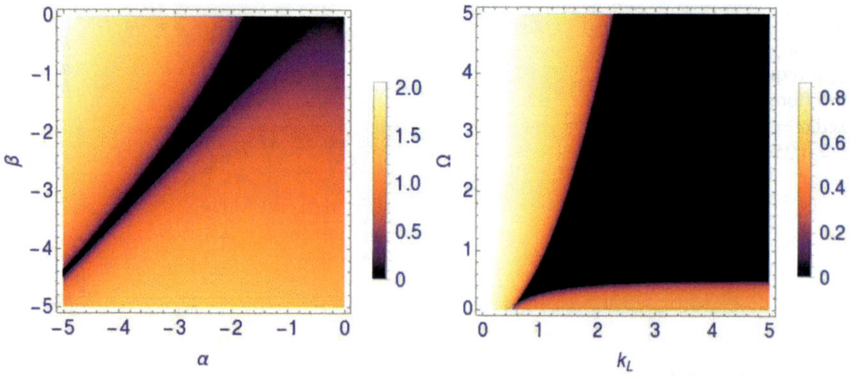

Fig. 5 Stability phase diagram of the BdG spectrum: (left) in the $\alpha - \beta$ plane the parameters are fixed as $\Omega = k_L = 1$, (right) in the Ω-k_L plane with fixed $\alpha = \beta = -1$. Note that the phase diagrams are obtained for the mode $k_x = 1$. $\alpha = \beta = -1$

6 Conclusion

In this paper, we have presented an analytical investigations of the effect of the interaction strengths and coupling parameters on the excitation spectrum of the attractive SO coupled binary BECs. We have used the Bogoliubov-de Gennes theory to analyze the excitation spectrum. The dynamical instability is characterized using the imaginary part of the negative eigenspectrum. We have obtained that the increase of intraspecies interaction leads to the formation of multiple instability bands while the increment in the interspecies interaction show the presence of only one instability band. The increase in the Rabi coupling for a fixed SO coupling (k_L) brings multi-band instability to single-band instability in the condensates. However, the effect of k_L is opposite those of Ω. Increase in k_L makes the condensates more stable when $\Omega \gtrsim 0.3$. However, for $\Omega \lesssim 0.3$ we observed multi-band instability from single-band instability upon increase in k_L and we do not have any stable state. In this paper, we have not analyzed the nature of dynamics and the eigenevector of the eigenspectrum which give the information about the detailed correlation between the spin and density of the condensate. The study is underway.

Acknowledgements Supported by DST-SERB (Department of Science and Technology - Science and Engineering Research Board) for the financial support through Project No. ECR/2017/002639.

References

1. Lin, Y.-J., Jiménez-García, K., Spielman, I.B.: Spin-orbit-coupled Bose-Einstein condensate. Nature **471**, 83 (2011). https://doi.org/10.1038/nature09887

2. Stanescu, T.D., Anderson, B., Galitski, V.: Spin-orbit coupled Bose-Einstein condensates. Phys. Rev. A **78**, 023616 (2008). https://doi.org/10.1103/PhysRevA.78.023616
3. Galitski, V., Spielman, I.B.: Spin-orbit coupling in quantum gases. Nature **494**, 49 (2013). https://doi.org/10.1038/nature11841
4. Zhai, H.: Degenerate quantum gases with spin-orbit coupling: a review. Rep. Prog. Phys. **78**, 026001 (2015). https://doi.org/10.1088/0034-4885/78/2/026001
5. Bychkov, Y.A., Rashba, E.I.: Oscillatory effects and the magnetic susceptibility of carriers in inversion layers. J. Phys. C: Solid State Phys. **17**, 6039 (1984)
6. Dresselhaus, G.: Spin-orbit coupling effects in zinc blende structures. Phys. Rev. **100**, 580 (1955). https://doi.org/10.1103/PhysRev.100.580
7. Martone, G.I., Li, Y., Pitaevskii, L.P., Stringari, S.: Anisotropic dynamics of a spin-orbit-coupled Bose-Einstein condensate. Phys. Rev. A **86**, 063621 (2012). https://doi.org/10.1103/PhysRevA.86.063621
8. Li, Y., Martone, G.I., Pitaevskii, L.P., Stringari, S.: Superstripes and the excitation spectrum of a spin-orbit-coupled Bose-Einstein condensate. Phys. Rev. Lett. **110**, 235302 (2013). https://doi.org/10.1103/PhysRevLett.110.235302
9. Khamehchi, M.A., Zhang, Y., Hamner, C., Busch, T., Engels, P.: Measurement of collective excitations in a spin-orbit-coupled Bose-Einstein condensate. Phys. Rev. A **90**, 063624 (2014). https://doi.org/10.1103/PhysRevA.90.063624
10. Ha, L.-C., Clark, L.W., Parker, C.V., Anderson, B.M., Chin, C.: Roton-maxon excitation spectrum of Bose condensates in a shaken optical lattice. Phys. Rev. Lett. **114**, 055301 (2015). https://doi.org/10.1103/PhysRevLett.114.055301
11. Ji, S.-C., Zhang, L., Xu, X.-T., Wu, Z., Deng, Y., Chen, S., Pan, J.-W.: Softening of roton and phonon modes in a Bose-Einstein condensate with spin-orbit coupling. Phys. Rev. Lett. **114**, 105301 (2015). https://doi.org/10.1103/PhysRevLett.114.105301
12. Zhu, Q., Zhang, C., Wu, B.: Exotic superfluidity in spin-orbit coupled Bose-Einstein condensates. Europhys. Lett. **100**, 50003 (2012). https://doi.org/10.1209/0295-5075/100/50003
13. Zhang, Y., Zhang, C.: Bose-Einstein condensates in spin-orbit-coupled optical lattices: flat bands and superfluidity. Phys. Rev. A **87**, 023611 (2013). https://doi.org/10.1103/PhysRevA.87.023611
14. Yu, Z.-Q.: Ground-state phase diagram and critical temperature of two-component Bose gases with Rashba spin-orbit coupling. Phys. Rev. A **87**, 051606(R) (2013). https://doi.org/10.1103/PhysRevA.87.051606
15. Ravisankar, R., Fabrelli, H., Gammal, A., Muruganandam, P., Mishra, P.K.: Effect of Rashba spin-orbit and Rabi couplings on the excitation spectrum of binary Bose-Einstein condensates. Phys. Rev. A **104**, 053315 (2021). https://doi.org/10.1103/PhysRevA.104.053315
16. Ravisankar, R., Sriraman, T., Salasnich, L., Muruganandam, P.: Quenching dynamics of the bright solitons and other localized states in spin-orbit coupled Bose-Einstein condensates. J. Phys. B: At. Mol. Opt. Phys. **53**, 195301 (2020). https://doi.org/10.1088/1361-6455/aba661

The Dynamics of COVID-19 Pandemic

Mapping First to Third Wave Transition of Covid19 Indian Data via Sigmoid Function

Supriya Mondal and Sabyasachi Ghosh

Abstract Understanding first and second wave of covid19 Indian data along with its few selective states, we have realized a transition between two Sigmoid pattern with twice larger growth parameter and maximum values of cumulative data. As a result of those transition, time duration of second wave shrink to half of that first wave with four times larger peak values. Realizing first and second wave Sigmoid pattern due to covid19 virus and its mutated variant—δ virus respectively, third wave was mapped by another Sigmoid pattern with three times larger growth parameter than that of first wave. After understanding the crossing zone among first, second and third wave curves due to covid19, δ and omicron respectively, a hidden Sigmoid pattern due to mutated $\delta+$ virus is identified in between δ and omicron. It is really interesting that entire covid19 data of India can be easily (offcourse grossly) understood by simple algebraic expressions of Sigmoid function and we can identify 4 Sigmoid patterns due to covid19 virus and its 3 dominant variants.

Keywords Covid19 · Omicron · Sigmoid function · 3rd Wave

1 Introduction

Spreading of the novel coronavirus, covid19, from China to entire globe become so alarming that the World Health Organization (WHO) declared it as a pandemic disease on 11th March 2020 [1, 2]. The data of covid infected, recovered and death are

S. Mondal (✉)
MMI College of Nursing, Pachpedi Naka, Raipur 492001, Chhattisgarh, India

VY Hospital, Adjacent to Kamal Vihar (Sector 12), New Dhamtari Road, Raipur 492001, Chhattisgarh, India
e-mail: supriyamondal.07@gmail.com

S. Ghosh
Indian Institute of Technology Bhilai, GEC Campus, Sejbahar, Raipur 492015, Chhattisgarh, India
e-mail: sabya@iitbhilai.ac.in

S. Banerjee and A. Saha (eds.), *Nonlinear Dynamics and Applications*, Springer Proceedings in Complexity,
https://doi.org/10.1007/978-3-030-99792-2_117

maintained by different countries in their government based websites. Reference [3] is citing the corresponding website for India. From 2020 to now, a huge amount of works are attempted to fit the covid infected daily cases and cumulative data for predicting the their pattern. One of the successful model is SIR model [4–6], on which a large number of works can be found. Few selective works are Refs. [7–13], which focus on SIR model application for explaining Indian covid19 data as well as other countries, e.g. Ref. [14]. SIR type interaction based alternative methodologies can be found in recent Refs. [15–17]. Since the lock down is one of the preventive measurement of this epidemic spreading, so some Refs. [9, 18–23] are focused to explore this fact mathematically. Alternative methods [24–30] like regression analysis [24], population ecology [25], machine learning [26], deep learning [27] etc. This modeling estimation helps different preventive measurement related qualitative research like Refs. [31–35]. Recently, Refs. [10, 36] work on second wave data and third wave prediction. Present work is intended towards similar aims

Though SIR model is quite well cultivated model for epidemic prediction, but a simple logistic function description like Sigmoid function [37–39] can also be an easy-dealing tools to understand the same epidemic outbreak. In our earlier works [40–42], we have used this Sigmoid function framework for predicting first wave of India data. This framework is also used for understanding epidemic size of other countries, for example Refs. [38, 39, 43].

Covid19 first wave growing is initiated in India from March, 2020 and peak was noticed in Sep, 2020. Then around Jan, 2021 the daily case data quietly went down but after that a second wave growing started, whose peak was noticed around May, 2021. It was more deadly than first wave due to mutated variant - δ virus. It also went down around June, 2021. Next, another mutated variant $\delta+$ came into the picture with a mild spreading coverage, but another mutated variant omicron create the next level rapid growth from Jan, 2022, which can be considered as actual third wave. Present work is aimed to explain the these first, second and third waves covid19 infection data due to 3 different variants with the help of three different Sigmoid functions.

The article is organized as follows. Starting with brief formalism part of Sigmoid function, we have discussed about the steps of generating curves in the Sect. 2. Next, in results section, we have described the first and second wave curves including our third wave predicted curves. After analyzing those results, at the end we summarized our work.

2 Mathematical Framework

In this framework part, we will discuss quickly about the Sigmoid function which will be used to interpret covid 19 data. Then we will discuss about the steps, through which we proceed.

The form of Sigmoid function is

$$N(t) = N_0 e^{\lambda t} / \left(\frac{N_0 e^{\lambda t}}{N_m} + 1 \right) , \quad (1)$$

where N_0 is initial number of cases, λ is growth parameter, N_m is the maximum values, where cumulative case $N(t)$ will be saturated. Here t represents number of days. Now, the time derivative of Sigmoid function is

$$\frac{dN}{dt} = \lambda N_0 e^{\lambda t} / \left(\frac{N_o e^{\lambda t}}{N_m} + 1 \right)^2 , \quad (2)$$

which is the number of new cases per day as we see in covid 19 data. Sigmoid function shows exponential behaviour in low values of t but it will saturate to a maximum values (N_{max}) at high values of t. When we analyze its derivative or slope, then we will get first increasing and then decreasing trends after showing a peak. The peak structure of daily cases depends on three parameters N_m, N_0, λ. The peak time t_p, when daily cases reach its highest value, can be expressed as

$$t_p = \frac{ln(N_m/N_0)}{\lambda} . \quad (3)$$

At $t = t_p$ daily cases and cumulative cases are respectively

$$\left(\frac{dN}{dt} \right)_p = \frac{\lambda N_m}{4} . \quad (4)$$

and

$$N_p = N_m/2 . \quad (5)$$

Above simple formalism can be useful to describe covid 19 data pattern. In India we found two waves whose cumulative and daily cases data patterns can expressed in terms of two consecutive Sigmoid functions and their time derivatives.

From the first wave of Covid 19 Indian data [3] we find out the values of t_{p1}, $(dN/dt)_{p1}$ and N_{m1}. These values are used in Eqs. (3) and (4) to find out the values of λ_1 and N_{01}. Subscript 1 is added in the notations of different parameters to assign first wave case. For India and some selective states—(1) Maharashtra (MH), (2) Kerala (KL), (3) Karnataka (KA), (4) Tamil Nadu(TN), (5) Andhrapradesh (AP), (6) West Bengal(WB), those parameters are tabulated in Table 1.

Next, when we will go for corresponding second wave data, we will not get N_{m2} as it is not till finished and we can not see the second saturated cumulative values. However, we can see the N_{p2} values from data and by using Eq. (5), we can guess N_{m2} by making twice of N_{p2}. Here, subscript 2 is added in the notations of different parameters to assign second wave case. Another important point for cumulative data of second wave is that we will redefine it by subtracting first wave maximum values

Table 1 Different parameters of Sigmoid functions, which can grossly describe covid19 first wave data of India and selective states—MH, KL, KA, TN, AP, WB

State	λ_1	N_{m1}	N_{01}	t_{p1}	$\left(\frac{dN}{dt}\right)_{p1}$
MH	0.05	19×10^5	142	190	25×10^3
KL	0.027	10×10^5	549	278	9×10^3
KA	0.048	8.6×10^5	37	209	10×10^3
TN	0.035	8×10^5	1469	148	7×10^3
AP	0.04	8.8×10^5	606	182	10×10^3
WB	0.029	5.5×10^5	522	240	4×10^3
India	0.045	94×10^5	1000	192	94×10^3

Table 2 Same as Table 1 for second wave

State	λ_2	N_{m2}	N_{02}	t_{p2}	$\left(\frac{dN}{dt}\right)_{p2}$	$N_{p2} + N_{m1}$
MH	0.05	46×10^5	350	84	67×10^3	42×10^5
KL	0.078	20×10^5	648	103	39×10^3	20×0^5
KA	0.11	16.8×0^5	43	96	49×10^3	17×10^5
TN	0.085	16×10^5	151	109	34×10^3	16×10^5
AP	0.09	10×10^5	74	106	24×10^3	14×10^5
WB	0.08	9×10^5	257	102	20×10^3	10×10^5
India	0.08	2×10^7	8579	96	4×10^5	2×10^5

N_{m1}. It means that Eqs. (1) and (2) for second waves will be

$$N_2(t) = N_{02}e^{\lambda_2 t} / \left(\frac{N_{02}e^{\lambda_2 t}}{N_{m2}} + 1 \right), \tag{6}$$

and

$$\frac{dN_2}{dt} = \lambda_2 N_{02}e^{\lambda_2 t} / \left(\frac{N_{02}e^{\lambda_2 t}}{N_{m2}} + 1 \right)^2, \tag{7}$$

where $N_{m2} = (N_{p2} - N_{m1}) \times 2$. So knowing N_{m1}, N_{p2} from data, we can guess about N_{m2}. Although, we should keep in mind that $N_{m1} + N_{m2}$ is actual saturated values of second wave case, when we compare it with actual data. The parameters of second waves for India and the selective states are given in Table 2.

3 Results

We have described the steps, through which we have find the parameters of first and second wave of covid 19 spreading. Figure 1 shows the Sigmoid function nature of cumulative N and daily case dN/dt data of India. In the left and right panels of Fig. 1 represents the data points (squares and circles) and corresponding Sigmoid fitted curves (dotted and solid lines) in first and second waves respectively. We consider 3 main data points of daily and cumulative cases at $t = (t_p - 2/\lambda), t_p, (t_p + 2/\lambda)$, within which spreading become most dominant. We have taken three different values of $\lambda = 0.04$, 0.045 and 0.05 to fit the three data points of first wave. In another aspect, Second wave is well fitted with one $\lambda = 0.08$. First waves is saturated in $N_{m1} = 10^7$ and second waves is saturated in $N_{m2} = 2 \times 10^7$ which are already seen in covid 19 data. They are implemented as important inputs to build corresponding Sigmoid curves.

Next, we will generate similar graphs in Figs. 2 and 3 for 6 selective states - MH, AP, WB, TN, KL and KA. In Fig. 2 we noticed that the daily cases data of MH, AP and TN in first wave are well fitted by (time derivative of) Sigmoid function but the same for WB, KA and KL are not so well fitted by Sigmoid function. On the other hand in second wave all those data are favouring the Sigmoid function, which can be seen in Fig. 3. In first wave there was no sharp peak for few states where as peak was clearly seen during second wave almost in every states. This is most probably because of rapidly growing of daily cases in second wave which was lacking for few states in first wave.

In first wave, we find the range of growth parameter $\lambda_1 = 0.03 - 0.05$ and peak value $\left(\frac{dN}{dt}\right)_{p1} = (0.04 - 0.25) \times 10^5$. The state level range of growth parameter is quite close to the range of $\lambda_1 = 0.04 - 0.05$ for entire country. Being added of state level peak values, we find $\left(\frac{dN}{dt}\right)_{p1} \approx 1 \times 10^5$ for India. Different states

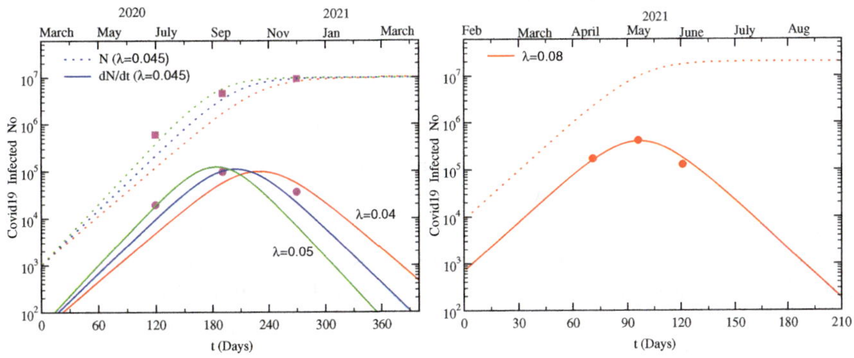

Fig. 1 Left panel: Cumulative (squares) and daily cases (circles) data, fitted in Sigmoid curves (dotted line) and their derivatives (solid lines) for first wave. Right panel: Same for second wave

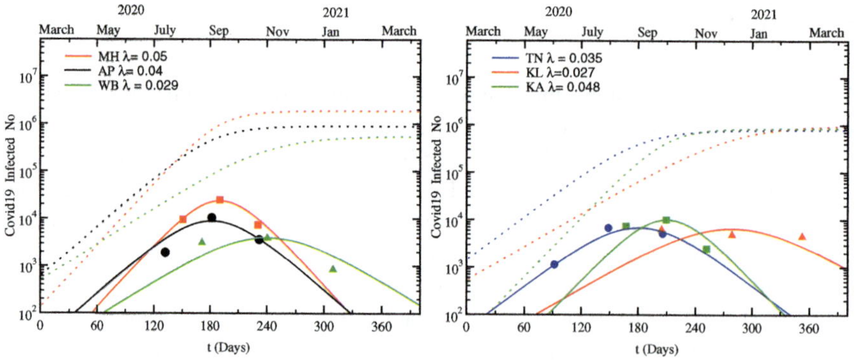

Fig. 2 Sigmoid curves (dotted lines) and their derivatives (solid lines) for first wave in MH, AP, WB, TN, KL, KA

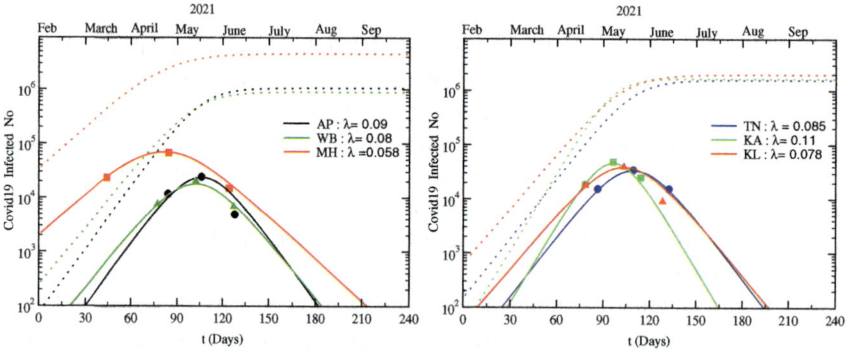

Fig. 3 Same as Fig. 2 for second wave

shown peak value at different time points t_{p1} which are in the range of $t_{p1} = 5 - 9$ months. India data shows the peak value around $t_{p1} = 6.5$ months. If we analyze second wave then state level ranges are $\lambda_2 = 0.078 - 0.1$ (excluding MH $\lambda_2 = 0.05$), $\left(\frac{dN}{dt}\right)_{p2} = (0.2 - 0.67) \times 10^5$, $t_{p2} = 2.8 - 3.5$ months and India data shows at $\lambda_2 = 0.08$, $\left(\frac{dN}{dt}\right)_{p2} \approx 4 \times 10^5$, $t_{p2} = 3.2$ months. If we compare first and second wave data of India and its different states, then we can notice their ratios as $\lambda_2/\lambda_1 \approx 2$, $t_{p2}/t_{p1} \approx 1/2$, $N_{m2}/N_{m1} \approx 2$. Although ratio between peak values of two waves for different states are not quite stable. As example, it is approximately 4 for India, 5 for WB, TN and 13 for KL etc. Considering India data as collective effect, we may conclude that first to second wave transition was just transition of parameters $\lambda_1 \rightarrow \lambda_2 = 2\lambda_1$, $t_{p1} \rightarrow t_{p2} = t_{p1}/2$, $N_{m1} \rightarrow N_{m2} = 2N_{m1}$ and $\left(\frac{dN}{dt}\right)_{p1} \rightarrow \left(\frac{dN}{dt}\right)_{p2} = 4\left(\frac{dN}{dt}\right)_{p1}$.

Now let us come to the question whether we can identify the reason for occurring second wave after the first wave? Single answer of this question is really difficult but

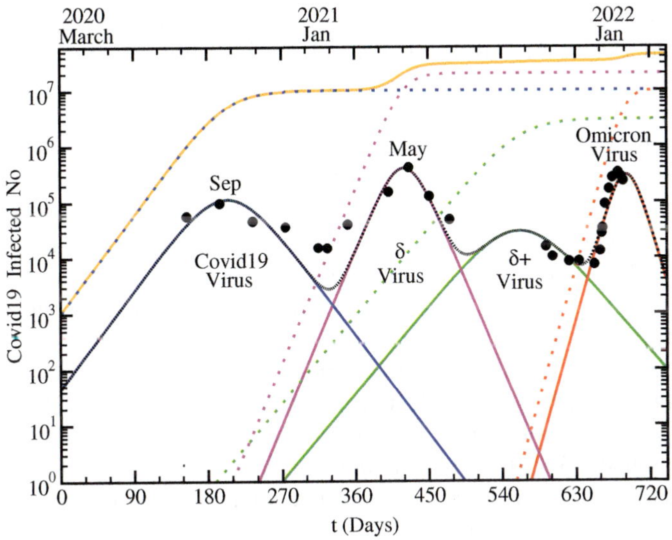

Fig. 4 Solid (daily cases) and dotted (cumulative cases) blue (covid19), magenta (δ), green (δ+) and red (omicron) lines are plotted in log scale against time (days) axis. Brown solid and black dotted lines denote the total cumulative and daily cases respectively

mutation of virus might be considered as a dominating point. Here, we will try to understand graphically but reader should considered that quantitative message with a very flexible way. In Fig. 4, we have drawn first and second wave daily cases curves in one portrait and we can see an overlapping/crossing zone of them around February, 2021. This is also observed in data (circles) as one can notice that second wave rising is started after Feb, 2021. We have put few selective daily cases data at $t_{p1} - \frac{2}{\lambda_1}, t_{p1}, t_{p1} + \frac{2}{\lambda_1}, t_{p1} + \frac{4}{\lambda_1}, t_{p1} + \frac{6}{\lambda_1}, t_{p2} - \frac{6}{\lambda_2}, t_{p2} - \frac{4}{\lambda_2}, t_{p2} - \frac{2}{\lambda_2}, t_{p2}, t_{p2} + \frac{2}{\lambda_2}$. One can notice that our Sigmoid curves for first and second waves are well fitted within $t_{p1,p2} \pm \frac{2}{\lambda_{1,2}}$. Outside those zones, the trends of data points and curves are going similar although bit of differences in numerical strengths are noticed. Now, assuming mutated virus as a dominating cause of second wave, let us try to describe the data as follows. After entering covid19 virus into India from March, 2020, first wave has received its peak around Sep, 2020 and then it went down until Feb, 2021. This imported covid19 virus are started to be mutated and among the different variant, delta virus [44] become more contagious than previous. We may consider this delta virus as a dominating reason for second wave, which is appeared to be started from Feb, 2021 and received peak around first week of May, 2021. Interestingly, first confirmed case of delta virus in India is observed around Oct, 2020, [45] which is quite earlier than Feb, 2021, from when second wave seems to grow. So it is quite interesting fact that the growing pattern of delta virus was hidden from Oct, 2020 to Feb, 2021. We can only see the decay pattern of first wave curve. If we extend our second wave Sigmoid curve before Feb, 2021, then we get $N_0 \approx 1$ around Oct, 2020. Recovering this empirical point by the simple logistic function is really very interesting fact. So if we crudely assign our first and second wave as the outcome of covid19 virus and mutated/delta virus spreading, then we can find recognize their overlapping zone.

After realizing the transition from first to second wave, we can use the idea for predicting third wave. After mutation of delta virus, recently delta plus ($\delta+$) virus is first reported in India around June, 2021. Initially it was expected to be $1.5 - 2$ times contagious [46, 47] than delta virus. On the basis of that speculation, a sudden grow was expected for $\delta+$ around mid Aug-Oct, 2021 and get a peak value within Sep-Nov, 2021. Although an opposite guess was suspected in parallel [48]. In real data, we see that $\delta+$ don't show any growing pattern within the duration June-Dec, 2021 and our hope was going toward end of the pandemic. A drastic change in data is appeared due to new variant **omicron**, which rapidly grows from first week of January, 2022 and reach its peak at the end of January, 2022 (guessed from data pattern). From June, 2021 to Dec, 2021, cumulative value was roughly saturated around 3.5×10^7 but due to omicron, at the end of Jan, 2022, when peak value in daily cases is reached, it cumulative value reaches around 4×10^7. So, $N_{p3} = (4 - 3.5) \times 10^7$ and $N_{m3} = 2N_{p3} \approx 1 \times 10^7$. Also, data show the peak value around $\left(\frac{dN}{dt}\right)_{p3} \approx 3.3 \times 10^5$, from where we can guess the $\lambda_3 \approx 0.12$ and $t_{p3} \approx 134$ days by using the relations

$$\left(\frac{dN}{dt}\right)_p = \frac{\lambda N_m}{4},$$
$$t_{p3} = \frac{ln(N_{m3}/N_{03})}{\lambda_3} \tag{8}$$

respectively. This omicron Sigmoid profiles along with those of earlier variants are sketched in single graph—Fig. 4. We are getting $t_{p3} \approx 134$ days assuming 10^3 daily cases in the beginning of Dec, 2021. Hence, our covid data with 3 waves can be expressed in terms of three Sigmoid functions for 3 variants—covid19, δ and omicron, which are nicely sketched in Fig. 4. Their parameter transformation can be expressed in a single equation:

$$\begin{aligned}
\lambda_1 \approx 0.04 &\to \lambda_2 \approx 0.08 \to \lambda_3 \approx 0.12 \\
N_{m1} \approx 1 \times 10^7 &\to N_{m2} \approx 2 \times 10^7 \to N_{m3} \approx 1 \times 10^7 \\
t_{p1} \approx 192 \text{ Days} &\to t_{p2} \approx 96 \text{ Days} \to t_{p3} \approx 134 \text{ Days} \\
\left(\frac{dN}{dt}\right)_{p1} \approx 1 \times 10^5 &\to \left(\frac{dN}{dt}\right)_{p2} \approx 4 \times 10^5 \to \left(\frac{dN}{dt}\right)_{p3} \approx 3.3 \times 10^5 .
\end{aligned} \tag{9}$$

with a hidden Sigmoid profile of $\delta+$ virus, whose parameters are roughly $\lambda_3^{\delta+} = 0.04$, $N_{m3}^{\delta+} = 0.3 \times 10^7$. We have used index 3 with $\delta+$ for assigning as pseudo 3rd wave by $\delta+$, which was actually hidden in between δ and omicron variants due to its mild properties (probably). However, reader should notice that its presence keep the daily case data as constant during June to Dec, 2021.

4 Summary

In summary, present work is intended to explain the existing first, second and third waves of covid19 infection data with the help of simple logistic function, called Sigmoid function. From the data points of peak values and peak positions for daily cases of India and its selective states MH, KL, KA, TN, WB, AP, we have found the required input parameters of the Sigmoid functions. Our results grossly indicates a transition between two Sigmoid pattern with twice larger growth parameter and maximum values of cumulative data during first to second wave transition. In parallel, time duration of second wave shrink to half of that first wave and peak values of daily cases becomes four times larger. From the basic properties of Sigmoid functions, those changes can be easily realized. Realizing these first and second waves are because of Sigmoid-type spreading of coviod19 and its mutated variant—δ virus, there was a speculation of third wave due to next level mutated variant $\delta+$ virus but it shows a mild growing Sigmoid profile, for which daily cases data from June to Dec, 2021 remain almost constant. In the beginning of January, a rapid growing of third wave profile is initiated due to another mutated variant—omicron. Our guess Sigmoid profile for third wave carry 3 times larger growth parameter than first wave but its saturate cumulative value remain almost same as that of first wave. Novelty of present work may be the finding 4 different Sigmoid profiles for 4 variants to explain roughly the actual covid19 data.

Acknowledgements SM and SG thank to their daughter Adrika Ghosh for allowing to work in home for this project.

References

1. WHO, Novel coronavirus (2019-nCoV) situation report - 11 (2020). https://apps.who.int/iris/handle/10665/330776
2. WHO, Coronavirus disease 2019 (COVID-19) situation report - 51 (2020). https://apps.who.int/iris/handle/10665/331475
3. Covid19 India. https://www.covid19india.org
4. Kermack, W.O., McKendrick, A.G.: A contribution to the mathematical theory of epidemics. Proc. R. Soc. Lond. Ser. A Contain. Papers Math. Phys. Character **115**(772), 700721 (1927). https://doi.org/10.1098/rspa.1927.0118
5. Hethcote, H.W.: The mathematics of infectious diseases. SIAM Rev. **42**(4), 599653 (2000). https://doi.org/10.1137/S0036144500371907
6. Anderson, R.M., May, R.M.: Infectious Diseases of Humans: Dynamics and Control. Oxford University Press (1992)
7. Tiwari, A.A.: Temporal analysis of covid19 peak outbreak. Epidemiol. Int. J. **4**(5), 000163, 2020. https://doi.org/10.23880/eij-16000163
8. Jakhar, M., Ahluawalia, P.K., Kumar, A.A.: COVID 19 Epidemic Forecast in Different States of India using SIR Model. https://doi.org/10.1101/2020.05.14.20101725
9. Rajesh, A., Pai, H., Roy, V., Samonta, S., Ghosh, S.: Covid 19 prediction for India from the existing data and SIR(D) model study. https://doi.org/10.1101/2020.05.05.20085902
10. Ranjan Aryan, R., Mahendra, S., Verma, K.: Characterization of the second wave of COVID-19 in India. Curr. Sci. **121**(1), 85–93 (2021). https://doi.org/10.1101/2021.04.17.21255665

11. Ranjan, R.: Temporal dynamics of COVID-19 outbreak and future projections: a data-driven approach. Trans. Indian Natl. Acad. Eng. **5**, 109–115 (2020). https://doi.org/10.1007/s41403-020-00112-y
12. Ranjan, R.: Covid-19 spread in India: dynamics, modeling, and future projections. J. Indian Stat. Assoc. **58**(2), 47–65 (2020)
13. Bagal, D.K., Rath, A., Barua, A., Patnaik, D.: Estimating the parameters of susceptible-infected-recovered model of COVID-19 cases in India during lockdown periods. Chaos Solitons Fractals **140**, 110154 (2020). https://doi.org/10.1016/j.chaos.2020.110154
14. Hussain, N., Li, B.: Using R-studio to examine the COVID-19 patients in Pakistan implementation of SIR model on cases. Int. J. Sci. Res. Multidiscip. Stud. **6**(8), 54–59, (2020). https://doi.org/10.13140/RG.2.2.32580.04482
15. Singh, R.R., Dhar, A.K., Kherani, A.A., Jacob, N.V., Misra, A., Bajpai, D.: Network based framework to compare vaccination strategies. In: International Conference on Computational Data and Social Networks, vol. 1311, pp. 218–230. Springer, Cham (2021)
16. Kherani, A.A., Kherani, N.A., Singh, R.R., Dhar, A.K., Manjunath, D.: On modeling of interaction-based spread of communicable diseases. In: International Conference on Computational Science and Its Applications, pp. 576–591. Springer, Cham
17. Kherani, N.A.: If the virus respected queues, submitted in Probability in Engineering anf Informational Science
18. Bhattacharyya, A., Bhattacharyya, D., Mukherjee, J.: The connection of growth and medication of covid-19 affected people after 30 days of lock down in India. Int. J. Sci. Res. **9**(7) (2020). https://doi.org/10.1101/2020.05.21.20107946, https://doi.org/10.36106/ijsr
19. Bhattacharyya, A., Bhowmik, D., Mukherjee, J.: Forecast and interpretation of daily affected people during 21 days lockdown due to covid19 pandamic in India. Indian J. Appl. Res. **10**(5) (2020). https://doi.org/10.36106/ijar , https://doi.org/10.1101/2020.04.22.20075572
20. Adhikari, A., Pal, A.: A six compartments with time-delay modelSHIQRD for the COVID-19 pandemic in India: during lockdown and beyond. Alex. Eng. J. **61**(2), 1403–1412 (2022). https://doi.org/10.1016/j.aej.2021.06.027
21. Pai, C., Bhaskar, A., Rawoot, V.: Investigating the dynamics of COVID-19 pandemic in India under lock-down. Chaos Solitons Fractals **138**, 109988 (2020). https://doi.org/10.1016/j.chaos.2020.109988
22. Sahoo, B.K., Sapra, B.K.: A data driven epidemic model to analyse the lockdown effect and predict thecourse of COVID-19 progress in India. Chaos Solitons Fractals **139**(110034), 19 (2020). https://doi.org/10.1016/j.chaos.2020.110034
23. Padhi, A., et al.: Studying the effect of lockdown using epidemiological modelling of COVID19 and a quantum computational approach using the Ising spin interaction. Sci. Rep. **10**, 21741 (2020). https://doi.org/10.1038/s41598-020-78652-0
24. Chauhan, P., Kumar, A., Jamdagni, P.: Regression analysis of COVID 19 Spread in India and its different states. https://doi.org/10.1101/2020.05.29.20117069
25. Biswas, D., Roy, S.: Analyzing COVID-10 Pandemic with a new growth model for population ecology. https://doi.org/10.13140/RG.2.2.34847.92324/1
26. Meghana, B.S.K., Kakulapati, V.: State-wise prevalence of covid 19 in India by machine learning approach. Int. J. Pharm. Res. **12**, 2 (2020). https://doi.org/10.31838/ijpr/2020.SP2.295
27. Arora, P., Kumar, H., Panigrahi, B.K.: Prediction and analysis of COVID-19 positive cases using deep learning models: A descriptive case study of India. Chaos Solitons Fractals **139**(110017), 18 (2020). https://doi.org/10.1016/j.chaos.2020.110017
28. Rafiq, D., Suhail, S.A., Bazaz, M.A.: Evaluation and prediction of COVID-19 in India: a case study of worst hit states. Chaos Solitons Fractals **139**(110014), 16 (2020). https://doi.org/10.1016/j.chaos.2020.110014
29. Easwaramoorthy, D., Gowrisankar, A., Manimaran, A., Nandhini, S., Lamberto, R., Santo, B.: An exploration of fractal-based prognostic model and comparative analysis for second wave of COVID-19 diffusion. Nonlinear Dyn. **106**, 1375–1395 (2021). https://doi.org/10.1007/s11071-021-06865-7

30. Gowrisankar, A., Lamberto, R., Banerjee, S.: Can India develop herd immunity against COVID-19? Eur. Phys. J. Plus **135**(6), 526 (2020). https://doi.org/10.1140/epjp/s13360-020-00531-4
31. Roy, S.: An algebraic interpretation of the spread of COVID-19 in India and an assessment of the impact of social distancing, World J. Adv. Res. Rev. **06**(3), 245–256 (2020). https://doi.org/10.30574/wjarr
32. Movsisyan, A., Burns, J., Biallas, R. et al.: Travel-related control measures to contain the COVID-19 pandemic: an evidence map. BMJ Open (2021) **11**(4) (2021). https://doi.org/10.1136/bmjopen-2020-041619
33. Kosfeld, R., et al.: The Covid 19 containment effects of public health measures a spatial difference-in-difference approach. J. Reg. Sci. **61**(4), 799–825 (2020). https://doi.org/10.1111/jors.12536
34. Kotwal, A., et al.: Predictive models of COVID19 in India: a rapid review. Med. J. Armed Forces India. **76**(4), 377–386 (2020). https://doi.org/10.1016/j.mjafi.2020.06.001
35. Bag, R., Ghosh, M., Biswas, B., Chatterjee, M.: Understanding the spatio-temporal pattern ofCOVID-19 outbreak in India using GIS and India'sresponse in managing the pandemic. Reg. Sci. Policy Pract. **12**(6), 1063–1103 (2020). https://doi.org/10.1111/rsp3.12359
36. Kavitha, C., Gowrisankar, A.1., Santo, B.: The second and third waves in India: when will the pandemic be culminated? Eur. Phys. J. Plus **136**(5), 596 (2021). https://doi.org/10.1140/epjp/s13360-021-01586-7
37. Wikipedia : Sigmoid function
38. Batista, M.: Estimation of the final size of the second phase of the coronavirus epidemic by the logistic model, medRxiv 2020.03.11.20024901, https://doi.org/10.1101/2020.03.11.20024901
39. Batista, M.: Estimation of the final size of the COVID-19 epidemic, medRxiv 2020.02.16.20023606. https://doi.org/10.1101/2020.02.16.20023606
40. Mondal, S., Ghosh, S.: Fear of exponential growth in Covid19 data of India and future sketching. Int. J. Creat. Res. Thoughts **8**(5), 2422 (2020). https://doi.org/10.1101/2020.04.09.20058933
41. Mondal, S., Ghosh, S.: Possibilities of exponential or Sigmoid growth of Covid19 data in different states of India. Indian J. Appl. Res. **10**(6), 53–56 (2020). https://doi.org/10.1101/2020.04.10.20060442
42. Mondal, S., Ghosh, S.: Searching the Sigmoid-type trend in lock down period covid19 data of India and its different states. J. Clin./Pharmaco-Epidemiol. Res. **2**(2) (2020). https://doi.org/10.46610/jcper.2020.v02i02.006
43. Merzoukia, M., Bentahirb, M., Najimia, M., Chigra, F., Gala, J.-L.: The Modeling of the capacity of the Moroccan health care system in the context of COVID-19: the relevance of the logistic approach. Bull. World Health Organ. (2020). https://doi.org/10.2471/BLT.20.258681
44. https://www.cdc.gov/coronavirus/2019-ncov/variants/variant-info.html, https://www.cdc.gov/coronavirus/2019-ncov/variants/variant-info.html
45. https://www.thehindu.com/news/international/who-says-covid-variant-in-india-of-concern/article34529654.ece, https://www.thehindu.com/news/international/who-says-covid-variant-in-india-of-concern/article34529654.ece
46. Hindustantimes News: https://www.hindustantimes.com/india-news/delta-plus-in-india-40-cases-1st-specimen-found-in-april-sample-what-we-know-so-far-101624448444003.html
47. Hindustantimes News: https://www.hindustantimes.com/india-news/govt-says-delta-plus-a-variant-of-concern-identifies-3-characteristics-101624405991131.html
48. https://www.bbc.com/news/world-asia-india-57564560, https://www.bbc.com/news/world-asia-india-57564560
49. Worldometer Coronavirus: https://www.worldometers.info/coronavirus/

Progression of COVID-19 Outbreak in India, from Pre-lockdown to Post-lockdown: A Data-Driven Statistical Analysis

Dipankar Mondal and Siddhartha P. Chakrabarty

Abstract In order to analyze the progression of COVID-19 outbreak in India, we present a data-driven analysis, by the consideration of four different metrics, namely, reproduction rate, growth rate, doubling time and death-to-recovery ratio. The incidence data of the COVID-19 (during the period of 2nd March 2020 to 11th September 2021) outbreak in India was analyzed, based on the estimation of time-dependent reproduction. The analysis suggested effectiveness of the lockdown, in arresting the growth and this continued for the post-lockdown period, except for the period of the setbacks resulting for the second wave. The approach adopted here would be useful in future monitoring of pandemics, including its progression.

Keywords Lockdown · Effective reproduction number · Estimation · COVID-19

1 Introduction

As of 31st October 2021, the coronavirus disease 2019 (COVID-19), first reported in Wuhan, China [1], has resulted in more than 246 million confirmed cases and nearly 4 million causalities [2]. The global pandemic resulting from the COVID-19 outbreak, was preceded by two other outbreaks of human coronavirus, in the 21st century itself, namely, the Severe Acute Respiratory Syndrome Coronavirus (SARS-CoV) and the Middle East Respiratory Syndrome Coronavirus (MERS-CoV) infections [3]. The index case for the COVID-19 outbreak in India was reported on 30th January 2020, in case of an individual with a travel history from Wuhan, China [4]. The data available on [4], suggests that during the early stages, the COVID-19 positive cases in India, was limited to individuals with a travel history involving the global hotspots of the outbreak. However, subsequently, the detected cases indicated the possibility of

D. Mondal · S. P. Chakrabarty (✉)
Indian Institute of Technology Guwahati, Guwahati 781039, Assam, India
e-mail: pratim@iitg.ac.in

D. Mondal
e-mail: dipankarcmi@gmail.com

S. Banerjee and A. Saha (eds.), *Nonlinear Dynamics and Applications*, Springer Proceedings in Complexity,
https://doi.org/10.1007/978-3-030-99792-2_118

community outbreak, which resulted in the Government of India announcing a lockdown across the country, driven by the necessity of crucial step towards curbing the growth of COVID-19 in densely populated country, like India. However, given the cost of the lockdown, from both the epidemiological as well as the economic perspective, the lockdown was withdrawn in a phased manner (contingent on the situation at the local level). Accordingly, this paper presents a data-driven analysis to examine the effectiveness of the lockdown, and the dynamics of post-lockdown spread of the pandemic. In order to accomplish this, we empirically analyze four different metrics, namely, reproduction number, growth rate, doubling time and death-to-recovery ratio, which quantify the transmission rate, the growth rate, the curvature of epidemic curve, and the improvement of health care capacity, respectively.

We now give a brief summary of some of the available literature on quantitative approaches to the modeling of transmission of COVID-19 outbreak. A system of ordinary differential equation (ODE) driven model for phasic transmission of COVID-19, was analyzed for calculating the transmissibility of the virus in [5]. Kucharski et al. [1] considered a stochastic transmission model on the data for cases in Wuhan, China (including cases that originated there) to estimate the likelihood of the outbreak taking place in other geographical locations. A literature survey by Liu et al. [6], summarized that the reproductive number (and hence the infectivity) in case of COVID-19, exceeded that of SARS. A Monte-Carlo simulation approach to assess the impact of the COVID-19 pandemic in India, was carried out in [7]. In a series of recent articles, the modeling of various aspects of the outbreak in India, have been studied from the perspective of fractal based prognosis assessment [8], a prediction approach for the second and third waves [9], and the vital question of achieving herd immunity in India [10].

A key identifier for the transmissibility of epidemiological diseases such as COVID-19 is the basic reproduction number R_0, which is defined as the average number of secondary infections resulting from an infected case, in a population whose all members are susceptible. However, we seek to determine the (data-driven) time-dependent reproduction number R_t, for better clarity on the time-variability of the reproduction number, particularly in the paradigm of its dynamics, both during the phases of nationwide pre-lockdown, lockdown and post-lockdown in India. In addition, we also estimate and analyze the statistical performance of growth rate, doubling time and death-to-recovery ratio. Therefore, the paper is organized as follows. In Sect. 2, we detail the source of the data as well as the statistical approaches used for the estimation of R_0 and R_t. This will be followed by the presentation of the data-driven analysis of the pre-lockdown to the post-lockdown period in Sect. 3. And finally, in the concluding remarks in Sect. 4, we highlight the main takeaways for this analysis.

2 Methodology for Estimating Reproduction Rate

The data of incidences used for the analysis reported in this paper was obtained from the website of India COVID 19 Tracker [4], and used for the purpose of estimation of R_0. This estimation was carried out making use of the `R0` package [11] of the statistical package `R`. The standardized approach included in the `R0` package, includes the implementation of the Exponential Growth (EG), Maximum Likelihood (ML) estimation, Sequential Bayesian (SB) method, and estimation of time dependent reproduction (TD) numbers, used during the H1N1 pandemic of 2009. The package is designed for the estimation of both the "initial" reproduction number, as well as the "time-dependent" reproduction number. Accordingly, we present a brief summary of the four approaches used in the paper.

1. *Exponential Growth (EG):* As observed in [12], the reproduction number can be indirectly estimated from the rate of the exponential growth. In order to address the disparity in the different differential equation models, the authors observed that this disparity can be attributed to the assumptions made about the shape of the generation interval distribution. Accordingly, the choice of the model, used for the estimation of the reproduction number, is driven by the shape of the generation interval distribution. Based on the assumption that the mean is equal to the generation intervals, the authors obtained the important result of determining an upper bound, on the possible range of values of the reproduction number, for an observed rate of exponential growth, which manifests into the worst case scenario for the reproductive number. Let the function $g(a)$ be representative of the generation interval distribution. If the moment generating function $M(z)$ of $g(a)$ is given by $M(z) = \int_0^\infty e^{za} g(a) da$, then the reproduction number is given by $R = \frac{1}{M(-r)}$, subject to the condition that $\frac{1}{M(-r)}$ exists. In particular, the Poisson distribution can be used in the analysis of the integer valued incidence data [13, 14], for (discretized) generation time distribution. An important caveat is that this approach is applicable to the time window in which the incidence data is observed to be exponential [11].
2. *Maximum Likelihood (ML) Estimation*: The maximum likelihood model as proposed in [15] is based on the availability of incidence data $N_0, N_1, \ldots, N_T$, with the notation N_t, $t = 0, 1, 2, \ldots, T$ denoting the count of new cases at time t. In practice, we take the index t in days, while noting that this indexing is applicable for other lengths of time intervals. This approach is driven by the assumption that the Poisson distribution models the number of secondary infections from an index case, with the average providing the estimate for the basic reproduction number. If we denote the number of observed incidences for consecutive time intervals by $n_1, n_2, \ldots, n_T$ and if p_i denotes the probability of the serial interval of a case in i days (which can be estimated apriori), then the likelihood function is the thinned Poisson: $L(R_0, \mathbf{p}) = \prod_{t=1}^{T} \frac{e^{-\mu_t} \mu_t^{n_t}}{n_t!}$. Note that

here $\mu_t = R_0 \sum_{i=1}^{\min(k,t)} n_{t-i} p_i$ and $\mathbf{p} = (p_1, p_2, \ldots, p_k)$. The absence of data from the index case can lead to an overestimation of the initial reproduction number, and accordingly a correction needs to be implemented [11].

3. *Sequential Bayesian (SB) Method*: A SIR model driven sequential estimation of the initial reproduction number was carried out by the sequential Bayesian method in [16]. It is based on the Poisson distribution driven estimate of incidence n_{t+1} at time $t+1$, with the mean of $n_t e^{\gamma(R-1)}$. In particular, the probability distribution for the reproduction number R, based on the observed temporal data is given by $P\left[R|n_1, n_1, \ldots, n_{t+1}\right] = \dfrac{P\left[n_1, n_1, \ldots, n_{t+1}|R\right] P[R]}{P\left[n_1, n_1, \ldots, n_{t+1}|R\right]}$, where $P[R]$ is the prior distribution of R and $P\left[n_1, n_1, \ldots, n_{t+1}\right]$ is independent of R.
4. *Estimation of Time Dependent (TD) Reproduction:* The TD method is amenable to the computation of the reproduction numbers through the averaging over all networks of transmission, based on the observed data [17]. Let i and j be two cases, with the respective times of onset of symptoms being t_i and t_j. Further, let p_{ij} denote the probability of i being infected by j. If $g(a)$ denotes the distribution of the generation interval, then $p_{ij} = \dfrac{g\left(t_i - t_j\right)}{\sum_{i \neq k} w\left(t_i - t_k\right)}$. Accordingly, the effective reproduction number is given by $R_j = \sum_i p_{ij}$, whose average is then given by $R_t = \dfrac{1}{n_t} \sum_{t_j = t} R_j$. In absence of observed secondary cases, a correction can be made to the time dependent estimation [18].

3 Data Driven Analysis of the Outbreak

The nationwide lockdown was imposed, on 25th March, 2020, with the goal of arresting the spread of infection, through strict restrictions on mass movement and encouraging social distancing, and it was expected that the spread rate would come down, along with the reduction in the possibility of community transmission. Thus, the first phase of lockdown until 14th April 2020, was extended to two more phases till 31st May, 2020. Subsequently, taking into account the economic cost of the lockdown, against the backdrop of community spread of the outbreak, a phasewise withdrawal of the lockdown was implemented, by taking into account the state of the pandemic at the local level. Accordingly, this section discusses various epidemiological aspects of the progression of the pandemic in India, from the pre-lockdown to post-lockdown, by analyzing four metrics, namely, the effective reproduction rate, the growth rate, the doubling time and the death-to-recovery ratio.

3.1 Analysis of Effective Reproduction Rate

One of the key mathematical indicator relied upon, in the paradigm of the spread of COVID-19 pandemic and consequent policy decisions is the effective reproduction rate (ERR) or the time-varying reproduction number. As ERR provides the information of time varying transmission rate, it would be a natural choice to measure its behaviour from the pre-lockdown, lockdown and the post-lockdown period. An analysis showed that the TD model fits well, for the Indian epidemic curve and accordingly, we discuss the results in the context of the TD-based R_t.

Figure 1a depicts the seven-day rolling ERR. It is clearly observed that, before the lockdown, the R_t was unsteady and high, but it started to decline after the commencement of the lockdown. In the first lockdown period, the overall average seven-day ERR was 1.64, which means that, during this period, if 100 individuals had COVID-19, they would have infected 164 people on an average. In the second lock-down period, the average ERR came down to 1.29, and then in the third lockdown period, the ERR furthers dropped to 1.21. Therefore, from the first lockdown to the end of third lockdown, the overall rate of reduction of ERR was 26%. However, after the lockdown period, the number of cases decreased gradually at first, and then increased unexpectedly. Accordingly, a sharp rise of ERR (of upto 1.14) was observed during the early March 2021 to early June 2021 (coinciding with the devastating second wave). Figure 1b displays the phase-wise[1] average R_t. The descriptive statistics of R_t and the corresponding confidence intervals are presented in Table 1. From these results, it can be observed that, there was a significant reduction of ERR, which continued to the early days after the lockdown was eased. But eventually, when after most of the restriction were lifted, a rise in ERR (consistent with the second wave) was observed, followed by a gradual decline.

3.2 Analysis of Growth Rate

The reduction of ERR should further reduce the growth rate of daily incidences. In order to see the growth rate, in a particular time period, we calculate the seven-day rolling growth rate in that period, and then take the average. Suppose that we have the daily incidence numbers, $D(t),\ t = 1, 2, 3, \ldots, 20$, for a period of 20 days. We first compute the seven-day rolling growth rates, $\dfrac{D(i+7) - D(i)}{D(i)}$, where $i = 1, 2, 3, \ldots, 13$, and we get a dataset of 13 points. Finally, the simple mean of the dataset is calculated. If the seven-day average growth is 30% in a month, then the average weekly number of positive cases would have increased from 100 to 130 in that month.

[1] L0, L1, L2 and L3 imply pre-lockdown, lockdown 1, lockdown 2 and lockdown 3, respectively. PL means post-lockdown period and each PL consists of approximately 100 days.

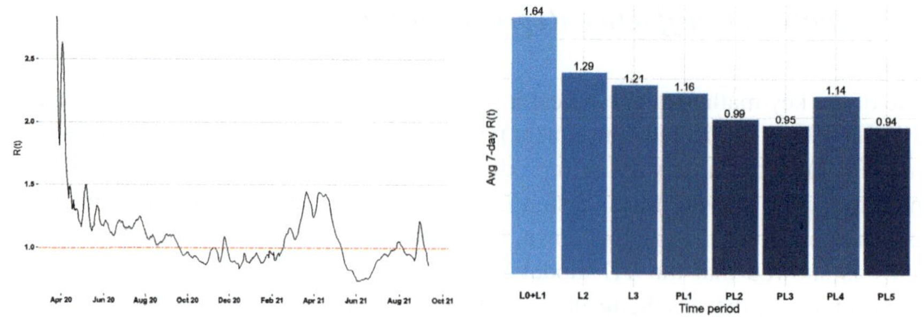

(a) R_t using TD (b) Average seven-day R_t

Fig. 1 Analysis of effective reproduction rate. **a** R_t using TD, **b** Average seven-day R_t

Table 1 Effective reproduction rate during different periods

	Period	Max	Min	Mean
1	L0 + L1	2.40	1.21	1.64
2	L2	1.43	1.16	1.29
3	L3	1.30	1.16	1.21
4	PL1	1.23	1.08	1.16
5	PL2	1.10	0.87	0.99
6	PL3	1.14	0.86	0.95
7	PL4	1.43	0.74	1.14
8	PL5	1.16	0.74	0.94

The seven-day moving average of the growth rate during the period considered is presented in Fig. 2a. Figure 2b illustrates the average weekly growth rate in different time periods. In the first lockdown period (L0 + L1), the growth rate was 97%. It means that the weekly number of positive cases, increased drastically from 100 to 197 in the pre-lockdown period. The growth rate then decreased to 42% in lockdown 2 (L2). It further reduced to 22% and 32% in lockdown 3 (L3) and post-lockdown 1 (PL1), respectively. Therefore, we can conclude, that the implementation of nationwide lockdown has resulted in slowing down the growth rate of COVID-19 positive cases. With the easing of lockdown curbs, this trend continued, but then in concurrence with the second wave there was period (PL4), when the rate again showed an upward trend.

3.3 *Analysis of Doubling Time*

One of the key indicator to observe the spread of any pandemic is the doubling time. It is referred to as the time (usually counted in number of days) it takes for the total

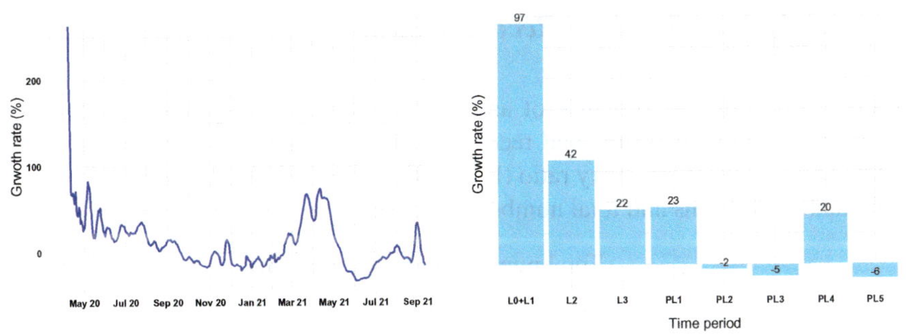

(a) 7-day moving average (b) Weekly growth rate

Fig. 2 Analysis of growth rate. **a** 7-day moving average, **b** Weekly growth rate

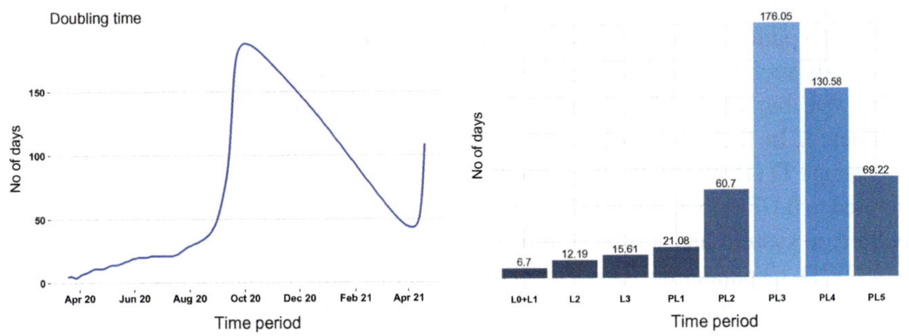

(a) Doubling time (b) Average doubling time in different periods

Fig. 3 Analysis of doubling time. **a** Doubling time, **b** Average doubling time in different periods

number of cases to double. The doubling time of n days means that if there were 100 cases at day 0, then, on day n, the number of cases would be 200. The more the doubling time is, the more the possibility of achieving a flattened epidemic curve.

Figure 3a displays the doubling rate for five-day moving averages. The escalation in doubling time is easily seen from the figure. The doubling time during the third lockdown period is about 15.6 days, up from 6.7 days at the beginning of lockdown. The phase-wise average doubling timings are shown in Fig. 3b. It further increases (upto 176) in the second post-lockdown period and then falls down to 69 at the end. The increment in doubling time is clearly visible from this figure. Therefore, from these results, we infer that the doubling time has improved significantly after the enforcement of nationwide lockdown, a trend which continued until the arrival of the second wave, when the doubling time showed a steady trend of decline.

3.4 Analysis of Death-to-Recovery Ratio

In a pandemic, the performance of any nation's health care system, is measured ultimately in terms of deaths and recoveries. This segment discuses the effect of lockdown on death-to-recovery ratio (DTR). The DTR is defined as a ratio between total number of deaths and total number of recoveries:

$$DTR_t = \frac{\text{Total number of deaths upto time } t}{\text{Total number of recoveries upto time } t}.$$

The DTR stipulates the clinical management ability or the efficiency of health system. It is highly important to keep the value of the DTR as low as possible. Mathematically, the closer this value is to zero, the better the efficiency of healthcare system, in dealing with the pandemic. For example, $DTR_t = 0.5$ implies that, for every 100 recoveries, 50 infected patients would have died. The seven-day rolling DTR is plotted in Fig. 4a. It is clearly seen that the DTR has declined significantly as time has progressed. The phase-wise bar chart (Fig. 4b) also depicts the reduction of DTR over the period considered. In the pre-lockdown (L0) and the lockdown 1 (L1) periods together, the average DTR was 0.28. It was reduced to 0.12 in lockdown 2 (L2) and further declined to 0.07 in lockdown 3 (L3), which shows that, in this short period, the Indian health care system has shown significant improvement in its preparedness to tackle the COVID-19 pandemic. This trend continued well into the post-lockdown periods, not withstanding the second wave, which is a highly encouraging sign.

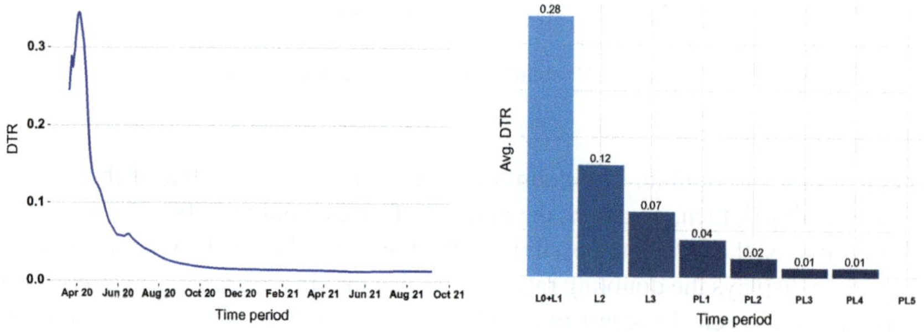

(a) Death-to-recovery ratio

(b) Average death-to-recovery ratio

Fig. 4 Analysis of death-to-recovery ratio. **a** Death-to-recovery ratio, **b** Average death-to-recovery ratio

4 Conclusion

In this paper, we have discussed the statistical analysis of the progression of COVID-19 pandemic in India, through a data-driven analysis. The goal was to ascertain the impact of the lockdown and the progression in the post-lockdown period, in terms of the intensity of the infection. Accordingly, we empirically analyzed different metrics that mainly measure the spread of infectious disease, like COVID-19. The metrics are effective reproduction rate, growth rate, doubling time and death-to-recovery ratio. For case of effective reproduction rate, it is seen that the lockdown has reduced the reproduction rate by more than 20%. The growth rate has also substantially decreased from the initial period to the end of lockdown. On the other hand, the doubling time has largely improved over the three month period. The rate of increment from pre-lockdown to lockdown 3 is nearly 123%. Finally, we described the impact on death-to-recovery ratio, which quantifies the number of death against the number of recoveries. We observed significant downfall of death-to-recovery ratio from the month of April. On average, the initial death-to-recovery of 0.28 has dipped downward to 0.08 at the third phase of lockdown. Therefore, despite rising cases of COVID-19 infection in India, the lockdown has managed to curb the spread to a great extent. Further, for the post-lockdown period, it is observed that the desirable trend for all the four metrics continued, except an adversarial trend observed, concurrently, with the devastating second wave.

Acknowledgements Siddhartha P. Chakrabarty was supported by Grant. No. MSC/2020/000049 from the Science and Engineering Research Board, India.

References

1. Kucharski, A.J., Russell, T.W., Diamond, C., Liu, Y., Edmunds, J., Funk, S., Davies, N.: Early dynamics of transmission and control of COVID-19: a mathematical modelling study. The Lancet Infectious Diseases (2020)
2. Weekly operational update on COVID-19-3 November 2021. https://www.who.int/publications/m/item/weekly-operational-update-on-covid-19---3-november-2021. Retrieved 8 November 2021
3. Gralinski, L.E., Menachery, V.D.: Return of the Coronavirus: 2019-nCoV. Viruses **12**(2), 135 (2020)
4. India COVID-19 Tracker. https://www.covid19india.org/
5. Chen, T.M., Rui, J., Wang, Q.P., Zhao, Z.Y., Cui, J.A., Yin, L.: A mathematical model for simulating the phase-based transmissibility of a novel coronavirus. Infect. Dis. Poverty **9**(1), 1–8 (2020)
6. Liu, Y., Gayle, A.A., Wilder-Smith, A., Rocklov, J.: The reproductive number of COVID-19 is higher compared to SARS coronavirus. J. Travel Med. (2020)
7. Chatterjee, K., Chatterjee, K., Kumar, A., Shankar, S.: Healthcare impact of COVID-19 epidemic in India: a stochastic mathematical model. Med. J. Armed Forces India (2020)
8. Easwaramoorthy, D., Gowrisankar, A., Manimaran, A., Nandhini, S., Rondoni, L., Banerjee, S.: An exploration of fractal-based prognostic model and comparative analysis for second wave of COVID-19 diffusion. Nonlinear Dyn. **106**(2), 1375–1395 (2021)

9. Kavitha, C., Gowrisankar, A., Banerjee, S.: The second and third waves in India: when will the pandemic be culminated? Eur. Phys. J. Plus **136**(5), 1–12 (2021)
10. Gowrisankar, A., Rondoni, L., Banerjee, S.: Can India develop herd immunity against COVID-19? Eur. Phys. J. Plus **135**(6), 526 (2020)
11. Obadia, T., Haneef, R., Boelle, P.Y.: The R0 package: a toolbox to estimate reproduction numbers for epidemic outbreaks. BMC Med. Inf. Decis. Mak. **12**(1), 147 (2012)
12. Wallinga, J., Lipsitch, M.: How generation intervals shape the relationship between growth rates and reproductive numbers. Proc. R. Soc. B: Biol. Sci. **274**(1609), 599–604 (2007)
13. Boelle, P.Y., Bernillon, P., Desenclos, J.C.: A preliminary estimation of the reproduction ratio for new influenza A (H1N1) from the outbreak in Mexico, March-April 2009. Eurosurveillance **14**(19), 19205 (2009)
14. Hens, N., Van Ranst, M., Aerts, M., Robesyn, E., Van Damme, P., Beutels, P.: Estimating the effective reproduction number for pandemic influenza from notification data made publicly available in real time: a multi-country analysis for influenza A/H1N1v 2009. Vaccine **29**(5), 896–904 (2011)
15. Forsberg White, L., Pagano, M.: A likelihood-based method for real-time estimation of the serial interval and reproductive number of an epidemic. Stat. Med. **27**(16), 2999–3016 (2008)
16. Bettencourt, L.M., Ribeiro, R.M.: Real time Bayesian estimation of the epidemic potential of emerging infectious diseases. PLoS One 3(5) (2008)
17. Wallinga, J., Teunis, P.: Different epidemic curves for severe acute respiratory syndrome reveal similar impacts of control measures. Am. J. Epidemiol. **160**(6), 509–516 (2004)
18. Cauchemez, S., Boelle, P.Y., Donnelly, C.A., Ferguson, N.M., Thomas, G., Leung, G.M., Hedley, A.J., Anderson, R.M., Valleron, A.J.: Real-time estimates in early detection of SARS. Emerg. Infect. Dis. **12**(1), 110 (2006)

Analysis of Fuzzy Dynamics of SEIR COVID-19 Disease Model

B. S. N. Murthy, M N Srinivas, and M A S Srinivas

Abstract The objective of this article is to build an SEIR epidemic system for episode COVID-19 (novel crown) with fuzzy numbers. Mathematical models might assist with investigating the transmission elements, forecast and control of Covid-19. The fuzziness in the infection rate, increased death owing to COVID-19, and recovery rate from COVID-19 were all deemed fuzzy sets, and their member functions were used as fuzzy parameters in the SEIR system. The age lattice technique is used in the SEIR system to calculate the fuzzy basic reproduction number and the system's stability at infection-free and endemic equilibrium points. Computer simulations are provided to comprehend the subtleties of the proposed SEIR COVID-19 model.

Keywords SEIR · Fuzzy parameter · Fuzzy basic reproduction number · Transmission rate · Stability

1 Introduction

COVID-19 infection has turned into a worldwide irresistible disease and more individuals are influenced. It is spread by a COVID-19-infected person through direct interaction with another individual or through minute precipitations from a COVID-19-infected individual's mouth that are moved by another person. Almost 2.27 mil-

B. S. N. Murthy (✉)
Department of Mathematics, Aditya College of Engineering and Technology, Surampalem, Andhra Pradesh, India
e-mail: bsn3213@gmail.com

M. N. Srinivas
Department of Mathematics, School of Advanced Sciences, Vellore Institute of Technology, Vellore, Tamilnadu, India
e-mail: mnsrinivaselr@gmail.com

M. A. S. Srinivas
Department of Mathematics, Jawaharlal Nehru Technological University, Hyderabad, Telangana, India
e-mail: massrinivas@gmail.com

S. Banerjee and A. Saha (eds.), *Nonlinear Dynamics and Applications*, Springer Proceedings in Complexity,
https://doi.org/10.1007/978-3-030-99792-2_119

lion individuals have been diseased with the virus around the biosphere, with 0.46 million of them succumbing to it in approximately 193 nations. The best and troublesome assignment for people is to control a similar climate which they have occupied. For this reason, notwithstanding, a few rules have been given or provided and cut off points have been static that past that landscape/climate shouldn't be upset. The execution of strategies to stop the communication of the infection has been viewed as a significant test. Consequently, as we realize that numerical models are useful assets to comprehend the communication elements of irresistible infections and to kind future arranging. Various universal and virus models have been examined in previous writing, allowing us to better examine and manage the feast of transferrable diseases [1–3].

Epidemiology, the SIR model is notable, and numerous extraordinary accomplishments have been complete on that [4–6]. However, the COVID-19 disease exists an inert stage throughout which the individual is contaminated however not so far suggestive. Another portion E is acquainted in the SEIR model with portray the non-suggestive yet contaminated stage. The SEIR model has numerous renditions, and numerical medicines can be found, for example, in Diekmann [7], Hethcote [8], among others. The scientific arrangement of the SEIR model is read for the unrestricted feast of the COVID-19 contagion in [9]. A multi draining SEIR pestilence system is set up for the COVID-19 contagion with universal prevalence charges in [10]. The objective of this analysis was to create an altered SEIR compartmental numerical classical for forecast of novel corona pandemic elements considering dissimilar intercession situations which may stretch experiences on the finest mediations to lessen the plague hazard. The COVID-19 model introduced in this paper depends on a past pandemic model announced in [11–13]. Mwalili et al. [14] investigated on SEIR pandemic system for COVID-19 contamination where the resistance of the populace assumes a significant part in recovering from hatching stage to the defenseless compartment. Shikha et al. [15] concentrated on the impact of regular resistance in the SEIS pestilence system. The above examinations has made some incredible accomplishments in exploring the COVID-19 disease.

The parameters used to communicate natural systems in a numerical manner are for the most part taken as certain. A couple of endeavours have been finished by the analysts to consider the natural vulnerability in their exploration works. Because of the reasonable circumstance, fuzzy parameters in natural demonstrating ought to be utilized more every now and again than the new endeavors. The usage of fuzzy sense and fuzzy sets in organic classifications has a lot of probables, but there aren't many of them [16–19] contains a few examples of fuzzy science applications. Mishra and Pandey [20] devised and implemented a plague model with fuzzy parameters to a PC network system. A few researchers [21–23] were concentrated on the fuzzy scourge models for human irresistible disease. As of late, fuzzy hypothesis stands out enough to be noticed. Therefore, the vulnerability in this SEIR system is portrayed by fuzzy statistics.

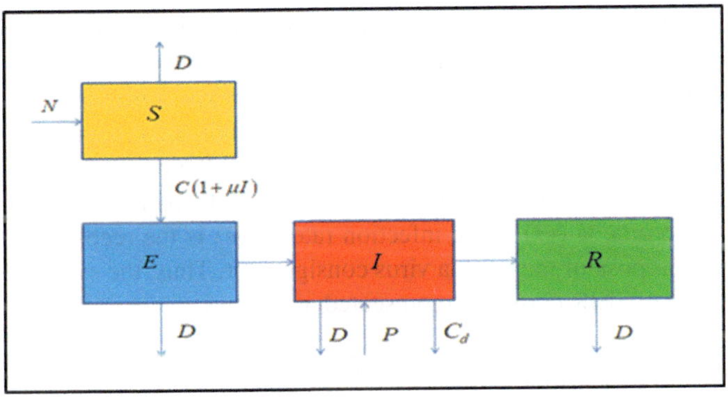

Fig. 1 A Schematic diagram of SEIR epidemic disease COVID-19

In this article, we construct a SEIR mathematical model in regularized system in Sect. 2, we try to investigate the spread of the COVID-19 virus from a fuzzy powerful structure perspective. Because of COVID-19, the infection rate, recovery rate, and death rate due to Covid-19 are all viewed as fuzzy parameters that are based on specific infection loads. In Sect. 3, we derive fuzzy basic reproduction number and existence conditions of fuzzy SEIR were discussed. In Sect. 4, we intended the stability of SEIR fuzzy system at these equilibrium points. In Sect. 5, numerical recreations have been accessible to exemplify the investigative results. Finally, a brief conversation and inferences have been assumed in Sect. 6.

2 Mathematical Formulation of Proposed SEIR Model

In the present article, we recommend a SEIR Covid-19 disease system to describe the following differential equations

$$\begin{cases} \frac{dS}{dt} = N - CIS(1 + \mu I) - DS \\ \frac{dE}{dt} = CIS(1 + \mu I) - (D + \psi)E \\ \frac{dI}{dt} = P + \psi E - (D + C_d + \delta)I \\ \frac{dR}{dt} = \delta I - DR \end{cases} \tag{1}$$

where $\Delta = N - P$is the total population. Let S is proportion of susceptible class, E is the proportion of exposed class, I is the proportion of infected class, R is the proportion of recovered class, Let N, P is the total population, who test is negative and positive; C is the proportionality constant; D is the natural death rate of individual class; μ is the individual lose of immunity. Figure 1 depicts transmission flow of SEIR epidemic COVID-19 disease model.

Assuming that the individuals are infected according to their Covid-19 virus load. So, in this manner the infection rate, death rate due to virus and recovery rate has been considered as fuzzy sets which rest on the corona virus consignment. Therefore, the Covid-19 virus consignment in an individual, the advanced the fortuitous of the Covid-19 virus spread in a contact interface. Let Θ is the corona virus load class. By considering the corona virus load Θ in each class, the parameters $C_d(\Theta)$ is the death due to corona, $\psi(\Theta)$ is the infection rate, $\delta(\Theta)$ is the recovery rate can be regarded as a purpose of the corona virus consignment. Thus, the system (1) can be stretched to fuzzy SEIR classical, represented as follows

$$\begin{cases} \frac{dS}{dt} = N - CIS(1+\mu I) - DS \\ \frac{dE}{dt} = CIS(1+\mu I) - (D+\psi(\Theta))E \\ \frac{dI}{dt} = P + \psi(\Theta)E - (D + C_d(\Theta) + \delta(\Theta))I \\ \frac{dR}{dt} = \delta(\Theta)I - DR \end{cases} \tag{2}$$

3 Analysis of Fuzzy SEIR Covid-19 Model

Only two levels of headings should be numbered. Lower level headings remain unnumbered; they are formatted as run-in headings.

3.1 Fuzzy Basic Reproduction Number

The fuzzy basic reproduction number $R_0(\Theta)$for system (2), is calculated using the next generation matrix technique [21].

$$R_0(\Theta) = \frac{C\psi(\Theta)N}{D^2(C_d(\Theta) + D + \delta(\Theta))}$$

3.2 Existence of Equilibrium Points

The system (2) has two points of equilibrium, such as (i) $E_{df}(S_0, 0, 0, 0)$(Infection free equilibrium point), (ii) $E_{ee}(S_*, E_*, I_*, R_*)$ (Endemic equilibrium point)

(i) Infection free equilibrium point $E_{df}(S_0, 0, 0, 0)$

The disease-free equilibrium point of system (2) is found by putting $I = I_0 = 0$. As a result, the SEIR fuzzy system (2) has a disease-free equilibrium point $E_{df}\left(\frac{N}{D}, 0, 0, 0\right)$

(ii) **Endemic equilibrium point** $E_{ee}(S_*, E_*, I_*, R_*)$
The possibility of spread of COVID-19 at the endemic equilibrium point $E_{ee}(S_*, E_*, I_*, R_*)$
where $S_* = \frac{N}{C(1+\mu I_*)I_* - D}; E_* = \frac{NC(1+\mu I_*)I_*}{[C(1+\mu I_*)I_* - D][\psi(\Theta)+D]}; R_* = \frac{\delta(\Theta)}{D} I_*$

$$I_* - -\frac{(D + C_d(\Theta) + \delta(\Theta) + P(\Theta)(\psi(\Theta) + D) + \psi(\Theta) NC) + \sqrt{\Phi}}{2N(\psi(\Theta) + D)(P(\Theta) + 1)};$$

where
$$\Phi = [D + C_d(\Theta) + P(\psi(\Theta) + D) + \psi(\Theta) NC]^2$$
$$-4\mu(\psi(\Theta) + D)(P + 1)\begin{pmatrix} C_d(\Theta) + \delta(\Theta) - D- \\ P(\psi(\Theta) + D) D \end{pmatrix}$$

4 Stability Analysis

Theorem 1 *At the disease-free equilibrium point $E_{df}(S_0, 0, 0, 0)$, system (2) is asymptotically stable locally, if $(C_d(\Theta) + \delta(\Theta) + D) D^2 - CN\psi(\Theta) > 0$ i.e., $R_0 < 1$ then the $E_{df}(S_0, 0, 0, 0)$ and if $(C_d(\Theta) + \delta(\Theta) + D) D^2 - CN\psi(\Theta) < 0$ i.e., $R_0 > 1$, then the disease free equilibrium point $E_{df}(S_0, 0, 0, 0)$ of the system (2) is unstable.*

Proof The Jacobian matrix (J) is written as follows based on system (2)

$$J = \begin{bmatrix} -CI(1+\mu I) - D & 0 & -CS - 2\mu CSI & 0 \\ CI(1+\mu I) & -D & CS + 2\mu CSI & 0 \\ 0 & \psi(\Theta) & -C_d(\Theta) - \delta(\Theta) - D & 0 \\ 0 & 0 & \delta(\Theta) & -D \end{bmatrix}$$

At infection free equilibrium point $E_{df}(S_0, 0, 0, 0)$, the characteristic equation is in the form of

$$(D + \lambda)^2 \left(\lambda^2 + \lambda(C_d(\Theta) + \delta(\Theta) + 2D) + (C_d(\Theta) + \delta(\Theta) + D) D - \frac{NC}{D}\psi(\Theta)\right) = 0 \tag{3}$$

Obviously, all the roots of an equation (3) are negative if $(C_d(\Theta) + \delta(\Theta) + D) D^2 - CN\psi(\Theta) > 0$
(i.e., if $R_0(\Theta) < 1$). As a result, at the disease-free equilibrium point $E_{df}(S_0, 0, 0, 0)$, system (2) is locally asymptotically stable if $\frac{C\psi(\Theta)N}{D^2(C_d(\Theta)+D+\delta(\Theta))} < 1$. Or else, the system (2) is unstable.

Theorem 2 *At the endemic equilibrium point* $E_{ee}(S_*, E_*, I_*, R_*)$*, system (2) is asymptotically stable locally if* $\frac{C\psi(\Theta)N}{D^2(C_d(\Theta)+D+\delta(\Theta))} > 1$*(i.e.,* $R_0(\Theta) > 1$*)*

Proof At endemic equilibrium point $E_{ee}(S_*, E_*, I_*, R_*)$, the characteristic equation is

$$\lambda^4 + \zeta_1\lambda^3 + \zeta_2\lambda^2 + \zeta_3\lambda + \zeta_4 = 0 \quad (4)$$

Here

$\zeta_1 = -\left(\sigma_{11} + \sigma_{22} + \sigma_{33}\right)$; $\zeta_2 = \sigma_{11}\sigma_{22} + \sigma_{11}\sigma_{33} + \sigma_{11}\sigma_{44} + \sigma_{22}\sigma_{33} + \sigma_{22}\sigma_{44} + \sigma_{33}\sigma_{44} - \sigma_{23}\sigma_{32}$; $\zeta_3 = -\sigma_{11}\sigma_{22}\sigma_{33} - \sigma_{11}\sigma_{22}\sigma_{44} - \sigma_{11}\sigma_{33}\sigma_{44} - \sigma_{22}\sigma_{33}\sigma_{44} - \sigma_{13}\sigma_{21}\sigma_{32} + \sigma_{11}\sigma_{23}\sigma_{32} + \sigma_{23}\sigma_{32}\sigma_{44}$; $\zeta_4 = \sigma_{11}\sigma_{22}\sigma_{33}\sigma_{44} + \sigma_{13}\sigma_{21}\sigma_{32}\sigma_{44} - \sigma_{11}\sigma_{23}\sigma_{32}\sigma_{44}$ where $\sigma_{11} = -CI_*(1+\mu I_*)$; $\sigma_{13} = -CS_*(1+\mu I_*)$; $\sigma_{21} = CI_*(1+2\mu I_*)$; $\sigma_{22} = -D - \psi(\Theta)$; $\sigma_{23} = CS_*(1+2\mu I_*)$; $\sigma_{32} = \psi(\Theta)$; $\sigma_{33} = -D - \psi(\Theta) - \delta(\Theta)$; $\sigma_{43} = \delta(\Theta)$; $\sigma_{44} = -D$. Clearly it is evident that $\zeta_1 > 0$, $\xi_3 > 0$, $\xi_4 > 0$, $\zeta_1\zeta_2 - \zeta_3 > 0$ and $\zeta_1\zeta_2\zeta_3 - \zeta_3^2 - \zeta_1^2\zeta_4 > 0$ if $NC\psi(\Theta) - (C_d(\Theta) + \delta(\Theta) + D)D^2 < 0$ (i.e., $R_0(\Theta) > 1$). The roots of an equation (4) have negative roots or negative real parts as per Routh-Hurwitz criteria [24]. Hence the system (2) is locally asymptotically stable at $E_{ee}(S_*, E_*, I_*, R_*)$

5 Numerical Simulation

The parameters that define the rates at which individuals progress from one stage to the next, such as infection rate, death rate owing to corona virus, and recovery rate, are directly dependent on the numerical analysis of the fuzzy SEIR system. The novel corona eruption in the inhabitants will never go away (i.e. $R_0(\Theta) < 1$), but it will be smaller than those who have been infected with COVID-19. The numerical simulation of Covid-19 model usages statistics on the number of Covid-19 cases in India. The initial population $(S_0, E_0, I_0, R_0) = (1405, 821, 0.034, 0.008609)$ (in millions) and the corresponding parameters in the system (2) are $N = 0.76$; $D = 0.02$; $C = 0.08960$; $P = 0.07112$; $\mu = 0.00009$; $C_d = 0.0004 - 0.0009$; $\Psi = 0.00002 - 0.00004$; $\delta = 0.01 - 0.05$

The following are the observations from the above figures: Fig. 2a shows that the time series evaluation of susceptible, exposed, infected, recovered population. Figure 3a represent the variations in susceptible populations of india along with time for various values of δ. we conclude that variation in susceptible population is more as δvaries and which says that δ plays a significant role on susceptible population of India. Figure 3b shows the variations in exposed populations of India for various values of δ. We can conclude that variation in exposed population of India is more as δ varies. So δ variations effect the exposed of India. Figure 3c represents the infected population of variations along time for India. From this figure we conclude that variations in infected populations in India is less as δ varies. So δ plays a significant role as dynamic sensitive parameter on the infected populations of India. Figure 3d represents the recovered populations of India for various values of δ. we conclude

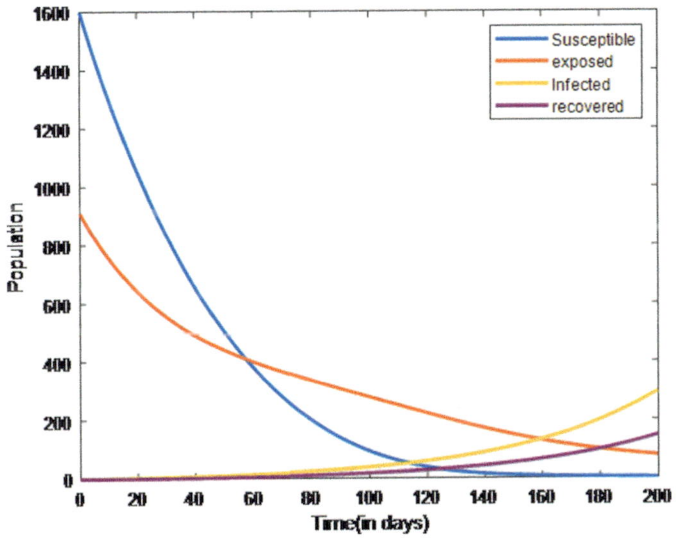

Fig. 2 Time series evaluation of population w.r.t the above set of parameter values

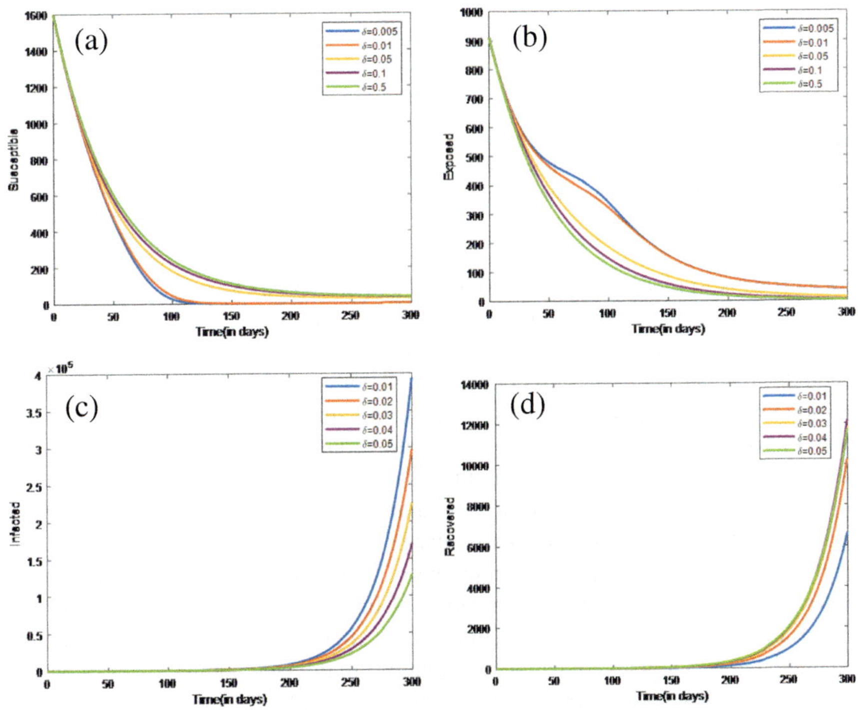

Fig. 3 a–d Time series evaluation of susceptible, exposed infected and recovered population of India where delta is varying with initial values from the above set of parameter values

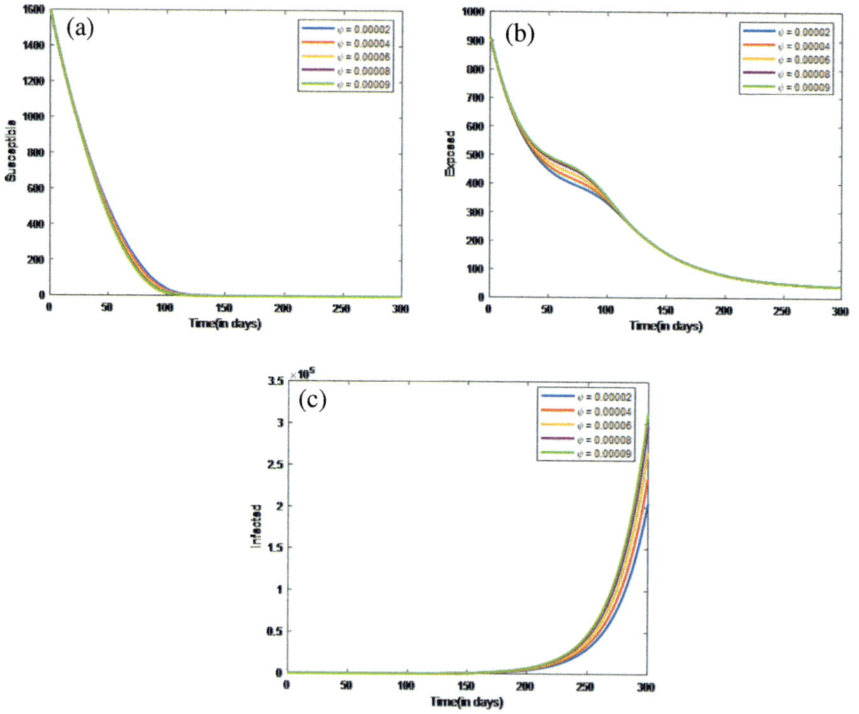

Fig. 4 **a** Time series evaluation of susceptible population of India where Ψ (Θ) is varying with initial values of above parameters. **b**, **c** Time series evaluation of exposed and infected population of India where Ψ (Θ) is varying with initial values of above parameters

that variations in recovered populations of India are less as δ varies. So δ acts as a dynamic attribute on recovered population of India. Figure 4a shows the variations in susceptible populations in India as ψ varies from 0.00003 to 0.00009. We can conclude that variations in susceptible population in India is very less. We also conclude that the attribute ψ is less dynamic in nature as it effects the susceptible populations of India. Figure 4b shows that the variation in exposed populations of India as ψ varies. We can conclude that variations in the exposed population in India almost high variations over a particular period of time in the population as we saw rapid spread of virus in a particular period. So, we conclude that ψ acts as highly dynamic on both exposed populations of India over a period of time. Figure 4c shows the variations in infected population of India as ψ varies. We conclude that the variations in infected population in India are slightly less as ψ varies. Hence we conclude that acts as a less dynamic on variations in infected population of India. As δ is increasing susceptible population is increasing slightly but infected and exposed are decreasing because of awareness in society about precautions such as social distancing, sanitizing and usage of herbs or home medicines. Vaccination awareness and drives also place a major role in reducing number of infected population. Also as there is less exposure of infection in society, recovered is effectively improving. Figure 4a–c shows the dynamics of susceptible, exposed and infected against time for different values of ψ. Also, it is observed that susceptible is increasing as ψ increases whereas infected and exposed are reducing because number of recoveries are going up with the help of growing immunities due to vaccination and prevention measures.

6 Conclusions and Remarks

The transmission dynamics of the COVID-19 outbreak are modeled with a SEIR fuzzy system in this paper. The parameters Ψ, δ and C_d are represented as fuzzy parameters in this study and are considered as association purposes of fuzzy numbers. These factors are dependent on the corona virus load Θ, and their fuzzy membership functions are specified by them. For $R_0(\Theta) < 1$ and $R_0(\Theta) > 1$, respectively, together the infection-free and the endemic equilibrium points are asymptotically stable locally.

Acknowledgements Supported by Organization Aditya College of Engineering and Technology, Surampalem, AP. and Vellore Institute of Technology, Vellore, Tamilnadu.

References

1. Alzahrani, E., Khan, M.A.: Modeling the dynamics of Hepatitis E with optimal control. Chaos Solitons Fract **116**, 287–301 (2018)
2. Barros, L.C., Bassanezi, R.C., Leite, M.B.F: The SI epidemiological models with a fuzzytransmission parameter. Comput. Math. Appl. **45**, 1619–1628 (2003)
3. Zhou, L., Fan, M.: Dynamics of an SIR epidemic model with limited resources visited. Nonlinear Anal. Real World Appl. **13**, 312–324 (2012)
4. Mccluskey, C.C.: Complete global stability for an SIR epidemic model with delay- distributed or discrete. Nonlinear Anal. **11**(1), 55–59 (2010)
5. Bjornstad, O.N., Finkenstadt, B.F., Grenfell, B.T.: Dynamics of measles epidemics: estimating scaling of transmission rates using a time series SIR model. Ecol. Monogr. **72**(2), 169–184 (2002)
6. Hu, Z., Ma, W., Ruan, S.: Analysis of SIR epidemic models with nonlinear incidence rate and treatment. Math. Biosci. **238**(1), 12–20 (2012)
7. Diekmann, O., Heesterbeek, H., Britton, T.: Mathematical tools for understanding infectious disease dynamics. In: Princeton Series in Theoretical and Computational Biology. Princeton University Press, Princeton (2013)
8. Hethcote, H.W.: The mathematics of infectious disease. SIAM Rev. **42**, 599–653 (2000)
9. He, S., Peng, Y., Sun, K.: SEIR modeling of the COVID-19 and its dynamics. Nonlinear Dyn. **101**, 1667–1680 (2020)
10. Overton, C.E.: Using statistics and mathematical modeling to understand infectious disease outbreaks: COVID-19 as an example. Infect. Dis. Model. **5**, 409–441 (2020)
11. Das, P., Upadhyay, R.K., Mishra, A.K.: Mathematical model of COVID-19 with comorbidity and controlling using non-pharmaceutical interventions and vaccination. Nonlinear Dyn. **106**, 1213–1227 (2021)

12. Haitao, S., Zhongwei, J., Zhen, J., Shengqiang, L.: Estimation of COVID-19 outbreak size in Harban. China. Nonlinear Dyn. **106**, 1229–1237 (2021)
13. Shidong, Z., Guoqiang, L., Huang, T.: Vaccination control of an epidemic model with time delay and its application to COVID-19. Nonlinear Dyn. **106**, 1279–1292 (2021)
14. Mwalili, S., Kimathi, M., Ojiambo, O., Gathungu, D.: Seir model for COVID-19 dynamics incorporating the environment and social distancing. BMC Res. Notes (2020)
15. Shikha, J., Sachin, K.: Dynamical analysis of SEIS model with nonlinear innate immunity and saturated treatment. Eur. Phys. J. Plus **136** (2021)
16. Jafelice, R., Barros, L.C., Bassanezei, R.C., Gomide, F.: Fuzzy modeling in symptomatic HIV virus infected population. Bull. Math. Biol. **66**, 1597–1620 (2004)
17. Massad, E., Burattini, M.N., Ortega, N.R.S.: Fuzzy logic and measles vaccination: designing a control strategy. Int. J. Epidemiol. **28**, 550–557 (1999)
18. Mondal, P.K., Jana, S., Halder, P., Kar, T.K.: Dynamical behavior of an epidemic model in a fuzzy transmission. Int. J. Uncertain. Fuzziness Knowl-Based Syst. **23**, 651–665 (2015)
19. Nagarajan, D., Lathamaheswari, M., Broumi, S., Kavikumar, J.: A new perspective on traffic control management using triangular interval type-2 fuzzy sets and interval neurosophic sets. Oper. Res. Perspect. **6**, 100099 (2019)
20. Mishra, B.K., Pandey, S.K.: Fuzzy epidemic model for the transmission of worms in computer network. Nonlinear Anal. Real World Appl. **11**(5), 4335–4341 (2010)
21. Gakkhar, S., Chavda, N.C.: Impact of awareness on the spread of dengue infection in human population. Appl. Math. **4**(8), 142–147 (2013)
22. Phaijoo, G.R., Gurung, D.B.: Mathematical model of dengue disease transmission dynamics with control measures. J. Adv. Math. Comput. Sci. **23**(3), 1–2 (2017)
23. Arqub, O.A., El-Ajou, A.M., Shawagfeh, N.: Analytical solutions of fuzzy initial value problem by HAM. Appl. Math. Inform. Sci. **7**, 1903–1919 (2013)
24. Brauer, F., Castillo-Chavez, C.: Mathematical models in population biology and epidemiology. Texts Appl. Math. **2** (2012)

Covid-19 Vaccination in India: Prophecy of Time Period to Immune 18+ Population

Anand Kumar, Agin Kumari, and Rishi Pal Chahal

Abstract In the present paper, we have prophesied how much time will be required to vaccinate 18+ population of India with at least one dose of COVID-19 vaccines. We have used non-linear extrapolation technique to prophecy, for this polynomial function is used for extrapolation. We have Fitted a non-linear polynomial of degree six to the cumulative vaccination data from 16 January 2021 to 24 July 2021 to estimate the required time period. Non-linear extrapolation results are depicted through the graphs, shows that the entire 18+ population will be vaccinated with at least 1 dose by mid of December of this year and 25% population will be fully vaccinated.

Keywords Covid-19 · Vaccination · India · Prophecy · Extrapolation

1 Introduction

Indian government has started vaccination drive for COVID-19 on 16 January 2021, as of 24 July 2021 over 3403.87 lacs (24.9%) population is vaccinated with at least 1 dose and over 927.63 lacs (6.8%) population is fully vaccinated with presently permitted vaccines.

Non-linear extrapolation is more reliable and inexpensive for statistical forecasts by using previous trend of data. This methodology estimates the dependent function for independent variable by interpolating a smooth nonlinear curve through all values of independent variable, using this nonlinear curve, extrapolates dependent values beyond available data. Polynomial function or rational function are used in this methodology.

A. Kumar (✉)
Department of Physics, Chaudhary Ranbir Singh University, Jind 126102, India
e-mail: anandkumar@crsu.ac.in

A. Kumari
Department of Mathematics, Chaudhary Bansi Lal University, Bhiwani 127021, India

R. Pal Chahal
Department of Physics, Chaudhary Bansi Lal University, Bhiwani 127021, India

S. Banerjee and A. Saha (eds.), *Nonlinear Dynamics and Applications*,
Springer Proceedings in Complexity,
https://doi.org/10.1007/978-3-030-99792-2_120

Gowrisankar et al. [1] used multifractal formalism to analyse COVID-19 data, assuming that infection rates in different countries follow a power law growth pattern. Radiom and Berret [2] constructed two models to explain the epidemic's fast-growth phase and to interpret the complete data set. Han et al. [3] studied a two-part framework, one to show repeated worldwide breakouts of COVID-19 and the other to investigate the underlying causes of recurrent outbreaks. Kavitha et al. [4] studied the SIR and fractal models on daily positive COVID-19 cases in India in order to forecast the outbreak's future trajectory. Moghadas et al. [5] constructed an agent-based model of SARS-CoV-2 transmission, and was parameterized using US statistics and COVID-19 age-specific results.

In the present paper, we have estimated how much time will be required to vaccinate 18 + population of India with at least one dose of COVID-19 vaccines. We have used non-linear extrapolation technique to prophecy, for this polynomial function is used for extrapolation. Nonlinear extrapolation method is used to prophecy as casual factors related to this kind of situation are expected to remain constant [6,7]. A non-linear polynomial of degree six is fitted to the cumulative vaccination data (vaccinated with at least 1 dose) from 16 January 2021 to 24 July 2021. A non-linear polynomial of degree five is fitted to the cumulative vaccination data (fully vaccinated) from 16 January 2021 to 24 July 2021. Non-linear extrapolation results are depicted through the graphs, shows that the entire 18+ population will be vaccinated with at least 1 dose by mid of December of this year and 25% population will be fully vaccinated.

Because COVID-19 vaccines stimulate a wide immune response including a variety of antibodies and cells, they should give some protection against new viral types. Vaccines should not be rendered fully ineffective due to changes or mutations in the virus. If any of these vaccinations prove ineffective against one or more variants, the vaccines' composition can be changed to defend against these variants. Even with minimal protection against infection, vaccination can have a significant influence on preventing COVID-19 outbreaks. To accomplish this benefit, however, ongoing adherence with non-pharmaceutical measures is required.

2 Modeling of Data and Fitting of Polynomial

In this section, we have constructed a table of vaccination data and fitted a suitable nonlinear polynomial for extrapolation. Table-1 contains cumulative vaccination data [8] of 18+ population of India observed on weekends for study. Figure 1 shows that a six-degree polynomial suited to the cumulative vaccination data (vaccinated with at least 1 dose) from 16 January 2021 to 24 July 2021 and a five-degree polynomial is suited to the cumulative vaccination data (fully vaccinated) from 16 January 2021 to 24 July 2021. Fitted polynomials are as follows;

$$y(x) = -0.0005x6 + 0.0437x5 - 1.4101x4 + 21.153x3 - 142.29x2$$

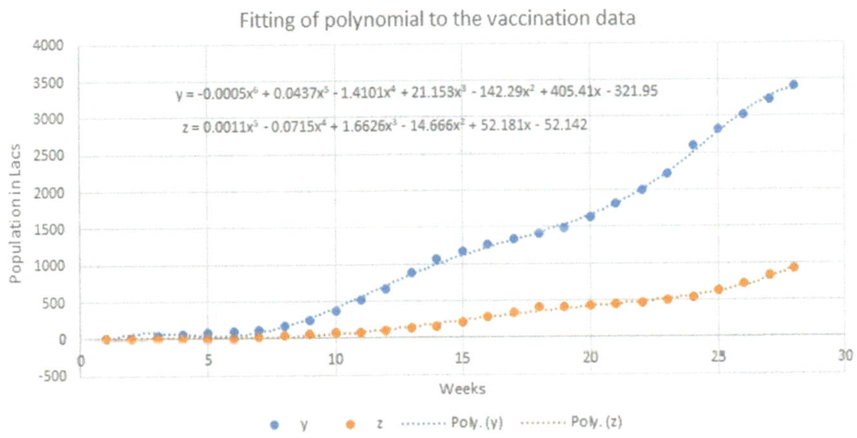

Fig. 1 Fitting of polynomial to the vaccination data

$$+405.41x - 321.95$$

$$z(x) = 0.0011x5 - 0.0715x4 + 1.6626x3 - 14.666x2 + 52.181x - 52.142$$

With the help of Table 1 and Fig. 1 it is observed that in the 2nd weekend of vaccination drive the data of population vaccinated with one dose is 15 times of 1st weekend. Gradual increment in 3rd to 7th weekends, moderate increment in 8th to 10th weekends, major increment in 11th–19th weekends, significantly great increment in 20th–28th weekends, in the data of population vaccinated with one dose is observed. Gradual increment in the data of Population fully vaccinated is observed.

3 Extrapolation Method

In this section, we have prophesied how much population will be vaccinated with 1 or 2 dose by method of nonlinear polynomial extrapolation. We have extrapolated the values of y and z with respect to x with confidence interval 95%, i.e. weekly estimation of population vaccinated with at least 1 dose and fully vaccinated.

With the help of vaccination data given in Table 1, The extrapolated values of population vaccinated with at least 1 dose and 2 dose are obtained. From Figs. 2 and 3 and Table 2, it is observed that up to 11th December over 80 crores (almost all 18+ population) population will be vaccinated with at least 1 and about 20 crores (25% of 18+ population) population will be fully vaccinated.

Table 1 Vaccination data at a glance in India (cumulative data observed at weekends from 16 January 2021 to 24 July 2021)

Data observed on weekends (date) (x)	Population vaccinated with at least 1 dose		Population fully vaccinated	
	in lacs (y)	in %	in lacs (z)	in %
1 (16 January)	1.99	0.0	0	0.0
2 (23 January)	15.82	0.1	0	0.0
3 (30 January)	37.44	0.3	0	0.0
4 (6 February)	57.75	0.4	0	0.0
5 (13 February)	80.45	0.6	0.07	0.0
6 (20 February)	99.64	0.7	8.73	0.1
7 (27 February)	117.88	0.9	24.54	0.2
8 (6 March)	171.68	1.3	37.54	0.3
9 (13 March)	243.07	1.8	54.31	0.4
10 (20 March)	371.25	2.7	74.79	0.5
11 (27 March)	514.41	3.8	88.28	0.6
12 (3 April)	657.39	4.8	102.40	0.7
13 (10 April)	888.86	6.5	127.09	0.9
14(17 April)	1064.31	7.8	161.91	1.2
15(25 April)	1177.95	8.6	213.90	1.6
16 (1 May)	1263.28	9.2	272.97	2.0
17 (8 May)	1333.66	9.8	341.27	2.5
18 (15 May)	1411.32	10.3	404.12	3.0
19 (22 May)	1492.19	10.9	416.22	3.0
20 (29 May)	1641.58	12.0	429.30	3.1
21 (5 June)	1809.72	13.2	445.99	3.3
22 (12 June)	1996.55	14.6	465.32	3.4
23 (19 June)	2214.93	16.2	494.15	3.6
24 (26 June)	2602.53	19.0	545.88	4.0
25 (3 July)	2822.31	20.7	620.69	4.5
26 (10 July)	3031.71	22.2	728.61	5.3
27 (17 July)	3218.93	23.6	830.37	6.1
28 (24 July)	3403.87	24.9	927.63	6.8

4 Discussion

In the present paper, we have obtained estimated time period to vaccinate the 18+ population in India in view of current vaccination rate. Extrapolated values of population vaccinated with at least 1 dosage and 2 dose are determined using vaccination data from Table 1 by a six-degree polynomial suited to the cumulative vaccination

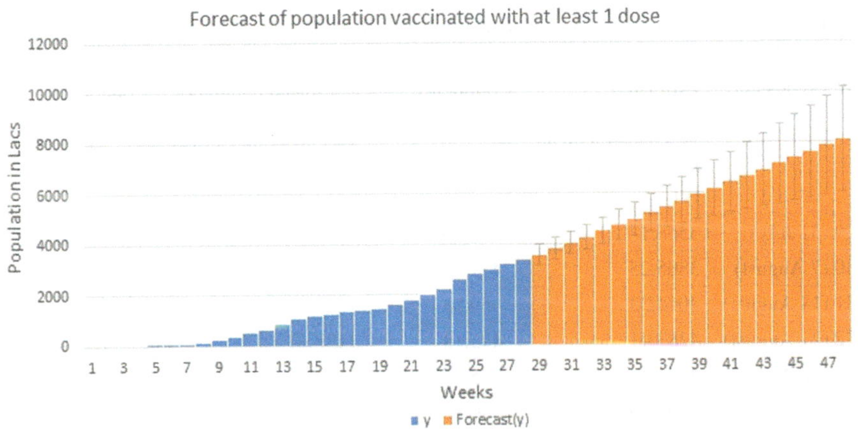

Fig. 2 Weekly Forecast of population vaccinated with at least 1 dose from 31 July 2021 to 11 December 2021

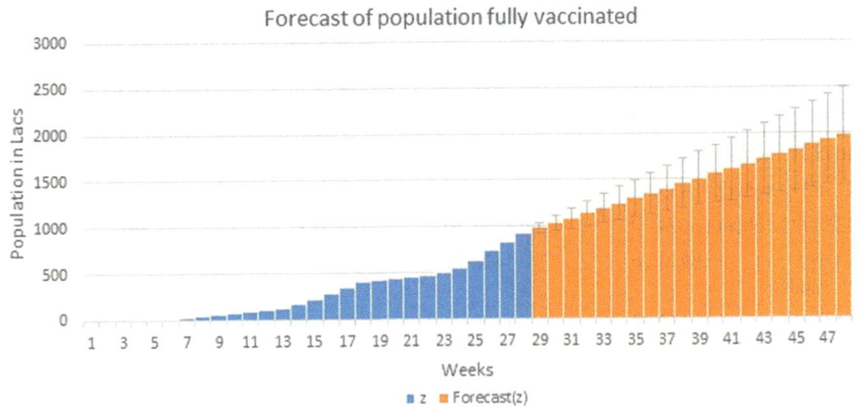

Fig. 3 Weekly Forecast of population fully vaccinated from 31 July 2021 to 11 December 2021

data (vaccinated with at least 1 dose) from 16 January 2021 to 24 July 2021 and a five-degree polynomial is suited to the cumulative vaccination data (fully vaccinated). Figures 2 and 3 represents the weekly extrapolated forecast values of population vaccinated with 1 or full dose from 31 July 2021 to 11 December 2021 along with confidence interval. From the Table-2, It is observed that over 80 crore population will be vaccinated with at least 1 dose up to 11 December 2021 i.e. almost all population will be covered and 25% population will be fully vaccinated. COVID-19 vaccinations will provide some resistance against new virus strains since they stimulate a broad immune response that includes a variety of antibodies and cells. Vaccines should not be rendered completely ineffective as a result of virus mutations or modifications. If any of these vaccines are shown to be ineffective against one or

Table 2 Weekly estimation of vaccinated population from 31 July 2021 to 11 December 2021 along with confidence interval

Weekends (x)	Estimated population will be vaccinated with at least 1 dose in lacs (y)	Confidence Interval (y)	Estimated population will be fully vaccinated in lacs (z)	Confidence Interval (z)
29 (31 July)	3566.17	424.00	980.31	57.15
30 (7 August)	3805.25	437.05	1033.10	84.83
31 (14 August)	4044.33	459.33	1085.88	108.94
32 (21 August)	4283.41	492.13	1138.67	131.70
33 (28 August)	4522.48	535.85	1191.45	153.93
34 (4 September)	4761.56	590.18	1244.24	175.99
35 (11 September)	5000.64	654.38	1297.02	198.12
36 (18 Sept)	5239.72	727.55	1349.81	220.44
37 (25 September)	5478.80	808.81	1402.59	243.02
38 (2 October)	5717.87	897.35	1455.38	265.92
39 (9 October)	5956.95	992.48	1508.16	289.18
40 (16 October)	6196.03	1093.62	1560.95	312.81
41 (23 October)	6435.11	1200.29	1613.73	336.84
42 (30 October)	6674.19	1312.10	1666.52	361.27
43 (6 November)	6913.26	1428.71	1719.30	386.10
44 (13 November)	7152.34	1549.84	1772.09	411.36
45 (20 November)	7391.42	1675.26	1824.87	437.03
46 (27 November)	7630.50	1804.77	1877.66	463.11
47 (4 December)	7869.58	1938.19	1930.44	489.60
48 (11 December)	8108.65	2075.36	1983.23	516.51

more variants, the vaccine's composition can be altered to protect against these variants. Vaccination, even if it provides only rudimentary protection against infection, can help avoid COVID-19 epidemics.

References

1. Gowrisankar, A., Rondoni, L., Banerjee, S.: Can India develop herd immunity against COVID-19? Eur. Phys. J. Plus **135**, 526 (2020)
2. Radiom, M., Berret, J.F.: Common trends in the epidemic of Covid-19 disease. Eur. Phys. J. Plus **135**, 517 (2020)
3. Han, C., Li, M., Haihambo, N., et al.: Mechanisms of recurrent outbreak of COVID-19: a model-based study. Nonlinear Dyn. **106**, 1169–1185 (2021)
4. Kavitha, C., Gowrisankar, A., Banerjee, S.: The second and third waves in India: when will the pandemic be culminated? Eur. Phys. J. Plus **136**, 596 (2021)
5. Moghadas, S.M., Vilches, T.N., Zhang, K. et al.: The impact of vaccination on COVID-19 outbreaks in the United States. medRxiv: the preprint server for health sciences, 2020.11.27.20240051 (2021)
6. Brezinski, C., Redivo Zaglia, M.: Extrapolation Methods-Theory and Practic. North Holand (1991)
7. Brezinski, C., Redivo Zaglia, M.: Extrapolation and Rational Approximation. Springer Nature, Switzerland (2020). ISBN 9783030584177
8. http://www.cowin.gov.in

COVID-19 Detection from Chest X-Ray (CXR) Images Using Deep Learning Models

Mithun Karmakar, Koustav Chanda, and Amitava Nag

Abstract Due to the tremendous rise in COVID cases around the world, early detection of Covid-19 has become critical. Deep learning technology has recently sparked a lot of attention as a means of detecting and classifying diseases quickly, automatically, and accurately. The goal of this study is to develop a deep learning based automatic COVID-19 detection system for better, faster, and more accurate COVID-19 detection from chest X-Ray (CXR) images. In our work, we have used pre-trained deep learning models such as VGG16, ResNet50, DenseNet201, InceptionV3 and Xception utilizing openly accessible dataset. Experimental results show that the DenseNet201 model performs the best with more than 97% accuracy. Moreover, in terms of size, DenseNet121 is beating the rest of the models. As a results, DenseNet201 is most suitable Deep Convolutional neural networks (CNN) architecture for developing an automatic covid-19 detection tool.

Keywords COVID-19 · Deep learning · Deep CNN · Chest X-ray images

1 Introduction

In recent years, a novel coronavirus (COVID-19) arising from the coronavirus SARS-COV2 has become a global epidemic. COVID-19 was declared a pandemic by the World Health Organization (WHO) on March 11, 2020, after it had spread to over a hundred countries [1]. A novel SARS-CoV-2 variant was reported to the WHO on November 24, 2021, from South Africa. As a variant of concern, the new variant (B.1.1.529) has been officially called Omicron. As compared to the other SARS-CoV-2 variants: Alpha, Beta, Gamma, and Delta, Omicron emerges as the most noticeable and different variant among the millions of SARS-CoV-2 genomes [2].

M. Karmakar (✉) · A. Nag
Department of CSE, CIT Kokrajhar, Kokrajhar, Assam, India
e-mail: m.karmakar@cit.ac.in

K. Chanda
L&T Infotech, Ranaghat, India

S. Banerjee and A. Saha (eds.), *Nonlinear Dynamics and Applications*,
Springer Proceedings in Complexity,
https://doi.org/10.1007/978-3-030-99792-2_121

Timely and accurate detection of Coronavirus diseases is of great importance for controlling COVID-19. Because of its high transmissibility, COVID-19 can easily be transmitted from asymptomatic to vulnerable groups. Fever, dry cough, myalgia, dyspnea, and headache are the most common symptoms of the COVID-19 [3]. In rare cases, no symptoms are also visible (asymptomatic) which makes the disease an even greater hazard to public health. The diagnosis of coronavirus is performed by conducting a reverse-transcription polymerase chain reaction (RT-PCR) test with a patient's respiratory tract or blood samples [4]. The RT-PCR is a laboratory-based COVID-19 detection procedure and takes 2–3 days to get the results, which is a long period in comparison to COVID-19's quick spread rate [5].

An alternative to RT-PCR is the rapid antigens test (RAT) which is mostly employed for large-scale testing. However, both RT-PCR and RAT have false negative rates in some cases. A useful supplement to both RT-PCR test and RAT is chest X-Rays (CXR) for the diagnosis of COVID-19 infection. The accuracy of CXR-based COVID-19 diagnosis is dependent on manual analysis and interpretation by radiologists, which can lead to inaccurate analyses due to expert radiologists and doctors who can accurately detect COVID-19 infection from chest X-Rays are in short supply in the context of the pandemic.

Based on the foregoing facts, it is essential to develop alternative, complementary, and low-cost technologies for speedy and accurate diagnosis of COVID-19. Artificial Intelligence (AI), a rapidly evolving technology, is now being used for speedy and precise diagnosis of a variety of ailments, including brain tumour detection, breast cancer detection, and so on. Deep learning, a type of artificial intelligence, has been a natural choice for use in healthcare applications such as medical image analysis in recent years. As a result, building deep learning-based computer-aided diagnostic (CAD) tools for better, faster, and more accurate COVID-19 diagnosis is worthwhile.

Advances in image processing provide an opportunity to expand its application in all areas of healthcare [6]. Applications of deep learning to medical imaging for automatic diagnosis of various diseases are growing rapidly [7, 8]. As a result, these methodologies are widely used for COVID-19 research [9]. The aim of this work is to develop a deep learning-based system for better, faster, and more accurate detection of COVID-19 from chest X-Ray (CXR) images.

The rest of the paper is organized as follows: The state-of-the-art deep Convolutional networks and the other materials and methodologies including dataset collection and preparation, required to accomplish this work are described in Sect. 2. The experimental setup as well as the results and performance analysis are provided in Sect. 3. Section 4 concludes the paper.

2 Materials and Methods

Deep Convolutional Neural Network (CNN) models are used in this study to reveal patterns in chest X-ray images that are imperceptible to the naked eye. The CNN is a class of Deep Learning (DL) techniques that is used to identify useful and

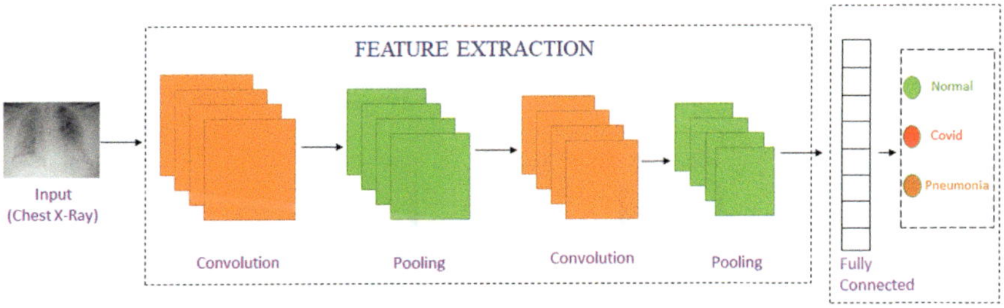

Fig. 1 A typical Convolution Neural Network (CNN) architecture

distinctive representations of images. A typical CNN is depicted in Fig. 1. However, the fundamental issue that a deep CNN model encountered during training was a large volume of image data. This problem has been addressed using a technique known as transfer learning (TL) which is designed for the CNNs. Several pre-trained models with transfer learning that have already trained on a huge annotated image library have been designed.

Recently, a number of CNN Architectures have been proposed which are used for COVID-19 diagnosis such as VGG16 [10], ResNet50 [11], DenseNet201 [12], InceptionV3 [13] and Exception[14], etc.

We used three different classes in our work (i.e., COVID-19, pneumonia, and healthy). Figure 2 depicts the entire detection process for COVID-19. In this work, first images of chest X-rays are collected from public datasets [15]. Then, image preprocessing was done. The only preprocessing used in this study was a simple rotation of the X-ray images (from 0 to 12° clockwise or anticlockwise). Finally, pre-trained CNN models with weights from ImageNet and with the proper fine-tuning are used for classification (Fig. 3).

2.1 X-ray Image DataSet

In our study, we have used anterior-to-posterior (PA)/posterior-to-anterior (AP) view of CXR images as this view of radiography is widely used by radiologists in clinical diagnosis. The dataset used in our work is collected from Kaggle [15]. It consists of 12,443 images and these images are divided into three different classes- 'Normal', 'Covid positive' and 'Viral Pneumonia'. There are 906 Covid positive images, 1345 viral pneumonia images and 10,192 normal images. These images are further split into training and testing data with a percentage ratio of 80% training data and 20% testing data.

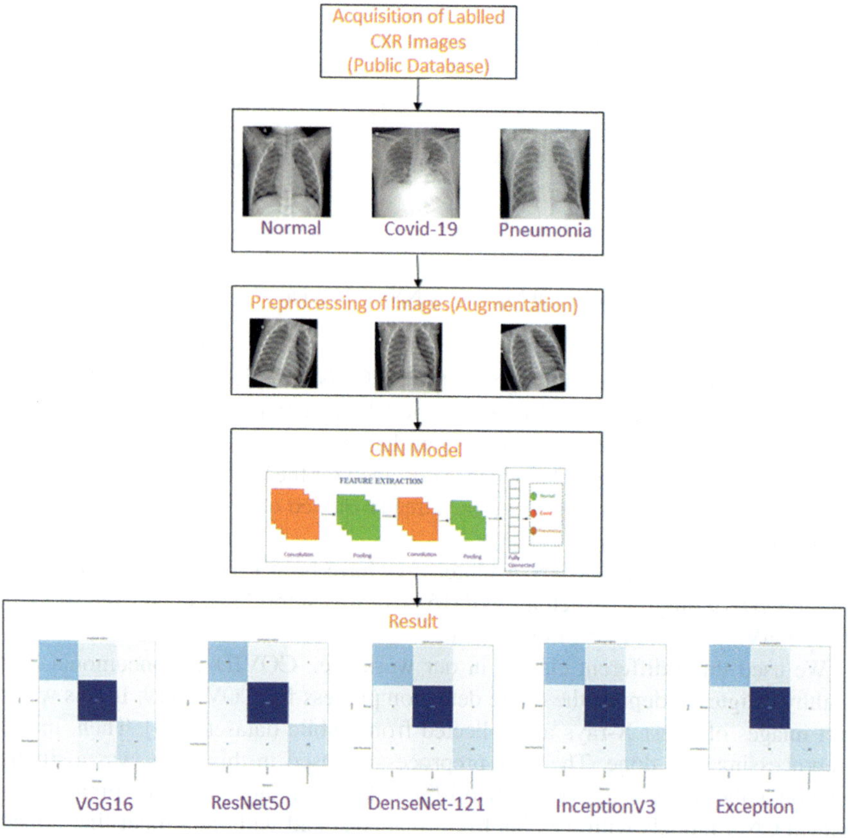

Fig. 2 Architecture of deep learning based COVID-19 diagnosis system

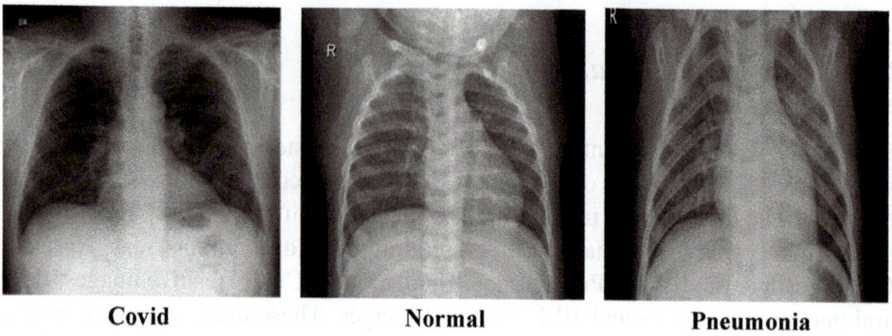

Fig. 3 Sample of X-ray images from dataset

Since the dataset is not uniform, therefore, first we have to resize the image to 224 × 224 for Vgg16, ResNet50, DenseNet201 architecture, and 299 × 299 for InceptionV3 and Xception architecture.

2.2 *Pretrained CNN Models with Fine-Tuning*

Training deep CNN models from scratch is complex as well as demand large amount of data in order to converge the model. Fine-tuning on a pre-trained CNN model can be an alternate solution. In this work, fine-tuning was performed on pre-trained CNN models such as VGG16, ResNet50, DenseNet201, Inception V3 and Xception architecture with CRX dataset [15]. The models were pre-trained with weights from ImageNet. Furthermore, the top layer of all models are truncated and a new fully-connected softmax layer is added on the top layer that used stochastic gradient descent (SGD) algorithm.

3 Results

Using the datasets given in Sect. 2.1, the classification of COVID-19 against the pneumonia and normal classes is performed. The CNN models described in the Sect. 2.2 was evaluated for COVID-19 detection from CRX images. In the experiment, the dataset was divided into two halves for training and testing: 80% for training and 20% for testing.

3.1 *Tools Used*

We used TensorFlow 2.2.0, Python 3.7, and Google Colab graphics processing units (GPU). The CNN models used in this work are implemented using the TensorFlow 2.2.0 deep learning framework, and the training and testing procedures are carried out on the Google Colab platform.

3.2 *Performance Evaluation*

The performance has been evaluated in terms of classification accuracy (CA). The classification accuracy is defined as follows:

Classification accuracy (CA): It is defined as the proportion of right predictions to the total number of predictions made on a given set of data:

$$Classification\ Accuracy\ (CA) = \frac{Number\ of\ unerring\ predictions}{Total\ number\ of\ predictions\ made}$$

3.3 Results Analysis

In this work, various state-of-the-art deep CNN models such as VGG16, ResNet50, DenseNet201, InceptionV3 and Xception are used for COVID-19 identification. The results for the performance of all evaluated models are provided in Table 1. As shown in Table 1, the size of the VGG16 trained dataset weight is the largest, which is 512 MB. As a result, the VGG16 model consumes a significant amount of storage space and bandwidth, making it inefficient. The DenseNet201 model was evaluated to be the most efficient in terms of both size and performance among all the five models we trained and tested. The training and test detection accuracy are also shown in Fig. 4.

Table 1 Experimental results of different models

	Size (MB)	Training accuracy	Test accuracy	Training loss	Test loss
VGG16	512	0.9667	0.9537	0.0920	0.1072
ResNet50	98	0.7998	0.9086	0.5369	0.2484
DenseNet201	80	0.9702	0.9704	0.2371	0.3435
InceptionV3	92	0.9398	0.9408	0.6222	0.7738
Xception	88	0.9592	0.9459	0.5204	1.0678

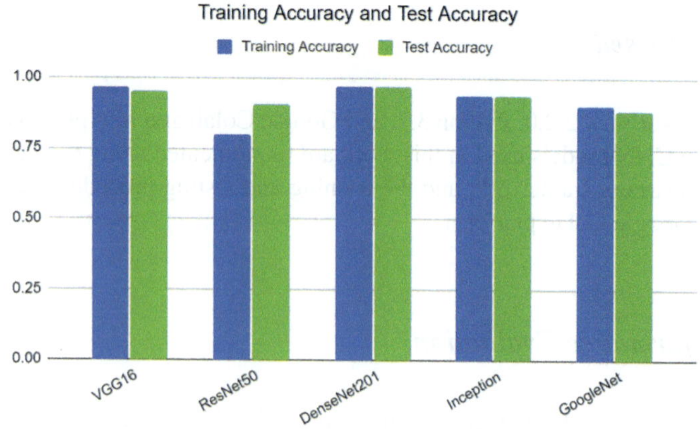

Fig. 4 COVID-19 detection accuracy

4 Conclusions

The wave of Covid-19 and the continued struggle to test the patients have led to increased cases. Effective screening and early medical attention for infected patients are required to combat the covid-19. The methods that are being used are mostly manual, and thus there is a delay in testing and inaccuracy of data, and as a result, most of the time, the patient will interact with, leading to the spread of the disease. The use of deep learning to help diagnose the diseases is a mechanism that would help ensure the diagnosis are more accurate and on time. In this work, various state-of-the-art deep CNN models for COVID-19 detection classification are used. The models evaluated include VGG 16, ResNet50, DenseNets121, InceptionV3 and Xception. The experimental results show that the best performing model is DenseNets121 followed by VGG16, Xception, InceptionV3 and Resnet50. Moreover, in terms of size, DenseNet121 is beating the rest of the models. DenseNets is, therefore, a good architecture for the development of computer-aided diagnostic tools.

References

1. Sohrabia, C., Alsafi, Z., O'Neill, N., Khan, M., Ahmed, K., Al-Jabir, A., Iosifidis, C., Agha, R.: World health organization declares global emergency: a review of the 2019 novel coronavirus (COVID-19). Int. J.Surg. **76**, 71e76 (2020)
2. Gowrisankar, A., Priyanka, T.M.C., Banerjee, S.: Omicron: a mysterious variant of concern. Eur. Phys. J. Plus **137**, 100 (2022)
3. Islam, M.M., Karray, F., Alhajj, R., Zeng, J.: A review on deep learning techniques for the diagnosis of novel coronavirus (covid-19). IEEE Access **9**, 30551–30572 (2021)
4. Islam, M.Z., Islam, M.M., Asraf, A.: A combined deep CNN-LSTM network for the detection of novel coronavirus (COVID-19) using X-ray images. Informat. Med. Unlocked **20**, 100412 (2020)
5. Naronglerdrit, P., Mporas, I., Sheikh-Akbari, A.: COVID-19 detection from chest X-rays using transfer learning with deep convolutional neural networks. In: Data Science for COVID-19, pp. 255–273. Academic Press (2021)
6. Altaf, F., Islam, S.M.S., Akhtar, N., Janjua, N.K.: Going deep in medical image analysis: concepts, methods, challenges, and future directions. IEEE Access **7**, 99540–99572 (2019)
7. Ghassemi, N., Shoeibi, A., Rouhani, M.: Deep neural network with generative adversarial networks pre-training for brain tumor classification based on MR images. Biomed. Signal Process. Control **57**, 101678 (2020)
8. Chowdhury, M.E.H., Rahman, T., Khandakar, A., Mazhar, R., Kadir, M.A., Mahbub, Z.B., Islam, K.R. et al.: Can AI help in screening viral and COVID-19 pneumonia? IEEE Access **8**, 132665–132676 (2020)
9. Bhattacharya, S., Maddikunta, P.K.R., Pham, Q.-V., Gadekallu, T.R., Chowdhary, C.L., Alazab, M., Piran, M.J.: Deep learning and medical image processing for coronavirus (COVID-19) pandemic: a survey. Sustain. Cities Soc. **65**, 102589 (2021)
10. Iandola, F., Moskewicz, M., Karayev, S., Girshick, R., Darrell, T., Keutzer, K.: Densenet: implementing efficient convent descriptor pyramids (2014). arXiv:1404.1869
11. He, K., Zhang, X., Ren, S., Sun, J.: Deep residual learning for image recognition. In: Proceedings of the IEEE conference on computer vision and pattern recognition, pp. 770–778 (2016)

12. Zhang, X., Zou, J., He, K., Sun, J.: Accelerating very deep convolutional networks for classification and detection. IEEE Trans. Pattern Anal. Mach. Intell. **38**(10), 1943–1955 (2015)
13. Szegedy, C., Vanhoucke, V., Ioffe, S., Shlens, J., Wojna, Z.: 'Rethinking the inception architecture for computer vision. In: Proceedings of the IEEE Conference on Computer Vision and Pattern Recognition, pp. 2818–2826 (2016)
14. Xception, C.F.: Deep learning with depthwise separable convolutions. In: Proceedings of the IEEE Conference on Computer Vision and Pattern Recognition, pp. 1251–8 (2017)
15. Dataset Link. https://www.kaggle.com/tawsifurrahman/covid19-radiography-database

Pre-covid and Post-covid Situation of Indian Stock Market-A Walk Through Different Sectors

Antara Roy, Damodar Prasad Goswami, and Sudipta Sinha

Abstract Sudden and unexpected outbreak of covid-19 has left a serious impression on Indian as well as global economy. A simple way of investigating and verifying this impact is to mind the movement of stock values and consequent market swing. In this piece of work, we tried to figure out the movement pattern of stock prices in different sectors. We carefully picked some representative stocks from each of the sectors and tried to perceive their beat to beat and overall movement in the pandemic period and express through mathematical language. This study offered some interesting, valuable and to some extent 'counter-intuitive' insights.

Keywords Covid · Indian stock market · Pandemic · Stock values

1 Introduction

Recently, the entire world has gone through a very tough situation due to the sudden outbreak of deadly virus covid-19. India is not an exception. We have witnessed two consecutive waves till now and a third wave is likely to occur according to the scientists. More or less 190 countries have been affected by the pandemic situation. Economic structure around the world has abruptly been impacted for the sudden advent of corona virus. Indian stock exchange has also been more or less affected during the pandemic situation. A simple way of measuring this impact on economy is to study the stock market behavior indifferent sectors. Some similar studies have already been performed. Buszko M studied utilization of an asymmetric exponential generalized autoregressive model to reflect the asymmetric effect on conditional

A. Roy (✉)
Asansol Institute of Engineering and Management-Polytechnic, Asansol, India
e-mail: royantara793@gmail.com

D. P. Goswami
Asutosh Mookerjee Memorial Institute, Sivotosh Mookerjee Science Centre, Kolkata, India

S. Sinha
Burdwan Raj College, University of Burdwan, Burdwan, India

S. Banerjee and A. Saha (eds.), *Nonlinear Dynamics and Applications*,
Springer Proceedings in Complexity,
https://doi.org/10.1007/978-3-030-99792-2_122

volatility [1]. Sahoo, M. collected evidence from the Indian stock market of covid-19 impact from Nifty 50, Nifty 50 Midcap, Nifty 100, Nifty 100 Midcap, Nifty 100 small cap, and Nifty 200 [2]. Awadhi et al. investigates whether infectious diseases affect stock market returns [3]. Khaled studied the impact of the COVID-19 pandemic on retailer performance [4]. Stefano R. et al. illustrated how anticipated real effects from the health crisis, a rare disaster were amplified through financial channels [5]. Md Akhtaruzzaman examined the impact of financial contagion through financial and non-financial firms between China and G7 countries during the COVID–19 period [6]. Naveen Donthu's paper covers different industrial sectors (e.g., tourism, retail, higher education) investigating the changes in consumer behavior, businesses, ethical issues, and aspects related to employees and leadership [7]. Dayong Zhanga. et al. [8] aimed to the mapping the general patterns of country-specific risks and systemic risks in the global financial markets. It also analyzed the potential consequences of policy interventions. The article developed by Okorie et al. [9] investigated the fractal contagion effect of the COVID-19 pandemic on the stock markets. Martin Scheicher [10] studied the regional and global integration of stock markets in Hungary, Poland, and the Czech Republic.

In this paper we have considered five different sectors like automobile sector, banking sector, construction sector, finance sector, tourism sector and some corresponding stocks picked up from these sectors which are mentioned below.

2 Materials and Methods

We have collected dataset of individual stocks from four different sectors, the Automobile, Banking, Finance and the construction sector starting from 1st November, 2019 up to 28th December, 2021. 'Yahoo finance' has been used as the data source. The opening value of the stocks on each day has been used in this study we plotted the stock values sector-wise and observed the nature. Finally, to compare these graphs with the covid situation, we also kept an eye on the covid graph of India built from count of daily infected persons. We demonstrate the graphs of each of the sectors one by one and quantify them with Pearson's Product Moment Correlation.

2.1 Automobile Sector

Figure 1 shows six representative stocks in the automobile sector like Ashok Layland, TVS motors, Mahendra & Mahendra, Hero Motor and Maruti. Here, x-axis denotes time and y-axis denotes the opening share price on each day. Being in different scale, they are not clearly visible in a single graph. So, we have normalized the values to [0,1] with the formula $\frac{x - x_{min}}{x_{max} - x_{min}}$ where x represents the time series values of stocks and plotted individual graphs in Fig. 2.

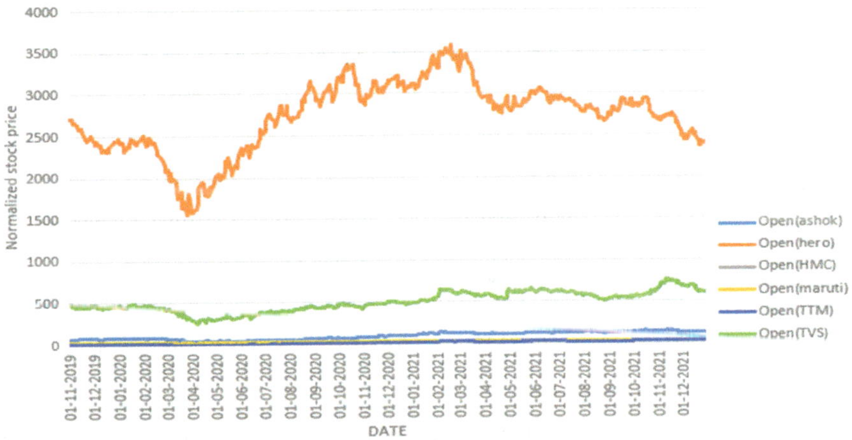

Fig. 1 Normal graph of automobile companies

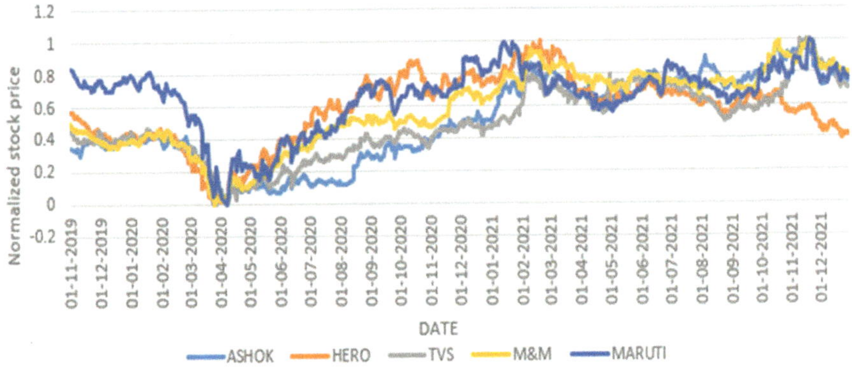

Fig. 2 Normalized graph of automobile sector

Now the graphs become comparable. This graph shows that different companies we chose, behave in a similar fashion to some extent.

To quantify this behavior, we calculated the correlation coefficients between each pair and tabulated them in Table 1.

We can observe that most of the coefficients are highly positive which explains their similar movements over time.

2.2 *Banking Sector*

The following graph presents the performance of different banks like Axis bank, Canara bank, Indian Bank, PNB, SBI and UCO bank during pandemic. Again,

Table 1 Correlation values between automobile companies

	Ashok Layland	TVS motors	HeroMoto crop	Mahindra & Mahindra	Maruti
Ashok Layland	1	0.935084	0.5475	0.922055	0.722622
TVS motors	0.935084	1	0.561567	0.921826	0.733054
HeroMoto crop	0.5475	0.561567	1	0.731019	0.683864
Mahendra & Mahendra	0.922055	0.921826	0.731019	1	0.760863
Maruti	0.722622	0.733054	0.683864	0.760863	1

we normalized the graph as before for visual comparison and quantified through correlation coefficients presented in Table 2 (Fig. 3).

Table 2 shows a good agreement among the stocks except PNB and UCO. They also have positive correlation with others.

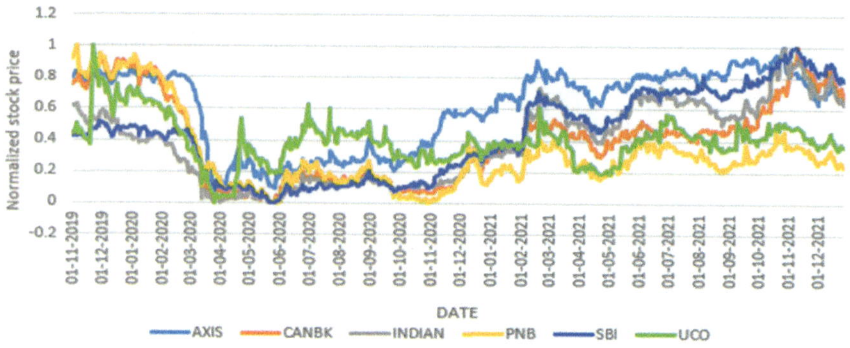

Fig. 3 Normalized graph in banking sector

Table 2 Correlation values between banks

	Axis bank	Canara bank	Indian bank	SBI	PNB	UCO
Axis bank	1	0.809175	0.869678	0.881935	0.591333	0.453348
Canara bank	0.809175	1	0.797292	0.767385	0.820716	0.654859
Indian bank	0.869678	0.797292	1	0.970295	0.43443	0.402588
SBI	0.881935	0.767385	0.970295	1	0.364925	0.310351
PNB	0.591333	0.820716	0.43443	0.364925	1	0.744481
UCO	0.453348	0.654859	0.402588	0.310351	0.744481	1

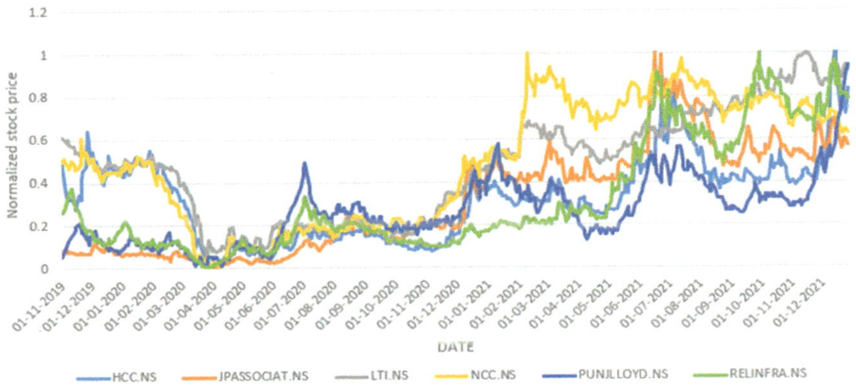

Fig. 4 Normalized graphs in construction sector

2.3 *Construction Sector*

We follow the same protocol here and plot the normalized values of different stocks like Larsen & Toubro ltd, NCC, Jaiprakash Associated ltd, Punj Lloyd, Hindusthan Construction Company and Reliance infrastructure Ltd and finally calculate the correlation coefficients for the quantification purpose (Fig. 4 and Table 3).

High positive correlation values indicate their similar behavior.

2.4 *Finance Sector*

Here, we plot the normalized values of different stocks like Bajaj, LIC, Muthoot, Rhfl.NS and finally calculate the correlation coefficients for the quantification purpose (Fig. 5).

Correlation values between stocks of different finance companies are given in Table 4.

Correlation values show that except the Muthoot Finance all are in good agreement. Muthoot Finance also have positive correlation with others.

2.5 *Travel and Tourism Sector*

We follow the same protocol here and plot the normalized values of different stocks like Expedia, Mahindra Holidays and Resorts India ltd, Thomas cook India ltd, Tripand finally calculate the correlation coefficients for the quantification purpose.

Correlation values between stocks of different travel companies are given in Table 5

Table 3 Correlation values between construction companies

	HCC	JPASSOCIATE	NCC	LTI.NS	PUNJLLOYED	RELINFRA.NS
HCC	1	0.663704	0.75423	0.812862	0.578739	0.701046
JPASSO-CIATE	0.663704	1	0.854078	0.76025	0.773943	0.830481
NCC	0.75423	0.854078	1	0.890897	0.564206	0.719297
LTI.NS	0.812862	0.76025	0.890897	1	0.568498	0.811581
PUNJLL-OYED	0.578739	0.773943	0.564206	0.568498	1	0.682547
RELINFRA.NS	0.701046	0.830481	0.719297	0.811581	0.682547	1

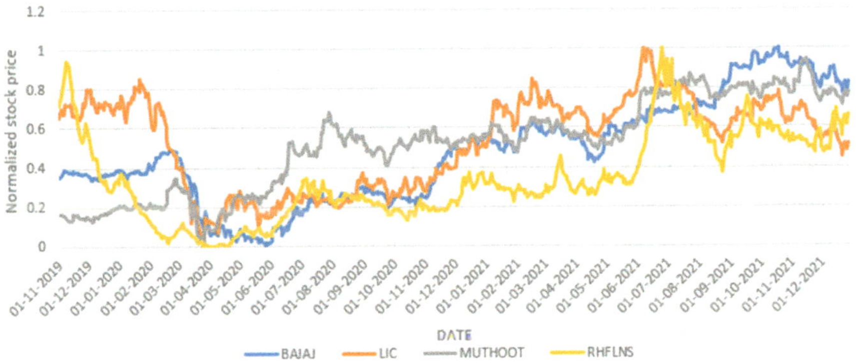

Fig. 5 Normalized graph of finance sector

Table 4 Correlation values between finance companies

	Bajaj	LIC	Muthoot	Rhfl
Bajaj	1	0.714902	0.760984	0.712494
LIC	0.714902	1	0.36199	0.647185
Muthoot	0.760984	0.36199	1	0.58674
Rhfl	0.712494	0.647185	0.58674	1

Correlation values are all positive but the correlation between EXPE and MHRILNS, THOMAS and EXPE, TRIP and THOMAS, MHRILNS and TRIP are less than 0.5 (Fig. 6).

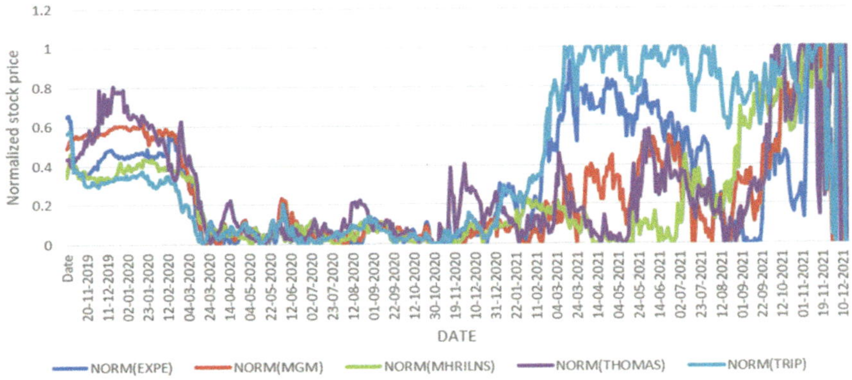

Fig. 6 Normalized graph of travel and tourism sector

Table 5 Correlation values between tourism companies

	EXPE	MGM	MHRILNS	THOMAS	TRIP
EXPE	1	0.6699114	0.354058	0.442404	0.773126
MGM	0.6699114	1	0.702177	0.773681	0.550526
MHRILNS	0.354058	0.702177	1	0.7666008	0.404562
THOMAS	0.442404	0.773681	0.766008	1	0.317964
TRIP	0.773126	0.550526	0.404562	0.317964	1

3 Covid Graph

Finally, we look at the covid graph of India to have a comparison.

Figure 7 represents the typical covid curve representing daily confirmed cases, familiar to all of us having two peaks corresponding to the first and second wave of the pandemic. The first peak occurs in the middle of September and these condone occurs at the first week of May. The pandemic though started a little bit earlier, becomes clearly visible from March, 2020.

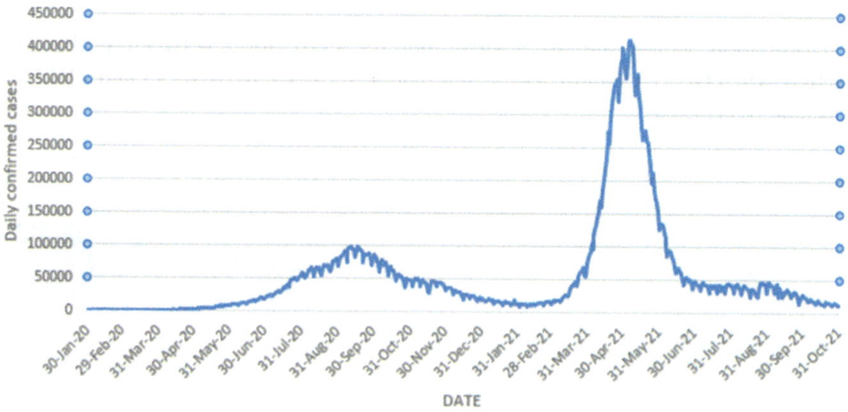

Fig. 7 Covid graph of India

4 Result and Discussion

In the automobile sector, all the stock values move in a similar fashion and reaches a low value in the April, 2020. Then they start to increase and peaks in February, 2021. After that, the curves become nearly flat. This indicates that the pandemic has no impact on the automobile sector. The initial dip is probably due to ongoing recession in Indian economy which is not connected with the pandemic. The bank stocks are low in April, 2020 and journey towards this low value started long before the pandemic. They reach high in April, 2021 and then continue to increase very slowly. Both the Finance sector and the construction sector reach at minimum in April, 2020 and then start to increase. So, pandemic effect could not be found here.

5 Conclusion

Among the five sectors covered in this study, the Automobile, Banking, Finance and the construction sector have not been grossly influenced by the ongoing pandemic. A one-line inference can be drawn as 'the second wave has no impact and the first wave also has a very little impact' on these sectors. On the contrary, the travel and tourism sector have been grossly influenced in this time period. From 2020 march throughout the year tourism sector faced a challenging time as it experienced the lowest stock market growth. This small and illustrative study shows that different sectors behave differently in this time period. Some of them win the battle and some lose. Deeper and extensive research is necessary to find out the economic, social and behavioral counterpart behind this behavior and these parameters should be linked with the observation.

6 Future Research Directions

A through, elaborate and comprehensive study is necessary with large number of sample stocks chosen from diverse sectors to have an insight on the movement of stock values during the pandemic and before it. A detailed and explorative study in this regard may reveal the reasons behind the ups and downs in these sectors. If any pattern can be determined, this may reveal a causality connection among observations.

Acknowledgements The authors thank the editor and the reviewers for their insightful comments and constructive suggestions that substantially improved this article.

Conflicts of Interest The authors declares that there is no conflict of interest.

References

1. Buszko, M., Orzeszko, W., Stawarz, M.: COVID-19 pandemic and stability of stock market-A sectoral approach. PLOS ONE **16**(5), e0250938 (2021)
2. Sahoo, M.: COVID-19 impact on stock market: evidence from the Indian stock market. J. Public Aff. **21**(4), e2621 (2021)
3. Al-Awadhi, A.M., Alsaifi, K., Al-Awadhi, A., Alhammadi, S.: Death and contagious infectious diseases: impact of the COVID-19 virus on stock market returns. J. Behav. Exp. Fin. **27**(1), 100326 (2020)
4. Khaled, A.S.D., Alabsy, N.B., Al-Homaidi, E.A., Saee, A.M.M.: The impact of the COVID-19 pandemic on retailer performance: empirical evidence from India. Innovat. Market. **16**(4), 129–138 (2020)
5. Ramelli, S., Wagner, A.F.: Feverish stock price reactions to COVID-19. Rev. Corpor. Fin. Stud. **9**(3), 622–655 (2020)
6. Akhtaruzzaman, M., Boubaker, S., Sensoy, A.: Financial contagion during COVID-19 crisis. Fin. Res. Lett. **38**, 101604 (2021)
7. Donthu, N., Gustafsson, A.: Effects of COVID-19 on business and research. J. Bus. Res. **117**, 284–289 (2020)
8. Zhang, D., Hu, M., Ji, Q.: Financial markets under the global pandemic of COVID-19. Financ. Res. Lett. **36**, 101528 (2020)
9. Okorie, D.I., Lin, B.: Stock markets and the COVID-19 fractal contagion effects. Fin. Res. Lett. **38**(4), 101640 (2021)
10. Martin Scheicher, F.: The co-movements of stock markets in Hungary, Poland and the Czech Republic. Int. J. Fin. Econ. **6**(1), 27–39 (2021)

A Mathematical Analysis on Covid-19 Transmission Using Seir Model

Sandip Saha, Apurba Narayan Das, and Pranabendra Talukdar

Abstract The current work describes the scenario of Covid-19 wave by SEIR model with the aid of mathematical analysis. The SEIR model describes the present scenario using a stability point of view, namely Disease-free equilibrium (DFE) and endemic (EE) equilibrium with the aid of the next-generation matrix, to predict the possible outcomes of recovery rate, infectious growth rate, and death rate and reproduction number.

Keywords COVID-19 epidemic · Equilibrium state · Stability analysis

1 Introduction

For the first time, the novel corona virus is found in Wuhan, Hubei Province, China, in the month of December 2019 [1, 2]. The transmission of the SARS-COV-2 virus caused the pandemic and it has become the most serious issue in the present world. It spreads rapidly across the globe in a very short span of time. Now, COVID-19 become a major research object in different branches, including Biology, chemical sciences, applied mathematics, economics, and even the living space, far outside the reach of medical science. In January 30, 2020, WHO [3, 4] declared an outbreak with an international public health crisis and finally, announced a pandemic situation on March 11, 2020. Some common symptoms of COVID-19 patient are fever, shortness of breath, exhaustion and lack of smell. However, some COVID-19 patients developed pneumonia and in some cases, acute respiratory distress syndrome has been found.

S. Saha
Department of Mathematics, Madanapalle Institute of Technology & Science, Madanapalle 517325, Andhra Pradesh, India

School of Advance Sciences (SAS) Mathematics, Vellore Institute of Technology Chennai, Chennai 600127, Tamilnadu, India

A. N. Das · P. Talukdar (✉)
Department of Mathematics, Alipurduar University, Alipurduar 736121, West Bengal, India
e-mail: dinhatapt@gmail.com

S. Banerjee and A. Saha (eds.), *Nonlinear Dynamics and Applications*,
Springer Proceedings in Complexity,
https://doi.org/10.1007/978-3-030-99792-2_123

According to the medical research, the general symptoms of sick people appear after five to six days. Some COVID-19 patient suffered shortness of breath quickly due to the lack of oxygen, hence chest X-rays, and CT scans become incompatible. The entire research community is still now trying to reveal the original fact that is going on inside the body of affected patients. Some recent investigations reported that Corona is a particular type of hypercoagulable state, which allows blood clotting in blood vessels. Kermack et al. [5], mathematically studied the characteristics of infectious diseases by employing the epidemiological statistical models and following in the works of Kermack and McKendrick [5], some eminent models to present the transmission mechanisms of some particular diseases have been proposed, such as the HIV model [6], Heathcote and Yorke [7] model to disseminate gonorrhea [8], Ronald Ross model for controlling malaria [9], Glass network [10, 11] etc. Currently, researchers are extensively involved to develop a statistical model for identifying COVID-19. Easwar Moorthy et al. [12] analyze the fractal-based prognostic model for second wave of COVID-19. In addition, [13, 14] researchers analyzed the predication of "when will the pandemic be culminated and immunity against covid-19 in India" At this situation, it is difficult to forecast the negative impact of COVID19 by applying a single model. Shahrear et al. [15] mathematically analyzes the pandemic of COVID transmission in Bangladesh using SEQIRP model and concluded that the infection rate is proportional to the number of infected populations. In addition, it was also studied that if an effective vaccine is not available to the populations, then the rate of death percentage will rise sharply, consequently recovery rate will diminish remarkably. Based on the Russia real-time data, Tomchin et al. [16] studied the pandemic scenario of COVID-19 using SEIR model. The calculations show that the peak in the number of infected in Russia should occur no earlier than the end of May. In 2021, Hamdi et al. [17] mathematically performed the pandemic of COVID transmission in Saudi Arabia using the SEIR model with the aid of next generation matrix and stated that if the rate of transmission, $\beta \leq 0.0000000112$, then the number of infected cases will reduce moderately. The devastating COVID-19 pandemic situation in the whole world claims a need to develop scientific models for improving a medical preparation and monitoring the pandemic in a long-term course. From the literature survey, it has been clear that in the view of mathematical perspectives, the number of research work on COVID-19 is very less. Most of the works performed are based on SEIR model, which provides us the enough confidence to find the research gap in that field. In the present article, the impact of Covid-2nd wave is analyzed using SEIR model. For numerical understanding, several authors have utilized different software [18–21], but here, we only use the symbolic computational MATLAB software. Presently, the COVID-2nd wave not only causes the falling of health issues but also affect the social, economic, and cultural issues. The prime objective of this investigation is to narrate the pandemic situation due to COVID in the whole World. Here, SEIR model forecast some prediction with the aid of stable and unstable equilibrium states and local and global stability analysis. Based on the reproduction number [21–23], the study of the model would reveal the stable and unstable situations. At the end, sensitivity analysis for the SEIR model have been introduced with some basic phenomena of Covid-19 such as low growth face, moderate growth face, transition face, steady and unsteady equilibrium face.

2 SEIR Model

The SEIR model can be treated as an extended version of the SIR model that are utilized by several authors [14–19, 24] to investigate the epidemic outbreaks. SEIR (susceptible-exposed-infectious-removed) model (Fig. 1) assumes the following relations,

a. Susceptible (S) persons have never been infected by the microorganism but considered as infected person,
b. Exposed (E) persons is infected by a microorganism that causes to create disease, but they are not considered as infectious, due to the passing over of the latent period of the disease,
c. Infectious (I) persons are infected and they can transmit the disease to others,
d. Removed (R) persons are not able to transmit that microorganism. This group contains the recovered individuals as well as the fatalities.

COVID-19 is one of the devastating pandemic that the entire human civilization has never been experienced before. In this model, it is assumed that the entire population initially belongs to a susceptible compartment. Contact rates in between the infected and suspected individuals in the population are utilized to compute the probability that a susceptible and an infectious people is in contact for a finite time and get infected. The prime difference between the SIR model and the SEIR model is that the inclusion of the exposed group to the SEIR model. The exposed group actually is a class that lies in between the susceptible group and the infectious group. It includes the people who are exposed to be infected but still are not infectious. In this model, four groups have been considered instead of three and thus, four differential equations required to describe the spread of pandemic. Susceptible class means those individuals who are at risk to be infected with the COVID-19 wave virus. The exposed group means group of those people who have been already infected with the COVID-19 virus but they have no symptoms (asymptomatic). Infectious groups are the individuals who are infected with the COVID-19 wave and also able to pass infection to susceptible persons. Finally, the recovered group are the individuals who have been already recovered. In this model, we consider that births and deaths occurred at the same rate

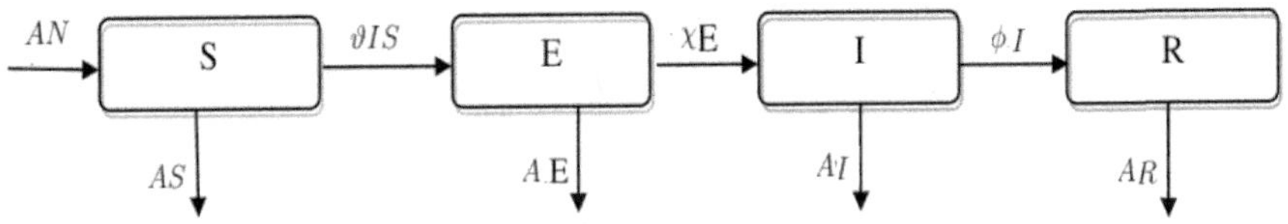

Fig. 1 Schematic diagram of SEIR model

and all the newborn are susceptible (that is, they have no inherited immunity). It is confirmed that the total population, N means that the sum of all the variables. Thus, we write

$$S(t) + E(t) + I(t) + R(t) = N \tag{1}$$

The Eq. (1) can be written in the following form:

$$s(t) + e(t) + i(t) + r(t) = 1, \tag{2}$$

where

$$s(t) = \frac{S(t)}{N}, e(t) = \frac{E(t)}{N}, i(t) = \frac{I(t)}{N}, r(t) = \frac{R(t)}{N}. \tag{3}$$

Now, we assume that the population blends homogeneously, irrespective of age, mobility or the other social factors. Thus, the following set of ODE^s is considered to represent this model:

$$\frac{dS}{dt} = AN - AS - \vartheta IS \tag{4}$$

$$\frac{dE}{dt} = \vartheta IS - (A + \chi)E \tag{5}$$

$$\frac{dI}{dt} = \chi E - (A + \varphi)I \tag{6}$$

$$\frac{dR}{dt} = \varphi I - AR \tag{7}$$

We consider the parameters $A, AS, \vartheta IS, \chi E, \vartheta > 0, \chi > 0, \varphi > 0, \psi > 0$, which respectively denote average birth and death rate, rate at which individuals are born into the susceptible class with no passive, rate at which susceptible enters into the exposed class without being infected, the rate at which an exposed person becomes infectious, rate at which an infected individual may recover fully transmission coefficient, latency coefficient, recovery coefficient and capital death rate respectively. Initially, we assume $(S(t = 0), E(t = 0), I(t = 0), R(t = 0)) \in \{(S, E, I, R) \in [0, N]^4 : S > 0, E > 0, I > 0, R > 0\}$.

The equations (4)-(7) formed as

$$\frac{dS}{dt} + \frac{dE}{dt} + \frac{dl}{dt} + \frac{dR}{dt} = AN - AS \tag{8}$$

Equations (2) and (3) are plug in the equations ((4)–(7)), we find the following ODEs.

$$\frac{ds}{dt} = A - (A + \vartheta i)s \tag{9}$$

$$\frac{de}{dt} = \vartheta Ie - (A + \chi)e \tag{10}$$

$$\frac{di}{dt} = \chi e - (A + \varphi)i \tag{11}$$

The above ODEs possess a disease-free equilibrium (DFE) and an endemic equilibrium (EE).

2.1 Calculation of R_0

The Jacobian method employed for the SEIR model gives a biologically reasonable result but this method cannot provide good result in a complex compartmental model, i.e., model having large number of infected compartments, as it depends on the algebraic Routh Hurwitz conditions for stability of the Jacobian matrix. Diekmann et al. [16] proposed an alternative method and later on, it is modified by van den et al. [17], which gives a new way to determine R_0 in the case of ODE compartmental model employing the next generation matrix. In this work, the method is briefly introduced and for details, readers are referred Van den et al. [19].

Let $x = (x_1, x_2, \cdots, x_n)^T$ denotes people present in each of the compartments with first $m < n$ compartments containing infected people. We assume the existence and stability of the DFE x_0 in absence of disease and the equations for $x_1, x_2, \cdots, x_n$ at the DFE effectively disassociates from the rest of the equations. For detailed discussions on the basic assumptions, see the references mentioned above. Consider the equation

$$\frac{dx_i}{dt} = F_i(x) - V_i(x), \; for \; i = 1, 2, \cdots, m \tag{12}$$

where $F_i(x)$ denotes the rate of increase infections in the ith compartment and $V_i(x)$ expresses the rate of transitions between the ith compartment and other infected compartments in other way with the assumption that F_i and $V_i \in \zeta^2$, and $F_i = 0$ if $i \in [m + 1, n]$. Now we define

$$F = \left[\frac{\partial F_i(x_0)}{\partial x_j}\right] \tag{13}$$

$$V = \left[\frac{\partial V_i(x_0)}{\partial x_j}\right] for 1 \le i, j \le m \tag{14}$$

Biologically, F and V represent the entry wise non-negativity and a non-singular M-matrix (see Berman and Plemmons (1994)) respectively, so V^{-1} also shows the entry wise non-negativity. If the number of infectious at the initial level be denoted by $\Omega(0)$, then FV^{-1} $\Omega(0)$ defines the entry wise non-negative vector and gives the expected number of newly infected people. The (i, j)th entry of the matrix FV^{-1} denotes the expected number of secondary infections in the ith compartment supposing that infected person is introduced in j th compartment. Thus, FV^{-1} is claimed to be the next generation matrix

$$R_0 = \rho(FV^{-1}), \tag{15}$$

where ρ signifies the spectral radius. The linear stability of the DFE is confirmed by $s(F - V)$, where s is the maximum of the real part of the eigenvalues of the Jacobian matrix and s often known as spectral bound. Considering the notations stated above, the interrelationship between this quantity and R_0 is described below in brief (for details, see the above cited references).

Theorem 1 *If x_0 is a DFE of the system*

$$\frac{dx_i}{dt} = F_i(x)V_i(x),$$

then x_0 is locally asymptotically stable if $R_0 = \rho(FV^{-1})$, but unstable if R_0>1.

The next generation matrix is expressed using the Eqs. (9)–(11) and (18) of the SEIR Model. Here E and I represent the infected compartments. F and V are described in the DFE matrices as

$$F = \begin{bmatrix} 0 & \vartheta s_0 \\ 0 & 0 \end{bmatrix}, and V = \begin{bmatrix} A + \chi & 0 \\ -\chi & A + \phi \end{bmatrix},$$

So FV^{-1} has the eigenvalues 0 and R_0, where

$$R_0 = \frac{\chi \vartheta s_0}{(A + \chi)(A + \varphi)} \tag{16}$$

as derived biologically by the Eq. (18). Here ϑs_0 is the rate of infection population of susceptible, $\frac{\chi}{A+\chi}$ represents the fraction transferring from E to I, $\frac{1}{A+\varphi}$ is the mean time in I, thus the (1, 1) entry of FV^{-1} shows expected number of the secondary infections created in compartment E by an infected people in E.

2.2 Local Stability

Considering the transformed sub-systems in ((4)-(6)), the local-stability is analyzed to find the disease-free equilibrium (DFE):

$$\text{LDFE} = (s, e, i), \tag{17}$$

and endemic equilibrium (DFE):

$$\text{LEE} = (s_1, e_1, i_1). \tag{18}$$

2.3 Disease-Free Equilibrium (DFE)

The conditions for constructing DFE is given by

$$(s_0, e_0, i_0, r_0) = \left(\frac{\gamma}{\alpha}, 0, 0, 0\right). \tag{19}$$

There exist two infected compartments (e and i), and it is found that the equations in two variables decide the stability of the DFE. The Eqs. (10) and (11), the Jacobian matrix at the DFE provides the characteristic equation:

$$l^2 + (2A + \chi + \varphi)l + ((A + \chi)(A + \varphi) - \chi\vartheta s_0) = 0. \tag{20}$$

All the roots of the equation will contain negative real parts (thus the DFE is linearly stable) if and only if each of the coefficient is positive, i.e., $\frac{\chi\vartheta s_0}{(A+\chi)(A+\varphi)} < 1$. Here, $\frac{\chi}{A+\chi}$ is showing the progression from exposed to infectious, and $\frac{1}{A+\varphi}$ defines the average infectious time for taking death. In the biological understanding, the fraction is considered as reproduction number (R_0),

$$R_0 = \frac{\chi\vartheta s_0}{(A + \chi)(A + \varphi)}. \tag{21}$$

If $R_0 \leq 1 \Rightarrow \lim_{t\to\infty}(\text{s(t), e(t), i(t), r(t)}) = \text{DFE}$, and for $R_0 > 1 \Rightarrow \lim_{t\to\infty}(\text{s(t), e(t), i(t), r(t)}) = \text{EE}$. From the Eq. (9)–(11) we find,

$$A - (A + \vartheta i)s = 0, \tag{22}$$

$$\vartheta is - (A + \chi)e = 0, \tag{23}$$

$$\chi e - (A + \varphi)i = 0. \tag{24}$$

Therefore, the Jacobian matrix is defined as follows:

$$J = \begin{bmatrix} -(\mathrm{A} + \vartheta \mathrm{i}) & 0 & \vartheta \mathrm{s} \\ \vartheta \mathrm{i} & -(\mathrm{A} + \chi) & \vartheta \mathrm{s} \\ 0 & \chi & -(\mathrm{A} + \varphi) \end{bmatrix}$$

.

Using Eq. (22)–(24), we find the DFE $(s, e, i) = (1, 0, 0)$, if we set $i = 0$, and the Jacobian matrix can be written as,

$$J_{DFE} = \begin{bmatrix} \mathrm{A} & 0 & \vartheta \mathrm{s} \\ \vartheta \mathrm{i} & -(\mathrm{A} + \chi) & \vartheta \mathrm{s} \\ 0 & \chi & -(\mathrm{A} + \varphi) \end{bmatrix}$$

.

Therefore from,

$$det(J_{DFE} - \lambda I_d) = \lambda^3 + b_1\lambda^2 + b_2\lambda + b_3$$

,

where

$$b_1 = 3A + \varphi + \chi, \tag{25}$$

$$b_2 = [(A + \chi)(A + \varphi) - \vartheta\chi + A(2A + \varphi + \chi)], \tag{26}$$

$$b_3 = A[(A + \chi)(A + \varphi) - \vartheta\chi]. \tag{27}$$

The stability criteria, given in [17, 18] suggest that if $b_1, b_3 > 0$, and $b_1b_2 - b_3 > 0$, then every roots of the characteristic equation will have negative real part that signifies a stable equilibrium. Therefore, $R_0 < 1$ signifies a stable DFE otherwise, it becomes unstable.

2.4 Endemic Equilibrium (EE)

To conclude the endemic equilibrium state from the Eq. (20), we find

$$e = \frac{(A + \varphi)i}{\chi}. \tag{28}$$

Putting the above value of e in Eq. (19), we get

$$s_1 = \frac{(A+\chi)(A+\varphi)}{\vartheta \chi s_0} = \frac{1}{R_0}. \tag{29}$$

Putting the value of s in Eq. (18), we find

$$i_1 = \frac{A}{\vartheta}(R_0 - 1), \tag{30}$$

and considering the Eqs. (24)–(26), we find

$$e_1 = \frac{(R_0 - 1)A}{R_0(A+\chi)} \tag{31}$$

Therefore, we find the endemic equilibrium points as,

$$(s_1, e_1, i_1) = \left(\frac{1}{R_0}, \frac{(R_0-1)A}{R_0(A+\chi)}, \frac{A}{\vartheta}(R_0 - 1)\right).$$

The Jacobian matrix is written as

$$J_{EE} = \begin{bmatrix} -AR_0 & 0 & -\frac{(A+\chi)(A+\varphi)}{\vartheta} \\ A(R_0-1) & -(A+\chi) & \frac{(A+\chi)(A+\varphi)}{\vartheta} \\ 0 & \chi & -(A+\varphi) \end{bmatrix}$$

Thus,

$$det(J_{EE} - \lambda I_d) = \lambda^3 + b_1\lambda^2 + b_2\lambda + b_3,$$

where

$$b_1 = (2 + R_0)A + \varphi + \chi, \tag{32}$$

$$b_2 = AR_0(2A + \varphi + \chi), \tag{33}$$

$$b_3 = A(R_0 - 1)\left[A^2 + A(\varphi + \chi) + \vartheta\varphi\right]. \tag{34}$$

From the stability criteria of Routh-Hurwitz (2013), if $b_1, b_3 > 0$, and $b_1 b_2 - b_3 > 0$, then the roots posses negative real part which means the occurrence of stable equilibrium. The first two conditions remain true for $R_0 > 1$ as $b_1, b_3 > 0$, and $b_1 b_2 - b_3 > 0$. Therefore, endemic state equilibrium becomes stable if $R_0 > 1$, otherwise unstable. It causes the occurrence of an outbreak and the epidemic grows.

The epidemic criteria in R_0, given in (21) presents the basic reproduction value, which is the most important factor for analyzing any epidemiological model. R_0, especially shows whether an epidemic appears due to the spread of infectious diseases because R_0 represents the average number of secondary infections created by one infected people during the mean period of infection in a susceptible population. If $R_0 \leq 1$, then on average, new infections created by one infected people over the mean course of the pandemic is less than unity, which implies the disease will disappear in no time. On the other hand, when $R_0 > 1$, the number of new infectious created by infected people becomes greater than unity, that causes the spread of infectious disease as an epidemic.

Algebraic analysis of R_0: If a plan for controlling the disease is considered for entire population, then Herd immunity is found. On the other hand, when the population contains several host types, then the strategy to control pandemic only found for one host population. For example, we consider a vector-host model of malaria where spray is applied to the mosquito vectors. To propose this Roberts et al. [18, 19] introduced a particular type of reproduction number, R_0. This type of controlling strategy influences single column/row of the next generation matrix, depending on control of impressing susceptibility or infectivity. Shuai et al. [20] recentralized this idea by singling out the entries to control and defined a target reproduction number. For example, shorten of contacts among the children can be taken as a reduction strategy for breaking out Cholera. Let us assume $K = \left[k_{ij}\right]$ be an nth order next generation matrix (i.e., $K = FV^{-1}$) and the entries for a set s are taken as a control strategy (this may be treated either in a decreasing or an increasing manner in the entries of s. The (i, j)th element of the target matrix Ks is defined as k_{ij}, if entry $(i, j) \in s$, and 0 otherwise. For $\rho(K - Ks) < 1$, the target reproduction number is denoted as F_s and defined by

$$F_s = \rho\left(Ks(I_d K + Ks)^{-1}\right) \tag{35}$$

where I_d represents an nth order unit matrix. When all the entries in a particular row/column or more rows/columns of K are selected, then the selected reproduction number is reduced to the reproduction number as narrated by Roberts et al. [14, 15]. If a fraction,$1 - \frac{1}{F_s}$ of total population can be prevented then the disease will no longer exist.

Theorem 2 *The solutions of SEIR model together with the initial condition become a subset in the interval* $[0, \infty)$ *and* $\{s(t), e(t), i(t), r(t)\} \geq 0$ *for* $0 \leq t < \infty$.

The right-hand sides of the SEIR model is totally continuous and locally Lipschitzian on t. The solutions $\{s(t), e(t), i(t), r(t)\}$ together with its initial conditions also exist and become unique in the interval $[0, \infty)$. From the Eq. (9), we have $[A - \vartheta i(t)s(t)] \geq 0$, then, we find valid inequality

$$\frac{ds}{dt} \geq -As(t). \tag{36}$$

By solving the above inequality, we find

$$s(t) \geq s_0 e^{-As(t)} \geq 0. \tag{37}$$

Hence, $s(t)$ becomes a non-negative function for all t such that, $t \in [0, \infty)$.
In the similar way, form the Eqs. (10) and (11), we find

$$e(t) \geq e_0 e^{-(A+\chi)t} \geq 0, \tag{38}$$

$$i(t) > i_0 e^{-(A+\varphi)t} \geq 0, \tag{39}$$

$$r(t) \geq r_0 e^{-At} \geq 0. \tag{40}$$

Thus, $e(t)$, $i(t)$ and $r(t)$ are all non-negative functions for all values of t such that, $t \in [0, \infty)$ and hence the proof.

Lemma 1 *All of the solutions of SEIR model that initiate in the zone* $[0, \infty)$ *are bounded inside the domain* ϖ, *given by* $\varpi = \{(s, e, i, r) \in [0, \infty) : 0 \leq N(t) \leq A\vartheta\}$ as$t \to \infty$.

2.5 *Uniqueness Theorem for DFE and EE Conditions:*

The general condition of DFE is given by

$$(s_0, e_0, i_0, r_0) = \left(\frac{\gamma}{\alpha}, 0, 0, 0\right) \tag{41}$$

Therefore, $|J_{DFE}| \neq 0$ means that the equilibrium is isolated, which means there is a disk around it that does not contain other equilibrium. $|J_{EE}| \neq 0$, where $R_0 = \frac{\chi\vartheta s_0}{(A+\chi)(A+\varphi)}$, Therefore, R_0 is unique, hence the proof.

2.6 *Global Stability of Equilibrium of the SEIR Model (Lyapunov's Stability Theorem)*

The Lyapunov functions represent the scalar functions that may be utilized in order to prove the state of global stability of equilibrium. Lyapunov reports that if the function $V(x)$ is globally positive definite as well as radially unbounded, and time derivative of $V(x)$ is globally negative, $V(x) < 0$ for all $x = x^*$ then the equilibrium, x^* becomes globally stable in an autonomous system

$$\frac{dx}{dt} = f(x), \tag{42}$$

and $V(x)$ is called the Lyapunov's function.

Lemma 2 *In the SEIR model, DFE* $(e_0) = \left(\frac{A}{\vartheta}, 0, 0, 0\right)$ *stands for globally stable of the DFE under the state* $R_0 < 1$.

Lemma 3 *When* $R_0 \leq 1$ *the DFE*(e_0) *becomes globally attractive.*

3 Conclusion

1. From the SEIR epidemiological model, it has been concluded that if the reproduction number $R_0 > 1$, then the disease will spread like an outbreak. The sensitivity analysis revealed that whenever the transmission rate is increased or the recovery rate is reduced, the disease will spread, but whenever the transmission rate is reduced or the recovery rate is increased, the disease will dies out.
2. Every roots of the characteristic equation contains negative real part that signifies a stable equilibrium and for R_0<1 signifies a stable DFE, otherwise it becomes unstable. If $R_0 \leq 1$, then on average, new infections created by one infected people over the mean course of the pandemic is less than unity that implies the infectious disease will vanish. On the contrary, when $R_0 > 1$, the number of new infections people created by one infected people becomes greater than unity, that causes the spread of infectious disease as an epidemic.

References

1. Paul, A., Chatterjee, S., Bairagi, N.: Prediction on Covid-19 epidemic for different countries: focusing on South Asia under various precautionary measures. MdRxiv (2020). https://doi.org/10.1101/2020.04.08.20055095
2. Lin, Q., Zhao, S., Gao, D., Lou, Y., Yang, S., Musa, S., He, D.: A conceptual model for the coronavirus disease 2019 (COVID-19) outbreak in Wuhan, China with individual reaction and governmental action. Int. J. Infect. Dis. **93**, 211–216 (2020)
3. Fang, Y., Nie, Y., Penny, M.: Transmission dynamics of the COVID19 outbreak and effectiveness of government interventions: a datadriven analysis. J. Med. Virol. **92**(6), 645–659 (2020)
4. Bajardi, P., Poletto, C., Ramasco, J., Tizzoni, M., Colizza, V., Vespignani, A.: Human mobility networks, travel restrictions, and the global spread of 2009 H1N1 pandemic. PloS One **6**(1) (2011)
5. Kermack, W.O., McKendrick, A.G.: A contribution to the mathematical theory of epidemics. Proc. R. Soc. Lond. Ser. A **115**(772), 700–721 (1927)

6. Anderson, R.M., May, R.M.: Population biology of infectious diseases, Part 1. Nature **820**, 361–367 (2005)
7. Diekmann, O., Heesterbeek, P.A.: Mathematical epidemiology of infectious diseases: model building, analysis and interpretation. Math. Comput. Biol. **15**, 1–13 (2000)
8. Driesschea, P.V., Watmough, J.: Reproduction numbers and sub-threshold endemic equilibria for compartmental models of disease transmission. Math. Biosci. **180**(1–2), 29–48 (2002)
9. Driesschea, P.V., Watmough, J.: Further notes on the basic reproduction number. Math. Epidemiol. **180**(1–2), 159–178 (1945)
10. Wang, F., Yang, Y., Zhao, D., Zhang, Y.: A worm defending model with partial immunization and its stability analysis. J. Commun. **10**(4), 276–283 (2015)
11. Shahrear, P., Glass, L., Edwards, R.: Chaotic dynamics and diffusion in a piecewise linear equation. Chaos **25**(3), 033103 (2015)
12. Easwara Moorthy, D., Gowrisankar, A., Manimaran, A., Nandhini, S., Rondoni, L., Banerjee, S.: An exploration of fractal-based prognostic model and comparative analysis for second wave of COVID-19 diffusion. Nonlinear Dyn. **106**(2), 1375–1395 (2021)
13. Kavitha, C., Gowrisankar, A., Banerjee, S.: The second and third waves in India: when will the pandemic be culminated? Eur. Phys. J. Plus **136**(5), 1–12 (2021). J. Plus **135**
14. Gowrisankar, A., Rondoni, L., Banerjee, S.: Can India develop herd immunity against COVID-19? Eur. Phys. **6**, 1–9 (2020)
15. SMS, R., Shahrear, P., Islam, M.S.: Mathematical model on branch canker disease in Sylhet, Bangladesh. Int. J. Math. **25**(3), 80–87 (2017)
16. Tomchin, D., Fradkov, A.: Partial prediction of the virus COVID-19 spread in Russia based on SIR and SEIR models. https://doi.org/10.1101/2020.07.05.20146969
17. Hamdy, M., Youssef, A.N., Alghamdi, A.N., Alghamdi, A.M., Ezzat, A.M.: Modified SEIR model with global analysis applied to the data of spreading COVID-19 in Saudi Arabia. AIP Adv. **10**(12), 125210 (2020)
18. Roberts, M.G., Heesterbeek, P.J.A.: A new method for estimating the effort required to control an infectious disease. Proc. Bio. Sci. **270**(1522), 1359–1364 (2017)
19. Roberts, M.G., Heesterbeek, P.J.A.: Model-consistent estimation of the basic reproduction number from the incidence of an emerging infection. J. Math. Biol. **55**(5–6), 803–816 (2017)
20. Shuai, Z., Heesterbeek, P.J.A., Driessche, V.D.: Extending the type reproduction number to infectious disease control targeting contacts between types. J. Math. Biol. **67**(5), 1067–1082 (2013)
21. Blower, M.S., Dowlatabadi, H.D.: Sensitivity and uncertainty analysis of complex models of disease transmission: an HIV model. Int. Statist. Rev. **62**(2), 229–243 (1994)
22. Shahrear, P., Rahman, S.M.S., Nahid, M.H.: Prediction and mathematical analysis of the outbreak of coronavirus (COVID-19) in Bangladesh. Results Appl. Math. **10**, 1–12 (2021)
23. Ruiz Estrada, M.A.: COVID-21. Available at SSRN 3686440 (2020)
24. Sandip, S., Biswas, P., Nath, S.: Numerical prediction for spreading novel coronavirus disease in india using logistic growth and SIR models. Eur. J. Med. Educ. Technol. **14**(2), 1–09 (2021)

Dynamics of Coronavirus and Malaria Diseases: Modeling and Analysis

Attiq ul Rehman and Ram Singh

Abstract COVID-19 has been declared a pandemic by the WHO on the 11th of Mar. 2020. This virus is believed to be born in China in 2019. The study of this disease is very complicated and challenging. In this manuscript, a fractional-order epidemic model to study the impact of COVID-19 and malaria disease has been proposed and analysed. The model is formulated with the help of fractional order by using the Caputo-Fabrizio derivative. The model is solved numerically with the help of the ABM method. The parameters which characterize the disease transmission are taken from real data of India [1]. The qualitative and quantitative behaviour of the proposed model is examined. The numerical work is performed to authenticate the analytic solutions. It is observed that the malaria disease acts as a launching pad for the COVID-19 dynamics as it weakens of humans immune system.

Keywords Coronavirus · Malaria · Modeling · Analysis

Abbreviations

COVID-19	Coronavirus Diseases-19
SARS-CoV-2	Severe Acute Respiratory Syndrome Coronavirus 2
WHO	World Health Organization
GHE	Global Health Emergency
DDT	Dichloro Diphyenyl Trichloroethane
ARDS	Acute Respiratory Distress Syndrome
CQ	Chloroquine
HCQ	Hydroxychloroquine
ABM	Adams Bashforth Moulton

A. Rehman (✉) · R. Singh
Department of Mathematical Sciences, BGSB University, Rajouri 185234, India
e-mail: attiq@bgsbu.ac.in

R. Singh
e-mail: drramsinghmaths@gmail.com

S. Banerjee and A. Saha (eds.), *Nonlinear Dynamics and Applications*,
Springer Proceedings in Complexity,
https://doi.org/10.1007/978-3-030-99792-2_124

C	Caputo
CF	Caputo-Fabrizio
ABC	Atangana Baleanu in Caputo sense
DFE	Disease Free Equilibrium
EE	Endemic Equilibrium
LAS	Locally Asymptotically Stable
GAS	Global Asymptotically Stable
ODE	Ordinary Differential Equation
FDE	Fractional Differential Equation
VIE	Volterra Integral Equation
PE	Predict Evaluate
CE	Correct Evaluate
PV	Predictor Value
CE	Corrector Value

1 General Introduction

The mathematical modeling of disease has been attracting the attention of many epidemiologists. COVID-19 was revealed to be supervised for a substantial number of pneumonia cases from China, in Dec. 2019. The number of cases has exploded in China, and then all over the world. On Dec. 31, 2019, the disease was first reported to the WHO, and on Jan. 30, 2020, this outbreak was declared a GHE, followed by a global pandemic on Mar. 11, 2020.

Fever, dry cough, and Fatigue are the most prevalent symptoms of COVID-19. Runny nose, nasal congestion, loss of taste, pains, rashes on the skin, and smell are all possible side effects. These symptoms usually appear and are modest over time. The majority of patients recover from the ailment without the need for medical treatment. All stages of people with underlying medical problems like respiratory or heart difficulties, diabetes, or weakened immune systems should be given special attention [2].

When someone with coughs, speaks, and minute droplets from the nose are transmitted from person to person, especially in improperly congested interior areas where the short-range aerosol transmission cannot be ruled out. It can also be contracted by contacting contaminated things or surfaces and then touching the eyes, nose, or mouth [3]. People who have no symptoms can to some accounts, spread the virus. But, the frequency of such transmissions is unknown.

Likewise COVID-19, malaria is a parasitic infection, transmitted by Anopheles vectors and caused by genus plasmodium parasites, that gives a notable worldwide health threat and leads to a long life-threatening illness. This disease increases mortality and mobility in these areas, causing a toxic impact on their economies and their health structures [4]. About 229 million cases of malaria and approx. 5,00,000 deaths have been reported by WHO, displaying the largest fraction of cases and the highest death [5].

It is critical to control the spread of malaria by implementing long-term and targeted interventions. These two actions should be sufficient to halt the spread of malaria [6]. Insecticide-treated fabrics, most likely vector nets, have been utilized to combat malaria. In addition, data suggest that anopheline vectors have begun to develop resistance to the pyrethroids used to treat vector nets. Indoor residual DDT pesticides and insect reflecting creams are two other options.

Drainage of standing waters is beneficial to human health. Chemoprophylaxis is applied as a prophylactic measure in endemic areas. This reduces the risk of severe maternal anemia and low birth weight in babies. However, due to a lack of Plasmodium falciparum tolerance, chemoprophylaxis is only marginally successful [7]. In circumstances of limited transmission, microgametocyte medicines may play a substantial role in malaria prevention. It lowers the rate of transmission. But, transgenic vectors become available, they may provide more benefits in the fight against malaria.

While both diseases can present similarly, the same symptoms they share include, but aren't controlled to, breathing problems, fever, head pain, and tiredness, that can give to a mistreat of both and vice versa, especially if the clinician data solely on signs [8]. Coronavirus had a big field of clinical signs from asymptomatic. Headache, myalgia, and vomiting are all possible symptoms. Malaria patients, on the other hand, commonly with fever, headache, and sweating, as well as fatigue, arthralgia, myalgia, nausea, and vomiting [9].

Malaria-affected people may be mistreated as COVID-19 due to the similarity of symptoms between both diseases, particularly difficulty breathing, fever, acute onset headache, and fatigue. Furthermore, both malaria and COVID-19 can cause complications such as ARDS, septic shock, as well as multi-organ failure. Dry cough, headache, and fever such as health protector patients and workers along with a long time of contact with a number of the COVID-19 peoples, is the first step in identifying a COVID-19 patient [10].

In terms of alertness locally, regionally, and internationally, a top-level of suspicion is demonstrated towards the COVID-19. Humans with fever are currently tested for COVID-19 and then return if the result is non-positive, siding the chance of malaria. The malaria consequences are toxic if a case of malaria is overlooked. Patients who are feverish and have a COVID-19 disease, on the other hand, may be screened for malaria. A single instance of the COVID-19 has the ability to show impact on 3.58 people [11]. The third situation is that a patient has both malaria and COVID-19, and the treatment, as well as diagnosis of one of these, leads to the other being missed.

The widespread utilization of CQ, HCQ, and other anti-malarial medications in malaria-positive regions, according to some researchers, explains the reverse connection between COVID-19 and malaria. It's worth noting that the ability of CQ and HCQ for the medication of COVID-19 disease has been examined since the initial SARS outbreak [12]. Some older studies elaborated on the usefulness of HCQ in the medication of SARS-CoV-2 and stated that a daily dose of four hundred milligrams of HCQ for ten days was the best treatment option. But, subsequent clinical data have recommended that HCQ for COVID-19 be given at a loading medicine of four hundred milligrams twice on the first day, followed by two hundred milligrams every 12 hours for the next 4 days.

In this manuscript, we have presented an epidemic model of COVID-19 and malaria. As there are very few papers in which the dynamics of both diseases are studied. It is important to study the dynamics of these epidemics.

The paper organization is as follows. Basic preliminaries are in Sect. 2. The mathematical epidemic model is described in Sect. 3. Existence and positiveness are proved in Sect. 4. Stability analysis is discussed in Sect. 5. The numerical work are presented in Sect. 6. At last, the conclusion is drawn in Sect. 7.

2 Basic Preliminaries

Since the fractional-order derivative is the generalization of the integer-order derivative. The basic concepts and definitions of Caputo, Caputo-Fabrizio, and Atangana-Baleanu-Caputo. The Caputo derivative has been derived with the power-law type of singular non-local kernel, CF with a non-singular kernel, and ABC with Mittag-Leffler kernel memory. But in this paper, we used only the CF fractional operator since it has a non-singular kernel.

Definition 1 For a function f from C^n space, the Caputo fractional derivative of order $\alpha \in (n-1, n]$ with $n \in \mathbb{Z}^+$ is given as follow:

$$ {}_0^C D_t^\alpha (f(t)) = \frac{1}{\Gamma(n-\alpha)} \int_0^t \frac{1}{(t-\varphi)^{\alpha-n+1}} f^{(n)}(\varphi) d\varphi, $$

The above Caputo fractional derivative approaches the ordinary derivative when order α is one.

Definition 2 For $a < 0$, $\alpha \in (0, 1]$, the fractional derivative with a non singular kernel, for a function $f(t)$ of C^1 space is given as follow:

$$ {}_0^{CF} D_t^\alpha f(t) = \frac{M(\alpha)}{1-\alpha} \int_a^t (f(t) - f(\varphi)) exp\left[-\alpha \frac{t-\varphi}{1-\alpha} \right] d\varphi. $$

Definition 3 If $f(t)$ is differentiable then, the new derivative is called the Atangana-Baleanu in the Caputo sense as follow:

$$ {}_a^{ABC} D^\alpha f(t) = \frac{B(\alpha)}{1-\alpha} \int_a^t D_t^\alpha f(\varphi) E_\alpha \left[-\alpha \frac{(t-\varphi)^\alpha}{1-\alpha} \right] d\varphi, $$

with $B(\alpha)$ is the normalization mapping s.t. $B(0) = B(1) = 1$, in which $B(\alpha) = 1 - \alpha + \alpha/\Gamma(\alpha)$.

Definition 4 For order $\alpha \in (0, 1)$, the unchangeable point x^* is called the critical point of the below Caputo fractional-order function:

$$ {}_0^C D_t^\alpha(x(t)) = f(t, x(t)) \iff f(t, x^*) = 0. \tag{1}$$

To show the stability of the non-linear fractional-order systems in the insight through Lyapunov way we firstly think of the afterward results from [13].

3 Fractional-Order Mathematical Model Formulation

To express the epidemic model, we split the COVID-19 affected host population into four sub-classes, namely susceptible, infectious, vaccinated, and discharged individuals. The complete host population is constituted by $\Sigma_1 = S_h + I_h + V_h + D_h$, where S_h is COVID-19 susceptible host, I_h is COVID-19 infectious host, V_h is the COVID-19 vaccinated host and D_h is COVID-19 discharge host population. Likewise, Σ_2 is the complete number of vectors which are further split into susceptible vectors S_m, and infected vectors I_m. So that $\Sigma_2 = S_m + I_m$. The mathematical model of six FDEs to relate the mechanism of the dynamical transmission of both disease as follow:

$$\begin{aligned}
{}_0^{CF} D_t^\alpha S_h &= (1-\sigma)Z_h - \beta_h S_h I_m + \rho L_0 h - x_h S_h, \\
{}_0^{CF} D_t^\alpha I_h &= \beta_h S_h I_m - (\eta + \kappa_h + x_h) I_h, \\
{}_0^{CF} D_t^\alpha V_h &= \sigma Z_h + \xi \kappa_h D_h - (\rho + x_h) V_h, \\
{}_0^{CF} D_t^\alpha D_h &= \eta I_h - (\xi \kappa_h + x_h) D_h, \\
{}_0^{CF} D_t^\alpha S_m &= Z_m - \beta_m S_m I_h - x_m S_m, \\
{}_0^{CF} D_t^\alpha I_m &= \beta_m S_m I_h - x_m I_m,
\end{aligned} \tag{2}$$

with i.c.'s $S_h(0) = S_{h0}$, $I_h(0) = I_{h0}$, $V_h(0) = V_{h0}$, $D_h(0) = D_{h0}$, $S_m(0) = S_{m0}$, and $I_m(0) = I_{m0} \geq 0$.

The exceeding proposed epidemic model Z_h and Z_m show the new upcoming rate of host and vector populations. The proportion of vaccination during the new coming host population is represented by σ. β_h is the transmission rate between humans to mosquitoes and β_m is the transmission rate between mosquito to human respectively. Also, In COVID-19 affected human population class η represent the discharge rate, κ_h represent disease-induced death rate, ρ represent the loss of vaccination-induced immunity, and ξ is the parameter with modifies disease death rate of discharge hosts. The natural mortality of the host individuals is denoted by x_h and for vector x_m respectively.

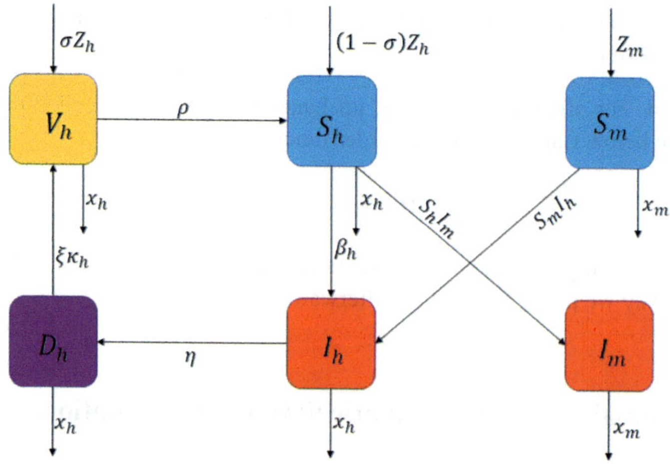

Fig. 1 Flowchart diagram of COVID-19 and malaria disease

In susceptible host population $(1-\sigma)Z_h$, vaccination host $\sigma\kappa_h$ and susceptible vector Z_m is the new upcoming rates. The rate $\beta_h S_h I_m$ is flowing from the susceptible human individual to the infected human individual, and $\beta_m S_m I_h$ is flowing from the susceptible mosquito individual to the infected mosquito individual, respectively. The rate ρ is the flow from vaccinated host to susceptible host, η is flow from I_h to D_h, and κ_h is from I_h to D_h. The natural death rate x_h is going out of each class of host population and x_m is going out of each class of vector population respectively. All these mentioned flowing rates of biological parameters are elaborated in Fig. 1.

4 Existence and Positiveness

To show the positivity of the dynamical system solution, we have $\mathbb{R}^6_+ = \{v \in \mathbb{R}^6 : v \geq 0\}$ and $v(t) = \left(S_h(t), I_h(t), V_h(t), D_h(t), S_m(t), I_m(t)\right)^T$. To move further, we have generalized mean value theorem.

Lemma 1 *Suppose that* $\phi \in [a, t]$, $\forall\ a < t < b$, $f(t) \in C[a, b]$ *and* $a <{}_0^{CF} D_t^{\alpha} f(t) \leq b$, *with* $\alpha(0, 1]$, *then* $f(t) = f(a) + \dfrac{1}{\Gamma(\alpha)}\left({}_0^{CF} D_t^{\alpha} f\right)(\phi)(t-a)^{\alpha}$.

Now we consider the following results.

Theorem 1 *The solution* $v(t)$ *of the system (2) is exists, non-negative and will stay in* $\mathbb{R}^6_+$.

Proof The existence as well as uniqueness of the fractional model can be easily demonstrated. Now to show the positive solution, it is compulsory that on each hyperplane bounding the positive orhant space to $\mathbb{R}^6_+$. Therefore, the proposed system (2), gives

$$\begin{aligned}
{}^{CF}_{0}D^{\alpha}_{t}S_h|_{S_h=0} &= (1-\sigma)Z_h + \rho L_0 h \geq 0,\\
{}^{CF}_{0}D^{\alpha}_{t}I_h|_{I_h=0} &= \beta_h S_h I_m \geq 0,\\
{}^{CF}_{0}D^{\alpha}_{t}V_h|_{V_h=0} &= \sigma Z_h + \xi\kappa_h D_h \geq 0,\\
{}^{CF}_{0}D^{\alpha}_{t}D_h|_{D_h=0} &= \eta I_h \geq 0,\\
{}^{CF}_{0}D^{\alpha}_{t}S_m|_{S_m=0} &= Z_m \geq 0,\\
{}^{CF}_{0}D^{\alpha}_{t}I_m|_{I_m=0} &= \beta_m S_m I_h \geq 0.
\end{aligned} \tag{3}$$

Hence, by Lemma 1, the solution will stay in $\mathbb{R}^6_+$, and so the biological feasible solution is given as:

$$\Sigma = \left\{ \left(S_h, I_h, V_h, D_h, S_m, I_m\right) \in \mathbb{R}^6_+ : S_h, I_h, V_h, D_h, S_m, I_m \geq 0 \right\}.$$

Afterward, we will evaluate the equilibrium points and basic reproduction number $\mathcal{R}_0$ of the proposed model (2) in the below subsection.

4.1 Equilibrium Points and Estimation of $\mathcal{R}_0$

The equilibria of the system (2) are acquired by solving the system as follows.

$${}^{CF}_{0}D^{\alpha}_{t}S_h = {}^{CF}_{0}D^{\alpha}_{t}I_h = {}^{CF}_{0}D^{\alpha}_{t}V_h = {}^{CF}_{0}D^{\alpha}_{t}D_h = {}^{CF}_{0}D^{\alpha}_{t}S_m = {}^{CF}_{0}D^{\alpha}_{t}I_m = 0.$$

Thus, we evaluated the proposed epidemic model and found two types of critical points. The DFE point is obtained as

$$E^1 = \left(S^1_h, I^1_h, V^1_h, D^1_h, S^1_m, I^1_m\right) = \left(\frac{(1-\sigma)Z_h}{x_h}, 0, \frac{\sigma Z_h}{\rho + x_h}, 0, \frac{Z_m}{x_m}, 0\right).$$

Next, the EE point is obtained as $E^2 = \left(S^2_h, I^2_h, V^2_h, D^2_h, S^2_m, I^2_m\right)$, with

$$
\begin{aligned}
S_h^2 &= \frac{(1-\sigma)Z_h + \rho V_h^2}{\beta_h I_m^2 + x_h}, \\
I_h^2 &= \frac{(1-\sigma)Z_h Z_m \beta_h \beta_m - x_h\big(\eta + \kappa_h + x_h\big)x_m^2}{\beta_m\big(\eta + \kappa_h + x_h\big)\big(Z_m \beta_h + x_h x_m\big)}, \\
V_h^2 &= \frac{\sigma Z_h + \xi \kappa_h D_h^2}{\rho + x_h}, \\
D_h^2 &= \frac{\eta I_h^2}{\big(\xi Z_h + x_h\big)}, \\
S_m^2 &= \frac{Z_m}{\beta_m I_h^2 + x_m}, \\
I_m^2 &= \frac{\beta_m Z_m I_h}{(\beta_m I_h^2 + x_m)x_m}.
\end{aligned} \tag{4}
$$

The EE point E^*, exist only if $\mathcal{R}_0 > 1$. The $\mathcal{R}_0$ for the non-integer COVID-19 and malaria disease models is determined by using the next-generation technique. $\mathcal{R}_0$ is biologically very essential and determines the global dynamical transmission of the model. The corresponding diseases matrices F(without infection) and V(with infection) are given by

$$
F = \begin{bmatrix} 0 & \frac{(1-\sigma)Z_h\beta_h}{x_h} \\ \frac{Z_m\beta_m}{x_m} & 0 \end{bmatrix}, \; V = \begin{bmatrix} \eta + \kappa_h + x_h & 0 \\ 0 & x_m \end{bmatrix}.
$$

The inverse of infected matrix V is

$$
V^{-1} = \begin{bmatrix} \frac{1}{\eta + \kappa_h + x_h} & 0 \\ 0 & \frac{1}{x_m} \end{bmatrix}, \; FV^{-1} = \begin{bmatrix} 0 & \frac{(1-\sigma)Z_h\beta_h}{x_h x_m} \\ \frac{Z_m\beta_m}{x_m\big(\eta + \kappa_h + x_h\big)} & 0 \end{bmatrix}.
$$

This above spectral radius is called the basic reproduction number of the model, and after some simplification, we have

$$
\mathcal{R}_0 = \sqrt{\frac{(1-\sigma)Z_h Z_m \beta_h \beta_m}{x_h x_m^2\big(\eta + \kappa_h + x_h\big)}}.
$$

5 Stability Analysis

In this section, we have presented the stability analysis of DFE results in both local as well as global stability. The Variational of the model (2) is as follows.

$$J_{E^1} = \begin{bmatrix} -x_h & 0 & \rho & 0 & 0 & \dfrac{-(1-\sigma)\beta_h Z_h}{x_h} \\ 0 & \eta + \kappa_h + x_h & 0 & 0 & 0 & \dfrac{(1-\sigma)\beta_h Z_h}{x_h} \\ 0 & 0 & -(\rho + x_h) & \xi\kappa_h & 0 & 0 \\ 0 & \eta & 0 & \xi\kappa_h - x_h & 0 & 0 \\ 0 & \dfrac{-\beta_m Z_m}{x_m} & 0 & 0 & -x_m & 0 \\ 0 & \dfrac{\beta_m Z_m}{x_m} & 0 & 0 & 0 & -x_m \end{bmatrix}.$$

Theorem 2 *For $q_1, q_2 \geq 0$ such that q_1 and q_2 are primitive root. Let $m_1 = \alpha = \frac{q_1}{q_2} > 0$, and $Q = q_2$, then E^1 is LAS if $|arg(\lambda)| > \frac{\pi}{2Q}$, with λ is the possible roots of the characteristic equation of the Variational matrix at E^1 as follows:*

$$det\big(diag[\lambda^{m_1}, \lambda^{m_1}, \lambda^{m_1}, \lambda^{m_1}, \lambda^{m_1}, \lambda^{m_1}] - J_{E^1}\big) = 0. \tag{5}$$

Proof By the expansion of Eq. (5) we get the below equation in terms of λ,

$$(-x_m - \lambda)(-\rho - x_h - \lambda)(\xi\kappa_h - x_h - \lambda)(-x_h - \lambda)\big[\lambda^2 + r_1\lambda + r_2\big], \tag{6}$$

with the coefficents are as follows:

$$r_1 = (\eta + \kappa_h + x_h + x_m),\, r_2 = x_m(\eta + \kappa_h + x_h)(1 - \mathcal{R}_0).$$

The arguments for the roots of the equation $(-x_m - \lambda) = 0$ are as below:

$$arg(\lambda_l) = \frac{\pi}{m_1} + l\frac{2\pi}{m_1} > \frac{\pi}{Z} > \frac{\pi}{2Z}, \quad with\ l = 0, 1, 2, \ldots, (m_1 - 1). \tag{7}$$

Likewise, the pattern, it can be demonstrated by the arguments for the roots of other equations, and all are greater than $\frac{\pi}{2Q}$. Also, if $\mathcal{R}_0 < 1$, then the condition $|arg(\lambda)| > \frac{\pi}{2Q}$ is proved for all the roots of the polynomial Eq. (6). But if $\mathcal{R}_0 > 1$, then by the Descartes sign's rule, there exists single root of characteristic equation with $|arg(\lambda)| < \frac{\pi}{2Q}$. Hence, if $\mathcal{R}_0 < 1$ then DFE point is LAS and if $\mathcal{R}_0 > 1$ then DFE point is unstable.

For the case of global stability of the DFE point, we have the following theorem.

Theorem 3 *If $\mathcal{R}_0 < 1$ then DFE point is GAS and if $\mathcal{R}_0 > 1$ then DFE point is unstable.*

Proof In order to obtain our result, we have the following Lyapunov function:

$$V(t) = N_1\left(S_h - S_h^1 - S_h^1 \frac{S_h}{S_h^1}\right) + N_2 I_h + N_3\left(L_0 h - V_h^1 - V_h^1 \frac{L_0 h}{V_h^1}\right)$$
$$+ N_4 D_h + N_5\left(S_m - S_m^1 - S_m^1 \frac{S_m}{S_m^1}\right) + N_6 I_m, \tag{8}$$

with $N_1, N_2, N_3, N_4, N_5, N_6$ are arbitrary positive constant. By Eqs. (5) and (8), along with the system of Eq. (2) we have

$$ {}_0^{CF}D_t^\alpha V(t) = N_1\left(\frac{S_h - S_h^1}{S_h}\right) {}_0^{CF}D_t^\alpha S_h + N_2\, {}_0^{CF}D_t^\alpha I_h + N_3\left(\frac{V_h - V_h^1}{V_h}\right) {}_0^{CF}D_t^\alpha L_0 h$$
$$+ N_4\, {}_0^{CF}D_t^\alpha D_h + N_5\left(\frac{S_m - S_m^1}{S_m}\right) {}_0^{CF}D_t^\alpha S_m + N_6\, {}_0^{CF}D_t^\alpha I_m$$
$$= x_m(\eta + \kappa_h + x_h)(\mathcal{R}_0 - 1). \tag{9}$$

Thus, if $\mathcal{R}_0 < 1$ then ${}_0^{CF}D_t^\alpha V(t) < 0$. Hence, E^1 is GAS in Σ.

6 Numerical Work

In this section, we employed the ABM method to solve the dynamical system (2). The main mathematical explanation steps that occur in the mechanism are demonstrated here for the order $\alpha \in (0, 1]$. This is the generalization of the ordinary ABM method applied to solve numerically the first order ODEs. This appears in the category of the so-called PE and CE type because it contains calculation of the PV which is further utilization to find the CV. This aforesaid numerical method and its behaviour are well known in the area of fractional calculus as well as in applied areas [14]. To find the approximate solution to this analogy, take the below non-linear FDE.

$$D^\alpha f(t) = H(t, f(t)), t \in [0, T]\ \&\ f^{(l)} = f_0^l, l = 0, 1, \ldots, n-1,\ for\ n = \lceil \alpha \rceil \tag{10}$$

The Eq. (10) is equivalent to the VIE:

$$f(t) = \sum_{l=0}^{n-1} f_0^{(l)} \frac{t^l}{l!} + \frac{1}{\Gamma(\alpha)} \int_0^t \frac{1}{(t-s)^{1-\alpha}} H(s, f(s))ds. \tag{11}$$

By utilizing this algorithm on the proposed model (2), Eq. (11) can be discretized as below [15].

$$
\begin{aligned}
K_{n+1} &= K_0 + \frac{h^\alpha}{\Gamma(\alpha+2)}\Bigg[\big((1-\sigma)Z_h - \beta_h K^p_{n+1} P^p_{n+1} + \rho M^p_{n+1} - x_h K^p_{n+1}\big) \\
&\quad + \sum_{j=0}^{n} a_{j,n+1}\big((1-\sigma)P_h - \beta_h K_j P_j + \rho M_j - x_h K_j\big)\Bigg], \\
L_{n+1} &= L_0 + \frac{h^\alpha}{\Gamma(\alpha+2)}\Bigg[\big(\beta_h K^p_{n+1} P^p_{n+1} - (\eta + \kappa_h + x_h) L^p_{n+1}\big) \\
&\quad + \sum_{j=0}^{n} a_{j,n+1}\big(\beta_h K_j P_j - (\eta + \kappa_h + x_h) L_j\big)\Bigg], \\
M_{n+1} &= M_0 + \frac{h^\alpha}{\Gamma(\alpha+2)}\Bigg[\big(\sigma Z_h + \xi \kappa_h N^p_{n+1} - (\rho + x_h) M^p_{n+1}\big) \\
&\quad + \sum_{j=0}^{n} a_{j,n+1}\big(\sigma Z_h + \xi \kappa_h N_j - (\rho + x_h) M_j\big)\Bigg], \\
N_{n+1} &= N_0 + \frac{h^\alpha}{\Gamma(\alpha+2)}\Bigg[\big(\eta L^p_{n+1} - (\xi \kappa_h + x_h) X^p_{n+1}\big) \\
&\quad + \sum_{j=0}^{n} a_{j,n+1}\big(\eta L_j - (\xi \kappa_h + x_h) N_j\big)\Bigg], \\
O_{n+1} &= O_0 + \frac{h^\alpha}{\Gamma(\alpha+2)}\Bigg[\big(Z_m - \beta_m O^p_{n+1} L^p_{n+1} - x_m O^p_{n+1}\big) \\
&\quad + \sum_{j=0}^{n} a_{j,n+1}\big(Z_m - \beta_m O_j L_j - x_m O_j\big)\Bigg], \\
P_{n+1} &= P_0 + \frac{h^\alpha}{\Gamma(\alpha+2)}\Bigg[\big(\beta_m O^p_{n+1} L^p_{n+1} - x_m P^p_{n+1}\big) \\
&\quad + \sum_{j=0}^{n} a_{j,n+1}\big(\beta_m O_j L_j - x_m P_j\big)\Bigg], \qquad (12)
\end{aligned}
$$

where

$$
\begin{aligned}
K^p_{n+1} &= K_0 + \frac{1}{\Gamma(\alpha)} \sum_{j=0}^{n} b_{j,n+1}\big((1-\sigma)Z_h - \beta_h K_j P_j + \rho M_j - x_h K_j\big), \\
L^p_{n+1} &= L_0 + \frac{1}{\Gamma(\alpha)} \sum_{j=0}^{n} b_{j,n+1}\big(\beta_h K_j P_j - (\eta + \kappa_h + x_h) L_j\big), \\
M^p_{n+1} &= M_0 + \frac{1}{\Gamma(\alpha)} \sum_{j=0}^{n} b_{j,n+1}\big(\sigma P_h + \xi \kappa_h N_j - (\rho + x_h) M_j\big),
\end{aligned}
$$

$$N_{n+1}^{p} = N_0 + \frac{1}{\Gamma(\alpha)} \sum_{j=0}^{n} b_{j,n+1}\big(\eta L_j - (\xi \kappa_h + x_h) N_j\big),$$

$$O_{n+1}^{p} = O_0 + \frac{1}{\Gamma(\alpha)} \sum_{j=0}^{n} b_{j,n+1}\big(P_m - \beta_m O_j L_j - x_m O_j\big),$$

$$P_{n+1}^{p} = P_0 + \frac{1}{\Gamma(\alpha)} \sum_{j=0}^{n} b_{j,n+1}\big(\beta_m O_j L_j - x_m P_j\big), \tag{13}$$

in which $a_{j,n+1} = \begin{cases} n^{\alpha+1} - (n+1)(n-\alpha), & j = 0, \\ (n-j)^{\alpha+1} - 2(n-j+1)^{\alpha+1} + (n-j+2)^{\alpha+1}, & j = [1, n], \\ 1, & j = n+1, \end{cases}$

and $b_{j,n+1} = \frac{h^{\alpha}}{\alpha}\big[(n-j+1)^{\alpha} - (n-j)^{\alpha}\big], \quad j \in [0, n].$

Figures 2 and 3 demonstrate that $I_h(t)$ decreases significantly with time t. In Fig. 2 it has been observed that for $\alpha = 0.75$, the graph of I_h increase from $t = 10$ to $t = 50$ and then decline significantly. Likewise for $\alpha = 0.65$ and $\alpha = 0.55$ we have the behaviour of I_h versus time t. This graph indicates that for large value of α have shown good impact on state variables. Since α has a reciprocal impact on $I_h(t)$ and $I_m(t)$. In order to find a stable and convergent solution, the value of α should be as small as possible and should be greater than zero. Afterwards, Fig. 4 shows the effects of η on $I_h(t)$. Since, as we increase the value of η the graph of $I_h(t)$ also increases. In Fig. 5 we observe the effect of ρ on $S_h(t)$ class. On increase of ρ, $S_h(t)$

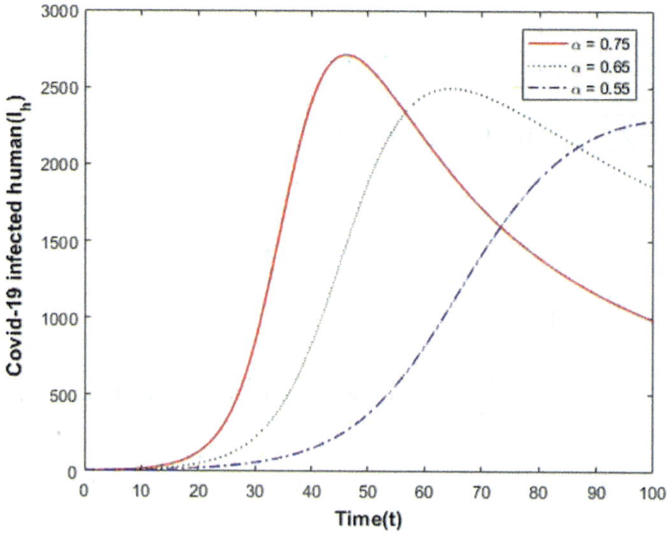

Fig. 2 Effect of α on I_h

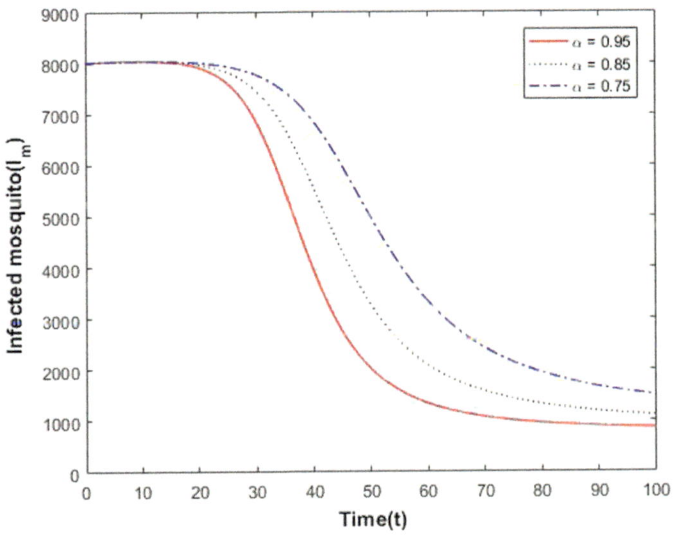

Fig. 3 Effect of α on I_m

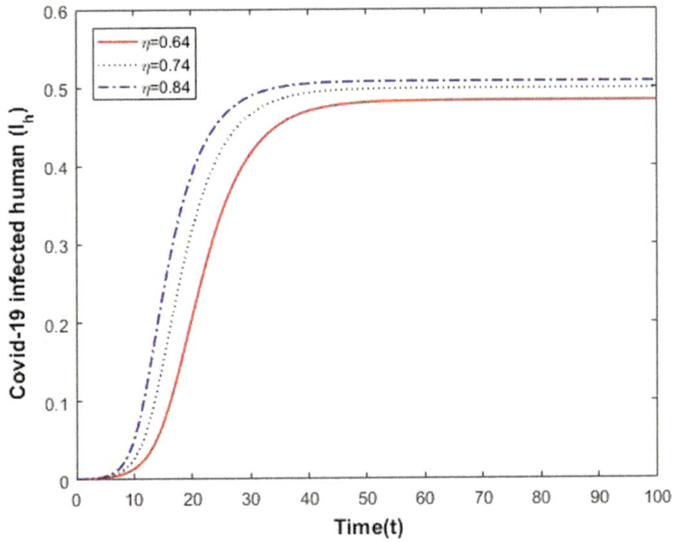

Fig. 4 Effect of η on I_h

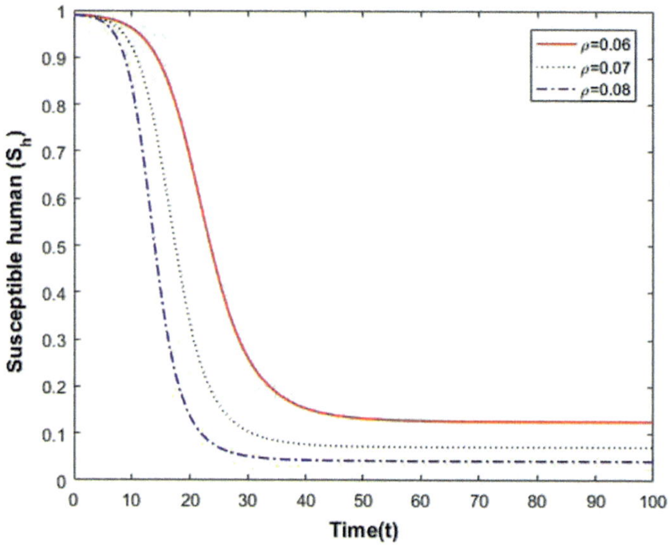

Fig. 5 Effect of ρ on S_h

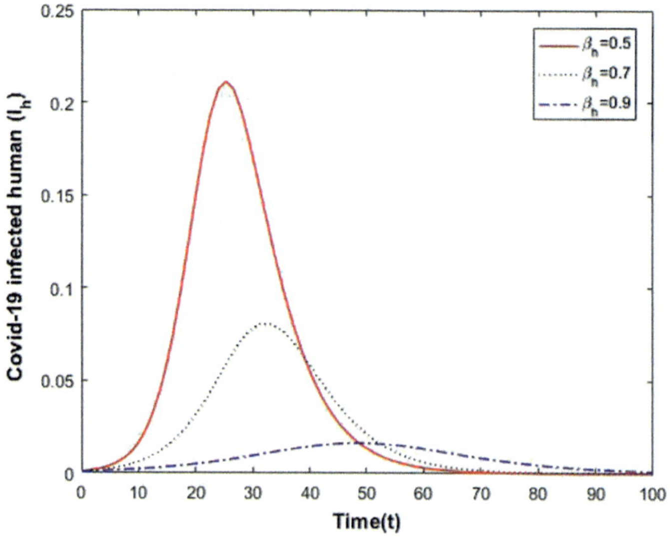

Fig. 6 Effect of β_h on I_h

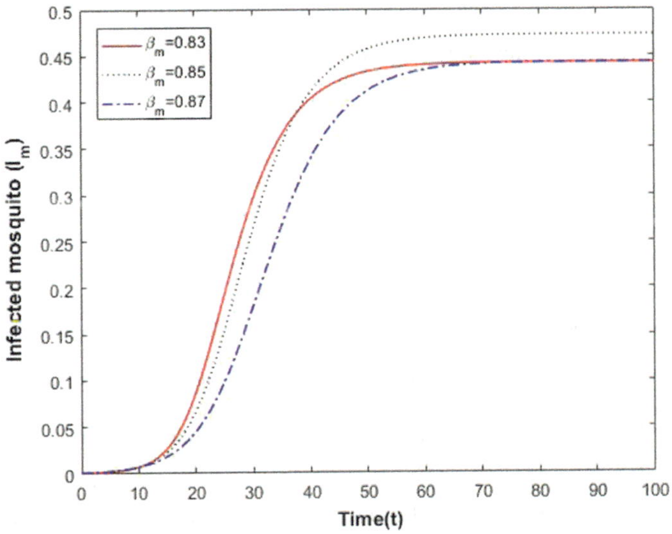

Fig. 7 Effect of β_m on I_m

get decrease. Next, in Fig. 6 demonstrates the impact of β_h on the $I_h(t)$ class. As we increase the value of β_h, the curves of I_h are increased. Finally, Fig. 7 shows the effect of β_m on $I_m(t)$. On the increase of β_m, the curves of the $I_m(t)$ class are decreased. In Fig. 7 it has been seen that for $\beta_m = 0.83, 0.85, 0.87$, the graph of I_m increase from $t = 10$ to $t = 40$. This show the realistic fact. As the transmission rate increase the number of infectious mosquitoes get increased in the population.

7 Conclusion

Malaria and Coronavirus are odd infectious diseases that wreak havoc over the world, and their route of transmission is still a mystery. We studied the dynamics of malaria and COVID-19 disease mathematically in this investigation. We computed the $\mathcal{R}_0$ and established a stability analysis using several key theorems. Figures demonstrate the impact of non-integer order α and various transmission rates graphically. Our findings are consistent with those that suggest malaria serves as a launchpad for the SARS-CoV-2 virus and causes death [16]. In the spirit of CF, we compared these several models. In the numerical work, CF is found to have even better outcomes in terms of stability than the other operators like Riemann-Liouville and Caputo.

Concluded remark:

- It is observed that DFE is LAS when $\mathcal{R}_0 < 1$, and unstable if $\mathcal{R}_0 > 1$.
- It is also observed that for the disease model, the traditional condition of $\mathcal{R}_0 < 1$ is not sufficient for eradication of the COVID-19 and malaria.

Future direction:
The future direction of the study will incorporate the optimization of medical treatment for COVID-19 and malaria.

References

1. Government of India, my gov.in/COVID-19
2. Gowrisankar, A., Rondoni, L., Banerjee, S.: Can India develop herd immunity against COVID-19 ?. Eur. Phys. J. Plus **135**, 526 (2020)
3. Kavitha, C., Gowrisankar, A., Banerjee, S.: The second and third waves in India: when will the pandemic be culminated ?. Eur. Phys. J. Plus **136**, 596 (2021)
4. Pinto, C.M.A., Machado, J.A.T.: Fractional model for malaria transmission under control strategies. Comput. Math. Appl. **66**(5), 908–916 (2013)
5. WHO, World Malaria Reports. MMV (2020)
6. Manda, S., Sarkar, R.P., Sinha, S.: Mathematical models of malaria - a review. Malar. J. **10**(1), 2–19 (2002)
7. Tumwiine, J., Mugisha, J.Y.T., Luboobi, L.S.: A mathematical model for the dynamics of malaria in a human host and mosquito vector with temporary immunity. Appl. Math. Comput. **189**(2), 1953–1965 (2007)
8. Chanda-Kapata, P., Kapata, N., Zumla, A.: COVID-19 and malaria: a symptom screening challenge for malaria endemic countries. Int. J. Infect. Dis. **94**(2020), 151–153 (2020)
9. Hussein, M.I.H., Albashir, A.A.D., Elawad, O.A.M.A., Homeida, A.: Malaria and COVID-19: unmasking their ties. Malar. J. **19**(457), 1–11 (2020)
10. Yu, X., Qi, G., Hu, J.: Analysis of second outbreak of COVID-19 after relaxation of control measures in India. Nonlinear Dyn. **106**, 1149–1167 (2021)
11. Sher, M., Shah, K., Khan, Z.A., Khan, H., Khan, A.: Computational and theoretical modeling of the transmission dynamics of novel COVID-19 under Mittag-Leffler power law. Alex. Eng. J. **59**(5), 3133–3147 (2020)
12. Ray, M., Vazifdar, A., Shivaprakash, S.: Co-infection with malaria and coronavirus disease-2019. J. Glob. Infect. Dis. **12**(3), 162–163 (2020)
13. Abdeljawad, T., Banerjee, S., Wu, G.C.: Discrete tempered fractional calculus for new chaotic systems with a short memory and image encryption. Optik **203**, 163698 (2020)
14. Rehman, A.U., Singh, R., Abdeljawad, T., Okyere, E., Guran, L.: Modeling, analysis and numerical solution to malaria fractional model with temporary immunity and relapse. Adv. Differ. Equ. **2021**, 390 (2021)
15. Li, C., Tao, C.: On the fractional Adams method. Comput. Math. Appl. **58**(8), 1573–1588 (2009)
16. Thakare, P., Sharma, S., Gupta, A., Lilare, N., Utpat, K., Desai, U., JM, J., Bharmal, R.N.: A Case Series on Early Virological Clearance in Malaria COVID-19 Co-infection. Int. Clin. Case Rep. 1(1), 1-3 (2021)

Design of Imidazole-Based Drugs as Potential Inhibitors of SARS-Cov-2 of the Delta and Omicron Variant

An *in Silico* Approach

Peter Solo and M. Arockia Doss

Abstract 30 imidazole-based drugs were designed, prepared and optimized using Gaussian 09, and were screened for drug-likeness with SWISS-ADME server. Molecular docking was performed using MOE 09, where the designed drugs were docked with the spike glycoprotein of the Delta (B.1.617.2) and Omicron (B.1.1.529) variant of SARS CoV-2. Nafamostat and Hydroxychloroquine were used as standards in comparing the docking results. Among the designed drugs, those drugs which used Benzil and Hyroxy/methoxy-benzaldehyde as the starting compound exhibited good binding scores, and can be potential inhibitors of the Delta and Omicron variant of SARS CoV-2.

Keywords SARS-CoV-2 · Imidazole-based drug design · Molecular docking

1 Introduction

The outbreak of severe acute respiratory syndrome corona virus 2 (SARS CoV-2) continues to prevail. The detection of the new omicron variant (B.1.1.529) in south Africa has ignited fresh concern over the already contagious delta variant (B.1.617.2). On 26 November 2021, WHO has designated omicron as a variant of concern which has now spread across the entire globe [1]. Unlike the other variants, omicron has large number of mutation with 32 mutations in the spike glycoprotein [2] and 15 mutations in the receptor-binding domain (RDB) alone [3]. Computational studies have affirmed that the Omicron variant had a higher affinity for human angiotensin-converting enzyme 2 (ACE2) as compared to the delta variant due to the significant

Supplementary Information The online version contains supplementary material available at https://doi.org/10.1007/978-3-030-99792-2_125.

P. Solo (✉) · M. A. Doss
Department of Chemistry, St. Joseph University, Dimapur 797115, India
e-mail: Solopeter82@gmail.com

P. Solo
Department of Chemistry, St. Joseph's College (Autonomous), Jakhama 797001, India

S. Banerjee and A. Saha (eds.), *Nonlinear Dynamics and Applications*, Springer Proceedings in Complexity,
https://doi.org/10.1007/978-3-030-99792-2_125

of mutations in the RBD regions of the spike protein [4]. Initial modelling suggests that the omicron variant is 2.8 times more infectious than the delta variant [5]. Most monoclonal antibodies are not effective against the omicron variant as they target the spike protein which has significant mutations [6].

Imidazole and its derivatives forms a prominent class of hetero cyclic compounds, displaying diverse applications especially with regards to its biological and pharmacological activities. Many substituted Imidazole are found to exhibit anti-inflammatory [7, 8], anti-fungal [9, 10], anti-bacterial [11], anti-epileptic, anti-microbial [12], anti-cancer [13, 14] and anti-tubercular activities [15]. They are also found to have analgesic properties [16]. Imidazole is the main structural feature in many drugs found in the markets, namely isoconazole, ketoconazole, clotrimazole, miconazole, econazole, fluconazole, sertaconazole, eprosartan, losartan, olmesartan, metronidazole, etc. [17]. The study aims at designing imidazole-based drug in search for potential inhibitor against SARS CoV-2 using *in silico* approach.

2 Materials and Methods

2.1 Designing of Ligands

The Ligands were designed based on a published work [18] where the heterocyclic imidazole ring forms an important structural feature with regard to its pharmacological activity as antimicrobial and antifungal (Fig. 1). Alpha-diketone, aromatic aldehyde, 2-(piperazin-1-yl) ethanamine and ammonium acetate are the reactants in the synthesis of the drugs. From a synthetic point of view various derivatives of the drug were designed by replacing various substituents at R1, R2 and R3 positions. R1 and R2 are replaced by alkyl or aromatic groups using diacetyl or benzyl as reactants respectively. R3 groups can have the most number of variations as it is formed by an aromatic aldehyde. All the drugs/ligands were designed using Gaussian 09 [19]. Optimization was performed with DFT/B3LYP method using 6-31G(d,p) level of theory. The optimized structures were saved in both MOL and SDF format.

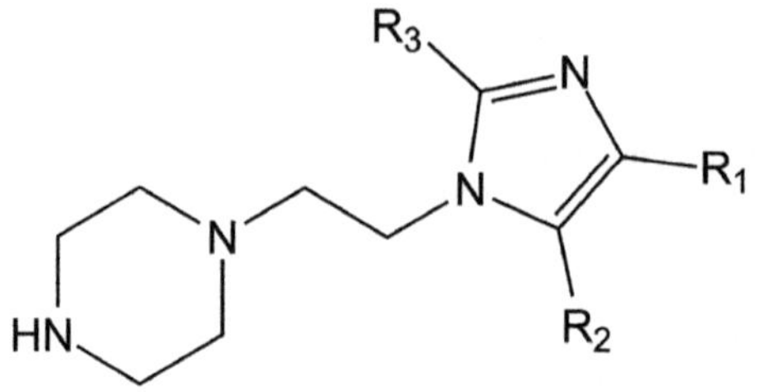

Fig. 1 Framework for imidazole-based drug design

2.2 Screening of Ligands for Drug-Likeness

The designed ligands were screened for drug-likeness with SWISS-ADME server [20], which analyses many important aspects of the drug, namely Lipophilicity, Water Solubility, Pharmacokinetics, Drug-likeness and medicinal properties of the drug. The screening for drug-likeness with Lipinski, Ghose, Veber, Egan and Muegge are all included in the SWISS-ADME server. The SDF structures of the designed-drugs were fed into the SWISS-ADME server and were converted into SMILES format which were then screened for drug-likeness.

2.3 SARS CoV-2 Spike Glycoprotein Target Selection and Homology Modelling

The 3D structure for the spike glycoprotein of the delta variant (B.1.617.2), with PDB ID:7so9, was downloaded from RCSB protein data bank (https://www.rcsb.org/structure/7SO9). All other chains and water molecules were deleted, and only chains A, F and K of the spike glycoprotein was retained. The 3D structure for the spike glycoprotein for the Omicron variant (B.1.1.529) was not available at RCSB protein data bank, therefore, it was constructed with homology modelling using Swiss Model server [21]. The target sequence of the Omicron variant with GenBank ID UFO69279.1 (https://www.ncbi.nlm.nih.gov/protein/UFO69279.1) was downloaded in FASTA format from NCBI website and was pasted in Swiss Model server for the construction of 3D model. Blast and HHblits methods were used to search for template structure in the SWISS-MODEL template library. The template with PDB ID 7n1u.1.A was selected for building the 3D model. The details of the template and homology modelling results are given in Table 1.

2.4 Protein and Ligand Preparation

MOE 09 docking tools [22] was used to prepared both the target proteins and all the ligands. Structure preparation of the target proteins was executed and all corrections were performed for the PDB structures of the two target proteins, i.e., Delta variant and Omicron variant. Energy minimization of the structures were performed using Amber12:EHT Force field where hydrogens and charges were added. A database of ligand containing all the 30 compounds were prepared in MDB format. Each ligand in SDF format was entered into the database after 3D protonation and energy minimization with Amber12:EHT Force field.

Table 1 Details of template and homology modelling

SARS-CoV-2	GenBank IDs	Template	Seq identity	Method	Seq similarity	coverage	description	Qmean score
B.1.1.529	UFO69279.1	7n1u.1.A	97.00%	EM	0.61	1.0	Spike glycoprotein	−1.67

3 Results and Discussion

In this study, 30 imidazole-based drugs were designed in search for potential antiviral candidate for the prevailing SARS CoV-2 pandemic using in silico method. The drugs were designed based on published literature where the imidazole ring forms an essential structural feature for its pharmacological activity as antimicrobial and antifungal drugs. Designing and optimization of the 3D structures of the drugs were performed efficiently using Gaussian 09.

Most compounds performed very well in the screening for drug-likeness with SWISS-ADME server and there were no serious violations or alerts (S.1). The alerts were due to higher molecular mass and higher lipophilicity of some compounds to the presence of aromatic rings. All compounds showed high GI absorption and there was no Lipinski's violation. All the 30 compounds were selected to be docked with the target proteins.

The docking analysis with MOE 09 reveals that most of the designed drugs binds quite well to the target spike glycoprotein as compared to the standards (S. 1). The top 3 ligand with the best binding energy are displayed in Table 2. The designed drugs with alkyl substituents at R1 and R2 Positions (Fig. 1) did not top the list, which suggest that aromatic substituents at these positions increases ligand affinity for the target spike proteins. Ligand number 28 binds very well to both the target spike proteins exhibiting a binding affinity of −8.06753635 kcal/mol for the Delta variant and −7.19685173 kcal/mol for the Omicron variant.

In most of the interactions of the ligands with the active site, the imidazole ring is actively involved either through Conventional Hydrogen bond, Carbon Hydrogen bond, alkyl interactions or pi-alkyl interactions with the residues of the target site. In the docking with omicron spike protein the imidazole ring is involved in pi-alkyl interaction with ILE A:192 (ligand 20) and ILE B:192 (ligand 22) residues (Fig. 2). In the docking with delta spike protein the imidazole ring is involved in alkyl interaction with PHE K:377 residue of the target protein (ligand 23). It is also involved in convention hydrogen bond and amide pi-stacked interactions with ARG A:408 and SER K:375 residues respectively (Fig. 3). Interaction of ligand 20 with omicron variant is mainly stabilized with three conventional hydrogen bonding with THR B:390, SER B:511 and ASN B:391, and carbon hydrogen bond with THR B:390 (Fig. 2). On the other hand, interaction of ligand 28 with the delta variant is mainly stabilized by five carbon hydrogen bonds with GLY A:404, ASP F:405, GLU F:406 and PHE A:374 (Fig. 3).

Drugs which were prepared using Benzil and hydroxy/methoxy-benzaldehyde as starting compound aided the binding of the drugs with multiple favorable interactions. The hydroxyl group (ligand 20) in complex with the omicron variant is involved in two conventional hydrogen bond with SER B:511 and ASN B:391 residues (Fig. 2). In the complex with delta variant, the methoxy-groups of ligand 28 is involved in two carbon hydrogen bond with GLU F:406 and PHE A:374 residues, and alkyl interaction with LYS F:417 (Fig. 3).

Table 2 Docking scores of the top 3 ligands with the standards

Top hits ligands docking with Omicron variant (B.1.1.529)	Docking score (kcal/mol)	Top hits ligands Docking with delta variant (B.1.617.2)	Docking score (kcal/mol)
Ligand no: 20 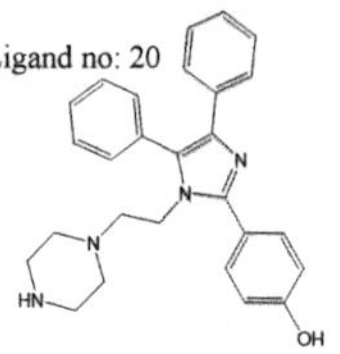	−7.22973061	Ligand no: 28 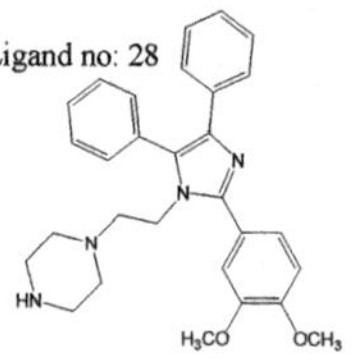	−8.06753635
Ligand no: 28 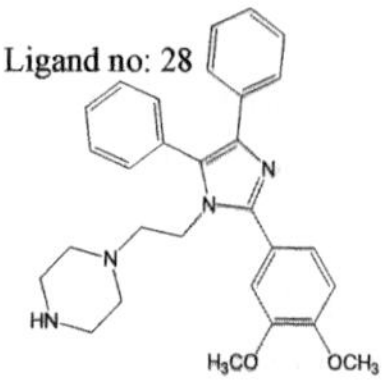	−7.19685173	Ligand no: 23 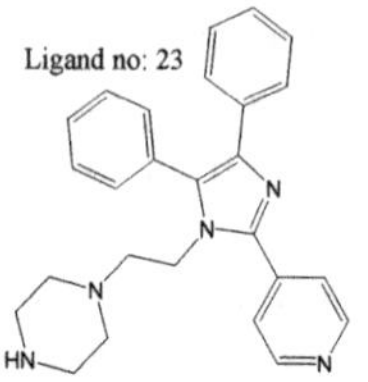	−7.96867609
Ligand no: 22 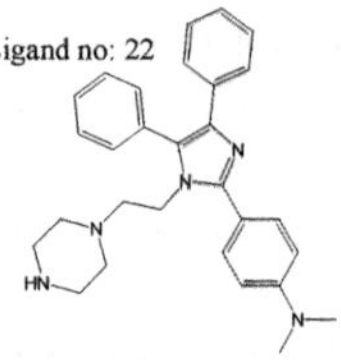	−7.17171955	Ligand no: 25 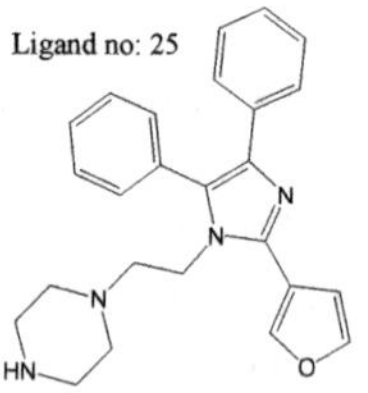	−7.96448946
Hydroxychloroquine	−6.51112795	Hydroxychloroquine	−6.40599966
Nafamostat	−7.13362694	Nafamostat	−6.89814949

The common amino acids at the binding site of the ligands with the Omicron variant are ASP B976, LEU C543, LEU C514 and LEU C515 (Fig. 2 and S. 2). At the active site of the binding of the ligands with the Delta variant, ARG A408, LYS A417, GLU A406, PHE K374, PHE K377, SER K375 and ARG K408 (Fig. 3 and S. 2), are the common amino acids. The difference in the common amino acid residues at the binding sites of the two spike protein variants reflects the enormous difference in their structure at the active site.

All common amino acids residues, accept for ASP B976, are in the receptor binding domain (RBD) which initiates viral entry into the host by binding to the cell receptor ACE2 in the aminopeptidase N [23]. Therefore, we can theoretically conclude that the designed drugs can inhibit the viral binding to ACE2 receptor.

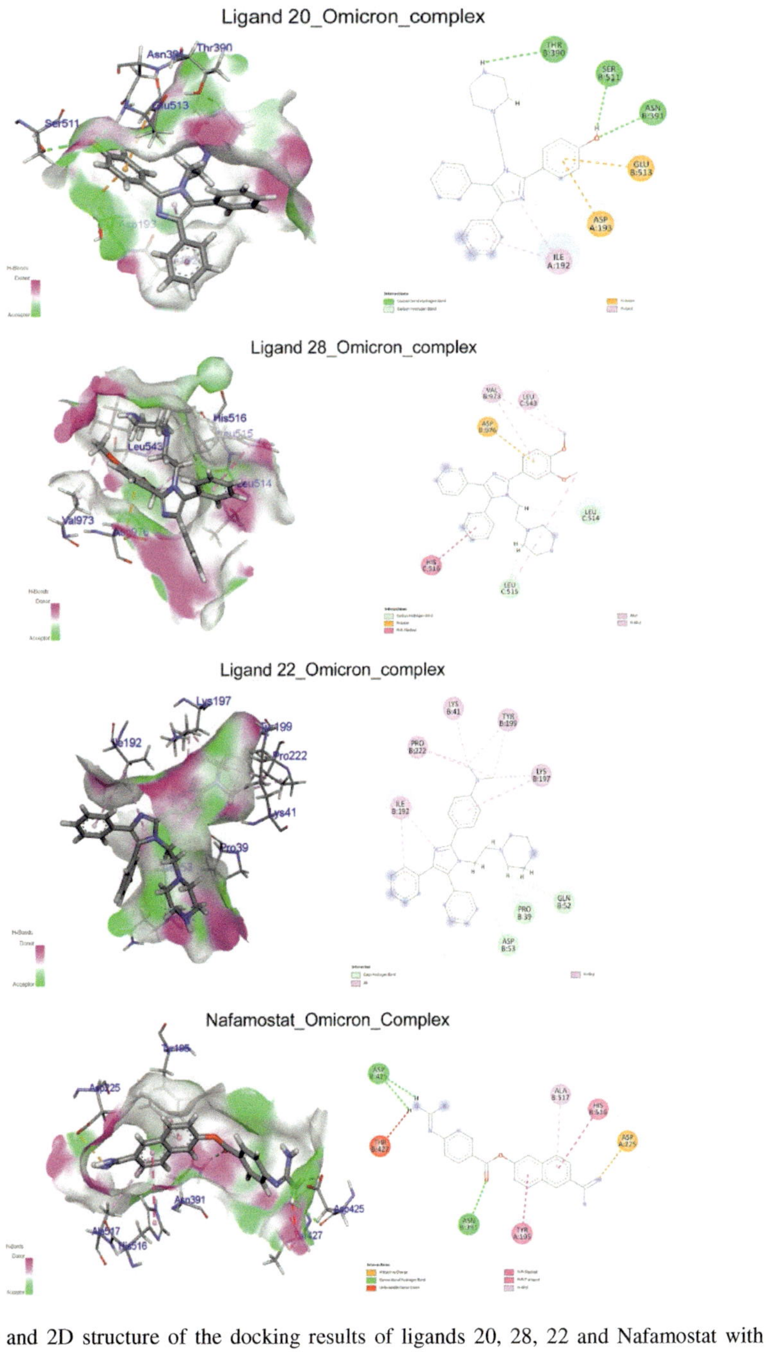

Fig. 2 3D and 2D structure of the docking results of ligands 20, 28, 22 and Nafamostat with Omicron (B.1.1.529) spike glycoprotein

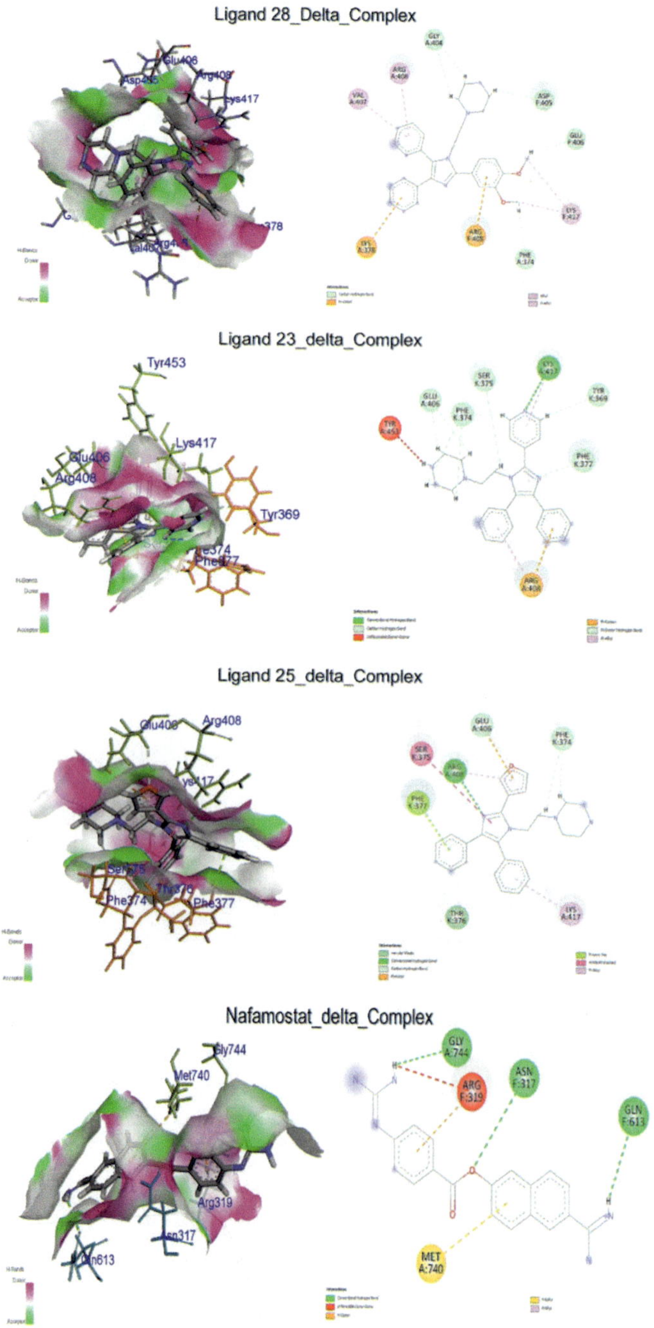

Fig. 3 3D and 2D structure of the docking results with ligands 28, 23, 25 and Nafamostat with Delta (B.1.617.2) spike glycoprotein

4 Conclusion

The study reports potential inhibitors for the SARS-CoV-2 main protease, through designing of Imidazole-based drugs and computational analysis. All the designed drugs were screened for drug-likeness using SWISS-ADME web based server. Gaussian 09 was used to design the drugs and optimization was performed using DFT/B3LYP method with 6-31G(d,p) level of theory. All the designed drugs were docked with the target spike proteins at the receptor binding domain using MOE 09. The docking results identified ligand 28 with good binding affinity for both the target proteins. The study also reports the active participation of the imidazole in the binding pockets of all the docked structures.

In silico approach for new drug discovery as potentials inhibitor against the prevailing contagious SARS CoV-2, can be a great contribution in the search for antiviral drugs for the virus, as the world is pushed to its limit with time and resources. Computational resources and software with user friendly interface are available in many open sources which can be easily utilized by those interested.

In conclusion the study reports that Imidazole-based drugs are found to bind quite effectively to the target spike proteins of Delta and Omicron variant. More Imidazole-based drugs can be designed in search for more efficient inhibitor for SARS CoV-2. In general, most heterocyclic-based drugs would display amazing pharmacological activities and hence, there are huge avenues in heterocycle-based drug designing. Further in vivo and in vitro analysis can be performed to support the conclusion made by the study.

Acknowledgements The authors gratefully acknowledge the support from St. Joseph University Dimapur (India) and St. Joseph's College (autonomous) Jakhama (India).

References

1. WHO Homepage. https://www.who.int/news/item/26-11-2021-classification-of-omicron-(b.1.1.529)-sars-cov-2-variant-of-concern
2. Gao, S.J., Guo, H., Luo, G.: Omicron variant (B.1.1.529) of SARS-CoV-2, a global urgent public health alert!. J. Med. Virol. (2022). https://doi.org/10.1002/jmv.27491
3. Thakur, V., Radha R.K.: OMICRON (B.1.1.529): A new SARS-CoV-2 variant of concern mounting worldwide fear. J. Med. Virol. 1–4 (2021). https://doi.org/10.1002/jmv.27541
4. Kumar, S., et al.: Omicron and Delta variant of SARS-CoV-2: a comparative computational study of spike protein. J. Med. Virol. 1–9 (2021). https://doi.org/10.1002/jmv.27526
5. STATPEARLS Homepage. https://www.statpearls.com/ArticleLibrary/viewarticle/132707
6. Li, P., et al.: SARS-CoV-2 Omicron variant is highly sensitive to molnupiravir, nirmatrelvir, and the combination. Cell Res. (2022). https://doi.org/10.1038/s41422-022-00618-w
7. Lombardino, J.G., Wiseman, E.H.: J. Med. Chem. **17**(11), 1128–1188 (1974). https://doi.org/10.1021/jm00257a011
8. Puratchikody, A., Mukesh, D.: Antinociceptive and anti-inflammatory activities and QSAR studies on 2-substituted-4,5-diphenyl-1H-imidazoles. Bioorg. Med. Chem. **15**(2) (2007). https://doi.org/10.1016/j.bmc.2006.10.025

9. Vita, D.D., et al.: Synthesis and antifungal activity of a new series of 2-(1H-imidazol-1-yl)-1-phenylethanol derivatives. J. Med. Chem. **49**, 334–342 (2012). https://doi.org/10.1016/j.ejmech.2012.01.034
10. Heeres, J., et al.: Antimycotic imidazoles. Part 4. Synthesis and antifungal activity of ketoconazole, a new potent orally active broad-spectrum antifungal agent. J. Med. Chem. **22**(8), 1003–1005 (1979). https://doi.org/10.1021/jm00194a023
11. Ganguly, S., et al.: Synthesis, antibacterial and potential anti-HIV activity of some novel imidazole analogs. Acta Pharm. **61**(2), 187–201 (2011). https://doi.org/10.2478/v10007-011-0018-2
12. Amir, M. et al.: Design and synthesis of some azole derivatives containing 2,4,5-triphenyl imidazole moiety as anti-inflammatory and antimicrobial agents. Indian J. Chem. **B50**, 207–213 (2011). http://www.nopr.niscair.res.in/handle/123456789/11029
13. Özkay, Y., et al.: Synthesis of 2-substituted-N-[4-(1-methyl-4,5-diphenyl-1H-imidazole-2-yl)phenyl]acetamide derivatives and evaluation of their anticancer activity. Eur. J. Med. Chem. **45**(8), 3320–3328 (2010). https://doi.org/10.1016/j.ejmech.2010.04.015
14. Yadav, S. et al.: Synthesis and evaluation of antimicrobial, antitubercular and anticancer activities of 2-(1-benzoyl-1Hbenzo[d]imidazole-2-ylthio)-N-substituted acetamides. Chem. Central J. **12**(66) (2018). https://doi.org/10.1186/s13065-018-0432-3
15. Shankar, G., et al.: Novel imidazo[2,1-b][1,3,4]thiadiazole carrying rhodanine-3-acetic acid as potential antitubercular agents. Bioorg. Med. Chem. Lett. **22**(5), 1917–1921 (2012). https://doi.org/10.1016/j.bmcl.2012.01.052
16. Uçucu, Ü., et al.: Synthesis and analgesic activity of some 1-benzyl-2-substituted-4,5-diphenyl-1H-imidazole derivatives. Il Farmaco **56**(4), 285–290 (2001). https://doi.org/10.1016/S0014-827X(01)01076-X
17. Bastide, M., et al.: A comparison of the effects of several antifungal imidazole derivatives and polyenes on Candida albicans: an ultrastructural study by scanning electron microscopy. Can. J. Microbiol. **28**(10), 1119–1126 (1982). https://doi.org/10.1139/m82-166
18. Rajkumar, R., et al.: Synthesis, spectral characterization and biological evaluation of novel 1-(2-(4,5-dimethyl-2-phenyl-1H-imidazol-1-yl)ethyl)piperazine derivatives. J. Saudi Chem. Soc. **18**(5), 735–743 (2014). https://doi.org/10.1016/j.jscs.2014.08.001
19. Gaussian 09, Revision A.02, M. J. Frisch, Gaussian, Inc., Wallingford CT, 2016. https://www.gaussian.com
20. Daina, A. et al.: SwissADME: a free web tool to evaluate pharmacokinetics, drug-likeness and medicinal chemistry friendliness of small molecules. Sci. Rep. **7**(42717) (2017). https://doi.org/10.1038/srep42717
21. Waterhouse, A., et al.: SWISS-MODEL: homology modelling of protein structures and complexes. Nucl. Acids Res. **41**(W1), W296–W303 (2018). https://doi.org/10.1093/nar/gky427
22. Chemical Computing Group ULC, 1010 Sherbooke St. West, Suite #910, Montreal, QC, Canada, H3A 2R7, 2021. https://www.chemcomp.com/Products.htm
23. Huang, Y., et al.: Structural and functional properties of SARS-CoV-2 spike protein: potential antivirus drug development for COVID-19. Acta Pharmacol. Sin. **41**, 1141–1149 (2020). https://doi.org/10.1038/s41401-020-0485-4

9783030997946VOL02